Hazardous Chemicals

Safety Management and Global Regulations

Hazardous Chemicals

Safety Management and Global Regulations

T.S.S. Dikshith

CRC Press
Taylor & Francis Group
Boca Raton   London   New York

CRC Press is an imprint of the
Taylor & Francis Group, an **informa** business

CRC Press
Taylor & Francis Group
6000 Broken Sound Parkway NW, Suite 300
Boca Raton, FL 33487-2742

First issued in paperback 2017

No claim to original U.S. Government works
Version Date: 20121120

ISBN 13: 978-1-138-07736-2 (pbk)
ISBN 13: 978-1-4398-7820-0 (hbk)

Library of Congress Cataloging-in-Publication Data

Dikshith, T. S. S., 1932-
  Hazardous chemicals : safety management and global regulations / T.S.S. Dikshith.
  pages cm
  Includes bibliographical references and index.
  ISBN 978-1-4398-7820-0
  1. Hazardous substances--Safety measures--Standards. I. Title.

T55.3.H3D55 2013
363.17'91--dc23   2012036757

Visit the Taylor & Francis Web site at
http://www.taylorandfrancis.com

and the CRC Press Web site at
http://www.crcpress.com

To my parents,

Gowramma and Turuvekere Subrahmanya Dikshith

and my wife, Saroja Dikshith

A hundred times every day I remind myself, that my inner and outer life depended on labours of other men, living and dead, and I must exert myself in order to give in the same measure as I have received and am still receiving.

Albert Einstein

Contents

Foreword

This outstanding work, *Hazardous Chemicals: Safety Management and Global Regulations*, discusses various aspects of potentially toxic and hazardous chemicals in simple comprehensive language. It is a welcome boon and much needed publication for students, scientists, and occupational health professionals. Frequent use, storage and disposal of hazardous chemicals without proper adequate knowledge and understanding of the possible consequences can result in health disorders, fatalities and disasters that can have an impact not only in the workplace, but also on a whole generation of human beings. Exposure to a wide variety of chemicals even in household environments is a harsh reality. Transportation of chemical substances by road, rail, air and sea with adequate safety measures needs careful study and attention. Generation of reliable and reproducible data, protocols required to evaluate chemical safety, and good laboratory practices will soon occupy centre stage. Fortunately the learned author, Prof. T.S.S. Dikshith, drawing our attention to the management of hazardous chemicals and related global safety standards, states that these standards are fast approaching uniformity. This is a big step towards preserving and protecting our fragile environment.

The ending of human adventure on earth is a possibility we have to reckon with. Faced by such a challenge, our differences of race and religion, class or colour, nation and ideology, become irrelevant. We have to devise a realistic way by which humanity does not bring about its own destruction.

Phosphene gas was used in the First World War first by the Germans on February 25–26, 1918. American troops had retaliated with the same gas in June 1918 near Bosis De Mort Mare in France. Phosphene is a colourless, very poisonous gas and has a characteristic strong odour which can be described as similar to that of freshly mown hay. It is one of the most dangerous gases used in industry. Brief inhalation of air containing only five parts per million (PPM) is fatal. Chemical warfare during the Vietnam war is still fresh in our memory, as are hazardous chemical leaks from nuclear plants positioned in different parts of the world. Bhopal, the capital city of the state of Madhya Pradesh, on the morning of Monday, December 3, 1984, virtually turned into a gas chamber, when deadly methyl isocyanate (MIC) poisonous gas leaked from the pesticide factory of a multinational company killing more than 500 people, mostly children, and affecting about 25,000 others. The tragedy described by experts as the worst environmental disaster in India struck around 1:00 AM when the entire town was asleep. The gas quickly spread to an area of 40 sq. km.

The victims woke up coughing with complaints of giddiness, burning of eyes, and obstruction of throat. Many could hardly talk and several complained of temporary blindness. The Bhopal gas tragedy was more serious than what happened in Italy, in 1983 when gas leakage from a chemical plant turned the town into a gas chamber. The MIC leakage in Bhopal occurred when the valve of the storage tank broke when pressure had built up inside. International safety management and global regulations were not in place, resulting in the disaster, the side effects of which are still being felt with babies being born with various birth defects and deformities. Genetic deformities that have damaged the human immune system, tumor growth and many other disturbances have already been passed from generation to another.

The population and ecosystem of the United States according to Prof. Dikshith are subject to possible exposure to more than 75,000 synthetic chemicals, most of which are poorly tested or untested for adverse health effects. Of this large list more than 70,000 such substances are in commercial use, and only a small fraction is known to be tested and evaluated. Disasters resulting from chemical explosions, contaminant seepage into soil, run off into rivers and sources of drinking water have assumed global proportions and they are bound to cause unimaginable disaster if serious attention is not given. Agriculture, forestry, industry and even the entire human environment is in great danger.

Chemical weapons of defence have now become weapons of universal destruction. We must have a clear understanding of the facts surrounding chemical weapons and their implications. We must work for the abolition of chemical weapons that threaten death to millions and diseases to more. Chemical warfare is a catastrophe in which everyone loses. Ending the threat of war would be a victory for mankind.

The world today is a madhouse where individuals exaggerate their racial superiority, religious pride, or national egotism and thus become victims of moral and spiritual blindness. We have to view the world as a whole, a single community, a fellowship of human beings who have the same instincts of hunger and sex, the same aspiration of generosity and fellow feeling, the same faith in the unseen. We have to march towards this goal of a world community in spite of blind alleys and setbacks.

Human folly in politics has no limit. Some years ago, T. S. Eliot envisioned the future of our civilization as, "internecine fighting ... people killing one another in the streets". Civilization seemed to him a crumbling edifice destined to fall to pieces and perish in the flames of war. This despair of life cannot be the end of things. The tragedy of the human condition imposes on us the opportunity and the obligation to give meaning and significance to life, to develop human dignity, work for the future, and believe in the young. This is the cause we have to serve and the cause which must win if humanity is worth its name and if it is to survive.

Dr. Dikshith, who has published over 70 research papers and has to his credit half a dozen books all published by leading world publishers and has participated in a number of national and international conferences, seminars and workshops, has in a unique way brought to bear the timeless wisdom of Indian sages and saints in his writings. We are living in the age of information explosion. Ancient Indian sages and saints constantly cautioned their students and society at large that information should not be mistaken for "knowledge." Information must be transformed into knowledge by due process of filtration. Knowledge in turn should be distilled and matured into wisdom. It is this wisdom that issues forth as action. A lasting society can be built upon the good of the people without any discrimination, based on wisdom and not on the basis of information alone.

Most of us in today's world think we are rational, have a scientific bent of mind, and are all scientists because we press a button to access the internet, then press another button to prepare food in a microwave oven in minutes, and then press a third button by which we can control the lighting in our house even though we are nowhere near the house. But we do not know how these things function. What knowledge is enshrined in these instruments or gadgets? We know only how to press a button and get something done. That is, we live on the surface of human life. We are at best mechanics or technicians. We have no idea of the depth that these things indicate and stand for. If we knew that, we would never make the mistake to say that science is a different discipline and not a human discipline, and that the other things, history, literature, philosophy, etc., make a different kind of man. We must study science not from the surface but from its very depth. The very transformation that science has brought about should make one question the meaning of existence. That is part of the quest for knowledge. One should want to know why knowledge is what it is, what is the knowledge that science contributes, and what its relation is to the knowledge that other branches of discipline give us.

Ancient Indian sages more than 10,000 years ago were very clear in their minds when they stated all the branches of discipline have only one end. *Sarva Shashtra proyajanam Atma darshanam*, tells us that reality is the end of all kinds of discipline. We must recognize that knowledge cannot be divided against itself. All truth must be regarded as a whole. Whether one practices a discipline or another, one should practice the one supreme pursuit of truth. This makes the human being civilized in the true sense of the term. The need of the hour is to be able to do it, so that one becomes a truly civilized human being. In the ultimate analysis, a society maintains its cohesion among the huge majority of its members by holding in common certain guiding principles. Civilization is not built with brick and mortar, steel and machinery, but with men and women, clarity of mind, charity of heart, and spirit of cooperation.

This excellent book deserves careful study so that one can make the right choices at theright time and not just get carried away by so called "industrial development". Having been myself a postgraduate student of chemistry at the Indian Institute of Technology Kanpur and later during my career when I had to deal with the infamous unfortunate Bhopal gas tragedy, I had an opportunity to meet and interact with some of the top-most technical people across the globe, scientists of international repute and social welfare workers with proven records, and I can with reasonable certainty say that this meticulously researched book in a very valuable publication that speaks volumes of the author's dedication and scholarship in the field of chemicals and toxicology.

George Hellman, in his biography of Benjamin Cardozo, recalls a passage from T. E. Shaw's translation of Homer's *Odyssey* to describe Cardozo: "God will so crown what he says with a plume of beauty that all who look on Him are moved. When He holds forth in public, it is with assurance, yet with so honey-sweet a modesty that makes him shine out above the ruck of men who gaze at him whenever He walks their city". This aptly applies to Prof. T. S. S. Dikshith.

A. Ramaswamy
Indian Administrative Service
Bangalore, India

Preface

Let noble thoughts come to us from every side

Rig Veda (I-89-I)

Asato ma sadgamaya: meaning, lead me from untruth to the truth;
Tamaso ma jyotirgamaya: meaning, lead me from darkness to light;
Mrtyorma amrtam gamaya: meaning, lead me from death to immortality;
Om shanthi shanthi shanthihi: meaning, Let there be peace, peace, peace.

Brhadaranyaka Upanishad – I.iii.28

Anityaani shareerani: meaning, human life is very temporary;
Vibhavo niva shaswataha: meaning, luxuries are never permanent;
Mrutyuhu sannihito nityam: meaning, death approaches fast and is certain;
Kartavyo dharma sangrahaha: meaning, purpose of human life/
society is to be virtuous.

Science of Ethics

Nahe gyanena sadrusham: There is nothing equal to Knowledge/
Knowledge has NO Equivalence…

Bhagavadgita

We have received the most valuable and Priceless Asset – the Human body. This is like a strong boat to cross across the Ocean of Ignorance. It is propelled by the favorable wind – the Grace of Almighty and steered by the Spiritual Master and Knowledge. With all this rich facility and the priceless asset, and as humans if we do not cross this Ocean of Ignorance, then this Golden opportunity is lost.

Value of Human Life

Science without religion is lame; Religion without science is blind

Albert Einstein

Chemical substances are present everywhere in one form or another. Many of them are used to improve the quality of life. The management of different kinds of chemical substances used in industries and in daily life has been categorised in brief as hazardous, toxic, and dangerous from the viewpoint of health and safety. Most chemical substances are not harmful to the environment or to human health. However, some of them have the potential to produce adverse health effects in humans and animals, while some others cause damage to the living environment. With the production of huge amounts of different chemical substances worldwide, their proper and safe management has become essential for all countries. It is imperative for them to improve the degree of protection against hazardous chemicals and the handling of chemical substances as well as to safeguard against the implications of improper use and negligence during use. In fact, a number of new and proactive measures are essential to ensure the proper management of chemical substances. Scientific data and societal norms influence and set acceptable standards for the use and handling of chemical substances to improve the quality of life.

This book describes a large number of chemical substances of different classes and kinds produced and handled by equally large groups of people around the globe. Improper use and negligence in their handling would cause great concern to the general public and to regulatory agencies in different countries of the world. This is of special significance in the context of occupational safety and health. The nature and properties of different chemical substances and their carcinogenicity, mutagenicity, and toxicity to the reproductive system, as well as their persistence, bioaccumulation in food and body tissues of humans and animals, and corrosive properties – as highly toxic fumigants, gases, mists, and vapours and as endocrine disrupters – have all demanded the availability of proper and quality data, especially with regard to their toxicity and to the safety of humans, animals, and the living environment.

Studies have indicated that several incidents and chemical disasters associated with hazardous chemical substances have led to human exposures and have posed an important public health and regulatory challenge for different countries around the world. These chemical disasters and workplace incidents range from chemical release, leakage and spillage of hazardous chemicals to the release of contaminated chemical products in the living environment.

Preventing exposure to toxic chemicals is of primary concern at hazardous waste sites. Most sites contain a variety of chemical substances in gaseous, liquid, or solid form. They can enter the unprotected body by inhalation, skin absorption, ingestion, or through a puncture wound (injection). A contaminant can cause damage at the point of contact or can act systemically, causing a toxic effect at a part of the body distant from the point of initial contact. Chemical exposures are generally divided into two categories: acute and chronic. Symptoms resulting from acute exposures usually occur during or shortly after exposure to a sufficiently high concentration of a contaminant. The concentration required to produce such effects varies widely from chemical to chemical. The term 'chronic exposure' generally refers to exposures to 'low' concentrations of a contaminant over a long period of time.

The 'low' concentrations required to produce symptoms of chronic exposure depend upon the chemical, the duration of each exposure, and the number of exposures. For a given contaminant, the symptoms of an acute exposure may be completely different from one resulting from chronic exposure. For either chronic or acute exposure, the toxic effect may be temporary and reversible, or may be permanent (disability or death). Some chemicals may cause obvious symptoms such as burning, coughing, nausea, tearing eyes, or rashes. Other chemicals may cause health damage without any such warning signs (this is of particular concern for chronic exposures to low concentrations). Severe adverse health effects such as cancer or respiratory disease may not become manifested and identified in workers until after several years or after a prolonged period of exposure. Also, several hazardous chemical substances may not produce any type of immediate or obvious physiological sensations, and the sensations and feelings of exposed workers cannot be relied upon in all cases to warn of potential toxic exposure. The effects of exposure not only depend on the chemical, its concentration, route of entry, and duration of exposure but may also be influenced by other factors such as personal habits, for example, smoking, chewing of tobacco, alcohol consumption, overuse of and addiction to drugs, irregular eating habits of workers/individuals as well as physiological factors such as age, sex, and many other modulatory conditions.

Inhalation is another important manner of exposure to hazardous chemical substances. The pulmonary system, more specifically the lungs, is extremely vulnerable to different types of adverse actions and interactions of hazardous chemical substances. Even chemical substances that may not directly affect the lungs may pass through the lung tissue and get absorbed into the blood vascular system, from where they are transported to other vulnerable areas of the body.

Some toxic chemical substances that are present in the atmosphere may not be detected by human senses, that is, they may be colourless, odourless, and their toxic effects may not produce any immediate symptoms. Respiratory protection is therefore extremely important if there is a possibility that the work-site atmosphere may contain such hazardous substances. Chemicals can also enter the respiratory tract through punctured eardrums. Where this is a hazard, individuals with punctured eardrums should be medically evaluated specifically to determine if such a condition would place them at unacceptable risk and preclude their working at the task in question. Direct contact of the skin and eyes to hazardous substances is another important route of exposure. Some chemicals directly affect the skin. Some pass through the skin into the bloodstream, where they are transported into vulnerable organs. Skin absorption is enhanced by abrasions, cuts, heat, and moisture. The eye, in particular, is vulnerable because airborne chemicals can dissolve in its moist surface and be carried to the rest of the body through the bloodstream (capillaries are very close to the surface of the eye). Wearing protective equipment, not using contact lenses in contaminated atmospheres (since they may trap chemicals against the eye surface), keeping hands away from the face, and minimising contact with liquid and solid chemicals can help avoid skin and eye contact. Although ingestion should be the least significant route of exposure at a site, it is important to be aware of how this type of exposure can occur. Deliberate ingestion of chemicals is unlikely; however, personal habits such as chewing gum or tobacco, alcohol consumption, eating, smoking, and applying cosmetics on site may provide a route for the entry of chemicals. The last primary route of chemical exposure is injection, whereby chemicals are introduced into the body through puncture wounds (e.g., by stepping or tripping and falling onto sharp, contaminated objects). Safety management of hazardous chemical substances is an important aspect and involves the total commitment of occupational workers, workplace managers, and the organisation as a whole. In short, a safety management system involves myriad safety challenges and complexities for workers, workers' training, workplace environment, and management skills to maintain and protect work records and many other documents.

Regular training of occupational workers and regular supervision of their habits at the workplace and of the work culture, specifically during handling of hazardous chemical substances, provide a meaningful solution to contain accidents and hazards. For example, use of protective equipment, such as face masks, safety goggles, safety gloves, and shoes and use of a fume hood to avoid chemical/physical hazards, as well as taking common sense precautions without being negligent, are some important measures that each occupational worker, skilled or semiskilled, should adhere to in order to achieve health safety and to contain possible accidents and disasters associated with hazardous chemical substances.

There are many potential causes of explosions and fires at hazardous waste sites, as follows:

- Explosion, fire, or heat due to chemical reactions
- Ignition of explosive or flammable chemicals
- Ignition of materials due to oxygen enrichment
- Agitation of shock- or friction-sensitive compounds
- Sudden release of materials under pressure

Explosions and fires may arise spontaneously. However, more commonly, they result from site activities, such as moving drums, accidentally mixing incompatible chemicals, or introducing an ignition source (such as a spark) into an explosive or flammable environment. At hazardous waste sites, explosions and fires not only pose the obvious hazards of intense heat, open flame, smoke inhalation, and flying objects but may also cause the release of toxic chemicals into the environment. Such releases can threaten both personnel on site and members of the general public living or working nearby. To protect against the hazard, have qualified personnel field monitor for explosive atmospheres and flammable vapours; keep all potential ignition sources away from an explosive or flammable environment; use non-sparking, explosion-proof equipment; and follow safe practices when performing any task that might result in the agitation or release of chemical substances. This book attempts to discuss, in brief, the elements of toxicology with regard to the effects of chemical substances and human use.

Chemical substances are required for health, progress, and societal development. In the very close linkage with an array of chemical substances and societal development, human health cannot be ignored. Therefore, scientists worldwide have framed regulations about the manners and methods of use of chemical substances. There are no safe chemical substances; all are toxic in some way or another. In fact, the safety of a chemical substance depends on the concentration and manner of exposure and use. No chemical substance is absolutely safe. This is important and should be heeded by students, occupational workers, and household users who handle, store, transport, and dispose of different chemical substances. Improper use and negligence in the management of chemical substances cause injury, death, or disaster.

Therefore, chemical substances as and when they are marketed for human use in the form of drugs, food additives, cosmetics, and many other forms require safety data and detailed evaluation. To generate quality data about candidate chemical substances, different countries and international regulatory agencies have framed elaborate procedures. By understanding the basics in toxicology, adhering correctly to the regulations, and observing precautions, the benefits of chemicals would enrich human society and free it from hunger and disease.

It has become mandatory for industries that market chemical substances to apply for an authorisation for use of every chemical. The regulatory agencies of different countries worldwide consider and grant an authorisation if the risk to human health or the environment from the use of the substance is adequately controlled. Further, even if the health and safety risks cannot be adequately controlled, a time-limited authorisation could also be granted if it was shown that the socioeconomic benefits outweigh the risks and if there were no suitable alternative substances or technologies. In such a case, it is recommended that the applicant submit a plan to develop alternative substances or technologies. Downstream users could use a substance provided that they have obtained the substance from a company to which an authorisation has been granted and that they meet the conditions of that authorisation. Such downstream users would also be required to notify the chemical control agency that they are using an authorised substance.

This book offers easy and quick access to information on safety management of chemical substances with appropriate recent scientific information for students and various types of occupational workers. Adequate and appropriate safety precautions must be followed by occupational workers and by the general public to protect themselves from the health risks and hazards of chemical substances. The aim of this book is to offer additional knowledge to students, occupational workers, and management as a whole about the importance of complying with laws and regulations set by regulatory bodies. The book also offers guidance to occupational workers about the importance of proper management of chemical substances and to avoid possible chemical injuries and hazards when handling them. The organisation and workplace management should establish and implement written procedures to maintain safety details and precautions required during handling of hazardous chemical substances. This information helps workers prevent or minimise the consequences of catastrophic releases of toxic, reactive, flammable, or explosive chemical substances which may result in toxic or explosive hazards. The information also provides guidance to occupational workers during transportation of chemical substances by different routes such as by road, rail, air, or sea.

Chemical substances have been used extensively over the years around the world and more so in recent years with newer compounds. To prevent the possible hazards of chemical substances, an effective safety information management system needs to be made available to students and occupational workers. Also a hazard communication tool should be maintained among management, workplace managers, users, and occupational workers. This book provides students and skilled or semi-skilled workers an understandable information tool to become aware of, to be alert, and to avoid negligence during use and handling of hazardous chemical substances. This book hopefully covers these basic elements that are closely linked with the safety to students and workers in a comprehensive manner. The application of this knowledge would be useful for workers to improve management of hazardous chemical substances. This information will also be useful for the general public to recognise the risks to the community posed by the use, transport, storage, and disposal of specific chemical substances.

Information regarding the safety management of hazardous chemical substances is very essential to all personnel and occupational workers associated with the purchase, storage, transport, use, and disposal of chemical wastes and materials.

Several kinds of chemical substances are in use in abundance in different countries worldwide, and their varieties and number are ever growing. Many chemical substances that are in use have not been fully assessed for their toxicity, safety, and their hazardous impacts on human health and the living environment. Chemical industries worldwide should take responsibility to declare in simple and unequivocal terms the safety data and quality of each candidate chemical substance that enters the global market. The quality data should indicate whether the chemical substance in question is dangerous or not before it is used by the general public and qualified or semi-trained occupational workers with necessary safety precautions set by global regulatory agencies.

Occupational workers newly engaged in teams/employees in chemical laboratories and the chemical industry should be made aware of their responsibilities and should have basic knowledge about proper handling of chemical substances. Workers should follow standard methods of use, storage, and disposal and should immediately report to the workplace supervisors/management of any kind of uneventful chemical reaction that may occur during handling of the chemical substance at the workplace. The responsibility for proper and judicious management of chemical substances at any workplace, be it a chemical laboratory, industry, or organisation, to achieve health and safety includes, but is not limited to, the following:

- Students/skilled and semi-skilled occupational workers
- Industrial and workplace managers
- Laboratory workers and supervisors

- Laboratory co-ordinators
- Students in research laboratory
- Safety officers and health unit
- Workers associated with disposal of chemical wastes

It goes without saying that the toxicological profile of chemical substances becomes much more complicated because of the presence of a mixture of contaminants, that is aggregated chemical substances, that often leads to risks. Occupational workers must be well aware of the consequences of the synergist effects of each chemical substance and many contaminants in a workplace, which is yet to be studied properly and understood. Also, in spite of the efforts to present an up-to-date and comprehensive compilation, many gaps still remain in this book. This book has attempted to describe only a very small section/area of the huge wealth of global knowledge. Readers should note that the descriptions of many hazardous chemical substances may appear as repetitions or as a revision of my earlier books. *Users, occupational workers, and readers of this book should always remember that repetitions of this subject and the text material suggest, both directly and indirectly, the importance of the candidate hazardous chemical substance, alone and in different combinations/formulations, and the care and precaution that must strictly be observed by each user, occupational worker – skilled and or semi-skilled – to maintain safety and protection during handling, with no negligence in use.* The very purpose of repetitions of text regarding use, handling of chemical substances, and the strict precautions to be observed is to make a deeper impression on students, semi-skilled occupational workers, and the general public about the judicious use and proper management of hazardous chemical substances. The book hopes to provide an integrated, yet simple, description of several hazardous chemical substances used, handled, stored, and transported by householders, occupational workers, and workplace management. For more information, readers can refer to the following books:

1. *The Merck Index: An Encyclopedia of Chemicals, Drugs, and Biologicals.* Merck & Co., Inc. Rahway, NJ.
2. R.P. Pohanish. 2011. *Sittig's Handbook of Toxic and Hazardous Chemicals and Carcinogens,* 6th edn. CRC Press, Boca Raton, FL.
3. C.B. Strong and T.R. Irvin. 1996. *Emergency Response and Hazardous Chemical Management: Principles and Practices.* CRC Press, Boca Raton, FL.
4. R.J. Lewis, Sr. 2001. *Hawley's Condensed Chemical Dictionary,* 14th edn. Wiley-Interscience, New York.
5. NRC. *Prudent Practices in the Laboratory: Handling and Management of Chemical Hazards.* National Research Council, National Academy Press, Washington, DC.

It has become very common and well documented in recent years that societal life is regularly exposed to varieties of chemical substances that include, but are not limited to, industrial solvents, metals and metal compounds, pesticides, dusts, fumigants and fumes, vapours, toxic gases, corrosive chemical substances, asphyxiates, irritants, and oxidising agents. A number of chemical substances, depending on their concentration, have the capacity and potency to cause adverse health effects that are comparable to those of neurotoxicants, nephrotoxicants, carcinogens, mutagens, and teratogens. Different chemical substances that are broadly included are industrial solvents, metals and metal compounds, pesticides, gases, dusts, fumes, fibres, gas, mists, asphyxiants, corrosives, irritants, and flammables.

There is an urgent need for students, occupational workers, householders, and the general public to be aware of the true potential of each hazardous chemical substance and the immediate and delayed implications thereof, if any, as well as of the consequences of negligent and improper use and handling and management. This knowledge is essential for all of us and especially for occupational workers to maintain health safety and to protect the living environment.

Societal life requires and demands that all chemical substances be *safe*. This requirement is exercised through legislation concerned with the safety of chemicals, usually introduced post hoc, in response either to workplace chemical accidents or global chemical disasters or to perceived inadequacies in the proper implementation of earlier legislation. The most important purpose of safety evaluation of chemical substances is to understand and identify the possible toxicity, hazards, benefits, and overall safety and to have the correct knowledge of the management of the candidate chemical substance and of their combinations. In short, every occupational worker and workplace management team member who uses different kinds of hazardous chemical substances at work must have access to basic information about chemical safety and sufficient information to take quick, preventive safety measures. This handbook has been developed to help occupational workers/ personnel to strictly follow the globally accepted acts and regulations step by step during use and handling of hazardous chemical substances at their workplace. Workers should be educated and trained regarding the use and handling of hazardous chemical substances. At the cost of sounding repetitive, while there is a wealth of chemical substances, I have attempted here *to describe and discuss only selected chemical substances and their safety management*. I have generously drawn information from a large number of published materials – for instance, from several material safety data sheets (MSDSs) of common industrial and household products; the hazardous materials table from the United States 'Code of Federal Regulations' (title 49 section 172.101); the National Institute for Occupational Safety and Health Pocket Guide to Chemical Hazards; the U.S. DOT 1996, 2000 and 2004 Emergency Response Guidebooks; the U.S. National Library of Medicine; as well as my own earlier publications and other related resources.

This book hopefully promotes the knowledge of procedures and methods that are essential for proper and judicious management of hazardous chemical substances. It is important to mention here that I have purposely not included in-depth information and/or technical data of different hazardous chemical substances. Readers and specialists who are interested in more information should refer to the literature. In the compilation of this book, I have drawn relevant information and the recommended guidelines from different regulatory agencies of different countries, for instance, the U.S. Environmental Protection Agency (U.S. EPA), the National Institutes of Health (NIH), and the Occupational Safety and Health Administration (OSHA). With specific and common chemical substances as examples, the book describes appropriate methods of use, handling, storage, chemical waste disposal, workplace equipment, and precautions, all of which have to be strictly observed by each user and occupational worker.

Turuvekere Subrahmanya Shanmukha Dikshith
Bangalore, India

Acknowledgements

This book contains information drawn from a large number of published scientific literature/materials as well as from my own previous publications. The compilation of this book could not have been possible without the generous permission granted by different agencies, publishers, and international organizations. I gratefully acknowledge the copyright permission granted by the U.S. Environmental Protection Agency (U.S. EPA), Agency for Toxic Substances and Disease Registry (ATSDR), National Institute for Occupational Safety and Health (NIOSH), International Registry on Potentially Toxic Chemicals (IRPTC), International Program on Chemical Safety (IPCS), World Health Organization (WHO), Occupational Safety and Health Administration (OSHA), National Library of Medicine (NLM), Hazardous Substances Data Bank (HSDB), Centers for Disease Control and Prevention (CDCP), Central Insecticides Board (CIB), Ministry of Agriculture and Cooperation (Government of India), and many others that allowed me to cite, refer to, and use their published scientific information.

I would like to express my sincere thanks to administrative authorities of the United States Food and Drug Administration (Joan G. Lytle, Jill Smith); Agency for Toxic Substances and Disease Registry (ATSDR), Division of Toxicology, Atlanta, Georgia; and the National Center for Environmental Health, Atlanta, Georgia; as well as to Dr. Chris Blackwell, chemistry area safety officer, Department of Chemistry, University of Oxford, Oxford, United Kingdom; Dr. Hugh Cartwright, Department of Chemistry, University of Oxford, Oxford, England; Carl J. Foreman, director, EH&S, Davis, California; International Labor Office, Geneva; NIOSH, Cincinnati, Ohio; California Department of Health Services; U.S. Geological Survey; and the Canadian Centre for Occupational Health and Safety (CCOHS).

It is my pleasure to acknowledge the many scientists, authorities, and individuals who provided support and copyright permission to cite, refer to, quote, and use the already published scientific information of a large number of chemical substances listed in this handbook. If the contribution of any particular individual has not been appreciated, it is due more to an inadvertent miss than to any other reason.

I would like to express my deep gratitude to Narasimha Kramadhati and Pratibha Narasimha as well as to Dr. Deepak Murthy and Prerana Murthy for their encouragement and cooperation in the work and for sharing their thoughts about the subjects covered in this book. It is my pleasure to acknowledge Dr. Steven G. Gilbert of the Institute of Neurotoxicology and Neurological Disorders, Seattle, Washington, for granting permission to cite his published work in the book. My thanks also to several of my friends for providing technical support in the completion of this work. Finally, it has been a pleasure to work with the staff of CRC Press/Taylor & Francis, namely, Mindy Rosenkrantz, permissions coordinator, who granted me copyright permission to cite selected literature from my earlier book, *Safe Use of Chemicals: A Practical Guide*. Also, thanks to the following people for their cooperation and coordinating the publication of this book: Kari Budyk, senior project coordinator, editorial project development; Cindy Renee Carelli, senior acquisitions editor; Michele Smith, editorial assistant; Richard Tressider, project editor; and Alfred Samson, project manager at SPi Global.

Author

Turuvekere Subrahmanya Shanmukha Dikshith, PhD, was responsible for the establishment of the Pesticide Toxicology Laboratory at Industrial Toxicology Research Centre, Council of Scientific and Industrial Research (CSIR), Government of India, Lucknow, India. He was director of the Toxicology Research Laboratory, VIMTA Labs Ltd., Hyderabad, India. He also worked as a consultant, Pharmacology Division, Indian Institute of Chemical Technology (IICT), Hyderabad, India. Dr. Dikshith served on the committees and expert panels constituted by different ministries of the Government of India. As an expert committee member, he also served the Central Insecticides Board (CIB), Directorate of Plant Protection, Quarantine and Storage, Ministry of Agriculture, Government of India, India. He is a member of the World Health Organization Task Group on Environmental Health Criteria Documents and the International Program on Chemical Safety. Dr. Dikshith is the technical specialist for Standards Australia Quality Assurance Services (SAQAS), Australia, and Lloyd Register Quality Assurance Ltd. (LRQA), London, for the quality management of the laboratory and good laboratory practice. He is also the recipient of the Chandra Kanta Dandiya prize in pharmacology.

As a fellow of the World Health Organization, Dr. Dikshith has worked at the Institute of Comparative and Human Toxicology, Albany Medical College, Albany, New York, and also at the International Center of Environmental Safety, Holloman, New Mexico. He has worked in several laboratories in France, Germany, and Canada. Dr. Dikshith has edited *Toxicology of Pesticides in Animals* published by CRC Press, Boca Raton, Florida, and has authored a book chapter published by Plenum Press, New York. He has also authored several other books, including *Safety Evaluation of Environmental Chemicals* published by New Age International Publishers, India; *Industrial Guide to Chemical and Drug Safety* published by John Wiley & Sons, Inc., Hoboken, New Jersey; and *Safe Use of Chemicals: A Practical Guide* and *Handbook of Chemicals and Safety* published by CRC Press, Boca Raton, Florida.

1

Introduction

Chemical substances are very essential and important to human society. They are available in the form of solids, liquids, and gases, and their judicious use, disposal, and overall management have offered health, wealth, and prosperity to human society. Avoiding the risks associated with the use of hazardous chemical substances requires elementary knowledge of different chemicals, their proper application/use, handling, and methods of safety precautions.

Modern society in general and the lives of humans in particular have been greatly improved by the development of a variety of chemical substances such as pharmaceuticals, building materials, housewares, pesticides, and industrial chemicals. Proper and judicious use of chemical substances helped us to increase food production, cure diseases, build more efficient houses, travel globally and make scientific discoveries. To gain these benefits, great care, caution, and total absence of negligence in the management of chemical substances are very essential. Improper use of hazardous chemical substances causes danger to the living environment and human safety.

Just as all humans are exposed to radiation in their daily routine, they are also exposed to different chemical substances. Some potentially hazardous chemical substances exist in the natural environment. In many areas of the country, soils contain naturally elevated concentrations of metals such as selenium, arsenic, or molybdenum, which now have been proven hazardous to humans or animals. Even some of the food items that we have consumed for decades contain natural toxins such as aflatoxin – a known toxin found in peanuts – and cyanide in apple seeds. Also exposures to many more hazardous chemical substances result from the direct or indirect activities of humans. Building materials used for the construction of homes and other structures do contain hazardous materials, for example, formaldehyde in some insulation materials, asbestos in insulation and ceiling tiles, and lead in paints and gasoline. Besides these kinds of exposures, many more chemical substances are present in our living environment as a result of industrialisation, application of different kinds of pest control products and pesticides (crop pest control, agriculture), control of vector-borne diseases, and drugs and health products. Unexpected releases of toxic, reactive, or flammable liquids and gases in processes involving highly hazardous chemical substances have been reported for many years in various industries around the world. Regardless of the industry that uses these highly hazardous chemicals, there is a potential for an accidental release any time they are not properly controlled, creating the possibility of disaster.

Now studies have shown that persistent chemical substances have been transported from long distances through the atmosphere from industrial sources and deposited on soil, air, or water.

Chemical substances released in the air remain suspended for long periods of time, or they may be rapidly deposited on plants, soil, and water and released as liquid wastes called effluents, which subsequently enter streams and rivers. Accumulation of scientific data and information about the properties and management of different chemical substances has increased over the decades and around the globe. It is important that students, occupational workers, individuals, and the general public be aware of the benefits of proper management of chemical substances and the possible dangers of improper use of different chemical substances. The majority of students, semi-skilled workers, householders, and many others, require essential information about chemicals and safety than those individuals who are qualified, trained, and well aware of the basic knowledge of chemical substances and safety management. The workplace chemical accidents and disasters that have occurred, and have been occurring repeatedly, in different countries clearly confirm the paucity of appropriate scientific information.

Various regulatory agencies and governments around the world have set rules that are to be observed – mandatory – during use, handling, and the overall management of hazardous chemical substances. The laws and regulations on the use and handling of hazardous materials differ depending on the activity and status of the chemical substance/material. For instance, the requirements during the use of chemical substances in the workplace are different in comparison during the response and management of chemical wastes, spills, and fire hazards due to flammable and corrosive chemical substances.

The very purpose of this book is to provide ready information on hazardous chemical substances and methods of safe use and handling to workers. The cornerstone of chemical safety is the availability of ready, understandable, and quick information to occupational workers at the workplace. The workplace hazardous materials data system or the material safety data sheets (MSDSs) should contain information of labelling, toxicological profile and possible and associated health hazards, and the important steps of precautions to be observed by the user. The book also provides readily available information on personal protective equipment (PPE) during handling chemicals.

To contain such chemical disasters, it is important that a scientific, clear, non-technical, concise, pragmatic, yet comprehensive guide should be readily available at all workplaces.

The author hopes that this book provides guidance to all categories of occupational workers and students and helps to appropriately comply with local, state, and global laws and regulations regarding hazardous chemical substances and management.

Bibliography

Burgess, W. A. 1995. *Recognition of Health Hazards in Industry: A Review of Materials and Processes*, 2nd edn. John Wiley & Sons, Inc., New York, p. 338.

Chan, P. K., G. P. O'Hara, and A. W. Hayes. 1982. Principles and methods for acute and subchronic toxicity. In: *Principles and Methods of Toxicology*. Raven Press, New York.

Daugherty, J. (ed.). 1997. *Assessment of Chemical Exposures: Calculation Methods for Environmental Professionals*. CRC Press, Boca Raton, FL.

Dikshith, T. S. S. (ed.). 2011. *Handbook of Chemicals and Safety*. CRC Press, Boca Raton, FL.

Pradyot, P. (ed.). 2007. *A Comprehensive Guide to the Hazardous Properties of Chemical Substances*. John Wiley & Sons, Inc., Hoboken, NJ.

Scott, R. M. (ed.). 1989. *Chemical Hazards in the Workplace*. CRC Press, Boca Raton, FL.

Sittig, M. (ed.). 1985. *Handbook of Toxic and Hazardous Chemicals and Carcinogens*, 2nd edn. Noyes Publications, Park Ridge, NJ.

Sullivan, J. B. and G. R. Krieger. 1992. *Hazardous Materials Toxicology: Clinical Principles of Environmental Health*. William & Wilkins, Baltimore, MD, p. 418.

2

Hazardous Chemical Substances: Characterisation

Introduction

Chemical substances are everywhere around us, in the environment; in our food, drinks, and clothes; and in the body. Many of these chemical substances are used to improve the quality of our lives. Most of these chemical substances are not harmful to the environment or human health. However, some have the potential to cause harm, in certain doses, and should only be used when the risks are appropriately managed. Chemical substances' classes and categories are large and numerous. A vast number of different chemical substances are in commercial use over these years, and their numbers are increasing. Many of these chemical substances are considered toxic or otherwise hazardous to animals, humans, and other living beings. Toxic chemicals are associated with a variety of serious health problems, including cancer, brain and nervous system disorders, reproductive disorders, organ damage, as well as asthma. Toxic chemicals that are persistent in the environment and bioaccumulate through the food chain can make exposure during childhood dangerous. Chemicals also can irritate the skin, eyes, nose, and throat. Some chemicals pose significant safety hazards, such as fire or explosion risks.

Safety management of hazardous chemical substances includes several components as follows:

- Regulations and implementation/enforcement
- Rules and regulations for proper/judicious use, storage, transport, and waste disposal
- Proper identification, labels, and precautions during use and handling
- Quality control of chemical substances, health implications, hazards, and research
- Workplace management, regular inspection/audit, and implementation of safety regulations
- Identification through categorisation via successive rounds of assessment regulatory action and data generation on the uses and health effects of chemical substances

Different chemical substances have been grouped/classified based on the type of hazard they pose. Understanding the different types of chemicals in a school is important for developing an effective chemical management policy. Hazardous substances in schools may fall into one or more of the following categories: flammables/explosives, corrosives (acids and bases), oxidisers/reactive, toxics, and compressed gases. Hazardous materials (HMs) are products that pose a risk to health, safety, and property during transportation. The term often is shortened to hazmat in government regulations. HMs include explosives; various types of gas, solids, and flammable and combustible liquid; and other materials. Because of the risks involved and the potential consequences these risks impose, all levels of government regulate the handling of HMs.

It is very important to know and remember that the regulations are intended to protect the user, the workplace, and the living environment around us all.

Federal agencies and legislative authorities have developed specific definitions for each of those categories; however, the hazards can be described generally as follows:

Carcinogens: Chemical substances/agents that cause cancer. Cancer is caused by changes in a cell's DNA – its genetic 'blueprint'. Some of these changes may be inherited from the parents, while others may be caused by different modulating factors/environmental factors, lifestyle factors, nutrition, use of tobacco, exposures to ultraviolet light, radon gas, infectious agents, chemotherapy, radiation, suppression of the immune system due to drugs, etc.

The International Agency for Research on Cancer (IARC) is part of the World Health Organization (WHO). Its major goal is to identify causes of cancer. Carcinogens are agents that can cause cancer. In industry, there are many potential exposures to carcinogens. Generally, workplace exposures are considered to be at higher levels than public exposures. Material safety data sheets (MSDSs) should always contain an indication of carcinogenic potential. The most widely used system for classifying carcinogens comes from the IARC that over past 30 years, evaluated the cancer-causing potential of a large number of chemical substances and placed them into one or the other of the following groups:

- Group 1: Carcinogenic to humans
- Group 2A: Probably carcinogenic to humans
- Group 2B: Possibly carcinogenic to humans
- Group 3: Unclassifiable as to carcinogenicity in humans
- Group 4: Probably not carcinogenic to humans

Perhaps not surprisingly, based on how hard it can be to test these candidate carcinogens, most are listed as being of probable, possible, or unknown risk. Only a little over 100 are classified as 'carcinogenic to humans'. Listed are the selected chemical substances that have been proven scientifically to cause cancer in humans The examples include, but are not limited to, acrylonitrile, arsenic, asbestos, benzene, beryllium, cadmium and cadmium compounds, chromium, cobalt, ethylene oxide, formaldehyde, nickel, radon, sulphur trioxide, thorium, and vinyl chloride.

Colours, Dyes and Pigments

Colours, dyes and pigments are chemical substances. These are solid, opaque particles and provide colour. Pigment molecules typically link together in crystalline structures. The origins of pigments and dyes have a long history and in the beginning came from common natural products such as minerals, berries, roots, and insects. For instance, the origin of India ink – *masi* – dates back to the early fourth century BC. The composition has been identified to contain pitch, tar, burnt bony material, and other substances. In fact, available historical documents and records indicate that India and other Asian countries used sharp-pointed needle *masi*, the coloured ink for writing documents in original languages, Sanskrit (Samskruta, *Prakrita*), and other Indo-Aryan languages.

In subsequent decades, the organic chemical industry gave new direction, and a host of synthetic chemical dyes and pigments has been created since then. Qualities such as blue, green, and yellow vary depending on the source and type of pigment. Pigments containing lead or arsenic have long been recognised as being dangerous. Both inorganic and organic pigments are used as colorants. Dry pigments are especially hazardous because they are easily inhaled and ingested. They are used in encaustic, paper marbleising and in the fabrication of paint products. The long-term health hazards of the different kinds of modern synthetic organic pigments have not been well studied. In common, the dye based inks are generally much stronger than pigment-based inks. The reason being that dye-based printer ink stick to papers stronger and use of a water-solution enables the colours on the prints better. Also the dye-based printer inks tend to produce prints with more vibrant and mixed colours, carefully placed and embedded on the print-outs. Dye-based prints use heat and pressure to evaporate the excess water from the printer ink, leaving the finished product with colours set in the papers. Also these are known to produce much more colour of a given density per unit of mass. Dyes are dissolved in the liquid phase and have a tendency to soak into paper, making the ink less efficient and potentially allowing the ink to bleed at the edges of an image. The possible dangerous effects of pigments containing cadmium, chromium, manganese, and mercury, and the colours include the cadmium reds, cadmium yellows, cadmium orange, viridian and chrome oxide opaque, manganese blue, manganese violet, burnt and raw umber, and vermilion/mercuric sulphide. Chromium hexavalent (CrVI) compounds, often called hexavalent chromium, exist in several forms. Industrial uses of hexavalent chromium compounds include chromate pigments in dyes, paints, inks, and plastics; chromates added as anticorrosive agents to paints, primers, and other surface coatings; and chromic acid electroplated onto metal parts to provide a decorative or protective coating. Hexavalent chromium is also formed during melting chromium, metal processing, fabrication, and welding on stainless steel. In these situations, the chromium is not originally hexavalent, but the high temperatures involved in the process result in oxidation that converts the chromium to a hexavalent state.

There are hundreds of organic pigments used in art materials. Most of the natural organic pigments are not particularly toxic. Only a small percentage of the synthetic pigments have been studied for toxicity or long-term hazards. The synthetic pigments are hazardous because they contain highly toxic impurities such as cancer-causing PCBs. Dry pigments are known to cause dangers because they carry pigment dust from the work area into the workplace, living environment, and food and water. Some pigments are related to the chemical 'benzidine', known to cause bladder cancer. Benzidine pigments and dyes may also cause this disease. Recent epidemiological studies of artist painters and industrial painters found elevated incidence of diseases, especially bladder cancer. Paints are pigments mixed with a vehicle or binder. There is a misconception that ink is non-toxic even if swallowed. Once ingested, ink can be hazardous to one's health. Certain inks, such as those used in printers and even those found in a common pen, can be harmful. Though ink does not easily cause death, inappropriate contact can cause effects such as severe headaches, skin irritation, or nervous system damage.

These effects can be caused by solvents or by the pigment ingredient *p*-anisidine, which helps create some inks' colour and shine.

The highly toxic pigments include, but are not limited to, antimony white (antimony trioxide); barium yellow (barium chromate); burnt or raw umber (iron oxides, manganese silicates or dioxide); cadmium red, orange, or yellow (cadmium sulphide, cadmium selenide); chrome green (Prussian blue, lead chromate); chrome orange (lead carbonate); chrome yellow (lead chromate); cobalt violet (cobalt arsenate or cobalt phosphate); cobalt yellow (potassium cobalt nitrate); lead or flake white (lead carbonate); lithol red, the reddish synthetic azo dye (sodium, barium, and calcium salts of azo pigments); manganese violet (manganese ammonium pyrophosphate); molybdate orange (lead chromate, lead molybdate, lead sulphate); Naples yellow (lead antimonate); strontium yellow (strontium chromate); vermilion (mercuric sulphide, zinc sulphide); and zinc yellow (zinc chromate). The moderately toxic pigments include, but are not limited to, alizarin crimson, carbon black, cerulean blue (cobalt stannate), cobalt blue (cobalt stannate), cobalt green (calcined cobalt, zinc and aluminium oxides), chromium oxide green (chromic oxide), phthalo blue and greens (copper phthalocyanine), manganese blue (barium manganate, barium sulphate), Prussian blue (ferric ferrocyanide), toluidine red and yellow (insoluble azo pigment), viridian (hydrated chromic oxide), and zinc white (zinc oxide).

Safe Handling and Precautions

During use and handling of different kinds of pigments and as a rule, workers should be careful, observe set regulations, and be aware of the fact that waste disposal of dyes and pigments has caused accidents and disasters to the streams and water bodies and posed hazards both to human health and ecological risks. These chemical substances are highly hazardous, toxic, and poisonous, examples being black colour antimony sulphide, logwood extract, and manganese dioxide. Therefore, users and occupational workers should always wear protective clothing, and the workplace must have proper ventilation. At workplaces, occupational workers must strictly wear personal protective equipment (PPE) such as a face shield for eye protection and acid-/solvent-resistant protective quality gloves to avoid contamination of colours, pigments, paints, varnishes, and binders; workers should not be negligent to wash their hands before they eat, drink, and smoke. The workplace should have an approachable eye-wash station, a shower, etc., as a safety precaution.

Compressed Gas and Gases

Any material that is normally a gas that is placed under pressure or chilled and contained in a cylinder is considered to be a compressed gas. These materials are dangerous because they are under pressure. If the cylinder is broken, the container can 'rocket' or 'torpedo' at great speeds, and this is a danger to anyone standing too close. If the cylinder is heated (by fire or rise in temperature), the gas may try to expand and the cylinder will explode. Leaking cylinders are also a danger because the gas that comes out is very cold and it may cause frostbite if it touches your skin (e.g. carbon dioxide or propane).

Common examples of compressed air include carbon dioxide, propane, oxygen, ethylene oxide, and welding gases. The hazard symbol is a picture of a cylinder or container of compressed gas surrounded by a circle.

Gases are stored under high pressure such that cracks or damage to the tanks and valves used to control these gases could cause significant physical harm to those in the same room. The examples include acetylene, helium, and nitrogen. Additional dangers may be present if the gas has other hazardous properties. For example, propane is a compressed gas and it will burn easily. Propane would have two hazard symbols – the one for a compressed gas and another to show that it is a flammable material.

Corrosive Chemical Substances

Acids and alkalis: Chemical substances that are capable of destroying materials on very contact as it happens to skin, or any other biological tissues. Corrosive is the name given to materials that can cause severe burns to skin and other human tissues, such as the eye or lung, and can attack clothes and other materials including metal. Corrosives are grouped in this special class because their effects are permanent (irritants whose effects may be similar but temporary are grouped in Class D-2). Common corrosives include acids such as sulphuric acid and nitric acids, bases such as ammonium hydroxide and caustic soda, and other materials such as ammonia gas, chlorine, and nitrogen dioxide.

Combustible Materials

Combustible substances are those that ignite and burn readily. A combustible material can be a solid or liquid. These materials – wood, paper, oil, and gas – can catch fire when exposed to sufficient heat and oxygen. Combustible substances/combustible liquids are known to be present in almost every workplace. Fuels and many common products like solvents, thinners, cleaners, adhesives, paints, waxes, and polishes may be flammable or combustible liquids. Users and occupational workers who handle these substances/liquids must be aware of the hazards – fire, workplace explosion, chemical reactivity, and health – and how to handle them safely. Proper storage and use of combustible materials is absolutely critical in maintaining a safe workplace. Workers should completely and strictly avoid placing or using combustible materials near sources of heat or flame – direct sunlight, furnaces, pilot lights, etc. Also workers must be cautious during waste disposal of combustible materials and must remember that static electricity poses a very real threat, and hence, workers should observe all regulatory standards of bonding and grounding practices. Workers/workplace supervisor should properly label/mark all containers used to store combustible materials and liquids. Workers should correctly name the combustible material/liquid and its hazard on the container and mark it with a DANGER label.

Explosive Chemical Substances

These are chemical substances that have the potential to catch fire rapidly and burn in the air. Liquids, gases, and solids (in the form of dusts) can be flammable and/or explosive. Examples include paint thinner and laboratory solvents, such as acetone, alcohols, acetic acid, and hexane, and adhesives.

Flammables: Flammable means that the material will burn or catch fire easily at normal temperatures (below 37.8°C or 100°F). Combustible materials must usually be heated before they will catch fire at temperatures above normal (between 37.8°C and 93.3°C or 100°F and 200°F). Reactive flammable materials are those that may suddenly start burning when they touch air or water or may react with air or water to make a flammable gas. The material may be a solid, liquid, or gas that makes up the different divisions that fall under this class. Common examples include acetone, propane, butane, acetylene, ethanol, turpentine, toluene, kerosene, Stoddard solvent, spray paints, and varnish. It also includes flammable gases, flammable liquids, flammable solids, and flammable aerosols.

Fumigants: These are chemical substances/compounds used in the gaseous state as a pesticide or disinfectant. Chemical formulations are designed to increase toxicity, reduce flammability, release/give off warning odours, and provide a sort of absorption of the substance at different rates. Physical types of fumigants include gases, liquids, and solids. There are several chemical types of fumigants. Fumigants are used in space fumigation operations related to the management of crop pests and stored grain pests and to disinfest food-processing plants, warehouses, grain elevators, boxcars, ship holds, stores, and households, and in spot fumigations within those structures. They are used in atmospheric vaults and vacuum chambers and are applied extensively to stacked bags of grain or stored foods under polyethylene sheets, to trees under tents to control scale insects, and to areas of land to destroy crop weeds, soil-infesting insects, and nematodes. The list of fumigant chemicals includes, but not limited to, many hydrocarbons, carbon tetrachloride (CCl_4), calcium cyanide, cyanides, chloropicrin, 1,3-dichloropropene, chloroform, ethylene dibromide (EDB), ethylene dioxide, formaldehyde, hydrogen cyanide, iodoform, phosphine, sulphur-containing compounds, methyl bromide, and methyl isocyanate. In brief, fumigation is the act of introducing a pesticide into an enclosed space in such a manner that it disperses quickly and acts in a gaseous state on the target organism. Pesticides formulated as fumigants have physical characteristics that cause them to occupy all air spaces within an enclosed area and to penetrate commodities within these areas. Aluminium and magnesium phosphide fumigants are generally used in space and commodity fumigation, when they are applied to properly sealed structures, containers, or rodent burrows.

Fumigants come in gas, liquid, and solid forms. Gases such as hydrogen peroxide remain in gas form when put in a room at normal temperature. Liquids such as carbon disulphide evaporate into gas on contact with the air. Solids such as methyl bromide and phosphorus trihydride are exposed to oxygen for a chemical reaction that produces hydrogen phosphate gas. Lighter fumigants such as hydrogen cyanide and aluminium chloride diffuse at a very fast rate. These types of fumigants often are used to destroy pests that are emerging but have not yet invaded the entire area. Other factors that affect which fumigant to choose include the moisture in the area, the presence of outside materials like dust, and the temperature. The process and operation of fumigation is very harmful and risky. The operation using hazardous chemical substances should be carried out under strict supervision by trained workers. Also generally, fumigation operation requires a legal permission along with the operator/

management and is conducted by designated persons who hold official certification to perform the fumigation operation because the input of the operation involves highly hazardous, toxic chemical substances that are known poisons to most forms of life, including humans.

Also only the designated persons shall enter hazardous atmospheres and are aware of the nature of the hazard, the precaution to be taken, and the use of protective and emergency equipment. The workplace supervisor/standby observers, similarly equipped and instructed, shall continuously monitor the activity of employees within such a space. The signs and symptoms of fumigant poisoning shall be clearly posted where fumigants, pesticides, or hazardous preservatives have created a hazardous atmosphere. These signs shall note the danger, identify specific chemical hazards, and give appropriate information and precautions including instructions for the emergency treatment of employees affected by any chemical in use.

Irritant Chemical Substances

These are chemical substances that are capable of irritating or inflaming the skin, eyes, or respiratory system like dermatitis and bronchitis.

Mutagens: Mutagenic chemical substances: These are chemical substances that are capable of causing damage to genes.

Organic peroxides: An organic peroxide is any organic (carbon-containing) compound having two oxygen atoms joined together (–O–O–). Organic peroxides are available as solids (usually fine powders), liquids, or pastes. Organic peroxide has extensive industrial applications, for instance, in plastics and rubber industries. Organic peroxides and mixtures containing an organic peroxide are used as accelerators, activators, catalysts, cross-linking agents, curing agents, hardeners, initiators, and promoters. The most common example of an organic peroxide is methyl ethyl ketone peroxide (also known as 2-butanone peroxide, ethyl methyl ketone peroxide, or MEKP). It is used as a polymerisation catalyst in the manufacture of polyester and acrylic resins and as a hardening agent for fibreglass-reinforced plastics. Regulatory agencies including the U.S. National Fire Protection Association (NFPA) have classified organic peroxide and formulations for their known hazards. The NFPA classification system provides the fire and explosion hazards of organic peroxide and the formulations. This information is very essential and mandatory during transport, normal shipping, storage, and management of containers. In fact this has to be approved by the regulatory agencies associated with global transport systems (DOT, Center for Devices and Radiological Health, Consumer Product Safety Commission, Defense Nuclear Facilities Safety Board, Nuclear Regulatory Commission (NRC), Department of Agriculture (USDA), Department of Health and Human Services (HHS), Department of Energy (DOE), Federal Emergency Management Agency (FEMA), Food and Drug Administration (FDA), Occupational Safety and Health Administration (OSHA), USEPA Radiation Protection Division). The following are the details of chemicals and formulation classification: (i) Class I, formulations are capable of deflagration but not detonation; (ii) Class II, formulations burn very rapidly and are a severe reactivity hazard; (iii) Class III, formulations burn rapidly and have a moderate reactivity hazard; (iv) Class IV, formulations burn in the same manner as ordinary combustibles and have a minimal reactivity hazard; and (v) Class V, formulations burn with less intensity than ordinary combustibles or they do not support combustion and present no reactivity hazard.

Oxidisers/Oxidising Agents/Materials

Oxygen is necessary for a fire to occur. Some chemicals can cause other materials to burn by supplying oxygen. Oxidisers do not usually burn themselves but they will either help the fire by providing more oxygen or cause materials that normally do not burn to suddenly catch fire (spontaneous combustion). In some cases, a spark or flame (source of ignition) is not necessary for the material to catch fire but only the presence of an oxidiser. Oxidisers can also be in the form of gases (the most common example is oxygen and ozone), in the form of liquids (the examples are nitric acid and perchloric acid solutions), and in the form of solids (examples are potassium permanganate and sodium chlorite). Some oxidisers such as the organic peroxide family are extremely hazardous because they will burn (they are combustible) as well as have the ability to provide oxygen for the fire. They can have strong reactions that can result in an explosion.

Pesticides

Pesticides are chemical compounds that are used to kill pests, including insects, rodents, fungi, and unwanted plants (weeds). Pesticides are used in public health to kill vectors of disease, such as mosquitoes, and in agriculture to kill pests that damage crops. By their nature, pesticides are potentially toxic to other organisms, including humans, and need to be used safely and disposed of properly. Proper use of different kinds of pesticides helped the societal needs by the management of crop pests and plant disease and vector-borne diseases and helped in the increased production of food grains, improvement of crops yields, agriculture economy, and a safe and diverse food supply. For example, pesticides are used to kill pests and vectors of human diseases like mosquitoes, that can transmit potentially deadly diseases such as malaria and yellow fever. Exposure to pesticides causes acute, delayed, and various adverse health effects. These effects can range from simple irritation of the skin and eyes to more severe effects such as affecting the nervous system, mimicking hormones causing reproductive problems, and also causing cancer. Insecticides can protect animals from illnesses that can be caused by parasites like fleas. Pesticides can prevent sickness in humans that could be caused by mouldy food or diseased produce. Herbicides can be used to clear roadside weeds, trees, and brush. They can also kill invasive weeds that may cause environmental damage. Herbicides are commonly applied in ponds and lakes to control algae and plants such as water grasses that can interfere with activities like swimming and fishing and cause the water to look or smell unpleasant. Uncontrolled pests such as termites and mould can damage structures such as houses. Pesticides are used in grocery stores and food storage facilities to manage rodents and insects that infest food such as grain. Each use of a pesticide carries some associated risk. Proper pesticide use decreases these associated risks to a level deemed acceptable by pesticide regulatory agencies such as the U.S. Environmental Protection Agency (U.S. EPA) and the Pest Management Regulatory Agency (PMRA) of Canada. There are many kinds and classes of chemical pesticides, and they are grouped as follows:

1. *Organophosphate pesticides*: These pesticides affect the nervous system by disrupting the enzyme that regulates acetylcholine, a neurotransmitter. Most organophosphates are insecticides. They were developed during the early nineteenth century.

Some are very poisonous and had been used in World War II as nerve agents. The parent chemical substances are several and include the following: acephate, azinphos-methyl, bensulide, chlorethoxyfos, chlorpyrifos, chlorpyrifos-methyl, diazinon, dichlorvos (DDVP), dicrotophos, dimethoate, disulfoton, ethoprop, fenamiphos, fenthion, malathion, methamidophos, methidathion, methyl parathion, mevinphos, naled, oxydemeton-methyl, phorate, phosalone, phosmet, phostebupirim, pirimiphos-methyl, profenofos, terbufos, tetrachlorvinphos, tribufos, and trichlorfon. For details of each OP compound, readers should refer to their corresponding text pages and literature.

2. *Carbamate pesticides*: Pesticides of the carbamate group are produced from carbamic acid and are in extensive use in gardens, agriculture, and homes for the control of insects and pests. Their mode of action is like that of the organophosphates and their inhibition of cholinesterase enzymes, affecting nerve impulse transmission. Carbamates are extremely toxic to insects of the Hymenoptera class. It is important that users and workers avoid exposure to foraging bees or parasitic wasps. The most well known and the first carbamate chemical used as a pesticide in 1956 is carbaryl with its wide use in lawn and garden settings. These pesticides cause adverse effects on the nervous system and disrupt the important enzyme that regulates acetylcholine, a neurotransmitter. The enzyme effects are usually reversible. The signs and symptoms of carbamate poisonings are similar to those caused by the organophosphate pesticides. Carbamate pesticides include aldicarb, carbaryl, carbofuran, fenoxycarb, methiocarb, methomyl, oxamyl, and thiodicarb.

3. *Organochlorine insecticides*: The use of insecticides has been restricted/removed from the market due to their health and environmental effects and their persistence in the living environment. The examples include chlordane, DDT, DDE, dieldrin, endosulfan, endosulfan epoxide, heptachlor epoxide, hexachlorobenzene, mirex, and oxychlordane. Chlordane and heptachlor are structurally related organochlorine pesticides. For details, refer to other pages of the text and references.

4. *Pyrethroid pesticides*: These pesticides have been developed as a synthetic version of the naturally occurring pesticide pyrethrin, an active ingredient found in chrysanthemums. The synthetic pyrethroids are toxic to the nervous system.

The other groups of pesticides include the following:

Microbial pesticides: Pesticide is a broad term and comprises substances or mixture of substances intended for preventing, destroying, repelling or mitigating any pest. The target pests include, insects, plant, weeds, molluscs, birds, mammals, fish, nematodes/roundworms, and microbes that destroy property and cause nuisance, spread diseases. Pesticides are of different types such as, algicides, antifouling agents, antimicrobials, attractants, biopesticides, biocides, disinfectants and sanitizers, fungicides, fumigants, herbicides, insecticides, miticides (also called acaricides), microbial pesticides, molluscicides, nematicides, ovicides, pheromones, repellents, rodenticides, defoliants, desiccants, and insect growth regulators.

Algicides are used to control algae in lakes, canals, swimming pools, water tanks, and other sites.

Antifouling agents are used to destroy or repel organisms that attach to underwater surfaces, such as boat bottoms.

Antimicrobial chemicals are used to control or kill microorganisms (such as bacteria and viruses).

Attractants are used to attract pests (e.g. to lure an insect or rodent to a trap). (However, food is not considered a pesticide when used as an attractant.)

Biopesticides are certain types of pesticides derived from such natural materials as animals, plants, bacteria, and certain minerals.

Biocides are used to kill microorganisms.

Disinfectants and sanitisers are used to inactivate disease-producing microorganisms on inanimate objects.

Fungicides are used to destroy/kill fungi (including blights, mildews, moulds, and rusts).

Fumigants are gases or vapours intended to destroy pests in buildings or soil.

Herbicides are used to control/destroy weeds and other unwanted plants that grow where they are not wanted.

Insecticides are used to control/kill insects and other arthropods.

Miticides (also called acaricides) are used to control mites that feed on plants and animals.

Microbial pesticides are microorganisms that kill, inhibit, or outcompete pests, including insects or other microorganisms.

Molluscicides are used to control/kill snails and slugs.

Nematicides are used to destroy nematodes (microscopic, worm-like organisms that feed on plant roots).

Ovicides are used to destroy eggs of insects and mites.

Pheromones are biochemicals used to disrupt the mating behaviour of insects.

Repellents are used to repel pests, insects, mosquitoes, and birds.

Rodenticides are used to control mice and other rodents.

Defoliants are that cause leaves or other foliage to drop from a plant, usually to facilitate harvest.

Desiccants are used to promote the process of drying of living tissues, such as unwanted plant tops.

Insect growth regulators are used to disrupt the moulting, maturity from pupal stage to adult, or other life processes of insects.

Plant growth regulators (excluding fertilisers or other plant nutrients) alter the expected growth, flowering, or reproduction rate of plants.

Pyrophoric chemical substances: Pyrophoric chemical substances are liquids or solids that will ignite spontaneously in air below 130°F (54.4°C). These materials must be stored in an atmosphere of inert gas or under kerosene. White phosphorus and titanium dichloride are pyrophoric chemicals.

Reactive agents/chemical substances: Reactivity of a chemical substance or the combination of chemical substances is the tendency of chemical substances. Reactivity is very essential and useful for a chemical substance since the property permits a wide variety of useful materials to be synthesised. Chemical substances/materials that are considered dangerously reactive exhibit three different properties or abilities: reacts very strongly, quickly, and 'vigorously' with water to make a toxic gas; reacts within itself when it gets shocked and 'bumped or dropped' or whenever temperature or pressure increases. If a material is dangerously reactive, it will most likely be described as 'unstable'. Most of these materials can be extremely hazardous if they are not handled properly because they can react in such a quick manner very easily. Examples of these products are ethyl acrylate, vinyl chloride, ethylene oxide, picric acid, and anhydrous aluminium chloride.

Radioactive chemicals: The technical papers presented here cover nuclear material safety topics on the storage and disposal of excess plutonium and highly enriched uranium fissile materials, including vitrification, mixed oxide fuel fabrication, plutonium ceramics, reprocessing, geological disposal, transportation, and the Russian regulatory process. The book, and the meeting from which it is derived, helps to provide a sound basis for maintaining and developing a continuing dialogue between Russian, European, and U.S. experts, improving the safety of future nuclear material operations in all countries involved. The common objective must be the safe and secure storage and disposal of excess fissile nuclear materials.

Sensitising chemical substances, sensitisers are capable of causing allergic reactions, such as asthma and skin allergies.

Teratogens are capable of causing birth defects.

Toxic alcohols can damage the heart, kidneys, and nervous system, such as methanol (wood alcohol) or ethylene glycol (antifreeze).

Water-reactive substances react with water to release a gas that is either flammable or a health hazard. Sodium metal and many of the metal hydrides are water reactive.

Appendix: Categories of Hazardous Chemical Substances

Anticoagulants: Poisons that prevent the blood from clotting properly.

Biotoxins: Poisons that come from plants or animals.

Blister agents/vesicant: Chemicals that severely blister the eyes and skin.

Blood agents: Poisons that affect the body by being absorbed into the blood.

Caustics (acids): Chemicals that burn on contact or corrode the skin, eyes, and mucous membranes.

Choking agents: Chemicals that cause severe irritation or swelling of the respiratory tract and lining of the nose, throat, and lungs.

Incapacitating agent: Chemicals that make it difficult to think clearly or that lead to semi-consciousness or unconsciousness.

Metallic poisons: Toxic compounds made from metals like arsenic or mercury.

Nerve agents: Chemicals that prevent the nervous system from functioning.

Organic solvents: Chemicals that damage the skin and other tissues by dissolving fats and oils.

Riot control/tear gas: Highly irritating agents normally used by law enforcement.

Vomiting agents: Chemical substances that cause nausea and vomiting.

Bibliography

Ali, M. and B. Ali (eds.). 2004. *Handbook of Industrial Chemistry: Organic Chemicals*. McGraw-Hill Handbooks, New York.

Cheremisinoff, N. P. (ed.). 2003. *Industrial Solvents Handbook*, 2nd edn. Marcel Dekker, Inc., New York.

Costa, L. G. 2006. Current issues in organophosphate toxicology. *Clin. Chim. Acta* 366(1–2): 1–13.

Dikshith, T. S. S. (ed.). 1991. *Toxicology of Pesticides in Animals*. CRC Press, Boca Raton, FL.

Dikshith, T. S. S. (ed.). 2011. *Handbook of Chemicals and Safety*. CRC Press, Boca Raton, FL.

Ecobichon, D. J. 1990. Toxic effects of pesticides. In: *Casarett and Doull's Toxicology*, 4th edn., C. D. Klaasen, M. O. Amdur, and J. Doull (eds.). Macmillan Publishing, New York.

Findley, E. 1986. *Procedures for Handling the Disposal of Dangerous Chemical, Biological and Radioactive Waste*. Findley & Co., Avondale, GA.

Gallo, M. A. and N. J. Lawryk. 1991. Organic phosphorus pesticides. In: *Handbook of Pesticide Toxicology*, W. J. Hayes, Jr. and E. R. Laws, Jr. (eds.). Academic Press, New York.

Gosselin, R. E., R. P. Smith, and H. C. Hodge (eds.). 1984. *Clinical Toxicology of Commercial Products*. Williams & Wilkins, Baltimore, MD.

Hayes, W., Jr. (ed.). 1992. *Pesticides Studied in Man*. Williams & Wilkins Publishers, Baltimore, MD.

ICON Group International, Inc. 2009. The 2009 import and export market for radioactive chemical elements and isotopes (and fissile or fertile elements and isotopes) and their compounds including mixtures and residues in Asia. San Diego, CA.

International Agency for Research on Cancer (IARC). 2004. *IARC Monographs on the Evaluation of Carcinogenic Risks to Humans*, Vol. 83. IARC, Lyon, France.

International Agency for Research on Cancer (IARC). 2011. *Agents Classified by the IARC Monographs*, Vols. 1–100. IARC, Lyon, France.

Jardine, L. J. and Moshkov, M. M. (eds.). 1999. *Nuclear Materials Safety Management*, NATO Science Partnership Sub-Series: Vols. 1 and 2. Springer Publishers, Dordrecht, the Netherlands.

Kidd, H. and D. R. James (eds.). 1991. *The Agrochemicals Handbook*, 3rd edn. Royal Society of Chemistry Information Services, Cambridge, U.K.

Klaasen, C. 2007. *Casarett & Doull's Toxicology: The Basic Science of Poisons*, 7th edn. McGraw-Hill Professional, New York.

Lewis, R. J., Sr. (ed.). 1992. *Sax's Dangerous Properties of Industrial Materials*, 8th edn. Van Nostrand Reinhold, New York.

Lewis, R. J., Sr. (ed.). 2007. *Hawley's Condensed Chemical Dictionary*, 15th edn. John Wiley & Sons, Inc., New York.

Lotti, M. and A. Moretto. 2005. Organophosphate-induced delayed polyneuropathy. *Toxicol. Rev.* 24(1): 37–49.

Meister, R. T. (ed.). 1995. *Farm Chemicals Handbook '95*. Meister Publishing Company, Willoughby, OH.

Meister, R. T. (ed.). 2005. *Crop Protection Handbook*, vol. 91. Meister Publishing Co., Willoughby, OH.

Patnaik, P. (ed.). 1992. *A Comprehensive Guide to the Hazardous Properties of Chemical Substances*. Van Nostrand Reinhold, New York.

Patnaik, P. (ed.). 2007. *A Comprehensive Guide to the Hazardous Properties of Chemical Substances*. John Wiley & Sons, Inc., Hoboken, NJ.

Sircar, D. C. 1996. *Indian Epigraphy*. Motilal Banarsidass, New Delhi, India.

Sittig, M. (ed.). 1985. *Handbook of Toxic and Hazardous Chemicals and Carcinogens*, 2nd edn. Noyes Publications, Park Ridge, NJ.

Sittig, M. (ed.). 1991. *Handbook of Toxic and Hazardous Chemicals*, 3rd edn. Noyes Publications, Park Ridge, NJ.

Solomon, R. 1998. *Indian Epigraphy*. Oxford University Press, Oxford, U.K.

Tomlin, C. D. S. (ed.). 2006. *The Pesticide Manual—A World Compendium*, 14th edn. British Crop Protection Council (BCPC), Farnham, Surrey, U.K.

U.S. Department of Health and Human Services, Public Health Service (U.S. DHHS, PHS). Agency for Toxic Substances and Disease Registry (ATSDR). 1994. *Toxicological Profile for Chlordane*. U.S. DHHS, PHS, Atlanta, GA.

U.S. Department of Health and Human Services, Public Health Service (U.S. DHHS, PHS), Agency for Toxic Substances and Disease Registry (ATSDR). 2007. *Toxicological Profile for Heptachlor and Heptachlor Epoxide*. U.S. DHHS, PHS, Atlanta, GA.

U.S. Department of Health and Human Services, Public Health Service (U.S. DHHS, PHS). Agency for Toxic Substances and Disease Registry (ATSDR). 2003. *Toxicological Profile for Pyrethrins and Pyrethroids*. U.S. DHHS, PHS, Atlanta, GA.

U.S. Department of Health and Human Services, Public Health Service, National Toxicology Program. 2011. *Report on Carcinogens*, 12th edn. DHHS, Washington, DC.

U.S. Environmental Protection Agency (U.S. EPA). 2005. *Guidelines for Carcinogen Risk Assessment*. 51 FR 33992-34003 (updated 2011).

3

Hazardous Chemical Substances:
A

Abamectin
Abate
Acacia Powder
Acenaphthene
1-Acenaphthenol
Acenaphthylene
Acephate
Acetal
Acetaldehyde
Acetaldehyde Ammonia
Acetamide
Acetamide Acid (Methomyl)
Acetic Acid
Acetic Anhydride
Acetone Cyanohydrin
Acetonitrile
Acetophenone
Acetophenone Oxime
p-Acetotoluidide (4'-Methylacetanilide)
Acetylacetone
Acetyl Bromide
Acetyl Chloride
Acetylene
Acetylene Dichloride
2-Acetylfluorene
2-Acetylfuran
Acetyl Iodide
Acrylonitrile
Adipamide
Adipic Acid
Adipamide

Abamectin (CAS No. 71751-41-2)

Molecular formula: $C_{48}H_{72}O_{14}$ (Avermectin B1a); $C_{47}H_{70}O_{14}$ (Avermectin B1b)
Synonyms and trade names: Avermectin; Affirm; Avermectin B1; Agri-mek; Agrimek; Vertimec; Zephyr
Chemical name: Avermectin B1

Avermectin is a colourless to yellowish crystalline powder. It is soluble in acetone, methanol, toluene, chloroform, and ethanol but insoluble in water. It is stable and incompatible with strong oxidising agents. Abamectin is a mixture of avermectins containing about 80% avermectin B1a and 20% avermectin B1b. These two components, B1a and B1b, have very similar biological and toxicological properties. The avermectins are insecticidal/miticidal compounds derived from the soil bacterium *Streptomyces avermitilis*. Abamectin is used to control insect and mite pests of citrus, pear, and nut tree crops, and it is used by homeowners for control of fire ants. It acts on the nervous system of insects and causes paralysing effects to insects. Abamectin is a general use pesticide (GUP). It is grouped as toxicity class IV, meaning practically non-toxic, requiring no precautionary statement on its label.

Avermectin is very toxic and causes adverse health effects if swallowed and/or inhaled. Emulsifiable concentrate formulations of avermectin cause slight to moderate eye irritation and mild skin irritation. The symptoms of poisoning observed in laboratory animals include pupil dilation, vomiting, convulsions and/or tremors, and coma. At very high doses, laboratory mammals develop symptoms of nervous system depression, incoordination, tremors, lethargy, excitation, and pupil dilation. Very high doses have caused death from respiratory failure in animals. Also, avermectin has been reported to cause reproductive hazard. Laboratory studies have indicated that abamectin may affect the nervous system in experimental animals. A 1 year study with dogs given oral doses of abamectin (0.5 and 1 mg/kg/day) caused adverse health effects such as pupil dilation, weight loss, lethargy, tremors, and recumbency.

Safe Handling and Precautions

Exposure to avermectin and its absorption through skin cause moderate eye irritation. Workers should avoid contact with skin, eyes, or clothing and avoid breathing spray mist. Wash thoroughly with soap and water after handling and before eating, drinking, or using tobacco. Remove contaminated clothing and wash it before reuse. Keep children or pets away from treated area until dry. Occupational workers, during use/handling of avermectin, should work in a well-ventilated area and use safety glasses, gloves, and protective clothing to prevent prolonged skin contact.

Abate (CAS No. 3383-96-8)

Molecular formula: $C_{16}H_{20}O_6P_2S_3$
Synonyms and trade names: Abat; Abate; Abathion; Biothion; Bithion; Ecopro; Difennthos; Lypor; Nimitox; Swebate; Temephos
IUPAC name: O,O'-(thiodi-4,1-phenylene) bis(O,O-dimethyl phosphorothioate)

Abate is the trade name for Temephos, the known organophosphate larvicide. Abate is used to treat water infested with disease-carrying fleas and control mosquito, midge, and black fly larvae. Abate quickly controls mosquito and other insect populations because it kills insect larvae before they mature, and the residual activity of abate prevents insect populations. On decomposition, abate produces oxides of carbon, phosphorous, and sulphur.

Exposures to abate in workplaces cause adverse health effects (refer to Temephos for details).

Safe Handling and Precautions

Abate should be properly stored to avoid contamination of water, food, or feed. It should be stored in a secure, dry, well-ventilated room, building, or covered area. Similarly, the methods of disposal should be in accord with local, state, and federal regulations. During use and handling of abate, occupational workers/applicators should wear long-sleeved shirt and long pants, shoes and socks, chemical-resistant gloves, and protective eyewear, goggles, or safety glasses. Flaggers must wear chemical-resistant headgear and protective eyewear.

Acacia Powder (CAS. No. 9000-01-5)

Synonyms and trade names: Gum arabic; Acacia gum; Acacia dealbata gum; *Acacia senegal*; Acacia syrup; Australian gum; Gum acacia; Gum ovaline; Indian gum; Senegal gum; Wattle gum

Acacia gum is an odourless white to yellow-white powder. It is soluble in water and incompatible with alcohol and oxidising agents and precipitates. It gels on addition of solutions of ferric salts, borax, lead subacetate, alcohol, sodium silicate, gelatin, and ammoniated tincture of guaiac. It is non-toxic and non-hazardous. It is a water-soluble gum from several species of the acacia tree, especially *Acacia senegal* and *A. arabica*, and used in the manufacture of adhesives and ink and as a binding medium for marbling colours.

Gum arabic is also known as gum acacia and is a natural gum made of hardened sap taken from two species of the Acacia tree – *Acacia senegal* and *Acacia seyal*. Gum arabic is a natural product of the *Acacia senegal* tree, occurring as an exudate from the trunks and branches. It is used primarily in the food industry as a stabiliser but has had more varied uses. It is normally collected by hand when dried, when it resembles a hard, amber-like resin normally referred to as 'tears'. Gum arabic is widely used in the food industry as an emulsifier, thickener, and flavouring and thickening agent. It is employed as a soothing agent in inflammatory conditions of the respiratory, digestive, and urinary tracts and is useful in diarrhoea and dysentery. It exerts a soothing influence upon all the surfaces with which it comes in contact. Gum acacia is an ingredient of all the official Trochisci and various syrups, pastes, and pastilles or jujubes. During the time of the gum harvest, the Moors of the desert are said to live almost entirely on it, and it has been proved that 6 oz. is sufficient to support an adult for 24 h.

Gum acacia is highly nutritious, is a mixture of saccharides and glycoproteins, and provides the properties of a glue and binder suitable for human edibility. In many cases of disease, it is considered that a solution of gum arabic may for a time constitute the exclusive drink and food of the patient. Gum arabic reduces the surface tension of liquids, which leads to increased fizzing in carbonated drinks.

Safe Handling and Precautions

Exposures to gum arabic dust produce a weak allergen reaction. Prolonged period of inhalation of dust may cause allergic respiratory reaction, headache, coughing, dizziness, and respiratory symptoms such as asthma, watery nose and eyes, cough, wheezing, nausea, vomiting, dyspnoea, and urticaria. Hives, eczema, and swelling may also occur. Ingestion and inhalation of gum acacia are considered non-toxic, but sensitive individuals may develop symptoms of mild toxicity. Workers should avoid breathing its dust, avoid getting it in eyes or on skin, and wash thoroughly after handling the material. Acacia powder should be kept stored in a dry place away from direct sunlight, heat, and incompatible materials.

Acenaphthene (CAS No. 83-32-9)

Molecular formula: $C_{12}H_{10}$
Synonyms and trade names: 1,2-Dihydroacenaphthylene; 1,8-Ethylenenaphthalene; peri-Ethylenenaphthalene; Naphthyleneethylene

Acenaphthene is a tricyclic aromatic hydrocarbon and crystalline solid at ambient temperature. Acenaphthene does not dissolve in water but is soluble in many organic solvents. Acenaphthene is a component of crude oil and a product of combustion. Acenaphthene occurs in coal tar produced during the high-temperature carbonisation or coking of coal. It is used as a dye intermediate in the manufacture of some plastics and as an insecticide and fungicide. Acenaphthene is a component of crude oil and a product of combustion which may be produced and released to the environment during natural fires. Emissions from petroleum refining, coal tar distillation, coal combustion, and diesel-fuelled engines are major contributors of acenaphthene to the environment. Acenaphthene is an environmental pollutant and has been detected in cigarette smoke, automobile exhausts, and urban air; in effluents from petrochemical, pesticide, and wood preservative industries; and in soils, groundwater, and surface waters at hazardous waste sites. This compound is one among a number of polycyclic aromatic hydrocarbons (PAHs) on U.S. EPA's (Environmental Protection Agency) priority pollutant list.

Safe Handling and Precautions

Exposures to acenaphthene cause poisoning and include symptoms such as irritation to the skin, eyes, mucous membranes, and upper respiratory tract. Studies on laboratory animals orally exposed to acenaphthene showed loss of body weight, peripheral blood changes, increased aminotransferase levels in blood serum, and mild morphological damage to the liver and kidneys. Chronic exposure to acenaphthene is known to cause damage to the kidney and liver. Acenaphthene is irritating to the skin and mucous membranes of humans and animals. Oral exposure of rats to acenaphthene for 32 days produced peripheral blood changes, mild liver and kidney damage, and pulmonary effects. However, detailed studies with acenaphthene in humans are limited.

1-Acenaphthenol (CAS No. 6306-07-6)

Molecular formula: $C_{12}H_{10}O$
Synonym: 1-Hydroxyacenaphthene

1-Acenaphthenol is a white to cream solid in appearance. It is almost insoluble in water. 1-Acenaphthenol is a stable and combustible chemical substance. It is incompatible with strong oxidising agents. It is used in the syntheses of organic compounds.

Exposures to 1-acenaphthenol cause irritant effects. It is harmful by ingestion, inhalation, as well as through skin absorption. There are no complete data on the toxicology of 1-acenaphthenol.

Safe Handling and Precautions

Students and workers should use safety glasses and avoid breathing dust of 1-acenaphthenol.

Acenaphthylene (CAS No. 208-96-8)

Molecular formula: $C_{12}H_8$
Synonym: Cyclopenta(de)naphthalene

Acenaphthylene is a PAH with three aromatic rings. It is an intermediate chemical for the manufacture of dyes, soaps, pigments, pharmaceuticals, insecticides, fungicides, herbicides, plant growth hormones, naphthalic acids, naphthalic anhydride (pigments), and acenaphthylene (resins) and is used to manufacture plastics. The largest emissions of PAHs result from incomplete combustion of organic materials during industrial processes and other human activities. These include (1) processing of coal, crude oil, and natural gas, including coking, coal conversion, and petroleum refining; (2) production of carbon blacks, creosote, coal tar, and bitumen; (3) aluminium, iron, and steel production in plants and foundries; (4) heating in power plants and residences and cooking; (5) combustion of refuse; (6) motor vehicle traffic; and (7) environmental tobacco smoke.

Safe Handling and Precautions

Acenaphthylene is irritating to the skin and mucous membranes of rabbits. Sub-chronic oral doses of acenaphthylene caused adverse effects to the kidneys, liver, blood, reproductive system, and lungs of experimental animals. Prolonged period of inhalation to low doses caused pulmonary effects like bronchitis, pneumonia, and desquamation of the bronchial and alveolar epithelium in rats.

Acephate (CAS No. 30560-19-1)

Molecular formula: $C_4H_{10}NO_3PS$

Synonyms and trade names: Acetamidophos; Asataf; Pillarthene; Kitron; Aimthane; Orthene; Ortran

Acephate is an organophosphate foliar spray insecticide of moderate persistence with residual systemic activity. It is a contact and systemic insecticide and very effective against a large number of crop pests such as alfalfa looper, aphids, armyworms, bagworms, bean leaf beetle, bean leafroller, black grass bugs, bollworm, budworm, and cabbage looper. Acephate is a colourless to white solid organophosphate insecticide. Exposures to acephate cause poisoning to animals and humans. Acephate inhibits acetylcholine esterase (AchE), the essential nervous system enzyme, and causes characteristic organophosphate poisoning. The symptoms of toxicity include, but are not limited to, headache, nervousness, blurred vision, weakness, nausea, fatigue, stomach cramps, diarrhoea, difficulty in breathing, chest pain, sweating, pinpoint pupils, tearing, salivation, clear nasal discharge and sputum, vomiting, muscle twitching, muscle weakness, and, in severe poisoning, convulsions, respiratory depression, coma, and death. Acephate causes cholinesterase inhibition leading to overstimulation, respiratory paralysis, and death. The U.S. EPA classified acephate as Group C, meaning a possible human carcinogen, and therefore requires judicious handling and management.

Acetal (CAS No. 105-57-7)

Molecular formula: $C_6H_{14}O_2$

Synonyms and trade names: 1,1-Diethoxyethane; Diethylacetal; Ethylidene diethyl ether

Acetal is a clear, colourless, and extremely flammable liquid with an agreeable odour. The vapour is susceptible to cause flash fire. Acetal is sensitive to light and, on storage, may form peroxides. In fact, it has been reported to be susceptible to autoxidation and should, therefore, be classified as peroxidisable. Acetal is incompatible with strong oxidising agents and acids.

Safe Handling and Precautions

Exposures to acetal cause irritation to the eyes, skin, and gastrointestinal tract; nausea; vomiting; and diarrhoea. In high concentrations, acetal produces narcotic effects in workers. Acetal, under normal storage conditions, forms peroxidisable compounds that accumulate and may explode when subjected to heat or shock. This chemical substance becomes very hazardous when peroxide levels are concentrated during the process of distillation or evaporation.

Acetaldehyde (CAS No. 75-07-0)

Molecular formula: C_2H_4O

Synonyms and trade names: Acetic aldehyde; Aldehyde; Ethanol; Ethylaldehyde

Acetaldehyde is a highly flammable, volatile colourless liquid. It has a characteristic, pungent, and suffocating odour and is miscible in water. Acetaldehyde is ubiquitous in the ambient environment. It is an intermediate product of higher plant respiration and formed as a product of incomplete wood combustion in fireplaces and woodstoves, burning of tobacco, vehicle exhaust fumes, coal refining, and waste processing. Exposures to acetaldehyde occur during the production of acetic acid and various other industrial chemical substances, for instance, manufacture of drugs, dyes, explosives, disinfectants, phenolic and urea resins, rubber accelerators, and varnish.

Safe Handling and Precautions

Exposures to acetaldehyde liquids and vapours for a prolonged period in work areas cause irritation to the eyes, skin, upper respiratory passages, and bronchi. Continued exposure is known to damage the corneal epithelium, nasal mucosa, and trachea and cause dermatitis, oedema, necrosis, photophobia, foreign body sensations, and persistent lacrimation or discharge of tears. Acetaldehyde causes bronchitis and reduction in the number of pulmonary macrophages. The severity of lung damage increases with the build-up of fluid in the lungs (pulmonary oedema) and the respiratory distress in the worker.

Laboratory animal studies in rats indicated that exposures to acetaldehyde through its vapours cause health effects such as nasal tumours and laryngeal tumours in hamsters. However, there are no adequate data available regarding acetaldehyde as a human carcinogen. The U.S. EPA has classified acetaldehyde as Group B2, meaning possible human carcinogen.

Acetaldehyde Ammonia (CAS No. 75-39-8)

Molecular formula: C_2H_7NO
Synonyms and trade names: 1-Aminoethanol

Acetaldehyde ammonia is a white crystalline solid and very soluble in water. It is used to make other chemicals and vulcanise rubber. On long exposure to air, acetaldehyde ammonia oxidises, hardens, and turns yellow or brown in colour and reacts exothermically with water to evolve gaseous ammonia. Acetaldehyde ammonia reacts with strong oxidising agents and halogens, and when heated, it readily decomposes into acetaldehyde and ammonia. It attacks copper, aluminium, zinc, and their alloys and reacts with mercury and silver oxides to form shock-sensitive compounds.

Safe Handling and Precautions

Exposure to acetaldehyde ammonia causes harmful effects, eye and skin irritation, and digestive and respiratory tract irritation. The toxicological properties of the chemical substance have not been fully investigated.

Acetaldehyde ammonia is hygroscopic – absorbs moisture from the air – and workers should handle the chemical in a chemical fume hood. Users and occupational workers should wear appropriate chemical protective gloves, boots, and goggles and should avoid breathing vapours or dusts. To clean up chemical spills at the workplaces, occupational workers should vacuum or sweep up the material and place it into a suitable disposal container.

Acetamide (CAS No. 60-35-5)

Molecular formula: C_2H_5NO
Synonyms and trade names: Acetic acid amide; Ethanamide; Methane carboxamide; Acetimidic acid

Acetamide occurs as hexagonal colourless deliquescent crystals with a musty odour. It is incompatible with strong acids, strong oxidising agents, strong bases, and triboluminescent materials. Acetamide is used primarily as a solvent, a plasticiser, and a wetting and penetrating agent. Workplace exposures to acetamide are associated with the plastic and chemical industries.

Safe Handling and Precautions

Repeated ingestion of acetamide may cause liver tumours. Animals, after oral exposures to acetamide, developed liver tumours. However, no information is available on the carcinogenic effects of acetamide in humans. Acetamide causes mild skin irritation in humans from acute exposure. The U.S. EPA has not classified acetamide for carcinogenicity. The International Agency for Research on Cancer (IARC) has, however, classified acetamide as a Group 2B, meaning a possible human carcinogen and requires careful use and handling. During use and handling of acetamide, workers should wear special protective equipment. After leaving the work areas, workers should wash hands, face, forearms, and neck; shower; dispose of outer clothing; and change to clean garments at the end of the day. Acetamide should be kept stored in a tightly closed container, in a cool, dry, ventilated area. It should be kept protected against physical damage and away from any source of heat or ignition and oxidising materials.

Acetamide Acid (Methomyl) (CAS No. 16752-77-5)

Molecular formula: $C_6H_{12}N_2O_3S$
Synonyms and trade names: Methomyl; Acinate; Agrinate; DPX-x1179; DuPont 1179; S-methyl N-(methylcarbamoyloxy) thioacetimidate; Flytek; Kipsin; Lannate; Lannate lv; Lanox; Metomil; Mesomile; Memilene; Methavin; Methomex; Nudrin; NuBait; Pillarmate

Acetamide acid (methomyl) is a carbamate broad-spectrum insecticide. Methomyl is classified as a restricted use pesticide (RUP). It is a crystalline solid with a slight sulphurous odour and very soluble in methanol, acetone, ethanol, and isopropanol. It decomposes with heat and releases hazardous gases/vapours, such as sulphur oxides, methyl isocyanate, and HCN. Acetamide acid (methomyl) is used both as a contact insecticide and as a systemic insecticide. It is used as an acaricide to control ticks and spiders; as fly bait; for foliar treatment of vegetable, fruit, field crops, cotton, and commercial ornamentals; and in and around poultry houses and dairies.

Safe Handling and Precautions

Acetamide acid (methomyl) is potentially a highly poisonous material to humans. Exposures to acetamide acid cause adverse health effects. It is highly toxic and causes inhibition of cholinesterase activity. The symptoms of toxicity include, but are not limited to, weakness, lack of appetite, blurred vision, pupillary constriction, corneal injury, headache, nausea, abdominal cramps, burning sensation, coughing, wheezing, laryngitis, shortness of breath, vomiting, and irritation. It may be harmful if absorbed through the skin of the mucous membranes and may cause respiratory tract and chest discomfort, constriction of pupils, sweating, muscle tremors, and decreased pulse. Occupational workers, after severe poisoning, show symptoms of twitching, giddiness, confusion, muscle incoordination, heart irregularities, loss of reflexes, slurred speech, paralysis of muscles of the respiratory system, and death. The target organs of methomyl toxicity include nerves, cardiovascular system, liver, and kidneys.

Workers should keep acetamide acid stored in a tightly closed container in a cool (below 0°C.), dry, well-ventilated area. The chemical should be kept away from incompatible chemical substances, strong oxidising agents, strong bases, other pesticides, and food or feed. Occupational workers should avoid contact of acetamide acid (methomyl) with skin, eyes, or clothing and avoid exposures to its vapour or mist. Workers should be careful in storage and or disposal of acetamide acid and avoid the dust from contaminating water, food, or feed by storage or disposal. Occupational workers during use of acetamide acid should use appropriate personal protective equipment (PPE) safety workplace dress such as, approved respirator, chemical-resistant gloves, and chemical safety goggles.

Acetic Acid (CAS No. 64-19-7)

Chemical formula: $C_2H_4O_2$
Synonyms and Trade names: Acetic acid; Ethanoic acid; Glacial acetic acid (pure compound); Methanecarboxylic acid
Chemical name: Acetic acid, Glacial

Acetic acid is a colourless liquid or crystal with a sour, vinegar-like odour and is one of the simplest carboxylic acids and is an extensively used chemical reagent. Acetic acid has wide application as a laboratory reagent, in the production of cellulose acetate mainly for photographic film and polyvinyl acetate for wood glue, synthetic fibres, and fabric materials. Acetic acid has also been of large use as a descaling agent and acidity regulator in food industries.

Safe Handling and Precautions

Acetic acid is a very hazardous chemical substance. Exposure to acetic acid cause severe health effects such as, irritation of the skin, eyes, lacrimation, nose, conjunctivitis, throat, skin burns, skin sensitization black skin, hyperkeratosis, dental injury/erosion. Also exposures to acetic acid as a spray or mist may produce tissue damage particularly of mucous membranes of eyes, mouth, and respiratory tract. Liquid or spray mist may produce tissue damage particularly on mucous membranes of eyes, mouth, and respiratory tract. Skin contact may produce burns. Inhalation of the spray mist may produce severe irritation of respiratory tract, characterised by coughing, choking, or shortness of breath. Inflammation of the eye is characterised by redness, watering, and itching. Skin inflammation is characterised by itching, scaling, reddening, and/or, occasionally, blistering. Repeated and prolonged exposure to acetic acid causes deleterious effects to kidneys, mucous membranes, skin, and teeth. Repeated or prolonged exposure to acetic acid in the form of spray, mist, and fumes produces severe eye irritation and skin irritation and produces respiratory tract irritation leading to frequent attacks of bronchial infection. Repeated or prolonged skin contact may cause thickening, blackening, and cracking of the skin. Acetic acid vapours may form explosive mixtures with air. Reactions between acetic acid and the following materials are potentially explosive: 5-azidotetrazole, bromine pentafluoride, chromium trioxide, hydrogen peroxide, potassium permanganate, sodium peroxide, and phosphorus trichloride.

Accidental contact with concentrated acetic acid at workplaces produces damage to the skin, and inhalation causes lung oedema, permanent eye damage, and irritation to the mucous membranes. These burns or blisters may not appear until hours after exposure. Similarly, accidental ingestion of acetic acid at workplaces causes sore throat, burning sensation, abdominal pain, vomiting, shock, and/or collapse. Therefore, acetic acid must be handled by users and occupational workers with appropriate care and precaution.

Experiments have shown that exposures to acetic acid is known to cause reproductive effects in laboratory animals. May cause adverse liver effects. May cause adverse kidney effects. Chronic exposure to corrosive fumes/gases may cause erosion of the teeth followed by jaw necrosis. Bronchial irritation with chronic cough and frequent attacks of pneumonia is common. Gastrointestinal disturbances may also be seen.

Precautions

During use and handling of acetic acid, students and occupational workers should be careful and should avoid skin contact with formic acid and/or acetic acid. Workers must use full-length protective gloves made of Butyl rubber/Viton rubber (flour rubber) and polyethylene. Workers must be aware that the liquid may penetrate the gloves. Frequent change is advisable.

Also, workers must wear suitable protective clothing as protection against splashing or contamination. Workplace manager should provide eyewash station and safety shower.

The spilled acetic acid should be absorbed with an inert material and disposed of in an appropriate waste disposal. Workers should neutralise the chemical residue with a dilute solution of sodium carbonate and should keep acetic acid stored in a segregated and approved, cool, well-ventilated area. Keep container tightly closed and sealed until ready for use. Avoid all possible sources of ignition (spark or flame) since, as stated earlier, acetic acid is a flammable and a corrosive liquid. Workers should keep acetic acid away from heat and other sources of ignition.

Users and occupational workers, during use of acetic acid, should use splash goggles, quality and impervious gloves, synthetic apron, and vapour respirator of an approved/certified respirator or equivalent. Workers should AVOID ALL types of CONTACT with acetic acid!

Acetic Anhydride (CAS No. 108-24-7)

Molecular formula: $(CH_3CO)_2O/C_4H_6O_3$
Synonyms and trade names: Acetic acid anhydride, Acetic oxide, Ethanoic anhydride, Acetyl ether, Acetyl oxide, Ethanoic anhydrate

Acetic anhydride is a colourless liquid and appreciably soluble in water. It is of ethanoic acid smell. Acetic acid anhydride is flammable, moisture sensitive, and incompatible with strong oxidising agents, water, strong bases, alcohols, metals, reducing agents, amines, ammonia, nitrates, nitric acid, permanganates, phenols, sodium hydroxide, hydrogen peroxide, chromium trioxide, potassium hydroxide, perchloric acid, and ethanol. Acetic anhydride is mainly used for the acetylation of cellulose to cellulose acetate for photographic film and other applications. Upon burning, acetic anhydride decomposes and produces toxic gases and toxic fumes including acetic acid fumes. It attacks many metals with or without the presence of water.

Safe Handling and Precautions

Acetic acid anhydride is very poisonous and harmful on exposure. Exposure to acetic anhydride causes conjunctivitis, lacrimation (discharge of tears), corneal oedema, opacity, photophobia (abnormal visual intolerance to light), nasal and pharyngeal irritation, cough, dyspnoea (breathing difficulty), bronchitis, skin burns, vesiculation, and sensitisation dermatitis. It is corrosive, and improper use and handling causes severe burns. When workers use it without care and precautions, it may cause harmful effects. When accidentally swallowed or inhaled at workplaces, acetic acid anhydride causes severe effects on mucous membranes, respiratory irritation, and burns. Users and occupational workers should observe strict precautions: the workplace should have adequate ventilation, and workers must use personal protective equipment (PPE), chemical protection suit, face shield, respirators, and safety glasses. Occupational workers should immediately remove the contaminated clothing and wash it before reuse. Any kind of spillage of acetic acid anhydride during use and or handling must be prevented. Workers should not allow water to get into the chemical container and should avoid the possible violent chemical reactions of acid anhydride. Workers should ground and bond containers during transferring the chemical and use spark-proof tools and explosion-proof equipments. Workers should collect leaking and spilled liquid in sealable containers as much as possible and absorb remaining spilled liquid in sand or inert absorbent and remove to a safe place. While handling acetic acid anhydride, users and occupational workers should strictly avoid breathing the harmful dust, mist, or vapour and avoid chances of the chemical substance getting in eyes, on skin, or on clothing. Workers should keep away the used/empty containers because these may still retain the chemical residue as liquid and/or vapour leading to further dangerous situations at the workplace. Users and workers should always remember to handle acetic anhydride away from heat, sparks, and flame and work with adequate ventilation. Occupational workers should not store acetic anhydride in direct sunlight; it should be kept away from contact with water, oxidising materials, and flammable areas. Occupational workers should use alcohol foam or carbon dioxide as preferred materials for fire suppression. Because acetic anhydride is an irritant and flammable chemical substance on its reactivity towards water, alcohol foam or carbon dioxide is preferred for fire suppression. The vapour of acetic anhydride is harmful. Workers should keep the container tightly closed in a cool, well-ventilated place. Workers should be aware of the hazards of retained residue in the empty containers.

Alert! Danger! Acetic anhydride is a combustible and corrosive liquid. It reacts with water, causes digestive and respiratory tract burns, and causes severe eye and skin irritation with possible burns. During handling, acetic anhydride users and workers should observe strict precautions. Keep away from water and away from sources of ignition.

Acetone Cyanohydrin (CAS No. 75-86-5)

Molecular formula: **(CH₃)₂C(OH)(CN)O**

Synonyms and trade names: ACH, 2-Hydroxy-2-methyl propionitrile; 2-Methyl-lactonitrile; 2-Cyano-2-propanol; *p*-Hydroxyisobutyronitrile, α-Hydroxyisobutyronitrile

Acetone cyanohydrin is an organic compound used in the production of methyl methacrylate, the monomer of the transparent plastic polymethyl methacrylate (PMMA), also known as acrylic. Acetone cyanohydrin is a flammable colourless liquid with a faint odour of bitter almond. It is incompatible with sulphuric acid. Acetone cyanohydrin readily undergoes decomposition by water to form hydrogen cyanide and acetone. Cassava tubers contain linamarin, a glucoside of acetohydrin, and the enzyme linamarinase for hydrolysing the glucoside. Crushing the tubers releases these compounds and produces acetone cyanohydrin, which is potentially lethally toxic. Acetone cyanohydrin is classified as an extremely hazardous substance in the U.S. Emergency Planning and Community Right-to-Know Act. The principal hazards of acetone cyanohydrin arise from its ready decomposition on contact with water, which releases highly toxic hydrogen cyanide.

Safe Handling and Precautions

Acetone cyanohydrin is extremely toxic. Exposures to acetone cyanohydrin cause adverse health effects. The symptoms of toxicity include, but are not limited to, irritation to the eyes, skin, and respiratory system, cough, nausea, laboured breathing, shortness of breath, unconsciousness, vomiting, irregular heartbeat, tightness in the chest, dizziness, lassitude (weakness, exhaustion), headache, confusion, pulmonary oedema, asphyxia, abdominal cramps, burning sensation, convulsions, and liver and kidney injury. The target organs include eyes, skin, respiratory system, central nervous system, cardiovascular system, liver, kidneys, and gastrointestinal tract. The toxicity of acetone cyanohydrin is believed to be predominantly attributable to dissociation of the cyanide molecule with the resultant formation of molecular (undissociated) hydrocyanic acid. Hydrocyanic acid, by virtue of its small size and lack of charge, readily penetrates the external membranes of aquatic organisms and inhibits respiration. Any potential environmental problems would be caused by cyanide rather than the parent compound. The rapid formation of hydrogen cyanide from acetone cyanohydrin is of concern, and the critical adverse health effect is acute lethality. However, at anticipated levels of human exposure, no systemic effects are likely to occur. Acetone cyanohydrin is not genotoxic or toxic to development or the reproductive system. Users and occupational workers, during handling of acetone cyanohydrin, should strictly follow safety regulations and should use face shield or eye protection in combination with breathing protection and completely avoid negligence, eating, drinking, or smoking during work and at the workplace.

Acetonitrile (CAS No. 75-05-8)

Molecular formula: **C₂H₃N**

Synonyms and trade names: Cyanomethane; Ethanenitrile, ethyl; Ethyl nitrile; Methyl cyanide; Methanecarbonitrile

Acetonitrile is a colourless liquid, with an ether-like odour, and a polar solvent. Acetonitrile is predominantly used as a solvent in the manufacture of pharmaceuticals, for spinning fibers and for casting and molding of plastic materials, in lithium batteries, for the extraction of fatty acids from animal and vegetable oils, and in chemical laboratories for the detection of materials such as pesticide residues. Acetonitrile is also used in dyeing textiles and in coating compositions as a stabilizer for chlorinated solvents and in perfume production as a chemical intermediate. It is a by-product from the manufacture of acrylonitrile, and acetonitrile has, in fact, replaced the acrylonitrile. It is used as a starting material for the production of acetophenone, alpha-naphthaleneacetic acid, thiamine, and acetamidine. It has been used as a solvent in making pesticides, pharmaceuticals, batteries, rubber products, and formulations for nail polish remover despite its low but significant toxicity. Acetonitrile has been banned in cosmetic products in the European Economic Area (EEA) since early 2000, and acetone and ethyl are often preferred as safer for domestic use. Acetonitrile has a number of uses: primarily as an extraction solvent for butadiene; as a chemical intermediate in pesticide manufacturing; as a solvent for both inorganic and organic compounds to remove tars, phenols, and colouring matter from petroleum hydrocarbons not soluble in acetonitrile; in the production of acrylic fibres; and in pharmaceuticals, perfumes, nitrile rubber, and acrylonitrile butadiene styrene (ABS) resins; in high-performance liquid and gas chromatographic analysis; and in extraction and refining of copper.

Safe Handling and Precautions

Acetonitrile liquid or vapour is irritating to the skin, eyes, and respiratory tract. Acetonitrile itself has only a modest toxicity, but it can be metabolised in the body to hydrogen cyanide and thiocyanate. Acetonitrile causes delayed symptoms of poisoning (several hours after the exposure) that include, but are not limited to, salivation, nausea, vomiting, anxiety, confusion, hyperpnoea, dyspnoea, respiratory distress, disturbed pulse rate, unconsciousness, convulsions, and coma. Cases of acetonitrile poisoning in humans (or, more strictly, of cyanide poisoning after exposure to acetonitrile) by inhalation, ingestion, and (possibly) by skin absorption are rare but not unknown. Repeated exposure to acetonitrile may cause headache, anorexia, dizziness, weakness, and macular, papular, or vesicular dermatitis.

Acetophenone (CAS No. 98-86-2)

Molecular formula: C_8H_8O
Synonyms and trade names: 1-Phenyl-1-ethanone; 1-Phenylethanone; Acetophenon; Acetylbenzene; Hypnon; Hypnone; Methyl phenyl ketone; Phenyl methyl ketone

Acetophenone is the simplest aromatic ketone and is a clear liquid/crystal and very slightly soluble in water with a sweet pungent taste and odour resembling oranges. It is used as a polymerisation catalyst for the manufacture of olefins. Acetophenone is used in perfumery as a fragrance ingredient in soaps, detergents, creams, lotions, and perfumes; as a flavouring agent in foods, non-alcoholic beverages, and tobacco; as a specialty solvent for plastics and resins; as a catalyst for the polymerisation of olefins; and as a photosensitiser in organic syntheses. Acetophenone is a raw material for the synthesis of some pharmaceuticals and is also listed as an approved excipient by the U.S. FDA. Acetophenone occurs naturally in many foods such as apple, apricot, banana, and beef. Occupational exposure to acetophenone occurs during its manufacture. Acetophenone has been detected in ambient air and drinking water; exposure of the general public may occur through the inhalation of contaminated air or the consumption of contaminated water. It is highly flammable and will get easily ignited by heat, sparks, or flames, and the vapours may form explosive mixtures with air.

Safe Handling and Precautions

Acute exposure of humans to acetophenone vapour may produce skin irritation and transient corneal injury. One study noted a decrease in light sensitivity in exposed humans. Acute oral exposure has been observed to cause hypnotic or sedative effects, haematological effects, and a weakened pulse in humans. Congestion of the lungs, kidneys, and liver was reported in rats acutely exposed to high levels of acetophenone via inhalation. Studies have been reported that acetophenone caused moderate acute effects after oral or dermal exposures in experimental rats, mice, and rabbits. There is no information indicating that acetophenone caused chronic (long-term), reproductive, developmental carcinogenic effects in humans or animals. The U.S. EPA classified acetophenone as a Group D, meaning not classifiable as carcinogenic to humans.

Acetophenone Oxime (CAS No. 613-91-2)

Molecular formula: C_8H_9NO
Synonyms and trade names: Phenylethan-1-one oxime; 1-Phenyl-1-ethanone oxime; Acetophenone oxime; Acetylbenzene oxime; 4-Acetophenone oxime; Benzoyl methide oxime; Hypnone oxinie; Methyl phenyl ketone oxime

Acetophenone oxime is a solid but slightly soluble substance. Acetophenone oxime is incompatible with strong oxidising agents, moisture, acids, and metal and alkali compounds. Acetophenone oxime on decomposition releases irritating and toxic fumes and gases. There is no published data about the hazardous polymerisation of acetophenone oxime.

Safe Handling and Precautions

Exposures to acetophenone oxime cause irritation to the eyes, skin, and respiratory and digestive tract. The toxicological properties of acetophenone oxime are not been fully investigated. Acetophenone oxime should be kept stored in a cool, dry place and in closed container when not in use. Upon exposure to acetophenone oxime, occupational workers should immediately flush eyes with plenty of water. Workers should use proper PPEs and wear appropriate gloves to prevent skin exposure and appropriate protective eyeglasses or chemical safety goggles as described.

---

p-Acetotoluidide (4'-Methylacetanilide) (CAS No. 103-89-9)

Molecular formula: $C_9H_{11}NO$

Synonyms and trade names: 4'-Methylacetanilide; *p*-Acetotoluidide; *p*-cetamidotoluene; Acetyl-*p*-toluidine; 1-Acetamido-4-methylbenzene; 4-(Acetylamino)toluene; 4-Acetotoluide; Acet-*p*-toluidide; *p*-Acetotoluidine; N-(4-Methylphenyl)acetamide; N-Acetyl-*p*-toluidide; 4-Acetotoluidide; *p*-Methylacetanilide; *p*-Acetotoluide; *n*-Acetyl-*p*-toluid

p-Acetotoluidide is an off-white to brown flake, odourless, solid or crystalline powder (pure form), which is soluble in hot water, alcohol, ether, chloroform, acetone, glycerol, and benzene. It is combustible and undergoes self-ignition (at 5450°C) but is otherwise stable under most conditions. *p*-Acetanilide on decomposition releases carbon monoxide, oxides of nitrogen, carbon dioxide, and toxic fumes. *p*-Acetanilide is used as an inhibitor of peroxides and stabiliser for cellulose ester varnishes. It is used as an intermediate for the synthesis of rubber accelerators, dyes and dye intermediate, and camphor. It is also used as a precursor in penicillin synthesis and other pharmaceuticals including painkillers and intermediates. Phenylacetamide structure shows analgesic and antipyretic effects. But acetanilide is not used directly for this application due to causing methemoglobinemia (the presence of excessive methemoglobin which does not function reversibly as an oxygen carrier in the blood).

Safe Handling and Precautions

Exposures to *p*-acetotoluidide, through ingestion and inhalation, cause potential health effects. The symptoms of toxicity include, but are not limited to, irritation to the eyes and redness of the skin; wheezing; cough; shortness of breath; burning in the mouth, throat, or chest; and respiratory tract irritation. *p*-Acetanilide should be kept stored in a cool, dry place and the container closed when not in use. During handling and use of *p*-acetotoluidide, workers should wear proper PPE and strictly avoid contact of the chemical with eyes, skin, and clothing and avoid ingestion and inhalation. On accidental exposure at the workplace, workers should immediately flush the eyes and exposed part of the body with plenty of water. Workers should avoid using the *p*-acetotoluidide-contaminated clothing and shoes.

Caution! Exposures to p-acetotoluidide may cause eye and skin irritation, and, respiratory and digestive tract irritation. The toxicological properties of this material have not been fully investigated.

Acetylacetone (CAS No. 123-54-6)

Molecular formula: $C_5H_8O_2$

Synonyms and trade names: Hacac; Acetoacetone; 2,4-Pentadione; Pentane; Acetyl 2-propanone; Diacetylmethane; Pentan-2,4-dione

Acetylacetone (2,4-pentanedione) is a clear or slightly yellowish liquid with a putrid odour. It is readily soluble in water and in organic solvents and incompatible with light, ignition sources, excess heat, oxidising agents, strong reducing agents, and strong bases.

On decomposition, acetylacetone releases hazardous products such as carbon monoxide, irritating and toxic fumes and gases, and carbon dioxide. Acetylacetone is used in the production of anti-corrosion agents and its peroxide compounds for the radical initiator application for polymerisation. It is used as a chemical intermediate for drugs (such as sulphamethazine, nicarbazine, vitamin B6, and vitamin K) and pesticides sulfonylurea herbicides and pesticides. It is used as an indicator for the complexometric titration of Fe (III), for the modification of guanidino groups and amino groups in proteins, and for the preparation of metal acetylacetonates for catalyst application.

Safe Handling and Precautions

Exposures to acetylacetone cause eye irritation, chemical conjunctivitis and corneal damage, and skin irritation (harmful if absorbed through the skin). Long periods of inhalation/ dermal absorption of acetylacetone at low concentration cause irritation and dermatitis, cyanosis of the extremities, pulmonary oedema, and burning sensation in the chest. On ingestion/accidental ingestion in workplace, it results in gastrointestinal irritation, nausea, vomiting and diarrhoea, and CNS depression. Inhalation of high concentrations may cause central nervous system effects characterised by nausea, headache, dizziness or suffocation, unconsciousness, and coma. The target organ of acetylacetone poisoning has been identified as the central nervous system.

Acetylacetone should be kept away from heat, sparks, flame, and sources of ignition. It should be stored in a tightly closed container, in a cool, dry, well-ventilated area away from incompatible substances and from flammable work area. Occupational workers should use/handle acetylacetone only in a well-ventilated area with spark-proof tools and explosion-proof equipment. Workers during work should not cut, weld, braze, solder, drill, grind, pressurise, or expose empty containers to heat, sparks, or open flames.

Acetyl Bromide (CAS No. 506-96-7)

Molecular formula: **CH$_3$COBr**
Synonyms and trade names: Ethanoyl bromide; Acetic acid; Bromide

Acetyl bromide is a colourless fuming liquid with pungent odour, is combustible, and turns yellow on exposure to air. It is used as an acetylating agent in the synthesis of fine chemicals, agrochemicals, and pharmaceuticals. It is also used as an intermediate for dyes. Acetylation, a case of acylation, is an organic synthesis process whereby the acetyl group is incorporated into a molecule by substitution for protecting –OH groups.

Safe Handling and Precautions

Exposures to acetyl bromide cause pain to the eyes, loss of vision, shock, collapse, abdominal pain, sore throat, cough, burning sensation, shortness of breath, and respiratory distress. Contact to skin causes pain, redness, blisters, dermatitis, and skin burns. The occupational worker may show delayed symptoms and lung oedema. The vapour is corrosive to the eyes, the skin, and the respiratory tract. The irritation caused by acetyl bromide may lead to chemical pneumonitis and pulmonary oedema and may cause burns to the respiratory tract. The target organs include eyes, skin, and mucous membranes.

Acetyl bromide should be stored in a tightly closed container, in a cool, dry, well-ventilated area away from water and incompatible substances. Acetyl bromide is combustible and gives off irritating or toxic fumes (or gases) in a fire. It should not have any contact with open flames and water. During use, the occupational workers should use protective gloves, protective clothing, face shield, or eye protection in combination with breathing protection, and should not eat, drink, or smoke. Acetyl bromide decomposes on heating and produces toxic and corrosive fumes such as hydrogen bromide and carbonyl bromide. It reacts violently with water, methanol, or ethanol to form hydrogen bromide. Acetyl bromide attacks and damages many metals in the presence of water.

Acetyl Chloride (CAS No. 75-36-5)

Molecular formula: CH_3COCl
Synonyms and trade names: Acetic chloride; Ethanoyl chloride; Acetic acid chloride

Acetic chloride is a colourless to light yellow liquid with a pungent and choking odour. Acetic chloride is highly flammable and reacts violently with DMSO, water, lower alcohols, and amines to generate toxic fumes. Along with air, acetic chloride may form an explosive mixture. It is incompatible with water, alcohols, amines, strong bases, strong oxidising agents, and most common metals. On decomposition when heated, acetic chloride produces carbon monoxide, carbon dioxide, hydrogen chloride, and phosgene.

Safe Handling and Precautions

Exposure to acetic chloride causes severe health effects. It is corrosive and causes severe skin burns. On contact with eyes and skin and accidental ingestion, acetic chloride causes permanent eye damage and serious burns to mouth and stomach. The spray mist or liquid produces tissue damage (mucous membranes of eyes, mouth, and upper respiratory tract). Inhalation of the spray mist produces severe irritation of the respiratory tract, with symptoms of burning sensation, coughing, wheezing, laryngitis, shortness of breath, headache, nausea, and vomiting. Prolonged period of inhalation of acetic chloride may become fatal as a result of spasm, inflammation, oedema of the larynx and bronchi, chemical pneumonitis, and pulmonary oedema/characterised by coughing, choking, or shortness of breath. However, there is no published information about the carcinogenicity, mutagenicity, teratogenicity, and developmental toxicity of acetic chloride in animals and humans.

During handling and use of acetic chloride, care is needed to keep the container dry, and it should be kept away from heat and sources of ignition. Avoid contact with skin and eyes, ingestion, and breathing of gas/fumes/vapour/spray. During use, the workers should never add water to acetic chloride, and they should wear suitable respiratory equipment. Acetic chloride should be kept away from incompatible materials such as oxidising agents, alkalis, and moisture; ground all equipment containing the material and store in a segregated and approved area. Keep the container of acetic chloride in a cool, well-ventilated area tightly closed and sealed until ready for use. Users should avoid all possible sources of ignition, for instance, spark or flames and smoking area.

Acetylene (CAS No. 74-86-2)

Molecular formula: C_2H_2
Synonyms and trade names: Ethyne; Acetylen; Ethine; Narcylen; Vinylene; Welding gas

Acetylene (100% purity) is odourless, but commercial purity has a distinctive garlic-like odour and is very soluble in alcohol and almost miscible with ethane. Acetylene is a flammable gas and kept under pressure in gas cylinders. Under certain conditions, acetylene can react with copper, silver, and mercury to form acetylides, compounds which can act as ignition sources. Brasses containing less than 65% copper in the alloy and certain nickel alloys are suitable for acetylene. Acetylene is not compatible with strong oxidisers such as chlorine, bromine pentafluoride, oxygen, oxygen difluoride and nitrogen trifluoride, brass metal, calcium hypochlorite, heavy metals such as copper, silver, mercury, and their salts, bromine, chlorine, iodine, fluorine, sodium hydride, caesium hydride, ozone, perchloric acid, and potassium.

Safe Handling and Precautions

Acetylene is highly explosive; it must be stored and handled with great care. Acetylene may cause anaesthetic effects. It is highly flammable under pressure and spontaneously combustible in air at pressures above 15 psig. Acetylene liquid is shock sensitive. Prolonged period of exposures to acetylene causes symptoms which include headaches, respiratory difficulty, ringing in ears, shortness of breath, wheezing, dizziness, drowsiness, unconsciousness, nausea, vomiting, and depression of all the senses. The skin of a victim of overexposure may have a blue colour. There are currently no known adverse health effects associated with chronic exposure to the components of this compressed gas. Lack of sufficient oxygen may cause serious injury or death. The target organs include kidneys, central nervous system, liver, respiratory system, and eyes.

Acetylene reacts vigorously and explosively when combined with oxygen and other oxidisers including all halogens and halogen compounds and with trifluoromethyl hypofluorite. The presence of moisture, certain acids, or alkaline materials tends to enhance the formation of copper acetylides. Do not attempt to dispose of residual or unused quantities. Return cylinder to supplier. Unserviceable cylinders should be returned to the supplier for safe and proper disposal. To extinguish fire, only water spray, dry chemical, carbon dioxide, or chemical foam should be used. Acetylene should be kept stored in a cool, dry place in a tightly closed container and should be used only in a well-ventilated area. Cylinders should be separated from oxygen and other oxidisers by a minimum of 20 ft or by a barrier of non-combustible material at least 5 ft high having a fire resistance rating of at least 1/2 h. Storage in excess of 2500 ft³ is prohibited in buildings with other occupancies. Cylinders should be stored upright with a valve protection cap in place and firmly secured to prevent falling or being knocked over. The cylinders should be protected from physical damage, and avoid dragging, rolling, sliding, or dropping the cylinder. During transport, workers should use a suitable hand truck for cylinder movement. Care should be taken to label 'No Smoking or Open Flames' signs in the storage or use areas. There should be no sources of ignition. All electrical equipment should be explosion-proof in the storage and use areas.

Acetylene is shipped in a cylinder packed with a porous mass material and a liquid solvent, commonly acetone. Acetylene is dissolved in the acetone solution and dispersed throughout the porous medium. When the valve of a charged acetylene cylinder is opened, the acetylene comes out of the solution and passes out in the gaseous form. It is crucial that fuse plugs in the tops and bottoms of all acetylene cylinders be thoroughly inspected whenever handled. Remove and quarantine in a safe location any defective cylinder. Acetylene cylinders should be properly secured at all times. Movement of cylinders should be done with care by a qualified and trained worker, and the cylinders should be protected from flame or heat.

Acetylene Dichloride (CAS No. 540-59-0)

Molecular formula: $C_2H_2Cl_2$
Synonyms and trade names: *cis*-Acetylene dichloride; *trans*-Acetylene dichloride; sym-Dichloroethylene; 1,2-Dichloroethylene

Acetylene dichloride is a colourless liquid (usually a mixture of the *cis* and *trans* isomers) with a slightly acrid, chloroform-like odour. Acetylene dichloride is a chemical used mainly in the production of perfumes, dyes, and thermoplastics. The type and severity of symptoms varies depending on the amount of chemical involved and the nature of the exposure. It is incompatible with strong oxidisers, strong alkalis, potassium hydroxide, and copper. Acetylene dichloride is highly flammable and in a fire gives off irritating or toxic fumes/gases. (Acetylene dichloride usually contains inhibitors to prevent polymerisation.)

Safe Handling and Precautions

Exposures to acetylene dichloride cause irritation of the eyes and respiratory system and central nervous system depression with symptoms of cough, sore throat, dizziness, nausea, drowsiness, weakness, unconsciousness, and vomiting. Prolonged period of exposures to acetylene dichloride damages the skin and may have effects on the liver. Exposure to more than 200 ppm of acetylene dichloride is dangerous to human health. Acetylene dichloride decomposes on heating or, under the influence of air, light, and moisture, produces toxic and corrosive fumes including hydrogen chloride. Acetylene dichloride reacts with copper bases and attacks plastic and strong oxidants.

2-Acetylfluorene (CAS No. 781-73-7)

Molecular formula: $C_{15}H_{12}O$
Synonyms and trade names: 2-Fluorenyl methyl ketone; 1-(9H-Fluoren-2-yl)-1-ethanone

2-Acetylfluorene is a crystalline powder of light yellow and is stable under normal temperatures and pressures. 2-Acetylfluorene is incompatible with strong oxidising agents. 2-Acetylfluorene on decomposition releases carbon monoxide and carbon dioxide.

Safe Handling and Precautions

Exposures to 2-acetylfluorene cause irritations to the eyes, skin, respiratory tract, and digestive tract. The toxicological properties have not been fully investigated. 2-Acetylfluorene should be kept stored in a cool, dry place and in a tightly closed container. Occupational workers should avoid exposures to 2-acetylfluorene; breathing its dust, vapour, mist, or gas; contact with skin and eyes; and ingestion and inhalation. Workers should wear appropriate protective gloves and chemical splash goggles to prevent skin exposure.

2-Acetylfuran (CAS No. 1192-62-7)

Molecular formula: $C_6H_6O_2$
Synonyms and trade names: 2-Furyl methyl ketone; Acetyl furan; Methyl 2-furyl ketone; 1-(2-Furanyl) ethanone
IUPAC name: 1-(furan-2-yl)ethanone

2-Acetylfuran is a gold-coloured solid with the odour of sweet balsam almond cocoa caramel coffee. It is incompatible with strong oxidising agents, strong reducing agents, and strong bases. It is combustible. 2-Acetylfuran on decomposition releases carbon monoxide and carbon dioxide. It is insoluble in water but soluble in alcohol, dipropylene glycol, and ethyl ether.

Safe Handling and Precautions

Exposures to 2-acetylfuran by ingestion, inhalation, and skin absorption cause harmful effects. It causes irritation to the skin and eyes, coughing, respiratory tract irritation, and respiratory distress. Accidental ingestion causes irritation of the digestive tract. There are no data about the chronic exposure and adverse effects. 2-Acetylfuran should be kept away from sources of ignition, heat, sparks, flame, and light. It should be kept stored in a tightly closed container in a dry area/refrigerated (below 4°C/39°F). During handling and use of 2-acetylfuran, occupational workers should use appropriate and proper PPE and wear a self-contained breathing apparatus. Occupational workers should avoid generating dusty conditions, minimise generation and accumulation of dust, remove all sources of ignition, and use a spark-proof tool. Workers should use 2-acetylfuran only in a chemical fume hood.

Acetyl Iodide (CAS No. 507-02-8)

Molecular formula: C_2H_3IO
Synonyms and trade names: Acetyl iodide; Ethanoyl iodide; Acetyliodide95+%; Acetic acid iodide

Acetyl iodide is a colourless, fuming liquid with a pungent odour. It is soluble in benzene and ether. Acetyl iodide is toxic, corrosive, and reacts with water or steam to produce toxic and corrosive fumes. The vapours of acetyl iodide are irritating to the eyes and mucous membranes. Vapour causes pulmonary oedema. It is corrosive to metals and skin. Acetyl iodide turns brown on exposure to air as it reacts exothermically with moisture in the air to give hydrogen iodide (hydroiodic acid), a strong irritant. It decomposes in water to give acidic products. Inhalation, ingestion, or contact of the skin and the eyes with vapours, dusts, or substance may cause severe injury, burns, or death. Contact with molten form of acetyl iodide causes severe burns to skin and eyes. Reaction with water or moist air will release toxic, corrosive, or flammable gases. Reaction with water may generate much heat that will increase the concentration of fumes in the air. Fire will produce irritating, corrosive, and/ or toxic gases. Run-off from fire control or dilution water may be corrosive and/or toxic and cause pollution.

Safe Handling and Precautions

Acetyl iodide may burn but does not ignite readily. Substance will react with water (some violently) releasing flammable, toxic, or corrosive gases and run-off. When heated, vapours may form explosive mixtures with air: indoors, outdoors, and sewers explosion hazards. Most vapours are heavier than air. They will spread along the ground and collect in low or confined areas (sewers, basements, tanks). Vapours may travel to source of ignition and flash back. Contact with metals may evolve flammable hydrogen gas. Containers may explode when heated or if contaminated with water.

Contact with molten substance may cause severe burns to skin and eyes. Reaction with water or moist air will release toxic, corrosive, or flammable gases. Reaction with water may generate much heat that will increase the concentration of fumes in the air. Fire will produce irritating, corrosive, and/or toxic gases. Run-off from fire control or dilution water may also be corrosive and/or toxic and cause pollution. Acetyl iodide decomposes exothermically in water or alcohol. It reacts vigorously and exothermically with bases and may react vigorously or explosively if mixed with diisopropyl ether or other ethers in the presence of trace amounts of metal salts.

In case of chemical accidents, occupational workers should use dry chemical or carbon dioxide and should not use water on the material itself. If large quantities of combustibles are involved, workers should use water in flooding quantities as spray and fog and use water spray to knock down vapours. Users and occupational workers should cool all affected chemical containers with flooding quantities of water. Apply water from as far a distance as possible. Occupational workers during handling of the chemical should wear suitable protective clothing and wear suitable gloves.

During use and handling of acetyl iodide, occupational workers should keep sparks, flames, and other sources of ignition away. Keep material out of water sources and sewers. Build dikes to contain flow as necessary. Use water spray to knock down vapours. Do not use water on the material itself. Neutralise spilled material with crushed limestone, soda ash, or lime. During handling of acetyl iodide, occupational workers should avoid breathing toxic chemical vapours. Workers should wear positive pressure self-contained breathing apparatus and must avoid bodily contact with the material, Wear appropriate chemical protective gloves, boots, and goggles. Workers should not handle broken packages unless wearing appropriate PPE and must wash away any kind of material which may have contacted the body with soap and plenty of water. Users and occupational workers should be never negligent to wear appropriate chemical protective clothing during handling of acetyl iodide.

Danger – Alert! Acetyl iodide is a highly flammable liquid and vapour and corrosive to metals. Negligence during use and handling causes severe skin burns and eye damage.

Acrylonitrile (CAS No. 107-13-1)

Molecular formula: **$H_2C=CHCN$**
Synonyms and trade names: AN; Cyanoethylene; 2-Propenenitrile; Propenenitrile, Vinyl cyanide

Acrylonitrile is a colourless, flammable liquid. Its vapours may explode when exposed to an open flame. Acrylonitrile does not occur naturally. It is produced in very large amounts by several chemical industries in the United States, and its requirement and demand are increasing in recent years. Acrylonitrile is a heavily produced, unsaturated nitrile. It is used to make other chemicals such as plastics, synthetic rubber, and acrylic fibres. It has been used as a pesticide fumigant in the past; however, all pesticide uses have been discontinued. This compound is a major chemical intermediate used in creating products such as pharmaceuticals, antioxidants, and dyes, as well as in organic synthesis.

The largest users of acrylonitrile are chemical industries that make acrylic and modacrylic fibres and high-impact ABS plastics. Acrylonitrile is also used in business machines, luggage, construction material, and manufacturing of styrene-acrylonitrile (SAN) plastics for automotive, household goods, and packaging material. Adiponitrile is used to make nylon, dyes, drugs, and pesticides.

Safe Handling and Precautions

Acrylonitrile is a carcinogen and mutagen. Accidental ingestion or inhalation of acetonitrile at workplaces can be fatal. A full assessment of all risks must be made before using acetonitrile in the laboratory. Eye contact of acetonitrile may lead to serious damage. Acrylonitrile is harmful to the environment, and exposures may cause long-term adverse effects to workers. Occupational workers must use most protective respirators and avoid inhalation of acrylonitrile at concentrations above 1 ppm. The Occupational Safety and Health Administration (OSHA) currently requires in 29 CFR 1919.1045 that workers be provided with and required to wear fully encapsulating, vapour-protective clothing for spills and leaks and use the 'most protective' respirators. Occupational workers, during use and handling of acetonitrile, should wear safety glasses and gloves. Effective ventilation is essential at workplaces. Before starting work, consider whether a less hazardous chemical can be used.

Adipamide (CAS No. 628-94-4)

Molecular formula: **$C_6H_{12}N_2O_2$**
Synonyms and trade names: Adipic acid amide; Adipic acid diamide; Adipic diamide; 1,4-Butanedicarboxamide; Hexanediamide; NCI-C02095

Adipamide is a chunky white powder. It is slightly soluble in water and incompatible with strong oxidising agents. Adipamide on combustion and decomposition produces hazardous products of toxic fumes of carbon monoxide, carbon dioxide, and nitrogen oxides. Exposures to adipamide by inhalation, ingestion, or skin absorption cause adverse health. The symptoms include irritation of the eyes, skin, mucous membranes, and upper respiratory tract. There is no complete information about the toxicological properties of the chemical.

Safe Handling and Precautions

Workers should avoid repeated exposures to adipamide. Workers during use should wear suitable protective clothing, self-contained breathing apparatus, chemical safety goggles, rubber boots, and heavy rubber gloves. Occupational workers should wash thoroughly after handling adipamide. Workers should remove contaminated clothing and wash it before reuse. Use with adequate ventilation. Minimise dust generation and accumulation. Avoid contact with eyes, skin, and clothing. Keep container tightly closed. Avoid breathing dust of adipamide, which should be kept stored in a tightly closed container, in a cool, dry, well-ventilated area away from incompatible substances.

Adipic Acid (CAS No. 124-04-9)

Molecular formula: $C_6H_{10}O_4$
Synonyms and trade names: Hexanedioic acid; 1,4-Butane dicarboxylic acid

Adipic acid is a white crystalline solid/crystalline granule, and its odour has not been characterised. It is stable and incompatible with ammonia and strong oxidising agents. It may form combustible dust concentrations in air. The likely routes of exposure to workers are by skin contact and inhalation at workplaces. It is used in the manufacture of nylon, plasticisers, urethanes, adhesives, and food additives.

Safe Handling and Precautions

Exposures to adipic acid cause pain, redness of the skin and eyes, tearing, or lacrimation. Adipic acid has been reported as a non-toxic chemical. Excessive concentrations of adipic acid dust are known to cause moderate eye irritation, irritation to skin, dermatitis, and respiratory tract irritation. It may be harmful if swallowed or inhaled. It causes respiratory tract irritation with symptoms of coughing, sneezing, and blood-tinged mucous. Occupational workers should keep stored adipic acid in a tightly closed container, in a cool, dry, ventilated area and protected against physical damage. In storage, adipic acid should be isolated from incompatible substances, and workers should avoid dust formation and ignition sources. Occupational workers should avoid contact of the adipic acid with eyes, avoid breathing its dust, and keep the container closed. Workers should use adipic acid only with adequate ventilation. *Workers should wash* thoroughly after handling adipic acid and keep it away from heat, sparks, and flame. Also, the workers should use rubber gloves and lab coat, apron, or coveralls and avoid creating a dust cloud in handling, transfer, and clean-up.

Bibliography

Berg, G. L., C. Sine, S. Meister, and J. Poplyk. (eds.). 1984. *Farm Chemicals Handbook*, 70th edn. R. T. Meister, G. L. Meister Publishing Co., Willoughby, OH.

Bevelacqua, A. S. and R. H. Stilp. 2006. *Hazardous Materials Field Guide*, 2nd edn. Delmar Cengage Learning, Clifton Park, NY.

Budavari, S. (ed.). 1996. *The Merck Index—An Encyclopedia of Chemicals, Drugs, and Biologicals*. Merck and Co. Inc., Whitehouse Station, NJ.

Burgess, W. A. 1995. *Recognition of Health Hazards in Industry: A Review of Materials and Processes*, 2nd edn. John Wiley & Sons, Inc., New York.

Chemical Data Sheet. Gum arabic. Chemeo chemicals. Ocean Service, National Oceanic and Atmospheric Administration (NOAA).

Cheremisinoff, N. P. (ed.). 1999. *Handbook of Hazardous Chemical Properties*. Butterworth Heinemann, Boston, MA.

Dikshith, T. S. S. 2008. *Safe Use of Chemicals: A Practical Guide*. CRC Press, Boca Raton, FL.

Environmental Health & Safety, 2007a. Acetyl chloride. MSDS no. A0806. Mallinckrodt Baker, Inc., Phillipsburg, NJ.

Environmental Health & Safety. 2007b. Acacia powder. MSDS no. A0012. Mallinckrodt Baker, Inc., Phillipsburg, NJ.

EXTOXNET. 1995. Acephate. Pesticide Information Profiles. Extension Toxicology Network. Cornell University, Ithaca, NY (updated 2001).

Faust, R. A. 1994. Toxicity summary for acenaphthylene. Chemical Hazard Evaluation Group. (DE-AC05–84OR21400). Oak Ridge National Laboratory. U.S. Department of Energy, Oak Ridge, TN.

Gosselin, R. E., R. P. Smith, and H. C. Hodge (eds.). 1984. *Clinical Toxicology of Commercial Products*, 5th edn. Williams & Wilkins, Baltimore, MD.

Hayes, W. J. and E. R. Laws. (ed.). 1990. *Handbook of Pesticide Toxicology, Volume 3, Classes of Pesticides*. Academic Press, Inc., New York.

International Agency for Research on Cancer (IARC). 1999. *Acetamide, Summaries and Evaluations*. IARC, Lyon, France.

International Programme on Chemical Safety (IPCS). 1997. Acetyl bromide. ICSC Card No. 0365 (updated 2005).

International Programme on Chemical Safety and the Commission of the European Communities (IPCS CEC). 1994. Acetyl bromide. ICSC Card No. 0365 (updated 1997).

International Programme on Chemical Safety and the Commission of the European Communities (IPCS-CEC). 1995. Acetyl chloride. International Chemical Safety Cards. ICSC Card No. 0210. NIOSH, Centers for Disease Control and Prevention, Atlanta, GA (updated 2005).

International Programme on Chemical Safety and the Commission of the European Communities (IPCS-CEC). 1998. Acetone cyanohydrin. ICSC Card No. 0611. IPCS, Geneva, Switzerland (updated 2005).

International Programme on Chemical Safety and the Commission of the European Communities (IPCS-CEC). 2003. 1,2-Dichloroethene. ICSC no. 0436. IPCS CEC, Geneva, Switzerland.

International Programme on Chemical Safety and the Commission of the European Communities (IPCS-CEC). 2005. Acenaphthene. IPCS Card no. ICSC: 1674. IPCS, Geneva, Switzerland (updated 2006).

International Programme on Chemical Safety and the Commission of the European Communities (IPCS-CEC). 2010. Acetic acid. ICSC Card No. 0363. IPCS-CEC, Geneva.

Knobloch, K., S. Szedzikowski, and A. Slusarcyk-Zablobona. 1969. Acute and subacute toxicity of acenaphthene and acenaphthylene. *Med. Pracy.*, 20: 210–222 (in Polish).

Lewis, R. J., Sr. 1996. *Sax's Dangerous Properties of Industrial Materials*, 9th edn. John Wiley & Sons, New York.

Lewis, R. J., Sr. 2004. *Sax's Dangerous Properties of Industrial Materials*, 8th edn. Van Nostrand Reinhold Company, New York.

Lewis, R. J., Sr. (ed.). 2007. *Hawley's Condensed Chemical Dictionary*, 15th edn. Wiley-Interscience. John Wiley & Sons, New York.

Material Safety Data Sheet (MSDS). Acetyl iodide. CAMEO Chemicals, National Oceanic and Atmospheric Administration.

Material Safety Data Sheets (MSDS). 1991. Methomyl. DuPont Agricultural Products. DuPont, Wilmington, DE.

Material Safety Data Sheets (MSDS). 1996. Methomyl. Extension Toxicology Network, Pesticide Information Profiles. (EXTOXNET). USDA/Extension Service. Oregon State University, Corvallis, OR.

Material Safety Data Sheet. 2002. Acetyl bromide. Laboratory Reagents.

Material Safety Data Sheet (MSDS). 2003. Acetylacetone. Physical Chemistry at Oxford University, Oxford, U.K.

Material Safety Data Sheet (MSDS). 2005a. Acetic anhydride. International Chemical Safety Cards (ICSC). ICSC Card No. 0209. International Programme on Chemical Safety (IPCS), Geneva, Switzerland.

Material Safety Data Sheet (MSDS). 2005b. P-Acetotoluidide. Acros Organics N.V., Fair Lawn, NJ.

Material Safety Data Sheet (MSDS). 2006a. Acetamide. Environmental Health & Safety. MSDS no. A0120. Mallinckrodt Baker, Inc., NJ.

Material Safety Data Sheet (MSDS). 2006b. Acetic anhydride. International Programme on Chemical Safety (IPCS). International Labour Organization (ILO) (updated 2007).

Material Safety Data Sheet (MSDS). 2007a. Acetophenone oxime. MSDS ACC no. 02208. Acros Organics N.V., Fair Lawn, NJ.

Material Safety Data Sheet (MSDS). 2007b. Adipamide. Department of Physical and Theoretical Chemistry, Oxford University, Oxford, U.K.

Material Safety Data Sheet (MSDS). 2007c. Safety data for acetic anhydride. Department of Physical and Theoretical Chemistry Laboratory, Oxford University, Oxford, U.K.

Material Safety Data Sheet (MSDS). 2007d. Adipamide. The Physical & Theoretical Chemistry Lab., Oxford University, Oxford, U.K. (updated 2009).

Material Safety Data Sheet (MSDS). 2008a. Acetal. Physical Chemistry at Oxford University, Oxford, U.K.

Material Safety Data Sheet (MSDS). 2008b. Adipic acid. MSDS No.A1716. Environmental Health & Safety. Mallinckrodt Baker, Inc, NJ.

Material Safety Data Sheet (MSDS). 2009a. 1-Acenaphthenol. Department of the Physical and Theoretical Laboratory, Oxford University, Oxford, U.K.

Material Safety Data Sheet (MSDS). 2009b. 2-Acetylfluorene. The Physical and Theoretical Chemistry Laboratory, Oxford University, Oxford, U.K.

Material Safety Data Sheet (MSDS). 2009c. Safety data for acrylonitrile. Department of Physical Chemistry, Oxford University, Oxford, U.K.

Material Safety Data Sheets (MSDS). 2009d. 2-Actylfuran. Physical Chemistry at Oxford University, Oxford, U.K.

Material Safety Data Sheet (MSDS). 2010a. Acetic acid. Fisher Scientific, Fair Lawn, NJ.

Material Safety Data Sheet (MSDS). 2010b. Acetic anhydride. NIOSH Pocket Guide to Chemical Hazards. Centers for Disease Control and Prevention (CDCP), Atlanta, GA. (updated 2011).

Material Safety Data Sheet (MSDS). 2010c. Safety data for acetylene. Department of Physical Chemistry, Oxford University, Oxford, U.K.

National Institute for Occupational Safety and Health (NIOSH). 2005. Acetonitrile. NIOSH Pocket Guide to Chemical Hazards. Centers for Disease Control and Prevention, Atlanta GA (updated 2008).

National Oceanic and Atmospheric Administration (NOAA). *n*-Acetyl-*p*-Toluidine. NOAA, USA.

National Toxicology Program (NTP). 1996. Toxicology and carcinogenesis studies of acetonitrile (CAS No. 75-05-8) in F344/N rats and B6C3F1 mice (inhalation studies) (NTP TR-447). Department of Health and Human Services, Public Health Service, National Institutes of Health, Bethesda, MD.

NIOSH Pocket Guide to Chemical Hazards. 2010a. Acetic acid. Centers for Disease Control and Prevention. Atlanta, GA.

NIOSH Pocket Guide to Chemical Hazards. 2010b. Acetone cyanohydrin. National Institute for Occupational Safety and Health (NIOSH). Centers for Disease Control and Prevention (CDCP), Atlanta, GA.

Patnaik, P. (ed.). 1999. *A Comprehensive Guide to the Hazardous Properties of Chemical.* McGraw Hill, New York.

Pohanish, R. P. and S. Greene. 2009. *Wiley Guide to Chemical Incompatibilities.* John Wiley & Sons, New York.

Pradyot, P. (ed.). 2007. *A Comprehensive Guide to the Hazardous Properties of Chemical Substances.* John Wiley & Sons., Inc., Hoboken, NJ.

Richard, P. P. (ed.). 2008. *Sittig's Handbook of Toxic and Hazardous Chemicals and Carcinogens*, 6th edn. William Andrew Publishing, Norwich, NY.

Routt, R. J. and J. R. Roberts. 1999. Organophosphate insecticides. Recognition and Management of Pesticide Poisonings. U.S. Environmental Protection Agency, National Technical Information Service (U.S. EPA 735-R-98–003; pp. 55–57), Washington, DC.

Sax, N. I. and R. J. Lewis. (eds.). 2004. *Dangerous Properties of Industrial Materials*, 11th edn. Van Nostrand Reinhold Company, New York.

Schwanecke, R. 1966. Safety hazards in the handling of acrylonitrile and methacrylonitrile. *Zentralbl Arbeitsmed Arbeitsschutz* 16(1):1–3 (in German).

Sinclair, J. S., D. T. McManus, M. D. O'Hara, and R. Millar. 1994. Fatal inhalation injury following an industrial accident involving acetic anhydride. *Burns* 20(5):469–470.

Sittig, M. (ed.). 1985. *Handbook of Toxic and Hazardous Chemicals and Carcinogens*, 2nd edn. Noyes Publications, Park Ridge, NJ.

Sittig, M. (ed.). 1991. *Handbook of Toxic and Hazardous Chemicals*, 3rd edn. Noyes Publications, Park Ridge, NJ.

Tomlin, C. (ed.). 2006. *The Pesticide Manual*, 14th edn. British Crop Protection Council, a Blackwell Publications, Cambridge, U.K.

Tomlin, C. (ed.). 2008. *The Pesticide Manual*, 15th edn. British Crop Protection Council (BCPC). Blackwell Scientific Publications, Hampshire, U.K.

U.S. Department of Health and Human Services (U.S. HHS). 1993. Hazardous substances data bank (HSDB). National Library of Medicine, Bethesda, MD.

U.S. Department of Labor. 2011. Acetylene. Mine Safety and Health Administration (MSHA). Arlington, VA.

U.S. Environmental Protection Agency (U.S. EPA). 1987a. Health effects assessment for acenaphthylene. Environmental Criteria and Assessment Office, Office of Health, Cincinnati, OH.

U.S. Environmental Protection Agency (U.S. EPA). 1987b. Acephate. Pesticide Fact Sheet. National Technical Information Service. U.S. EPA. Washington, DC.

U.S. Environmental Protection Agency (U.S. EPA). 1987c. Health and environmental effects document for acetophenone. Environmental Criteria and Assessment Office, Cincinnati, OH.

U.S. Environmental Protection Agency (U.S. EPA). 1997. Acenaphthene. The Risk Assessment Information System. U.S. EPA, Washington, DC.

U.S. Environmental Protection Agency (U.S. EPA). 2000a. Acetaldehyde hazard summary. Technology Transfer Network. Air Toxics Web Site. U.S. EPA (updated 2007).

U.S. Environmental Protection Agency (USEPA). 2000b. Acetonitrile hazard summary. Technology Transfer Network. Air Toxics Web Site U.S. EPA, NC (updated 2007).

U.S. Environmental Protection Agency (U.S. EPA). 2002. Acetamide hazard summary. Technology Transfer Network Air Toxics (updated 2007).

Wagner, F. S., Jr. 1991. Acetic anhydride. In: *Kirk-Othmer Encyclopedia of Chemical Technology*, 4th edn., Vol. 1. John Wiley & Sons, New York, pp. 142–154.

William, B., Jr. (ed.). 2002. *Methacrylic Acid and Derivatives, Ullmann's Encyclopedia of Industrial Chemistry*. Wiley-VCH, Weinheim, Germany.

World Health Organization (WHO). International Programme on Chemical Safety (IPCS). 1983. Acrylonitrile. Environment health criteria. Document No. 28. IPCS, WHO, Geneva. (updated 2011).

4

Hazardous Chemical Substances:
B

Barium
Barium Compounds
   Barium Chloride (Anhydrous)
   Barium Dichloride Dihydrate
   Barium Fluoride
   Barium Nitrate (as Ba)
   Barium Sulphate
Bendiocarb
Benfuracarb
Benomyl
Benzaldehyde
Benzamide
1,2-Benzanthracene
Benzene
Benzidine
Benzotrichloride
2,3-Benzofuran
Benzyl Acetate
Benzyltrimethylammonium Dichloroiodate
Beryllium
Beryllium Chloride
Bifenthrin
1,1'-Biphenyl
Bis(chloromethyl) Ether
Bismuth
Boron
Boron Compounds
   Boric Acid
   Boron Tribromide
   Boron Carbide
   Boron Nitride
   Borazole/Borazine
   Boron Trichloride
   Boron Trifluoride
   Boron Trifluoride Diethyl Ether Complex
   Sodium Tetraborate Decahydrate/Borax (Anhydrous)
   Sodium Tetraborate Decahydrate/Borax (Decahydrate))
   Sodium Tetraborate Pentahydrate

1,3-Butadiene
Butane
1,3-Butanediol
Butanesultone
Butanethiol
2-Butanone
1-Butene
2-Butoxyethanol
Butyl Acetates
   Isobutyl Acetate
   n-Butyl Acetate
   sec-Butyl Acetate
   tert-Butyl Acetate
n-Butyl Alcohol
n-Butylamine

Barium (CAS No. 7440-39-3)

Molecular formula: **Ba**

Barium is a silvery-white metal. It exists in nature only in ores containing mixtures of elements. The important combinations are the peroxide, chloride, sulphate, carbonate, nitrate, and chlorate. The pure metal oxidises readily and reacts with water emitting hydrogen. It combines with other chemicals such as sulphur or carbon and oxygen to form barium compounds. Barium compounds are used by the oil and gas industries to make drilling muds. Barium attacks most metals with the formation of alloys; iron is the most resistant to alloy formation. Barium forms alloys and intermetallic compounds with lead, potassium, platinum, magnesium, silicon, zinc, aluminium, and mercury. Barium compounds exhibit close relationships with the compounds of calcium and strontium, which are also alkaline earth metals. Twenty-five barium isotopes have been identified, ^{138}Ba being the most abundant, and the others are unstable isotopes with half-lives ranging from 12.8 days for ^{140}Ba to 12 s for ^{143}Ba. Two of these isotopes, ^{131}Ba and ^{139}Ba, are used in research as radioactive tracers. The general population is exposed to barium through air, drinking water, and food.

Safe Handling and Precautions

The health effects of barium compounds depend on how well the compound dissolves in water or in the stomach contents. Barium compounds that do not dissolve well, such as barium sulphate, are considered as not harmful. Barium carbonate dust and barium oxide dust have been reported to be a bronchial irritant. While barium carbonate is a dermal irritant, barium oxide is a nasal irritant. Occupational workers exposed to barium dust, usually in the form of barium sulphate or carbonate, often develop a benign pneumoconiosis also called as 'baritosis'. The effect of baritosis has been shown to be a reversible one and has not caused any kind of severe pulmonary adverse effect.

However, health effects of the different barium compounds depend on the degree of their water solubility. The compounds that dissolve well in water are known to cause harmful health effects when ingested in high levels. The symptoms of poisoning include stomach irritation, brain swelling, muscle weakness, liver and kidney damage, adverse effects to the heart, increased blood pressure, changes in heart rhythm, effects on spleen, and difficulties in breathing.

Exposures to high concentrations of barium through food and drinking water cause GI and muscular disturbances. Barium causes vomiting, abdominal cramps, diarrhoea, difficulties in breathing, increased or decreased blood pressure, numbness around the face and muscle weakness, changes in heart rhythm, or paralysis and possibly death. Animals exposed to barium over long periods showed kidney damage, decreased body weight, and fatal injury.

Ingestion of large amount of barium chloride (2 and 4 g) causes fatal injury because barium ions paralyse the heart. Acute poisoning of barium causes nausea and diarrhoea, cardiac problems, and muscular spasms as well as cardiac arrest. Thus, barium, at concentrations normally found in our environment, does not pose any significant risk for the general population. However, for specific subpopulations and under conditions of high barium exposure, the potential for adverse health effects should be taken into account.

Barium has not been classified as a carcinogen by the Department of Health and Human Services (DHHS) and the International Agency for Research on Cancer (IARC). Similarly, the U.S. Environmental Protection Agency (U.S. EPA) has observed that barium on ingestion or inhalation has not caused any carcinogenic effects to humans. The IARC has indicated that barium chromate (VI) is the only barium compound for which there is sufficient evidence that it is a human carcinogen.

For barium, the U.S. EPA has set a limit of 2 ppm (2.0 mg/L) in drinking water. The Occupational Safety and Health Administration (OSHA) has set the permissible exposure limit (PEL) at 0.5 mg/m^3 in workplace air. National Institute of Occupational Safety and Health (NIOSH) has set the recommended exposure limit (REL) at 500 mg/m^3 for soluble barium compounds, 10 mg/m^3 for total barium dust, and 5 mg/m^3 for barium sulphate. The OSHA has set a limit for barium sulphate dust at 15 mg/m^3 of total dust and 5 mg/m^3 as desirable fraction. Barium is a strong reducing agent and reacts violently with oxidants and acids, reacts violently with halogenated solvents, and reacts with water, forming flammable/explosive gas. Barium is known to spontaneously ignite on contact with air if in powder form, causing fire and explosion hazard.

Barium Compounds

Barium compounds are of extensive use in the manufacture of alloys for nickel barium parts found in ignition equipment for automobiles and in the manufacture of glass, ceramics, and television picture tubes. Barite (BaSO$_4$), or barium sulphate, is primarily used in the manufacture of lithopone, a white powder containing 20% barium sulphate, 30% zinc sulphide, and less than 8% zinc oxide. Lithopone is widely employed as a pigment in white paints. Chemically precipitated barium sulphate – blanc fixe – is used in high-quality paints, in x-ray diagnostic work, and in the glass and paper industries. It is also used in the manufacture of photographic papers, artificial ivory, and cellophane. Crude barite is used as a thixotropic mud in oil well drilling.

Barium hydroxide (Ba(OH)$_2$) is found in lubricants, pesticides, the sugar industry, corrosion inhibitors, drilling fluids, and water softeners. It is also used in glass manufacture, synthetic rubber vulcanisation, animal and vegetable oil refining, and fresco painting. Barium carbonate ($BaCO_3$) is obtained as a precipitate of barite and is used in the brick, ceramics, paint, rubber, oil-well drilling, and paper industries. It also finds use in enamels, marble substitutes, optical glass, and electrodes. Barium oxide (BaO) is a white alkaline powder that is used to dry gases and solvents. At 450°C, it combines with oxygen to produce barium peroxide (BaO_2), an oxidising agent in organic synthesis and a bleaching material for animal substances and vegetable fibres. Barium peroxide is used in the textile industry for dyeing and printing, in powder aluminium for welding, and in pyrotechnics. Barium chloride ($BaCl_2$) is obtained by roasting barite with coal and calcium chloride and is used in the manufacture of pigments, colour lakes, and glass and as a mordant for acid dyes. It is also useful for weighting and dyeing textile fabrics and in aluminium refining. Barium chloride is a pesticide, a compound added to boilers for softening water, and a tanning and finishing agent for leather. Barium nitrate ($Ba(NO_3)_2$) is used in pyrotechnics and the electronics industries.

Barium metal has only limited use and presents an explosion hazard. The soluble compounds of barium (chloride, nitrate, hydroxide) are highly toxic; the inhalation of the insoluble compounds (sulphate) may give rise to pneumoconiosis. Many of the compounds, including sulphide, oxide, and carbonate, may cause local irritation of the eyes, nose, throat, and skin. Certain compounds, particularly peroxide, nitrate, and chlorate, present fire hazards in use and storage. When the soluble compounds enter by the oral route, they are highly toxic, with a fatal dose of the chloride thought to be 0.8–0.9 g. However, although poisoning due to the ingestion of these compounds occasionally occurs, very few cases of industrial poisoning have been reported. Poisoning may result when workers are exposed to atmospheric concentrations of the dust of soluble compounds such as may occur during grinding. These compounds exert a strong and prolonged stimulant action on all forms of muscle, markedly increasing contractility. In the heart, irregular contractions may be followed by fibrillation, and there is evidence of a coronary constrictor action. Other effects include intestinal peristalsis, vascular constriction, bladder contraction, and an increase in voluntary muscle tension. Barium compounds also have irritant effects on the mucous membranes and the eye (Table 4.1).

Barium carbonate, an insoluble compound, does not appear to have pathological effects from inhalation; however, it can cause severe poisoning from oral intake, and in rats, it impairs the function of the male and female gonads; the foetus is sensitive to barium carbonate during the first half of pregnancy. Barium sulphate is characterised by its extreme insolubility, a property that makes it non-toxic to humans. For this reason and due to its high radio-opacity, barium sulphate is used as an opaque medium in x-ray examination

TABLE 4.1

Toxicity of Barium Compounds to Humans

Barium Compound	Exposure Data	Effect
Barium carbonate	Lowest lethal dose (57 mg/kg)	Death
	Lowest toxic dose (29 mg/kg)	Flaccid paralysis without anaesthesia; paraesthesia; muscle weakness
Barium chloride	Lowest lethal dose (11.4 mg/kg)	Death
Barium polysulphide	Lowest toxic dose (226 mg/kg)	Flaccid paralysis without anaesthesia; muscle weakness; dyspnoea

of the GI, respiratory, and urinary systems. It is also inert in the human lung, as has been demonstrated by its lack of adverse effects following deliberate introduction into the bronchial tract as a contrast medium in bronchography and by industrial exposure to high concentrations of fine dust.

Inhalation of barium sulphate results in its deposition in the lungs in sufficient quantities to produce baritosis (a benign pneumoconiosis, which principally occurs in the mining, grinding, and bagging of barite, but has been reported in the manufacture of lithopone). The first reported case of baritosis was accompanied by symptoms and disability, but these were associated later with other lung diseases. Subsequent studies have contrasted the unimpressive nature of the clinical picture and the total absence of symptoms and abnormal physical signs with the well-marked x-ray changes, which show disseminated nodular opacities throughout both lungs. The opacities are discrete but sometimes so numerous as to overlap and appear confluent. No massive shadows have been reported. The outstanding feature of the radiographs is the marked radio-opacity of the nodules, which is understandable in view of the substance's use as a radio-opaque medium. The size of the individual elements may vary between 1 and 5 mm in diameter, although the average is about 3 mm or less, and the shape has been described variously as 'rounded' and 'dendritic'. In some cases, a number of very dense points have been found to lie in a matrix of lower density.

In one series of cases, dust concentrations of up to 11,000 particles/cm^3 were measured at the workplace, and chemical analysis showed that the total silica content lays between 0.07% and 1.96%, quartz not being detectable by x-ray diffraction. Men exposed for up to 20 years and exhibiting x-ray changes were symptomless, had excellent lung function, and were capable of carrying out strenuous work. Years after the exposure has ceased, follow-up examinations show a marked clearing of x-ray abnormalities. Reports of post-mortem findings in pure baritosis are practically non-existent. However, baritosis may be associated with silicosis in mining due to contamination of barite ore by siliceous rock and in grinding, if siliceous millstones are used.

Adequate washing and other sanitary facilities should be provided for workers exposed to toxic soluble barium compounds, and rigorous personal hygiene measures should be encouraged. Smoking and consumption of food and beverages in workshops should be prohibited. Floors in workshops should be made of impermeable materials and frequently washed down. Employees working on such processes as barite leaching with sulphuric acid should be supplied with acid-resistant clothing and suitable hand and face protection. Although baritosis is benign, efforts should still be made to reduce atmospheric concentrations of barite dust to a minimum. In addition, particular attention should be paid to the presence of free silica in the airborne dust.

Barium Chloride, Anhydrous (CAS No. 10361-37-2)

Molecular formula: **BaCl$_2$**
Synonyms and trade names: Barium dichloride; Hydrochloric acid; Barium salt

Barium dichloride is a white solid, odorless, hygroscopic chemical substance. Barium dichloride is used in the manufacture of pigments, in the manufacture of other barium salts and in fireworks to give a bright green color. It is one of the most common water-soluble salts of barium. Like other barium salts, it is toxic and imparts a yellow-green coloration to a flame. Barium chloride has wide application in the laboratory.

Safe Handling and Precautions

Occupational workers exposed to barium chloride, anhydrous has been reported to suffer with health disorders. Accidental exposures at workplaces through ingestion, or inhalation and or of skin contact barium chloride, anhydrous cause irritation, tearing, and burning pain of the eyes, allergic reactions, sensitization, causes necrosis to the wet or moist skin and ingestion or swallowing causes kidney damage. The barium ion has been reported as a muscle poison causing stimulation and paralysis. The symptoms of poisoning include, but not limited to, nausea, vomiting, colic, and diarrhea, abdominal cramps, convulsions, myocardial and general muscular stimulation with tingling in the extremities, dullness, and unconsciousness. Barium chloride affects the muscles – the smooth muscles of the cardiovascular and respiratory systems, salivation, tingling of the mouth or face, convulsions, numbness, muscle paralysis, respiratory failure, slow pulse rate, pulmonary edema. Workers on overexposure suffer with severe abdominal pain, violent purging with watery and blood y stools. Barium chloride, anhydrous has not been listed by ACGIH, IARC as a human carcinogen.

Warning! Precautions!!

Barium chloride, along with other water-soluble barium salts, is highly toxic. Repeated exposures to barium chloride, anhydrous may cause respiratory tract irritation, lung damage, cardiac disorders and muscular disorders, kidney damage and could cause death.

Barium chloride is water-soluble and workers should keep it stored in dry containers in a cool, place.

Regulations: Based on laboratory animal studies Occupational Safety and Health Administration (OSHA) (Hazard Communication Standard, 29 CFR 1910.1200) Barium chloride is listed as a hazardous chemical.

Barium Dichloride Dihydrate (CAS No. 10326-27-9)

Molecular formula: $BaCl_2\ 2H_2O$
Synonyms: Barium chloride

Barium chloride is a white/colourless solid and stable under ordinary conditions of use and storage. It is incompatible with bromine trifluoride and 2-furan percarboxylic acid (anhydrous).

Safe Handling and Precautions

Exposures to barium chloride cause sore throat, coughing, and laboured breathing and become harmful and fatal if swallowed and inhaled. Prolonged exposures cause irritation of skin, eyes, and respiratory tract and involve the heart, respiratory system, and central nervous system (CNS). On accidental ingestion, barium chloride causes severe digestive tract irritation with abdominal pain, gastroenteritis, vomiting, diarrhoea, haemorrhaging of the digestive tract, and kidney damage. Barium dichloride dihydrate affects the muscles, especially the smooth muscles of the cardiovascular and respiratory systems. Severe effects of exposure also include tingling of the mouth or face, numbness, muscle paralysis (paralysis of arms and legs), respiratory failure, slow pulse rate, pulmonary oedema, tremors, faintness, and slow or irregular heartbeat. In severe cases, barium chloride may cause respiratory failure leading to collapse and death.

The PELs for soluble barium compounds have been set by OSHA at 0.5 mg (Ba)/m^3; the American Conference of Governmental Industrial Hygienists (ACGIH) has set the threshold limit value (TLV) at 0.5 mg (Ba)/m^3 and the estimated lethal dose in humans.

Barium chloride is grouped by IARC and NTP as A4, meaning not classifiable as a human carcinogen. However, barium chromate (VI) is the only barium compound for which there is sufficient evidence that it is a human carcinogen (IARC, 1980). Workers should be careful during handling and management of barium dichloride dihydrate. It should be kept in a tightly closed container, protected from physical damage, and stored in a cool, dry, ventilated area away from sources of heat, moisture, and incompatibilities. Containers of this material may be hazardous when empty since they retain product residues (dust, solids); observe all warnings and precautions listed for the product.

Occupational workers with pre-existing skin disorders or impaired respiratory function may be more susceptible to the effects of barium dichloride dihydrate. During handling and use of barium chloride, it is important to avoid contact of the substance with eyes, skin, and clothing and avoid ingestion and inhalation. After handling/use, workers should wash the exposed parts, eyes, and skin and change clothes. After use and handling, workers of barium chloride should wash hands thoroughly before eating. Workers should handle the chemical in areas with adequate ventilation, should minimise dust generation and its accumulation.

Occupational workers during handling of barium dichloride dihydrate should use appropriate personal protective equipment (PPE). Workers should wear appropriate protective eyeglasses or chemical safety goggles as described by OSHA's eye and face protection regulations in 29 CFR 1910.133, appropriate protective gloves to prevent exposure, protective clothing to prevent skin exposure, and a NIOSH-approved respirator when necessary.

Special Precautions: Warning

Barium dichloride dehydrate is incompatible with interhalogens and furan-2-percarbonic acid. It is harmful if swallowed.

Barium Fluoride (CAS No. 7787-32-8)

Molecular formula: **BaF$_2$**
Synonym and trade name: Barium difluoride

Barium fluoride appears as colourless, odourless crystals/white powder and is slightly soluble in water but soluble in ethanol and methanol. When heated, barium fluoride decomposes and emits toxic fumes of fluorine and barium. Barium fluoride is used as a preopacifying agent and in enamel and glazing frits production. Its other use is in the production of welding agents (an additive to some fluxes, a component of coatings for welding rods and in welding powders). It is also used in metallurgy as a molten bath for refining aluminium.

Safe Handling and Precautions

Barium compounds including barium fluoride are known to cause severe gastroenteritis, including abdominal pain, vomiting, and diarrhoea; tremors; faintness; paralysis of the arms and legs; and slow or irregular heartbeat. In severe cases, it leads to collapse and death due to respiratory failure. Inhalation of fumes of barium fluoride causes sore throat, coughing, laboured breathing, irritation of the respiratory tract, salivation, nausea, vomiting, diarrhoea, and abdominal pain, followed by weakness, tremors, shallow respiration, convulsions, brain and kidney damage, and coma. Chronic poisoning and inhalation of barium fluoride cause severe bone changes, bone fluorosis, loss of weight, anorexia, anaemia, dental defects, irritation of the respiratory tract and mucous membranes, asthma attacks and lung damage, lung granulomas and pulmonary oedema, pulmonary fibrosis, hyperaemia, cellular eosinophilia and vascular granulomata, acute chemical pneumonitis, subacute bronchitis, and focal hypertrophic emphysema. Repeated and high concentrations of the chemical substance are known to cause poisoning with multiple symptoms such as immediate defecation, writhing, loss of muscle coordination, laboured respiration, sedation, hypotension, dyspnoea, hyperaemia, liver oedema and necrosis, portal congestion, pleural effusion and granulomatous peritonitis with serous and hemorrhagic ascites, and respiratory and cardiac failure.

Occupational workers during handling of barium fluoride must strictly avoid eating, drinking, or smoking while handling this material and must wash hands and face thoroughly after working with material. Workers should store barium fluoride at room temperature (15°C–20°C) in a tightly sealed/closed container in a cool, dry, well-ventilated area, away from incompatible products. The container must be kept well protected from direct sunlight and moisture and away from combustibles (wood, paper, etc.) as it may ignite them. Occupational workers during use and handling of barium fluoride must also wear safety glasses; rubber or plastic gloves, adequate ventilation, full-face mask, self-contained certified breathing apparatus with full protective clothing to prevent contact with skin and eyes. Fumes from fire are hazardous. Isolate run-off to prevent environmental pollution.

Barium Hydroxide Octahydrate (CAS No. 12230-71-6)

Molecular formula: $Ba(OH)_2$
Synonyms and trade names: Barium hydroxide octahydrate; Barium hydrate; Barium hydroxide; 8-Hydrate

Barium hydroxide is available as colourless or white crystals. It is odourless and soluble in water. It is stable under ordinary conditions of use and storage. It is incompatible with acids, oxidisers, and chlorinated rubber. Barium hydroxide is corrosive to metals such as zinc. Barium hydroxide is very alkaline and rapidly absorbs carbon dioxide from air, becoming completely insoluble in water. Barium hydroxide is used in analytical chemistry for the titration of weak acids. Barium hydroxide is used in organic synthesis as a strong base. Barium hydroxide decomposes to barium oxide when heated to 800°C.

Exposures to barium hydroxide through inhalation of its dust cause harmful adverse health effects, blood abnormalities, and toxicity. Barium hydroxide is corrosive and affects the respiratory tract, digestive track, CNS, and kidneys. Barium hydroxide presents the same health hazards like other strong bases and other water-soluble barium compounds. The signs and symptoms of toxicity include, but are not limited to, irritation of the nose, throat, and respiratory tract. Workers without proper protection suffer from irritation of the eyes and skin, skin burns, sore throat, coughing, and shortness of breath. Repeated exposures by ingestion to high concentrations of barium hydroxide cause systemic poisoning with symptoms of severe irritation of the GI tract, tightness in the muscles of the face and neck, vomiting, diarrhoea, abdominal pain, muscular tremors, anxiety, weakness, laboured breathing, cardiac irregularity, convulsions, cardiac and respiratory failure, and death. Exposures to very high doses (1–15 g) cause kidney damage and death within hours or up to a few days. Occupational workers who suffer with pre-existing health disorders of skin and nervous system and impaired respiratory or kidney function have been shown to be more susceptible to the effects of barium hydroxide. Barium hydroxide is grouped as A4, meaning not classifiable as a human carcinogen.

The U.S. OSHA has set the PEL for the soluble barium compounds at 0.5 mg (Ba)/m³ at the workplace air, and the ACGIH has set the TLV at 0.5 mg (Ba)/m³. Barium hydroxide should be stored in a tightly closed container, in a cool, dry, ventilated area. The chemical substance should be kept protected against physical damage, in isolation and away from incompatible chemical substances. The containers of barium hydroxide may be hazardous when empty because they retain residues as dust and solids. Occupational workers should be careful during handling of barium hydroxide. Workers should wear impervious protective clothing, including boots, gloves, lab coat, apron, or coveralls, as appropriate, to prevent skin contact. Workers should use chemical safety goggles and full-face shield. Workplace should maintain eyewash fountain and quick-drench facilities. Occupational workers should observe precautions of not releasing chemical spill and waste products to drains but vacuuming or sweeping up material and placing it into a suitable disposal container. The waste spills should immediately be cleaned up observing precautions. It is important that during use of barium hydroxide, workers should minimise dust generation and accumulation at the work area, which should have adequate ventilation.

Barium Nitrate (as Ba) (CAS No. 10022-31-8)

Molecular formula: **Ba (NO$_3$)$_2$**
Synonyms and trade names: Barium (II) nitrate; Barium dinitrate; Nitric acid; Barium salt

Barium nitrate is a stable, strong oxidiser. It is incompatible with combustible material, reducing agents, acids, acid anhydrides, and moisture-sensitive substance. Barium nitrate is poisonous, is a respiratory irritant, and is hazardous if mixed with flammable materials. Barium oxide plus zinc, aluminium and magnesium alloys are combustibles (paper, oil, wood), acids, and oxidisers and is hazardous. Mixtures with finely divided aluminium–magnesium alloys are easily ignitable and extremely sensitive to friction or impact.

Barium nitrate is a severe poison and becomes fatal if accidentally swallowed during improper use and handling. The symptoms of poisoning include, but are not limited to, coughing and shortness of breath. Systemic poisoning and the symptoms of ingestion include tightness of the muscles of the face and neck, vomiting, diarrhoea, abdominal pain, muscular tremors, anxiety, weakness, laboured breathing, cardiac irregularity, convulsions, and death from cardiac and respiratory failure. Chronic exposure to barium nitrate causes skin burns and ulcerations and damage to the CNS, liver, spleen, kidney, and bone marrow system. Workers during use and handling of barium nitrate should use a PPE such as a dust mask since the mixtures of metal powders and barium nitrate sometimes heat up spontaneously and may ignite, especially when moist. Barium nitrate mixed with aluminium powder, a formula for flash powder, is highly explosive. However, barium nitrate is non-corrosive in presence of glass. It is used in military thermite grenades, in the manufacturing process of barium oxide, in the vacuum tube industry, and in pyrotechnics for green fire.

Safe Handling and Precautions

Barium nitrate has not been listed by the National Toxicology Program (NTP), IARC, and OSHA as a human carcinogen. Barium nitrate should be stored in a tightly closed container, in a cool, dry, well-ventilated area, protected against physical damage. It should be completely separated from heat, sources of ignition, incompatible substances, combustibles, organic or other readily oxidisable materials, acids, alkalis, and reducing agents. Barium nitrate should be kept for later disposal as solid waste.

Workers should ensure that the container is properly labelled with the name of the material and shows that it is a strong oxidising agent. Barium nitrate should not be kept on wood floors or along with food and beverages. Occupational workers after accidental exposure by ingestion, swallow, inhalation of barium nitrate, should get immediate medical attention and support. Never give anything by mouth to an unconscious person and get medical attention immediately.

Barium Sulphate (CAS No. 7727-43-7)

Molecular formula: $BaSO_4$
Synonyms and trade names: Barite; Barium sulphate; Sulphuric acid; Barium salt; Barytes; Blanc fixe; Barite; Enamel white

Barium sulphate is available as odourless, tasteless, white or yellowish crystals or powder or polymorphic crystals. It is stable and insoluble or negligibly soluble in water, and on burning, it may produce sulphur oxides. It reacts violently with aluminium powder. It occurs naturally as mineral barite, barytes. It has wide use as inert filler and pigment extender in paints, primers, inks, plastics, floor tiles, paper coatings, polymer fibres, and rubber. It is used as the semi-transparent base (lake) for organic pigments and as a thixotropic weighting mud in oil well drilling. Barium sulphate is a contrast agent that is used to help x-ray diagnosis of problems in areas of the upper GI tract, like the oesophagus, the stomach, and/or the small intestine. The raw barium sulphate has wide applications such as in the manufacture of lithopone, a white pigment in the manufacture of photographic paper, wallpaper and glassmaking, in battery plate expanders, and in heavy concrete for radiation shield.

Safe Handling and Precautions

Barium sulphate is also known to cause several side effects, which include stomach cramps, diarrhoea, nausea, vomiting, constipation, weakness, pale skin, sweating, ringing in the ears, itching, skin redness, swelling or tightening of the throat, difficulty breathing or swallowing, hoarseness, agitation, confusion, fast heartbeat, and bluish skin colour.

Exposures to barium sulphate cause irritation of the eyes, lacrimation, redness, scaling, and itching, characteristic of skin inflammation. Repeated and/or prolonged exposure has been reported to cause adverse effects to the lungs and mucous membranes and can produce target organ damage. Although barium sulphate has been identified as a nontoxic dust, long-term inhalation of dust in high concentrations has caused benign pneumoconiosis (baritosis) and deposition of dust in lungs in sufficient quantities to produce adverse effects like rhinitis, frontal headache, wheezing, laryngeal spasm, salivation, and anorexia. This produces a radiological picture even though symptoms and abnormal signs may not be present. Long-term effects include nervous disorders and adverse effects on the heart, circulatory system, and musculature. There are, however, no reports indicating that barium sulphate has potential occupational hazards or carcinogenicity. The exposure limits for barium sulphate have been determined as follows: OSHA PEL as 15 mg/m³ total dust and 5 mg/m³ respirable dust and ACGIH TLV as 10 mg/m³ (TWA).

Barium sulphate should be used only by or under the direct supervision of a qualified supervisor or doctor. During use and handling of barium sulphate, workers should avoid contact of the chemical substance with the skin and should not breathe in its dust. After use, the workers should wash their hands with soap and water. While not expected to be a health hazard, chronic exposure and long-term inhalation of the chemical's dust may lead to deposition in lungs in sufficient quantities to produce baritosis – a benign pneumoconiosis. This produces a radiological picture even though symptoms and abnormal signs may not be present. While dispensing, care is needed to properly and correctly label the container to avoid confusion with poisonous barium sulphide, sulphite, or carbonate. Barium sulphate should be kept in its original sealed glass container with proper security when not in use. Barium sulphate should be stored in a cool, dry, ventilated area, away from sources of heat, moisture, incompatibilities, and foodstuff. During handling of barium sulphate, occupational workers should use splash goggles, full suit, dust respirator, boots, gloves, and self-contained breathing apparatus (SCBA) to avoid inhalation of the product.

Bendiocarb (CAS No. 22781-23-3)

Molecular formula: $C_{11}H_{13}NO_4$
Synonyms and trade names: Bencarbate; Bendiocarbe; 1,3-Benzodioxol-4-ol; 2,2-Dimethyl-, methylcarbamate; Ficam; Ficam 80 W; Garvox; NC 6897; Seedox; Dycarb; Multamat; Niomil; Rotate; Seedox; Tattoo; Turcam
IUPAC name: 2,3-isopropylidenedioxyphenyl methylcarbamate
Toxicity class: U.S. EPA, II; WHO, II

Bendiocarb is an odourless, non-corrosive white crystalline solid. Bendiocarb is a carbamate ester. Carbamates are chemically similar to but more reactive than amides. Like amides, they form polymers such as polyurethane resins. Some of the formulations of bendiocarb are classified as general use pesticides (GUP), while Turcam and its 2.5 G formulation have been classified as restricted use pesticides (RUP). Bendiocarb is stable under normal temperatures and pressures, but should not be mixed with alkaline preparations. Thermal decomposition products may include toxic oxides of nitrogen. It is non-corrosive. Flammable gaseous hydrogen is produced by the combination of active metals or nitrides with carbamates. Strongly oxidising acids, peroxides, and hydroperoxides are incompatible with carbamates.

Bendiocarb as a carbamate insecticide is effective against a wide range of insects that cause nuisance and act as disease vectors. It is used to control mosquitoes, flies, wasps, ants, fleas, cockroaches, silverfish, ticks, and other pests in homes, industrial plants, and food storage sites. Bendiocarb is also used as a seed treatment on sugar beets and maize and against snails and slugs. Pesticides containing bendiocarb are formulated as dusts, granules, ultra-low volume sprays, and wettable powders. It gets hydrolysed rapidly in alkali media and slows under acid and neutral conditions. Flammable gaseous hydrogen is produced by the combination of active metals or nitrides with carbamates.

Safe Handling and Precautions

Bendiocarb belongs to a carbamate class of systemic insecticides and its contact results in a rapid knock-down effect of the organism. The early symptoms associated with bendiocarb exposure include, but are not restricted to, headache, malaise, muscle weakness, nausea, GI cramps, sweating, and restlessness. Bendiocarb also causes pinpoint pupils, tearing, excessive salivation, nasal discharge, vomiting, diarrhoea, muscle twitching, and ataxia. Severe poisonings can result in convulsions, CNS depression, coma, and death. Ingestion and/or absorption through the skin may cause hepatic and renal diseases and respiratory problems.

Absorption through the skin is the most likely route of exposure. Bendiocarb is absorbed through all the normal routes (oral, dermal, and inhalation) of exposure, but dermal absorption is especially rapid. Carbamates generally are excreted rapidly and do not accumulate in mammalian tissue. If exposure does not continue, cholinesterase inhibition and its symptoms reverse rapidly. In non-fatal cases, the illness generally lasts less than 24 h. Bendiocarb is moderately toxic to birds. The LD50 in mallard ducks is 3.1 mg/kg and in quail is 19 mg/kg. Bendiocarb is moderately to highly toxic to fish. The LC50 (96 h) for bendiocarb in rainbow trout is 1.55 mg/L.

Bendiocarb is a mild irritant to the skin and eyes. Prolonged exposures to high concentrations of bendiocarb cause severe poisoning, and symptoms of irritation and pain, blurred vision, tearing, muscle spasms, and unresponsive pupils (to changes in light) may all occur if bendiocarb gets in the eyes. Symptoms also include twitching, giddiness, confusion, muscle incoordination, slurred speech, low blood pressure, heart irregularities, and loss of reflexes. Death can result from discontinued breathing, paralysis of muscles of the respiratory system, intense constriction of the openings of the lung, or all three. These effects are due to anti-cholinesterase activity. In one case of exposure while applying bendiocarb, the victim experienced symptoms of severe headache, vomiting, and excessive salivation, and his cholinesterase level was depressed by 63%. He recovered from these symptoms in less than 3 h with no medical treatment and his cholinesterase level returned to normal within 24 h. In another case, poisoning occurred when an applicator who was not wearing protective equipment attempted to clean contaminated equipment. The victim experienced nausea; vomiting; incoordination; pain in his arms, hands, and legs; muscle spasms; and breathing difficulty. Bendiocarb is readily absorbed by the GI tract and is rapidly metabolised. Bendiocarb causes disruption of the nervous system by adding a carbamyl moiety to the active site of the acetylcholinesterase enzyme. This prevents acetylcholine from reaching the active site and renders the enzyme inactive. However, the carbamyl group is released by spontaneous hydrolysis, thus reversing the disruption and restoring the activity.

A 2 year study with rats exposed to high doses of bendiocarb (10 mg/kg/day) showed a wide range of adverse effects in organ weights, blood, and urine characteristics, as well as an increased incidence of stomach and eye lesions. In a three-generation study with rats, fertility and reproduction were not affected by bendiocarb at dietary doses of up to 12.5 mg/kg/day. Very high doses (40 mg/kg/day) during prenatal and post-natal periods caused toxic effects to rat dams and reduced pup weight and survival rates. No effects were seen at 20 mg/kg/day. Thus, no reproductive effects are likely to occur in humans at expected exposure levels.

Laboratory studies have indicated that bendiocarb is not carcinogenic, teratogenic, and/or mutagenic in species of animals. The acceptable daily intake (ADI) for bendiocarb has been set at 0.004 mg/kg/day. Published data regarding recommended limits/levels of bendiocarb in drinking water, surface water, or foods or other items and in daily diets are not available from the ACGIH, NIOSH, and/or OSHA.

The preliminary risk assessment of U.S. EPA showed that applicators, including home owners, were at risk when mixing or applying the pesticide. All bendiocarb-containing products in the United States recently had their registrations cancelled due to concerns of overexposure of those applying the products. Regulatory agencies in the United Kingdom have still not restricted the use of bendiocarb as a biocide. As a precaution, bendiocarb should be purchased and used only by certified and trained applicators. Its registration was voluntarily cancelled in September 1999, and all products containing bendiocarb lost registration in December 2001 with cancelled registrations cannot be purchased after that date, but existing stocks have been permitted for use according to labelled directions. It has been reported that products of bendiocarb (as granular, dust, or liquid spray formulations) are used in gardens, turf, soil, and ornamental plants.

Benfuracarb (CAS No. 82560-54-1)

Molecular formula: $C_{20}H_{30}N_2O_5S$
Chemical name: Ethyl 3-[[(2,2-dimethyl-3H-benzofuran-7-yl)oxycarbonyl-methyl-amino] sulfanyl-propan-2-yl-amino]propanoate

Safe Handling and Precautions

Benfuracarb cause poisoning on exposure by ingestion, inhalation, and or by subcutaneous route. When heated, it decomposes and emits toxic vapours of NOx and SOx. Benfuracarb is moderately toxic by skin contact and is harmful by inhalation, and it is toxic to aquatic organisms. Occupational workers, after repeated and long periods of exposure to benfuracarb, may develop health effects. Workers during handling of benfuracarb should use personal protective equipment (PPE), should wear self-contained equipment for firefighters, and respirators. Suitable extinguishing include water spray, carbon dioxide (CO_2), dry powder, and alcohol-media-resistant foam. Workers should avoid using media with high-volume waterjet for safety reasons and collect the contaminated fire-extinguishing water separately, avoiding discharging it into drains. Also, workers should store benfuracarb in their original chemical container, tightly closed in a cool, well-ventilated place and protected from exposures to frost.

Regulations: The review report for benfuracarb (active grade) was finalised in the Standing Committee on the Food Chain and Animal Health at its meeting on 16 March 2007, in support of a decision concerning the non-inclusion of benfuracarb in Annex I of Directive 91/414/EEC and the withdrawal of authorisations for plant protection products containing this active substance. The information available was found to be insufficient to meet the requirements set out in Annex II and Annex III Directive 91/414/EEC in particular with regard to the following:

- The risk to groundwater, especially with regard to a number of relevant metabolites.
- The consumer exposure, the risk to birds and mammals, the risk to aquatic organisms, and the risk to earthworms, non-target arthropods, and other soil organisms.
- The concerns were identified with regard to the toxicity of the substance and the high toxicity of some of its metabolites; the presence of carcinogenic impurities in the technical substance; the consumer exposure that is regarded inconclusive and that indicated, mainly due to certain metabolites, a potential acute risk to certain vulnerable groups of consumers; the possible contamination of groundwater, especially by a number of relevant metabolites; and the substantial lack of data for almost all groups in the ecotoxicological field.

Assessments of bendiocarb suggest that it should, not be included in Annex I of Directive 91/414/EEC. Benfuracarb should, therefore, not be included in Annex I of Directive 91/414/EEC.

Benomyl (CAS No. 17804-35-2)

Molecular formula: $C_{14}H_{18}N_4O_3$
Synonyms and trade names: Benlate; Tersan; Fungicide 1991; Fundazol
Chemical name: Carbamic acid, ((1-(butylamino)carbonyl)-1H-benzimidazol-2-yl), methyl ester

Benomyl, a tan-coloured crystalline solid/powder, is a systemic fungicide with a characteristic odour. It belongs to the benzimidazole family. Benomyl decomposes at high temperature. Benomyl is essentially insoluble in water. It is stable under normal storage conditions but will decompose to carbendazim in water. On decomposition by heat, benomyl produces toxic fumes including nitrogen oxides.

Benomyl is a systemic and broad-spectrum fungicide that is currently registered for use in more than 50 countries on more than 70 crops for the control of diseases in fruit trees, nut crops, vegetables, cereals, tropical crops and ornamentals, turf, and many field crops. Benomyl is marketed as a wettable powder and as a dry flowable formulation (dispersible granules).

Safe Handling and Precautions

Benomyl has a very low acute toxicity. Benomyl causes contact dermatitis and dermal sensitisation in some farm workers. It is only a mildly irritant to the skin and eyes but sensitises skin. Repeated or prolonged contact may cause skin sensitisation. Animal tests show that this substance possibly causes toxic effects upon human reproduction. Benomyl poisoning in the general population has not been reported in scientific literature. Recent data used to estimate dietary exposure based on food consumption patterns within the United States indicate exposures well below the no observable effects levels (NOELs) in animal toxicity tests. Also no inadvertent poisoning of agricultural or forestry workers has been documented. Further occupational exposures during manufacture or crop application are below the established. Benomyl is rapidly converted to carbendazim in various environmental compartments.

Benomyl should be stored in a well-ventilated area. Occupational workers should keep the benomyl containers tightly closed and away from food, drink, and children. Workers should never allow benomyl to become wet during storage. This may lead to certain chemical changes that could increase its toxicity (lacrimation because of the formation of butyl isocyanate) and reduce the effectiveness of benomyl as a fungicide.

Liquid formulations of benomyl-containing organic solvents may be flammable. Workers, during use of benomyl, should observe rules about the treatment, storage, transportation, and disposal that follow local regulations. Workers, during handling of large quantities (2 kg bags or greater) of solid formulations of benomyl, should use a dust mask and protective clothing.

Benzaldehyde (CAS No. 100-52-7)

Molecular formula: C_7H_6O
Synonyms and trade names: Artificial almond oil; Benzoic aldehyde; Benzene carbaldehyde; Benzenecarbonal; Benzene carboxaldehyde; Benzenemethylal; Phenylmethanal; Synthetic bitter almond oil

Benzaldehyde is a colourless to yellow, oily liquid with an odour of bitter almonds. Benzaldehyde is commercially available in two grades: (1) pure benzaldehyde and (2) double-distilled benzaldehyde. The latter has applications in the pharmaceutical, perfume, and flavour industries. Benzaldehyde may contain trace amounts of chlorine, water, benzoic acid, benzal chloride, benzyl alcohol, and/or nitrobenzene. Benzaldehyde gets ignited relatively easily like contact with hot surfaces. This has been attributed to the property of very low autoignition temperature. Benzaldehyde also undergoes autoxidation in air and is liable to self-heat. Benzaldehyde exists in nature, occurring in combined and uncombined forms in many plants. Benzaldehyde is also the main constituent of the essential oils obtained by pressing the kernels of peaches, cherries, apricots, and other fruits. Benzaldehyde is released to the environment in emissions from combustion processes such as gasoline and diesel engines, incinerators, and wood burning. It is formed in the atmosphere through photochemical oxidation of toluene and other aromatic hydrocarbons. Benzaldehyde is corrosive to grey and ductile cast iron (10% solution) and all concentrations of lead. However, pure benzaldehyde is not corrosive to cast iron. Benzaldehyde does not attack most of the common metals, like stainless steel, aluminium, aluminium bronze, nickel and nickel-base alloys, bronze, naval brass, tantalum, titanium, and zirconium. On decomposition, benzaldehyde releases peroxybenzoic acid and benzoic acid. Benzaldehyde is used in perfumes, soaps, foods, drinks, and other products and as a solvent for oils, resins, some cellulose ethers, cellulose acetate, and cellulose nitrate. The uses of benzaldehyde in industries are extensive as in the production of derivatives that are employed in the perfume and flavour industries like cinnamaldehyde, cinnamyl alcohol, cinnamic acid, benzylacetone, and benzyl benzoate, in the production of triphenylmethane dyes and the acridine dye, benzoflavin; as an intermediate in the pharmaceutical industry, for instance, to make chloramphenicol, ephedrin, and ampicillin, as an intermediate to make benzoin, benzylamine, benzyl alcohol, mandelic acid, and 4-phenyl-3-buten-2-one (benzylideneacetone), in photochemistry, as a corrosion inhibitor and dyeing auxiliary, in the electroplating industry, and in the production of agricultural chemicals.

Safe Handling and Precautions

Exposures to vapour of benzaldehyde cause irritation of the upper respiratory tract, intolerable irritation of the nose and throat, headache, nausea, dizziness, drowsiness, and confusion. It is a CNS depressant. Exposures to benzaldehyde cause moderate to severe eye irritation and prolonged period of exposures cause corrosive effects to skin like burns, scarring, and skin injury; fatigue; headache; nausea; dizziness; and loss of coordination. At higher concentrations, benzaldehyde produces more severe effects such as sore throat, abdominal pain, nausea, CNS depression, convulsions, and respiratory failure. The estimated lethal dose of benzaldehyde has been reported as 2 oz. There are no data with humans and the information and conclusions are based on evidence obtained from animal studies. The NTP or the IARC, the OSHA, and the ACGIH have not reported benzaldehyde as a human carcinogen.

Benzaldehyde should be kept in a tightly closed container and protected against physical damage. Storage of the chemical substance outside or in a detached area is preferred, whereas inside storage should be in a standard flammable liquids storage room or cabinet. Benzaldehyde should be stored away from sources of heat, sparks and flame, and ignition and separated from oxidising materials.

Occupational workers, during handling of benzaldehyde, should avoid breathing gas/fumes/vapour/spray of the chemical. Workers should wear suitable protective clothing, and work area should have proper and sufficient ventilation and suitable respiratory equipment. Also the storage and use areas should be strictly labelled as 'NO SMOKING'. Containers of benzaldehyde, when empty, may pose a hazard since they retain product residues (vapours, liquid); therefore, observe all warnings and precautions listed for the product.

Workers should be careful while using benzaldehyde because there is a risk of spontaneous combustion. It may ignite spontaneously if it is absorbed onto rags, cleaning clothes, clothing, sawdust, diatomaceous earth (kieselguhr), activated charcoal, or other materials with large surface areas in workplaces as well. Workers should avoid handling of the chemical substance and not cut, puncture, or weld on or near the container.

Benzamide (CAS No. 55-21-0)

Molecular formula: $C_6H_5CONH_2$
Synonyms and trade names: Benzoylamide; Benzoic acid amide; Phenylcarboxyamide

Benzamide appears as off-white crystals or powder. It is combustible and incompatible with strong oxidising agents and strong bases. On combustion and thermal decomposition, it emits nitrogen oxides, carbon monoxide, and carbon dioxide.

Safe Handling and Precautions

Exposure to benzamide causes harmful health effects. Accidental ingestion or swallowing of benzamide at workplaces causes effects of irritation of the skin and eyes and GI irritation, nausea, vomiting, and diarrhoea.

During handling of benzamide, workers should wear an SCBA and full protective gear. During use and handling of benzamide, occupational workers should be careful and must have a well-ventilated area and minimise dust generation and its accumulation. Workers should avoid contact of the chemical substance with eyes, skin, and clothing and keep the chemical substance in a tightly closed container. Benzamide must be stored in a cool, dry, well-ventilated area away from non-compatible substances and children.

1,2-Benzanthracene (CAS No. 56-55-3)

Molecular formula: $C_{18}H_{12}$
Synonyms and trade names: 1,2-Benz(a)anthracene; Benzanthrene; Benzo(a)anthracene; Benzo(b)phenanthrene; 2,3-Benzophenanthrene; Naphthanthracene; Tetraphene

1,2-Benzanthracene is available as colourless to yellow-brown fluorescent flakes or powder. It is stable, combustible, and incompatible with strong oxidising agents. On decomposition, 1,2-benzanthracene releases carbon monoxide, carbon dioxide, acrid smoke, and fumes. During work, 1,2-benzanthracene can be absorbed into the body of occupational workers by inhalation, through the skin, and by ingestion. Exposures may cause irritation of the eyes, skin, and respiratory tract.

Safe Handling and Precautions

Occupational workers, during use and handling of 1,2-benzanthracene, should be careful and avoid exposures since the chemical substance is known to cause kidney damage. However, published data on the neurotoxicity, teratogenicity, reproductive toxicity, and mutagenicity of 1,2-benzanthracene are not available. Benz[a]anthracene has been shown to be carcinogenic to experimental animals. 1,2-Benzanthracene has been grouped as an A2, meaning suspected human carcinogen by the American Conference of Industrial Hygienists (ACGIH), the International Agency for Research on Cancer (IARC) and the National Toxicology Program (NTP) have classified 1,2-Benzanthracene as A2, meaning probable human carcinogen. The OSHA has not listed the exposure limit to 1,2-benzanthracene.

Users and occupational workers, after the use of benzanthracene, should wash their hands thoroughly, and workplace must have a well-ventilated facility. Workers should minimise dust generation and accumulation and avoid contact of the chemical substance with the eyes, skin, and clothing. Workers should keep the container of the chemical tightly closed, stored in a cool, dry, well-ventilated area away from incompatible substances.

Benzene (CAS No. 71-43-2)

Molecular formula: C_6H_6
Synonyms and trade names: Benzine; Benzol; Aromatic hydrocarbon

Benzene is a colourless, flammable liquid with a pleasant odour. It is used as a solvent in many areas of industries, such as rubber and shoe manufacturing, and in the production of other important substances such as styrene, phenol, and cyclohexane. It is essential in the manufacture of detergents, pesticides, solvents, and paint removers. It is present in fuels such as in gasoline up to the level of 5%. Uses of benzene are several.

Safe Handling and Precautions

Exposure to low concentrations of benzene vapour or to the liquid causes dizziness, light-headedness, headache, loss of appetite and stomach upset, and irritation of the nose and throat. Prolonged exposure to high concentrations of benzene leads to functional irregularities in the heartbeat and, in severe cases, to death. Benzene is a known carcinogen to humans. It causes leukaemia and blood disorders such as aplastic anaemia. The major types of leukaemia related to benzene exposure are (i) acute myelogenous leukaemia (AML); (ii) acute lymphocytic leukaemia (ALL); (iii) chronic myelogenous leukaemia, also called chronic myeloid leukaemia (CML); and (iv) chronic lymphocytic leukaemia (CLL) and hairy cell leukaemia (HCL).

Occupational exposure to benzene is frequent such as in road tanker drivers and Chinese glue and shoemaking factory workers. AML (known as acute myeloid leukaemia or acute non-lymphocytic leukaemia) is a blood cancer that develops in specific types of white blood cells (granulocytes or monocytes). White blood cells are used by the body to fight infections. The blood cells affected, granulocytes and monocytes, are created from stem cells (haematopoietic stem cells that will turn into different blood cells). These blood stem cells originate in a person's bone marrow that creates blood cells. With the development of AML, the normal development of white blood cells becomes disturbed and they do not grow properly. Possibly due to some sort of change or damage to their genetic material or DNA, the cells are prevented from growing beyond a certain point.

ALL is a malignant cancer that develops in a person's white blood cells called lymphocytes. Under normal circumstances, ALL is rare among adults; only about 1500 adults get the disease each year in the United States. However, ALL is the prevalent form of leukaemia in children. Nearly 85% of leukaemia in children is ALL. In adults, the disease may be related to genetics or exposure to solvents containing benzene.

Long-term exposure to benzene increases the risks of getting cancer; however, cancer linked to benzene has been discovered in people exposed for less than 5 years. Workers exposed for decades are at increased risk for these rare forms of leukaemia and long-term exposure may also adversely impact bone marrow and blood production. Still other workers have been diagnosed with aplastic anaemia, a group of disorders that prevent bone marrow from producing all three types of blood cells: red blood cells, white blood cells, and platelets.

The U.S. EPA has set the maximum permissible level of benzene in drinking water at 5 ppb. The OSHA has set a limit of 1 ppm of benzene in workplace air for 8 h (TWA). The NIOSH recommends that benzene should be treated as a potential human carcinogen. The NIOSH usually recommends that occupational exposures to carcinogens be limited to the lowest feasible concentration: 0.1 ppm for 10 h (TWA) and 1 ppm for 15 min (short-term exposure limit [STEL]). The TLVs for benzene are as follows: 0.5 ppm for 8 h (TWA) and 2.5 ppm for 15 min (STEL) on skin. More information on the toxicity and health effects of benzene among occupational workers may be found in the literature.

Benzidine (CAS No. 92-87-5)

Molecular formula: $C_{12}H_{12}N_2$
Synonyms and trade names: 4,4'-Diaminophenyl; 4,4'-Diphenylenediamine; 4,4'-Biphenyldiamine; 4,4'-Biphenylenediamine; p-Benzidine; 1,1'-Biphenyl-4; 4'-Diamine; p-Diaminodiphenyl; p,p'-Bianiline; 4,4'-Bianiline; p,p'-Diaminobiphenyl; Fast corinth base b; Benzidine base

Benzidine is a white, greyish-yellow, or slightly reddish crystalline solid or powder. The major use for benzidine is in the production of dyes, especially azo dyes in the leather, textile, and paper industries and as a synthetic precursor in the preparation and manufacture of dyestuffs. It is also used in the manufacture of rubber, as a reagent, and as a stain in microscopy. It is slightly soluble and slowly changes from a solid to a gas.

Published reports have indicated that a large number of workers involved in the production of dyes, textile, paper, and leather goods potentially are exposed to dyes based on benzidine, o-tolidine, and o-dianisidine. Because more than one of these dyes may be found concurrently in the same industry, it is difficult to count exposed workers and to define the extent of exposure to any specific dye.

Safe Handling and Precautions

Exposure to benzidine causes irritation of the eyes. Laboratory studies of animals exposed through food suggest that as low as 0.01%–0.08% in food may cause a decrease in liver, kidney, and body weight; an increase in spleen weight; swelling of the liver; and blood in the urine. Exposure may cause an increase in urination, blood in the urine, and urinary tract tumours.

Benzidine is considered to be very acutely toxic to humans by ingestion, with an estimated oral lethal dose between 50 and 500 mg/kg. The symptoms of acute ingestion exposure include cyanosis, headache, mental confusion, nausea, and vertigo. Dermal exposure may cause skin rashes and irritation. Prolonged exposure to benzidine causes bladder injury in humans.

Laboratory animals exposed to benzidine via oral, inhalation, and injection developed various tumour types at multiple sites. Epidemiologic studies have indicated that occupational exposure to benzidine causes an increased risk of bladder cancer. The U.S. EPA has classified benzidine as Group A, meaning human carcinogen. Users and occupational workers should be well aware of the safety of benzidine, which at high temperatures breaks down and releases highly poisonous fumes. Benzidine decomposes on heating and on burning, producing toxic fumes including nitrogen oxides. It reacts violently with strong oxidants, especially nitric acid. During use and handling of benzidine, occupational workers should wear butyl rubber gloves, goggles, and full-body plastic coveralls and ensure that no skin is exposed. Workers should keep benzidine in a cool, well-ventilated area; in closed, sealed containers; and out of sunlight and away from heat.

Workplace regulations should be strict, and any kind of access to the workplace should be restricted to authorised workers only. Workers should completely avoid eating, drinking, or smoking in the regulated areas.

Dangers and Precautions

Occupational workers, during handling of benzidine, are likely to be exposed in the workplace by inhalation of its dust or vapours. There is less likelihood to be exposed through the skin or by ingestion. However, it can still be fatal if inhaled, swallowed, or absorbed through the skin. Acute symptoms include nausea, vomiting, and painful and irregular urination. Benzidine is an OSHA-regulated carcinogen covered under CFR Title 29 Part 1910.1010. Inhalation or absorption through the skin of the dust has been recognised as a cause of bladder tumours. Target organs include the liver, kidney, urethra, bladder, blood, and skin. During use and handling of benzidine, students and occupational workers should be very careful. Use of benzidine is banned in many countries. Benzidine should be used only in special circumstances with strict measures of precautions and workplace risk assessment. Workers handling benzidine must have complete PPE and must wear and use a half-face filter-type respirator with filters for dusts, mists, and fumes or air-purifying canisters or cartridges.

Benzotrichloride (CAS No. 98-07-7)

Molecular formula: $C_7H_5Cl_3$

Synonyms and trade names: Trichloromethylbenzene; (Trichloromethyl)benzene; alpha,alpha,alpha-Trichlorotoluene; omega,omega,omega-Trichlorotoluene; Benzenyl chloride; Benzyl trichloride; Benzylidene chloride; Benzoic trichloride; Phenylchloroform; Phenyltrichloromethane; Toluene trichloride; Trichlorophenylmethane; 1-1-Trichloromethyl)benzene

Benzotrichloride is a colourless to light yellow oily liquid with penetrating odour and is a stable substance. It is sensitive to moisture and incompatible with strong oxidising agents and strong acids, alkali metals, alkaline earth metals, aluminium, amines, and water. Benzotrichloride hydrolyses in contact with moisture/water and releases toxic and corrosive fumes of hydrogen chloride and aqueous hydrochloric acid. Benzotrichloride is very toxic, and exposed occupational workers suffer harmful effects such as sore throat, coughing, shortness of breath and delayed lung oedema, and irritation effects to the skin and respiratory system. It causes lacrimation, and repeated exposure causes eye burns. Accidental ingestion (swallowing) of benzotrichloride at workplaces causes burning effects of GI tract, burns, and CNS depression.

Safe Handling and Precautions

Occupational workers should be very careful during the use and handling of benzotri-chloride and must use proper PPE. Exposure to benzotrichloride causes cancer in humans. Laboratory studies in mice have shown that inhalation exposures of benzotrichloride for 30 min/day for a period of 12 months produce bronchitis, bronchial pneumonia, and tumours at multiple sites of the experimental animal. Skin irritation and liver and tes-ticular effects were reported in rabbits exposed dermally to benzotrichloride for 3 weeks.

During use and handling of benzotrichloride, workers should have good and adequate ventilation. Workers should wear chemical splash goggles, quality gloves, clothing to pre-vent skin exposure, and standard-quality respirator. During handling of the chemical sub-stance, workers should avoid light, moisture, and excess heat. Workers should thoroughly wash after handling the chemical substance and remove the contaminated clothing.

Users and occupational workers, during handling of benzotrichloride, should strictly remember these simple rules:

- Workers must obtain permission before use and handling of the chemical substance.
- Do not handle until all safety precautions have been read and understood.
- Do not breathe mist/vapours/spray.
- Wash hands thoroughly after handling.
- Do not eat, drink, or smoke when using this product.
- Should be used only outdoors or in a well-ventilated area.
- Wear protective gloves/protective clothing/eye protection/face protection.
- Use ventilation system or PPE as required and respiratory protection.

2,3-Benzofuran (CAS No. 271-89-6)

Molecular formula: C_8H_6O
Synonyms and trade names: Benzofurfuran; NCI-C56166; Coumarone; Cumarone; 1-Oxindene; Benzo(b)furan

2,3-Benzofuran is a colourless, sweet-smelling, oily liquid made by processing coal into coal oil. It may also be formed during other uses of coal or oil. 2,3-Benzofuran is not used for any commercial purposes, but the part of the coal oil that contains 2,3-benzofuran is made into a plastic called coumarone–indene resin. This resin resists corrosion and is used to make paints and varnishes. The resin also provides water resistance and is used in coatings on paper products and fabrics. It is used as an adhesive in food containers and some asphalt floor tiles. The resin has been approved for use in food packages and as a coating on citrus fruits. We do not know how often the resin is used or whether any 2,3-benzofuran in the coating or packaging gets into the food. 2,3-Benzofuran may enter the air, water, and soil during its manufacture, use, or storage at hazardous waste sites.

Safe Handling and Precautions

Exposure to 2,3-benzofuran causes adverse health effects. On exposure to (ingestion) high levels of 2,3-benzofuran, laboratory animals developed liver and kidney damage. Those animals exposed over a long time to moderate levels showed liver, kidney, lung, and stomach damage. Similar health disorders due to 2,3-benzofuran in humans require further studies. Laboratory mice and rats exposed to 2,3-benzofuran for long periods of time developed cancer of the kidneys, lungs, liver, or stomach. There are no studies on 2,3-benzofuran's potential to cause cancer in humans, but studies report that this chemical substance may be a possible human carcinogen. The DHHS has not classified 2,3-benzofuran as a human carcinogen. The IARC and the U.S. EPA have also not classified 2,3-benzofuran as to its human carcinogenicity.

During use and handling of 2,3-benzofuran, occupational workers must be very careful since it is highly flammable and sensitive to prolonged exposure to air. 2,3-Benzofuran should be handled by workers only in a chemical fume hood and in a well-ventilated area away from incompatible substances. Workers should ground and bond containers when transferring the chemical substance. Workers should use spark-proof tools and explosion-proof equipment and should avoid inhaling/ingesting the chemical and/or contact of the chemical with eyes, skin, and clothing. Empty containers retain product residue (liquid and/or vapour) and can be dangerous. Keep chemical away from heat, sparks, and flame. 2,3-Benzofuran should be kept in a tightly closed container under an inert atmosphere, at refrigerated temperatures, and away from sources of ignition, incompatible substances, and flammable area.

Benzyl Acetate (CAS No. 140-11-4)

Molecular formula: $CH_3COOCH_2C_6H_5$
Synonyms and trade names: Acetic acid; Phenylmethyl ester; Acetic acid; Benzyl ester; Benzyl ethanoate

Benzyl acetate is a colourless liquid with fruity odour. On burning and decomposition, it produces irritating fumes. Benzyl acetate reacts with strong oxidants causing fire and explosion hazards.

Safe Handling and Precautions

Workers exposed to benzyl acetate develop adverse health effects. The symptoms of toxicity and poisoning include, but are not limited to, irritation of the skin and eyes, burning sensation, confusion, dizziness, drowsiness, laboured breathing, sore throat, nausea, vomiting, and diarrhoea. Benzyl acetate also causes adverse health effects to the respiratory tract and CNS with neurological effects.

Occupational workers, during use and handling of benzyl acetate, should be very careful. The chemical substance should be handled with good industrial practices. Workers should avoid contact with the chemical, and the workplace must have adequate ventilation and cleanliness. Workers should store benzyl acetate in a cool, dry place with container closed when not in use.

Workers, after handling and use of benzyl acetate, should wash thoroughly the contaminated clothing before reuse, and should avoid any kind of contact of benzyl acetate with eyes and skin. Workers should wear appropriate personal protective equipment such as face shield, gloves and chemical-resistant apron and safety dress to prevent skin exposure.

During handling of benzyl acetate, occupational workers should not ignore that prolonged inhalation of the vapour of the chemical substance results in serious health disorders of the user/worker. In fact, breathing high concentrations of chemical vapour causes anaesthetic effects and related health disorders. Therefore, workers should handle the chemical at areas with adequate local exhaust ventilation and should use NIOSH-approved respiratory protection. Workers should observe good personal hygiene practices and wash hands before eating, drinking, or smoking, or using toilet facilities.

Benzyltrimethylammonium Dichloroiodate (CAS No. 114971-52-7)

Molecular formula: $C_6H_5CH_2N(ICl_2)(CH_3)_3$

Benzyltrimethylammonium dichloroiodate is a slightly brown solid chemical and hygroscopic (absorbs moisture from the air).

Safe Handling and Precautions

Benzyltrimethylammonium dichloroiodate is a skin, eye, and respiratory irritant. Accidental ingestion causes GI irritation with nausea, vomiting, and diarrhoea, Users and occupational workers should observe reasonable precautions. Benzyltrimethylammonium dichloroiodate is flammable and liberates toxic gases on contact with acid.

The IARC, NTP, and OSHA have not listed benzyltrimethylammonium dichloroiodate as a human carcinogen.

Occupational workers should use and handle the chemical under the fume hood and away from flame at all times. During handling of benzyltrimethylammonium dichloroiodate, occupational workers should wear approved protective clothing, SCBA, appropriate protective gloves, and eye and face protection mask to prevent exposure to the chemical.

Benzyltrimethylammonium dichloroiodate should be disposed by an approved contractor and according to the local regulations.

Not much information on chronic toxicity of benzyltrimethylammonium dichloroiodate is available in the literature.

Beryllium (CAS No. 7440-41-7)

Molecular formula: **Be**

Beryllium is a brittle, steel-grey metal found as a component of coal, oil, certain rock minerals, volcanic dust, and soil. It reacts with strong acids and strong bases forming flammable/explosive gas. It has several applications in the aerospace, nuclear, and manufacturing industries. In addition, beryllium is amazingly versatile as a metal alloy where it is used in dental appliances, golf clubs, non-sparking tools, wheelchairs, and electronic devices. Beryllium is used in alloys with a number of metals including steel, nickel, magnesium, zinc, and aluminium, the most widely used alloy being beryllium copper – properly called 'a bronze' – which has a high tensile strength and a capacity for being hardened by heat treatment.

The inhalation of beryllium dust or vapours at work can cause disease in susceptible individuals. The effects usually develop slowly, often over years after contact with beryllium has stopped.

One of the largest uses of beryllium is as a moderator of thermal neutrons in nuclear reactors and as a reflector to reduce the leakage of neutrons from the reactor core. A mixed uranium–beryllium source is often used as a neutron source. As a foil, beryllium is used as window material in x-ray tubes. Its lightness, high elastic modulus, and heat stability make it an attractive material for the aircraft and aerospace industry. Beryllium ores are used to make special ceramics for electrical and high-technology applications. Beryllium alloys are used in automobiles, computers, sports equipment (golf clubs and bicycle frames), and dental bridges. It is used in nuclear reactors as a reflector or moderator for it has a low thermal neutron absorption cross section. It is used in gyroscopes, computer parts, and instruments where lightness, stiffness, and dimensional stability are required. The oxide has a very high melting point and is also used in nuclear work and ceramic applications. In general, the public is exposed to low levels of beryllium normally present in air, food, and water. Occupational workers associated with multiple industries where beryllium is mined, processed, machined, or converted into metal, alloys, and other chemicals may be exposed to high levels of beryllium. People living near these industries may also be exposed to higher than normal levels of beryllium in air.

Safe Handling and Precautions

Beryllium and its compounds are highly toxic and hazardous substances to users and occupational workers. When exposed to it, at or above toxic the threshold values, it can lead to chronic beryllium disease (CBD) (i.e. berylliosis) or acute beryllium disease. Beryllium combines with a protein and is deposited in the liver, spleen, and kidneys, but beryllium when bound with a biological protein, a hapten, can result in the chronic form of the disease that is believed to be a delayed hypersensitivity immune response. The major toxicological effects of beryllium are on the respiratory tract, specifically the lungs and their alveoli. Thus, beryllium is known to cause adverse effects and affect all organ systems, although the primary organ involved is the lung. The symptoms of CBD include shortness of breath with physical activity, continuous dry cough, fatigue, night sweats, chest and joint pain, and loss of appetite. CBD primarily affects the lungs, causing granulomas, inflammation, and sometimes scarring. Beryllium causes systemic disease by inhalation and can distribute itself widely throughout the body after absorption from the lungs. Ingestion and breathing of beryllium is harmful. Acute exposures to high levels of beryllium cause mild inflammation of the nasal mucous membranes and pharynx, rhinitis and pharyngitis, tracheobronchitis, and pneumonitis.

The signs and symptoms of chronic beryllium poisoning include, but are not limited to, acute pneumonitis, with cough, respiratory distress, substernal discomfort or pain, loss of appetite, weakness, tiredness, chest pain, fatigue, dyspnoea, anorexia, cyanosis, clubbing, hepatomegaly, and splenomegaly with complications of cardiac failure, renal stone, and pneumothorax.

Beryllium is also known to cause skin irritation, and its traumatic introduction into subcutaneous tissue causes local irritation and granuloma formation. Levels of absorption of beryllium through GI tract are very small, but it is a potent inhibitor of various enzymes of phosphate metabolism, particularly of alkaline phosphatase.

The health hazards associated with beryllium are almost exclusively confined to inhalation exposure and skin contact. Users and occupational workers should therefore be careful during handling of beryllium and its salts since the chemical substance is toxic and should observe precautions with no negligence. Beryllium and its compounds should not be tasted to verify its sweetish nature. Beryllium can be very harmful when humans breathe it in because it can damage the lungs and cause pneumonia. In about 20% of all cases, people die of this disease. Breathing in beryllium in the workplace is what causes berylliosis. People that have weakened immune systems are most susceptible to this disease. Beryllium can also cause allergic reactions with people that are hypersensitive to this chemical. The symptoms are weakness, tiredness, and breathing problems. Some people that suffer from CBD will develop anorexia and blueness of hands and feet. Sometimes people can even be in such a serious condition that CBD can cause their death. Next to causing berylliosis or CBD, beryllium can also increase the chances of cancer development and DNA damage. The latency of the disease can be from 1 to 30 years, most commonly occurring 10–15 years after first exposure. It has been reported that from the use pattern of beryllium, it can be deduced that toxicologically relevant exposure to beryllium is largely confined to the workplace. Only a few exposure situations have been reported for the general population. Because epidemiological studies have indicated that beryllium and its compounds could be carcinogenic to humans, workers should strictly observe precautions during the management of beryllium and beryllium compounds. Based on sufficient evidence for animals, but inadequate evidence for humans, the U.S. EPA has grouped beryllium as B2, meaning probable human carcinogen. Similarly, the DHHS and the IARC have determined that beryllium is a human carcinogen. In view of this, the OSHA sets a limit of 0.002 mg/m^3 (2 µg/m^3) and 8 h TWA of beryllium at the workplace air. In contrast, NIOSH has set a REL at 0.5 µg/m^3 for beryllium.

Beryllium Chloride (CAS No. 7787-47-5)

Molecular formula: **BeCl$_2$**

Beryllium chloride is available as colourless to yellow crystals. It decomposes rapidly on contact with water, producing hydrogen chloride, and attacks many metals in presence of water. Beryllium chloride gives off irritating or toxic fumes (or gases) in a fire.

Exposure to beryllium occurs during routine handling, material transfer, chemical processing, or further processing. If this material is converted or becomes part of a solid shape, exposure can occur when machining, melting, casting, dross handling, pickling, welding, grinding, sanding, polishing, milling, crushing, or otherwise heating or abrading the surface of this material in a manner that generates particulate materials of the compound. The particulate deposits on hands, gloves, and clothing and also gets transferred to the surfaces of the body and enters the system during normal course of breathing/inhalation, during normal hand to face motions, rubbing of the nose or eyes, sneezing, and coughing.

Safe Handling and Precautions

Exposures to beryllium chloride cause redness, blurred vision, nausea, vomiting, and abdominal pain. Inhalation of beryllium chloride causes cough, sore throat, and shortness of breath, but the worker may feel that the symptoms are delayed. Inhaling particulate containing beryllium chloride is known to cause a serious, chronic lung disease called chronic beryllium chloride disease (CBD) in some individuals. Over time, lung disease can be fatal. CBD is a hypersensitivity or allergic condition in which the tissues of the lungs become inflamed. The health disorder and inflammation, sometimes with accompanying fibrosis (scarring), may restrict the exchange of oxygen between the lungs and the bloodstream. Medical science suggests that CBD may be related to genetic factors. The IARC and the NTP have listed beryllium compounds as a Group 1 carcinogen, meaning a known human carcinogen. Occupational workers involved in refining, machining, and producing of beryllium metal are therefore normally prone to health disorders associated with an increased risk of lung cancer; 'the greater excess was in workers hired before 1950 when exposures to beryllium in the workplace were relatively uncontrolled and much higher than in subsequent decades'. The highest risk for lung cancer was observed among workers and individuals diagnosed with acute beryllium-induced pneumonitis, who represent a group that had the most intense exposure to beryllium. The IARC also reported that 'prior to 1950, exposure to beryllium in working environments was usually very high, and concentrations exceeding 1 mg/m^3 (1000 micrograms/cubic meter) were not unusual'. The ACGIH has reported beryllium chloride as A1, meaning confirmed human carcinogen. Keeping these factors in view, the TLV of beryllium chloride (as Be) has been set as low as 0.002 mg/m^3 as TWA and the STEL as 0.01 mg/m^3.

Special Precautions

Workplace management should make strict regulations that each worker should practise and read the material safety data sheet (MSDS) of beryllium and beryllium compounds before working/handling of the chemical substance.

Occupational workers should be very careful and strict to observe precautions during the handling and management of beryllium compounds. Workers should keep storage container of beryllium compounds tightly sealed and transfer material in closed systems or within a completely hooded containment with local exhaust ventilation. Workers should avoid spillage, prevent contact with clothing, and flush container clean before discarding. During use and if and when any kind of chemical spillage happens, the workplace manager should evacuate all persons from the area immediately. Workers should use local exhaust ventilation or other engineering controls to control exposure to airborne particulate material. Workers should use local exhaust ventilation or other engineering controls to control exposure to airborne particular material and avoid any kind of disruption of the airflow in the area of a local exhaust inlet by equipment such as a cooling fan. Workplace manager/management should provide proper training on the use and operation of ventilation to all workers and involve qualified professionals to design and install ventilation systems. Workplace management should develop work practices and procedures that prevent particulate from coming in contact with workers' skin, hair, and/ or personal clothing and provide appropriate cleaning/washing facilities to avoid deposition of the particulate material on skin, hair, or clothing.

Users and occupational workers must be strict to use protective overgarments or work clothing during workplace activities such as machining, furnace rebuilding, air cleaning, equipment filter changes, maintenance, and furnace tending. Contaminated work clothing and overgarments must be managed in a controlled manner to prevent secondary exposure to workers of third parties, to prevent the spread of particulate to other areas, and to prevent particulate from being taken home. Workers must wear gloves to prevent contact with particulate or solutions and to prevent metal cuts and skin abrasions during handling.

Bifenthrin (CAS No. 82657-04-3)

Molecular formula: $C_{23}H_{22}ClF_3O_2$

Synonyms and trade names: Bifenthrin; Bifenthrin (ANSI); Bifenthrin (Talstar); Bifenthrin (trans isomer); Bifentrin; Bifentrina; Biflex; Biphenthrin; Brigade; Capture; Cyclopropanecarboxylic acid; 3-(2-Chloro-3,3,3-trifluoro-1-propenyl)-2,2-dimethyl-; (2-Methyl{1,1'-biphenyl}-3-yl)methyl ester; FMC 54800

Bifenthrin is a synthetic pyrethroid insecticide/miticide/acaricide. Bifenthrin is off-white to pale tan waxy solid granules with a faint, musty odour and a slightly sweet smell. Bifenthrin is soluble in methylene chloride, acetone, chloroform, ether, and toluene and slightly soluble in heptane and methanol. It is slightly combustible and support combustion at elevated temperatures. Thermal decomposition and burning may form toxic by-products such as carbon monoxide, carbon dioxide, hydrogen chloride, and hydrogen fluoride. Bifenthrin treatment affects the nervous system and causes paralysis in insects.

Safe Handling and Precautions

Bifenthrin is moderately toxic to species of mammals when ingested. Exposures to large doses of bifenthrin cause poisoning with symptoms that include, but are not limited to, incoordination, tremor, salivation, vomiting, diarrhoea, and irritability to sound and touch. Exposures to bifenthrin by skin absorption and/or inhalation of the dust cause adverse health effects. Occupational workers on contact with bifenthrin develop adverse health effects that include skin sensations, rashes, numbness, and burning and tingling type of effects. As a pyrethroid poison, bifenthrin disturbs the electrical impulses in nerves, overstimulating nerve cells causing tremors and eventually causing paralysis. The skin-related health effects were found to be reversible and subside after a brief period of time and stoppage of further exposures to bifenthrin. Workers and individuals often get exposed to bifenthrin through ingestion or skin contact, although dermal absorption has little to no risks beyond mild discomfort. However, in the body, bifenthrin is broken down and excreted quickly. In a 7 day study on rats, excess bifenthrin was found accumulated in high-fat tissues, including skin and ovaries of females. The U.S. EPA listed bifenthrin as a developmental toxicant in the toxic release inventory. The EPA has also identified bifenthrin as a class C carcinogen, meaning that it is a possible human carcinogen.

Although bifenthrin causes no inflammation or irritation on human skin, it can cause a tingling sensation that lasts about 12 h. Bifenthrin causes no symptoms of irritation to rabbit eyes. The U.S. EPA classified bifenthrin as toxicity Class II, meaning a moderately toxic chemical substance. While a 2 year study of bifenthrin in laboratory rats showed no evidence of carcinogenicity, higher doses of bifenthrin caused an increased tumour incidence in the urinary bladder and lung of laboratory mice. However, the IARC, NTP, OSHA, and ACGIH have not listed bifenthrin as a carcinogen.

Occupational workers, during handling of bifenthrin, should be careful to store the chemical in a cool, dry, well-ventilated place and away from heat, open flame, or hot surfaces. Also bifenthrin should be stored only in original containers. Workers should be cautious and avoid any kind of contamination of bifenthrin with other substances, pesticides, fertilisers, water, food, or feed.

Occupational workers, during handling of bifenthrin, should be careful and avoid splash, mist, or spray exposure and must wear chemical protective goggles or a face shield, coveralls or long-sleeved shirt and long pants, chemical protective gloves (of nitrile, neoprene, or Viton brand), head covering, and shoes plus socks. Workers, during increased exposures, must wear a full-body cover barrier suit, such as a PVC rain suit. Contaminated leather articles, such as shoes, belts, and watchbands, should be removed and destroyed, and clothing should be separated from household materials. Occupational workers should avoid all kinds of workplace exposures of bifenthrin through accidentals, by ingestion/swallowing, inhalation and or/skin absorption. Workers should avoid breathing of dust, vapor or spray mist of bifenthrin. Care should be observed to store bifenthrin in a cool, dry place and away from heat, food, feedstuffs, and out of reach of children'.

1,1'-Biphenyl

Molecular formula: $C_{12}H_9NO_2$
Synonyms and trade names: 4-Nitro-1,1'-diphenyl; 4-Nitrodiphenyl; 4-Nitrobiphenyl; 4-Phenyl-nitrobenzene; 1-Nitro-4-phenylbenzene

1,1'-Biphenyl is a clear colourless liquid with a pleasant odour and is the most thermally stable organic compound. It is combustible at high temperatures, producing carbon dioxide and water when combustion is complete. Partial combustion produces carbon monoxide, smoke, soot, and low molecular weight hydrocarbons. It is used extensively in the production of heat-transfer fluids, for example, as an intermediate for polychlorinated biphenyls and dye carriers for textile dyeing. It is also used sometimes as a mould retardant in citrus fruit wrappers, formation of plastics, optical brighteners, and hydraulic fluids.

Safe Handling and Precautions

On exposures to 1,1'-biphenyl, occupational workers develop many acute and chronic health effects. The symptoms of health effects include, but are not limited to, polyuria, accelerated breathing, lacrimation, anorexia, weight loss, muscular weakness and coma, fatty liver cell degeneration, and severe nephrotic lesions. Exposure to biphenyl fumes for short periods of time also causes nausea, vomiting, irritation of the eyes and respiratory tract, and bronchitis. Breathing small amounts of 1,1'-biphenyl over long periods of time causes damage to the liver and nervous system of exposed workers. Breathing of the mists, vapours, or fumes may irritate the nose, throat, and lungs. Depending on the concentration and duration of exposure, the symptoms include, but are not limited to, sore throat, coughing, laboured breathing, sneezing and burning sensation, effects of CNS depression, headache, excitation, euphoria, dizziness, in-coordination, drowsiness, light-headedness, blurred vision, fatigue, tremors, convulsions, loss of consciousness, coma, respiratory arrest, and death.

The finely dispersed particles of 1,1'-biphenyl form explosive mixtures in air.

The OSHA recommends the PEL for biphenyl at 0.2 ppm (TWA). Similarly, the ACGIH recommends the TLV at 0.2 ppm (TWA).

1,1'-Biphenyl should be stored in tightly closed containers in a cool, dry, isolated, properly ventilated area, away from heat, sources of ignition, incompatibles, and contact with strong oxidisers.

Occupational workers, during use and handling of 1,1'-biphenyl, should be careful because the chemical substance is a known mutagen and an experimental carcinogen. During the use of the chemical substance, occupational workers should use proper PPE, protective eyeglasses, or chemical safety goggles as described by OSHA's eye and face protection regulations. Workers should clean up chemical spills and the protective equipment immediately, strictly observing precautions. Workers should sweep up or absorb the spilled chemical substance, placing it into a suitable, clean, dry, closed container and properly putting it to the disposal unit. During the use of 1,1'-biphenyl, workers should avoid generating dusty conditions, and the workplace should have appropriate ventilation facilities.

Danger and Precautions

4-Nitro-1,1'-biphenyl is a known mutagen. It is an experimental carcinogen in studies with laboratory animals and probable human carcinogen. This chemical substance has been included in Schedule 2 of the Control of Substances Hazardous to Health Regulations (COSHH) 1999, suggesting its prohibition for all kinds of 'manufacture and use for all purposes'.

Bis(chloromethyl) Ether (CAS No. 542-88-1)

Molecular formula: **(CH₂Cl)₂O**

Molecular formula: **$(CH_2Cl)_2O$**

Synonyms and trade names: BCME; Chloro(chloromethoxy)methane; Dichloromethyl ether; Dimethyl-1,1'-dichloroether

Bis(chloromethyl) ether is a clear liquid with a strong unpleasant odour. It does not occur naturally. It dissolves easily in water but degrades rapidly and readily evaporates into air. During earlier years, bis(chloromethyl) was used to make several types of polymers, resins, and textiles, but its use is now highly restricted. Only small quantities of bis(chloromethyl) ether are produced in the United States. The small quantities that are produced are only used in enclosed systems to make other chemicals. However, small quantities of bis(chloromethyl) ether may be formed as an impurity during the production of another chemical, chloromethyl methyl ether. Along with other chemicals, rain, and sunlight, it undergoes chemical reactions and breaks down as formaldehyde and hydrochloric acid.

Safe Handling and Precautions

Exposures to bis(chloromethyl) ether cause irritation of the skin, eyes, throat, and lungs and, in cases of severe exposures, cause damage to the lungs (swelling and bleeding) and death.

Breathing low concentrations will cause coughing and nose and throat irritation. Chloromethyl methyl ether, after subcutaneous administration, produced local sarcomas in mice and was found to be an initiator of mouse skin tumours. Reports have indicated that bis(chloromethyl) ether causes lung cancer and other tumours in occupational workers. The DHHS observes that bis(chloromethyl) ether is a known human carcinogen. Bis(chloromethyl)ether and chloromethyl methyl ether (technical grade) are classified as Group 1, meaning human carcinogens. The U.S. EPA recommends that the levels of bis(chloromethyl) ether in drinking water and fish in lakes and streams should be limited to 0.0000038 ppb to prevent possible health effects. Also the OSHA has set a limit of 1 ppb as the highest acceptable level in workplace air.

Bismuth (CAS No. 7440-69-9)

Molecular formula: **Bi**

Bismuth is a white, crystalline, brittle metal with a pinkish tinge. Bismuth is the most diamagnetic of all metals, and the thermal conductivity is lower than any metal. It occurs naturally in the metallic state and in minerals such as bismite. The most important ores are bismuthinite or bismuth glance and bismite, and countries such as Peru, Japan, Mexico, Bolivia, and Canada are major producers of bismuth. It is found as crystals in the sulphide ores of nickel, cobalt, silver, and tin. Bismuth is mainly produced as a by-product from lead and copper smelting. It is insoluble in hot or cold water. Bismuth explodes if mixed with chloric or perchloric acid. Molten bismuth explodes and bismuth powder glows red-hot on contact with concentrated nitric acid. It is flammable in powder form.

Bismuth is used in the manufacture of low melting solders and fusible alloys; as key components of thermoelectric safety appliances, such as automatic shut-offs for gas and electric water-heating systems and safety plugs in compressed gas cylinders; in the production of shot and shotguns; in pharmaceuticals; in the manufacturing of acrylonitrile; and as the starting material for synthetic fibres and rubbers. Bismuth oxychloride is sometimes used in cosmetics. Also bismuth subnitrate and bismuth subcarbonate are used in medicine. Bismuth subsalicylate is used as an anti-diarrhoeal and as a treatment of some other gastrointestinal diseases.

Safe Handling and Precautions

Exposures to bismuth salts are associated primarily by ingestion. Bismuth is known to cause adverse health effects. The symptoms include, but are not limited to, irritation of the eyes, skin, respiratory tract and lungs; foul breath; metallic taste; and gingivitis. On ingestion, bismuth causes nausea, loss of appetite and weight, malaise, albuminuria, diarrhoea, skin reactions, stomatitis, headache, fever, sleeplessness, depression, and rheumatic pain, and a black line may form on gums in the mouth due to deposition of bismuth sulphide. Prolonged exposure to bismuth causes mild but deleterious effects on kidneys, and high concentrations of bismuth result in fatalities.

Occupational exposures to bismuth occur during the manufacture of cosmetics, industrial chemicals, and pharmaceuticals. Bismuth also causes neurotoxicity. Bismuth pentafluoride is highly toxic and irritating to the skin, eyes, and respiratory tract, while bismuth subnitrate causes blurred vision. There are no reports about the carcinogenicity of bismuth to humans, and bismuth has not been grouped as a human carcinogen.

Boron (CAS No. 7440-42-8)

Molecular formula: **B**

Boron was discovered by Sir Humphry Davy and J.L. Gay-Lussac in 1808. It is a trivalent non-metallic element that occurs abundantly in the evaporite ores borax and ulexite. Boron is never found as a free element on Earth. Boron appears as charcoal-grey pieces or black powder or as crystalline; is a very hard, black material with a high melting point; and exists in many polymorphs. Boron has several forms, and the most common one is amorphous boron, a dark powder, non-reactive to oxygen, water, acids, and alkalis. It reacts with metals to form borides. Boron is an essential plant micronutrient. Sodium borate is used in biochemical and chemical laboratories to make buffers. Boric acid is produced mainly from borate minerals by the reaction with sulphuric acid. Boric acid is an important compound used in textile products. The most economically important compound of boron is sodium tetraborate decahydrate or borax, used for insulating fibreglass and sodium perborate bleach. Compounds of boron are used in organic synthesis, in the manufacture of a particular type of glasses, and as wood preservatives. Boron filaments are used for advanced aerospace structures, due to their high strength and light weight.

It is used as an antiseptic for minor burns or cuts and is sometimes used in dressings. Boric acid was first registered in the United States as an insecticide in 1948 for control of cockroaches, termites, fire ants, fleas, silverfish, and many other insects. It acts as a stomach poison affecting the insect's metabolism, and the dry powder is abrasive to the insect's exoskeleton. Boric acid is generally considered to be safe to use in household kitchens to control cockroaches and ants. The important use of metallic boron is as boron fibre. Borate-containing minerals are mined and processed to produce borates for several industrial uses, for instance, glass and ceramics, soaps and detergents, fire retardants, and pesticides.

Industries and workplaces where boron compounds are found in abundance include borate mines and processing plants. The boron exposure for the general population is mostly through the ingestion of food and, to a lesser extent, water.

The fibres are used to reinforce the fuselage of fighter aircraft, for example, the B-1 bomber. The fibres are produced by vapour deposition of boron on a tungsten filament. Pyrex is a brand name for glassware, introduced by Corning Incorporated in 1915. Originally, Pyrex was made from thermal shock-resistant borosilicate glass. The common borate compounds include boric acid, sodium tetraborates (borax), and boron oxide.

Safe Handling and Precautions

Boron is harmful by inhalation, ingestion, and skin absorption. It is highly inflammable. Exposure to boron causes irritant effects and disturbances of the CNS in exposed workers. Boron is potentially toxic, although humans tend not to accumulate high levels of boron due to the ability to rapidly excrete it. Large doses of boron cause acute poisoning. There are fatal case reports of infants who have been exposed to boron by mouth or on the skin. Boron toxicity may cause skin rash, nausea, vomiting (may be blue green in colour), diarrhoea (may be blue green in colour), abdominal pain, and headache.

Boron has been studied extensively for its nutritional importance in animals and humans. There is a growing body of evidence that suggests that boron may be an essential element in animals and humans. Many nutritionists believe that more boron and many popular multivitamins such as Centrum in the diet would be beneficial. The adverse health effects of boron on humans are limited. However, ingestion/inhalation cause irritation of the mucous membrane and boron poisoning.

Short-term exposures to boron in work areas are known to cause irritation of the eye, the upper respiratory tract, and the nasopharynx, but the irritation disappears with the stoppage of further exposure. Ingestion of large amounts of boron (about 30 g of boric acid) over short periods of time is known to affect the stomach, intestines, liver, kidneys, and brain and can eventually lead to death in exposed people. Boron has not been listed as a human carcinogen by the IARC, NTP, and OSHA. The concentrations of boron in workers is known to increase to levels that can cause health problems because of consumption of large amounts of boron-containing food. When exposure to small amounts of boron takes place, irritation of the nose, throat, or eyes may occur. As little as 5 g of boric acid causes adverse health effects and illness to any individual, while 20 g or more leads to fatal poisoning and danger.

The U.S. EPA has determined the limit of boron in drinking water at 4 mg/L/day. The OSHA has set a legal limit of 15 mg/m^3 for boron oxide in air averaged over an 8 h work day. Elemental boron is non-toxic and common boron compounds such as borates and boric acid have low toxicity (approximately similar to table salt with the lethal dose being 2–3 g/kg) and therefore do not require special precautions while handling. Some of the more exotic boron hydrogen compounds, however, are toxic as well as highly flammable and do require special care when handling.

Boron Compounds

There are several commercially important borates, including borax, boric acid, sodium perborate, and many boron compounds. The following list includes a few selected compounds:

Boric Acid (CAS No. 10043-35-3)

Molecular formula: H_3BO_3
Synonyms and trade names: ortho-Boric acid; Boracic acid; Borofax; Boric acid; Hydrogen orthoborate

Boric acid is a white powder or granule and is odourless. It is incompatible with potassium, acetic anhydride, alkalis, carbonates, and hydroxides. Boric acid has uses in the production of textile fibreglass, flat-panel displays, and eye drops. Boric acid is recognised for its application as a pH buffer and as a moderate antiseptic agent and emulsifier.

Boron Tribromide (CAS No. 10294-33-4)

Molecular formula: BBr_3
Synonyms and trade names: Borane; Tribromoboron; Boron bromide; Tribromoborane

Boron tribromide is commercially available and is a strong Lewis acid. It is an excellent demethylating or dealkylating agent for ethers, often in the production of pharmaceuticals.

Additionally, it also finds applications in olefin polymerisation and in Friedel–Crafts chemistry as a Lewis acid catalyst. The electronics industry uses boron tribromide as a boron source in pre-deposition processes for doping in the manufacture of semiconductors. Boron tribromide is a colourless, fuming liquid compound containing boron and bromine. It is usually made by heating boron trioxide with carbon in the presence of bromine: this generates free boron that reacts vigorously with the bromine. It is very volatile and fumes in air because it reacts vigorously with water to form boric acid and hydrogen bromide. Boron tribromide is used extensively in industries associated with pharmaceutical manufacturing, image processing, semiconductor doping, plasma etching, and photovoltaic manufacturing and as a reagent for different chemical processes.

Boron tribromide is slightly reactive to moisture. It undergoes hazardous decomposition, condensation, or polymerisation; it may react violently with water to emit toxic gases; or it may become self-reactive under conditions of shock or increase in temperature or pressure. Boron tribromide is a colourless to amber liquid. Boron tribromide is an extremely hazardous chemical substance, and on contact with the skin and eyes, it causes severe irritation effects and skin damage such as scaling, reddening, or, occasionally, blistering. Occupational workers exposed to boron tribromide suffer with harmful in lungs, damage of the mucous membranes of eyes, mouth, and respiratory tract. It causes cough, choking, or shortness of breath among exposed workers.

Safe Handling and Precautions

Occupational workers should avoid contact with boron tribromide since exposures are very toxic (may be fatal if swallowed or inhaled) and corrosive (may cause severe burns to skin or eyes). Chronic exposure may lead to liver or kidney damage. Workers during use and handling of boron tribromide should be careful. The chemical substance should be kept locked up in a safe room, and the container should be kept dry, away from heat, away from sources of ignition, and away from direct sunlight. Workers should avoid breathing in gas/fumes/vapour/spray of boron tribromide. Workers should never add water to boron tribromide. Occupational workers should handle boron tribromide in areas with sufficient ventilation. Workers should wear suitable respiratory equipment. Users should follow strict regulations and precautions. Precautions should be observed by a technically qualified person when opening the container of boron tribromide.

Boron Carbide (CAS No. 12069-32-8)

Molecular formula: B_4C

Boron carbide is used as a ceramic material and is used to make armour materials, especially in bulletproof vests for soldiers and police officers.

Sodium Tetraborate Decahydrate/Borax (Anhydrous) (CAS No. 1330-43-4)

IUPAC name: Disodium tetraborate anhydrous
Molecular formula: $Na_2B_4O_7$
Synonyms and trade names: Anhydrous borax; Borax dehydrated; Boron sodium oxide; Disodium salt of boric acid; Disodium tetraborate; Fused borax; Sodium borate (anhydrous); Sodium tetraborate.

Sodium tetraborate decahydrate/borax (anhydrous) is a clear, colorless or pale yellow hygroscopic substance with a faint odor of detergent. It is stable and is incompatible with powdered metalsand slightly soluble in water. It is extensively used in the industrial manufacturing of metallurgical fluxes, fiberglass, ceramics, fertilizers, enamels, heat-resistant glass (e.g., Pyrex), and other chemicals. It decomposes on heating or on burning producing toxic fumes including sodium oxide, reacts with strong oxidants, and in fire gives off irritating or toxic fumes (or gases).

Safe Handling and Precautions

Accidental ingestion/swallowing of large amounts of sodium tetraborate decahydrate/borax (anhydrous) may cause gastrointestinal disorders with symptoms such as nausea, vomiting, and diarrhea; these may be followed by weakness, depression, headaches, skin rashes, drying skin. Sodium borate decahydrate powder or solutions on eye contact have been reported to cause serious eye irritation. Prolonged or repeated ingestion or skin absorption may cause anorexia, weight loss, vomiting, mild diarrhea, skin rash, convulsions, and anemia. Inhalation of the vapor of sodium tetraborate decahydrate is known to cause weakness, cough, shortness of breath, sore throat, irritation eyes, skin, upper respiratory system, dermatitis, epistaxis (nosebleed), and dyspnea (breathing difficulty). Ingestion of high doses or absorption through broken skin have been reported to cause ill effects on the central nervous system, kidneys and gastrointestinal tract.

No chronic effects of sodium tetraborate decahydrate have been recognized, but continued contact of the chemical substance should be avoided. Local skin irritation may result from the contact of powder. Strong solutions of the chemical may get absorbed through broken skin and cause poisoning. Repeated or prolonged contact of the chemical substance with skin may cause dermatitis. Inhalation of the dust in high concentrations has been reported to cause upper respiratory tract irritation. Workers should be aware of the incompatibility property of sodium tetraborate decahydrate with strong reducing agents such as metal hydrides or alkali metals and elemental zirconium. Alkali metals will generate hydrogen gas that could create an explosive hazard.

Precautions

Users and occupational workers during use and handling of tetraborate decahydrate should wear appropriate, adequate and workplace safety protective dress. They should avoid spilling and contact with skin and eyes, and keep the chemical away from heat, sparks and open flame. The workplace should have good ventilation and workers should avoid breathing vapours.

Alert: Warning! Precautions: Exposure to sodium tetraborate decahydrate is harmful. Whenever accidentally swallowed, inhaled or absorbed through skin at workplaces it causes irritation to skin, eyes and respiratory tract.

Sodium Tetraborate Decahydrate/Borax (Decahydrate) (CAS No. 1303-96-4)

Molecular formula: $Na_2B_4O_7, 10H_2O$

Borax is used in the production of adhesives and in anti-corrosion systems.

Sodium Tetraborate Pentahydrate (CAS No. 12179-04-3)

Molecular formula: $Na_2B_4O_7, 5H_2O$
Synonyms and trade names: Sodium tetraborate pentahydrate; Disodium tetraborate pentahydrate; Borax pentahydrate

Sodium tetraborate pentahydrate is used in large amounts in making insulating fibreglass and sodium perborate bleach.

Boron Nitride (CAS No. 10043-11-5)

Molecular formula: BN

Boron nitride is a material in which the extra electron of nitrogen (with respect to carbon) enables it to form structures that are isoelectronic with carbon allotropes.

Borazole/Borazine (CAS No. 6569-51-3)

Molecular formula: $B_3N_3H_6$

Borazole (borazine) is a colourless liquid with an aromatic smell. With water, it decomposes to form boric acid, ammonia, and hydrogen. The reaction product of boron and ammonia at high temperatures is also known as inorganic benzene.

Boron Trichloride (CAS No. 10294-34-5)

Molecular formula: BCl_3
Synonyms and trade names: Borane; Boron chloride; Trichloroborane; Chlorure; Trichloroboron

Boron trichloride is a colourless gas with a pungent odour. Boron trichloride reacts violently with water and, on decomposition and hydrolysis, yields hydrochloric and boric acid. Boron trichloride is extremely corrosive. It has a pungent, highly irritating odor. Workplace exposure to boron trichloride is known to destroy human skin or lung tissue or corrode metals. Occupational exposure to boron and boron compounds can occur in industries that produce special glass, washing powder, soap and cosmetics, leather, cement, etc.

Safe Handling and Precautions

Fumes of boron trichloride irritate the eyes and mucous membranes. It is corrosive to metals and living tissues. On inhalation, boron trichloride causes chemical pneumonitis, pulmonary oedema – a result of exposure to the lower respiratory tract and deep lung. Occupational workers exposed to boron trichloride show symptoms such as tearing of eyes, coughing, laboured breathing, and excessive salivation and sputum formation leading to pulmonary malfunction. Boron affects the CNS, causing depression of circulation as well as shock and coma.

Ingestion of boron trichloride in work areas leads to circulatory collapse of the worker. There are no reports indicating that boron trichloride is a human carcinogen.

Boron trichloride vigorously attacks elastomers and packing materials, and natural and synthetic rubbers. It also reacts energetically with nitrogen dioxide/dinitrogen tetraoxide, aniline, phosphine, triethylsilane, or fat and grease. It reacts exothermically with chemical bases such as amines, amides, and inorganic hydroxides. Occupational workers should use gloves of neoprene or butyl rubber, PVC, or polyethylene; safety goggles or glasses and face shield; and safety shoes.

Boron trichloride cylinders should be kept protected from physical damage. The cylinders should be kept upright and firmly secured to prevent falling or being knocked over and in a cool, dry, well-ventilated area of non-combustible construction away from heavily trafficked areas and emergency exits.

Alert! Warning!!

Boron trichloride is a pungent-smelling, an extremely corrosive, poisonous, flammable, high-pressure gas. Improper handling may cause injuries to the eyes/impaired vision or complete loss of vision, skin and mucous membranes. Inhalation of boron trichloride gas and mists has been reported to cause frostbite, ulceration of the skin blistering, and pain, damage the tissues of the mouth, throat, esophagus, and tissues of the digestive system and upper respiratory tract and kidney damage. Accidental ingestion of Boron trichloride cause chemical pneumonia and fatal injury. Workers should handle the Boron trichloride cylinders only with protective dress/clothing, self-contained breathing apparatus, only in well ventilated areas.

Workers should secure cylinder when using to protect from falling, use suitable hand truck to move cylinders and store in well ventilated areas, with valve protection cap on cylinders when not in use and protect from physical damage.

Exposure to boron trichloride without proper use and handling is known to cause respiratory tract burns, skin burns, eye burns, and mucous membrane burns. Containers of boron trichloride, when kept in places with severe heat, are known to rupture or explode and release toxic, corrosive, flammable, or explosive gas.

Boron Trifluoride (CAS No. 7637-07-2)

Molecular formula: BF_3
Synonyms and trade names: Trifluoroborane; Boron fluoride

Boron trifluoride is a colourless gas with an acrid suffocating odour. It forms thick acidic fumes in moist air. Dry boron trifluoride is used with mild steel, copper, copper–zinc and copper–silicon alloys, and nickel. Moist gas is corrosive to most metallic materials and some plastics. Therefore, Kel-F® and Teflon® are the preferred gasketing materials. Mercury-containing manometers should not be used since boron trifluoride is soluble in mercury. It decomposes in hot water, yielding hydrogen fluoride. Boron trifluoride is widely used as a catalyst for organic synthesis reactions.

Boron Trifluoride Diethyl Ether Complex (CAS No. 109-63-7)

Molecular formula: $C_4H_{10}BF_3O$
Synonyms and trade names: Boron fluoride diethyl etherate; Boron fluoride etherate; Boron fluoride monoetherate; Boron trifluoride etherate; Boron trifluoride diethyl etherate

Boron trifluoride diethyl ether complex is a stable, highly flammable, colourless to brown fuming, corrosive liquid with a sharp pungent odour. It forms explosive peroxides in contact with air or oxygen. It reacts exothermically with water to form extremely flammable diethyl ether and toxic, corrosive boron trifluoride hydrates. The chemical is incompatible with bases, amines, and alkali metals. It immediately gets hydrolysed by moisture in air to form hydrogen fluoride. Boron trifluoride diethyl ether has applications in chemical laboratory as a catalyst in chemical reactions.

Safe Handling and Precautions

Boron trifluoride etherate may be corrosive to skin, eyes, and mucous membranes. Boron trifluoride etherate may be toxic by inhalation. Upon exposure to water, boron trifluoride etherate may emit flammable and corrosive vapours. Prolonged exposure to boron trifluoride diethyl ether complex by inhalation, ingestion, and skin contact is harmful to users and occupational workers. Workers exposed to the vapours of boron trifluoride etherate develop dizziness or suffocation. On contact with the skin, the chemical substance causes severe burns, and any kind of contact/chemical splash to the eyes causes serious and permanent eye damage. Reduction with lithium aluminium hydride in the attempt to prepare diborane caused an explosion ignited by heat, sparks, or flames. Vapours may form explosive mixtures with air and are known to travel to source of ignition and flash back. The ether component is known to cause anaesthetic effects. Exposure to boron trifluoride etherate causes corrosive burns to the throat and stomach, rash, headache, tingling sensation, twitching, severe irritating effects to the respiratory tract, nausea, diarrhoea, stomach pain, bloody vomit, blood in the urine, shortness of breath, damage to kidneys and liver, CNS depression, paralysis, convulsions, shock, coma, heart failure, and death.

Users and occupational workers, during handling of boron trifluoride diethyl ether complex, should be careful and must use safety glasses, and the work area should have good ventilation, away from any kind of sources of sparks, ignition, and open flames. Workers should keep chemical containers closed and fully protected from physical damage. Workers should avoid breathing vapours (keep upwind), and avoid bodily contact with the material. They must wear positive pressure SCBA and appropriate chemical protective clothing. They must not handle broken packages unless they are wearing appropriate PPE.

On any kind of unusual fire accident at the workplace, workers should use carbon dioxide and dry chemical and avoid direct water stream and foams containing water.

1,3-Butadiene (CAS No. 106-99-0)

Molecular formula: C_4H_6
Synonyms and trade names: Biethylene; Bivinyl; Butadiene; Divinyl; Erythrene; Vinylethylene

1,3-Butadiene is a simple conjugated diene. It is a colourless gas with a mild aromatic or gasoline-like odour and incompatible with phenol, chlorine dioxide, copper, and crotonaldehyde. The gas is heavier than air and may travel along the ground; distant ignition is possible. It is an important industrial chemical used as a monomer in the production of synthetic rubber. Most butadiene is polymerised to produce synthetic rubber. While polybutadiene itself is a very soft, almost liquid, material, polymers prepared from mixtures of butadiene with styrene or acrylonitrile, such as ABS, are both tough and elastic. Styrene–butadiene rubber is the material most commonly used for the production of automobile tyres. Smaller amounts of butadiene are used to make nylon via the intermediate adiponitrile, other synthetic rubber materials such as chloroprene, and the solvent sulpholane. Butadiene is used in the industrial production of cyclododecatriene via a trimerisation reaction.

Safe Handling and Precautions

Exposures to 1,3-butadiene in high concentrations cause damage to the CNS. The symptoms of poisoning include distorted blurred vision, vertigo, general tiredness, decreased blood pressure, headache, and nausea. Several studies show butadiene exposure increases risk of cardiovascular diseases and cancer. Animal studies have shown breathing butadiene during pregnancy increase reproductive, developmental effects and birth defects while data on the effects of butadiene in humans is lacking.

Studies on laboratory animals have indicated that 1,3-butadiene causes carcinogenic effects on the bone marrow, resulting in leukaemia. While these data reveal important implications on the risks of human exposure to butadiene, more data are necessary to draw more conclusive risk assessments. Also it has been reported that women have a higher sensitivity over men when exposed to butadiene. The IARC, based on the human epidemiological evidence, observed that evidences for carcinogenicity are limited and categorised butadiene as Group 2A, meaning probable human carcinogen.

The ACGIH has set the TLV limit at 2 ppm (TWA) (ACGIH 2004). The OSHA has set the PEL at 1 ppm (TWA), and the NIOSH has set the immediately dangerous to life and health (IDLH) concentration at Ca 2000 ppm.

Butane (CAS No. 106-97-8)

Molecular formula: C_4H_{10}
Synonyms and trade names: *n*-Butane; Butyl hydride; Methylethylmethane

The main sources of butane are the refinery of crude oil and the processing of natural gas. It is commonly blended into motor vehicle gasoline to increase the fuel's volatility and to make engine starting easier. Butane contains mixtures of methane, ethane, propane, isobutane, and *n*-butane and is a colourless aliphatic hydrocarbon gas with a gasoline-like odour.

Butane is a component of liquefied petroleum gas (LPG) and as such is used in a wide variety of fuel applications for both recreational and leisure use, including heating and air conditioning, refrigeration, cooking, and lighters. Butane is commonly used alone or in mixtures as a propellant in aerosol consumer products, such as hairsprays, deodorants and antiperspirants, shaving creams, edible oil and dairy products, cleaners, pesticides, and coatings (e.g. automobile or household spray paint). Butane is used as a chemical intermediate in the production of maleic anhydride, ethylene, methyl *tert*-butyl ether (MTBE), synthetic rubber, and acetic acid and its by-products.

Butane is a simple asphyxiant with explosive and flammable potential. It is also a widely used substance of abuse. The main target organs are in the CNS and cardiovascular system. Improper use and handling cause poisoning. Exposure to high levels of butane vapors can result in asphyxia. The symptoms of butane poisoning include but not limited to, rapid breathing and pulse rate, headache, dizziness, visual disturbances, mental confusion, incoordination, mood changes, muscular weakness, tremors, cyanosis, narcosis and numbness of the extremities, and unconsciousness leading to central nervous system injury.

Safe Handling and Precautions

Exposures to butane (general and simple as well) cause excitation, blurred vision, slurred speech, nausea, vomiting, coughing, sneezing, and increased salivation. With increased period of exposure to high concentrations of butane, signs and symptoms of toxicity become more severe. For instance, the exposed worker demonstrates confusion, perceptual distortion, hallucinations (ecstatic or terrifying), delusions, behavioural changes, tinnitus, and ataxia.

Warning! Danger!!

Butane is extremely flammable gas, contains gas under pressure. If heated may cause explosion. Workers should keep butane away from ignition sources as heat/sparks/open flame – No smoking. Workers should use good personal hygiene practices and wear appropriate personal protective Equipment. Butane should be handled only in well ventilated areas.

1,3-Butanediol (CAS No. 107-88-0)

Molecular formula: $C_4H_{10}O_2$
Synonyms and trade names: 1,3-Butylene glycol; Beta-butylene glycol; Butane-1,3-diol; 1,3-Dihydroxybutane; 1-Methyl-1,3-propanediol; Methyl trimethylene glycol

1,3-Butanediol is a colourless, hygroscopic liquid and is soluble in water. It is a stable and flammable/combustible chemical substance. It is incompatible with strong oxidising agents.

1,3-Butanediol is a four-carbon glycol with a sweet flavour and a bitter aftertaste. It is a clear, viscous, low-volatility liquid that is miscible with water and most polar organic solvents but only slightly soluble in ether. It is insoluble in aliphatic hydrocarbons, benzene, and carbon tetrachloride. Its most extensive use is as an intermediate in the manufacture of polyester plasticisers and other chemical products. It finds some use as a solvent and humectant, a useful chemical intermediate. It has extensive application in the manufacture of structural materials for boats, custom mouldings, and sheets and boards for construction applications. 1,3-Butanediol imparts resistance to weathering plus flexibility and impact resistance. It is also used in the manufacture of saturated polyesters for polyurethane coatings, where the glycol imparts greater flexibility to the polyester molecule. 1,3-Butanediol is currently used in many personal care products.

Exposure to 1,3-butanediol causes stinging sensation and irritates the eyes, the skin, and the respiratory tract. Several studies have been conducted to determine the potential health effects of 1,3-butanediol. However, results of several studies have shown very little potential for adverse health effects from administration of even large quantities of 1,3-butanediol to experimental animals or humans. A principle of toxicology is that if the body can handle a non-pharmacologically active compound using normal metabolic and excretion mechanisms, then few compound-related adverse effects will occur. This principle is superbly illustrated in the case of 1,3-butanediol, which can actually be used as a nutritional caloric source by mammals.

During handling of 1,3-butanediol, occupational workers should use safety glasses, adequate ventilation and should be kept protected from air and moisture.

Butanesultone (CAS No. 1633-83-6)

Molecular formula: $C_4H_8O_3S$
Synonyms and trade names: 1,4-Butane sultone; 1,2-Oxathiane; 2,2-Dioxide; 1-Butanesulphonic acid; 4-hydroxy-

Butanesultone is a viscous, clear, colourless – or yellowish/brown – and odourless liquid. When heated, vapours of butanesultone are known to form explosive mixtures with air and are reported to cause explosion hazards.

Safe Handling and Precautions

Exposure to butanesultone causes eye, skin, and respiratory tract irritation; it is harmful if absorbed through the skin and harmful if swallowed, causing irritation of the digestive tract. Chronic exposure causes possible risk of irreversible effects. Occupational workers should use and handle butanesultone with precautions, and the workplace should have adequate ventilation. Workers should avoid contact with eyes, skin, and clothing and avoid ingestion and/or inhalation of the chemical substance. During use and handling of butanesultone workers should avoid contact and exposure of the chemical, should wear appropriate protective eye glasses or chemical safety goggles as described by OSHA's protection regulations. Butanesultone should be stored in a tightly closed container in a cool, dry, well-ventilated area away from incompatible substances, heat, sparks, and flame. In case of accidental spillage of butanesultone at the workplace, workers should absorb spill with inert material, such as dry sand or earth; place it into a waste chemical or waste container; and avoid run-off of the waste into storm sewers and ditches that lead to waterways. Occupational workers should clean up chemical spills immediately, using the appropriate protective equipment, and remove all sources of ignition.

Butanesultone has not been listed by the ACGIH, IARC, or NTP as a human carcinogen.

Butanethiol (CAS No. 109-79-5)

Molecular formula: $C_4H_{10}S$
Synonyms and trade names: Butyl mercaptan; Butanethiol; 1-Butanethiol; *n*-Butyl mercaptan; Thiobutyl alcohol; 1-Mercaptobutane; Butyl sulfhydrate

Butanethiol is also known as butyl mercaptan. It is a volatile, clear to yellowish liquid with an extremely foul smell, with a strong, garlic-, cabbage-, or skunk-like odour. It is slightly oily in nature. Butanethiol is used as a chemical intermediate in the production of insecticides and herbicides. It is also used as a gas odorant.

Safe Handling and Precautions

Exposures to butanethiol cause adverse health effects and poisoning. The symptoms of poisoning in exposed workers include, but are not limited to, asthenia, muscular weakness, malaise, sweating, nausea, vomiting, headache, restlessness, increased respiration, incoordination, skeletal muscle paralysis in most cases, heavy to mild cyanosis, lethargy and/or sedation, and respiratory depression followed by coma. Severe cases of poisoning lead to death of the exposed worker.

In laboratory studies, animals given intraperitoneal and oral exposures of butanethiol survived near-lethal single doses. However, after 20 days posttreatment, they showed pathological changes involving liver and kidney damage. The pathomorphological changes in the liver included cloudy swelling, fatty degeneration, and necrosis. Pathological changes in the kidneys included cloudy swelling, and the lungs displayed capillary engorgement, patchy oedema, and occasional haemorrhage.

2-Butanone (CAS No. 78-93-3)

Molecular formula: C_4H_8O
Synonyms and trade names: Ethyl methyl ketone; 2-Butanone; Methyl acetone; Methyl ethyl ketone (MEK); Methylpropanone

2-Butanone is a stable, highly flammable chemical. It is incompatible with oxidising agents, bases, and strong reducing agents. It is a colourless liquid with a sharp, sweet odour. 2-Butanone is produced in large quantities. It is used as a solvent and nearly half of its use is in paints and other coatings because it will quickly evaporate into the air and it dissolves many substances. It is also used in glues and as a cleaning agent.

Safe Handling and Precautions

Occupational workers get exposed to 2-butanone by breathing contaminated air in workplaces associated with the production or use of paints, glues, coatings, or cleaning agents. Prolonged exposures to 2-butanone cause symptoms of poisoning such as cough, dizziness, drowsiness, headache, nausea, vomiting, dermatitis, irritation of the nose, throat, skin, and eyes and at very high levels cause drooping eyelids, uncoordinated muscle movements, loss of consciousness, and birth defects. Chronic inhalation studies in animals have reported slight neurological, liver, kidney, and respiratory effects. However, information on the chronic (long-term) effects of 2-butanone (MEK) in humans is limited. The DHHS has not classified 2-butanone as a human carcinogen. The IARC and the U.S. EPA have also not classified 2-butanone as a human carcinogen. The OSHA, the ACGIH, and the NIOSH have set an occupational exposure limit (OEL) of 200 ppm at workplace air for an 8 h workday, 40 h workweek.

2-Butanone vapour and air mixtures are explosive. It reacts violently with strong oxidants and inorganic acids causing fire and explosion hazard. 2-Butanone should be kept protected from moisture.

1-Butene (CAS No. 106-98-9)

Molecular formula: C_4H_8
Synonyms and trade names: 1-Butylene; alpha-Butene; alpha-Butylene; Butene-1; *N*-Butene; Normal butene; Ethylethylene; STCC 4905707

1-Butene is a colourless, stable but polymerises exothermically, extremely flammable liquefied gas with an aromatic odour. It is insoluble in water and is one of the isomers of butane.

1-Butene readily forms explosive mixtures with air. It is incompatible with strong oxidising agents, halogens, halogen acids, metal salts, boron trifluoride, fluorine, and nitrogen oxides.

1-Butene of high purity is made by cracking naphtha and separating it from other products by an extra-high-purity distillation column. It is an important organic compound in the production of several industrial materials – for instance, linear low-density polyethylene (LLDPE), a more flexible and resilient polyethylene, and a range of polypropylene resins – and in the production of polybutene, butylene oxide, and the C4 solvents secondary butyl alcohol (SBA) and MEK. The vapour of 1-butene is heavier than air and may travel long distances to an ignition source and flash back.

Safe Handling and Precautions

Exposures to 1-butene cause the effects of an asphyxiant and/or an anaesthetic (at high concentrations). Workers exposed to 1-butene develop eye irritation. The symptoms of poisoning include, but are not limited to, fatigue, dizziness, nausea, vomiting, disorientation, mood swings, tingling sensation, loss of coordination, suffocation, convulsions, possible respiratory collapse, unconsciousness, and coma. It should be noted that before adverse health effects or suffocation could occur, the lower flammability limits of the components of this gas mixture in air may be exceeded, possibly causing an explosive atmosphere as well as an oxygen-deficient environment. Workers/individuals in ill health should not be allowed to work with or handle 1-butene. 1-Butene is not listed in the IARC, NTP, or OSHA as a carcinogen or potential carcinogen.

While handling 1-butene, workers should select a workplace with good ventilation and away from sources of ignition. Workplace supervisor should be well aware of the fact that vapour cloud formation from 1-butene liquid or vapour leaks and subsequent contact with an ignition source can cause an explosion because 1-butene is like petroleum gases, which is heavier than air and can travel along the ground towards distant ignition sources, which may cause an explosive flashback. Therefore, take extreme care to prevent leakage. Workers should wear preferably approved full-face positive pressure-supplied air respirator or an SCBA. Workers should not wear contact lenses. Before sampling operations of 1-butene are undertaken workers training, methods of handling and of the workplace should be done. Several commercial websites provide access to the Code of Federal Regulations, NIOSH, and OSHA databases, which may help in answering questions and setting up safety programmes. All workers potentially exposed to 1-butene shall be provided with information and training in accordance with the requirements of OSHA Hazard Communication Standard 29 CFR 1910.1200. During transport, loading, or unloading of 1-butene, occupational workers must take special precautions to avoid contact with any source of ignition. Workers should shut down any equipment that may be an ignition source and workers must plan and control the loading and unloading of 1-butene in order to limit personnel exposure and environmental releases. Secure the area and inform personnel in the area of the operation being performed. Management should not permit any unauthorised worker/personnel to enter an empty tank that has been used for 1-butene. Also workers should be aware of the fact that strong oxidising agents such as chlorates, nitrates, and peroxides are known to react with 1-butene and should avoid contact with these types of materials. The ACGIH classifies 1-butene as a simple asphyxiant and a cryogenic fluid when contained under high pressures. When released to the atmosphere, 1-butene immediately vaporises (Table 4.2).

TABLE 4.2

Precautions and Emergency

Occupational workers during handling of 1-butene must know the following:

- DANGER! Flammable liquid and gas under pressure.
- Can form explosive mixtures with air.
- May cause frostbite.
- May cause dizziness and drowsiness.
- SCBA may be required by rescue workers.
- Odour: slightly aromatic.
- Extinguishing media: CO_2, dry chemical, water spray, or fog.
- Occupational workers should use piping and equipment adequately designed to withstand pressures.
- Occupational workers during handling and use of 1-butene must be in a closed system, use only spark-proof tools and explosion-proof equipment and avoid heat, sparks, and open flame.
- Occupational workers must protect 1-butene cylinders from direct sunlight and isolate it from cylinders of oxygen and chlorine. Gas can cause rapid suffocation due to oxygen deficiency and hence store and use of 1-butene require adequate ventilation. This product is heavier than air. It tends to accumulate near the floor of an enclosed space, displacing air and pushing it upwards. This creates an oxygen-deficient atmosphere near the floor. Ventilate space before entry. Verify sufficient oxygen concentration.
- Occupational workers should close cylinder valve when not in use; keep cylinder closed even when empty. Never work on a pressurised system. If there is a leak, blow the system down in an environmentally safe manner in compliance with all federal, state, and local laws.
- Workplace supervisor should prevent unauthorised access; workplace should be free from smoking.
- The container should be kept tightly closed in a dry and well-ventilated place.
- Observe label precautions. Cylinders should be stored upright and be firmly secured to prevent it from falling or being knocked over.
- Cylinders can be stored in the open, but in such cases, it should be protected against extremes of weather and from the dampness of the ground to prevent rusting.
- Occupational workers should check and comply with regulations for electrical installations/working materials with the technological safety standards.

Workers should dispose non-refillable cylinders of 1-butene in accordance with federal, state, and local regulations. Allow gas to vent slowly to the atmosphere in an unconfined area or exhaust hood. If the cylinders are of the refillable type, return cylinders to supplier with any valve outlet plugs or caps secured and valve protection caps in place. Workers should return cylinders with any residual product to MESA Specialty Gases & Equipment and should not be disposed of locally.

2-Butoxyethanol (CAS No. 111-76-2)

Molecular formula: $C_6H_{14}O_2$
Synonyms and trade names: Ethylene glycol monobutyl ether; Monobutyl glycol ether; 2-Butoxy-1-ethanol; Ethylene glycol butyl ether; Ethylene glycol *n*-butyl ether; Butyl cellosolve; Butyl glycol; and Butyl oxitol

2-Butoxyethanol is a clear colourless liquid with ether-like smell and belongs to the family of glycol ether/alkoxy alcohol. It is miscible in water and soluble in most organic solvents. 2-Butoxyethanol does not occur naturally. It is usually produced by reacting ethylene oxide with butyl alcohol. It is used as a solvent for nitrocellulose, natural and synthetic resins, and soluble oils; in surface coatings, spray lacquers, enamels, varnishes, and latex paints; as an ingredient in paint thinners, quick-dry lacquers, latex paint, and strippers; and as varnish removers and herbicides. It is also used in textile dyeing and printing, in the treatment of leather, in the production of plasticisers, as a stabiliser in metal cleaners and household cleaners, and in hydraulic fluids, insecticides, herbicides, and rust removers. It is also used as an ingredient in liquid soaps, cosmetics, industrial and household cleaners, dry-cleaning compounds, silicon caulks, cutting oils, and hydraulic fluids. It is a fire hazard when exposed to heat, sparks, or open flames.

Safe Handling and Precautions

2-butoxyethanol is readily absorbed following inhalation and oral and dermal exposure. 2-Butoxyethanol gets released into air or water by different industrial activities and facilities that manufacture, process, or use the chemical. Exposure to 2-butoxyethanol causes irritating effects to the eyes and skin, but it has not induced skin sensitisation in guinea pigs. Information on human health effects associated with exposure to 2-butoxyethanol is limited. However, case studies of individuals who had attempted suicide by ingesting 2-butoxyethanol-containing cleaning solutions showed that they suffered poisoning with symptoms such as haemoglobinuria, erythropenia, hypotension, metabolic acidosis, shock, non-cardiogenic pulmonary oedema, albuminuria, hepatic disorders, and haematuria.

The DHHS, the IARC, and EPA have not classified 2-butoxyethanol and 2-butoxyethanol acetate as to their human carcinogenicity. There are no available data on the carcinogenicity of 2-butoxyethanol and 2-butoxyethanol acetate on people or animals. On the basis of inadequate data, the IARC has observed that 2-butoxyethanol is grouped as 3, meaning not classifiable as a human carcinogen. For 2-butoxyethanol, the OSHA has set an exposure limit of 50 ppm in workplace air for an 8 h workday, 40 h workweek. Whenever possible, fire-resistant containers should be used. Workers must wear appropriate protective equipment to prevent skin and eye contact.

2-Butoxyethanol should be stored in tightly closed, grounded containers in a cool area with adequate ventilation, away from normal work areas and sources of heat and sparks and electrical equipment. At the storage and handling area, workers should use solvent-resistant materials.

Butyl Acetates

Isobutyl Acetate (CAS No. 110-19-0)

Molecular formula: $C_6H_{12}O_2$
Synonyms and trade names: Acetic acid; Isobutyl ester; Acetic acid; 2-Methylpropyl ester; 2-Methylpropyl acetate; IMIS21: 1534

Isobutyl acetate, also known as 2-methylpropyl ethanoate (IUPAC name) or β-methylpropyl acetate, is a common solvent. It is produced from the esterification of isobutanol with acetic acid. It is used as a solvent for lacquer and nitrocellulose. Like many esters, it has a fruity or floral smell at low concentrations and occurs naturally in raspberries, pears, and other plants. At higher concentrations, the odour can be unpleasant and may cause symptoms of CNS depression such as nausea, dizziness, and headache.

n-Butyl Acetate (CAS No. 123-86-4)

Molecular formula: $C_6H_{12}O_2$
Synonyms and trade names: Acetic acid; Butyl ester; 1-Butyl acetate; Butyl ethanoate; IMIS12: 0440

n-Butyl acetate, also known as butyl ethanoate, is an organic compound commonly used as a solvent in the production of lacquers and other products. It is also used as a synthetic fruit flavouring in foods such as candy, ice cream, cheeses, and baked goods. Butyl acetate is found in many types of fruit, where along with other chemicals, it imparts characteristic flavours. Apples, especially of the Red Delicious variety, are flavoured in part by this chemical. It is a colourless flammable liquid with a sweet smell of banana.

sec-Butyl Acetate (CAS No. 105-46-4)

Molecular formula: $C_6H_{12}O_2$
Synonyms and trade names: Acetic acid 1-methylpropylacetate; 1-Methylpropyl acetate; Acetic acid; 2-Butyl ester; Acetic acid; *sec*-Butyl ester; *sec*-Butylacetate; Butyl acetate, *sec*-

sec-Butyl acetate is a colourless liquid with a pleasant odour and is highly flammable. It reacts with strong oxidants, strong bases, strong acids, and nitrates, causing fire and explosion hazard.

Safe Handling and Precautions

sec-Butyl acetate, during prolonged occupational exposures, causes health effects. The symptoms of toxicity include irritation of the skin and eyes. Exposures to high concentrations of *sec*-butyl acetate cause adverse health effects. Ingestion and/or inhalation of *sec*-butyl acetate can cause CNS depression, producing symptoms of dizziness and disorientation, irritation of the nose and throat, coughing and respiratory distress, headache, nausea, vomiting, drowsiness, and coma. For *sec*-butyl acetate, the legal airborne PEL is 200 ppm averaged over an 8 h work shift. Similarly, the NIOSH and ACGIH also recommend the airborne exposure limit at 200 ppm (TWA) over a 10 h/8 h work shift, respectively. During handling of *sec*-butyl acetate, occupational workers should avoid use of open flames and sparks and remove all ignition sources and smoking. The vapour of *sec*-butyl acetate mixes well with air and easily forms explosive mixtures. Workers should avoid using compressed air for filling, discharging, or handling.

tert-Butyl Acetate (CAS No. 540-88-5)

Molecular formula: $C_6H_{12}O_2$
Synonyms and trade names: *tert*-Butyl acetate of acetic acid; *tert*-Butyl ester of acetic acid; 1,1-Dimethyl ethyl ester; Texaco lead appreciator; *tert*-butyl acetate

tert-Butyl acetate, or *t*-butyl acetate, is a colourless flammable liquid with a camphor- or blueberry-like smell. It is used as a solvent in the production of lacquers, enamels, inks, adhesives, thinners, and industrial cleaners. It is also used as an additive to improve the anti-knock properties of motor fuels. *tert*-Butyl acetate has three isomers: *n*-butyl acetate, isobutyl acetate, and *sec*-butyl acetate.

Safe Handling and Precautions

Exposure to *tert*-butyl acetate causes eye, skin, and respiratory irritation in workers. By analogy of the effects of exposure to similar esters, *tert*-butyl acetate may act as a CNS depressant at high concentrations. The signs and symptoms of acute exposure to *tert*-butyl acetate include, but are not limited to, itchy or inflamed eyes and irritation of the nose and upper respiratory tract. Exposures to *tert*-butyl acetate at high concentrations may cause headache, drowsiness, and other narcotic effects.

For *tert*-butyl acetate, the NIOSH has set the REL at 200 ppm (950 mg/m^3) TWA, and the U.S. OSHA has set the PEL at 200 ppm (950 mg/m^3) TWA. The IDLH concentration of *tert*-butyl acetate is reported at 500 ppm.

tert-Butyl acetate should be stored in a cool, dry, well-ventilated area in tightly sealed containers that are labelled in accordance with regulatory standards. Containers of *tert*-butyl acetate should be protected from physical damage and should be stored separately from nitrates, strong oxidisers, strong acids, strong alkalis, heat, sparks, and open flame. Because containers that formerly contained *tert*-butyl acetate may still hold product residues, they should be handled appropriately.

If *tert*-butyl acetate or a solution in work areas gets into the eyes, immediately flush the eyes with large amounts of water for a minimum of 15 min, lifting the lower and upper lids occasionally. Always use safety spectacles or eye protection in combination with breathing protection.

---

n-Butyl Alcohol (CAS No. 71-36-3)

Molecular formula: C_4H_{10}
Synonyms and trade names: *n*-Butanol; Butyl hydroxide; *n*-Propylcarbinol; Butyric hydroxybutane; Butanol; *n*-Butanol; Butan-1-ol; HemStyp; 1-Hydroxybutane; Primary butyl alcohol; *n*-Butyl alcohol; Propyl methanol; Butyl hydroxide

n-Butyl alcohol is a colourless flammable liquid with strong alcoholic odour. *n*-Butyl alcohol is a highly refractive liquid and burns with a strongly luminous flame. It is incompatible with strong acids, strong oxidising agents, aluminium, acid chlorides, acid anhydrides, copper, and copper alloys. *n*-Butyl alcohol has an extensive use in a large number of industries. For instance, it is used as solvent in industries associated with the manufacturing of paints, varnishes, synthetic resins, gums, pharmaceuticals, vegetable oils, dyes, and alkaloids. *n*-Butyl alcohol finds its use in the manufacture of artificial leather, rubber, plastic cements, shellac, raincoats, perfumes, and photographic films.

Safe Handling and Precautions

Exposures to *n*-butyl alcohol by inhalation, ingestion, and/or skin absorption are harmful. *n*-Butyl alcohol is an irritant, with narcotic effect, and a CNS depressant. Butyl alcohols have been reported to cause poisoning with symptoms that include but are not limited to irritation of the eyes, nose, throat, and respiratory system. Prolonged exposure results in symptoms of headache, vertigo, drowsiness, corneal inflammation, blurred vision, photophobia, and cracked skin. It is advised that workers coming in contact with *n*-butyl alcohol should use protective clothing and barrier creams. Occupational workers with pre-existing skin disorders or eye problems, or impaired liver, kidney, or respiratory function, may be more susceptible to the effects of the substance.

The typical PEL of *n*-butyl alcohol is 50–100 ppm (OSHA) and ACGIH TLV 20 ppm (TWA).

Occupational workers and users should store *n*-butyl alcohol with proper care in a cool, dry, well-ventilated location, away from smoking areas/fire hazard. Outside or detached storage is preferred. Separate chemical from incompatibles. Containers should be bonded and grounded for transfers to avoid static sparks.

n-Butylamine (CAS No. 109-73-9)

Molecular formula: $C_4H_{11}N$
Synonyms and trade names: 1-Aminobutane; Butylamine; Monobutylamine; 1-Butanamine; Mono-*n*-butylamine; Norvalamine

n-Butylamine is one of the four isomeric amines of butane, the others being *sec*-butylamine, *tert*-butylamine, and isobutylamine. It is a colourless to yellow liquid and is highly flammable. It is stable and incompatible with oxidising agents, aluminium, copper, copper alloys, and acids. *n*-Butylamine finds its uses in the manufacture of pesticides (such as thiocarbazides), pharmaceuticals, and emulsifiers. It is also a precursor for the manufacture of *N,N'*-dibutylthiourea, a rubber vulcanisation accelerator, and *n*-butylbenzenesulphonamide, a plasticiser of nylon.

Safe Handling and Precautions

Exposures to butylamine (inhalation, ingestion, and skin contact) are harmful. It is very destructive to mucous membranes and causes redness, severe deep burns, and loss of vision. The symptoms include, but are not limited to, sore throat, cough, burning sensation, headache, flushing of the face, vomiting, dizziness, abdominal pain, diarrhoea, nausea, shock or collapse, shortness of breath, laboured breathing, depression, convulsions, narcosis, and possibly unconsciousness. Exposure of this nature is unlikely, however, because of the irritating properties of the vapour. On catching fire, butylamine gives off irritating or toxic fumes (or gases). Repeated or prolonged contact with skin may cause dermatitis. The vapour is corrosive to the eyes, the skin, and the respiratory tracts and causes lung oedema.

The exposure limits of *n*-butylamine have been set as follows: TLV 5 ppm, 15 mg/m³ (ceiling values) (skin) (ACGIH 1997); OSHA PEL C 5 ppm (15 mg/m³) skin; NIOSH REL C 5 ppm (15 mg/m³) skin; and NIOSH IDLH 300 ppm.

Some metals in the presence of water and in contact with *n*-butylamine undergo deformation. Persons with pre-existing skin disorders or eye problems or impaired respiratory function may be more susceptible to the effects of *n*-butylamine.

n-Butylamine containers should be protected against physical damage. *n*-Butylamine should be stored in a cool, dry, well-ventilated location, away from any area where the fire hazard may be acute, preferably outside/detached area of storage, separated from incompatible chemical substances, and the containers should be bonded and grounded for transfers to avoid static sparks.

Bibliography

Agency for Toxic Substances and Disease Registry (ATSDR). 1992a. *Toxicological Profile for 2,3-Benzofuran*. U.S. Department of Health and Human Services, Public Health Service, Atlanta, GA (updated 2011).

Agency for Toxic Substances and Disease Registry (ATSDR). 1992b. *Toxicological Profile for 2-Butanone*. U.S. Department of Health and Human Services, Public Health Service, Atlanta, GA (updated 2008).

Agency for Toxic Substances and Disease Registry (ATSDR). 1995. *Toxicological Profile for Benzidine*. U.S. Public Health Service, U.S. Department of Health and Human Services, Atlanta, GA.

Agency for Toxic Substances and Disease Registry (ATSDR). 1998. *Toxicological Profile for 2-Butoxyethanol and 2-Butoxyethanol Acetate*. U.S. Department of Health and Human Services, Public Health Service, Atlanta, GA.

Agency for Toxic Substances and Disease Registry (ATSDR). 1999. Benzidine. *Toxicological Profile for Benzidine (Draft)*. U.S. Public Health Service, U.S. Department of Health and Human Services, Atlanta, GA.

Agency for Toxic Substances and Disease Registry (ATSDR). 1999. *Toxicological Profile for Bis(chloromethyl) Ether*. U.S. Department of Health and Human Services, Public Health Service, Atlanta, GA (updated 2008).

Agency for Toxic Substances and Disease Registry (ATSDR). 2002. *Toxicological Profile for Beryllium*. U.S. Department of Health and Human Services, Public Health Service, Atlanta, GA (updated 2011).

Agency for Toxic Substances and Disease Registry (ATSDR). 2007a. *Toxicological Profile for Barium and Compounds*. U.S. Department of Public Health and Human Services, Public Health Service, Atlanta, GA.

Agency for Toxic Substances and Disease Registry (ATSDR). 2007b. *Toxicological Profile for Barium*. U.S. Department of Health and Human Services, Public Health Service, Atlanta, GA (updated 2011).

Agency for Toxic Substances and Disease Registry (ATSDR). 2007c. *Toxicological Profile for Benzene*. U.S. Public Health Service, U.S. Department of Health and Human Service, Atlanta, GA.

Agency for Toxic Substances and Disease Registry (ATSDR). 2007d. *Toxicological Profile for Boron (Draft for Public Comment)*. U.S. Department of Health and Human Services, Public Health Service, Atlanta, GA.

Amweg, E. L., D. P. Weston, and N. M. Ureda. 2005. Use and toxicity of pyrethroid pesticides in the Central Valley, California, USA. *Environ. Toxicol. Chem.* 24(4):966–972.

Bauman, J. E., B. S. Dean, and E. P. Krenzelok. 1991. Myocardial infarction and neurodevastation following butane inhalation. *Vet. Hum. Toxicol.* 4:150.

Beliles, R. P. 1994a. The metals: Barium In: *Patty's Industrial Hygiene and Toxicology*, 4th edn., G. D. Clayton and F. E. Clayton, eds. John Wiley & Sons, New York, pp. 1925–1929.

Beliles, R. P. 1994b. The metals: Bismuth. In: *Patty's Industrial Hygiene and Toxicology*, Vol. 2, 4th edn., G. D. Clayton and F. E Clayton, eds. John Wiley & Sons, Inc., New York, pp. 1948–1954.

Bradberry, S. M. 1996. International Program on Chemical Safety (IPCS). INTOC (Data bank) CEHM Bismuth UKPID Monograph. *National Poisons*. Information Service. Birmingham, U.K.

Brühne, F. et al. 2005. Benzaldehyde. In: *Ullmann's Encyclopedia of Industrial Chemistry*, 7th edn. John Wiley & Sons, Inc., New York.

Budavari, S., M. J. O'Neil, A. Smith, and P. E. Heckelman (eds.). 1989. *The Merck Index: An Encyclopedia of Chemicals, Drugs and Biologicals*, 11th edn. Merck & Co., Inc., Rahway, NJ.

Canadian Centre for Occupational Health & Safety (CCOHS). 2007. Benzaldehyde. Chemin No. 232. CCOHS, Hamilton, Ontario, Canada.

Carpenter, C. P., U. C. Pozzani, C. S. Woil, H. Nair, G. A. Keck, and H. F. Smyth, Jr. 1956. The toxicity of butyl cellosolve solvent. *Am. Med. Assoc. Arch. Ind. Health* 14:114–131.

Cheremisinoff, N. P. (ed.). 2003. *Industrial Solvents Handbook*, 2nd edn. Marcel Dekker, Inc., New York.

Chu, and S. V. Church. Thirteen-week inhalation toxicity study of *n*-butyl and t-butyl mercaptan in rats.

Czerwinski, A. W. and H. E. Ginn. 1964. Bismuth nephrotoxicity. *Am. J. Med.* 37:969–975.

Dalvi, R. R. 1992. Effect of the fungicide benomyl on xenobiotic metabolism in rats. *Toxicology* 71:63–68.

Dean, B. S. and E. P. Krenzelok. 1992. Clinical evaluation of pediatric ethylene glycol monobutyl ether poisonings. *J. Toxicol. Clin. Toxicol.* 30(4):557–563.

Dietz, D. D., M. R. Elwell, W. E. Davis, Jr. et al. 1992. Subchronic toxicity of barium chloride dihydrate administered to rats and mice in the drinking water. *Fundam. Appl. Toxicol.* 19:527–537.

Dikshith, T. S. S. (ed.). 2009. *Safe Use of Chemicals—A Practical Guide.* CRC Press, Boca Raton, FL.

Dikshith, T. S. S. (ed.). 2011. *Handbook of Chemicals and Safety.* CRC Press, Boca Raton, FL.

Dikshith, T. S. S. and P. V. Diwan. (eds.). 2003. *Industrial Guide to Chemical and Drug Safety.* John Wiley & Sons, Inc., Hoboken, NJ.

Environmental Health & Safety. 2008. *n*-Butyl alcohol. MSDS No. B5860. Mallinckrodt Baker, Inc., Phillipsburg, NJ.

Extension Toxicology Network (EXTOXNERT). 1994. Bendiocarb. Pesticide Information Profile. USDA/Extension Service, Oregon State University, Corvallis, OR.

Extension Toxicology Network (EXTOXNET). 1995. Bifenthrin. Extension Toxicology Network, Pesticide Information Profiles. Oregon State University, USDA/Extension Service.

Flick, E. W. (ed.). 1998. *Industrial Solvents Handbook*, 5th edn. William Andrew Publishing/Noyes, Park Ridge, NJ.

Grant, W. M. and J. S. Schuman. 1993. *Toxicology of the Eye*, 4th edn. Charles C. Thomas, Springfield, IL.

Groth, D. H., C. Kommiheni, and G. R. MacKay. 1980. Carcinogenicity of beryllium hydroxide and alloys. *Environ. Res.* 21:63–84.

Haley, P. J., G. L. Finch, M. D. Hoover, and R. G. Cuddihy. 1990. The acute toxicity of inhaled beryllium metal in rats. *Fundam. Appl. Toxicol.* 15:767–778.

Hall, T. C., C. H. Wood, J. D. Stoeckle, and L. B. Tepper. 1959. Case data from the beryllium registry. *Am. Med. Assoc. Arch. Ind. Health* 19:100–103.

Health Organization (WHO). 2001. Barium and barium compounds. Concise International Chemical Assessment Document No. 33. WHO, Geneva, Switzerland.

Hudson, M., N. Ashley, and G. Mowat. 1989. Reversible toxicity in poisoning with colloidal bismuth subcitrate. *Br. Med. J.* 299:59.

Huwez, F., A. Pall, D. Lyons, and M. J. Stewart. 1992. Acute renal failure after overdose of colloidal bismuth subcitrate. *Lancet* 340:1298.

International Agency for Research on Cancer (IARC). 1980. *Some Metals and Metallic Compounds.* International Agency for Research on Cancer, Lyon, France. p. 205 (IARC Monographs on the Evaluation of the Carcinogenic Risk of Chemicals to Humans, Vol. 23).

International Agency for Research on Cancer (IARC). 1997. *Summaries & Evaluations, Beryllium and Beryllium Compounds.* IARC, World Health Organization, Geneva, Switzerland.

International Chemical Safety Cards (ICSC). 2004. Barium nitrate. ICSC Card No.1480 (updated 2010).

International Labour Organization, or the World Health Organization. 2005. Butyl acetates. CICAD no. 64. Geneva, Switzerland.

International Programme on Chemical Safety (IPCS). 1990. Beryllium. Environmental health criteria. 106. World Health Organization, Geneva, Switzerland.

International Programme on Chemical Safety (IPCS). 2006. 2-Butoxyethanol. *IARC Summary & Evaluation, IARC Monographs on the Evaluation of Carcinogenic Risks to Humans*, Vol. 88. IARC, Geneva, Switzerland.

International Programme on Chemical Safety and the Commission of the European Communities (IPCS CEC). 1994a. 1,3-Butadiene. ICSC no. 0017 (updated 2000).

International Programme on Chemical Safety and the Commission of the European Communities (IPCS CEC). 1994b. Beryllium. ICSC no. 0226. IPCS, Geneva, Switzerland.

International Programme on Chemical Safety (IPCS) and the Commission of the European Communities (CEC). 1994c. ICSC No. 1182. IPCS CEC.

International Programme on Chemical Safety and the Commission of the European Communities (IPCS CEC). 1994d. 2,3-Benzofuran. ICSC Card No. 0388 (updated 2002).

International Programme on Chemical Safety and the Commission of the European Communities (IPCS CEC). 1994e. Sodium tetraborate. ICSC no. 1229. (updated 1995).

International Programme on Chemical Safety and the Commission of the European Communities (IPCS CEC). 1998a. Methyl ethyl ketone. ICSC no. 0179. ICSC, Geneva, Switzerland.

International Programme on Chemical Safety and the Commission of the European Communities (IPCS-CEC). 1998b. *n*-Butylamine. ICSC no. 0374. IPCS, Geneva, Switzerland.

International Programme on Chemical Safety and the Commission of the European Communities (IPCS, CEC). 1999a. ICSC no. 0840. International Labor Organization (ILO), Geneva, Switzerland.

International Programme on Chemical Safety (IPCS) and the Commission of the European Communities (CEC). 1999b. ICSC no. 0615. IPCS, CEC, Geneva, Switzerland (updated 2005).

International Programme on Chemical Safety (IPCS) and the European Commission (EU). 2003. Butane. ICSC No. 0232, IPCS. WHO, Geneva, Switzerland.

International Programme on Chemical Safety and the Commission of the European Communities (IPCS CEC). 1999c. Barium chloride. ICSC no. 0614. IPCS CEC (updated 2010).

International Programme on Chemical Safety & the Commission of the European Communities (IPCS CEC). 1999d. Barium sulfate. ICSC no. 0827. IPCS CEC (updated 2010).

International Programme on Chemical Safety and the Commission of the European Communities (IPCS-CEC). 2000. Beryllium chloride. ICSC no. 1354. IPCS, Geneva, Switzerland (updated 2004).

International Programme on Chemical Safety and the Commission of the European Communities (IPCS-CEC). 2003. *sec*-Butyl acetate. ICSC no. 0840. IPCS, Geneva, Switzerland (updated 2009).

James, J. A. 1968. Acute renal failure due to a bismuth preparation. *Calif. Med.* 109:317–319.

Kidd, H. and D. R. James (eds.). 1991. *The Agrochemicals Handbook*, 3rd edn. Royal Society of Chemistry Information Services, Cambridge, U.K.

Lewis, R. J. 1993. *Hazardous Chemicals Desk Reference*, 3rd edn. Van Nostrand Reinhold, New York.

Lewis, R. J., Sr. (ed.). 2002. Benzaldehyde. *Hawley's Condensed Chemical Dictionary*, 14th edn. John Wiley & Sons, Inc., New York.

Material Safety Data Sheet (MSDS). 1991. Bendiocarb. Occupational Health Services (OHS), Inc., Secaucus, NJ.

Material Safety Data Sheet (MSDS). 1996. Benzaldehyde. MSDS No. B0696. Mallinckrodt Baker, Inc., Phillipsburg, NJ.

Material Safety Data Sheet (MSDS). 1997. Benzyltrimethylammonium dichloroiodate 99%. ACC No. 23605. Acros Organics N.V., Fair Lawn, NJ (updated 2006).

Material Safety Data Sheet (MSDS). 1999. 1,2-Benzanthracene. ACC# 50930, International Chemical Safety Cards. Benz(a)anthracene. ICSC card no. 0385 (updated 2007).

Material Safety Data Sheet (MSDS). 2002. Boron trichloride (updated 2006).

Material Safety Data Sheet (MSDS). 2003a. Safety data for 1,3-butanediol. Department of Physical and Theoretical Chemistry Laboratory, Oxford University, Oxford, U.K.

Material Safety Data Sheet (MSDS). 2003b. Safety data for 4-nitro-1,1'-biphenyl. Department of Physical and Theoretical Chemistry, Oxford University, Oxford, U.K.

Material Safety Data Sheet (MSDS). 2003c. Safety data for barium sulfate. Department of Physical and Theoretical Chemistry, Oxford University, Oxford, U.K.

Material Safety Data Sheet (MSDS). 2003d. Benzaldehyde. MSDS No. ACC# 02590. Fisher Scientific, Fair Lawn, NJ.

Material Safety Data Sheet (MSDS). 2004. Benzyl acetate. International Programme on Chemical Safety and the European Commission (IPCS-EC), 1999. ICSC Card no. 1331. IPCS, Geneva, Switzerland (updated 2004).

Material Safety Data Sheet (MSDS). 2005a. N,N'-diphenylbenzidine, 97%. Acros Organics N.V., Fair Lawn, NJ.

Material Safety Data Sheet (MSDS). 2005b. Safety data for barium chloride dehydrate. Department of Physical and Theoretical Chemistry, Oxford University, Oxford, U.K.

Material Safety Data Sheet (MSDS). 2005c. Safety data for barium hydroxide octahydrate. Department of Physical Chemistry, Oxford University, Oxford, U.K.

Material Safety Data Sheet (MSDS). 2005d. Safety data for benzamide. Department of Physical and Theoretical Chemistry, Oxford University, Oxford, U.K.

Material Safety Data Sheet (MSDS). 2005e. Safety data for boron. Department of Physical and Theoretical Chemistry, Oxford University, Oxford, U.K.

Material Safety Data Sheet (MSDS). 2005f. Safety data for 2,3-benzofuran. Department of Physical and Theoretical Chemistry, Oxford University, Oxford, U.K.

Material Safety Data Sheet (MSDS). 2005g. Safety data for benzaldehyde. Department of Theoretical and Physical Chemistry, Oxford University, Oxford, U.K.

Material Safety Data Sheet (MSDS). 2006a. Boron trifluoride ether complex. NOOA's Ocean Service, National Oceanic and Atmospheric Administration (NOAA), Washington, DC

Material Safety Data Sheet (MSDS). 2006b. Safety data for alpha-(trichloromethyl) benzyl acetate. Department of Physical and Theoretical Chemistry, Oxford University, Oxford, U.K.

Material Safety Data Sheets (MSDS). 2006c. Safety data for benzidine. Department of Theoretical and Physical Chemistry, Oxford University, Oxford, U.K.

Material Safety Data Sheet (MSDS). 2007a. 1,4-Butane sultone. ACC No. 94056. Acros Organics N.V., Fair Lawn, NJ.

Material Safety Data Sheet (MSDS). 2007b. Safety data for barium fluoride. Department of Physical Chemistry at Oxford University, Oxford, U.K.

Material Safety Data Sheet (MSDS). 2007c. Barium hydroxide. Environmental Health & Safety (EHS). MSDS No. B0372. Mallinckrodt Baker, Inc., Phillipsburg, NJ.

Material Safety Data Sheet (MSDS). 2007d. Barium nitrate. MSDS No. B0 432. Environmental Health & Safety, Mallinckrodt Baker, Inc., Phillipsburg, NJ.

Material Safety Data Sheet (MSDS). 2008a. Barium chloride. Environmental Health & Safety (EHS). MSDS No. B0372. Mallinckrodt Baker, Inc., Phillipsburg, NJ.

Material Safety Data Sheet (MSDS). 2008b. Safety data for benzotrichloride. Department of Physical and Theoretical Chemistry, Oxford University, Oxford, U.K.

Material Safety Data Sheet (MSDS). 2009a. Safety data for barium nitrate.

Material Safety Data Sheet (MSDS). 2009b. Safety data for 1-butene. Department of Physical and Theoretical Chemistry, Oxford University, Oxford, U.K.

Material Safety Data Sheet (MSDS). 2009c. Safety data for boric acid. Department of Physical and Theoretical Chemistry, Oxford University, Oxford, U.K.

Material Safety Data Sheet (MSDS). 2010a. Beryllium chloride. MSDS No. M 23. American National Standard for Hazardous Industrial Chemicals, Cincinnati, OH (updated 2011).

Material Safety Data Sheet (MSDS). 2010b. Safety data for boron tribromide. Department of Physical and Theoretical Chemistry, Oxford University, Oxford, U.K.

Material Safety Data Sheet (MSDS). 2010c. Safety data for boron trifluoride diethyl ether complex. Department of Physical and Theoretical Chemistry, Oxford University, Oxford, U.K.

Material Safety Data Sheet (MSDS). 2010d. Safety data for 2-butanone. Department of Physical and Theoretical Chemistry, Oxford University, Oxford, U.K.

Material Safety Data Sheet (MSDS). 2011. *sec*-Butyl acetate. Canadian Centre for Occupational Health & Safety, Hamilton, Ontario, Canada.

Mattawan, M. I. and International Research and Developmental Corporation. 1982. Confidential study, submitted to the committee by Chevron Phillips Chemical Co., Houston, TX.

Meister, R. T. (ed.). 1991. *Farm Chemicals Handbook '91*. Meister Publishing Company, Willoughby, OH.

National Center for Biotechnology Information, U.S. National Library of Medicine.

National Institute for Occupational Health and Safety (NIOSH). 1980. NIOSH special occupational hazard review of benzidine-based dyes. NIOSH Report DHEW (NIOSH) Publication No. 80–109.

National Institute for Occupational Safety and Health (NIOSH). 1976. Criteria for a Recommended Standard: Occupational Exposure to Boron Trifluoride. DHHS (NIOSH) Publication No. 77–122. Centers for Disease Control and Prevention (CDCP). Atlanta, GA.

National Institute for Occupational Safety and Health (NIOSH). International Programme on Chemical Safety (IPCS) and the Commission of the European Communities (CEC). 1994. Boron trichloride. ICSC card no. 0616. Atlanta, GA.

National Institute of Occupational Safety and Health (NIOSH). 1994. International Program on Chemical Safety (IPCS), and Commission of the European Communities (CEC). Benzidine. ICSC Card No. 0224. Centers for Disease Control and Prevention, Atlanta, GA.

National Institute for Occupational Safety and Health (NIOSH). 2005a. 1,3-Butadiene. *Pocket Guide to Chemical Hazards*. Centers for Disease Control and Prevention, Atlanta, GA.

National Institute for Occupational Safety and Health (NIOSH). 2005b. *tert*-Butyl acetate. NIOSH, Centers for Disease Control and Prevention, (CDCP), Atlanta, GA.

National Institute for Occupational Safety and Health (NIOSH). 2010a. Borates, tetra, sodium salts (anhydrous). *NIOSH Pocket Guide to Chemical Hazards*. NIOSH, Atlanta, GA.

National Institute for Occupational Safety and Health (NIOSH). 2010b. 2-Butoxyethanol. *NIOSH Pocket Guide to Chemical Hazards*. NIOSH, Centers for Disease Control and Prevention, Atlanta, GA (updated 2011).

National Toxicology Program (NTP). 1994. Technical report on the toxicology and carcinogenesis studies of barium chloride dihydrate in F344/N rats and B6C3F1 mice (drinking water studies). NIH Pub. No. 94-3163. NTIS Pub. PB94-214178. NTP TR 432. National Toxicological Program, Research Triangle Park, NC.

NIOSH. 2005. *sec*-Butyl acetate. *Pocket Guide to Chemical Hazards*. NIOSH Publication No. 2005-149, Cincinnati, OH.

NIOSH. 2011. *Pocket Guide to Chemical Hazards*. National Institute for Occupational Safety and Health (U.S.) and Centers for Disease Control and Prevention, Atlanta, GA (updated 2011).

Patnaik, P. (ed.). 1999. *A Comprehensive Guide to the Hazardous Properties of Chemicals*. McGraw Hill, New York.

Proctor, N. H., J. P. Hughes, and M. L. Fischman. 1988. *Chemical Hazards of the Workplace*. J.B. Lippincott Company, Philadelphia, PA.

Ramsey, J., H. R. Anderson, K. Bloor, and R. J. Flanagan. 1989. Mechanism of sudden death associated with volatile substance abuse. *Human Toxicol.* 8:261–269.

Richard, P. P. and A. G. Stanley. 2005. Hazardous chemicals safety and compliance. *Handbook for the Metalworking Industries*. Industrial Press, NY.

Robert, P., R. P. Ryan, C. E. Terry, and S. S. Leffingwell (eds.). 1997. *Toxicology Desk Reference: The Toxic Exposure and Medical Monitoring Index*, 4th edn. CRC Press, Boca Raton, FL.

Sakabe, H. 1973. Lung cancer due to exposure to bis(chloromethyl)ether. *Ind. Health* 11:145–148.

Sax, N. I. and R. J. Lewis, Sr. (eds.). 1989. *Dangerous Properties of Industrial Materials*, 7th edn. Van Nostrand Reinhold, New York.

Shepherd, R. T. 1989. Mechanism of sudden death associated with volatile substance abuse. *Human Toxicol.* 8:287–292.

Sittig, M. (ed.). 1985. *Handbook of Toxic and Hazardous Chemicals and Carcinogens*, 2nd edn. Noyes Publications, Park Ridge, NJ.

Snyder, R., G. Witz, and B. D. Goldstein. 1993. The toxicology of benzene. *Environ. Health Perspect.* 100: 293–306.

Sullivan, M. D., Jr., B. John, and R. K. Gary. *Hazardous Materials Toxicology: Clinical Principles of Environmental Health*. William & Wilkins, Baltimore, MD, p. 418.

The European Commission Health & Consumer Protection Directorate-General. 2007. Directorate E—Safety of the food chain Unit E.3—Chemicals, contaminants, pesticides. Code of Federal Regulations: Food and Drugs: Records and Reports of listed chemicals and certain machines-substances covered (21 CFR 1310.02). U.S. Government Printing Office, Washington, DC (updated 2009).

The National Institute for Occupational Safety and Health (NIOSH). 2010. Ethyl acetate. (updated 2011).

Tomlin, C. (ed.). 1994. *The Pesticide Manual*, 10th edn. British Crop Protection Council/Royal Society of Medicine, Surrey, UK.

U.S. Department of Health and Human Services (U.S. DHHS). 2001. *Barium and Chemicals, Facts on Toxicity*. DHHS, Washington, DC.

U.S. Department of Labor, Occupational Safety & Health Administration (OSHA). 1999. *tert*-Butyl acetate. OSHA, Atlanta, GA.

U.S. Environmental Protection Agency (U.S. EPA). 1984. Health effects assessment for barium.

U.S. Environmental Protection Agency (U.S. EPA). 1987. Health assessment document for beryllium. Prepared by the Office of Health and environmental Assessment, Environmental Criteria and Assessment Office, External Review Draft. EPA 600-8-84-026B, 1986.

U.S. Environmental Protection Agency (U.S. EPA). 1987. *Guidance for the Reregistration of Pesticide Products Containing Benomyl as the Active Ingredient*. U.S. EPA, OPP, Washington, DC.

U.S. Environmental Protection Agency (U.S. EPA). 1996. List of chemicals evaluated for carcinogenic potential. U.S. EPA Office of Pesticide Programs, Washington, DC.

U.S. Environmental Protection Agency (U.S. EPA). 1998. Beryllium and compounds. The Risk Assessment Information System (IRIS) (updated 2007).

U.S. Environmental Protection Agency (U.S. EPA). 2000. Technology transfer network air toxics web site. Benzidine. Hazard Summary-1992 (updated 2007).

U.S. Environmental Protection Agency (U.S. EPA). 2008. Toxicological Review of Beryllium and compounds (EPA/635/R-08/009A). U.S. EPA, Washington, DC.

U.S. Environmental Protection Agency (U.S. EPA). 2011a. Benzotrichloride. Integrated Risk Information System (IRIS), Washington, DC.

U.S. Environmental Protection Agency (U.S. EPA). 1,1'-Biphenyl. Integrated Risk Information System (IRIS). 2011b. IRIS, EPA, Washington, DC.

U.S. EPA Office of Pesticides and Toxic Substances. 1988. Bifenthrin. Fact Sheet No. 177. U.S. EPA, Washington, DC.

Weiss, W. and W. G. Figueroa. 1976. The characteristics of lung cancer due to chloromethyl ethers. *J. Occup. Med*. 18: 623–627.

Weiss, W., R. Moser, and O. Auerbach. 1979. Lung cancer in chloromethyl ether workers. *Am. Rev. Respir. Dis*. 120:1031–1037.

Winship, K. A. 1983. Toxicity of bismuth salts. *Adverse Drug React. Acute Poisoning Rev*. 2:103–121.

World Health Organization (WHO). 1990a. Barium. Environmental Health Criteria No. 107. WHO, Geneva, Switzerland.

World Health Organization (WHO). 1990b. Benomyl. Health and Safety Guide 81. WHO, Geneva, Switzerland.

World Health Organization (WHO). 1993. Benomyl. International Programme on Chemical Safety (IPCS). Health and Safety Guide No. 81. IPCS, Geneva, Switzerland.

World Health Organization (WHO). 1998. 2-Butoxyethanol. Concise International Chemical Assessment Document (CICAD) No. 10. WHO, Geneva, Switzerland.

World Health Organization (WHO), International Programme on Chemical Safety (IPCS). 1989. Allethrins: Allethrin, D-allethrin, bioallethrin, S-bioallethrin. Environmental health criteria no. 87. IPCS, WHO, Geneva, Switzerland.

World Health Organization (WHO), International Programme on Chemical Safety (IPCS). 1991. Barium. Health and Safety Guide No. 46. WHO, IPCS, Geneva, Switzerland.

5

Hazardous Chemical Substances:
C°

Cadmium and Its Compounds
    Cadmium
    Cadmium Acetate
    Cadmium Arsenide
    Cadmium Chloride Anhydrous
    Cadmium Compounds
    Cadmium Fume (as Cd)
    Cadmium Iodide
    Cadmium Oxide
    Cadmium Sulphate
Calcium Cyanide
Captafol
Carbaryl
Carbofuran
Carbophenothion
Cefadroxil
Cefixime
Chloramine-T
Chlordane
Chlorine
Chlorofluorocarbons
p-Chloronitrobenzene
Chlorpyrifos
Chlorpyrifos-Methyl
Cobalt and Its Compounds
    Cobalt Metal, Dust, and Fume (as Co)
    Cobalt Iodide
    Cobalt Sulphate
Copper and Its Compounds
    Copper
    Copper (Dust, Fume, and Mist, as Cu)
    Copper Fume (as Cu)
    Copper Sulphate
Coumaphos
Cyanide

Cyanide Compounds
   Cyanogen Bromide
   Cyanogen Chloride
   Potassium Cyanide
   Potassium Silver Cyanide
   Sodium Cyanide
Cyclohexane

Cadmium and Its Compounds

Cadmium (CAS No. 7440-43-9)

Molecular formula: **Cd**

Cadmium is a grey-white, soft, blue-white malleable, lustrous metal. It is insoluble in cold water, hot water, methanol, diethyl ether, and *n*-octanol. It is stable and incompatible with strong oxidising agents, nitrates, nitric acid, selenium, and zinc, and the powdered metal may be pyrophoric and flammable. Cadmium is associated with occupations such as industrial processes, metal plating, and production of nickel–cadmium batteries, pigments, plastics, and other synthetics. Cadmium metal is produced as a by-product from the extraction, smelting, and refining of the non-ferrous metals zinc, lead, and copper. In view of the unique properties, cadmium metal and cadmium compounds are used as pigments, stabilisers, coatings, specialty alloys, and electronic compounds.

Safe Handling and Precautions

Cadmium is hazardous in case of ingestion and inhalation, slightly hazardous in case of skin contact that causes allergic reactions, irritant to the skin and eyes, and sensitiser. Human exposures to cadmium cause nausea, vomiting, abdominal cramping, diarrhoea, increased salivation, haemorrhagic gastroenteritis, headache, dizziness, cough, dyspnoea, chills (metal fume fever), alopecia, anaemia, arthritis, cirrhosis of the liver, renal cortical necrosis, and cardiomyopathy. Acute inhalation of cadmium causes nasopharyngeal irritation, chest pain, enlarged heart, pulmonary oedema, pulmonary fibrosis, emphysema, bronchiolitis, alveolitis, and renal cortical necrosis, particularly of proximal tubule cell necrosis. Prolonged period of exposure to high concentrations of cadmium causes adverse effects to the skeletal system (arthritis) and cardiovascular system (hypertension), and severe overexposure can result in death.

The carcinogenicity of cadmium is still an area of controversy. While some jurisdictions have classified cadmium as a known human carcinogen and others have indicated that it is a possible or probable human carcinogen, the Department of Health and Human Services (DHHS) has determined that cadmium and cadmium compounds are known human carcinogens. The observation is based on sufficient evidence of carcinogenicity in humans, including epidemiological and mechanistic information that indicate a causal relationship between exposure to cadmium and cadmium compounds and human cancer. The American Conference of Governmental Industrial Hygienists (ACGIH) classified it as A2, meaning suspected human carcinogen. Laboratory animals exposed to cadmium develop cancer of the lung, prostate, testes, haematopoietic system, liver, and pancreas. Occupational workers on exposures to cadmium developed tumours of the lung and prostate.

The Occupational Health and Safety Administration (OSHA) set the permissible exposure limit (PEL) of 5 µg/m³ (dust and fume) at the workplace exposure for 8 h workday. The U.S. Food and Drug Administration (U.S. FDA) set 0.005 ppm limit of cadmium concentration in bottled drinking water, and the ACGIH has set the TLV at 0.2 mg/m³ (dust and salts). Cadmium should be stored in a tightly closed container in a cool place. It should be kept in a separate locked safety storage cabinet. On exposures to cadmium, wash the skin immediately with plenty of water and gently with running water and non-abrasive soap. Workers should cover the exposed skin with an emollient.

Cadmium Acetate (CAS No. 543-90-8)

Molecular formula: **$C_4H_6CdO_4$**
Synonyms and trade names: Acetic acid; Cadmium salt; Bis(acetoxy)cadmium; Cadmium (II) acetate; Cadmium diacetate; Cadmium ethanoate

Cadmium acetate is a colourless crystal with a characteristic odour. It is not combustible, but it decomposes on heating, producing toxic fumes of cadmium oxide. It is incompatible with oxidising agents, metals, hydrogen azide, zinc, selenium, and tellurium. Occupational exposure to cadmium and cadmium compounds occurs in workplaces mainly in the form of airborne dust and fume. Occupations and workplaces include cadmium production and refining, nickel–cadmium battery manufacture, cadmium pigment manufacture and formulation, cadmium alloy production, mechanical plating, zinc smelting, soldering, and polyvinylchloride compounding. Cadmium and compounds enter the body mainly by inhalation and by ingestion.

Safe Handling and Precautions

Exposures to cadmium acetate cause cough, skin redness, abdominal pain, nausea, vomiting, salivation, choking, dizziness, and diarrhoea. On catching fire, cadmium acetate gives off irritating or toxic metal oxide fumes. Inhalation of dust produces perforation of the nasal septum, loss of smell, irritation, headache, metallic taste, and cough. Prolonged exposures to cadmium acetate cause and may produce shortness of breath, chest pain, flu-like symptoms, chills, weakness, fever, muscular pain, pulmonary oedema, liver and kidney damage, and death. Cadmium acetate causes effects on the kidneys and bones, leading to kidney impairment and osteoporosis (bone weakness), and liver. Accidental ingestion or inhalation of cadmium acetate may become fatal to the workers.

There is sufficient evidence in humans for the carcinogenicity of cadmium and cadmium compounds. The International Agency for Research on Cancer (IARC) has classified cadmium and cadmium compounds as Group 1, meaning human carcinogens. Cadmium and cadmium compounds are highly toxic and experimental carcinogens. Cadmium acetate has been listed by the National Toxicology Program (NTP) as A2, meaning suspected human carcinogen. The OSHA has set the PEL at 5 µg/m³ of cadmium and time-weighted average (TWA) at 2.5 µg/m³, and the ACGIH set the threshold limit value (TLV) at 0.01 mg/m³ total dust and 0.002 mg/m³ for cadmium and compounds.

Occupational workers during use and handling of cadmium acetate should be careful. Workers should use protective gloves and immediately remove the contaminated clothing and shoes. The workplace should provide the eyewash fountain and quick-drench facilities. During use of cadmium acetate, workers should avoid heat, flame, ignition sources, dusting, and incompatibles.

Cadmium Arsenide (CAS No. 12006-15-4)

Molecular formula: Cd_3As_2

Cadmium arsenide appears as a dark-grey powder or pieces in appearance with no odour. It is incompatible with acids and acid fumes. Cadmium arsenide, when heated to high temperatures, decomposes and emits toxic fumes of airborne cadmium, arsenic fumes, cadmium oxide, and hydrogen gas.

Safe Handling and Precautions

Occupational workers exposed to cadmium arsenide at workplaces suffer from poisoning. Accidental and direct workplace contact of cadmium arsenide with skin causes irritation, dermatitis, change in pigmentation, and cancerous changes, while contact with the eyes causes irritation, redness, pain, smarting, and conjunctivitis. Accidental ingestion of cadmium arsenide causes acute arsenic toxicity; irritation of mouth and throat; increased salivation; burning sensation and cramps in stomach; nausea; headache; vomiting; weakness; dizziness; diarrhoea; shock; convulsions; irreversible renal tubular dysfunctional; functional changes in the liver, pancreas, and adrenal glands; and chronic arsenic poisoning. The symptoms of poisoning due to inhalation of cadmium arsenide include irritation of the upper respiratory system, vertigo, constriction of the throat, metallic taste in the mouth, cough, dyspnoea, cyanosis, chest pain, flu-like symptoms, and pulmonary oedema. Severe and prolonged exposure to cadmium arsenide causes pulmonary fibrosis/hypertrophy of bronchial vessels, irreversible lung injury, damage to the olfactory nerve, and renal necrosis and/or liver damage. All routes of exposure of cadmium arsenide are known to cause kidney damage, osteomalacia, osteoporosis, spontaneous fractures, haemolytic and iron-deficiency anaemia, weight loss, irritability, renal tubular necrosis, cardiovascular effects, liver damage, and prostatic and respiratory cancers. The affected target organs include the respiratory system, kidneys, liver, prostate, blood, and skin.

Occupational workers and firefighters during handling of cadmium arsenide must wear full-face, self-contained breathing apparatus with full protective clothing to prevent contact with skin and eyes. Fumes from a fire are hazardous. Isolate run-off to prevent environmental pollution.

Occupational workers should keep cadmium arsenide stored in a cool, dry area, in tightly sealed containers, and away from incompatible materials. Workers should be alert and handle the chemical substance in a controlled, inert atmosphere and properly wash hands after handling and before eating and attending other duties.

Cadmium Chloride (Anhydrous) (CAS No. 10108-64-2)

Molecular formula: $CdCl_2$
Synonyms and trade names: Cadmium dichloride; Cadmium chloride; Hydrate (2:5); Cadmium chloride; Hemipentahydrate

Cadmium chloride is a colourless and odourless crystal. It is used for the preparation of cadmium sulphide, used as 'cadmium yellow', a brilliant-yellow pigment, which is stable to heat and sulphide fumes. Cadmium chloride has a high solubility in water and is a non-combustible solid, but the dust can be a moderate fire hazard when exposed to heat or flame or when reacted with oxidising agents. It is incompatible with bromine trifluoride, potassium oxidisers, zinc, selenium, tellurium, and hydrogen azide.

Safe Handling and Precautions

Exposures to cadmium chloride and cadmium salts cause headache, chills, diarrhoea, abdominal pains, choking, dizziness, sweating, nausea and muscular pain, irritations to the mucous membranes and upper respiratory tract, cough, shortness of breath, pulmonary fibrosis, emphysema, perforation of the nasal septum, loss of smell, chest pain, flu-like symptoms, pulmonary oedema, liver and kidney damage, and death. Cadmium chloride has been listed by IARC as category 1, meaning, known human carcinogen, and NTP listed it as a known carcinogen. It is well known that cadmium and cadmium compounds are potential carcinogens and care is required during use. Occupational workers during use of cadmium chloride should use proper protectives and avoid all kinds of direct contact with the body. Workers should avoid generation of cadmium dust in the workplace and use the chemical with adequate ventilation.

Cadmium Fume (Cd) (CAS No. 1306-19-0)

Molecular formula: **Cdo/Cd**
Synonyms and trade names: Cadmium oxide

Cadmium Iodide (CAS No. 10102-68-8)

Molecular formula: CdI_2

Cadmium iodide is a crystalline solid.
   Exposures to cadmium iodide cause adverse health effects on the skin, blood, kidneys, and lungs. It causes irritation to the eyes and skin and corrosive effects to the skin. It is extremely hazardous in case of ingestion. The ACGIH and the NTP classified cadmium iodide as A2, meaning suspected for human carcinogen, and as Group 2, meaning reasonably anticipated human carcinogen, respectively.

Cadmium Oxide (CAS No. 1306-19-0)

Molecular formula: **CdO**
Synonyms and trade names: Cadmium fume

Cadmium oxide as cadmium fume appears as dark-brown powder or crystals and is negligibly soluble in water. It is incompatible with magnesium.

Safe Handling and Precautions

Exposure to cadmium oxide fume during work is of main concern in workplaces in industries. The effects of such exposure cause poisoning with symptoms that include, but are not limited to, sore eyes, nose, throat, coughing, headache, dizziness and weakness, chill, fever, chest pains and breathlessness, nausea, vomiting, diarrhoea, muscular cramps, and salivation. Severe exposures to cadmium oxide cause poisoning and result in fatal injury. It is a severe respiratory irritant, an experimental carcinogen in animals, and a probable human carcinogen. Chronic exposure may cause irreversible lung injury, kidney damage, and other serious effects.
   Occupational workers should be very alert during handling of cadmium compounds/cadmium oxide. Cadmium powders must be handled with care since the material easily produces dust.

Workers should use the dust extraction system or respirators and follow the instructions. The respirators should be clean, should be in proper working condition, and should be changed regularly.

Occupational workers should avoid welding, brazing, or burning on cadmium-plated metals unless he/she is fully equipped with fume extraction system or respirators. In brief, long-term exposure to cadmium oxide fume and dust has caused severe chronic effects, kidney failure, and may, with longer exposure and/or higher concentrations, lead to severe respiratory disease and death. Inhalation of cadmium by smokers is known to accelerate the development of respiratory diseases. There is evidence that long-term exposure to cadmium produces lung cancer. The OSHA has defined cadmium as a carcinogen with no further categorisation, and thus the observations require more confirmatory data. The typical PEL is set at 0.3 mg/m^3, and the typical TLV at 0.05 mg/m^3 (TWA).

Cadmium Sulphate (CAS No. 10124-36-4)

Molecular formula: $CdSO_4$; $CdSO_4 \cdot H_2O$ (monohydrate); $3CdSO_4 \cdot 8H_2O$ (octahydrate)
Synonyms and trade names: Cadmium sulphate anhydrous; Sulphuric acid cadmium salt

Safe Handling and Precautions

Exposures to cadmium salts by absorption are most efficient via respiratory tract. The symptoms of health effects include but are not limited to irritation, headache, metallic taste, and/or cough. With severe exposures, it causes shortness of breath, chest pain, and flu-like symptoms with weakness, fever, headache, chills, sweating, nausea and muscular pain, pulmonary oedema, liver and kidney damage, and death. Prolonged exposures even at relatively low levels/concentrations may result in kidney damage, anaemia, pulmonary fibrosis, emphysema, perforation of the nasal septum, loss of smell, male reproductive effects, and an increased risk of cancer of the lung and of the prostate and may be toxic to blood, kidneys, lungs, liver, upper respiratory tract, bones, and teeth. Repeated or prolonged exposure to the substance can produce target organ damage. Decrease in bone density, renal stones, and other evidence of disturbed calcium metabolism have been reported. Cadmium compounds have been listed as A2, meaning suspected human carcinogens.

Cadmium compounds cause more health disorders to occupational workers and persons with pre-existing skin disorders, eye problems, blood disorders, prostate problems, or impaired liver, kidney, or respiratory function. These workers are more susceptible to the effects of the cadmium salts. On contact with cadmium, exposed workers should wash the skin and eyes immediately with plenty of water. Users and occupational workers during handling of cadmium sulphate should be careful and must use safety glasses and gloves, and the workplace should have good ventilation. Workers should handle cadmium sulphate as a possible carcinogen.

Captafol (CAS No. 2425-06-1)

Molecular formula: $C_{10}H_9Cl_4NO_2S$
Synonyms and trade names: Crisfolatan; Difolatan; Difosan; Folcid; Haipen; Kenofol; Merpafol; Pillartan; Sanseal; Santar-SM; Sanspor
Chemical name: cis-N-(I, 1,2,2,-tetrachloroethylthio)-4-cyclohexene-I,2-dicarboximide

Captafol appears as white, colourless to pale yellow, or tan (technical-grade) crystals or as a crystalline solid or powder, with a slight characteristic pungent odour. It is practically insoluble in water but is soluble or slightly soluble in most organic solvents. Captafol reacts with bases, acids, acid vapours, and strong oxidisers. Captafol is a broad-spectrum nonsystemic fungicide that is categorised as a phthalimide fungicide based on its tetra-hydrophthalimide chemical ring structure (other phthalimide fungicides include captan and folpet). It hydrolyses slowly in aqueous emulsions or suspensions but rapidly in acidic and basic aqueous alkaline media. Captafol will not burn, but when heated to decomposition, it emits toxic fumes, including nitrogen oxides, sulphur oxides, phosgene, and chlorine.

Captafol is a combustible, broad-spectrum protective contact fungicide belonging to the sulphanilamide group. Liquid formulations of captafol containing organic solvents may be flammable. On heating/fire, captafol emits irritating or toxic fumes (or gases) and corrosive fumes including hydrogen chloride, nitrogen oxides, and sulphur oxides and reacts violently with bases causing fire and explosion hazard.

Captafol is effective for the control of almost all fungal diseases of plants except powdery mildews and is widely used outside the United States for the control foliage and fruit disease on apples, citrus, tomato, cranberry, potato, coffee, pineapple, peanut, onion, stone fruit, cucumber, blueberry, prune, watermelon, sweet corn, wheat, barley, oilseed rape, leek, and strawberry. It is also used as a seed protectant in cotton, peanuts, and rice. Captafol is also used in the lumber and timber industries to reduce losses from wood rot fungi in logs and wood products. Formulations of captafol include dusts, flowables, wettables, water dispersibles, and aqueous suspensions. Mixed formulations include (captafol +) triadimefon, ethirimol, folpet, halacrinate, propiconazole, and pyrazophos. Captafol is compatible with most plant-protection products, with the exception of alkaline preparations and formulating material.

Safe Handling and Safety

Exposure to captafol by accidental workplace inhalation causes cough, sore throat, and wheezing; skin absorption causes redness, rash, and blisters; and ingestion causes burning sensation.

Safe Handling and Precautions

Exposure to captafol causes users and occupational workers to develop adverse health effects and poisoning. Exposure to captafol at workplaces and its direct contact with the skin have been reported to cause allergy, dermatitis, skin sensitisation, and conjunctivitis of the eyes and eye injury. The symptoms of poisoning include, but are not limited to, irritation of the eyes, skin, and respiratory tract; bronchitis; wheezing; diarrhoea; vomiting; and injury to liver and kidney.

Oral administration of captafol to mice produced eye and skin irritation and of the respiratory tract. Captafol also produced high incidence of adenocarcinomas of the small intestine, vascular tumours of the heart and spleen, and hepatocellular carcinomas. Oral administration in rats caused a dose-related increase in the incidence of renal carcinomas in males, benign renal tumours in females, and liver tumours in both sexes. Based on evidences of experimental animals, captafol has been listed as a reasonably anticipated human carcinogen.

Federal Insecticide, Fungicide, and Rodenticide Act

Captafol is classified as Group B, meaning a probable human carcinogen based on mammary gland and liver tumours in female Sprague–Dawley rats, kidney tumours in both male and female rats, and lymphosarcoma and haemangiosarcoma in both male and female CD-1 mice, with Harderian gland tumours in male mice.

Regulatory status: Captafol is no longer sold in the United States.

Bulgaria – Final Regulatory Action. In Bulgaria, captafol was excluded from the list of active substances authorised for use in plant-protection products in 1986 under the law on protection of plants against pests and blights. It has prohibited the production, use, and place on the market of all plant-protection products containing captafol according to the annual adopted list of active ingredients banned for use in plant-protection products under the Plant Protection Act. Captafol is designated as a person in charge (PIC) chemical (Annex I of the Regulation on the import and export of certain dangerous chemicals on the Bulgarian territory). The chemical is listed in Annex II of the Regulation as prohibited for export from and import in the country.

During handling of captafol, workers should avoid open flames and must use face shield and eye protection in combination with breathing protection if powder. The workplace should have adequate ventilation, local exhaust, or breathing protection.

Carbaryl (CAS No. 63-25-21)

Molecular formula: $C_{12}H_{11}NO_2$
IUPAC name: 1-Naphthyl methylcarbamate
Synonyms and trade names: Bugmaster; Carbamec; Carbamine; Crunch; Denapon; Dicarbam; Hexavin; Karbaspray; Rayvon; Septene; Sevin; Tercyl; Torndao; Thinsec; Tricarnam

Carbaryl is a colourless to light tan or white or grey solid crystal depending on the purity of the compound. The crystals are essentially odourless and stable to heat, light, and acids but are not stable under alkaline conditions. It is non-corrosive to metals, packaging materials, and application equipment. Carbaryl is classified as a general use pesticide (GUP). It is sparingly soluble in water, but soluble in dimethylformamide, dimethyl sulfoxide, acetone, cyclohexanone, isopropanol, and xylene. Carbaryl is a wide-spectrum carbamate insecticide, which controls over 100 species of insects on citrus, fruit, cotton, forests, lawns, nuts, ornamentals, shade trees, and other crops, as well as on poultry, livestock, and pets. It is also used as a molluscicide and an acaricide. Carbaryl works whether it is ingested into the stomach of the pest or absorbed through direct contact. It is available as bait, dusts, wettable powders, granules, dispersions, and suspensions.

Safe Handling and Precautions

Exposures to carbaryl cause a moderately to very toxic health disorder among workers. Carbaryl produces adverse effects in humans by skin contact, inhalation, or ingestion. The symptoms of acute toxicity are typical of the other carbamates. Direct contact of the skin or eyes with moderate levels of this pesticide can cause burns. Inhalation or ingestion of very large amounts can be toxic to the nervous and respiratory systems resulting in nausea, stomach cramps, diarrhoea, and excessive salivation. Exposures to high concentrations of carbaryl cause poisoning with symptoms such as excessive sweating, headache, weakness, giddiness, nausea, vomiting, stomach pains, slurred speech and muscle twitching, blurred vision, incoordination, and convulsions. The effects of carbaryl on the nervous system of rats, chicken, monkeys, and humans are primarily related to the inhibition of AChE, which under normal situations is transitory. The only documented fatality from carbaryl was through intentional ingestion.

Laboratory studies have indicated that the acute oral toxicity (LD50) of carbaryl ranges from 250 to 850 mg/kg in rats and from 100 to 650 mg/kg in mice. The inhalation toxicity (LC50) in rats is greater than 206 mg/L. Low doses of carbaryl cause minor skin and eye irritation in rabbits. The acute dermal toxicity (LD50) of carbaryl to rabbits is measured as greater than 2000 mg/kg. In a 90 day feeding study, carbaryl did not cause any significant adverse effects in rats. Carbaryl in high doses has caused no reproductive or foetal effects in a long-term feeding study of rats. Ingestion of carbaryl affected the lungs, kidneys, and liver of experimental animals. Inhalation of carbaryl caused adverse effect to lungs. High dose of carbaryl for a prolonged period caused nerve damage in rats and pigs. Several studies indicate that carbaryl can affect the immune system in animals and insects. The evidence for teratogenic effects due to chronic exposure is minimal in test animals. Birth defects in rabbit and guinea pig offspring occurred only at dosage levels that were highly toxic to the mother.

Carbaryl has been shown to affect cell division and chromosomes in rats. However, numerous studies indicate that carbaryl poses only a slight mutagenic risk. There is a possibility that carbaryl may react in the human stomach to form a more mutagenic compound, but this has not been demonstrated. The available information thus suggests the lack of evidence on the mutagenic potential of carbaryl and as a human mutagen. Long-term and lifetime studies in mice and rats exposed to technical-grade carbaryl did not show production of tumours. While *N*-nitrosocarbaryl, a possible by-product, has been shown to be carcinogenic in rats at high doses, this product has not been detected. The U.S. Environmental Protection Agency (U.S. EPA) has not classified carbaryl as a human carcinogen. The International Programme on Chemical Safety (IPCS) has classified carbaryl as Group 3, meaning not classifiable as a human carcinogen. The acceptable daily intake (ADI) for carbaryl has been set at 0.01 mg/kg/day.

Carbofuran (CAS No. 1563-66-2)

Molecular formula: $C_{12}H_{15}NO_3$
IUPAC name: 2,3-Dihydro-2,2-dimethylbenzofuran-7-yl methylcarbamate
Synonyms and trade names: Agrofuran; Carbodan; Carbosip; Cekufuran; Chinufur; Furacarb; Furadan; Terrafuran
Toxicity class: USEPA: I (Formulation, Furadan 4F); II (Furadan G); WHO: Ib

Carbofuran is a broad-spectrum carbamate insecticide and nematocide. It is an odourless, white crystalline solid. On heating, its breakdown can release toxic fumes and irritating or poisonous gases. It is sparingly soluble in water but very soluble in acetone, acetonitrile, benzene, and cyclohexone. The liquid formulations of carbofuran are classified as restricted use pesticides (RUP) because of their acute oral and inhalation toxicity to humans. Granular formulations are also classified as RUP. In fact, carbofuran was first registered in the United States in 1969 and classified as RUP. Exposure to heat breaks down carbofuran, with the release of toxic fumes. Carbofuran is used for the control of soil-dwelling and foliar-feeding insects. It is also used for the control of aphids, thrips, and nematodes that attack vegetable, ornamental plants, crops of sunflower, potatoes, peanuts, soybeans, sugar cane, cotton, rice, and variety of other crops.

Safe Handling and Precautions

The acute oral LD50 of carbofuran to male and female rats is about 8 mg/kg, while the acute dermal LD50 for rats is more than 3000 mg/kg. Carbofuran is mildly irritating to eyes and skin of rabbits. The acute inhalation toxicity (LC50, 4 h) is 0.075 mg/L to rats. As with other carbamate compounds, carbofuran's cholinesterase-inhibiting effect is short term and reversible.

The symptoms of carbofuran poisoning include, but are not limited to, nausea, vomiting, abdominal cramps, sweating, diarrhoea, excessive salivation, weakness, imbalance, blurring of vision, breathing difficulty, increased blood pressure, and incontinence. Death may result at high doses from respiratory system failure associated with carbofuran exposure. Complete recovery from an acute poisoning by carbofuran, with no long-term health effects, is possible if exposure ceases and the victim has time to regain his or her normal level of cholinesterase and to recover from symptoms. Reports have indicated that risks from exposure to carbofuran are especially high among occupational workers and common public suffering from asthma, diabetes, cardiovascular disease, and gastrointestinal (GI) or urogenital tracts disturbances. The available studies indicate carbofuran is unlikely to cause reproductive effects in humans at expected exposure levels. Studies indicate carbofuran is not teratogenic. No significant teratogenic effects have been found in offspring of rats given carbofuran (3 mg/kg/day) on days 5–19 of gestation.

Carbofuran and its carcinogenicity have not been listed by the NTP or IARC or OSHA or ACGIH. Published reports have also indicated that carbofuran does not pose a risk of cancer to animals and humans.

Carbofuran and Bird Mortalities

Carbofuran has been associated with the deaths of millions of wild birds since its introduction in 1967. For instance, bald and golden eagles, red-tailed hawks, and migratory songbirds, as well as other wildlife, have been reported to suffer from heavy poisonings. Reports have shown that carbofuran has a high potential for groundwater contamination and has been detected in aquifers and surface waters. Also, reports have indicated that carbofuran has caused mortality in large number of birds from direct spraying, ingestion of granules or contaminated drinking water, and the consumption of contaminated prey. The U.S. Environmental Protection Agency (U.S. EPA), after a prolonged review of ecological and human health risks associated with the use of carbofuran found, it is ineligible for reregistration and cancelled its registration. The tolerance limit value (TLV) for carbofuran has been set at 0.1 mg/m^3 (8 h, TWA), and the ADI of carbofuran has been set at 0.01 mg/kg/day.

Precautions

During use/handling of carbofuran, occupational workers should wear coveralls or long-sleeved uniform, head covering, and chemical protective gloves made of materials such as rubber, neoprene, or nitrile. Occupational workers should know that areas treated with carbofuran are hazardous. The run-off of carbofuran material and the fire control releases irritating or poisonous gases. It is advisable that workers should enter store houses or carbofuran-treated close spaces with caution. Carbofuran should be stored in a cool, dry, well-ventilated place, in the original containers only. It should not be stored or used near heat, open flame, or hot surfaces.

Carbon Disulphide (CAS No. 75-15-0)

Molecular formula: CS_2
Synonyms and trade names: Carbon bisulphide; Dithiocarbonic anhydride

Pure carbon disulphide is a colourless liquid with a pleasant odour similar to that of chloroform, while the impure carbon disulphide is a yellowish liquid with an unpleasant odour, like that of rotting radishes. Exposure to carbon disulphide occurs in industrial workplaces. Industries associated with coal gasification plants release more of carbon disulphide, carbonyl sulphide, and hydrogen sulphide. Carbon disulphide is used in large quantities as an industrial chemical for the production of viscose rayon fibres. In fact, the major source of environmental pollution both indoor and outdoor by carbon disulphide is caused by emission released into the air from viscose plants.

Safe Handling and Precautions

Laboratory animals exposed to carbon disulphide caused deleterious health effects for instance, developmental effects, skeletal and visceral malformations, embryotoxicity, and functional and behavioural disturbances. Studies have also shown that animals exposed to carbon disulphide indicate destruction of the myelin sheath and axonal changes in both central and peripheral neurons along with changes in the cortex, basal ganglia, thalamus, brainstem, and spinal cord. Neuropathy and myelopathy were extensively studied in rats and rabbits. In the muscle fibres, atrophy of the denervation type occurred secondary to the polyneuropathy. Studies have also shown that carbon disulphide causes vascular changes in various organs of animals as well as myocardial lesions. Occupational workers exposed to carbon disulphide showed symptoms of irritability, anger, mood changes, manic delirium and hallucinations, paranoic ideas, loss of appetite, GI disturbances, and reproductive disorders. The slowing down of nerve conduction velocity in the sciatic nerves preceded clinical symptoms. Studies have indicated that carbon disulphide can affect normal functions of the brain, liver, and heart. Occupational workers exposed to high concentrations of carbon disulphide have suffered with skin burns when the chemical accidentally touched people's skin. The U.S. EPA and also the IARC have not classified carbon disulphide as a human carcinogen. There are no confirmed reports indicating that carbon disulphide is carcinogenic to animals and humans. The OSHA has set a limit of 20 ppm of carbon disulphide for an 8 h workday (TWA), while the NIOSH has set a limit of 1 ppm for a 10 h workday (TWA).

Occupational workers during handling of carbon disulphide should be careful and require proper clothing, eye protection, and respiratory protection. Workers should use the chemical under trained management. On contact with eyes, immediately flush with large amounts of water. On skin contact, the worker should quickly remove contaminated clothing and call for medical attention immediately.

Carbon Monoxide (CAS No. 630-08-0)

Molecular formula: **CO**
Synonyms and trade names: Carbonic oxide; Flue gas; CO; Carbon oxide

Carbon monoxide (CO) is a colourless, odourless, tasteless gas and extremely hazardous. CO can be formed from incomplete burning of gasoline, wood, kerosene, or other fuels. CO is also found in cigarette smoke and vehicle exhaust. In homes, CO can build up from poorly vented or malfunctioning heater, furnace, range, or any appliance that runs on natural gas or oil.

Safe Handling and Precautions

Carbon monoxide is a highly toxic gas often called as a chemical asphyxiant. When inhaled, it combines with haemoglobin more readily than does oxygen, displacing oxygen from haemoglobin and thereby interfering with oxygen transport by the blood. The early symptoms of CO poisoning include headaches, nausea, and fatigue, which are often mistaken for the flu because CO is not detected in a home. Prolonged exposure to CO causes deleterious health effects, brain damage, and eventually death. CO poisoning can happen to anyone, anytime, almost anywhere. Depending upon the period of exposure and concentration of CO, poisoning may be severe, moderate, and mild. (i) Extreme exposures cause confusion, drowsiness, rapid breathing or pulse rate, vision problems, chest pain, convulsions, seizures, loss of consciousness, cardiorespiratory failure, and death. (ii) Moderate exposures cause severe throbbing headache, drowsiness, confusion, vomiting, and fast heart rate. (iii) Mild exposures cause slight headache, nausea, and fatigue (often described as 'flu-like' symptoms). The symptoms of CO poisoning include, but are not limited to, drowsiness, nausea, tiredness, vomiting, headaches, dizziness, visual changes, abdominal pain, chest pains, memory and walking problems, brain damage, and, in severe cases, death. Exposure to high concentrations of CO causes severe headache, weakness, dizziness, irregular heartbeat, seizures, coma, respiratory failure, and unconsciousness.

The toxicity of CO results from its very tight binding to haemoglobin, the species that carries oxygen from your lungs to your bodily tissues. For haemoglobin to work, it cannot bind oxygen very tightly (otherwise, it could not release it at its destination). Unfortunately, CO binds to haemoglobin 200 times more tightly than oxygen. Carboxyhaemoglobin (the molecule formed when CO binds to haemoglobin) does not perform oxygen transport, and it rapidly builds up. In essence, victims are slowly suffocated because their haemoglobin is consumed. The fatal concentration of CO depends on the length of the air exposure and exertion. CO also causes a decrease in heart oxygen supply and induces myocardial hypoxia. Levels above 300 ppm for more than 1–2 h can lead to death, and exposure to 800 ppm (0.08%) can be fatal after an hour (Table 5.1). It is alarming to note that each year more than 500 Americans die from unintentional CO poisoning and more than 2000 com-

TABLE 5.1

Carbon Monoxide: Safety Precautions

Install a CO alarm on each level of your home
Inspect home heating system, chimney
Flue must be inspected and cleaned by a qualified technician every year
Keep chimneys clear of bird and squirrel nests, leaves, and residue to ensure proper ventilation
Make sure that the furnace and other appliances, such as gas ovens, ranges, and cooktops, are inspected for adequate ventilation
Do not burn charcoal inside the house even in the fireplace
Do not operate gasoline-powered engines in confined areas such as garages or basements. Do not leave your car, mower, or other vehicle running in an attached garage, even with the door open
Do not block or seal shut exhaust flues or ducts for appliances such as water heaters, ranges, and clothes dryers

mit suicide by intentionally poisoning themselves. The OSHA has set the PEL for CO at 50 ppm for an 8 h TWA, and the NIOSH has set a standard of 35 ppm.

Carbophenothion (CAS No. 786-19-6)

Molecular formula: $C_{11}H_{16}ClO_2PS_3$
IUPAC name: S-4-chlorophenylthiomethyl O,O-diethyl phosphorodithioate
Synonyms and trade names: Acarithion; Dagadip; Endyl; Garrathion; Hexathion (3); Lethox (2); Nephocarp; Trithion
Toxicity class: U.S. EPA: I; WHO: Ib

Carbophenothion is an off-white- to light-amber-coloured liquid with a mild mercaptan-like odour. It is slightly soluble in water and miscible with many organic solvents such as hydrocarbons, alcohols, ketones, and esters. The U.S. EPA has grouped it under RUP. Carbophenothion is used as a non-systemic insecticide and acaricide used for preharvest treatments on deciduous, citrus and small fruits, field crops, and vegetables and for the control of aphids, mites, suckers, and other pests on fruit, nuts, vegetables, sorghum, and maize.

Safe Handling and Precautions

Carbophenothion is highly toxic both through ingestion and skin absorption. Carbophenothion affects the nervous system by inhibiting ChE activity. On heating or on burning, carbophenothion undergoes decomposition and produces toxic fumes such as phosphorus oxides, sulphur oxides, and hydrogen chloride. Exposures to carbophenothion cause poisoning with symptoms such as headache, blurred vision, weakness, nausea, discomfort in the chest, abdominal cramps, vomiting, diarrhoea, salivation, sweating, and pinpoint pupils. It is highly toxic when eaten and nearly as toxic when absorbed through the skin. Large single doses of carbophenothion potentiate the toxicity of malathion. Long-term feeding studies of laboratory rats to carbophenothion (4 mg/kg/day) for a period of 2 years did not produce any tumours. Thus, carbophenothion as a human carcinogen is not established.

Cefadroxil (CAS No. 50370-12-2)

Molecular formula: $C_{16}H_{17}N_3O_5S$

Safe Handling and Precautions

Cefadroxil is a light-orange-coloured powder. It is a cephalosporin antibiotic and used for the treatment of bacterial infections. Cefadroxil is stable under recommended storage conditions. Cefadroxil has not been considered hazardous when handled under normal medical health conditions and with good housekeeping. Exposures to cefadroxil cause certain common side effects. These include nausea; vomiting; stomach disorders; rashes; itching; unusual tiredness or weakness; yellowing of the skin or eyes; red, swollen, or blistered skin; unusual bruising or bleeding; sore throat; respiratory distress; tightness in the chest; swelling of the mouth, face, lips, or tongue; decreased urination; dark urine; vaginal itching, odour, or discharge; fever; chills; joint pain; and seizures. Prolonged or long-term use of cefadroxil should be avoided. Cefadroxil should be used only under proper medical health care since it has properties of penicillin allergy, renal function damage, and GI tract disturbances. Pregnant women and breast-feeding women should avoid exposures to cefadroxil.

Cefixime (CAS No. 79350-37-1)

Molecular formula: $C_{16}H_{15}N_5O_7S_2$

Cefixime is an oral third-generation cephalosporin antibiotic. It was sold under the trade name Suprax in the United States until 2003, and the oral suspension form of 'Suprax' was relaunched. Cefixime is prescribed for bacterial infections of the chest, ears, urinary tract, and throat (tonsilitis and pharyngitis) and for uncomplicated gonorrhoea, upper and lower respiratory tract infections, acute otitis media, and gonococcal urethritis.

Safe Handling and Precautions

Exposures to cefixime may cause side effects that include, but are not limited to, stomach and abdominal pain, diarrhoea, vomiting, mild skin rash, headache, skin rash, fever, urticaria, pruritus, eosinophilia, leucopenia, anaphylaxis, superinfection, haemolytic anaemia, dyspepsia, and flatulence. Also cefixime may cause transient elevation of SGOT, SGPT, alkaline phosphatase, BUN, and creatinine. Reports have indicated that cefixime is mainly excreted unchanged in bile and urine. Users of cefixime should be careful in health conditions such as neonates, pregnancy, lactation, and renal failure. Users should follow suitable and appropriate healthcare measures to control superinfection (if it happens during therapy). It is contraindicated in patients with known allergy to penicillin or any other ingredients of TAXIMAX. TAXIM-O is contraindicated in patients with known allergy to the cephalosporin group of antibiotics.

Chloramine-t (CAS No. 127-65-1)

Molecular formula: $C_7H_7ClNNaO_2S$
Synonyms and trade names: Sodium N-chloro-p-toluenesulfonamide; Sodium N-chloro-4-toluenesulfonamide; Sodium N-chloro 4-methylbenzenesulfonamide; Tosylchloramide sodium

Chloramine T (CAS No. 7080-50-4)

Molecular formula: $C_7H_{13}ClNNaO_5S$
Synonyms and trade names: Chloroaminum; ChloramineTGr; ChloramineTGr99%; ChloraMine T Chloramine T GR 99%; ChlorominE-T SOLUTION; Chloramine t 3-hydrate; Chloramine T; Chloroamine t trihydrate; Chloramine-T, Sodium salt

Chloramine T is a white to yellow crystalline powder with characteristic odor. It is an air sensitive and corrosive, cause eye and skin burns and chemical dangers: Chloramine T is known to explode on heating above 130°C (anhydrous). Chloramine T decomposes slowly under the influence of air producing chlorine (trihydrate). On heating or on contact with acids it decomposes and produce toxic gases.

Safe Handling and Precautions

Repeated or prolonged contact of chloramine T is known to cause skin sensitization. Repeated or prolonged inhalation exposure may cause asthma. It may also cause sensitization by inhalation. Methemoglobin former – can cause cyanosis, severe respiratory tract irritation with possible burns. May cause severe digestive tract irritation with possible burns. The target organs include, blood, respiratory system, eyes, skin. Repeated or prolonged contact of chloramine T is known to cause skin sensitization. Repeated or prolonged inhalation exposure may cause asthma.

Alert!! Warning!! DANGER!!

Occupational workers should be careful during handling of chloramine T. Workers must use appropriate personal protective equipment (PPE), spills/leakequipment, and workplace safety dress. Workers should clean up spills immediately, observing precautions, sweep up or absorb material, then place into a suitable clean, dry, closed container for disposal, flush spill area with water.

Workers should store chloramine T in a cool, dry place. Keep container closed when not in use. Keep away from water. Keep containers tightly closed. Do not expose to air. Store under an inert atmosphere.

Chlordane (CAS No. 57-74-9)

Molecular formula: $C_{10}H_6C_{18}$
Synonyms and trade names: Aspon-chlordane; Belt; gamma-Chlordane; Chlorindan; Chlor kil; Chlorodane; Corodane; Dichlorochlordene; Dowchlor; Kilex lindane; Kypchlor; Octachlor; Starchlor; Synklor; Tat chlor 4; Termi-ded; Topichlor; Topichlor 20; Toxichlor; Velsicol 1068
IUPAC name: 1,2,4,5,6,7,8,8-Octachloro-2,3,3a,4,7,7a-tetrahydro-4,7-methanoindane
Toxicity class: U.S. EPA: II; WHO: II

Chlordane is a viscous, amber-coloured liquid. Technical-grade chlordane is a mixture of many structurally related compounds including *trans*-chlordane, *cis*-chlordane, -chlordene, heptachlor, and trans-nonachlor. Chlordane was used as a broad-spectrum pesticide in the United States from 1948 to 1988. The uses included termite control in homes and pest control on agricultural crops such as maize and citrus, on home lawns, gardens, turf, and ornamental plants. Chlordane is a persistent organochlorine insecticide. It kills insects when ingested and on contact. Formulations include dusts, emulsifiable concentrates, granules, oil solutions, and wettable powder.

Safe Handling and Precautions

Exposures to chlordane cause adverse health effects and poisoning to animals and humans. The acute oral LD50 values of technical-grade chlordane for rat range from 137 to 590 mg/kg, and acute dermal LD50 for the rabbit is 1720 mg/kg. Signs of acute chlordane intoxication include ataxia, convulsions, and cyanosis followed by death due to respiratory failure. Rats treated by gavage with 100 mg/kg once a day for 4 days had increased absolute liver weights; fatty infiltration of the liver; and increased serum triglycerides, creatinine phosphokinase, and lactic acid dehydrogenase. Sheep treated by stomach tube with 500 mg/kg showed signs of intoxication but recovered fully within 5–6 days; a dose 1000 mg/kg resulted in death after 48 h. Ingestion of chlordane induces vomiting, dry cough, agitation and restlessness, hemorrhagic gastritis, bronchopneumonia, muscle twitching, convulsions, and death among humans. Non-lethal, but accidental, poisoning of children has resulted in convulsions, excitability, loss of coordination, dyspnoea, and tachycardia. Recovery however was complete. Ingestion of chlordane-contaminated water (1.2 g/L) caused symptoms of GI and neurological disorders. Chronic inhalation of chlordane produced symptoms of poisoning, which included, but are not limited to, sinusitis, bronchitis, dermatitis, neuritis, migraine, GI distress, fatigue, memory deficits, personality changes, decreased attention span, numbness or paresthesias, blood dyscrasias, disorientation, loss of coordination, dry eyes, and seizures. Chlordane-treated laboratory rats showed blood diseases including aplastic anaemia and acute leukaemia.

Chlordane can promote cancer and is an endocrine disrupter. Chlordane exposures for prolonged period of time and at high doses (30–64 mg/kg/day for 80 weeks) caused liver cancer in laboratory mice. The U.S. EPA classified chlordane as carcinogen of Group B 2, meaning a probable human carcinogen. The ADI for chlordane has been set at 0.0005 mg/kg/day and the PEL at 0.5 mg/m^3 (8 h).

Chlorine (CAS No. 7782-50-5)

Molecular formula: **Cl**

Chlorine is a yellow-green gas that is heavier than air and has a strong irritating odour. Chlorine is extensively used in the production of paper products, dyestuffs, textiles, petroleum products, medicines, antiseptics, insecticides, food, solvents, paints, plastics, and many other consumer products. Chlorine is mainly used as a bleach in the manufacture of paper and cloth and to make a wide variety of products. Most of the chlorine produced is used in the manufacture of chlorinated compounds for sanitation, pulp bleaching, disinfectants, and textile processing. Further use is in the manufacture of chlorates, chloroform, and carbon tetrachloride and in the extraction of bromine. Organic chemistry demands much from chlorine, both as an oxidising agent and in substitution. In fact, chlorine was used as a war gas in 1915 as a choking (pulmonary) agent. Chlorine itself is not flammable, but it can react explosively or form explosive compounds with other chemicals such as turpentine and ammonia.

Chlorine is slightly soluble in water. It reacts with water to form hypochlorous acid and hydrochloric acid. The hypochlorous acid breaks down rapidly. Chlorine gas is used to synthesise other chemicals and to make bleaches and disinfectants. Chlorine is a powerful disinfectant and in small quantities ensures clean drinking water. It is used in swimming pool water to kill harmful bacteria. Chlorine has a huge variety of uses, for instance, as a disinfectant and purifier, in plastics and polymers, solvents, agrochemicals, and pharmaceuticals, as well as an intermediate in manufacturing other substances where it is not contained in the final product. Also, a very large percentage of pharmaceuticals contain and are manufactured using chlorine. Thus, chlorine is essential in the manufacture of medicines to treat illnesses such as allergies, arthritis, and diabetes.

Safe Handling and Precautions

Chlorine is a respiratory irritant. It causes irritation to the mucus membranes, and the liquid burns the skin. The poisoning caused by chlorine depends on the amount of chlorine a person or an occupational worker is exposed to and the length of exposure time. Prolonged exposures to high concentrations of chlorine cause poisoning with symptoms, which include, but are not limited to, coughing; burning sensation in the nose, throat, and eyes; blurred vision; nausea; vomiting; pain, redness, and blisters on the skin; chest tightness; and pulmonary oedema. There is no information indicating that exposures to chlorine cause cancer in animals and humans. The DHHS, the IARC, and the U.S. EPA have not classified chlorine as a human carcinogen. The OSHA has set a PEL of 1 ppm for chlorine for an 8 h workday (TWA), while the ACGIH set the limit of 0.5 ppm as the TLV for an 8 h workday (TWA) and a short-term exposure level (STEL) of 1.0 ppm of chlorine.

Chlorofluorocarbons (CFCs)

Chlorofluorocarbons (CFCs) are the most important ozone-destroying chemicals. These have been used in many ways since they were first synthesised in 1928. They are stable, non-flammable, low in toxicity, and inexpensive to produce. Over the time, CFCs found uses as refrigerants, solvents, foam blowing agents, aerosols, and in other smaller applications. When released into the air, CFCs rise into the stratosphere. In the stratosphere, CFCs react with other chemicals and reduce the stratospheric ozone layer, which protects Earth's surface from the sun. Reducing CFC emissions and eliminating the production and use of ozone-destroying chemicals are very important to protecting Earth's stratosphere.

CFCs are a family of organic compounds containing chlorine, fluorine, and carbon and are also called Freons. CFCs entered the industrial scene in the late 1920s and early 1930s. This was for identifying safer alternatives to the sulphur dioxide and ammonia refrigerants used at the time. The CFCs are inert and volatile compounds with extensive uses as refrigerants, blowing agents for cleaning agents, in the production of plastic foams, as solvents to clean electronic components, and as propellants in air conditioners and aerosol sprays. These compounds are low in toxicity, non-flammable, non-corrosive, and non-reactive with other chemical species and have desirable thermal-conductivity and boiling-point characteristics. The primary chlorine-containing products on the market are denoted by the industry nomenclature such as CFC-11, CFC-12, CFC-113, CFC-114, CFC-115, and the hydrochlorofluorocarbon HCFC-22. CFCs are marketed under many different trade names, for instance, Algcon, Algofrene, Arcton, Eskimon, Flugene, Forane, Freon, Frigen, Genetron, Isceon, and Osotron.

Safe Handling and Precautions

CFCs (commercial) are persistent in the environment because of their chemical stability. Prolonged period of exposures and accumulation to the inert CFCs in the atmosphere lead to depletion of ozone layer and increased intensity of sunlight. This in turn is known to cause health complications such as skin cancer, eye cataract, and ecological disasters. At high concentrations, CFCs cause neurological disorders such as tingling sensation, humming in the ears, apprehension, EEG changes, slurred speech, and decreased performance in psychological tests. CFCs are known to cause health complications such as skin cancer, eye cataract, and ecological disasters.

p-Chloronitrobenzene (CAS No. 100-00-5)

Molecular formula: p-$NO_2C_6H_4Cl$
Synonyms and trade names: 4-Chloronitrobenzene; *p*-Chloronitrobenzene; *p*-Nitrochlorobenzene; 1-Chloro-4-nitrobenzene; 4-Nitrochlorobenzene; PCNB; PNCB
IUPAC name: 1-Chloro-4-nitrobenzene
4-Chloronitrobenzene: $C_6H_4ClNO_2$ (nitrobenzol, oil of mirbane)
Nitrobenzene: $C_6H_5NO_2$ (nitrobenzol, oil of mirbane)
o-Nitrotoluene: $CH_3C_6H_4NO_2$ (*m*-nitrotoluene, *p*-nitrotoluene)

p-Chloronitrobenzene is extensively used in different industries as an intermediate in the manufacture of dyes, rubber, and agricultural chemicals. It is incompatible with strong oxidisers and alkalis.

Safe Handling and Precautions

Repeated exposure to high levels of *p*-chloronitrobenzene causes adverse health effects. The symptoms of toxicity include, but are not limited to, anoxia; unpleasant taste; anaemia; methemoglobinaemia; haematuria (blood in the urine); spleen, kidney, and bone marrow changes; and reproductive effects. The target organs of *p*-chloronitrobenzene poisoning have been identified as blood, liver, kidneys, cardiovascular system, spleen, bone marrow, and reproductive system.

The evaluations on the carcinogenicity of *p*-chloronitrobenzene are inadequate and the evidence is insufficient and further testing is required to assess their potential to cause cancer. In view of this, the IARC classified *p*-chloronitrobenzene as Group 3, meaning not classifiable as a human carcinogen. The PEL for *p*-chloronitrobenzene set by the OSHA is 1 mg/m³ (skin) (TWA), and the immediately dangerous to life or health (IDLH) has been identified as approximately 100 mg/m³.

Chlorpyrifos (CAS No. 2921-88-2)

Molecular formula: $C_9H_{11}Cl_3NO_3PS$
Synonyms and trade names: Dursban; Empire; Eradex; Lorsban; Paqeant; Piridane; Scout; Stipend
IUPAC name: o,o-Diethyl o-(3,5,6-trichloro-2-pyridyl) phospothorothioate
Toxicity class: U.S. EPA: II; WHO: II

Chlorpyrifos belongs to a class of insecticides known as organophosphates. Technical chlorpyrifos is amber to white crystalline solid with a mild sulphur odour. It is insoluble in water but soluble in benzene, acetone, chloroform, carbon disulphide, diethyl ether, xylene, methylene chloride, and methanol. Formulations of chlorpyrifos include emulsifiable concentrate, dust, granular wettable powder, microcapsule, pellet, and sprays. Chlorpyrifos is widely used as an active ingredient in many commercial insecticides such as Dursban and Lorsban to control household pests, mosquitoes, and pests in animal houses. The U.S. EPA classified chlorpyrifos as GUP.

Safe Handling and Precautions

Exposures to chlorpyrifos cause adverse health effects and poisoning. The symptoms include, but are not limited to, headache, dizziness, respiratory problems, muscular and joint pains, numbness, tingling sensations, incoordination, tremor, nausea, abdominal cramps, vomiting, sweating, blurred vision, respiratory depression, slow heartbeat, nervousness, weakness, cramps, diarrhoea, chest pain, sweating, pinpoint pupils, tearing, salivation, clear nasal discharge and sputum, muscle twitching, and in severe poisonings convulsions, coma, and death. Exposures to chlorpyrifos cause adverse effects to the nervous system. The effects include phosphorylation of the active site and disturbance in the activity of the acetylcholinesterase (AchE) enzyme (inactivity). The AchE is necessary for stopping the transmission of the chemical neurotransmitter.

High concentrations of chlorpyrifos cause poisoning in occupational workers with symptoms of unconsciousness, convulsions, and/or fatal injury. Persons with respiratory ailments and disturbed liver function are known to be at increased health risk. Also, repeated exposures to chlorpyrifos have been reported to cause disturbances in the process of brain development.

The published reports of the NTP, the IARC, and the NIOSH indicate that there are no evidences suggesting that chlorpyrifos is carcinogenic. The U.S. EPA categorised chlorpyrifos as Group E, meaning no evidence as a human carcinogen. The ADI for chlorpyrifos is set at 0.003 mg/kg/day and the no observable effect level (NOEL) is set at 0.03 mg/kg/day.

Occupational workers should be careful during handling and use of chlorpyrifos. The workplace should have adequate washing facilities at all times and close to the site of the handling and use. Eating, drinking, and smoking should be prohibited during handling and before washing after handling. During use and handling of chlorpyrifos, workers should strictly avoid eating, drinking and smoking.

Chlorpyrifos-Methyl (CAS No. 5598-13-0)

Molecular formula: $C_7H_7Cl_3NO_3PS$
Synonyms and trade names: DOWCO 214; ENT 27520; OMS-1155; Reldan; Zertell
IUPAC name: O,O-Dimethyl O-3,5,6-trichloro-2-pyridyl phosphorothioate

Chlorpyrifos-methyl is a general use organophosphate insecticide registered in 1985. It is used for the control of stored grain pests, weevils, moths, borers, beetles and mealworms, red flour beetle, and grain moth; for seed treatment; and for warehouse. It is effective against rice stem borer, aphids, cutworms, plant and leaf hoppers, mole crickets and some moths, and stored grain pests. It is poorly soluble in water, moderately soluble in hexane and alcohols, and readily soluble in other organic solvents such as acetone, benzene, and chloroform.

Safe Handling and Precautions

Exposures to chlorpyrifos-methyl cause excessive salivation, sweating, rhinorrhoea and tearing, muscle twitching, blurred vision/dark vision, slurred speech, weakness, tremor, incoordination, headache, dizziness, nausea, vomiting, abdominal cramps, diarrhoea, respiratory depression, tightness in the chest, wheezing, productive cough, fluid in lungs, and pinpoint pupils. In cases of severe exposure and poisoning, chlorpyrifos-methyl causes seizures, incontinence, respiratory depression, loss of consciousness, respiratory paralysis, and death. Exposures to chlorpyrifos-methyl cause cholinesterase inhibition in workers, and the systemic toxicity includes body weight loss, decreased food consumption, and liver, kidney, and adrenal pathology. The workplace should have adequate washing facilities at all times during handling and should be close to the site of the handling. Occupational workers during handling of chlorpyrifos-methyl should avoid eating, drinking, and smoking. Containers of chlorpyrifos-methyl should be kept away from foodstuffs, animal feed and their containers and out of reach of children.

Cobalt and Its Compounds

Human exposures to cobalt and cobalt compounds cause cough, chest tightness, pain in chest on coughing, dyspnoea, malaise, ache, chilling, sweating, shivering, and aching pain in back and limbs. After some more days of exposures to high concentrations of cobalt, the worker develops more severe pulmonary responses such as severe dyspnoea, wheezing, chest pain and praecordial constriction, persistent cough, weakness and malaise, anorexia, nausea, diarrhoea, nocturia, abdominal pain, diffuse nodular fibrosis, respiratory hypersensitivity, asthma, abdominal pain, sensation of hotness, cardiomyopathy, lung damage, haemoptysis, prostration, and death.

Cobalt Metal, Dust, and Fume (as Co) (CAS No. 7440-48-4)

Molecular formula: **Co**

Cobalt is a silvery, bluish-white, odourless, and magnetic metal. The fume and dust of cobalt metal are odourless and black. The appearance and odour of cobalt compounds and their dusts and fumes vary with the compound. Cobalt metal in powdered form is incompatible with fused ammonium nitrate, hydrozinium nitrate, or strong oxidising agents and should be avoided. It ignites on contact with bromide pentafluoride. Powdered cobalt ignites spontaneously in air. Exposure to cobalt metal fume and dust can occur through inhalation, ingestion, and eye or skin contact. Acute exposure to cobalt metal, dust, and fume is characterised by irritation of the eyes and, to a lesser extent, irritation of the skin. In sensitised individuals, exposure causes an asthma-like attack, with wheezing, bronchospasm, and dyspnoea. Ingestion of cobalt may cause nausea, vomiting, diarrhoea, and a sensation of hotness.

Safe Handling and Precautions

Cobalt metal, dust, and fume are pulmonary toxins and respiratory and skin sensitisers. Inhalation of cobalt metal fume and dust may cause interstitial fibrosis, interstitial pneumonitis, myocardial and thyroid disorders, and sensitisation of the respiratory tract and skin. Chronic cobalt poisoning may also produce polycythemia and hyperplasia of the bone marrow. Myocardial disorders have also been observed in cobalt production workers. Chronic exposure to cobalt metal, dust, or fume is known to cause respiratory or dermatologic injury. On skin sensitisation, contact with cobalt workers develop skin eruptions, dermatitis, frictional surfaces of the arms, legs, and neck. Chronic cobalt poisoning may cause polycythemia, hyperplasia of the bone marrow and thyroid gland, pericardial effusion, and damage to the alpha cells of the pancreas.

The current OSHA PEL for cobalt metal, dust, and fume (as Co) is 0.1 mg per cubic metre (mg/m^3) of air as an 8 h TWA concentration. The NIOSH has established a recommended exposure limit (REL) for cobalt metal, dust, and fume of 0.05 mg/m^3 as a TWA for up to a 10 h workday and a 40 h workweek.

Cobalt metal dust (powdered metal) should be stored in a cool, dry, well-ventilated area in tightly sealed containers that are labelled in accordance with OSHA standard. Containers of cobalt metal dust should be protected from physical damage and ignition sources and should be stored separately from strong oxidisers.

Cobalt Iodide (CAS No. 15238-00-3)

Molecular formula: CoI_2

Cobalt iodide appears as black crystals with a slight 'iodine'-like odour. When heated to decomposition, it may emit toxic fumes of iodine and oxides of nitrogen. Anhydrous cobalt (II) iodide is sometimes used to test for the presence of water in various solvents. Cobalt (II) iodide is used as a catalyst, for example, in carbonylations. It catalyses the reaction of diketene with Grignard reagents, useful for the synthesis of terpenoids.

It has a low toxicity by ingestion. Ingestion of soluble salts produces nausea and vomiting by local irritation. In animals, administration of cobalt salts produces an increase in the total red cell mass of the blood. In humans, a single case of poisoning with liver and kidney damage has been attributed to cobalt. Locally, cobalt has been shown to produce dermatitis, and investigators have been able to demonstrate a hypersensitivity of the skin to cobalt. There have been reports of haematologic, digestive, and pulmonary changes in humans. Iodides are similar in toxicity to bromides. Prolonged absorption of iodides may produce 'iodism', which is manifested by skin rash, running nose, headache, and irritation to mucous membranes. In severe cases, the skin may show pimples, boils, redness, black and blue spots, hives, and blisters. Weakness, anaemia, loss of weight, and general depression may occur. Generally, it is very soluble in water and easily absorbed into the body (Sax, Dangerous Properties of Industrial Materials, eighth edition).

Cobalt Sulphate (CAS No. 10026-24-1)

Molecular formula: $CoSO_4 \cdot 7H_2O$
Synonyms and trade names: Cobalt sulphate 7-hydrate; Sulphuric acid; Cobalt (2+) salt; Heptahydrate; Bieberite

Cobalt sulphate appears as pink or red crystals. It is stable, non-flammable, and hygroscopic and undergoes dehydration at 41°C and 71°C.

Safe Handling and Precautions

Cobalt sulphate is harmful when accidentally swallowed or inhaled at workplaces. It is a respiratory irritant and causes an allergic skin reaction. Cobalt sulphate has been reported as a carcinogen if inhaled.

Occupational workers during use and handling of cobalt sulphate should be alert, avoid to inhale/breathe the contaminated dust at the workplace. Workers should use safety workplace dress, PPE, gloves, goggles etc.

Copper and Its Compounds

Copper (CAS No. 7440-50-8)

Molecular formula: **Cu**

Copper is a metal that occurs naturally throughout the environment, in rocks, soil, water, and air. Copper is an essential element in plants and animals (including humans), which means it is necessary for us to live. Therefore, plants and animals must absorb some copper from eating, drinking, and breathing.

Copper in very low levels is essential for good health. Copper is used to make many different kinds of products like wire, plumbing pipes, and sheet metal. The U.S. pennies made before 1982 are made of copper, while those made after 1982 are only coated with copper. Copper is also combined with other metals to make brass and bronze pipes and faucets. Copper compounds are commonly used in agriculture to treat plant diseases like mildew, for water treatment, and as preservatives for wood, leather, and fabrics.

Safe Handling and Precautions

Copper is an essential micronutrient for human health. While very low levels of copper are essential for good health, high levels of copper cause health effects and can be harmful. Occupational breathing of high levels of copper in workplaces is known to cause irritation of the nose and throat. Acute ingestion of excess copper in drinking water can cause GI tract disturbances. Ingestion of high levels of copper causes nausea, vomiting, and diarrhoea, and very high doses of copper cause damage to the kidneys and can become fatal. The U.S. EPA has not classified copper as a human carcinogen.

Copper (Dust, Fume, and Mist, as Cu) (CAS No. 7440-50-8)

Molecular formula: **Cu**

Safe Handling and Precautions

Occupational workers exposed to copper fumes, dust, and mists in work areas develop symptoms of poisoning. These include irritation to mucous membrane, nasal and pharyngeal irritation, nasal perforation, eye irritation, metallic or sweet taste, and dermatitis, and prolonged period of exposure to high concentrations causes anaemia, adverse effects to lung and liver, and kidney damage. The exposed worker also suffers from metal fume fever, chills, muscle aches, nausea, fever, dry throat, coughing, weakness, lassitude, irritation of eyes and the upper respiratory tract, discoloured skin and hair, and acute lung damage.

Occupational workers exposed to copper dust suffer from GI disturbances, headache, vertigo, drowsiness, and hepatomegaly. Vineyard workers chronically exposed to Bordeaux mixture (copper sulphate and lime) exhibit degenerative changes of the lungs and liver. Dermal exposure to copper may cause contact dermatitis in some individuals. Copper is required for collagen formation. Copper deficiency is associated with atherosclerosis and other cardiovascular conditions. Any kind of imbalance of copper in the body causes health disorders, which include, but are not limited to, arthritis, fatigue, adrenal burnout, insomnia, scoliosis, osteoporosis, heart disease, cancer, migraine headaches, seizures, gum disease, tooth decay, skin and hair problems, uterine fibroids, and endometriosis (in females). Copper deficiency is associated with aneurysms, gout, anaemia, and osteoporosis.

Exposures to copper in the form of dusts and mists cause irritation to the eyes, respiratory system, mucous membrane, nasal and pharyngeal irritation, cough, dyspnoea (breathing difficulty), and wheezing. Prolonged exposures are known to cause nasal perforation. It has caused anaemia and damage to the lung, liver, and kidney in experimental laboratory animals. Reports have indicated that copper dusts and fume are potential occupational carcinogen.

The OSHA, the ACGIH, and the NIOSH have set the PEL for copper in general industry at $1.0 \, mg/m^3$ (TWA).

Copper Fume (as Cu) (CAS No. 1317-38-0)

Molecular formula: **CuO**

Copper metal, metal compounds, and alloys are often used in 'hot' operations in the workplace. The workplace operations include, but are not limited to, welding, brazing, soldering, plating, cutting, and metallising. At the high temperatures reached in these operations, metals often form metal fumes, which have different health effects.

Exposures to copper fume cause fever; chills; muscle aches; nausea; dry throat; coughing; weakness; lassitude; irritation to the eyes, nose, throat, skin, and upper respiratory tract; chest tightness; cough; nose bleeding; oedema; and lung damage. Symptoms of copper fume poisoning also include metallic or sweet taste, skin itching, skin rash, skin allergy, and greenish colour to the skin, teeth, and hair. Workers have increased risk of Wilson's disease. There is evidence that workers in copper smelters have an increased risk of lung cancer, but this is thought to be due to arsenic trioxide and not copper. Occupational workers should use protective clothing such as suits, gloves, footwear, and headgear and promptly change the contaminated clothing/work dress. Workers should not eat, smoke, or drink where copper dust or powder is handled, processed, or stored. Workers should wash hands carefully before eating, drinking, smoking, or using the toilet. The workplace should have the vacuum or wet method facilities to reduce the metal dust during cleanup.

Copper Sulphate (CAS No. 7758-98-7)

Molecular formula: **$CuSO_4 \, 5H_2O$**
Synonyms and trade names: Copper sulphate; Copper sulphate; Cupric sulphate; Cupric sulphate; Bluestone; Blue vitriol; Copper sulphate pentahydrate; Hydrated copper sulphate

Copper sulphate (anhydrous form) is green or grey-white powder, whereas the pentahydrate, the most commonly encountered salt, is bright blue. The anhydrous form occurs as a rare mineral known as chalcocyanite. The hydrated copper sulphate occurs in nature as chalcanthite. Copper sulphate is made by the action of sulphuric acid with a variety of copper compounds. Copper sulphate is used in hair dyes, colouring glass, processing of leather, textiles, and pyrotechnics as a green colorant.

Copper sulphate pentahydrate is used as a fungicide, and a mixture with lime is called Bordeaux mixture, which is used to control fungus on grapes, melons, and other berries, and as a molluscicide for the destruction of slugs and snails, particularly the snail host of the liver fluke. Copper sulphate is a very versatile chemical with extensive uses in industry as it has in agriculture. The metal industry uses large quantities of copper sulphate as an electrolyte in copper refining, for copper coating steel. The paint industry uses it in anti-fouling paints and it plays a part in the coloring of glass. To test blood for anemia, Copper sulphate is a very versatile chemical with extensive uses in industry as it has in agriculture. The metal industry uses large quantities of copper sulphate as an electrolyte in copper refining, for copper coating steel. The paint industry uses it in anti-fouling paints and it plays a part in the coloring of glass. To test blood for anemia Copper sulfate of a particular specific gravity is used to blood for anaemia, in Fehling's and Benedict's solution to test reducing sugars

Safe Handling and Precautions

Workers who accidentally ingest copper sulphate develop abdominal pain, burning sensation, corrosive effects, nausea, vomiting, abdominal pain and cramps, loose bowel movement, and a metallic taste. Exposures to copper sulphate by ingestion or skin absorption cause severe irritating effects to the eyes and skin The aerosol is irritating to the respiratory tract and has effects on the blood, kidneys, and liver, resulting in haemolytic anaemia, kidney impairment, liver impairment, and shock or collapse. At large doses, accidental intake of copper sulphate causes renal failure, comatose, and even death. Long-term exposure to copper sulphate may lead to liver damage, lung diseases, and decreased female fertility. Workers should keep stored copper sulphate in cool, dry area with sufficient ventilation. It should be kept away from alkalis, magnesium, ammonia, acetylene, and sodium hypobromite.

During handling and use of copper sulphate, students and occupational workers should wear safety glasses and should not breathe the material in powder form. Copper sulphate is an environmental pollutant and must be carefully incorporated when used in its varied applications. Workers should wear protective clothing, goggle, impermeable gloves, and rubber boots to avoid skin contact.

Coumaphos (CAS No. 56-72-4)

Molecular formula: $C_{14}H_{16}ClO_5PS$
Synonyms and trade names: Agridip; Asunthol; Meldane; Muscatox; Umbethion; Asuntol; Baymix
IUPAC name: O-(3-chloro-4methyl-2-oxo-2H-Chromen-7-ylO,O-diethyl phosphorothioate)
Toxicity class: U.S. EPA: II; WHO: Ia

Technical coumaphos is a tan crystalline solid with a slight sulphur odour. It is insoluble in water; slightly soluble in acetone, chloroform, and ethanol; and soluble in organic solvents. Coumaphos when heated undergoes decomposition and releases very toxic fumes of sulphur oxides, phosphorous oxides, and chlorides. It is used for the control of a wide variety of livestock insects, including cattle grubs, screwworms, lice, scabies, flies, and ticks. It is used against ectoparasites, which are insects that live on the outside of host animals such as sheep, goats, horses, pigs, and poultry. Coumaphos is incompatible with alkalis, strong oxidising agents, pyrethroids, and piperonyl butoxide. The U.S. EPA has grouped coumaphos as RUP.

Safe Handling and Precautions

Coumaphos is highly toxic by inhalation and ingestion and moderately toxic by dermal absorption. Exposures to coumaphos cause signs of poisoning such as diarrhoea, drooling, difficulty in breathing, and leg and neck stiffness among occupational workers. Inhalation of coumaphos causes headaches, dizziness, and incoordination. Moderate poisoning causes muscle twitching and vomiting, while severe poisoning leads to fever, toxic psychosis, lung oedema, and high blood pressure. Repeated exposures cause irritability, confusion, headache, sweating, speech difficulties, effects on memory concentration, disorientation, severe depressions, laboured breathing, unconsciousness, pupillary constriction, muscle cramp, excessive salivation, and drowsiness or insomnia/sleepwalking among poisoned occupational workers. The important site of effect of coumaphos has been identified as the nervous system where coumaphos causes the inhibition of the cholinesterase enzyme very essential for the normal and proper nerve functioning. No organ effects were seen in acute or chronic studies of coumaphos. The toxic effects of coumaphos are limited to those related to cholinesterase inhibition. Coumaphos has been classified as Group A4, meaning not classifiable as a human carcinogen.

Cyanide (CAS No. 57-12-5)

Molecular formula: **CN**

Cyanides are fast-acting poisons that can be lethal. They were used as chemical weapons for the first time in World War I. Low levels of cyanides are found in nature and in products we commonly eat and use. Cyanides can be produced by certain bacteria, fungi, and algae. Cyanides are also found in cigarette smoke, in vehicle exhaust, and in foods such as spinach, bamboo shoots, almonds, lima beans, fruit pits, and tapioca. There are several chemical forms of cyanide. Hydrogen cyanide (HCN) is a pale blue or colourless liquid at room temperature and is a colourless gas at higher temperatures. It has a bitter almond odour. Sodium cyanide (NaCN) and potassium cyanide (KCN) are white powders, which may have a bitter almond-like odour. Other chemicals called cyanogens can generate cyanides. Cyanogen chloride is a colourless liquefied gas that is heavier than air and has a pungent odour. While some cyanide compounds have a characteristic odour, odour is not a good way to tell if cyanide is present. Some people are unable to smell cyanide. Other people can smell it at first but then get used to the odour.

Historically, HCN has been used as a chemical weapon. Cyanide and cyanide-containing compounds are used in pesticides and fumigants, plastics, electroplating, photo developing, and mining. Dye and drug companies also use cyanides. Some industrial processes, such as iron and steel production, chemical industries, and wastewater treatment, can create cyanides. During water chlorination, cyanogen chloride may be produced at low levels. The most common cyanide is HCN and its salts – NaCN and KCN. Cyanides are ubiquitous in nature, arising from both natural and anthropogenic sources. Cyanide is released to the environment from numerous sources. Metal finishing and organic chemical industries as well as iron and steel production are major sources of cyanide releases to the aquatic environment. More than 90% of emissions to the air are attributed to releases in automobile exhaust. Workers in a wide variety of occupations may be exposed to cyanides. The general population may be exposed to cyanides by inhalation of contaminated air, ingestion of contaminated drinking water, and/or consumption of a variety of foods. Cyanide process, also known as cyanidation, is the most widely used process for extracting gold and silver from ores. The ores are powdered grounds and can be concentrated by flotation. It is then mixed with dilute solutions of sodium (or potassium or calcium) cyanide, while air is bubbled through it to form the soluble complex ion, $Au(CN)2-1$. The precious metals are precipitated from solution by zinc. The precipitates are smelted to remove the zinc and treated with nitric acid to dissolve the silver.

Cyanide most commonly occurs as HCN and its salts – sodium and potassium cyanides. Cyanides are both man-made and naturally occurring substances. They are found in several plant species as cyanogenic glycosides and are produced by certain bacteria, fungi, and algae. In very small amounts, cyanide is a necessary requirement in the human diet. Cyanides are released to the environment from industrial sources and car emissions. Cyanides are readily absorbed by the inhalation, oral, and dermal routes of exposure. HCN and its simple soluble salts such as sodium and potassium cyanides are among the most rapidly acting poisons. The central nervous system (CNS) is the primary target organ for cyanide toxicity. Neurotoxicity has been observed in humans and animals following ingestion and inhalation of cyanides. Neurotoxic effects including convulsions and coma preceded death in guinea pigs dermally exposed to HCN. Exposed laboratory animals developed signs of toxicity and death in 3–12 min after the eyes were treated. Cardiac and respiratory effects, possibly CNS mediated, have also been reported. In tropical regions of Africa, a high incidence of ataxic neuropathy, goitre, amblyopia, and other health disorders has been associated with chronic ingestion of cassava, one of the dietary staples containing cyanogenic glycosides that release HCN when metabolised in vivo.

Safe Handling and Precautions

Upon exposure, cyanide quickly enters the bloodstream. The body handles small amounts of cyanide differently than large amounts. In small doses, cyanide in the body can be changed into thiocyanate, which is less harmful and is excreted in urine. In the body, cyanide in small amounts can also combine with another chemical to form vitamin B12, which helps maintain healthy nerve and red blood cells. In large doses, the body's ability to change cyanide into thiocyanate is overwhelmed. Large doses of cyanide prevent cells from using oxygen, and eventually these cells die. The heart, respiratory system, and CNS are most susceptible to cyanide poisoning.

The health effects from high levels of cyanide exposure can begin in seconds to minutes. The severity of health effects depends upon the route and duration of exposure, the dose, and the form of cyanide. Some signs and symptoms of such exposures are the following:

Weakness and confusion

Headache

Nausea/feeling 'sick to your stomach'

Gasping for air and difficulty breathing

Loss of consciousness/'passing out'

Seizures

Cardiac arrest

The U.S. EPA has set a limit of 0.2 ppm for cyanide in drinking water. The OSHA has set a limit of 10 ppm for HCN and most cyanide salts in the workplace.

Cyanide Compounds

Cyanide compounds are a group of chemical substances. These are based on a common structure formed when elemental nitrogen and carbon are combined. Cyanides are produced by certain bacteria, fungi, and algae and may be found in food and plants. Cyanide itself is an ion or combining form that carries a positive or negative charge. It is a powerful and rapid-acting poison. Cyanide, in combination with metals and organic compounds, forms simple and complex salts such as KCN and NaCN. Cyanide compounds most commonly occur as hydrogen cyanide and its salts – sodium and potassium cyanides. Cyanides are both man-made and naturally occurring substances. They are found in several plant species as cyanogenic glycosides and are produced by certain bacteria, fungi, and algae. In very small amounts, cyanide is a necessary requirement in the human diet. Cyanides are released to the environment from industrial sources and car emissions. In general, toxic effects of cyanides, whether acute or chronic, in animals and humans are believed to result primarily from inhibition of cellular respiration and consequent histotoxic anoxia. Effects on the thyroid resulting from long-term exposure are likely to be caused by thiocyanate. Cyanide salts are mainly used in electroplating, metallurgy, and production of organic chemicals; in photographic development; as anti-caking agents in road salts; in the extraction of gold and silver from ores; and in the making of plastics. Minor uses of cyanide salts include as insecticides and rodenticides, chelating agents, and in the manufacture of dyes and pigments.

Cyanide is released to the environment from numerous sources. Metal finishing and organic chemical industries as well as iron and steel production are major sources of cyanide releases to the aquatic environment. More than 90% of emissions to the air are attributed to releases in automobile exhaust. Workers in a wide variety of occupations may be exposed to cyanides. The general population may be exposed to cyanides by inhalation of contaminated air, ingestion of contaminated drinking water, and/or consumption of a variety of foods. Cyanogenic glycosides, producing HCN upon hydrolysis, are found in a number of plant species. Cyanide compounds are many and important.

Calcium Cyanide (CAS No. 592-01-8)

Molecular formula: $Ca(CN)_2$
Synonyms and trade names: Calcid; Calcium cyanide-solid; Calcyan; Calcyanide; Cianuro de calcio; Cyanide of calcium; Cyanogas; Cyanogas a-dust; Cyanogas g-fumigant; Cyanure de calcium

Calcium cyanide is used mainly for the extraction or cyanidation of gold and silver ores. It is also used in the production of prussiates or ferrocyanides, in the froth flotation of minerals, in processes where gold complexes are adsorbed on carbon, in the manufacture of stainless steel, as a fumigant and rodenticide, and as a cement stabiliser. The main users of cyanides are the steel, electroplating, mining, and chemical industries. The principal cyanide compounds used in industrial operations are potassium and sodium cyanide and calcium cyanide, particularly in metal leaching operations. Cyanides have been well established in uses as insecticides and fumigants; in the extraction of gold and silver ores; in metal cleaning;

in the manufacture of synthetic fibres, various plastics, dyes, pigments, and nylon; and as reagents in analytical chemistry. Calcium cyanide decomposes on heating above 350°C, producing toxic fumes including nitrogen oxides and HCN. It reacts violently with water, moist air, carbon dioxide, acids, and acid salts producing highly toxic and flammable HCN. It reacts violently when heated with oxidising substances causing fire and explosion hazard.

Safe Handling and Precautions

Exposure to calcium cyanide causes adverse health effects. Chronic exposure to cyanide in humans via inhalation results in effects such as headaches, dizziness, numbness, tremor, loss of visual acuity, cardiovascular and respiratory effects, an enlarged thyroid gland, irritation to the eyes and skin, chest tightness, confusion, convulsions, cough, dizziness, headache, laboured breathing, nausea, shortness of breath, unconsciousness, vomiting, weakness, and red coloration of the skin. Chronic exposure to cyanide in humans via inhalation results in effects on the CNS, such as headaches, dizziness, numbness, tremor, loss of visual acuity, irritation of the eyes and mucous membranes, and irritations to the skin. Accidental ingestion at workplaces causes lethal poisoning to users and workers. Calcium cyanide on contact with acids releases very poisonous gas. Calcium cyanide causes severe adverse health effects on blood, CNS, heart, and the cardiovascular system. It causes effects on the intracellular oxygen metabolism, resulting in seizures and unconsciousness, and may result in death.

Also, calcium cyanide is very toxic to aquatic organisms and causes long-term adverse effects in the aquatic environment. During handling of calcium cyanide, workers should avoid all kinds of contact with the chemical and must use full protective equipment (PPE) and protective gloves made of 'rubber or plastic' and must wear approved safety tight-fitting goggles to avoid dust if and when generated at the workplace. The workplace should have eyewash facilities and emergency shower during handling of calcium cyanide. Workers should observe precautions to store the chemical, and it should be kept in cool, dry, ventilated storage place and closed containers. The containers should be closed tightly and away from acids. Wear appropriate clothing to prevent any possibility of skin contact and wash at the end of each work shift and before eating, smoking, and using the toilet. Also, workers should promptly wash with soap and water if skin has become contaminated during work and promptly remove any clothing that becomes contaminated, and the contaminated clothing must be placed in closed container until proper disposals are done. Workers should avoid eating, smoking, and drinking at the prohibited and immediate work area. As a safety precaution, workplace manager should restrict the use of calcium cyanide to persons who understand its dangers and proper handling, because calcium cyanide on contact with moisture releases cyanide gas. During handling of calcium cyanide workers should not be negligent to use fume cupboard, appropriate respiratory protection, protective dress, gloves made of Rubber or plastic. Workplace manager should confirm the provision of good general ventilation throughout areas where cyanides are used or stored.

Cyanogen Bromide (CAS No. 506-68-3)

Molecular formula: **CBr**

Exposures to cyanogen bromide cause danger. The chemical substance is poisonous and causes fatal injury if swallowed, inhaled, or absorbed through skin. It is corrosive, and the vapours cause severe irritation to eyes and respiratory tract and cause burns to any area of contact. On contact with acids, cyanogen bromide liberates poisonous gas and affects blood, cardiovascular system, CNS, and thyroid.

Cyanogen Chloride (CAS No. 506-77-4)

Molecular formula: **CNCl**
Synonyms and trade names: Cyanogen chloride; Chlorine cyanide; Chlorine cyanide; Chlorocyan; Chlorocyanide; Chlorocyanide; Chlorocyanogen; Canochloride

Cyanogen chloride is a liquefied gas with an irritating odor. Cyanogen chloride has been reported a potentially hazardous chemical and fatal if swallowed, harmful if inhaled, respiratory tract burns, skin burns, eye burns, mucous membrane burns.

Hydrogen Cyanide (CAS No. 74-90-8)

Molecular formula: **HCN**
Synonyms and trade names: Carbon hydride nitride; Cyanane; Cyclon; Formic anammonide; Formonitrile; Hydrocyanic acid; Prussic acid

HCN is a colourless gas or bluish-white liquid with a bitter almond odour. HCN reacts with amines, oxidisers, acids, sodium hydroxide, calcium hydroxide, sodium carbonate, caustic substances, and ammonia. HCN was first isolated from a blue dye, Prussian blue, in 1704. HCN is obtainable from fruits that have a pit, such as cherries, apricots, and bitter almonds, from which almond oil and flavouring are made. HCN is used in fumigating, electroplating, mining, and producing synthetic fibres, plastics, dyes, and pesticides. It also is used as an intermediate in chemical syntheses. Exposures to cyanide occur in the workplaces such as the electroplating, metallurgical, firefighting, steel manufacturing, and metal cleaning industries. Human exposures to cyanide also occur from wastewater discharges of industrial organic chemicals, iron and steel works, and wastewater treatment facilities. An air odour threshold concentration for HCN of 0.58 part per million (ppm) parts of air has been reported.

Safe Handling and Precautions

HCN is particularly dangerous because of its toxic/asphyxiating effects on all life requiring oxygen to survive. HCN combines with the enzymes in tissue associated with cellular oxidation. The signs and symptoms of HCN poisoning are non-specific and very rapid. The symptoms include excitement, dizziness, nausea, vomiting, headache, weakness, drowsiness, gasping, thyroid, blood changes, confusion, fainting, tetanic spasm, lockjaw, convulsions, hallucinations, loss of consciousness, coma, and death. When oxygen becomes unavailable to the tissues, it leads to asphyxia and causes death. Children are more vulnerable to HCN gas exposure. HCN is readily absorbed in the lungs; symptoms of poisoning begin within seconds to minutes. Inhalation of HCN results in the most rapid onset of poisoning, producing almost immediate collapse, respiratory arrest, and death within minutes (Table 5.2).

Hydrogen cyanide and carcinogenicity: Information on the carcinogenicity of HCN in humans or animals for oral exposure is unavailable. Similarly, there are no reports that cyanide

TABLE 5.2

HCN and Human Poisoning: Possible Workplace Exposures

The manufacture and transportation of HCN
Use in fumigation of ships, structures, and agricultural crops and as a nematocide
Liberated during use of cyanide salts or solutions in metal treatment operations, blast furnace and coke oven operations, metal ore processing, and photoengraving operations
Use in production of intermediates in synthesis of resin monomers, acrylic plastics, acrylonitrile, Nylon 66, cyanide salts, lactic acid, nitrates, chelating agents, dyes, pharmaceuticals, and specialty chemicals
Liberated during petroleum refining and electroplating
Use in the manufacture of silver and metal polishes, and electroplating solutions, and as a reagent
Use as the instrument of execution for convicted criminals in prison gas chambers in some states
Occupational workers should not ignore to wear PPE and clothing during handling of CHN. In the event of any kind of accidental chemical spill or leak of HCN, workers should avoid the entry of the contaminated areas until cleanup has been completed. Workplace supervisor should immediately notify the persons of contaminated area
Remove all sources of heat and ignition; ventilate potentially explosive atmospheres
Workers should avoid to touch the spilled material; stop the leak if it is possible to do so without risk; use non-sparking tools
If source of leak is a cylinder and the leak cannot be stopped in place, remove the leaking cylinder to a safe place and repair leak or allow cylinder to empty. Use water sprays to protect personnel attempting to locate and seal the source of escaping HCN gas
For small liquid spills, take up with sand or other non-combustible absorbent material and place into closed containers for later disposal; prevent the HCN from accumulating in a confined space, such as a sewer, because of the possibility of an explosion

can cause cancer in animals and humans. The U.S. EPA has classified cyanide as a Group D, meaning not classifiable as to human carcinogenicity. Exposure limits: The OSHA set the PEL of HCN at 10 ppm at the workplace, and NIOSH set the IDLH concentration at 50 ppm.

Additional Steps of Precautions

Occupational workers should be very careful in the management of HCN since the gas in air is explosive at concentrations over 5.6%, equivalent to 56,000 ppm, and it does not provide adequate warning of hazardous concentrations. HCN concentration of 300 mg/m^3 in air becomes fatal within about 10 min, and HCN at a concentration of 3500 ppm (about 3200 mg/m^3) kills a human in about 1 min.

Workers should use appropriate personal protective clothing and equipment that must be carefully selected, used, and maintained to be effective in preventing skin contact with HCN. The selection of the appropriate PPE such as gloves, sleeves, and encapsulating suits should be based on the extent of the worker's potential exposure to HCN.

If HCN contacts the skin, workers should flush the affected areas immediately with plenty of water, followed by washing with soap and water.

Clothing contaminated with HCN should be removed immediately, and provisions should be made for the safe removal of the chemical from the clothing. Persons laundering the clothes should be informed of the hazardous properties of HCN, particularly its potential for severe systemic toxicity by dermal absorption or inhalation.

A worker who handles HCN should thoroughly wash hands, forearms, and face with soap and water before eating, using tobacco products, using toilet facilities, applying cosmetics, or taking medication.

Workers should not eat, drink, use tobacco products, apply cosmetics, or take medication in areas where HCN or a solution containing HCN is handled, processed, or stored.

Potassium Cyanide (CAS No. 151-50-8)

Molecular formula: **KCN**

KCN is a white solid or colourless water solution with a faint bitter almond odour. As a solution, it is slightly soluble in ethanol. It is a poison that reacts with acid or acid fumes to emit deadly HCN. When heated to decomposition, it emits very toxic fumes. As a solid, KCN is incompatible with nitrogen trichloride, perchloryl fluoride, sodium nitrite, acids, alkaloids, chloral hydrate, and iodine. A synonym for KCN is potassium salt of hydrocyanic acid.

Potassium Silver Cyanide (CAS No. 506-61-6)

Molecular formula: **AgK(CN)$_2$**
Synonyms and trade names: Potassium argentocyanide and potassium dicyanoargentate

Potassium silver cyanide is a poisonous, white solid made of crystals, which are light sensitive. It is soluble in water and acids and slightly soluble in ethanol. It emits very toxic fumes when heated to decomposition.

Safe Handling and Precautions

Silver KCN is an extremely hazardous and a corrosive chemical substance. Any kind of accidental contact to the eyes and skin is very harmful. It causes corneal damage or blindness to the eyes, skin contact produces inflammation and blistering, and the inhalation of dust causes irritation to GI and respiratory tract and causes burning, sneezing, and coughing. Severe overexposure to silver KCN produces lung damage, choking, unconsciousness, or death.

Occupational workers on poisoning of potassium silver cyanide is characterised by gasping for breath, loss of consciousness, develop violent convulsions, epileptiform, or tonic, and weak breathing, cardiac arrest and death. Opisthotonos and trismus may develop. Involuntary micturition and defecation occur. Paralysis follows the convulsive stage. The skin is covered with sweat. The eyeballs protrude, and the pupils are dilated and unreactive. The mouth is covered with foam, which is sometimes bloodstained, and the skin colour may be brick red. Cyanosis is not prominent in spite of weak and irregular gasping. In the unconscious patient, bradycardia and the absence of cyanosis may be key diagnostic signs. In death from respiratory arrest, as long as the heartbeat continues, prompt and vigorous treatment offers some promise of survival of the poisoned individual.

Sodium Cyanide (CAS No. 143-33-9)

Molecular formula: **NaCN**
Synonyms and trade names: Hydrocyanic acid, sodium salt; Cyanide of sodium; Prussiate of soda

NaCN is a white crystalline solid that is odourless when dry but emits a slight odour of HCN in damp air. It is slightly soluble in ethanol and formamide. It is very poisonous. It explodes if melted with nitrite or chlorate at about 450°F. It produces a violent reaction with magnesium, nitrites, nitrates, and nitric acid. On contact with acid, acid fumes, water, or steam, it will produce toxic and flammable vapours.

Safe Handling and Precautions

NaCN is very hazardous, and exposures and any kind of contact cause effects of irritation to the skin, local skin destruction, or dermatitis and effects to the eyes and corneal damage or blindness. Skin contact can produce inflammation, watering (tears), itching, scaling, reddening, and blistering. Inhalation of NaCN dust causes severe irritation of the GI or respiratory tract, and the symptoms of poisoning include weakness, headache, dizziness, confusion, burning, sneezing and coughing, lung damage, and choking. NaCN is known to inhibit cellular respiration causing metabolic tissue anoxia, cardiovascular effects, cyanosis (bluish skin due to deficient oxygenation of the blood), weak and irregular heartbeat, collapse, unconsciousness, convulsions, coma, and death.

Users and occupational workers should remember that NaCN is a corrosive and poisonous solid. Workers should keep the container of NaCN tightly closed and store the container in a cool, well-ventilated area away from incompatible substances. Keep it away from strong acids. Poison room should be locked. Keep containers tightly closed. Store in an area protected from moisture.

Workers during use and handling should use all personal protective equipment (PPE) safety dress, splash goggles, full suit, vapour and dust respirator, boots, and a self-contained breathing apparatus to avoid inhalation of the product (Table 5.3).

TABLE 5.3

Safe Handling of Cyanides

Cyanides shall only be sold to customers who have the knowledge and capability to handle, store, and dispose of the product or its packaging in a safe, responsible, and legal manner including accident prevention systems
All regulations regarding the sale and use of cyanides shall be observed
Cyanides shall only be supplied into applications, which do not, by their very nature, put people or the environment at any risk
Cyanide producers jointly condemn any misuse of cyanides and will endeavour to avoid such misuse of their product
Cyanides shall be transported in such a way as to minimise the possibility of accidental loss of containment or injury. Cyanides shall be transported in such a way as to minimise the possibility of accidental loss of containment or injury
Efforts shall be made to ensure that the end user or distributor is a bona fide company with a legitimate use for the product
Cyanides and cyanide samples/compounds shall be supplied with particular attention under controlled conditions
Much before the shipment, the member shall satisfy itself that the customer's facilities are adequate to safely handle cyanides according to the 'guidelines for the storage, handling and distribution for alkali cyanides'. If not, the member may provide recommendations for improvement within a defined timescale or decline to supply until the improvements are made
The producers reserve the right to carry out periodic follow-up inspections
Workers should be provided with a copy of the safety data sheet, 'guidelines for the storage, handling and distribution of alkali cyanides', and if needed additional handling instructions and training
Cyanides shall only be sold to customers who have the knowledge and capability to handle, store, and dispose of the product or its packaging in a safe, responsible, and legal manner including accident prevention systems
All regulations regarding the sale and use of cyanides shall be observed
Cyanides shall only be supplied into applications, which do not, by their very nature, put people or the environment at any risk

Cyclohexane (CAS No. 110-82-7)

Molecular formula: C_6H_{12}
Synonyms and trade names: Benzene, hexahydrobenzene, hexamethylene, hexanapthene.

Bibliography

Agency for Toxic Substances and Disease Registry (ATSDR). 1996. *Toxicological Profile for Carbon Disulfide*. U.S. Department of Health and Human Services, Public Health Service, Atlanta, GA (updated 2007).

Agency for Toxic Substances and Disease Registry (ATSDR). 1997a. *Toxicological Profile for Cyanide*. U.S. Department of Health and Human Services, Public Health Service ATSDR. Atlanta, GA.

Agency for Toxic Substances and Disease Registry (ATSDR). 1997b. *Toxicological Profile for Chlorpyrifos*. U.S. Department of Health and Human Services, Public Health Service, Atlanta, GA (updated 2007).

Agency for Toxic Substances and Disease Registry (ATSDR). 2002. Chlorine. *Managing Hazardous Materials Incidents. Volume III—Medical Management Guidelines for Acute Chemical Exposures*. U.S. Department of Health and Human Services, Public Health Service, Atlanta, GA (updated 2007).

Agency for Toxic Substances and Disease Registry (ATSDR). 2004a. *Toxicological Profile for Copper*. U.S. Department of Health and Human Services, Public Health Service, Atlanta, GA (updated 2007).

Agency for Toxic Substances and Disease Registry. 2004b. *Medical Management Guidelines for Hydrogen Cyanide*. Division of Toxicology, U.S. Department of Health and Human Services. Public Health Service: Atlanta, GA.

Agency for Toxic Substances and Disease Registry (ATSDR). 2006. *Toxicological Profile for Cyanide*. U.S. Department of Health and Human Services, Public Health Service, Atlanta, GA.

Agency for Toxic Substances and Disease Registry (ATSDR). 2008. *Toxicological Profile for Cadmium (Draft for Public Comment)*. U.S. Department of Health and Human Services, Public Health Service, Atlanta, GA.

American Conference of Governmental Industrial Hygienists (ACGIH). 2001. Hydrogen cyanide and cyanide salts. In: *Documentation of the Threshold Values and Biological Exposure Indices*, 8th edn. Cincinnati, OH.

Banerjee, K. K., B. Bishayee, and P. Marimuthu. 1997. Evaluation of cyanide exposure and its effect on thyroid function of workers in a cable industry. *J. Occup. Med.* 39:255–260.

Bhattacharya, R. and P. V. L. Rao. 1997. Cyanide induced DNA fragmentation in mammalian cell cultures. *Toxicology* 123:207–215.

Centers for Disease Control and Prevention. 2004. Cyanide. Emergency preparedness and response. U.S. Department of Health and Human Services. Public Health Service: Atlanta, GA.

Chandra, H., B. N. Gupta, S. K. Bhargava, S. H. Clerk, and P. N. Mahendre. 1980. Chronic cyanide exposure: A biochemical and industrial hygiene study. *J. Anal. Toxicol.* 4:161–165.

Chandra, H., B. N. Gupta, and N. Mathur. 1988. Threshold limit value of cyanide: A reappraisal in Indian context. *Indian J. Environ. Protect.* 8:170–174.

Clary, T. and B. Ritz. 2003. Pancreatic cancer mortality and organochlorine pesticide exposure in California, 1989–1996. *Am. J. Ind. Med.* 43(3):306–313.

Department of Health. 2004. Carbon monoxide. Chemical Fact Sheets, Department of Health, WI.

Department of Health and Human Services (DHHS). 2006. Facts about chlorine. DHHS, Center for Disease Control and Prevention, Atlanta, GA.

Department of Public Health. 2004. Carbon monoxide poisoning. Department of Public Health Review, IA.

Dikshith. T. S. S. 2009. *Safe Use of Chemicals: A Practical Guide*. CRC Press, Boca Raton, FL.

Dikshith, T. S. S. (ed.). 2011. *Handbook of Chemicals and Safety*. CRC Press, Boca Raton, FL.

Extension Toxicology Network (EXTOXNET). 1994. Coumaphos. USDA, Atlanta, GA.

Extension Toxicology Network (EXTOXNET). 1995. Captafol. USDA/Extension Service. Cornell University, Ithaca, NY.

Gosselin, R. E. et al. 1976. *Clinical Toxicology of Commercial Products*, 4th edn. Williams and Wilkins, Baltimore, MD.

Gosselin, R. E., R. P. Smith, and H. C. Hodge (eds.) 1984. *Clinical Toxicology of Commercial Products*, 5th edn. Williams & Wilkins, Baltimore, MD.

Hayes, W. J., Jr. 1971. *Clinical Handbook on Economic Poisons*. U.S. EPA Pesticides Programs, Public Health Service Publication 476.

Hayes, W. Jr. 1982. *Pesticides Studied in Man*. Williams & Wilkins, Baltimore, MD.

Hayes, W. J., Jr. and E. R. Laws, Jr. (eds.). 1991. *Handbook of Pesticide Toxicology. Volume 2. Classes of Pesticides*. Academic Press, Inc., New York, p. 106.

Henry, C. R., D. Satran, B. Lindgren, C. Adkinson, I. Caren, R. N. Nicholson, and T. D. Henry. 2006. Myocardial injury and long-term mortality following moderate to severe carbon monoxide poisoning. *JAMA* 295:398–402.

Hill, E. F. and M. B. Camardese. 1986. Lethal dietary toxicities of environmental contaminants to coturnix, Technical Report Number 2. U.S. Department of Interior, Fish and Wildlife Service, Washington, DC.

International Agency for Research on Cancer (IARC). 1990. *IARC Monographs on the Evaluation of Carcinogenic Risk to Humans—Cadmium and Cadmium Compounds*, Supplement 7 and Vols. 43–61.

International Agency for Research on Cancer (IARC). 1996a. 2-Chloronitrobenzene, 3-chloronitrobenzene and 4-chloronitrobenzene. In: *Printing Processes and Printing Inks, Carbon Black and Some Nitro Compounds*.

International Agency for Research on Cancer (IARC). 1996b. *IARC Monographs on the Evaluation of Carcinogenic Risks to Humans*, Vol. 65, pp. 263–296, Lyon, France.

International Agency for Research on Cancer (IARC). 1997. Cadmium and cadmium compounds. Summaries & Evaluations. IARC, Lyon, Paris.

International Chemical Safety (IPCS). 2002. Hydrogen cyanide, liquefied. IPCS Card No. 0492). IPCS, World Health Organization, Geneva, Switzerland.

International Programme on Chemical Safety (IPCS). 1990. Fully halogenated chlorofluorocarbons, Environ Health Criteria No. 113. World Health Organization, Geneva, Switzerland.

International Programme on Chemical Safety (IPCS). 1992. Carbaryl. Health and safety. Guide No. 78. IPCS, World Health Organization, Geneva, Switzerland.

International Programme on Chemical Safety (IPCS). 1993. Captafol. Poisons Information Monograph No. 097. IPCS, Geneva, Switzerland.

International Programme on Chemical Safety (IPCS). 1999. Calcium cyanide. ICSC Card No. International Chemical Safety Cards. 0407. WHO, Geneva, Switzerland (updated 2007).

International Programme on Chemical Safety and the Commission of the European Communities (IPCS-CEC). 1993. International Chemical Safety Cards. Captafol. ICSC Card No. 0119. IPCS-CEC, Geneva, Switzerland.

International Programme on Chemical Safety and the Commission of the European Communities, Carbophenothion (IPCS, CEC). 1997. ICSC Card No. 0410, WHO, Geneva, Switzerland (updated 2005).

International Programme on Chemical Safety and the Commission of the European Communities (IPCS, CEC). 1999. Cadmium. ICSC Card No. ICSC: 0020. IPCS-CEC, Geneva, Switzerland (updated 2005).

International Programme on Chemical Safety and the Commission of the European Communities (IPCS-CEC). 2005. Chloramine-T. International Chemical Safety Card. ICSC Card No. 1059. IPCS-CEC, Geneva, Switzerland.

International Programme on Chemical Safety, World Health Organization. 2005. Chlorpyrifos. International Chemical Safety Card No. 0851 (updated 2007).

Kennedy, G. L., Jr., D. W. Arnold, and M. L. Keplinger. 1975. Mutagenicity studies with captan, captofol, folpet and thalidomide. *Food Cosmet. Toxicol.* 13(1):55–61.

Kidd, H. and D. R. James (eds.). 1991. *The Agrochemicals Handbook,* 3rd edn. Royal Society of Chemistry Information Services, Cambridge, U.K.

Material Safety Data Sheet (MSDS). 1993a. Carbaryl. Extension Toxicology Network, USDA/ Extension Service, Oregon State University, OR.

Material Safety Data Sheet (MSDS). 1993b. Carbofuran. Extension Toxicology Network, USDA/ Extension Service, Oregon State University, OR.

Material Safety Data Sheet (MSDS). 1995. Carbophenothion. Extension Toxicology Network, Pesticide Information Profiles. USDA/Extension, Oregon State University, OR.

Material Safety Data Sheet (MSDS). 1996a. Chlordane. Extension Toxicology Network, Pesticide Information Profiles. USDA/Extension Service/Oregon State University, OR.

Material Safety Data Sheet (MSDS). 1996b. Chlorpyrifos. Extension Toxicology Network. Pesticide Information Profiles. USDA/Extension Service. Oregon State University, OR.

Material Safety Data Sheet (MSDS). 1997. Cumaphos. International Chemical Safety Cards. ICSC Card No. 0422 (updated in July 2007).

Material Safety Data Sheet (MSDS). 2003. Safety data for silver potassium cyanide hydrogen cyanide. Physical Chemistry at Oxford University, Oxford, U.K.

Material Safety Data Sheet (MSDS). 2005a. Safety data for cadmium oxide. Department of Physical and Theoretical Chemistry. Oxford University, Oxford, U.K.

Material Safety Data Sheet (MSDS). 2005b. Safety data for cadmium sulfate anhydrous. Department of Physical and Theoretical Chemistry, Oxford University, Oxford, U.K.

Material Safety Data Sheet (MSDS). 2005c. Safety data for chlordane. Department of Physical and Theoretical Chemistry, Oxford University, Oxford, U.K.

Material Safety Data Sheet (MSDS). 2005d. Safety data for coumaphos. Physical Chemistry, Oxford University, Oxford, U.K.

Material Safety Data Sheet (MSDS). 2005. Cefadroxi. U.S. National Library of Medicine. Rockville Pike, Bethesda, MD.

Material Safety Data Sheet (MSDS). 2006. Safety data for cobalt sulfate heptahydrate. Department of Physical and Theoretical Chemistry, Oxford University, Oxford, U.K.

Material Safety Data Sheet (MSDS). 2008a. Cadmium acetate. MSDS no. C0077. Environmental Health & Safety. Mallinckrodt Baker, Inc. Phillipsburg, NJ.

Material Safety Data Sheet (MSDS). 2008b. Cadmium chloride. MSDS no.C0105. Environmental Health & Safety. Mallinckrodt Baker, Inc. Phillipsburg, NJ.

Material Safety Data Sheet (MSDS). 2009a. Copper sulfate. Chemical Safety Data: Copper (II) sulfate. Physical Chemistry, Oxford University, Oxford, U.K.

Material Safety Data Sheet (MSDS). 2009b. Safety data for hydrogen cyanide. Physical Chemistry at Oxford University, Oxford, U.K.

Mathangi, D. C. and A. Namasivayam. 2000. Effect of chronic cyanide intoxication on memory in albino rats. *Food Chem. Toxicol.* 38:51–55.

Meister, R. T. (ed.) 1991. *Farm Chemicals Handbook.* Meister Publishing Co., Willoughby, OH.

National Institute for Occupational Safety and Health (NIOSH). 1976. Occupational exposure to hydrogen cyanide and cyanide salts. Publication no. 1-191.

National Institute for Occupational Safety and Health (NIOSH). 1994. Cadmium acetate. ICSC: 1075 NIOSH, Centers for Disease Control and Prevention, Atlanta, GA (updated 2007).

National Institute for Occupational Safety and Health (NIOSH). 2003. Sodium cyanide. International Chemical Safety Cards ICSC Card No.: 1118. NIOSH, Atlanta, GA.

National Institute for Occupational Safety and Health (NIOSH). 2010. Sodium cyanide. *NIOSH Pocket Guide to Chemical Hazards.* Centers for Disease Control and Prevention (CDCP). Atlanta, GA.

National Institutes of Health (NIH). 2003. Cefixime. Department of Health & Human Services (DHHS), Bethesda, MD (updated 2007).

National Library of Medicine (NLM). 1995. Hazardous substances data bank: Cobalt metal. NLM. Bethesda, MD.

New Jersey Department of Health and Senior Services. 1986. Copper (dust, fume or mist). Occupational Disease and Injury Services. Trenton, NJ.

Nolan, R. J., D. L. Rick, N. L. Freshour, and J. H. Saunders. 1984. Chlorpyrifos: Pharmacokinetics in human volunteers. *Toxicol. Appl. Pharmacol.* 73:8–15.

Occupational Health Services (OHS). 1991. MSDS for coumaphos. OHS Inc., Secaucus, NJ.

Occupational Health Services (OHS) Database. 1993 (December). Occupational Health Services, Inc. MSDS for captafol. OHS Inc., Secaucus, NJ.

Patnaik, P. 1999. *A Comprehensive Guide to the Hazardous Properties of Chemical Substances*, 2nd edn. John Wiley & Sons, New York, pp. 288–295.

Phin, N. 2005. Carbon monoxide poisoning (acute). *Clin. Evidence.* 13:1732–1743.

Pohanish, R. P. (ed.). 2011. *Sittig's Handbook of Toxic and Hazardous Chemicals and Carcinogens*, 6th edn. William Andrew, New York.

Richard, J., Sr. (ed.). 1992. *Sax's Dangerous Properties of Industrial Materials*, 8th edn. Van Nostrand Reinhold, New York, NY.

Richardson, R. J. 1995. Assessment of the neurotoxic potential of chlorpyrifos relative to other organophosphorus compounds: A critical review of the literature. *J. Toxicol. Environ. Health* 44(2):135–165.

Ryan, R. P. and C. E. Terry. (eds.). 1997. *Toxicology Desk Reference*, Vols. 1–3, 4th edn. Taylor & Francis, Washington, DC.

Satran, D., C. R. Henry, C. Adkinson, C. I. Nicholson, R. N. Yiscah Bracha, and T. D. Henry. 2005. Cardiovascular manifestations of moderate to severe carbon monoxide poisoning. *J. Am. Coll. Cardiol.* 45:1513–1516.

Sax, N. I. and R. J. Lewis Jr. 1989. *Dangerous Properties of Industrial Materials*, Vol. 3, 7th edn. Van Nostrand Reinhold, New York.

Sittig, M. 1991. *Handbook of Toxic and Hazardous Chemicals*, 3rd edn. Noyes Publications, Park Ridge, NJ.

Slotkin, T. A. 1999. Brain developmental damage occurs from common pesticide Dursban (chlorpyrifos), *Environ. Health Perspect.* 107 (Suppl. 1).

The National Toxicology Program (NTP). 2011. *12th Report on Carcinogens*. Research Triangle Park, NC.

Tomlin, C. D. S. (ed.). 2006. *The Pesticide Manual. A World Compendium*, 14th edn. British Crop Protection Council, Surrey, U.K.

U.S. Department of Labor. 2000. Copper dusts & mists (as Cu). OSHA, Washington, DC.

U.S. Department of Labor. 2008. Occupational safety and health guideline for cobalt metal, dust, and fume (as Co). OSHA, Washington, DC.

U.S. Environmental Protection Agency. 1984. Pesticide fact sheet number 35: Captafol. U.S. EPA, Office of Pesticide Programs, Registration Div., Washington, DC.

U.S. Environmental Protection Agency (U.S. EPA). 1985. US EPA Chemical profile: Coumaphos. Washington, DC.

U.S. Environmental Protection Agency (US EPA). 1988. Calcium cyanide: Tolerances for residues. Code of Federal Regulations 40 CFR 180.125, US EPA, Atlanta, GA.

U.S. Environmental Protection Agency (U.S. EPA). 1989a. Coumaphos. Pesticide fact sheet no. 207. Office of Pesticides and Toxic Substances, Washington, DC.

U.S. Environmental Protection Agency (U.S. EPA). 1989b. Health advisory summary for carbofuran. U.S. EPA, Washington, DC.

U.S. Environmental Protection Agency (U.S. EPA). 1990a. Calcium cyanide. Integrated Risk Information System (IRIS). Environmental Criteria and Assessment Office, Office of Health and Environmental Assessment, Cincinnati, OH.

U.S. Environmental Protection Agency (U.S. EPA). 1990b. Cyanogen. Integrated Risk Information System (IRIS). Environmental Criteria and Assessment Office, Office of Health and Environmental Assessment, Cincinnati, OH.

U.S. Environmental Protection Agency (U.S. EPA). 1990c. Silver cyanide. Integrated Risk Information System (IRIS). Environmental Criteria and Assessment Office, Office of Health and Environmental Assessment, Cincinnati, OH.

U.S. Environmental Protection Agency (U.S. EPA). 1990d. Potassium cyanide. Integrated Risk Information System (IRIS). Environmental Criteria and Assessment Office, Office of Health and Environmental Assessment, Cincinnati, OH.

U.S. Environmental Protection Agency (U.S. EPA). 1990e. Potassium silver cyanide. Integrated Risk Information System (IRIS). Environmental Criteria and Assessment Office, Office of Health and Environmental Assessment, Cincinnati, OH.

U.S. Environmental Protection Agency (U.S. EPA). 1991. Sodium cyanide. Integrated Risk Information System (IRIS). Environmental Criteria and Assessment Office, Office of Health and Environmental Assessment, Cincinnati, OH.

U.S. Environmental Protection Agency (U.S. EPA). 1992a. Carbaryl. Technology Transfer Network, Air Toxics Web Site (revised 2000, updated 2007).

U.S. Environmental Protection Agency (U.S. EPA). 1992b. Carbon disulfide. Hazard summary. Technology Transfer Network Air Toxics Web Site (Revised 2000, and updated 2007).

U.S. Environmental Protection Agency (U.S. EPA). 2000a. Chlordane. Technology Transfer Network, Air Toxics Web Site (updated 2007).

U.S. Environmental Protection Agency (U.S. EPA). 2000b. Cyanide compounds hazard summary. Technology Transfer Network Air Toxics Web Site (updated 2007).

U.S. Environmental Protection Agency (U.S. EPA). 2006. *Chemical Emergency Preparedness and Prevention Coumaphos, Emergency First Aid Treatment Guide*. U.S. EPA, Atlanta, GA.

World Health Organization (WHO). Chlorpyrifos-methyl. Data sheet on pesticides no. 33, WHO, Geneva, Switzerland.

World Health Organization (WHO). 1992. Cadmium. Environmental Health Criteria. No. 135. International Programme on Chemical Safety (IPCS), WHO, Geneva, Switzerland.

World Health Organization (WHO). 1999. Carbon monoxide, Environmental Health Criteria 213. WHO, Geneva, Switzerland.

World Health Organization (WHO). 2000. Carbon disulfide. WHO Regional Office for Europe, Copenhagen, Denmark.

World Health Organization (WHO). 2004. Hydrogen cyanide and Cyanides: Human health aspects. International Labour Organization (ILO), WHO, United Nations Environment Programme, Geneva, Switzerland.

6

Hazardous Chemical Substances:
D

DDT
Demeton-S-Methyl
Diazinon
Dichlorvos
Dicofol
Diethyl Phthalate
Dimethoate
Dimethyl Adipate
Dimethylamine
1,4-Dioxane
Diphenylamine
Diethylamine
Disulphoton
Diuron
Divinyl Benzene

DDT (CAS No. 50-29-3)

Molecular formula: $C_{14}H_9C_{15}$
Synonyms and trade names: Dichlorodiphenyltrichloroethane
IUPAC name: 1,1,1'-Trichloro-2,2-bis(4-chlorophenyl) ethane

The technical p,p'-DDT is a waxy solid but in its pure form appears as colourless crystals. It is a mixture of three isomers, namely, p,p'-DDT isomer (about ca. 85%); o,p'-DDT; and o,o'-DDT (in smaller levels). DDT is very soluble in cyclohexanone, dioxane, benzene, xylene, trichloroethylene, dichloromethane, acetone, chloroform, diethyl ether, ethanol, and methanol. The U.S. Environmental Protection Agency (U.S. EPA) grouped DDT as a restricted use pesticide (RUP). DDT is often contaminated with DDE (1,1-dichloro-2,2-bis(chlorophenyl) ethylene) and DDD (1,1-dichloro-2,2-bis(p-chlorophenyl) ethane).

Dichloro-diphenyl-trichloroethane (DDT) was first synthesised in 1873 by the German chemist Othmar Zeidler. Later in 1939, Paul Muller of Geigy Pharmaceutical in Switzerland discovered the effectiveness of DDT as an insecticide. For this discovery of DDT, Paul Muller was awarded the Nobel Prize in Medicine and Physiology in 1948. The use of DDT increased enormously on a worldwide basis after World War II, primarily because of its effectiveness against the mosquito that spreads malaria and lice that carry typhus. The World Health Organization (WHO) estimates that during the period of its use, approximately 25 million lives were saved.

DDT was extensively used during the World War II among Allied troops and certain civilian populations to control insect typhus and malaria vectors. Application of DDT became extensive because of easy control of a large number of crop pests and vectors of human diseases. Humans are exposed to DDT because of different activities and many factors. These include, but are not limited to, (i) consumption of eating contaminated foods, such as root and leafy vegetable, fatty meat, fish, and poultry, but levels are very low; (ii) eating contaminated imported foods from countries that still allow the use of DDT to control pests; (iii) breathing contaminated air or drinking contaminated water near waste sites and landfills that may contain higher levels of these chemicals; (iv) infants fed on breast milk from mothers who have been exposed; and (v) breathing or swallowing soil particles near waste sites or landfills that contain these chemicals. By the 1970s, overuse and misuse of DDT became obviously associated with environmental and health effects. Eventually, in June of 1972, the U.S. EPA cancelled all use of DDT on crops except in certain cases of disease control where the U.S. EPA allowed a limited use of DDT. However, many tropical countries are still using DDT for the control of malaria.

Use of DDT was banned for use in Sweden in 1970 and in the United States in 1972. In view of its large-scale use over the decades, many insect pests may have developed resistance to DDT. It is no longer registered for use in the United States barring public health emergency, for example, outbreak of malaria. The latest group meeting by 110 countries on DDT met in Geneva (September 17, 1999) to phase out the production and impose total ban on its use even for public health purposes resolved differently. The conference could not arrive at the conclusion for a global ban on DDT. Absence of a suitable substitute of DDT for the control of malaria and the absence of an anti-malaria vaccine emphasised for continuing DDT for malaria control.

Safe Handling and Precautions

Exposures in high concentrations to DDT cause adverse health effects and poisoning among occupational workers. The symptoms of toxicity include, but are not limited to, tremors, diarrhoea, dizziness, headache, vomiting, numbness, paresthaesias, hyperexcitability, and convulsions. Chronic exposures to DDT caused adverse effects on the nervous system, liver, kidneys, and immune systems in experimental animals. In laboratory rats and mice, DDT given at 16–32 and 6.5–13 mg/kg for about 26 weeks and 80–140 weeks, respectively, caused the tremors. Laboratory studies with non-human primates given DDT (10 mg/kg/day over 100 days) showed changes in cellular chemistry of the central nervous system (CNS) and at higher doses (50 mg/kg/day) caused loss of equilibrium in animals. Prolonged exposures to DDT caused adverse effects in species of animals, for instance, rats, mice, hamsters, and dogs, and monkeys showed pathomorphological changes in liver, kidney, CNS, and adrenal glands. Humans exposed to DDT have shown many adverse effects, for instance, nausea; diarrhoea; increased liver enzyme activity; irritation of the eyes, nose, and throat; disturbed gait; malaise; excitability; and, at higher doses, tremors and convulsions. It has been reported that the health effects of DDT in humans exposed to different doses and for different periods include poisoning, sweating, headache, nausea, convulsions, and, in some workers, changes in liver function as well. It is very important to remember that earlier findings on the toxicological effects of DDT in animals and humans, as reported earlier, require further conformity data since the studies did not observe the

GLP regulations. Therefore, the potential hazards of DDT to human health and environmental safety require careful evaluations.

DDT and Malaria Control

The U.S. government has finally begun to reverse policy on the insecticide DDT. Let's hope that this policy shift represents the beginning of the end of what can only be called a crime against humanity: the decades-old withholding of the world's most effective antimalarial weapon from billions of adults and children at risk of dying from the disease. The WHO has given the credit to DDT for helping one billion people live free from malaria and for saving millions of lives. After 30 years of worldwide use of DDT, the WHO, in 1973, concluded that the benefits of use of DDT were far greater than its possible risks. The United Nations Environment Program (UNEP), during its recent discussions, observed whether or not that DDT should have been totally banned together with 11 other persistent organic pollutants. The total ban of DDT was sharply criticised; in South Africa, a temporary total ban on the use of DDT for indoor spraying resulted in a sudden increase in malaria. Eleven countries in Africa, seven countries in Asia, and five countries in Latin America have been still using DDT for vector disease control. In fact, according to the general consensus, limited and strictly controlled and proper use of DDT should be allowed for public health purposes, in particular where other effective, safe, and affordable alternative pesticides are not available, and the benefits are clearly far superior to possible risks. In this context, it becomes important to note the observations that DDT can cause many toxicological effects, but the effects on human beings at likely exposure levels seem to be very slight. However, the perceived rather than the calculated risks from DDT use are an important consideration in maintaining public confidence. Thus, it would seem prudent that if its use was continued for anti-malarial campaigns and the benefits of use outweigh the risks, tight control should continue, and the effects of spraying DDT should be closely monitored (more information in literature). Studies in DDT-exposed workers did not show increases in cancer. Studies in laboratory animals exposed to DDT in food suggested that DDT can cause liver cancer. The Department of Health and Human Services (DHHS) indicated that DDT may reasonably be anticipated to be a human carcinogen. The International Agency for Research on Cancer (IARC) determined that DDT may possibly cause cancer in humans. The U.S. EPA also observed that DDT, DDE, and DDD are probable human carcinogens.

In most studies in which the relationship between exposure to organochlorine compounds and breast cancer was examined, residues were measured in serum, although they are higher in breast adipose tissue, which represents cumulative internal exposure at the target side for breast cancer. In a large study in which DDT and its metabolites DDE and DDD were measured in breast adipose tissue, the concentrations of DDE were higher than those of DDT in both breast cancer patients and controls. After adjustment for age, no relationship was found between the concentration of either DDT, DDE, or DDT + DDE + DDD and breast cancer. Laboratory studies showed that DDT increased incidences of lung tumours and lymphomas in mice, incidences of liver tumours in rats, and incidences of adrenal adenomas in hamsters. Long-term oral administration of DDT to non-human primates caused hepatic toxicity and malignant and benign tumours at various sites, and the adverse health effects found in non-human primates were to be of borderline statistical significance. However, it was reported that no conclusion could be made about the carcinogenicity of DDT in monkeys on the basis of a 130-month study at one dose. The reports on DDT and cancer have become contradictory and inconclusive.

For instance, vant't Veer et al. did not support the hypothesis that DDE increases risk of breast cancer in post-menopausal women in Europe. Thus, several workers in large and well-designed study found no evidence that exposures to DDT and DDE caused an increased risk of breast cancer. The evidences suggest generation of more adequate and quality data on the toxicity profile of DDT. The Occupational Safety and Health Administration (OSHA) has set the limit for DDT at 1.0 mg/m^3 at the workplace for an 8 h time-weighted average (TWA).

Demeton-S-Methyl (CAS No. 919-86-8)

Molecular formula: $C_6H_{15}O_3PS_2$
Synonyms and trade names: Metasystox; Metaphor; Meta-isosystox; Azotox; Duratox; Mifatox; Persyst
IUPAC name: S-2-Ethylthioethyl-O, O-dimethyl phosphorothioate
Toxicity class: U.S. EPA: I; WHO: Ib

Demeton-S-methyl is pale yellow in colour and oily and has a sulphur-like odour. It is sparingly soluble in water but very rapidly soluble in common polar organic solvents such as dichloromethane, 2-propanol, toluene, and n-hexane. It is a highly toxic, systemic, and contact insecticide and acaricide and classified as Category I. It is used for the control of insects, aphids, sawflies, and spider mites on cereals, fruits, vegetables, and ornamental plants. On heating, demeton-S-methyl undergoes decomposition and emits very toxic fumes, phosphorus oxides, and sulphur oxides.

Safe Handling and Precautions

Demeton-S-methyl is highly toxic to animals and humans. Careless occupational exposures to demeton-S-methyl cause severe symptoms of poisoning which include, but are not limited to, headaches, nausea, vomiting, diarrhoea, sweating, dizziness, tremors, lack of coordination, hiccough, and memory loss. Prolonged exposures to high concentrations cause pupillary constriction, blurred vision, muscle cramp, excessive salivation, weakness, sweating, abdominal cramps, unconsciousness, respiratory distress, convulsions, respiratory failure, and death. There are no reports indicating that demeton-S-methyl has embryotoxic or teratogenic potential or causes adverse effects on reproduction or development, or carcinogenic action to animals and humans. Demeton-S-methyl is grouped as A4, meaning not classifiable as a human carcinogen. The acceptable daily intake (ADI) of demeton-S-methyl has been set at 0.0003 mg/kg, the tolerance level value at 0.5 mg/m^3, the short-term exposure limit (STEL) should not exceed 1.5 mg/m^3 for and not more than a total of 30 min, and the no observable effect level (NOEL) at 1.0 mg/kg. Demeton-S-methyl should be stored separately and away from food and feedstuffs. Keep it in a well-ventilated room. During handling and use of demeton-S-methyl, occupational workers should use face shield or eye protection in combination with breathing protection, protective gloves, and protective clothing. During use and handling of the chemical substance, workers should avoid eating, drinking, or smoking during work and wash hands before eating. Workers should prevent generation of chemical mists and observe strict hygiene.

Diazinon (CAS No. 333-41-5)

Molecular formula: $C_{12}H_{21}N_2O_3PS$
Synonyms and trade names: Basudin; Knox-out; Dazzel; Gardentox; Kayazol; Nucidol
IUPAC name: O, O-Diethyl O-2-isopropyl-6-methyl (pyrimidin-4-yl) phosphorothioate
Toxicity class: U.S. EPA: II or III; WHO: II

Diazinon is available in the form of a colourless or dark brown liquid. It is sparingly soluble in water but very soluble in petroleum ether, alcohol, and benzene. Diazinon is used for the control of a variety of agriculture and household pests. These include pests in soil, on ornamental plants, fruit, vegetable, and crops and household pests like flies, fleas, and cockroaches. Diazinon undergoes decomposition on heating above 120°C and produces toxic fumes such as nitrogen oxides, phosphorus oxides, and sulphur oxides. It reacts with strong acids and alkalis with possible formation of highly toxic tetra ethyl thiopyrophosphates. Diazinon is classified as a RUP. Depending on the type of formulation, diazinon is classified as toxicity class II, meaning moderately toxic, or toxicity class III, meaning slightly toxic.

Safe Handling and Precautions

Human exposures to diazinon occur in workplaces of manufacture and professional applications. Diazinon causes poisoning with symptoms such as headache, dizziness, nausea, weakness, feelings of anxiety, vomiting, pupillary constriction, convulsions, dizziness, respiratory distress or laboured breathing, unconsciousness, muscle cramp, excessive salivation, convulsions, respiratory failure, and coma.

There are no evidences indicating that exposures to diazinon cause mutagenic, teratogenic, and/or carcinogenic effects in laboratory animals or humans. The DHHS, the IARC, the OSHA, and the U.S. EPA have not reported that diazinon causes cancer in humans or in animals and listed diazinon as Group 3, meaning not classifiable as a human carcinogen. Workers should avoid eye contact with diazinon and wear chemical safety glasses or goggles, protective clothing or equipment, waterproof boots, long-sleeved shirt, long pants, and a hat. Workers should avoid contamination of food and feed and wash thoroughly after handling and before eating or smoking. In fact, occupational workers should avoid eating, drinking, and smoking in areas of work with the chemical.

Dichlorvos (CAS No. 62-73-7)

Molecular formula: $C_4H_7Cl_2O_4P$
Synonyms and trade names: DDVP; Vapona; Phosvit; Vantaf; Uniphos; Swing; Nuvon
IUPAC name: 2,2-Dichlorovinyl dimethyl phosphate
Toxicity class: U.S. EPA: I; WHO: Ib

Dichlorvos is available as an oily colourless to amber liquid with an aromatic chemical odour. It is slightly soluble in water but soluble in kerosene, ethanol, chloroform, and acetone and miscible with alcohol, aromatic and chlorinated hydrocarbon solvents, aerosol propellants, and other non-polar solvents. Dichlorvos in contact with strong acids or alkalis undergoes decomposition and releases hazardous products toxic gases and vapours such as phosphorus, chlorinated oxides, and carbon monoxide. Dichlorvos is used as an agricultural insecticide for the control of crop pests such as flies, aphids, spider mites, caterpillars, and thrips; pests in store grains; Dichlorvos is also used for the control of parasitic worms in animals, and in flea collars for dogs. Occupational workers and general public get exposed to dichlorvos while working with the manufacture, formulation, and application in agriculture and household and when used as a fumigant and as pest strips. Human exposures also occur through food contamination.

Safe Handling and Precautions

Exposures to dichlorvos through all routes, namely, oral, dermal, and respiratory, cause adverse effects to species of laboratory animals such as rats, mice, and rabbits. The symptoms of poisoning include perspiration, nausea, salivation, vomiting, diarrhoea, drowsiness, fatigue, headache, and, in severe cases, tremors, ataxia, convulsions, coma, slow heartbeat, impaired memory concentration, disorientation, and severe depressions. Occupational workers with intake of alcoholic beverages have been reported to demonstrate much severe effects of poisoning of dichlorvos. Laboratory animals exposed to dichlorvos by gavage showed an increased incidence of tumours of the pancreas and leukaemia in male rats, tumours of the pancreas and mammary gland in female rats, and tumours of the forestomach in both sexes of mice. Also, accidental ingestion and/or inhalation exposure to dichlorvos at workplaces has been reported to cause poisoning in occupational workers. The symptoms and signs have been reported to include, but are not limited to, a feeling of tightness in the chest, allergic contact dermatitis, wheezing due to bronchospasm, blurring of vision, nausea, tearing, runny nose, irritability, confusion, frontal headache, twitching, muscle fasciculations, speech difficulties, sweating, blurred vision, drowsiness or insomnia, numbness, tingling sensations, incoordination, tremor, abdominal cramps, vomiting, diarrhoea, cramps, anorexia, difficulty in breathing or respiratory distress, and paralysis of the respiratory muscles leading to the fatal injury. There is no information regarding the carcinogenic effects of dichlorvos in humans. The U.S. EPA classified dichlorvos as Group B 2, meaning a probable human carcinogen.

Occupational workers during use and handling of dichlorvos should strictly observe workplace safety regulations. Workers should wear approved and appropriate workplace safety dress and personal protective equipment (PPE) to avoid dermal absorption. Any occupational worker who handles dichlorvos should thoroughly wash hands, forearms, and face with soap and water before handling food and drinks and eatables.

Workers should not store food and drinks in areas where dichlorvos or a solution containing dichlorvos is handled, processed, or stored. Dichlorvos should be stored in a cool, dry, well-ventilated area in tightly sealed containers, properly labelled in accordance with regulations, and stored separately and away from strong alkalis and/or strong acids.

Regulations:

- The American Conference of Governmental Industrial Hygienists (ACGIH) has set the threshold limit value (TLV) for dichlorvos at 0.1 ppm (0.90 mg/m^3) (TWA for a normal 8 h workday).
- The OSHA set a permissible exposure limit (PEL) for dichlorvos at 1 mg/m^3 (8 h TWA).

- Similarly, U.S. EPA set the limit of dichlorvos in food products at 0.02 ppm and the immediately dangerous to life and health limit (IDLH) at 100 mg/m^3.

Caution! Alert! Occupational workers should know that dichlorvos causes damage to some forms of plastic, rubber, and coatings. It is corrosive to iron and mild steel.

Dicofol (CAS No. 115-32-2)

Molecular formula: $C_{14}H_9Cl_5O$
Synonyms and trade names: Acarin; Cekudifol; Decofol; Dicaron; Dicomite; Difol; Hilfol; Kelthane; Mitigan
IUPAC name: 2,2,2-Trichloro-1,1-bis(4-chlorophenyl)ethanol
Toxicity class: U.S. EPA: II or III (depending upon the formulation); WHO: III

Pure dicofol is available as a white or grey powder or colourless crystal solid, while the technical dicofol is a red-brown or amber viscous liquid with an odour like fresh-cut hay. Dicofol undergoes decomposition on burning or on contact with acids, acid fumes, or bases producing toxic and corrosive fumes including hydrogen chloride. Dicofol is a persistent OCP used as acaricide and miticide. It is structurally similar to DDT. Dicofol is combustible and incompatible with strong oxidising agents. Dicofol is soluble in most aliphatic and aromatic solvents and most common organic solvents but practically insoluble in water and hydrolyses in basic solution. It is used on a wide variety of fruit, vegetable, ornamental, and field crops. Dicofol is manufactured from DDT. Dicofol is corrosive to some metals. Dicofol is used for foliar applications, mostly on cotton, apples, and citrus crops. Other crops include strawberries, mint, beans, peppers, tomatoes, pecans, walnuts, stonefruit, cucurbits, and non-residential lawns/ornamentals.

Safe Handling and Precautions

Exposures to dicofol cause adverse health effects and poisoning. Occupational workers who use and handle dicofol improperly get exposed to it either through inhalation or by ingestion and through skin and suffer with harmful effects and poisoning. The symptoms of poisoning include, but are not limited to, nausea, dizziness, weakness, and vomiting from ingestion or respiratory exposure and skin irritation or rash from dermal exposure. Dicofol-poisoned occupational workers show skin sensitisation, conjunctivitis of the eyes, and pathomorphological changes in liver, kidneys, and the CNS. Workers after exposures to high concentrations of dicofol show nervousness, hyperactivity, headache, nausea, vomiting, unusual sensations, fatigue, convulsions, coma, respiratory failure, and death. However, published literature is limited, and more data are required in occupational workers as well as on general population.

There is limited evidence that it may cause cancer in laboratory animals, but there is no evidence that it causes cancer in humans. This classification was based on animal test data that showed an increase in the incidence of liver adenomas (benign tumour) and combined liver adenomas and carcinomas in male mice. Dicofol is an experimental carcinogen and a human mutagen. It has been reported that the results of the experiment in mice provide limited evidence, and dicofol is carcinogenic to experimental animals. No data on humans were available. The available data are insufficient and not adequate to evaluate the carcinogenicity of dicofol as a human carcinogen. The U.S. EPA classified dicofol as Group C, meaning a possible human carcinogen. Similarly, the IARC listed dicofol as Group 3, meaning not classifiable as carcinogen to humans (Kelthane). The ADI for dicofol has been set at 0.002 mg/kg/day.

Diethylamine (CAS No. 109-89-7)

Molecular formula: $CH_3CH_2NH_2$
Synonyms and trade names: Aminoethane; DEA; Monoethylamine

Diethylamine is a colourless, strongly alkaline, fish odour liquid, and highly inflammable. It has an ammonia-like odour and is completely soluble in water. On burning, diethylamine releases ammonia, carbon monoxide, carbon dioxide, and nitrogen oxides. It is incompatible with several chemical substances such as strong oxidisers, acids, cellulose nitrate, some metals, and dicyanofuroxan. N-nitrosamines, many of which are known to be potent carcinogens, may be formed when diethylamine comes in contact with nitrous acid, nitrates, or atmospheres with high nitrous oxide concentrations. The applications of Diethylamine are numerous. Diethylamine is used in the production of pesticides. It is used in a mixture for the production of DEET which goes into the repellents that are found readily in supermarkets for general use. Diethylamine is also mixed with other chemicals to form Diethylaminoethanole, which is used mainly as a corrosion inhibitor in water treatment facilities as well as production of dyes, rubber, resins, and pharmaceuticals. Diethylamine is also used in manufacture of basic chemicals and pharmaceuticals.

Safe Handling and Precautions

Exposures to diethylamine cause adverse health effects. The symptoms of toxicity include irritation of skin, eyes, and mucous membrane. The acute oral LD_{50} and acute dermal LD_{50} in rat and rabbit are 540 and 580 mg/kg, respectively, and the acute inhalation LC_{50} (4 h) to rats is 4000 ppm. The pathomorphological changes caused by diethylamine include lungs, liver, and kidneys; cellular infiltration; bronchopneumonia; parenchymatous degeneration; and nephritis. The IARC, U.S. EPA, and National Toxicology Program (NTP) have not listed diethylamine as a human carcinogen and classified as Group A 4, meaning not classifiable as a human carcinogen.

Occupational workers during handling of diethylamine should be alert at the workplace. Diethylamine should be kept protected against physical damage and should be stored in a cool, dry well-ventilated location, away from incompatible chemical substances and away from fire hazard and smoking areas. The containers should be bonded and grounded for transfers to avoid static sparks. Occupational workers and users should be very careful during the use and chemical management of diethylamine. Workers should wear impervious protective clothing, including boots, gloves, lab coat, apron, or coveralls, as appropriate, to prevent skin contact. The chemical is very hazardous, corrosive, and harmful and is a very flammable liquid and vapour. Exposures to vapour may cause flash fire. It causes burns and causes adverse effects to the cardiovascular system. Workers should use chemical safety goggles and a full-face shield to avoid splashing of the chemical substance. An eyewash fountain and quick-drench facilities in work area should be maintained by the chemical management unit.

Danger! Workers should handle diethylamine with care. Improper use and negligence cause corrosive effects to the mucous membranes and eyes; corneal damage; and permanent visual impairment.
Regulations:

- The ACGIH has set the TLV for diethylamine at 5 ppm (15 mg/m^3) TWA for 8 h workday and a short-term exposure limit (STEL) of 15 ppm (45 mg/m^3) for a periods of 15 min.
- The U.S. OSHA has set the PEL for diethylamine in workplace air at 25 ppm (76 mg/m^3).
- The National Institute for Occupational Safety and Health (NIOSH) recommended the exposure limit for diethylamine at 10 ppm (30 mg/m^3) TWA for 10 h workday, and the STEL at 30 mg/m^3.
- The use of diethylamine in adhesives used in packaging, transporting, and/or holding food is regulated by the Food and Drug Administration (FDA) under 21 CFR 175.105.

Diethyl Phthalate (CAS No. 84-66-2)

Molecular formula: $C_{12}H_{14}O_4$
Synonyms and trade names: Anozol; DEP; Diethyl 1,2-benzene dicarboxylate; 1,2-Benzenedicarboxylic acid diethyl ester; Diethyl o-phthalate; Ethyl phthalate; Neantine; o-Benzenedicarboxylic acid diethyl ester; Palatinol A; Phthalol; Solvanol
Chemical name: Phthalic acid, diethyl ester

Diethyl phthalate is a colourless, odourless, and oily liquid. It is miscible with ethanol and ethyl ether. Diethyl phthalate is soluble in acetone, benzene, carbon tetrachloride, alcohols, ketones, esters, and aromatic hydrocarbons and partly miscible with aliphatic solvents. Diethyl phthalate, when exposed to heat, decomposes and emits acrid smoke and irritating fumes. It is incompatible with strong oxidisers, strong acids, nitric acid, permanganates, and water and attacks some forms of plastics. Diethyl phthalate is produced in the reaction of phthalic anhydride with ethanol in the presence of concentrated sulphuric acid catalyst.

Safe Handling and Precautions

The use of diethyl phthalate in consumer products and intake from contaminated foods, however, are likely to be the primary sources of human exposure. Occupational exposure may occur in industrial facilities where diethyl phthalate is used in the manufacture of plastics or consumer products. Diethyl phthalate on exposure is readily absorbed in the skin, intestinal tract, peritoneal cavity, and lung. In rats, rabbits, and dogs, oral and intravenous administration of diethyl phthalate caused stimulated respiration, lethargy, imbalance, cramps, and respiratory arrest. Dermally applied diethyl phthalate penetrates the skin and can be widely distributed in the body, but it does not accumulate in tissue. Diethyl phthalate is hydrolysed in the body to the monoester derivative. Hydrolytic metabolism of diethyl phthalate is qualitatively similar in rodents and humans.

Exposure to diethyl phthalate causes poisoning with symptoms that include, but not limited to, pain, numbness, lassitude (weakness, exhaustion), and spasms in arms and legs, irritation of the eyes, skin, nose, and throat, headache, dizziness, and nausea, lacrimation (discharge of tears), possible polyneuropathy and vestibular dysfunction.

In animals diethyl phthalate causes poisoning witheffects on reproductive system, respiratory system, central nervous system, peripheral nervous system.

Users and occupational workers during handling of diethyl phthalate should wear appropriate chemical protective workplace dress, solvent-resistant protective clothing, gloves, splash-proof chemical goggles, and face shield to prevent any kind of contact with the skin and body surfaces.

Regulations:

- The ACGIH has set the TLV of dicofol at 5 mg/m^3 (TWA).
- The NIOSH has set the recommended exposure limit (REL) for dicofol at 5 mg/m^3 (TWA).
- The OSHA had not set any PEL for diethyl phthalate. However, recent reports have indicated that OSHA is establishing PEL of 5 mg/m^3 (an 8 h TWA) for diethyl phthalate. The OSHA observed that the PEL should protect occupational workers against the possible significant risks of polyneuritis and vestibular dysfunction that have been associated with occupational exposure to diethyl phthalate at workplaces.

Dimethoate (CAS No. 60-51-5)

Molecular formula: **$C_5H_{12}NO_3PS_2$**
Synonyms and trade name: Daphene; Devigon; Dicap; Dimet; Rogodan; Rogodial; Rogor; Sevigor; Trimetion
IUPAC name: O,O-Dimethyl S-methylcarbamoylmethyl phosphorodithioate
Toxicity class: U.S. EPA: II; WHO: II

Dimethoate is a grey-white crystalline solid at room temperature. It is sparingly soluble in water, soluble in methanol and cyclohexane, but very soluble in chloroform and benzene. It has been classified by the U.S. EPA under GUP. Dimethoate is used extensively for the control of crop pests such as mites, aphids, thrips, plant hoppers, white-flies, and a wide range of other insects that damage, crops, fruits, vegetables, and ornamental plants. Dimethoate is also used for the control of cattle grubs that infect livestock. Thermal decomposition of dimethoate is highly hazardous due to the release of fumes of dimethylsulphide, methyl mercaptan, carbon monoxide, carbon dioxide, phosphorus pentoxide, and nitrogen oxides.

Safe Handling and Precautions

Dimethoate is toxic to animals and humans. Occupational exposures cause poisoning with symptoms that include, but are not limited to, sweating, headache, weakness, giddiness, nausea, vomiting, stomach pains, blurred vision, pupillary constriction, slurred speech, and muscle twitching. Workers repeatedly exposed to dimethoate suffered with symptoms such as numbness, tingling sensations, incoordination, headache, dizziness, pupillary constriction, muscle cramp, excessive salivation, sweating, nausea, dizziness, laboured breathing, weakness, tremor, abdominal cramps, difficulty breathing or respiratory depression, slow heartbeat, and speech difficulties. Prolonged exposures cause severe poisoning with adverse effects on the CNS, leading to incoordination, slurred speech, loss of reflexes, weakness, fatigue, involuntary muscle contractions, twitching, tremors of the tongue or eyelids, and eventually paralysis of the body's extremities and the respiratory muscles; psychosis; irregular heartbeats; unconsciousness; convulsions; coma; and death caused by respiratory failure or cardiac arrest. Dimethoate is possibly a human mutagen, teratogen, and carcinogen. The evidences of carcinogenicity, even with high-dose, long-term exposure, are inadequate and inconclusive suggesting that carcinogenic effects of dimethoate in humans are unlikely.

Dimethyl Adipate (CAS No. 627-93-0)

Molecular formula: $C_8H_{14}O_4$
Synonyms and trade names: Dimethyl hexanedioate; Methyl adipate; Hexanedioic acid; Dimethyl ester

Dimethyl adipate (DMA) is a colourless and flammable liquid. It is soluble in alcohol and ether but sparingly soluble in water. DMA is incompatible with strong oxidising agents, and on decomposition, it emits carbon monoxide, irritating and toxic fumes and gases, and carbon dioxide. DMA reacts with acids, alkalis, and strong oxidants. DMA is synthesised by the esterification of adipic acid. DMA is part of a dibasic ester (DBE) blend used as a major ingredient in several paint strippers, and the DBE blends used in paint stripping formulations contain a major portion (about 90%) of DMA. DMA is used as a chemical intermediate (polymers, agrochemicals), cellulose resins, a speciality solvent (inks, coatings, adhesives), and an emollient and can also be utilised as a paint remover and plasticiser.

Safe Handling and Precautions

Exposures to DMA cause toxicity and adverse health effects in laboratory animals and humans. Workplace exposures to DMA by inhalation, ingestion, or skin absorption cause harmful and irritation effects to users. There is no published information indicating that DMA is a human carcinogen. The ACGIH, the IARC, and the NTP have not listed DMA as a human carcinogen.

Workers should avoid accidental workplace spillage of chemical waste and avoid ingestion and/or inhalation of DMA. The work area should have adequate ventilation, and workers should strictly avoid contact of the chemical substance with eyes, skin, and clothing. DMA should be stored in a tightly closed container, in a cool, dry, well-ventilated area away from incompatible substances. Also, during handling of DMA, occupational workers should be careful and use self-contained breathing apparatus, rubber boots, and heavy rubber gloves and avoid prolonged period of exposures. After use and handling of DMA, workers should not be negligent to wash hand and contaminated body surface thoroughly, to remove the contaminated clothing, etc.

Dimethylamine (CAS No. 124-40-3)

Molecular formula: C_2H_7N

Synonyms and trade names: Anhydrous dimethylamine; Dimethylamine (aqueous solution); DMA; N-methylmethanamine

Dimethylamine is a colourless flammable gas at room temperature. It has a pungent, fishy, or ammonia-like odour at room temperature and is shipped and marketed in compressed liquid form. It is very soluble in water and soluble in alcohol and ether. It is incompatible with oxidising materials, acrylaldehyde, fluorine, maleic anhydride, chlorine, or mercury. Dimethylamine is a precursor to several industrially important compounds. For instance, it used in the manufacture of several products, for example, for the vulcanisation process of rubber, as detergent soaps, in leather tanning, in the manufacture of pharmaceuticals, and also for cellulose acetate rayon treatment.

Safe Handling and Precautions

Exposures to dimethylamine cause adverse health effects. The symptoms include, but are not limited to, severe pain to the eyes, corneal oedema/injury, redness, irritation and burning of the skin, chemical burns, and dermatitis. Severe inhalation exposure causes runny nose, coughing, sneezing, burning of the nose and throat, shortness of breath, delayed pulmonary effects like tracheitis, bronchitis, pulmonary oedema, and pneumonitis.

Regulations:

- The ACGIH has set the TLV for dimethylamine at 5 ppm (9.2 mg/m^3) as a TWA for a normal 8 h workday and a STEL of 15 ppm (27.6 mg/m^3) for periods not to exceed 15 min.
- The U.S. OSHA has set the PEL for dimethylamine at 10 ppm (18 mg/m^3) as an 8 h TWA concentration (29 CFR 1910.1000).
- The NIOSH has set the REL for dimethylamine at 10 ppm (18 mg/m^3) as a TWA for up to a 10 h workday and a 40 h workweek.
- The European Union has set the occupational exposure limit (OEL) for dimethylamine at 2 ppm and 3.8 mg/m^3 as TWA.
- The air odour threshold concentration for dimethylamine is 0.34 ppm of air.

Occupational workers should strictly avoid contact of dimethylamine with the eyes and skin and should avoid breathing the chemical vapours. Occupational workers should keep stored dimethylamine in a cool, dry, well-ventilated area in tightly sealed containers away from heat, sparks, flames, and other sources of ignition. Use with adequate ventilation. Use in accordance with package instructions and labelled in accordance with the chemical safety regulations set by national and international agencies. Containers of dimethylamine should be protected from physical damage and ignition sources and should be stored separately from oxidising materials, acrylaldehyde, fluorine, maleic anhydride, chlorine, or mercury. Outside or detached storage is preferred. If stored inside, a standard flammable liquids cabinet or room should be used. Ground and bond metal containers and equipment when transferring liquids. Empty containers of dimethylamine should be handled appropriately.

During handling of dimethylamine, workers should use proper fume hood, personal protective clothing and equipment, gloves, sleeves, and encapsulating suits and avoid skin contact. Dimethylamine is extremely flammable and may get ignited by heat, sparks, or open flames. Liquid dimethylamine will attack some forms of plastic, rubber, and coatings and is flammable. The vapours of dimethylamine are an explosion and poison hazard. Containers of dimethylamine may explode in the heat of the fire and require proper disposal. Workers should use dimethylamine with adequate ventilation, and the containers must be kept closed properly.

Warnings! Occupational workers should know that dimethylamine reacts violently with acid, strong oxidants, and mercury, can cause fire, and is an explosion hazard; it is corrosive – attacks aluminium, copper, zinc alloys, galvanised surfaces, plastic, rubber, and coatings.

1,4-Dioxane (CAS No. 123-91-1)

Molecular formula: $C_4H_8O_2$
Synonyms and trade names: 1,4-Dioxacyclohexane; Diethylene dioxide; 1,4-Diethylene dioxide; -Dioxacyclohexane; Diethylene; Dioxide; Glycol ethylene ether 8; Ethylene glycol ethylene ether; Diox; Dioxane

Technical grade 1,4-dioxane is a clear liquid with ether-like odour. It is highly flammable and forms explosive peroxides in storage (rate of formation increased by heating, evaporation, or exposure to light). 1,4-Dioxane is incompatible with oxidising agents, oxygen, halogens, reducing agents, and moisture. Industrial applications of 1,4-dioxane are extensive, for instance, as solvent for cellulose acetate, ethyl cellulose, benzyl cellulose, resins, oils, waxes, and some dyes; as a solvent for paper, cotton, and textile processing; and for various organic and inorganic compounds and products. It is also used in automotive coolant liquid and in shampoos and other cosmetics as a degreasing agent and as a component of paint and varnish. Human exposures to 1,4-dioxane have been traced to multiple occupations and breathing of contaminated workplace air and drinking polluted water. Industrial uses of 1,4-dioxane are very many. For instance, it is used as solvent for celluloses, resins, lacquers, synthetic rubbers, adhesives, sealants, fats, oils, dyes, and protective coatings; as a stabiliser for chlorinated solvents and printing inks; and as a wetting and dispersing agent in textile processing agrochemicals and pharmaceuticals, in different processing of solvent-extraction processes, and in the preparation and manufacture of detergents.

Safe Handling and Precautions

Laboratory studies in experimental animals indicated that repeated exposures to large amounts of 1,4-dioxane in drinking water, in air, or on the skin cause convulsions, collapse, and damage to liver and kidney in animals. The occupational workers on inhalation of 1,4-dioxane suffered from severe poisoning. The symptoms of poisoning included coughing, irritation of eyes, drowsiness, vertigo, headache, anorexia, stomach pains, nausea, vomiting, irritation of the upper respiratory passages, coma, and death. 1,4-Dioxane also caused hepatic and renal lesions and demyelination and oedema of the brain among workers. Laboratory studies with animals exposed to 1,4-dioxane showed induction of nasal cavity and liver carcinomas in rats, liver carcinomas in mice, and gall bladder carcinomas in

guinea pigs. The NIOSH observed dioxane as a potential occupational carcinogen. Further, the U.S. EPA classified 1,4-dioxane as Group B 2, meaning probable human carcinogen.
Regulations:

- The U.S. OSHA has set the PEL for 1,4-dioxane at 100 ppm (skin) (TWA) and the IDLH limit at 500 ppm.
- The NIOSH has set the REL for 1,4-dioxane at 1 ppm for 30 min.

Diphenylamine (CAS No. 122-39-4)

Molecular formula: $(C_6H_5)_2NH$
Synonyms and trade names: Anilinobenzene; Benzenamine; *N*-phenyl-anilinobenzene; Big dipper; Biphenylamine; DFA; DPA; C.I. 10355; Difenylamin; *N,N*-diphenylamine; *N*-fenylanilin; No scald; *N*-phenylaniline; Phenyl aniline; *N*-phenylbenzeneamine; Scaldip

Diphenylamine is a colourless monoclinic leaflet substance. It is used in the manufacture of a variety of substances, for instance, dye stuffs and their intermediates, pesticides, anthelmintic drugs, and as reagents in analytical chemistry laboratories.

Safe Handling and Precautions

Diphenylamine is highly toxic and is rapidly absorbed by skin and through inhalation. It has caused anorexia, hypertension, eczema, and ladder symptoms. Experimental animals exposed to diphenylamine demonstrated cystic lesions but failed to demonstrate cancerous growth. Inhalation of diphenylamine dust may cause systemic poisoning. The symptoms of toxicity include, but are not limited to, anoxia, headache, fatigue, anorexia, cyanosis, vomiting, diarrhoea, emaciation, hypothermia, bladder irritation, and kidney, heart, and liver damage. The ACGIH, IARC, and NTP have not listed diphenylamine as a human carcinogen.
Regulations:

- The ACGIH has set the TLV for diphenylamine at 10 mg/m³ (TWA).
- The U.S. NIOSH has set the REL for diphenylamine at 10 mg/m³.
- The OSHA is establishing 10 mg/m³ (8 h PEL TWA) for diphenylamine. The agency concludes that this limit will protect workers against the significant risks of liver, kidney, cardiovascular, and other systemic effects, all of which constitute material health impairments that are potentially associated with diphenylamine exposures above the new PEL.

Occupational workers should be careful in handling diphenylamine, and it should be kept protected against physical damage and from light. Diphenylamine should be kept safe and outside or in detached area as it should be in a standard flammable liquids storage room or cabinet for chemicals. Also, diphenylamine should be kept separately stored from oxidising chemical substances/materials and incompatible chemical substances. The storage and work areas should be posted with a *NO SMOKING* sign. Students and occupational workers should be careful during use and handling of diphenylamine. Workers should wear impervious protective clothing, including boots, gloves, lab coat, apron, or coveralls, as appropriate, to prevent skin contact. Diphenylamine in the form of fine particles that are present in the chemical mixtures causes hazards of explosions. Diphenylamine is very harmful on exposures by swallowing, inhalation, and/or skin absorption. Diphenylamine causes irritation to skin, eyes, and respiratory tract; causes blood vascular changes; and leads to methaemoglobinemia.

Disulphoton (CAS No. 298-04-4)

Molecular formula: $C_8H_{19}O_2PS_3$
Synonyms and trade names: Disyston; Disystox; Dithiodemeton; Dithiosystox; Solvigram; Solvirex
IUPAC name: O,O-Diethyl S-2-ethylthioethyl phosphorodithioate
Toxicity class: U.S. EPA: I; WHO: Ia

Disulphoton is a dark yellowish oil with an aromatic, sulphurous odour. It is soluble in most organic solvents and fatty oils. Disulphoton is a selective, systemic insecticide and acaricide. It is used for seed coating and for soil application to protect from insect attacks and for the control of sucking insects, aphids, leafhoppers, thrips, beetflies, spider mites, and coffee-leaf miners. Disulphoton has been extensively used in pest control on a variety of crops such as cotton, tobacco, sugar beets, corn, peanuts, wheat, ornamentals, cereal grains, and potatoes. It is grouped by the U.S. EPA under RUP. Human exposures to disulphoton occur through breathing contaminated air, drinking contaminated water, eating contaminated food, and working in industries that manufacture and formulate the pesticide.

Safe Handling and Precautions

Disulphoton is highly toxic to animals and humans by all routes of exposures, namely, by dermal absorption, through ingestion, and inhalation by respiratory route. The symptoms of poisoning include blurred vision, fatigue, headache, dizziness, sweating, tearing, and salivation. It inhibits cholinesterase and affects nervous system function. It does not cause delayed neurotoxicity. Prolonged period of exposures to high concentrations of disulphoton causes harmful effects to the nervous system with symptoms such as narrowing of the pupils, vomiting, diarrhoea, drooling, difficulty in breathing, lung oedema, tremors, convulsions, coma, and death. Disulphoton causes no mutagenic or teratogenic in laboratory animals. There are no reports indicating that disulphoton causes cancer in animals and humans. The DHHS, the IARC, and the U.S. EPA have not classified disulphoton as to its ability to cause cancer. (More information in literature.)

Diuron (CAS No. 330-54-1)

Molecular formula: $C_9H_{10}Cl_2N_2O$
Synonyms and trade names: 3-(3,4-Dichlorophenyl)-1,1-dimethylurea; Direx®; Karmex.

Diuron is a white crystalline solid/wettable powder and used as a herbicide. Diuron is registered for pre- and post-emergent herbicide treatment of both crop and non-crop areas, as a mildewcide and preservative in paints and stains, and as an algaecide. Diuron is a substituted urea herbicide for the control of a wide variety of annual and perennial broad-leaved and grassy weeds on both crop and non-crop sites. Thus, the application of diuron is wide for vegetation control and weed control in citrus orchards and alfalfa fields. The mechanism of herbicidal action is the inhibition of photosynthesis. Diuron was first registered in 1967. Products containing diuron are intended for both occupational and residential uses. Occupational uses include agricultural food and non-food crops; ornamental trees, flowers, and shrubs; paints and coatings; ornamental fish ponds and catfish production; and rights-of-way and industrial sites. Residential uses include ponds, aquariums, and paints. Occupational exposure to diuron has been reported to occur through inhalation of dust and dermal contact with this compound at workplaces where diuron and its produced are used. Monitoring data indicate that the general population gets exposed to diuron through inhalation of ambient air, ingestion of food and drinking water, and dermal contact with the chemical substance and other products containing diuron.

Safe Handling and Precautions

Exposure to diuron repeatedly and for a prolonged period of time causes adverse health effects. The symptoms of adverse health effects and poisoning include, but are not limited to, irritation of the eyes, skin, nose, and throat and methaemoglobinemia. The reports have indicated that diuron causes effects on blood, bladder, and kidney as target sites. Laboratory studies indicated that chronic exposure to diuron caused gross pathological disorders such as increased incidences of urinary bladder swelling and wall thickening at high doses in rats and mouse. The microscopic evaluation indicated a dose-related increase in the severity of epithelial focal damage. Based on the data and examination of urinary bladder carcinomas in both sexes of the Wistar rat, the kidney indicated carcinomas in the male rat (a rare tumour), and mammary gland carcinomas in the female NMRI mouse. The U.S. EPA lists diuron as a *known and/or likely* carcinogen because diuron has caused bladder cancer, kidney cancer, and breast cancer in studies with laboratory animals.

Occupational workers during use and handling of diuron should be careful and avoid negligence to wear appropriate PPE and workplace safety dress to prevent skin contact and eye contact. Workers should keep diuron away from sources of water and sewers. According to regulatory methods, diuron should be, filled to a land spill, dig a pit/pond, and disposed of subsequently. Workers should cover solids of diuron with a plastic sheet to prevent dissolving in rain or firefighting water.

Regulations: The Federal Insecticide, Fungicide, and Rodenticide Act (FIFRA) was amended in 1988 to accelerate the reregistration of pesticides. The U.S. EPA issued the Reregistration Eligibility Document (RED) for diuron. This RED document includes guidance and time frames for complying with any required label changes for products containing diuron. With the addition of the label restrictions and amendments detailed in this document, the agency has determined that all currently registered uses of diuron are eligible for reregistration.

Divinyl Benzene (CAS No. 1321-74-0)

Molecular formula: $C_{10}H_{10}$
Synonyms and trade names: 1,3-Divinylbenzene; Benzene, Divinyl-; Benzene, *m*-Divinyl-Diethenylbenzene; Divinyl benzene; DVB 960; *m*-Divinylbenzene; *m*-Vinylstyrene; Vinylstyrene

Divinyl benzene is a clear yellow liquid with an aromatic odour. It is a highly flammable and reactive chemical, and the liquid and vapour are combustible. It is incompatible with strong acids and oxidising agents and peroxides. Divinyl benzene reacts vigorously with strong oxidising agents. It also reacts exothermically with both acids and bases, with reducing agents (such as alkali metals and hydrides), and release gaseous hydrogen. May react exothermically with both acids and bases or initiators undergo exothermic polymerisation. It is an extremely versatile chemical cross-linking agent used to improve polymer properties. Divinyl benzene has extensive industrial applications for example, in the manufacture of adhesives, plastics, elastomers, ceramics, biological materials, catalysts, membranes, pharmaceuticals, specialty polymers, and ion exchange resins, as coatings, divinyl benzene has been reported to improve surface properties of rubber goods for biomedical applications, used in photocurable protective coatings for wood, metal, glass, and plastic materials.

Safe Handling and Precautions

During use and handling of divinyl benzene, workers should be alert to avoid the contact of the chemical substance with skin and eyes since it causes pain, redness, skin burns, swelling, tissue damage, and irritation of the mucous membranes and respiratory system. Exposures to vapours of divinyl benzene for a prolonged period at workplaces have been reported to cause irritation to the nose, throat, and lungs; lethargy; narcotic effects; dizziness; drowsiness; sensitisation reactions; and damage to the liver and kidneys. During use, workers should avoid the contact of divinyl benzene with oxidising materials, acids, metal halides, peroxides, brass, and copper. Occupational workers while working with divinyl benzene should wear protective gloves/protective clothing/eye protection/face protection. Workers should use and handle divinyl benzene only in a well-ventilated area, and divinyl benzene should be stored in tightly sealed container in a cool, dry place and away from oxidising agents and acids.

Occupational workers and users during handling of divinyl benzene should be alert and completely avoid negligence. Workplace manager and the workers should know that divinyl benzene liquid and vapours are highly reactive and combustible, and proper handling and storage of the chemical is a must. Workers should know that polymerisation is generally the greatest workplace hazard concern during shipment and storage. Under certain conditions (increased heat, low inhibitor concentration, and low oxygen content), rapid, runaway polymerisation may occur with potentially serious consequences, such as excessive heat and pressure build-up, which underlines strict observance of set instructions during working with divinyl benzene.

Workers should dispose of the chemical waste of divinyl benzene according to set official regulations of the state, local, or national regulations.

Bibliography

Agency for Toxic Substances and Disease Registry (ATSDR). 1995. *Toxicological Profile for Disulfoton*. U.S. Department of Health and Human Services, Public Health Service, Atlanta, GA (updated, 2007).

Agency for Toxic Substances and Disease Registry (ATSDR). 1997a. *Toxicological Profile for Dichlorvos*. Public Health Service, U.S. Department of Health and Human Services, Atlanta, GA (updated 2006).

Agency for Toxic Substances and Disease Registry (ATSDR). 1997b. *Toxicological Profile for Dichlorvos.* U.S. Department of Health and Human Services, Public Health Service, Atlanta, GA (updated 2011).

Agency for Toxic Substances and Disease Registry (ATSDR). 2002. *Toxicological Profile for DDT, DDDE, and DDD.* U.S. Department of Health and Human Services, Public Health Service, Atlanta, GA (updated 2007).

Agency for Toxic Substances and Disease Registry (ATSDR). 2008a. *Toxicological Profile for Diazinon.* U.S. Department of Health and Human Services, Public Health Service, Atlanta, GA (updated 2009).

Agency for Toxic Substances Disease Registry (ATSDR). 2008b. *Toxicological Profile for Diethylphthalate.* ATSDR, U.S. Department of Health & Human Services, Atlanta, GA.

Ahrens, W. H. 1994. *Herbicide Handbook of the Weed Science Society of America,* 7th edn. Weed Science Society of America, Champaign, IL.

American Conference of Governmental Industrial Hygienists (ACGIH). 2003. *1,4-Dioxane. Threshold Limit Values for Chemical Substances and Physical Agents and Biological Exposure Indices.* ACGIH, Cincinnati, OH.

Black, R. E., F. J. Hurley, and D. C. Havery. 2001. Occurrence of 1,4-dioxane in cosmetic raw materials and finished cosmetic products. *J. AOAC Int.* 84:666–670.

Budavari, S., M. J. O'Neil, A. Smith, and P. E. Heckelman (eds.). 1989. *The Merck Index: An Encyclopedia of Chemicals, Drugs, and Biologicals,* 11th edn. Merck & Co. Inc., Rahway, NJ.

Cheremisinoff, N. P. (ed.). 1999. *Handbook of Hazardous Chemical Properties.* Butterworth, Heinemann, Boston, MA.

DeRosa, C. T., S. Wilbur, J. Holler et al. 1996. Health evaluation of 1,4-dioxane. *Toxicol. Ind. Health* 12:143.

Dikshith, T. S. S. (ed.). 1991. *Toxicology of Pesticides in Animals.* CRC Press, Boca Raton, FL.

Dikshith, T. S. S. (ed.). 2011. *Handbook of Chemicals and Safety.* CRC Press, Boca Raton, FL.

Dikshith, T. S. S. and P. V. Diwan. 2003. *Industrial Guide to Chemicals and Drug Safety.* John Wiley & Sons Inc., Hoboken, NJ.

Ernstgard, I., A. Iregren, B. Sjogren et al. 2006. Acute effects of exposure to vapours of dioxane in humans. *Hum. Exp. Toxicol.* 25(12):723–729.

Extension Toxicology Network (EXTOXNET). 1996. Diazinon. Pesticide Information Profiles, Extension Toxicology Network. Cornell University, Ithaca, NY.

Ferrucio, B., C. A. Franchi, N. F. Boldrin, and J. L. de Camargo. 2010. Evaluation of diuron (3-[3,4-dichlorophenyl]-1,1-dimethyl urea) in a two-stage mouse skin carcinogenesis assay. *Toxicol. Pathol.* 38(5):756–764.

Hazardous Substances Data Bank (HSDB). TOXNET. 2009. Diethyl phthalate. HSDB, Bethesda, MD.

International Agency for Research on Cancer (IARC). 1979. *IARC Monographs on the Evaluation of the Carcinogenic Risk of Chemicals to Humans: Some Halogenated Hydrocarbons,* Vol. 20. World Health Organization, Lyon, France.

International Agency for Research on Cancer (IARC). 1983. Dicofol. IARC Monographs on the Evaluation of the Carcinogenic Risk of Chemicals to Humans, Vol. 30, pp. 87–101 (updated 1998).

International Programme on Chemical Safety (IPCS). 2003. Diethyl phthalate. Concise International Chemical Assessment Document (CICADS) 52: (updated 2008).

International Programme on Chemical Safety and the Commission of the European Communities (IPCS-CEC). 1993. Diazinon. ICSC card no. 0137 (updated 2001).

International Programme on Chemical Safety and the Commission of the European Communities (IPCS-CEC). 1994a. Diphenylamine. International Chemical Safety Cards (ICSC). ICSC Card No. 0466. IPCS-CEC, Geneva, Switzerland (updated 2006).

International Programme on Chemical Safety and the Commission of the European Communities (IPCS-CEC). 1994b. 1,4-Dioxane. ICSC Card No. 004. IPCS-CEC, Geneva, Switzerland.

International Programme on Chemical Safety and the Commission of the European Communities (IPCS-CEC). 1997a. Demeton-S-methyl. ICSC Card no. 0705. IPCS, Geneva, Switzerland (updated 2005).

International Programme on Chemical Safety and the Commission of the European Communities (IPCS-CEC). 1997b. Divinylbenzene (mixed isomers). International Chemical Safety Card. ICSC Card No. 0885. IPCS-CEC, Geneva, Switzerland.

International Programme on Chemical Safety and the Commission of the European Communities (IPCS-CEC). 2001. Dimethoate. ICSC No. 0741. Geneva, Switzerland.

International Programme on Chemical Safety and the Commission of the European Communities (IPCS-CEC). 2003. Dicofol. ICSC card no. 0752. World Health Organization, Geneva, Switzerland.

Kidd, H. and D. R. James (eds.). 1991. *The Agrochemicals Handbook*, 3rd edn. Royal Society of Chemistry Information Services, Cambridge, U.K.

Lewis, R. J., Sr. (ed.). 2007. *Hawley's Condensed Chemical Dictionary*, 15th edn. John Wiley & Sons, Inc., New York.

Lide, D. R. (ed.). 2005. *CRC Handbook of Chemistry and Physics*, 86th edn. CRC Press, Inc., Boca Raton, FL.

Material Safety Data Sheet (MSDS). 1996a. DDT. Extension Toxicology Network (EXTOXNET), Pesticide Information Profiles (PIP). Oregon State University/USDA Extension Service, Corvallis, OR.

Material Safety Data Sheet (MSDS). 1996b. Dicofol. Extension Toxicology Network (EXTOXNET), Pesticide Information Profiles (PIP). USDA/Extension Service/Oregon State University, Corvallis, OR.

Material Safety Data Sheet (MSDS). 1996c. Dimethoate. Extension Toxicology Network (EXTOXNET). Pesticide Information Profiles (PIP). USDA/Extension/Oregon State University, Corvallis, OR.

Material Safety Data Sheet (MSDS). 1998. Dimethyl adipate. ACC No. 58733. Acros Organics N.V., Fair Lawn, NJ.

Material Safety Data Sheet (MSDS). 2003. Safety data for dicofol. Physical Chemistry, Oxford University, Oxford, U.K.

Material Safety Data Sheet (MSDS). 2005. Safety data for diethylamine. Physical Chemistry, Oxford University, Oxford, U.K.

Material Safety Data Sheet (MSDS). 2007. Safety data for 1,4 dioxane. Department of Physical and Theoretical Chemistry, Oxford University, Oxford, U.K.

Material Safety Data Sheet (MSDS). 2008. Diethylamine. MSDS no. D 3056. Environmental Health & Safety, Mallinckrodt Baker, Inc., Phillipsburg, NJ.

Material Safety Data Sheet (MSDS). 2009. Diphenylamine. MSDS No. D7728. Environmental Health & Safety (EHS), Mallinckrodt Baker, Inc., Phillipsburg, NJ.

Material Safety Data Sheet (MSDS). 2011a. Diethyl phthalate. NIOSH, Institute for Occupational Safety and Health. Centers for Disease Control and Prevention, Atlanta, GA.

Material Safety Data Sheet (MSDS). 2011b. Boron trifluoride. National Institute for Occupational Safety and Health (NIOSH). Centers for Disease Control and Prevention, Atlanta, GA.

Material Safety Data Sheet (MSDS). U.S. Department of Labor. 1999. Dimethylamine. MSDS, Occupational Safety & Health Administration, Washington, DC.

Meister, R. T. (ed.). 1992. *Farm Chemicals Handbook '92*. Meister Publishing Company, Willoughby, OH.

National Institute of Occupational Safety and Health (NIOSH). 1997. Dichlorvos. IDLH Documentation. NIOSH, Atlanta, GA.

National Institute for Occupational Safety and Health (NIOSH). 2004. Dioxane. *NIOSH Pocket Guide to Chemical Hazards*. NIOSH, Washington, DC.

National Institute of Environmental Health Sciences (NIEHS). 2006. Developmental toxicity of diethyl phthalate. Department of Health & Human Services, NIEHS, Research Triangle Park, NC.

National Institutes of Health (NIH). National Toxicology Program (NTP). 2006. *Toxicology and Carcinogenesis Studies of Divinylbenzene-hp.in f344/n Rats and b6c3f1 Mice (Inhalation Studies)*. NTP, NIH, DHHS, Public Health Service, Research Triangle Park, NC.

National Library of Medicine (NLM). 1995a. Dichlorvos. Hazardous Substances Data Bank, NLM, Bethesda, MD.

National Library of Medicine (NLM). 1995b. Diethylamine. Hazardous Substances Data Bank, NLM, Bethesda, MD.

National Library of Medicine (NLM). 2004. 1,4-Dioxane. Environmental Standards & Regulations. Hazardous Substances Databank (HSDB), NLM, HSDB, Bethesda, MD.

National Oceanic and Atmospheric Administration (NOOA). 1999. Dimethyl adipate.

National Toxicology Program (NTP). 2002. 1,4-Dioxane. Report on carcinogens. NTP, U.S. Department of Health and Human Services, Public Health Service, Atlanta, GA.

New Jersey Department of Health (NJDH). 1987. Hazardous substance fact sheet: dichlorvos. NJDH, Trenton, NJ.

NIOSH. 2010a. Diuron. *Pocket Guide to Chemical Hazards*. Centers for Disease Control and Prevention, Atlanta, GA (updated 2011).

NIOSH. 2010b. Divinylbenzene. *Pocket Guide to Chemical Hazards*. NIOSH, Centers for Disease Control and Prevention, Atlanta, GA.

NIOSH. 2010. Diethyl phthalate. *Pocket Guide to Chemical Hazards*. Centers for Disease Control and Prevention, Atlanta, GA.

O'Neil, M. J. (ed.). 2006. *The Merck Index—An Encyclopedia of Chemicals, Drugs, and Biologicals*. Merck & Co., Inc., Whitehouse Station, NJ.

Patnaik, P. 1992. *A Comprehensive Guide to the Hazardous Properties of Chemical Substances*. Van Nostrand Reinhold, New York.

Pohanish, R. P. (ed.). 2002a. Diethyl phthalate. In: *Sittig's Handbook of Toxic and Hazardous Chemicals and Carcinogens*, 4th edn., Vol. 1. Noyes Publications, William Andrew Publishing, Norwich, NY, pp. 880–882.

Pohanish, R. P. (ed.). 2002b. Diuron. In: *Sittig's Handbook of Toxic and Hazardous Chemicals and Carcinogens*, 4th edn., Vol. 1. Noyes Publications, William Andrew Publishing, Norwich, NY.

Pohanish, R. P. (ed.). 2008. Divinylbenzene. In: *Sittig's Handbook of Toxic and Hazardous Chemicals and Carcinogens*, 5th edn., Vol. 1. Noyes Publications, William Andrew Publishing, Norwich, NY.

Ray, D. E. 1991. Pesticides derived from plants and other organisms. In: *Handbook of Pesticide Toxicology*, W. J. Hayes, Jr. and E. R. Laws, Jr. (eds.). Academic Press, New York.

Sax, N. I. (ed.). 1984. *Dangerous Properties of Industrial Materials*, 6th edn. Van Nostrand Reinhold Co., New York.

Sittig, M. (ed.). 1985. *Handbook of Toxic and Hazardous Chemicals and Carcinogens*, 2nd edn. Noyes Publications, Park Ridge, NJ.

Sittig, M. (ed.). 1991. *Handbook of Toxic and Hazardous Chemicals and Carcinogens*, 3rd edn. Noyes Publications, Park Ridge, NJ.

Sittig, M. (ed.). 2002. *Handbook of Toxic and Hazardous Chemicals and Carcinogens*, 4th edn. Noyes Publications, Norwich, NY.

Smith, A. G. 1991. Chlorinated hydrocarbon insecticides. In: *Handbook of Pesticide Toxicology*, W. J. Hayes, Jr. and E. R. Laws, Jr. (eds.). Academic Press Inc., New York.

The Agrochemicals Handbook. 1994. 3rd edn. Royal Society of Chemistry Information Systems, Unwin Brothers Limited, Surrey, U.K.

Tomatis, L. and J. Huff. 2000. Evidence of carcinogenicity of DDT in nonhuman primates. *J. Cancer Res. Clin. Oncol.* 126:246.

Tomlin, C. D. S. (ed.). 2000. *The Pesticide Manual—A World Compendium*, 12th edn. British Crop Protection Council, Farnham, Surrey, U.K.

Tomlin, C. D. S. (ed.). 2006. *The Pesticide Manual—A World Compendium*, 14th edn. British Crop Protection Council (BCPC), Surrey, U.K.

U.S. Department of Health and Human Services (U.S. DHHS). 1993. *Registry of Toxic Effects of Chemical Substances (RTECS)*. National Toxicology Information Program, National Library of Medicine, Bethesda, MD.

U.S. Department of Health and Human Services (U.S. DHHS), Agency for Toxic Substances and Disease Registry (ATSDR). 2007. *Toxicological Profile for 1,4-Dioxane*. U.S. DHHS, Public Health Service (PHS), ATSDR, Atlanta, GA.

U.S. Department of Labor (DOL). 1996a. *Occupational Safety and Health Guideline for Dichlorvos*. U.S. DOL, OSHA, Washington, DC.

U.S. Department of Labor. 1996b. *Occupational Safety and Health Guideline for Diethylamine*. Occupational Safety & Health Administration (OSHA), Washington, DC.

U.S. Environmental Protection Agency (U.S. EPA). 1999. *Integrated Risk Information System (IRIS) on Dichlorvos*. National Center for Environmental Assessment, Office of Research and Development, Washington, DC.

U.S. Environmental Protection Agency (U.S. EPA). 2000. Dichlorvos—Hazard summary. Technology Transfer Network Air Toxics Website, U.S. EPA (updated 2007).

U.S. Environmental Protection Agency (US EPA). 2007. *Disulfoton*. IRIS, U.S. EPA, Washington, DC.

U.S. Environmental Protection Agency (USEPA). 2003. Diuron. Office of Pesticide Programs; Reregistration Eligibility Decision Document. U.S. EPA, Atlanta, GA.

U.S. Environmental Protection Agency's (EPA). 2010. *Reregistration Eligibility Decision (RED) for Diuron*. U.S. EPA, Atlanta, GA.

U.S. Environmental Protection Agency (US EPA). Integrated Risk Information System (IRIS). 2010. *Toxicological Review of 1,4-Dioxane*. IRIS, U.S. Environmental Protection Agency, Washington, DC.

vant Veer, P. et al. 1997. DDT and post menopausal breast cancer in Europe—A case control study. *Brit. Med. J.* 12:315.

World Health Organization (WHO). 1988. Disulfoton. Data sheet on pesticides No. 68. WHO/VBC/DS/88.68. WHO, Geneva, Switzerland.

World Health Organization (WHO). 1997a. DDT and its derivatives. Environmental health criteria, no. 9. WHO, Geneva, Switzerland.

World Health Organization (WHO). 1997b. Dimeton-S-methyl. Environment health criteria no. 197. WHO, Geneva, Switzerland.

World Health Organization (WHO). International Programme on Chemical Safety (IPCS). 1989. Dimethoate. Environmental health criteria no. 90. IPCL, WHO, Geneva, Switzerland.

7

Hazardous Chemical Substances:
E

Endrin
S-Ethyl Dipropylthiocarbamate (EPTC)
Ethanol
Ethyl Alcohol Completely Denatured
Ethion
Ethylbenzol
Ethyl Carbamate
Ethylene Chlorohydrin
Ethylene Dichloride
Ethylene Oxide
Ethylenediamine
Ethyl Silicate
Ethyleneimine

Endosulfan (CAS No. 115-29-7)

Molecular formula: $C_9H_6Cl_6O_3S$
Synonyms and trade names: Acmaron; Agrosulphan; Endocel; Endomil; Endol; Endosol; Endotox 555; Fezdion; Hexa-sulfan; Kendan; Polydan; Thiodan; and many more names
Chemical name: 6,7,8,9,10,10-Hexachloro-1,5,5a,6,9,9a-hexahydro-6,9-methano-2,4,3-benzodioxathiepine-3-oxide
Toxicity class: U.S. EPA: IB; WHO: II

Endosulfan is a pesticide. It is a cream- to brown-coloured solid that may appear in the form of crystals or flakes. It has a smell like turpentine, but does not burn. It does not occur naturally in the environment. Endosulfan is used to control insects on food and non-food crops and also as a wood preservative. Endosulfan is used for the control of ticks and mites and the control of rice stem borers. It is a restricted use pesticide, meaning that it can only be used by professional applicators. Endosulfan, commonly known by its trade name Thiodan, is an insecticide and was first introduced in the 1950s. Endosulfan enters the air, water, and soil during its manufacture and use. It is often sprayed onto crops, and the spray may travel long distances before it lands on crops, soil, or water. Endosulfan on crops usually breaks down in a few weeks, but endosulfan sticks to soil particles and may take years to completely break down. Endosulfan does not dissolve easily in water. Endosulfan in surface water is attached to soil particles floating in water or attached to soil at the bottom. Endosulfan can build up in the bodies of animals that live in endosulfan-contaminated water. It is also extremely toxic to fish and other aquatic life.

Exposures to endosulfan occur among workers and people working in industries involved in making endosulfan or as pesticide applicators and by skin contact with soil containing endosulfan.

Safe Handling and Precautions

Endosulfan is readily absorbed by the stomach, by the lungs, and through the skin, meaning that all routes of exposure can pose a hazard. Exposures to endosulfan cause adverse health effects and poisoning. The symptoms of toxicity include, but are not limited to, hyperactivity, nausea, dizziness, headache, irritability, restlessness, muscular twitching, and convulsions and have been observed in adults exposed to high doses. Severe exposures cause poisoning and disturbances of the central nervous system (CNS) and may result in death. Laboratory studies in experimental animals indicated that long-term exposure to endosulfan can also damage the kidneys, testes, and liver and may possibly affect the body's ability to fight infection. More studies are needed to confirm similar situations in humans. Reports have indicated that the most prominent signs of acute exposure are hyperactivity, tremors, decreased respiration, difficulty in breathing, salivation, and convulsions. Long-term neurotoxic effects have been observed after high acute exposure. The National Poison Control Information Center of the Philippines recorded 278 poisonings including 85 deaths due to endosulfan in 1990. In Colombia, in 1993, at least 60 people were poisoned, and one person died as a result of exposure to Thiodan. There is no confirmatory data available about endosulfan as a human carcinogen. Studies in animals are inconclusive. The Department of Health and Human Services (DHHS), the National Toxicology Program (NTP), the International Agency for Research on Cancer (IARC), and U.S. Environmental Protection Agency (U.S. EPA) indicated that endosulfan is not classifiable as a human carcinogen.

Endosulfan and Human Poisonings

Many cases of accidental and suicidal poisonings related to endosulfan have been reported. In severe cases, death occurred within a few hours of ingestion of endosulfan. The poisoned worker showed signs and symptoms that included, but are not limited to, vomiting, restlessness, irritability, convulsions, pulmonary oedema, cyanosis, and EEG changes.

Regulations:

- The American Conference of Governmental Industrial Hygienists (ACGIH) has set the threshold limit value (TLV) at 0.1 mg/m^3.

- The Occupational Safety and Health Administration (OSHA) has set the permissible exposure limit (PEL) for endosulfan at 0.1 mg/m^3 in workroom air for 8 h time-weighted average (TWA).

- The National Institute for Occupational Safety and Health (NIOSH) has set the recommended exposure limit (REL) for endosulfan at 0.1 mg/m^3 and the acceptable daily intake (ADI) for endosulfan at 0.006 mg/kg/day.

- The U.S. Food and Drug Administration (U.S. FDA) has set the limit of endosulfan on dried tea below 24 ppm. The U.S. EPA recommends that levels of endosulfan in rivers, lakes, and streams should not be more than 74 ppm, and in other raw agricultural products should be below 0.1–2 ppm.

Endrin (CAS No. 72-20-8)

Molecular formula: $C_{12}H_8Cl_6O$ (stereoisomer of dieldrin)
Synonyms and trade names: 1,2,3,4,10,10-Hexachloro-6,7-epoxy-1,4,4a,5,6,7,8,8a-octahydro-1,4-endo-endo-5,8-dimethanonaphthalene; Endrin aldehyde; Mendrin; Nendrin; Hexadrin
Chemical name: 1,2,3,4,10,10-Hexachloro-6,7-epoxy-1,4,4a,5,6,7,8,8a-octahydro-1,4-endo,endo-5,8-dimethanonaphthalene

Endrin is an organochlorine compound. Endrin appears as a white or beige crystalline solid and is stable. It is incompatible with strong acids and strong oxidisers and corrodes some metals. Endrin decomposes on heating above 245°C, producing hydrogen chloride and phosgene. Endrin is incompatible with strong oxidisers, strong acids, and parathion and emits hydrogen chloride and phosgene when heated or burned. In fact, the U.S. EPA has sharply restricted the availability and uses of many organochlorine groups of pesticides. These include DDT, aldrin, dieldrin, endrin, heptachlor, mirex, chlordecone, and chlordane, while many other organochlorines, however, remain the active ingredients of various home and garden products and some agricultural, structural, and environmental pest control products.

Safe Handling and Precautions

Exposures to endrin cause toxicity and adverse health effects. Endrin is highly toxic to humans and animal species. The symptoms of poisoning include, but are not limited to, headache, dizziness, nausea, abdominal discomfort, vomiting, incoordination, insomnia, aggressiveness, confusion, tremor, mental confusion, anorexia, and hyperexcitable state. Exposures to very high concentrations of endrin can cause severe cases of poisoning symptoms that include convulsions, seizures, coma, and respiratory depression. In severe organochlorine compound poisoning, symptoms include myoclonic jerking movements, generalised tonic–clonic convulsions, and respiratory depression following the seizures and coma.

Endrin is identified as an experimental teratogen and produces reproductive hazard to laboratory animals. The epidemiological study carried out on occupationally exposed workers to endrin did not support cancer risk. The IARC classified endrin as Group 3, meaning not classifiable as a human carcinogen.

Regulations:

- The ACGIH has set the TLV for endrin at 0.1 m^3.
- The OSHA has set the PEL for endrin at 0.1 mg/m^3 (TWA, 8 h, skin).
- The NIOSH has set the REL for endrin at 0.1 mg/m^3 (TWA skin).

Students and occupational workers during use and handling of endrin should be careful. Workers should use personal protective equipment (PPE), appropriate and approved workplace safety dress to avoid exposure to the toxic chemical substance, and self-contained breathing apparatus. Workers should be aware of the fact that exposures to endrin cause effects on the CNS leading to convulsions and fatal injury. Workers should not wash away the chemical waste into sewer but sweep spilled chemical substance into sealable containers.

S-Ethyl Dipropylthiocarbamate (EPTC) (CAS No. 759-94-4)

Molecular formula: $C_9H_{19}NOS$

Synonyms and trade names: Carbamothioic acid, dipropyl-, S-ethyl ester; EPTC; S-Ethyl dipropylcarbamothioate; S-Ethyl-*N*,*N*-dipropylthiocarbamate

S-Ethyl dipropylthiocarbamate (EPTC) is a pale to dark yellow liquid with an aromatic odour/pleasant scent characteristic to thiocarbamates. EPTC is a commonly used herbicide. EPTC has a pleasant scent and at 20°C is miscible with most organic solvents, including acetone, ethyl alcohol, kerosene, methyl isobutyl ketone (MIBK), and xylene.

EPTC was the first thiocarbamate herbicide developed. EPTC formulations are used for the control of preemergent weeds, especially grasses. EPTC inhibits photosynthesis, respiration, and the synthesis of lipids, proteins, and RNA in these seedlings. Resistant plants are less sensitive to the herbicidal action apparently due to their ability to rapidly metabolise EPTC. Also known as EPTAM, EPTC is used to kill weeds and unwanted plants. Thiocarbamate, the selective herbicide is applied as a pre-plant and soil herbicide for the control of annual grasses, perennial weeds, broadleaf weeds in beans, forage legumes, potatoes; corn, sweet potatoes in some areas. The formulations include emulsifiable concentrate and granules. EPTC decomposes on heating or on burning, producing toxic fumes including nitrogen oxides and sulphur oxides. EPTC is a broad-spectrum herbicide that must be incorporated into soil to be effective. In California, EPTC is applied primarily to alfalfa, corn, sugar beets, and potatoes.

Safe Handling and Precautions

EPTC is a cholinesterase inhibitor. EPTC poisoning can cause harmful, even deadly, effects to the body.

Laboratory animals exposed to EPTC developed the most common clinical signs of acute toxicity which were lethargy, salivation, ataxia, red facial stains, bloody nasal discharge, anogenital stains, dyspnoea, laboured respiration, prostration, lacrimation, diarrhoea, convulsions, tremors, vocalisation, hyperactivity, and hypersensitivity to stimuli. Direct exposure to EPTC causes eye problems including restricted pupils, blurred vision, and severely irritated and reddened eyes. Accidental swallowing of EPTC is known to cause stomach cramps, diarrhoea, and nausea. Breathing EPTC has been reported to cause poisoning with symptoms such as headache, dizziness, and trouble breathing that can escalate to respiratory failure and death. Stomach and breathing problems usually start at the time of exposure. In children, the first symptom may be a convulsion, an intense, uncontrollable contraction of the muscles. Occupational workers exposed to EPTC develop symptoms of dizziness, headache and nausea, and the eyes and skin develop redness. EPTC has not undergone a complete evaluation and determination under the U.S. EPA's IRIS programme for evidence of human carcinogenic potential. The substance is harmful to aquatic organisms. This substance does enter the environment under normal use. Great care, however, should be given to avoid any additional release, for example, through inappropriate disposal. Thus, it is also very important to remember that detailed studies in species of animals indicated and led to the conclusion that the margins of safety (MOS) for the use of technical EPTC in herbicide formulations were greater than 100 for potential acute occupational exposure and acute or chronic dietary exposure. The MOS from the theoretical consumption of foods with the highest legal residues (tolerances) of EPTC were all greater than 100. Possible adverse effects associated with EPTC exposure in animals include neurotoxicity, nasal cavity degeneration/hyperplasia, blood coagulation abnormality, and neuromuscular degeneration in experimental animals. Dermal absorption studies conducted in rats indicated that EPTC is rapidly absorbed and eliminated. Percutaneous absorption of a dose comparable to field worker exposure was estimated to be 18.25% of the administered dose in 24 h.

During handling of EPTC, workers should completely avoid to eat any kind of food materials, to drink, and/or to smoke. Workers should be careful to wash hands before eating. During handling of EPTC, workers should not neglect use of PPE such as safety spectacles or eye protection in combination with breathing protection. Workers should avoid eye and skin contact, and EPTC should be stored in a cool, dry, well-ventilated area out of reach of children and should not be stored near food or feed. During use and handling, occupational workers must use protective clothing with long-sleeved shirt and long pants, chemical-resistant gloves, and chemical-resistant footwear plus socks.

Ethanol (CAS No. 64-17-5)

Molecular formula: C_2H_5OH/CH_3CH_2OH
Synonyms and trade names: Absolute alcohol; Ethyl alcohol; Ethyl hydrate; Ethyl hydroxide; Fermentation alcohol; Grain alcohol; Methylcarbinol; Molasses alcohol; Spirits of wine

Ethanol/ethyl alcohol is a colourless clear, highly flammable liquid, hygroscopic, and fully miscible in water. Ethanol is incompatible with a large number of chemicals such as strong oxidising agents, acids, alkali metals, ammonia, hydrazine, peroxides, sodium, acid anhydrides, calcium hypochlorite, chromyl chloride, nitrosyl perchlorate, bromine pentafluoride, perchloric acid, silver nitrate, mercuric nitrate, potassium tert-butoxide, magnesium perchlorate, acid chlorides, platinum, uranium hexafluoride, silver oxide, iodine heptafluoride, acetyl bromide, disulphuryl difluoride, acetyl chloride, permanganic acid, ruthenium (VIII) oxide, uranyl perchlorate, and potassium dioxide. Ethanol is extensively used as a solvent, in manufacture of chemicals, as a fuel additive, and for potable drink manufacture. Solvent use is mainly in paint and ink manufacture and in pharmaceutical production. Ethanol is widely used in consumer products and forms explosive mixtures with air.

Safe Handling and Precautions

Exposure to ethanol/ethyl alcohol is known to cause severe eye irritation, painful sensitisation to light, chemical conjunctivitis, and corneal damage. Ingestion of ethanol causes gastrointestinal irritation with symptoms of nausea, vomiting and diarrhoea, acidosis, and CNS depression, characterised by excitement, followed by headache, dizziness, drowsiness, and nausea. Users and workers after severe poisoning of ethanol develop symptoms of nausea, dizziness, respiratory tract irritation, narcotic effects, unconsciousness, collapse, coma, and possible fatal injury. Ethanol is readily absorbed by the oral and inhalation routes and, subsequently, metabolised and excreted in humans.

Upon repeat exposures at higher doses, male rats showed minor changes to organ weights and haematology/biochemistry, and the female rats showed minor biochemistry changes and increased length of oestrus cycle along with liver nodules; adverse liver effects were observed at concentrations of 3600 mg/kg/bw/day and above. Reports also indicate that prolonged exposure to ethanol/ethyl alcohol causes adverse effects and damages to the liver, kidney, heart, and CNS. Reports indicate that laboratory animals exposed to ethanol developed fetotoxicity in the embryo or foetus. Exposure to ethanol has been associated with a distinct pattern of congenital malformations – termed the *foetal alcohol syndrome*.

Epidemiological studies clearly indicate that drinking of alcoholic beverages is causally related to cancers of the oral cavity and pharynx (excluding the nasopharynx). There is no indication that the effect is dependent on type of beverage. Epidemiological studies clearly indicate that drinking of alcoholic beverages is causally related to cancer of the oesophagus. Also, the available information indicates that drinking of alcoholic beverages is causally related to liver cancer.

The ACGIH classified ethanol/ethyl alcohol as A4, meaning not classifiable as a human carcinogen. Similarly, the U.S. OSHA, the IARC, the NIOSH, and the NTP have not listed ethanol/ethyl alcohol as a human carcinogen.

Workers should keep ethanol/ethyl alcohol away from heat, sparks, flame, sources of ignition, and oxidising materials. Ethanol/ethyl alcohol should be stored in a tightly closed container, in a cool, dry, well-ventilated area away from incompatible substances, perchlorates, peroxides, chromic acid or nitric acid, and flammables area. Workers during handling of ethanol/ethyl alcohol should wear appropriate protective workplace dress, eye glasses or safety goggles.

Ethyl Alcohol Completely Denatured (CAS No. Not Applicable to Mixtures)

Molecular formula: Not applicable to mixtures
Synonyms and trade names: Alcohol; Spirits of wine; Potato alcohol; CDA Formula 19
Ingredients: Ethyl alcohol (CAS No. 64-17-5); 95%; MIBK (CAS No. 108-10-1); 4%; Kerosene (CAS No. 8008-20-6); less than 1.0%

Ethyl alcohol completely denatured is a mixture of many chemicals including MIBK. Ethyl alcohol completely denatured is incompatible with strong oxidising agents, perchlorates, aluminium, alkali metals, acetyl chloride, calcium hypochlorite, chlorine oxides, mercuric nitrate, hydrogen peroxide, nitric acid, bromine pentafluoride, chromyl chloride, permanganic acid, uranium hexafluoride, and acetyl bromide. It ignites on contact with phosphorus (III) oxide, platinum, disulphuric acid + nitric acid, and potassium tert-butoxide + acids. It will ignite and then explode on contact with acetic anhydride + sodium hydrogen sulphate. It forms explosive products in reaction with silver nitrate, ammonia + silver, and silver (I) oxide + ammonia or hydrazine. MIC is incompatible with aldehydes, nitric acid, perchloric acid, and strong oxidisers. Violent reaction occurs with potassium tert-butoxide.

Exposure of skin to completely denatured ethyl alcohol causes dryness with mild irritation and redness. Workers exposed to vapours of ethyl alcohol completely denatured show symptoms of depression, eye and upper respiratory tract irritation, burning sensation, headache, dizziness, tremors, and nausea. Ingestion of ethyl alcohol completely denatured with significant exposures causes dose-related CNS depression. The symptoms of poisoning include headache, tremor, fatigue, hallucinations, distorted perceptions, narcosis, convulsions, coma, respiratory failure, and death. Chronic exposures to high concentrations cause severe damage to the CNS, liver, blood, and reproductive system. Workers after chronic toxicity of the chemical substance show effects such as physical dependence, malnutrition, and neurological disorders like amnesia, dementia, and prolonged sleepiness. Chronic ingestion of ethyl alcohol completely denatured produces cancers of the oesophagus, liver, and the kidneys. Users and occupational workers should keep stored ethyl alcohol well protected against physical damage and away from children; free from smoking areas; in a cool, dry, well-ventilated location; and away from any area where the fire hazard may be acute. Storage of ethyl alcohol should be outside or preferably in a detached storage area. Storage of ethyl alcohol should be totally separated from incompatibles. Containers of ethyl alcohol completely denatured should be bonded and grounded for transfers to avoid static sparks. The ACGIH and OSHA have set the exposure levels of MIBK as follows: TLV (ACGIH) and the PEL (OSHA) at 50 ppm (TWA) and the short-term exposure limit (STEL) at 75 ppm.

Ethion (CAS No. 563-12-2)

Molecular formula: $C_9H_{22}O_4P_2S_4$
Synonyms and trade names: Cethion; Dhanumit; Ethanox; Ethiol; Hylmox; Nialate; Rhodiacide; Tafethion
IUPAC name: O,O,O',O'-tetraethyl S,S'-methylene bis (phosphorodithioate)
Toxicity class: U.S. EPA: II; WHO: II

Technical ethion is an odourless amber liquid. It is very sparingly soluble in water but soluble in most of the organic solvents. Ethion undergoes decomposition on heating or on burning and produces toxic and corrosive fumes including phosphorus oxides and sulphur oxides. It is used for the control of crop pests and household insects. These include, but are not limited to, aphids, mites, sticks, scales, thrips, leaf hoppers, maggots, leaf-feeding insects, foliar feeding larvae, and house flies. It may be used on a wide variety of food, fibre, and ornamental crops, including greenhouse crops, lawns, and turf. Ethion is often used on citrus and apples. It is mixed with oil and sprayed on dormant trees to kill eggs and scales. Occupational workers and general public get exposed to ethion while working in industries that manufacture ethion and during eating raw fruits or vegetables that have been treated with ethion. There are no residential uses for ethion.

Safe Handling and Precautions

Ethion is highly to moderately toxic by the oral route to animals and humans. It causes toxicity and poisoning with symptoms such as nausea, cramps, diarrhoea, excessive salivation, severe depression, irritability, confusion, headache, blurred vision, fatigue, tightness in chest, and abnormal heartbeat and breathing. Ethion on repeated exposures and in high concentrations caused severe symptoms of poisoning. The symptoms of toxicity and poisoning included, but are not limited to, pupillary constriction, muscle cramp, impaired memory and concentration, disorientation, speech difficulties, delayed reaction times, nightmares, sleepwalking, loss of coordination, convulsions, unconsciousness/coma, and death. Studies on laboratory animals and humans have not indicated any evidence of mutagenic, teratogenic, or carcinogenic effects of ethion.

Regulations:

- The ACGIH has set the TLV for ethion at 0.4 mg/m^3.
- The U.S. NIOSH has set the REL for ethion at 0.4 mg/m^3.

Ethylbenzol (CAS No. 100-41-4)

Molecular formula: C_8H_{10}
Synonyms and trade names: Ethylbenzene; Ethylbenzol; Phenylethane; EB

Ethylbenzol is a highly flammable chemical substance. It reacts with strong oxidants and attacks plastic and rubber.

Ethylbenzol is a colourless liquid found in a number of products including gasoline and paints. It is naturally found in coal tar and petroleum and is also found in manufactured products such as inks, pesticides, and paints. Ethylbenzene is used primarily to make another chemical, styrene. Other uses include as a solvent, in fuels, and to make other chemicals. Occupational exposure to ethylbenzol has been reported to occur in industrial workplaces, during the use of products containing it, for instance, gasoline, carpet glues, varnishes, and paints. Ethylbenzol has been reported to undergo ready biodegradation and is expected to undergo full mineralisation in the aquatic environment under aerobic conditions within days to a few weeks.

Safe Handling and Precautions

Exposure to ethylbenzol causes adverse health effects. Workers exposed to ethylbenzol develop symptoms of poisoning such as cough, sore throat, dizziness, drowsiness, and headache, and accidental ingestion causes burning sensation in the throat and chest. Ethylbenzol causes irritating effects to the eyes, to the skin, and to respiratory tract. Accidental swallowing of the liquid ethylbenzol at the workplace causes aspiration into the lungs with the risk of chemical pneumonitis. Ethylbenzol is also known to cause effects on the CNS, and ethylbenzol above the occupational exposure level (OEL) leads to the lowering of consciousness. Laboratory studies with animals indicated that ethylbenzene causes an increased weight of liver, kidney, and spleen. Repeated dermal application of ethylbenzene in rabbits produced erythema, oedema, and superficial necrosis. Aspiration of ethylbenzol causes extensive pulmonary oedema and haemorrhage. Repeated exposure to ethylbenzol has been reported to cause effects on the kidneys and liver and impairment and functional disorders. Repeated contact of ethylbenzol benzol with skin causes dryness, reddening, cracking, blistering of the skin, and dermatitis. Ethylbenzol has been reported to induce sister chromatid exchanges in human test systems.

The IARC concluded that there is inadequate evidence to classify ethylbenzene as a carcinogen in humans and sufficient evidence in experimental animals and list ethylbenzene as a possible human carcinogen. Also, the U.S. EPA classified ethylbenzene as Group D, meaning not classifiable as a human carcinogen.

Occupational workers and students should know that ethylbenzol rapidly gets absorbed through the skin, hands, and forearms. Workers during handling of ethylbenzol should use appropriate and adequate workplace safety dress and thoroughly wash hands, fore-arms, and face with soap and water before eating. Ethylbenzene should be stored in a cool, dry, well-ventilated area in tightly sealed containers that are labelled in accordance with regulations and kept protected from physical damage. Workers should know that ethyl-benzol forms combustible mixture with air, and hence they must be alert with appropri-ate precautions to prevent the formation of explosive and flammable vapour–air mixtures during handling, storage, and transport of ethylbenzol.

Regulations:

- The ACGIH has set the TLV for ethylbenzol at 100 ppm (TWA) and the STEL at 125 ppm.
- The U.S. OSHA has set the PEL of ethylbenzol for general industry at 435 mg/m^3 (TWA).
- The NIOSH has set the REL for ethylbenzol at 435 mg/m^3 (TWA) (100 ppm) and the STEL at 545 mg/m^3 (125 ppm).
- The European Union (E.U.) Dangerous Substances Directive has not classified eth-ylbenzol as hazardous for the environment.

Ethyl Carbamate (CAS No. 51-79-6)

Molecular formula: $C_3H_7NO_2$
Synonym and trade name: Carbamic acid ethyl ester; Ethyl aminoformate; Ethyl urethane; O-Ethylurethane; NSC 746; Leucethane; Uretano; Urethane

Ethyl carbamate occurs as colourless or white columnar crystals or granular powder. It is very soluble in water. With heat, ethyl carbamate undergoes decomposition and emits toxic fumes of carbon monoxide, carbon dioxide, and nitrogen oxides. Ethyl carbamate is used as an intermediate in the synthesis of a number of chemical products, for example, phar-maceuticals, in biochemical research and medicine, and as a solubiliser and co-solvent for pesticides and fumigants.

Prior to World War II, ethyl carbamate saw relatively heavy use in the treatment of multiple myeloma before it was found to be toxic, carcinogenic, and largely ineffective. However, due to U.S. FDA regulations, ethyl carbamate has been withdrawn from phar-maceutical use. However, small quantities of ethyl carbamate are also used in laboratories as an anaesthetic for animals.

Safe Handling and Precautions

Exposures to ethyl carbamate cause harmful effects. Occupational workers on short-term exposure to ethyl carbamate develop health disorders, and repeated exposure to high concentrations of the chemical substance for a long period of time (chronic exposure) has been known to cause severe poisoning. The symptoms of poisoning caused by ethyl car-bamate include vomiting, injury to the kidneys and liver, reproductive disorders. Ethyl carbamate cause damage to systems of bone marrow, CNS, and immune system and lead to cancer and coma. Laboratory animals on oral and inhalation exposures to ethyl carbamate developed increased incidence of lung tumours. While the U.S. EPA has not classified ethyl carbamate as a carcinogen, the IARC classified ethyl carbamate as a Group 2B, possibly carcinogenic to humans. Reports of IARC indicate that in 2007, the carcinogenicity grade of ethyl carbamate has been upgraded to as Group 2A carcinogen.

Occupational workers during handling of ethyl carbamate should strictly observe workplace regulations and safety management of hazardous chemical substances. Workers should keep stored ethyl carbamate in a cool dry place and away from strong oxidising agents, strong acids, or bases.

Ethylene Chlorohydrine (CAS No. 107-07-3)

Molecular formula: C_2H_5ClO
Synonyms and trade names: 2-Chloroethanol; 2-Chloroethyl alcohol; Ethylene chlorohydrine; Glycol chlorohydrine

Ethylene chlorohydrine is a clear, colourless liquid with mild, ethereal odour. It reacts with alkali metals. Ethylene chlorohydrine has applications as a laboratory reagent and as a pharmaceutical intermediate. Ethylene chlorohydrine is a building block in the production of pharmaceuticals, biocides, and plasticisers, used for manufacture of thiodiglycol, an important solvent for cellulose acetate and ethyl cellulose, textile-printing dyes, extraction of pine lignin, in dewaxing, refining of rosin, and the cleaning of machines. Several dyes are prepared by the alkylation of aniline derivatives with chloroethanol.

Safe Handling and Precautions

Users and occupational workers during handling of ethylene chlorohydrine should be very careful, because it is harmful and exposure causes adverse health effects. Like most organochlorine compounds, chloroethanol is combustible and releases hydrogen chloride and phosgene. The symptoms of poisoning include, but are not limited to, mucous membrane irritation, nausea, vomiting, vertigo, incoordination, numbness, vision disturbance, headaches, thirst, delirium, low blood pressure, possibility of shock, collapse, and coma. Reports indicate that the ACGIH, the IARC, and the NTP have not listed ethylene chlorohydrine as a human carcinogen.

Workers should handle the accidental chemical spills of ethylene chlorohydrine that occur at workplace with care, and the waste chemical spills should be appropriately contained, solidified, and placed in suitable containers for disposal according to the set workplace regulations. Workers should remember that the waste ethylene chlorohydrine should not be discharged into waterways or sewer systems without proper regulatory authorisation, and the workplace manager should be aware of the national, state, and local regulations. Ethylene chlorohydrine should be kept away from extreme heat, sources of ignition, and smoking, and segregated from alkalis and alkylating agents and incompatible substances. Workers should also note that ethylene chlorohydrine should be stored for a period of 6 months only.

Occupational workers during use and handling of ethylene chlorohydrine should strictly wear PPE, safety workplace dress, chemical resistant protective gloves, (butyl rubber, chloroprene rubber), chemical safety goggle, face shield to avoid chemical splashing, and standard, certified air purifying respirator/self-contained breathing apparatus.

Danger: Ethylene chlorohydrine is a combustible liquid, highly flammable, and harmful. Improper use and handling may cause fatal injury – when inhaled or absorbed through skin.

Regulations:

- The U.S. OSHA has set the PEL for ethylene chlorohydrine at 5 ppm (16 mg/m^3, skin).
- The U.S. NIOSH has set the REL for ethylene chlorohydrine at 1 ppm (3 mg/m^3, skin).
- The U.S. NIOSH has set the immediately dangerous to life or health (IDLH) concentrations for ethylene chlorohydrine at 7 ppm.
- The DSCL of EEC has listed ethylene chlorohydrine as very toxic by inhalation and skin contact.
- The Workplace Hazardous Materials Information System (WHMIS), Canada, has listed ethylene chlorohydrine as CLASS B-3, meaning combustible liquid; as CLASS D-1A, meaning chemical substance that causes immediate and serious toxic effects (VERY TOXIC); and CLASS D-2B, meaning material causing other toxic effects (TOXIC).

Ethylenediamine (CAS No. 107-15-3)

Molecular formula: $C_2H_8N_2$
Synonyms and trade names: alpha, omega-Ethanediamine; beta-Aminoethylamine; Diaminoethane; 1,2-Diaminoethane; 1,4-Diazabutane; Dimethylenediamine; 1,2-Ethanediamine; STC 4935628; NCI-C60402

Ethylenediamine is a colourless to light yellow liquid with an ammoniacal odour. It is air sensitive and hygroscopic and absorbs carbon dioxide from the air. It is incompatible with aldehydes, phosphorus halides, organic halides, oxidising agents, strong acids, copper, its alloys, and its salts.

Occupational workers have a wide range of chances of exposure to ethylenediamine in workplaces, for instance, in the various industrial or manufacturing facilities, during production, distribution, storage in closed systems, and while working with sampling, testing, or other procedures.

Safe Handling and Precautions

Ethylenediamine is an air-sensitive chemical substance. It absorbs carbon dioxide from the air and is flammable – note wide explosion limits. It should be protected from moisture.

Exposure to ethylenediamine causes harmful health effects. On accidental inhalation, swallowing, and/or absorption through the skin, it causes allergic reactions. Studies in laboratory animals exposed to ethylenediamine indicated that it is corrosive and severely irritating to the skin and eyes. Systemic toxicity from exposure to high vapour concentrations caused damage to the kidneys, liver, and lungs. Oral exposure to ethylenediamine causes adverse effects to the kidneys and liver of experimental rats. Ethylenediamine is a strong irritant of the eyes and respiratory tract and a skin and respiratory sensitiser. Also, ethylenediamine causes burns to the skin and is very destructive to mucous membranes.

During handling of ethylenediamine, occupational workers should protect themselves with appropriate personal protective equipment (PPE), good workplace ventilation, protective clothing, helmets, and goggles. Workers should keep stored and protected ethylenediamine away from moisture and from flammable and wide explosion limits.

Regulations:

- The ACGIH has set the TLV for ethylenediamine at 10 ppm (25 mg/m^3 as 8 h TWA).
- The OSHA has set the PEL for ethylenediamine at 10 ppm (25 mg/m^3 as 8 h TWA).
- The NIOSH has set the REL for ethylenediamine at 10 ppm (25 mg/m^3 as 10 h TWA).

Ethylene Dibromide (Dibromoethane) (CAS No. 106-93-4)

Molecular formula: $C_2H_4Br_{12}$
Synonyms and trade names: Bromofume; Ethylene bromide; EDB; Dowfume; Glycol bromide

Ethylene dibromide is a heavy, colourless liquid with a mild sweet odour, like chloroform. Ethylene dibromide is incompatible with strong oxidisers, magnesium, alkali metals, and liquid ammonia. It is also known as 1,2-dibromomethane. Ethylene dibromide is soluble in alcohols, ethers, acetone, benzene, and most organic solvents and slightly soluble in water. Ethylene dibromide was once of dominant use, although its use has faded as an additive in leaded gasoline. Ethylene dibromide (1,2-dibromoethane) reacts with lead residues to generate volatile lead bromides. It has been used as a pesticide in soil and various crops. Exposure to ethylene dibromide primarily occurs from its past use as an additive to leaded gasoline and as a fumigant. Most of the uses of ethylene dibromide have been stopped in the United States; however, it is still used as a fumigant for treatment of logs for termites and beetles, for the control of moths and beehives, and as a preparation for dyes and waxes. Ethylene dibromide was used as a fumigant to protect against insects, pests, and nematodes in citrus, vegetable, and grain crops and as a fumigant for turf, particularly on golf courses. Ethylene dibromide is a severe eye, mucous membrane, and skin irritant, and toxic to liver, kidneys and lungs. However, because of limitations in epidemiological study evidences for ethylene dibromide as a human carcinogen is inconclusive. In 1984, the U.S. EPA imposed a ban on its use as a soil and grain fumigant.

Safe Handling and Precautions

Exposures to ethylene dibromide cause adverse health effects and poisoning. Ethylene dibromide is extremely toxic to humans. Long-term exposures to ethylene dibromide caused deleterious effects to the liver, kidney, and the testis in the laboratory rats, irrespective of the route of exposure. Limited data on men occupationally exposed to ethylene dibromide indicate that long-term exposure to ethylene dibromide can impair reproduction by damaging sperm cells in the testicles. Several animal studies indicate that long-term exposure to ethylene dibromide increases the incidences of a variety of tumours in rats and mice in both sexes by all routes of exposure. The symptoms of toxicity include, but are not limited to, redness, inflammation, skin blisters, and ulcers on accidental swallowing/ingestion. Ethylene dibromide has also been reported to cause birth defects in exposed humans. There is sufficient evidence in experimental animals for the carcinogenicity of ethylene dibromide, while evidences in humans are inadequate. Ethylene dibromide is a severe skin irritant and can be absorbed through the skin as well as the respiratory tract. High concentrations can affect the lungs and injure liver and kidneys. Ethylene dibromide-exposed fumigated food grains fed to laying hens resulted in the decrease of egg size and number of eggs. Ethylene dibromide is more toxic to humans than methyl bromide.

Ethylene dibromide has been investigated for its carcinogenic effects and shown to be capable of producing cancer in laboratory animals. The IARC classified ethylene dibromide as Group 2A, meaning a probable human carcinogen. The U.S. EPA classified ethylene dibromide as Group B2, meaning probable human carcinogen. Also, on the basis of data on ethylene dibromide–induced tumours in multiple sites and by various routes of exposure in animals, the DHHS has listed that ethylene dibromide can reasonably be anticipated to be a human carcinogen. The results of epidemiological studies have been found inconclusive, and more studies are required. Elemental mercury has not been classified as a carcinogen or a non-carcinogen by IARC or by US Environmental Protection Agency, or by Department of Health Human Services (DHHS).

Workplace safety dress: Ethylene dibromide is absorbed through the skin. Occupational workers must be aware of the fact that ethylene dibromide readily penetrates most rubbers and barrier fabrics or creams. Workers should, therefore, use butyl rubber gloves for skin protection. Contact with concentrated ethylene dibromide vapour or liquid, cause delayed or immediate chemical/'thermal burns'.

Regulations:

- Ethylene dibromide was used in the past as an additive to leaded gasoline; however, since leaded gasoline is now banned, it is no longer used for this purpose.
- Republic of Korea: Ethylene dibromide and mixtures containing 50% or more of ethylene dibromide are banned for manufacture, import, and use as an industrial chemical.
- The United States banned ethylene dibromide as soil fumigant and insecticide.
- Use of EDB as agricultural chemical is banned in the following countries: Belize, Chile (1985), Colombia (1985), Cyprus (1987), Ecuador (1985), Kenya (1985), Argentina (1990), Sweden (1985; severely restricted), and United Kingdom (1981–1985).
- Chile prohibited use of EDB for fumigation of fruit and vegetables in 1985.
- The U.S. OSHA has set the PEL for ethylene dibromide at 20 ppm (8 h, TWA) and the acceptable ceiling concentration at 30 ppm not to exceed for 15 min.
- The U.S. NIOSH has set the REL for ethylene dibromide at 0.045 ppm and ceiling limit of 0.13 ppm not to exceed for 15 min.
- The IDLH for ethylene dibromide at 100 ppm.
- The ADI of ethylene dibromide (inorganic bromide) is set at 1.0 mg/kg/day.

Occupational workers associated with the management and fumigation work should receive thorough instructions on the properties of fumigants and training in safe methods of handling of the fumigant. During use and handling of ethylene dibromide, it is very important that no worker/person should work alone to avoid the possible serious consequences of the poisonous fumigant. Also, occupational workers should use approved and adequate respiratory protection; should be scrupulously careful to avoid any kind of contact of the fumigant with the skin, eyes, and body; and must wear appropriate clothing and footwear.

Ethylene Dichloride (CAS No. 107-06-2)

Molecular formula: $C_2H_4Cl_2$

Synonyms and trade names: 1,2-Dichloroethane; Dichloroethylene; 1,2-Bichloroethane; Ethylene chloride; Ethane dichloride; 1,2-Ethylene dichloride; Glycol dichloride; EDC; sym-Dichloroethane; alpha, beta-Dichloroethane; Borer sol; Brocide; Destruxol; Dichloremulsion; Dutch oil; Di-chlor-mulsion; Dutch liquid

Ethylene dichloride (1,2-dichloroethane) is a colourless oily liquid with a chloroform-like odour, detectable over the range of 6–40 ppm, with a sweet taste. Ethylene dichloride (1,2-dichloroethane), which has a carbon–carbon single bond, should be distinguished from 1,2-dichloroethene which has a carbon–carbon double bond. It is a skin irritant. Ethylene dichloride was also used as an extraction solvent, as a solvent for textile cleaning and metal degreasing, in certain adhesives, and as a component in fumigants for upholstery, carpets, and grain. Other miscellaneous applications include paint, varnish, and finish removers, soaps and scouring compounds, wetting and penetrating agents, organic synthesis, ore flotation, and as a dispersant for nylon, rayon, styrene-butadiene rubber, and other plastics. Reports indicate that ethylene dichloride has extensive industrial and other applications around the globe.

Safe Handling and Precautions

Exposures to ethylene dichloride cause depression of the CNS with symptoms that include, but are not limited to, dizziness, drowsiness, trembling, unconsciousness, nausea, vomiting, abdominal pain, skin irritation, dermatitis, eye irritation, corneal opacity, blurred vision, headache, sore throat, cough, bronchitis, pulmonary oedema (may be delayed), liver, kidney, cardiovascular system damage, cardiac arrhythmia, acute abdominal cramps, diarrhoea, internal bleeding (hemorrhagic gastritis and colitis), and respiratory failure. Ethylene dichloride involves kidneys, liver, eyes, skin, and cardiovascular system as the target organs. Ethylene dichloride is known to cause systemic effects and has been identified as a priority pollutant in many countries. Prolonged period of inhalation of vapours of ethylene dichloride irritates the respiratory tract. Severe symptoms of toxicity lead to CNS effects, liver damage, kidney damage, adrenal gland damage, cyanosis, weak and rapid pulse, unconsciousness, and respiratory and circulatory failure leading to fatal injury.

The acute effects of ethylene dichloride are similar for all routes of entry, namely, by ingestion, inhalation, and skin absorption. Acute exposures to ethylene dichloride result in nausea, vomiting, dizziness, internal bleeding, bluish-purple discoloration of the mucous membranes and skin (cyanosis), rapid but weak pulse, and unconsciousness. Acute exposures can lead to death from respiratory and circulatory failure. Autopsies in such situations have revealed widespread bleeding and damage in most internal organs. Repeated long-term exposures to ethylene dichloride have resulted in neurologic changes, loss of appetite and other gastrointestinal problems, irritation of the mucous membranes, liver and kidney impairment, and death.

The IARC classifies 1,2-dichloromethane (ethylene dichloride) as a Group 2B carcinogen, based on evidence in experimental animals. Studies in mice and rats have shown an increased incidence of tumours at various sites, including the stomach, lung, liver, mammary gland, and uterus. Although excesses of some cancers have been observed in epidemiological studies, analysis is complicated by potential exposure to multiple compounds. The U.S. EPA classified ethylene dichloride as a Group B2, meaning probable human carcinogen.

The 11th report on carcinogens of the NTP based on sufficient evidence in experimental animals lists 1,2-dichloroethane as 'reasonably anticipated to be a human carcinogen'. Further, ethylene dichloride is a probable human carcinogen and has been reported to cause liver damage and mutagenic effects and has been identified as an experimental transplacental carcinogen.

Regulations:

- The ACGIH has set the TLV for 1,2-dichloromethane (ethylene dichloride) at 10 ppm (TWA).
- The OSHA has set the PEL for1,2-dichloromethane (ethylene dichloride) at 50 ppm (TWA) with a ceiling limit of 100 ppm and a maximum limit of 200 ppm/5 min.
- The NIOSH has set the IDLH for 1,2-dichloromethane (ethylene dichloride) at 50 ppm.

Occupational workers should store and handle ethylene dichloride in accordance with set regulations and standards. Workers should keep the container tightly closed and properly labelled and stored in a cool, dry, well-ventilated area. Workers should not store the chemical in aluminium container or use aluminium fittings or transfer lines. Workers should strictly avoid heat, flames, sparks, and other sources of ignition and should keep ethylene dichloride separated from incompatible substances. Workers should avoid workplace negligence and contact of ethylene dichloride with eyes, skin, and clothing. Occupational workers should keep ethylene dichloride well protected against physical damage. Occupational workers should avoid use of ethylene dichloride along with oxidising agents, strong alkalis, strong caustics, magnesium, sodium, potassium, active amines, ammonia, iron, zinc, nitric acid, and aluminium. Ethylene dichloride is highly flammable and workers should keep it protected in air and light sensitive store house. On decomposition, ethylene dichloride emits toxic fumes of phosgene, hydrogen chloride, acetylene, and vinyl chloride. The NIOSH recommends that workers should use ethylene dichloride properly and with prudence in the workplace as if the chemical substance is a human carcinogen. Workplace management should limit the entry of other workers/individuals to the workplace. In other words, the work area with ethylene dichloride should be restricted to only those workers essential to the chemical management.

Alert! Ethylene dichloride is a highly flammable liquid and vapour. Negligence and prolonged/repeated exposure is harmful to the skin, eyes, CNS, liver, and kidney; it is suspected of causing genetic defects.

Ethyleneimine (CAS No. 151-56-4)

Molecular formula: C_2H_5N
Synonyms and trade names: Azacyclopropane; Aziridine; Aziran; Azirane; Dihydro-1H-azirine; Dihydroazirene; Dimethyleneimine; Ethylimine; Ethylenimine

Ethyleneimine is a colourless liquid with an ammonia-like smell or pungent odour. It is highly flammable and reacts with a wide variety of materials. Ethyleneimine is used in polymerisation products, as a monomer for polyethyleneimine and as a comonomer for polymers, for example, with ethylenediamine. Polymerised ethyleneimine is used in paper, textile chemicals, adhesive binders, petroleum, refining chemicals, fuels, lubricants, coating resins, varnishes, lacquers, agricultural chemicals, cosmetics, ion-exchange resins, photographic chemicals, colloid flocculants, and surfactants. Ethyleneimine readily polymerises, and it behaves like a secondary amine. Ethyleneimine is highly caustic, attacking materials such as cork, rubber, many plastics, metals, and glass except those without carbonate or borax. It polymerises explosively on contact with silver, aluminium, or acid. The activity of ethyleneimine is similar to that of nitrogen and sulphur mustards. Ethyleneimine is used as an intermediate in the production of triethylenemelamine.

Safe Handling and Precautions

Exposures to ethyleneimine cause adverse health effects and poisoning. Ingestion/swallowing or inhalation or absorption through exposures to skin causes severe irritation, blisters, severe deep burns, and effects of sensitisation. Ethyleneinime is corrosive to the eye tissue and may cause permanent corneal opacity and conjunctival scarring, severe respiratory tract irritation, and effects of inflammation in workers. Ethyleneimine is a severe blistering agent, causing third-degree chemical burns of the skin. The symptoms of toxicity include, but are not limited to, cough, dizziness, headache, laboured breathing, nausea, vomiting, tearing and burning of the eyes, sore throat, nasal secretion, bronchitis, shortness of breath, laryngeal oedema, and pronounced changes of the trachea and bronchi of lungs. Ethyleneimine with its corrosive effects causes injury on the mucous membranes, and acute oral exposure may cause scarring of the oesophagus in humans. The onset of symptoms and health effects caused by ethyleneimine depend on exposure concentration.

There are no reported data on the potential carcinogenicity of ethyleneimine in laboratory animals exposed by inhalation. The NIOSH reported ethyleneimine as a potential carcinogen, and the IARC classified it as Group 3, meaning not classifiable as a human carcinogen. However, ethyleneimine has not undergone a complete evaluation and determination under the U.S. EPA's IRIS programme for evidence of human carcinogenic potential. The U.S. EPA has not classified ethyleneimine as a carcinogen. The ACGIH has set the TLV of ethyleneimine at 0.5 ppm on skin exposure in workplace area. Also, there are no reports regarding the measurement of personal exposure to ethyleneimine. Students and occupational workers during use and handling of ethyleneimine should wear protective equipments such as gloves and safety glasses and have good ventilation, and ethyleneimine should be handled by workers as a carcinogen. Ethyleneimine vapour/air mixtures are explosive and pose risk of fire and explosion on contact with acids and oxidants.

Ethylene Oxide (CAS No. 75-21-8)

Molecular formula: C_2H_4O
Synonyms and trade names: EO; EtO; Alkene oxide; Ethylene oxide; 1,2-Epoxyethane; Dihydrooxirene; Oxacyclopropane; Dimethylene oxide; Oxane; Oxirane; Oxidoethane; Epoxyethane

Ethylene oxide is the simplest cyclic ether. It is a colourless gas or liquid and has a sweet, etheric odour. Etylene oxide is a flammable and very reactive and explosive chemical substance. On decomposition, vapours of pure ethylene oxide mix with air or inert gases and become highly explosive. Industrial use of ethylene oxide is extensive, as an intermediate in the production of monoethylene glycol, diethylene glycol, triethylene glycol, poly(ethylene) glycols, ethylene glycol ethers, ethanolamine, ethoxylation products of fatty alcohols, fatty amines, alkyl phenols, cellulose, and poly(propylene glycol). It is also used as a fumigant fungicide and insecticide, in the sterilisation of surgical instruments/equipments and heat-sensitive materials in the hospital, in the sterilisation/fumigation of some imported foods, in the fumigation of books and archival materials in museums, and in the fumigation of furs, textiles, and furniture.

Safe Handling and Precautions

Exposure to ethylene oxide causes adverse health effects. The symptoms of toxicity include headache, nausea, vomiting, dizziness, lethargy, behavioural disturbances, weakness, cyanosis, loss of sensation in the extremities, and reduction in the sense of smell and/or taste. Ethylene oxide is also an irritant to the skin and respiratory tract, and inhaling the vapours may cause the lungs to fill with fluid several hours after exposure. Ethylene oxide is a sensitising agent. Occupational workers exposed to ethylene oxide vapour develop irritation of the eyes and respiratory tract. Mild irritation of the skin has been reported after contact with aqueous solutions of ethylene oxide as low as 1%. The injuries to skin include oedema and erythema followed by the formation of vesicles. Breathing of ethylene oxide in low levels for a prolonged period of time is known to cause irritation of the eyes, skin, respiratory passages, nervous system disorders, memory loss, numbness, progressing with increasing exposure to convulsions, seizure, and coma. At higher levels of exposure for shorter periods, effects are similar but may be more severe. There is some evidence that exposure to ethylene oxide can cause a pregnant woman to have a miscarriage. Animal studies indicate that in addition to irritation of the respiratory passages, nervous system effects, and reproductive effects, the kidneys, adrenal gland, and skeletal muscles may be affected from long-term exposure to ethylene oxide.

Inhalation of ethylene oxide is known to cause dizziness or drowsiness among occupational workers. Liquid contact may cause frostbite and allergic skin reaction. Laboratory animals after oral exposure (ingestion) to ethylene oxide show adverse effects in blood, damage to liver and kidney, reproductive effects, miscarriages/spontaneous abortion, and cancer. Prolonged exposures of laboratory animals to ethylene oxide produced incidences of liver cancer. The OSHA classified ethylene oxide as a carcinogenic agent; the U.S. EPA classified ethylene oxide as a Group B1, meaning probable human carcinogen; and the ACGIH classified ethylene oxide as A2, meaning a suspected human carcinogen, while the NTP classified ethylene oxide as a known human carcinogen. Similarly, the NIOSH classified ethylene oxide as a potential human carcinogen, and the IARC classified ethylene oxide as Group 1, meaning human carcinogen. In brief, ethylene oxide is a directly acting alkylating agent and has been identified as a causative agent for malignancies of the lymphatic and haematopoietic system in both humans and experimental animals. Reports have indicated that ethylene oxide has limited evidence in humans for the carcinogenicity and sufficient evidence in experimental animals. The overall evaluation of ethylene oxide by the IARC which is part of the World Health Organization (WHO) suggested and classified it as a Group 1, meaning it is a known human carcinogen.

Ethylene oxide is dangerously explosive if exposed to fire; it is flammable over an extremely large range of concentrations in air and burns in the absence of oxygen. The OSHA has set a limit of 1.0 ppm over an 8 h workday, 40 h work week for ethylene oxide with a STEL (not to exceed 15 min) of 5 ppm; the NIOSH recommends that average workplace air should contain less than 0.1 ppm ethylene oxide averaged over a 10 h workday, 40 h workweek.

Ethyl Silicate (CAS No. 78-10-4)

Molecular formula: $(CH_2H_5O)_4$, Si
Synonyms and trade names: Ethyl orthosilicate; Ethyl silicate; Condensed silicon ethoxide; Silicic acid tetraethyl ester; Tetraethyl orthosilicate; Tetraethyl ester orthosilicic acid; Tetraethyl silicate; TEOS; Tetraethoxysilane

Ethyl silicate is a flammable, colourless liquid with a mild, sweet, alcohol-like odour. Exposure to ethyl silicate can occur through inhalation, ingestion, and eye or skin contact. It is practically insoluble in water, soluble in alcohol, and slightly soluble in benzene. Occupational workers get exposed to ethyl silicate at workplaces associated with the manufacture and transportation of ethyl silicate and during use as a bonding agent for industrial buildings and investment castings, ceramic shells, crucibles, refractory bricks, and other moulded objects, as a protective coating for heat- and chemical-resistant paints, lacquers, and films, in manufacture of protective and preservative coatings for protection from corrosion (primarily as a binder for zinc dust paints), chemicals, heat, scratches, and fire. During production of silicones, as a chemical intermediate in the preparation of soluble silica, as a gelling agent in organic liquids, as a coating agent inside electric lamp bulbs, and in the synthesis of fused quartz and during industrial use in textile industry workers get exposed to ethyl silicate; human exposure to the chemical substance is possible during various uses such as, aqueous emulsions, delustre, and fireproofing, as a component of lubricants, as a mould-release agent and as a heat-resistant adhesive. Users and occupational workers should know that ethyl silicate acts dangerously with strong oxidising agents, is incompatible with alkalis and mineral acids, and hydrolyses slowly and non-violently under moist alkaline or acidic conditions at ambient temperatures and atmospheric pressures to form silicon dioxide and ethanol. Ethyl silicate also reacts with water to form a silicone adhesive (a milky white mass) and cause swelling and hardening of some plastics.

Safe Handling and Precautions

Exposures to ethyl silicate cause adverse health effects. The symptoms of poisoning include, but are not limited to, irritation of the eye, mucous membrane, respiratory tract, respiratory difficulty, tremor, fatigue, narcosis, nausea, and vomiting. Prolonged period of skin contact may produce drying, cracking, inflammation, and dermatitis. As observed in laboratory animals, occupational workers exposed to the chemical substance may suffer from liver and kidney damage, CNS depression, and anaemia. At concentrations of 3000 ppm, ethyl silicate causes extreme and intolerable irritation of the eyes and mucous membranes; at 1200 ppm, it produces tearing of the eyes; at 700 ppm, it causes mild stinging of the eyes and nose; and at 250 ppm, it produces slight irritation of the eyes and nose. Occupational workers should avoid contact between ethyl silicate and strong oxidisers, water, mineral acids, and alkalis. Workers should use appropriate personal protective clothing and equipment that must be carefully selected, used, and maintained to be effective in preventing skin contact with ethyl silicate. Ethyl silicate should be kept stored in a cool, dry, well-ventilated area in tightly sealed containers that are labelled in accordance with OSHA's Hazard Communication Standard. Containers of ethyl silicate should be protected from physical damage and should be stored separately from strong oxidisers, water, mineral acids, and alkalis. The current OSHA PEL for ethyl silicate is 100 ppm as an 8 h TWA concentration. The NIOSH has established a REL for ethyl silicate of 10 ppm as a TWA for up to a 10 h workday and a 40 h workweek. The ACGIH has assigned ethyl silicate a TLV of 10 ppm as a TWA for a normal 8 h workday and a 40 h workweek.

Also, workers should select appropriate PPE, gloves, sleeves, and encapsulating suits on the basis of the worker's potential exposure to ethyl silicate. There are no published reports on the resistance of various materials to permeation by ethyl silicate. Workers should be careful during use and handling of ethyl silicate and avoid any kind of spill at the work area. Workers should absorb the accidentally spilled material with an inert material and appropriately manage waste disposal. Workers should be aware of the fact that ethyl silicate is a flammable liquid and insoluble in water and must be kept away from heat and all sources of ignition, and a large spill of ethyl silicate needs to be absorbed only with dry earth, sand, and/or other non-combustible materials without contact of water. Workers and others should not touch spilled material, and workplace management must not allow entry into sewers, basements, or confined areas.

Management of Ethyl Silicate: Display Precautions

- DANGER! Flammable liquid and vapour.
- Isolate hazard area and keep away unprotected and unnecessary workers.
- May form explosive mixtures with air; reaction with water releases flammable and toxic vapours.
- Avoid inhalation and eye contact and breathing vapour or mist.
- Improper use by workers causes health effects to the eyes, skin, respiratory tract, and damage to liver, and kidney, and blood.
- Anaesthetic effects in high concentrations.
- Workers MUST USE protective clothing and self-contained breathing apparatus.
- Workers should dispose of waste chemical container and unused contents in accordance with local, national, and international regulations.
- Strictly follow standard industrial hygiene practices.

Bibliography

Agency for Toxic Substances and Disease Registry (ATSDR). 1992. *Toxicological Profile for 1,2-Dibromoethane*. Public Health Service, U.S. Department of Health and Human Services, Atlanta, GA (updated 2007).

Agency for Toxic Substances and Disease Registry (ATSDR). 1996. *Toxicological Profile for Endrin*. U.S. Department of Health and Human Services, Public Health Service, Atlanta, GA (updated 2008).

Agency for Toxic Substances and Disease Registry (ATSDR). 1999. *Managing Hazardous Materials Incidents. Volume III—Medical Management Guidelines for Acute Chemical Exposures: Ethylene Oxide*. U.S. Department of Health and Human Services, Public Health Service, Atlanta, GA (updated 2011).

Agency for Toxic Substances and Disease Registry (ATSDR). 2000. *Toxicological Profile for Ethion*. U.S. Department of Health and Human Services, Public Health Service, Atlanta, GA (updated 2007).

Agency for Toxic Substances and Disease Registry (ATSDR). 2001a. *Toxicological Profile for 1,2-Dichloroethane*. U.S. Department of Health and Human Services, Public Health Service, Atlanta, GA (updated 2011).

Agency for Toxic Substances and Disease Registry (ATSDR). 2001b. *Toxicological Profile for Endosulfan*. U.S. Department of Health and Human Services, Public Health Service. Atlanta, GA (updated 2007).

Agency for Toxic Substances and Disease Registry (ATSDR). 2010. *Toxicological Profile for Ethyl Benzene*. U.S. Department of Health and Human Services, Public Health Service, Atlanta, GA.

Chaubey, R. C., B. R. Kavi, P. S. Chauhan, and K. Sundaram. 1977. Evaluation of the effect of ethanol on the frequency of micronuclei in the bone marrow of Swiss mice. *Mutat. Res.* 43:441–444.

Cheremisinoff, N. P. (ed.). 2003. *Industrial Solvents Handbook*, 2nd edn. Marcel Dekker, Inc. New York.

Department of Pesticide Regulation (DPR). 1991. Illness/injury associated with exposure to EPTC, 1984–1988. Worker Health and Safety Branch, California Department of Pesticide Regulation, Sacramento, CA.

Dikshith, T. S. S. (ed.). 1991. *Toxicology of Pesticides in Animals*. CRC Press, Boca Raton, FL.

Dikshith, T. S. S. (ed.). 2011. *Handbook of Chemicals and Safety*. CRC Press, Boca Raton, FL.

Dikshith, T. S. S. and P. V. Diwan. 2003. *Industrial Guide to Chemical and Drug Safety*. John Wiley & Sons, Hoboken, NJ.

Echobichon, D. J. 1996. Toxic effects of pesticides. In: *Casarett and Doull's Toxicology: The Basic Science of Poisons*, Klaassen, C. D. (ed.), 5th edn. McGraw-Hill, New York.

Environmental Protection Agency (U.S. EPA). Integrated Risk Information System (IRIS). 2011. S-Ethyl dipropylthiocarbamate (EPTC). IRIS. Washington, DC.

Ethylene diamine. IRIS, Washington, DC.

Field, K. J. and C. M. Lang. 1988. Hazards of urethane (ethyl carbamate): A review of the literature. *Lab. Anim.* 22:255–262.

Filser, J. G., P. E. Kreuzer, H. Greim, and H. M. Bolt. 1994. New scientific arguments for regulation of ethylene oxide residues in skin-care products. *Arch. Toxicol.* 68:401–405.

Fisher, A. 1988. Burns of the hands due to ethylene oxide used to sterilize gloves. *Cutis* 42:267–268.

Gardner, M. J., D. Coggon, B. Pannett, and E. C. Harris. 1989. Workers exposed to ethylene oxide: A follow up study. *Br. J. Ind. Med.* 46:860–865.

Gosselin, R. E., R. P. Smith, and H. C. Hodge (eds.). 1984. *Clinical Toxicology of Commercial Products*. Williams & Wilkins, Baltimore, MD.

Hathaway, G. J., N. H. Proctor, J. P. Hughes, and M. L. Fischman. 1991. *Proctor and Hughes' Chemical Hazards of the Workplace*, 3rd edn. Van Nostrand Reinhold, New York.

IARC Monographs. 1988. *Alcohol Drinking*, Vol. 44. Lyon, France.

International Chemical Safety Cards (WHO/IPCS/ILO). 2004. 1.2-Dichloroethane. Occupational Safety & Health Administration (OSHA), Ethylene dichloride. Washington DC.

International Programme on Chemical Safety (IPCS). 1996. Ethylbenzene. Environmental Health criteria No. 186. World Health Organization, Geneva, Switzerland.

International Programme on Chemical Safety and the Commission of the European Communities (IPCS-CEC). 2001. Endrin. ICSC no. 1023. 2001. IPCS, IARC, Geneva, Switzerland.

International Programme on Chemical Safety and the Commission of the European Communities (IPCS–CEC). 1996. S-Ethyl dipropylthiocarbamate (EPTC). International Chemical Safety Card. ICSC Card No. 0469. IPCS–CEC, Geneva, Switzerland.

International Programme on Chemical Safety and the European Commission (IPCS–EC). 2004. Ethion. ICSC Card No. 0888. IPCS, Geneva, Switzerland.

Kidd, H. and D. R. James. (eds.). 1991. *The Agrochemicals Handbook*, 3rd edn. Royal Society of Chemistry Information Services, Cambridge, U.K.

Knaak, J. B., M. Al-Bayati, F. Gielow, and O. G. Raabe. 1986. Percutaneous absorption of EPTC by the young male rat. California Department, Health Services and University of California, Davis, CA.

Korte, A. and G. Obe. 1981. Influence of chronic ethanol uptake and acute acetaldehyde treatment on the chromosomes of bone-marrow cells and peripheral lymphocytes of Chinese hamsters. *Mutat. Res.* 88:389–395.

Lewis, R. J., Sr. 2004. *Sax's Dangerous Properties of Industrial Materials*. Van Nostrand Reinhold Company, New York.

Lynch, D. W., T. T. Lewis et al. 1997. Carcinogenic and toxicologic effects of inhaled ethylene oxide and propylene oxide in F 344 rats. *Toxicol. Appl. Pharmacol.* 76:69–84.

Maier-Bode, H. 1968. Properties, effect, residues, and analytics of the insecticide endosulfan. *Residue Rev.* 22:10–44.

Material Safety Data Sheet (MSDS). 1986. Ethylenediamine. Hazardous substance fact sheet. New Jersey Department of Health (NJDH), Trenton, NJ.

Material Safety Data Sheet (MSDS). 1992. Ethylbenzene. New Jersey Department of Health, Trenton, NJ.

Material Safety Data Sheet (MSDS). 1996a. Endosulfan. Extension Toxicology Network (EXTOXNET). Pesticide Information Profiles (PIP). Oregon State University/USDA/Extension Service, Hall Corvallis, OR.

Material Safety Data Sheets (MSDS). 1996b. Ethion. Extension Toxicology Network (EXTOXNET). Pesticide Information Profiles (PIP). USDA/Extension Service, Oregon State University, OR.

Material Safety Data Sheet (MSDS). 2001. Tetraethyl orthosilicate. MSDS No. T113. Environmental Health and Safety, Mallinckrodt Baker, Inc., Phillipsburg, NJ.

Material Safety Data Sheet (MSDS). 2004. Safety data for ethyl alcohol, absolute. Department Physical and Theoretical Chemistry, Oxford University. Oxford, U.K.

Material Safety Data Sheet (MSDS). 2005a. Safety data for ethyleneimine. Physical Chemistry, Oxford University, Oxford, U.K.

Material Safety Data Sheet (MSDS). 2005b. Safety data for endrin. Physical Chemistry, Oxford University, Oxford, U.K.

Material Safety Data Sheet (MSDS). 2006a. Ethylene dichloride, MSDS No. E4700. Mallinckrodt Baker, Inc., Phillipsburg, NJ.

Material Safety Data Sheet (MSDS). 2006b. Safety data sheet ethylene chlorohydrine. BASF Corporation, Florham Park, NJ.

Material Safety Data Sheet (MSDS). 2007. Safety data for tetraethyl orthosilicate. Department of Physical Chemistry, Oxford University, Oxford, UK.

Material Safety Data Sheet (MSDS). 2008. Safety data for ethylenediaminein. Physical Chemistry, Oxford University, Oxford, U.K.

Material Safety Data Sheet (MSDS). 2009a. Ethyl alcohol completely denatured, MSDS No. E 2012. Environmental Health & Safety (EHS), Mallinckrodt Baker, Inc., Phillipsburg, NJ.

Material Safety Data Sheet (MSDS). 2009b. Safety data for ethylbenzene. Department of Physical Chemistry, Oxford University, Oxford, U.K.

Material Safety Data Sheet (MSDS). 2010. Ethyl alcohol. *NIOSH Pocket Guide to Chemical Hazards*. Centers for Disease Control and Prevention, Atlanta, GA.

Material Safety Data Sheet (MSDS). 2011a. Chemical safety data: Ethyl alcohol. Department of Physical Chemistry, Oxford University, Oxford, U.K.

Material Safety Data Sheet (MSDS). 2011b. Ethyl alcohol. International Chemical Safety Cards ICSC Card No. 0044. International Programme on Chemical Safety and the Commission of the European Communities (C) IPCS–CEC.

McClellan, P. P. 1950. Manufacture and uses of ethylene oxide and ethylene glycol. *Ind. Eng. Chem.* 42:2402–2407.

Meister, R. T. 2004. *Crop Protection Handbook*. Meister Media Worldwide, Willoughby, OH.

National Institute of Occupational Health (NIOH). 2003. Final report of the investigation of unusual illnesses allegedly produced by endosulfan exposure in Padre Village of Kasargod district (North Kerala). National Institute of Occupational Health. Indian Council for Medical Research (ICMR), Ahmadabad, India.

National Institute for Occupational Safety and Health (NIOSH). 1978a. Ethylene dichloride ((1,2-dichloroethane)). Cincinnati, OH (updated 1997).

National Institute for Occupational Safety and Health (NIOSH). 1978b. Ethylene dichloride. *NIOSH Pocket Guide to Chemical Hazards*. NIOSH, Atlanta, GA.

National Institute for Occupational Safety and Health (NIOSH). 1997. *Pocket Guide to Chemical Hazards*. U.S. Department of Health and Human Services, Public Health Service, Centers for Disease Control and Prevention, Cincinnati, OH.

National Institute for Occupational Safety and Health (NIOSH). 2005a. Ethylene oxide. *NIOSH Pocket Guide to Chemical Hazards*. DHHS, NIOSH Publication no. 2005-149, (updated 2008).

National Institute of Occupational Safety and Health (NIOSH). 2005b. Ethion, *Pocket Guide to Chemical Hazards*. NIOSH (No. 2005-149). Centers for Disease Control and Prevention, Atlanta, GA.

National Institute for Occupational Safety and Health (NIOSH). 2010. Endrin. *Pocket Guide to Chemical Hazards*. U.S. Department of Health and Human Services, Atlanta, GA.

National Library of Medicine (NLM). 1992a. Ethyl benzene. Hazardous Substances Data Bank. National Library of Medicine, Bethesda, MD.

National Library of Medicine (NLM). 1992b. Ethyl silicate. Hazardous Substances Data Bank. National Library of Medicine, Bethesda, MD.

National Library of Medicine (NLM). 1992c. Ethylenediamine. Hazardous Substances Data Bank. NLM, Bethesda, MD.

National Library of Medicine (NLM). 1995. Ethylene dibromide. NLM, Bethesda, MD.

Occupational Safety & Health Administration (OSHA). Occupational safety and health guideline for ethyl silicate. OSHA, Washington, DC.

Occupational Safety & Health Administration (OSHA). 1996. Ethyl benzene. Occupational Safety & Health Administration Washington, DC.

Organization for Economic Cooperation and Development (OECD). 2002. 1,2-Dichloroethane. SIDS Initial Assessment Profile for SIAM 15: United Nations Environment Programme (UNEP) Publications (October 22–25, 2002).

Patnaik, P. 1992. Ethylene chlorohydrine. *A Comprehensive Guide to the Hazardous Properties of Chemical Substances*. Van Nostrand Reinhold, New York.

Patnaik, P. (ed.). 2007. *A Comprehensive Guide to the Hazardous Properties of Chemical Substances*. John Wiley & Sons, Hoboken, NJ.

Plotnick, H. B. 1978. Carcinogenesis in rats of combined ethylene dibromide and disulfiram. *JAMA*, 239(16):1609.

Proctor, N. H., J. P. Hughes, and G. J. Hathaway (eds.). 2004. *Proctor and Hughes' Chemical Hazards of the Workplace*. Wiley-Interscience, Van Nostrand Reinhold, New York.

Rangaswamy, S. V. and M. Swaminathan. 1954. Ethylene dibromide: A fumigant for the 1954 food industry. Mysore Cent. *Food Technol. Res. Inst. Bull.* 4:3–4.

Rangaswamy, J. R., N. Vijayashankar, and M. Muthu. 1976. Colorimetric method for the determination of ethylene dibromide residues in grains and air. *J. Assoc. Off. Anal. Chem.* 59:1262–1265.

Rao, U. N., M. Aravindakshan, and P. S. Chauhan. 1994. Studies on the effects of ethanol on dominant lethal mutations in Swiss, C57Bl6 and CBA mice. *Mutat Res.* 311:69–76.

Roberts, D. M., A. Karunarathna, N. A. Buckley, G. Manuweera, M. H. Sheriff, and M. Eddleston. 2003. Influence of pesticide regulation on acute poisoning deaths in Sri Lanka. *Bull. World Health Organ.* 81(11):789–798.

Sax, N. I. and R. J. Lewis. 1989. *Dangerous Properties of Industrial Materials*, 7th edn. Van Nostrand Reinhold, New York.

Sittig, M. (ed.). 1985. *Handbook of Toxic and Hazardous Chemicals and Carcinogens*, 2nd edn. Noyes Publications, Park Ridge, NJ.

Sittig, M. 1991. *Handbook of Toxic and Hazardous Chemicals*, 3rd edn. Noyes Publications, Park Ridge, NJ.

Smith, A. G. 1991. Chlorinated hydrocarbon insecticides. In: *Handbook of Pesticide Toxicology*, Hayes, W. J., Jr. and E. R. Laws Jr. (eds.). Academic Press, New York.

Tates, A. D. 1980. Cytogenetic effects in hepatocytes, bone-marrow cells and blood lymphocytes of rats exposed to ethanol in the drinking water. *Mutat. Res.* 79:285–288.

Tomlin, C. D. S. (ed.). 2006. *The Pesticide Manual—A World Compendium*, 14th edn. British Crop Protection Council (BCPC), Farnham, Surrey, U.K.

U.S. Environmental Protection Agency (U.S. EPA). 1999. 1,2-dichloroethane. Integrated Risk Information System (IRIS). National Center for Environmental Assessment, Office of Research and Development, Washington, DC (updated 2007).

U.S. Environmental Protection Agency (U.S. EPA). 2002. Ethyleneimine(aziridine). Hazard summary, Technology Transfer Network Air Toxics Web Site (updated 2007).

U.S. Environmental Protection Agency (U.S. EPA). 2007. Pesticide: Reregistration, Ethion RED. U.S. EPA, Atlanta, GA.

U.S. Environmental Protection Agency (U.S. EPA). 2011. Integrated Risk Information System (IRIS).

United States Environmental Protection Agency (U.S. EPA). 1983. *Guidance for the Reregistration of Pesticide Products Containing EPTC as the Active Ingredient*. U.S. EPA, Washington, DC.

World Health Organization (WHO). 1986. Principles and methods for the assessment of neurotoxicity associated with exposure to chemicals. Environmental Health Criteria 60. WHO, Geneva, Switzerland.

Zariwala, M. B. A., V. S. Lalitha, and S. V. Bhide. 1991. Carcinogenic potential of Indian alcoholic beverage (country liquor). *Indian J. Exp. Biol.* 29:738–743.

8

Hazardous Chemical Substances:
F

Famotidine
Fenamiphos
Fensulphothion
Fenitrothion
Fenoxycarb
Fenthion
Fenvalerate
Ferbam
Ferrocene
Fipronil
Flocoumafen
Fluorine
Fluoroacetamide
Fluorobenzene
Fluoboric Acid
Fluorosulphuric Acid
Fluvalinate
Folpet
Fonophos
Formaldehyde
Formamide
Formic Acid
Formonitrile
Furan
Furfural

Famotidine (CAS No. 76824-35-6)

Molecular formula: $C_8H_{15}N_7O_2S_3$
Synonyms and trade names: Amfamox; Antodine; Apo-famotidine; Apogastine; Bestidine; Confobos; Dipsin; Dispromil; Fagastine; Famodine; Gaster; Gastridan; Muclox; Pepcid; Pepcidin; Ulceprax; Ulgarine

Famotidine is a competitive histamine H_2-receptor antagonist, and the main pharmaco-dynamic effect of famotidine is to cause the inhibition of gastric secretion. Famotidine on decomposition releases toxic products such as carbon oxides (CO, CO_2), nitrogen oxides

(NO, NO$_2$), and sulphur oxides (SO$_2$, SO$_3$). Famotidine is a medication that is available both in prescription and over-the-counter forms. It is used to treat conditions related to the oesophagus, stomach, and intestines. Some specific famotidine is used for the treatment of duodenal ulcers, gastric ulcers (stomach ulcers), gastroesophageal reflux disease (GERD), and pathological hypersecretory conditions that occur when stomach acid is secreted/produced in very large quantities, an abnormal health condition called 'Zollinger-Ellison syndrome'.

Safe Handling and Precautions

Famotidine, a competitive histamine H$_2$-receptor antagonist, is used to treat gastrointestinal disorders such as gastric or duodenal ulcer, GERD, and pathological hypersecretory conditions. Famotidine inhibits many of the isoenzymes of the hepatic CYP450 enzyme system. Other actions of famotidine include an increase in gastric bacterial flora such as nitrate-reducing organisms. Famotidine binds competitively to H$_2$ receptors located on the basolateral membrane of the parietal cell, blocking histamine affects. This competitive inhibition results in reduced basal and nocturnal gastric acid secretion and a reduction in gastric volume, acidity, and amount of gastric acid released in response to stimuli including food, caffeine, insulin, betazole, or pentagastrin. Reports have indicated that contact of intermediate products associated with famotidine synthesis exposure during synthesis of famotidine causes dermatitis among occupational workers with symptoms such as oedema, pruritus, erythema, and respiratory symptoms. Also, adverse nervous system effects such as headache and dizziness and gastrointestinal (GI) effects such as constipation and diarrhoea occur most frequently during famotidine therapy. It is important to remember that famotidine should be used with caution and the doses and/or frequency of administration decreased in patients with severe renal impairment, since the drug is excreted principally by the kidneys.

Individuals during handling of famotidine should be careful. Workers should keep famotidine away from heat and sources of ignition and avoid exposure to chemical dust or ingest. Workers handling famotidine should wear appropriate personal protective equipment (PPE), face mask, glove, and workplace dress to prevent chemical exposure.

Precautions: Improper use and handling of famotidine has been reported as hazardous. It is combustible at high temperature. Workers should avoid skin contact (irritant), eye contact (irritant), ingestion, and inhalation. Workers should strictly keep famotidine away from heat and away from sources of ignition.

Fenamiphos (CAS No. 22224-92-6)

Molecular formula: **C$_{13}$H$_{22}$NO$_3$PS**
Synonyms and trade names: Nemacur; Namiphos; Phenamiphos; Methaphenamiphos
IUPAC name: Ethyl-4-methylthio-m-tolyl isopropylphosphoramidate
Toxicity class: U.S. EPA: I; WHO: I

Fenamiphos is a colourless crystal or a tan, waxy solid. It is non-corrosive to metals and breaks down readily in strong acids and bases. Fenamiphos is used as a nematicide and an insecticide for use on a wide variety of field, vegetable, and fruit crops. Fenamiphos has been used primarily to control nematodes and thrips on various agricultural crops (i.e., citrus, grapes, peanuts, pineapples, tobacco, etc.) and non-agricultural (i.e., turf and ornamentals) sites. There are no residential uses for fenamiphos and is a Restricted Use Pesticide (RUP) due to high acute toxicity and toxicity to wildlife.

Safe Handling and Precautions

Occupational exposures to fenamiphos cause severe toxicity and adverse health effects. It inhibits the activity of the cholinesterase enzyme in humans leading to the overstimulation of the nervous system. The symptoms of poisoning include, but are not restricted to, nausea, dizziness, confusion, impaired memory, disorientation, severe depression, irritability, headache, speech difficulties, delayed reaction times, nightmares, sleepwalking, and drowsiness or insomnia. Exposures of fenamiphos at very high concentrations, due to accidental ingestion and/or major spillage cause respiratory paralysis and death in workers. Laboratory studies have indicated that teratogenic effects of fenamiphos occurred only at levels that caused overt maternal toxicity and are likely indirect consequences of this toxicity. Other studies have indicated that fenamiphos is non-mutagenic and also have no evidence suggesting that fenamiphos is a carcinogen to animals and humans.

Fenitrothion (CAS No. 122-14-5)

Molecular formula: $C_9H_{12}NO_5PS$
Synonyms and trade names: Accothion; Agrothion; Cytel; Dicofen; Fenstan; Folithion; Metathion; Novathion; Nuvano; Pestroy; Sumanone; and Sumithion
IUPAC name: O,O-dimethyl O-4-nitro-m-tolyl phosphorothioate
Toxicity class: U.S. EPA: II; WHO: II

Pure fenitrothion is a yellowish-brown liquid with an unpleasant odour. It is insoluble in water but readily soluble in common organic solvents such as acetone, alcohol, benzene, chlorinated hydrocarbons, dichloromethane, 2-propanol, toluene, ethers, methanol, and xylene. It decomposes explosively. Fenitrothion is a contact insecticide and selective acaricide of low ovicidal properties. Fenitrothion is effective against a wide range of pests, namely, penetrating, chewing, and sucking insect pests (coffee leaf miners, locusts, rice stem borers, wheat bugs, flour beetles, grain beetles, grain weevils) on cereals, cotton, orchard fruits, rice, vegetables, and forests. It may also be used as a fly, mosquito, and cockroach residual contact spray for farms and public health programmes. Fenitrothion is also effective against household insects and all of the nuisance insects. It belongs to the organophosphate family of insecticides. It is considered a cholinesterase inhibitor. Its effectiveness as a vector control agent for malaria is confirmed by the WHO. Fenitrothion is non-systemic and non-persistent. Fenitrothion was introduced in 1959. Fenitrothion comes in dust, emulsifiable concentrate, flowable, fogging concentrate, granules, ultralow volume (ULV), oil-based liquid spray, and wettable powder formulations. It is compatible with other neutral insecticides.

It is extensively used in other countries, including Japan, where parathion has been banned. Occupational workers get exposed to fenitrothion during mixing, loading/ transportation, and field applications.

Safe Handling and Precautions

Fenitrothion is toxic to animals and humans. Occupational workers, after prolonged period of exposures to high concentrations of fenitrothion, show poisoning. The symptoms include, but are not limited to, general malaise, fatigue, headache, loss of memory and ability to concentrate, anorexia, nausea, thirst, loss of weight, cramps, muscular weakness and tremors, and at sufficiently high dosage, produce typical cholinergic poisoning. The formulation product, sumithion 50EC, causes delayed neurotoxicity in adult rats as well as humans. Male and female rats fed with fenitrothion for as long as 2 years did not show any dose-related increase in tumour incidence. There are no published reports indicating that fenitrothion is a human carcinogen. The acceptable daily intake (ADI) of fenitrothion has been set at 0.003 mg/kg, and the threshold limit value (TLV) has not been established. Occupational workers should handle the technical fenitrothion and its formulations with care and should keep them stored in protected, locked, and well-ventilated area. Fenitrothion and its formulations should not be kept exposed to direct sunlight and should be out of reach of children and unauthorised personnel. Occupational workers should dispose of the formulation-contaminated containers and waste/surplus materials with care and burn in a proper incinerator at high temperatures and according to the set regulations. In short, workers should avoid contaminating the workplace, soil, water, and atmosphere by proper methods of storage, transport, handling, and waste disposal and comply with local and international legislation.

Fenoxycarb (CAS No. 79127-80-3)

Molecular formula: $C_{17}H_{19}NO_4$
Synonyms and trade names: Comply; Insegar; Logic; Pictyl; Torus; and Varikill
IUPAC name: Ethyl 2-(4-phenoxy-phenoxy)-ethyl carbamate
Toxicity class: U.S. EPA; IV

Fenoxycarb is a carbamate insect growth regulator. Fenoxycarb is a yellow granular solid, broad-spectrum insect growth regulator, and non-neurotoxic carbamate. Fenoxycarb is almost insoluble in water but soluble and very soluble in hexane, acetone, chloroform, diethyl ether, and methanol. Fenoxycarb is a general use pesticide (GUP), meaning the user or the pesticide applicator does not need a licence. It is used to control a wide variety of insect pests, such as fire ant bait and for flea and mosquito. It is also used for the control of cockroaches, butterflies, moths, beetles, scale, and sucking insects on olives, vines, cottons, and fruits. It is also used to control these pests on stored products. As a growth regulator, fenoxycarb blocks the ability of an insect to change into the adult stage from the juvenile stage (metamorphosis). Fenoxycarb interferes with larval moulting, the periodic shedding or moulting of the old exoskeleton, and production of a new exoskeleton. Although fenoxycarb is a carbamate insecticide, it exhibits no anti-cholinesterase activity and is thus considered non-neurotoxic. It mimics the action of the juvenile hormones (JHs) on a number of physiological processes, such as moulting and reproduction in insects.

Safe Handling and Precautions

Fenoxycarb is non-neurotoxic and does not have the same mode of action as other carbamate insecticides. Instead, it prevents immature insects from reaching maturity by mimicking JH. Fenoxycarb is practically non-toxic to mammals after oral ingestion. The oral LD50 for rats is greater than 10,000 mg/kg, and the dermal LD50 to the rat is greater than 2,000 mg/kg. Direct application of fenoxycarb on the skin of laboratory rats caused laboured breathing and diarrhoea. Although fenoxycarb does not irritate the skin, it is an eye irritant. The liver is the primary organ affected by fenoxycarb in long-term animal studies. Prolonged period of oral exposures to rats and dogs with very low doses of fenoxycarb caused no health effects, while high concentrations caused adverse effects to the liver of rats, mice, and dogs. Reports on the teratogenicity and mutagenicity of fenoxycarb are not available in literature. The observations of the U.S. EPA peer review and the observations of the Ciba-Geigy Corporation's cancer risk calculations have indicated that fenoxycarb is not a human carcinogen. However, results of animal toxicology lifetime feeding studies indicated that fenoxycarb was carcinogenic at high dose levels to male mice with regard to an increase in lung and harderian gland tumours. No significant increase in carcinogenic effects was observed in either female mice or the rat lifetime feeding study. There are no other reports on the carcinogenicity of fenoxycarb to animals and humans. According to the State of California report, fenoxycarb (Award fire ant bait) is known to cause cancer (EPA signal word: Caution). Also, the International Agency for Research on Cancer (IARC), the U.S. National Toxicology Program (U.S. NTP), and the U.S. Toxic Inventory Program (TRI) have not listed fenoxycarb as a carcinogen, while the U.S. EPA reports that fenoxycarb is a likely carcinogen, and the California Prop 65 reports that fenoxycarb is a known carcinogen.

Fensulphothion (CAS No. 115-90-2)

Molecular formula: $C_{11}H_{17}O_4PS_2$
Synonyms and trade names: O,O-diethyl O-(p-methylsulfonyl)phenyl)phosphorothioate, Dasanit®, Terracur

Fensulphothion is a brown liquid or yellow oily chemical substance. It is a combustible chemical substance. The liquid formulations containing organic solvents are flammable and release irritating or toxic fumes (or gases) in a fire.

Safe Handling and Precautions

Occupational workers on exposure to fensulphothion develop poisoning. The symptoms of poisoning include, but are not restricted to, pupillary constriction, muscle cramp, muscle fasciculation, excessive salivation, headache, nausea, vomiting, abdominal cramps, diarrhoea, sweating, dizziness, lassitude (weakness, exhaustion), rhinorrhea (discharge of thin nasal mucus), chest tightness, cardiac irregularity, blurred vision, miosis, dyspnoea (breathing difficulty), convulsions, and unconsciousness. Repeated and short-term exposure to fensulphothion causes adverse health effects on the nervous system, resulting in convulsions and respiratory failure.

During use and handling of fensulphothion, workers should use protective gloves, protective clothing, face shield, or eye protection in combination with breathing protection.

Also, workers should avoid to eat, drink, and smoke during work and wash hands before eating and other activities.

Fenthion (CAS No. 55-38-9)

Molecular formula: $C_{10}H_{15}O_3PS_2$
Synonyms and trade names: Lebaycid; Mercaptophos; Prentox; Queletox; Spotton; Talodex
IUPAC name: O,O-dimethyl O-4-methylthio-*m*-tolyl phosphorothioate
Toxicity class: U.S. EPA: II; WHO: II

Pure fenthion is a colourless liquid. Technical fenthion is a yellow or brown oily liquid with a weak garlic odour. It is insoluble or very sparingly soluble in water but soluble in all organic solvents, alcohols, ethers, esters, halogenated aromatics, and petroleum ethers. It is grouped by U.S. EPA under RUP and hence requires handling by qualified, certified, and trained workers. Fenthion is used for the control of sucking and biting pests, for instance, fruit flies, stem borers, mosquitoes, and cereal bugs. In mosquitoes, it is toxic to both the adult and immature forms (larvae). The formulations of fenthion include dust, emulsifiable concentrate, granular, liquid concentrate, spray concentrate, ULV, and wettable powder.

Safe Handling and Precautions

Fenthion is moderately toxic to mammals and highly toxic to birds. Exposures to fenthion cause poisoning with symptoms among occupational workers as observed with organophosphate pesticide-induced toxicity. These include, but are not limited to, numbness, tingling sensations, incoordination, headache, dizziness, tremor, nausea, abdominal cramps, sweating, blurred vision, difficulty breathing or respiratory depression, and slow heartbeat. Very high doses may result in unconsciousness, incontinence, and convulsions or fatality. Reports have indicated that exposures to fenthion cause adverse effects on the central and peripheral nervous systems and heart of exposed workers. Reports have indicated that evidences are neither sufficient nor adequate to draw conclusions on the mutagenicity, teratogenicity, and carcinogenicity of fenthion on animals and humans and to classify fenthion as a human carcinogen.

Fenvalerate (CAS No. 51630-58-1)

Molecular formula: $C_{25}H_{22}ClNO_3$
Chemical name: α-Cyano-3-phenoxybenzyl-2-(4-chlorophenyl)-isovalerate

Fenvalerate is a potent insecticide that has been in use since 1976. It is an ester of 2-(4-chlorophenyl)-3-methylbutyric acid and alpha-cyano-3-phenoxybenzyl alcohol but lacks a cyclopropane ring. However, in terms of its insecticidal behaviour, it belongs to the pyrethroid insecticides. Technical grade fenvalerate is a yellow or brown viscous liquid having a specific gravity of 1.175 at 25°C. The vapour pressure is 0.037 mPa at 25°C, and it is relatively non-volatile. It is practically insoluble in water (approximately 2 µg/L) but soluble in organic solvents such as acetone, xylene, and kerosene. It is stable to light, heat, and moisture but unstable in alkaline media.

Safe Handling and Precautions

As a pyrethroid insecticide, fenvalerate has been reported to cause moderate toxicity to humans. On exposure at workplaces, fenvalerate gets absorbed from the GI tract and from the lungs, and through the skin. Occupational workers and individuals on exposure to fenvalerate develop adverse health effects. The symptoms of poisoning include, but are not limited to, numbness, itching, tingling, cough, dizziness, headache, nausea, abdominal pain, burning sensations, and respiratory tract irritation. The symptoms in exposed workers develop after a latent period of approximately 30 min and disappear within 24 h. Repeated exposure to high concentrations of fenvalerate causes effects on heart, GI tract, skin, eyes, head, and muscle tissue. Some poisoning cases have resulted from occupational exposure because of the negligence of safety precautions and overexposure. The exposure of the general population to fenvalerate is expected to be very low. It is not likely to present a hazard provided it is used as recommended.

In brief, with reasonable work practices, hygiene measures, and safety precautions, fenvalerate is unlikely to present a hazard to those occupationally exposed to it. Occupational workers during use and handling, mixing of fenvalerate formulations, etc., should use protective workplace dress such as impermeable boots, clean overalls, gloves, and a suitable respirator to avoid spray mist.

Ferbam (CAS No. 14484-64-1)

Molecular formula: $C_9H_{18}FeN_3S_6$
Synonyms and trade names: Aafertis; Bercema fertam 50; Ferbam 50; iron salt; Fermate; Ferradow; Ferric dimethyldithiocarbamate; Hexaferb; Trifungol; Tris (dimethyldithiocarbamato)iron; Vancide FE-95

Ferbam is a carbamate fungicide. It is a stable, black powder and combustible and gives off irritating or toxic fumes (or gases) in a fire. Ferbam is incompatible with strong oxidants. It is used for foliar protectant against scab, rust, mould, and many fungal diseases on fruits, vegetables, melons, and ornamentals. Also, it works as a repellent towards Japanese beetles. The major uses of ferbam are in the control of apple scab and cedar apple rust, peach leaf curl, tobacco blue mould, and cranberry diseases. Ferbam is a broad-spectrum registered fungicide for the control of certain diseases in fruit trees, small fruit and berry crops, potatoes, ornamentals, conifers, and tobacco and for use on citrus, pome, stone fruits, and cranberries. Major areas of use include Florida, Massachusetts, New Jersey, and other countries. Ferbam is used as water-dispersible granule, 76WG.

Safe Handling and Precautions

Occupational workers on exposure to ferbam develop adverse health effects. The symptoms of poisoning include irritation of the eyes, the skin, and the respiratory tract. The substance may cause effects on the central nervous system (CNS). Repeated or prolonged contact of ferbam with skin is known to cause dermatitis, skin sensitisation, and adverse effects on the nervous system and thyroid. In acute studies, ferbam has low acute toxicity (toxicity category III) via the oral and dermal routes and moderate (toxicity category II) acute toxicity via the inhalation route of exposure. Exposures to ferbam cause mild health disorders such as irritation, slight eye and skin irritation, weak sensitisation, confusion, drowsiness, headache, and nausea. Repeated or prolonged contact may cause skin sensitisation. The substance may have effects on the nervous system and thyroid in high doses.

A substantially complete database has been assembled for ferbam as a result of bridging of data from the ziram and thiram databases. In longer-term studies, ferbam is toxic to the liver, kidneys, and lungs. There were no tumour effects observed in the ferbam studies; therefore, no cancer assessment was done. Based on information from the developmental neurotoxicity (DNT) study, the regulatory agency concluded that ferbam requires no DNT study. The U.S. EPA has determined that all products containing ferbam as the active ingredient are eligible for reregistration, provided changes specified in the ferbam reregistration eligibility decisions (RED) are incorporated into the label and additional data identified in Section V of the RED confirm this conclusion.

Ferbam has been tested by oral administration in mice and rats and by single subcutaneous injection in mice. Although no carcinogenic effect was observed in these tests, the available data are insufficient for an evaluation of the carcinogenicity of this compound to be made. However, ferbam can react with nitrite under mildly acid conditions, simulating those in the human stomach, to form *N*-nitrosodimethylamine, which has been shown to be carcinogenic in seven animal species.

Ferrocene (CAS No. 102-54-5)

Molecular formula: $C_{10}H_{10}Fe$
Synonyms and trade names: Dicyclopentadienyl iron; Bis(cyclopentadienyl)iron; Catane; Di-2,4-cyclopentadien-1-yliron; Ferrotsen; Iron bis(cyclopentadiene); Iron dicyclopentadienyl

Ferrocene has application as a catalyst, as anti-knock additive in gasoline, and as combustion supporting catalyst for solid propellants. It is a stable, highly flammable solid and heat-sensitive chemical substance. It is incompatible with strong oxidising agents.

Safe Handling and Precautions

Exposure to ferrocene causes harmful effects, very hazardous when swallowed, ingested, or inhaled. Ferrocene is very hazardous in case of ingestion. Inhalation of ferrocene is toxic to lungs and mucous membranes, and reports have indicated that repeated or prolonged exposure to ferrocene produces blood abnormalities and causes kidney and liver damage, delayed pulmonary oedema, and lung damage. Reports have indicated that the toxicological properties of ferrocene have not been fully investigated, and hence users should not be negligent during its handling and management.

Students and occupational workers during use and handling of ferrocene should be careful to use personal precautions; protective equipment to avoid breathing of vapours, mist, or gas; and flame-retardant anti-static protective clothing and strictly observe emergency procedures and good industrial hygiene and safety practice. Ferrocene is an organometallic chemical and reacts violently with tetranitromethane. Students and occupational workers should avoid/minimise generation, formation, and accumulation of dust at the workplace and avoid breathing vapours, mist, or gas. The workplace should be free from all sources of ignition and smoking and should have appropriate and adequate exhaust ventilation. Workers should keep the container of ferrocene tightly closed in a cool, dry, and well-ventilated place.

Fipronil (CAS No. 120068-37-3)

Molecular formula: $C_{12}H_4Cl_{12}F_6N_4OS$
Synonyms and trade names: 5-Amino-1-(2,6-dichloro-4-(trifluoromethyl)phenyl)-4-((trifluoromethyl)sulfinyl)-1H-pyrazole-3-carbonitrile, Cockroach gel
IUPAC name: 5-Amino-1-(2,6-dichloro-α,α,α-trifluoro-p-tolyl)-4-(trifluoromethyl)sulfinyl] pyrazole-3-carbonitrile

Fipronil is a white powder with a mouldy odour. It has a low solubility in water and is a slow-acting poison. It does not bind strongly with soil, and the half-life of fipronil–sulphone is 34 days. Fipronil is a broadspectrum insecticide of the phenylpyrazole group. Fipronil was first used extensively for the control of ants, beetles, cockroaches, fleas; ticks, termites, mole crickets, thrips, rootworms, weevils, flea of pets, field pest of corn, golf courses, and commercial turf, and other insects. Fipronil was first used in the United States in 1996. Fipronil is a topical insecticide. It kills adult fleas and larvae, ticks, and chewing lice. As a liquid, fipronil is applied on the back side skin of dogs and cats. Reports indicate that some animals do become hypersensitive (allergic) to fipronil. Fipronil was first used to control registered for use in the United States in 1996.

Safe Handling and Precautions

Fipronil is a broad-spectrum insecticide that disrupts the insect's CNS by blocking the passage of chloride ions through the GABA receptor and glutamate-gated chloride (GluCl) channels, components of the CNS. This causes hyperexcitation of contaminated insects' nerves and muscles. Reports have indicated that fipronil is more toxic to insects than humans and pets because it is more likely to bind to insect nerve endings.

Fipronil has been found to cause disruption of the normal function of the CNS in insects. Users and occupational workers when exposed to fipronil by inhalation, ingestion of contaminated food, and/or through skin absorption at workplaces showed adverse health effects. The symptoms of health disorders included, but are not limited to, restlessness, anxiety, sweating, nausea, vomiting, headache, stomach pain, dizziness, weakness, seizures, and tremor, and prolonged exposure is known to cause serious damage to health. While the American Conference of Governmental Industrial Hygienists (ACGIH), IARC, and NTP have not listed fipronil as a human carcinogen, the U.S. EPA classified fipronil as a possible human carcinogen.

Fipronil when used and handled according to prescribed instructions causes no hazardous reactions. It is highly toxic to fish and aquatic invertebrates. Its tendency to bind to sediments and its low water solubility may reduce the potential hazard to aquatic wildlife. It is toxic to bees and should not be applied to vegetation when bees are foraging. Fipronil has been found to be highly toxic to upland game birds. It is very toxic to aquatic organisms, to fauna, and to bees and causes long-term damage to the environment.

Flocoumafen (CAS No. 90035-08-9)

Molecular formula: $C_{33}H_{25}F_3O_4$

Synonyms and trade names: Amfamox; Antodine; Apo-famotidine; Apogastine; Bestidine; Confobos; Dipsin; Dispromil; Fagastine; Famodine; Gaster; Gastridan; Muclox; Pepcid; Pepcidin; Storm; Stratagem; WL-108366; Anti-coagulant rodenticide; Ulceprax; Ulgarine

IUPAC name: 2-Hydroxy-3-[3-[4-([4-(trifluoromethyl) phenyl] methoxy) phenyl]-1,2,3,4-tetrahydronaphthalen-1-yl] chromen- 4-one

Flocoumafen is a second-generation anti-coagulant used as a rodenticide. It has a very high toxicity and is restricted to indoor use and sewers (in the United Kingdom). This restriction is mainly due to the increased risk to non-target species. Studies have shown that rodents resistant to first-generation anti-coagulants can be adequately controlled with flocoumafen. The chemical substance is off-white solid, does not mix well with water (1.1 mg/L), and is sparingly soluble in acetone, ethanol, xylene, and octanol. Flocoumafen is stable to hydrolysis. Because of the acute toxicity of flocoumafen and its intended use as a rodenticide, chronic toxicity studies have not been reported. However, flocoumafen is known to cause adverse health effects and abnormal prothrombin.

Safe Handling and Precautions

Flocoumafen is a highly dangerous chemical substance and a poison. On exposure to flocoumafen, workers suffer with poisoning, and the symptoms include nausea, dizziness, drowsiness, vomiting, bleeding gums, easy bruising, blood in the urine, and excessive bleeding from minor wounds, shock, or collapse. Severe poisonings can cause shock, coma, and death. The effects of poisoning may be delayed for days. Flocoumafen causes effects on the blood and impairment of blood clotting. During use and handling of flocoumafen, workers should wear self-contained breathing apparatus and chemical-protective clothing. In case of fire and/or explosion, do not breathe fumes. Persons and occupational workers with impaired respiratory function, airway diseases, and conditions such as emphysema or chronic bronchitis may incur further disability if excessive concentrations of particulate are inhaled. The major symptom of poisoning is breathlessness. Repeated exposure to some coumarin derivatives may cause nosebleed, bleeding gut and pharynx, dark-red bleeding spots, widespread bruising, blood swelling, and blood in the phlegm, vomitus, urine, or stools. Bleeding into the organs, digestive tract, joints, and abdomen can cause localised pain. Exposure at work can cause anaemia with weakness, pallor, and shock. The lung shadows show on x-ray. Many coumarins cause mutations and cancer, and coumarins also inhibit tumour production by carcinogens and inhibit metastasis.

Workers should keep containers cool by spraying with water if exposed to fire, should collect contaminated extinguishing water separately, and should not allow to reach sewage or effluent systems; the fire debris and contaminated extinguishing water should be disposed of in accordance with official regulations and also should not be discharged into drains/surface waters or groundwater.

Workers during handling of flocoumafen should avoid any kind of contact of the chemical with the skin, eyes, and clothing. Workers should use protective workplace clothe/dress and remove contaminated clothes, undergarments, and shoes immediately.

Workers should have breathing protection to avoid solid and liquid particles whenever the ventilation is inadequate.

Flocoumafen should be used by or in accordance with directions of accredited pest control managers who have had training and are aware of the procedures for safe use of the chemical substance. Flocoumafen should be stored in original containers and securely sealed and kept in a cool, dry, well-ventilated area, away from incompatible materials and foodstuff containers. The containers should be protected against physical damage and should be checked regularly for leaks.

Fluorine (CAS No. 7782-41-4)

Molecular formula: F_2
Sodium fluoride (CAS No. 7681-49-4); Hydrogen fluoride (CAS No. 7664-39-3)

Fluorine is pale yellow gas with a pungent odour. It is a stable, extremely strong oxidant, which may react violently with combustible materials, including plastics, reducing agents, and organic material. It reacts with water to form corrosive acids. Fluorine is very toxic and may be fatal if inhaled. Exposure to fluorine causes severe burns and serious damage to eyes, skin, and respiratory system. Fluorine reacts violently with many oxidising agents (e.g. perchlorates, peroxides, permanganates, chlorates, nitrates, chlorine, bromine, and fluorine), strong acids (hydrochloric, sulphuric, and nitric), organic compounds, combustible materials like oil and paper, hydrogen, bromine, iodine, and chemically active metals like, potassium, sodium, magnesium, and zinc.

Safe Handling and Precautions

Exposure to fluorine and its contact cause severe eye and skin irritation and burns and lead to permanent eye damage, nosebleeds, nausea, vomiting, loss of appetite, diarrhoea, and constipation. Fluorine gas is corrosive to all exposed tissues including upper and lower respiratory tracts, eyes, nose, and any other exposed mucous membranes. Breathing of fluorine irritates the nose, throat, and lungs and causes coughing and shortness of breath. Repeated exposure to higher concentration of fluorine causes pulmonary oedema with a build-up of fluid in the lungs and severe shortness of breath. People exposed to fluorine for prolonged period of time suffer with bone pain, fractures, osteosclerosis, mottled teeth, and damage to the liver and kidneys.

Much before use, handling, and working with fluorine, workers should have adequate knowledge and training of proper handling and storage of the chemical substance. Workers during handling of fluorine fumes, gases, or vapours must be very careful and should wear nonvented, impact-resistant goggles. Workers should keep stored fluorine in tightly closed containers in a cool, well-ventilated area away from water and steam as ozone and hydrofluoric acid are produced.

Users and occupational workers during handling of fluorine should be very careful and observe strict methods of chemical management. Workers must use safety glasses and gloves (medium or heavyweight Viton, nitrile, or natural rubber gloves) and work only in a thoroughly, well-ventilated area and should not use open laboratory. Workers should use fluorine after the completion of a full risk assessment has been made and must confirm absence of oil or grease in the gas-handling system to avoid the violent chemical reaction or explosion. Fluorine gas reacts with H_2, and on release into the air, it readily becomes hydrogen fluoride (HF). It is lethal at very low levels, and the immediately dangerous to life or health (IDLH) concentration is 25 ppm.

Alert! Fluorine and Dangers

Fluorine is a very reactive non-flammable gas or liquid, which will enhance combustion of other materials. If possible, stop flow of gas to fire or remove cylinders from fire area.

Fluorine reacts violently with many materials, many at room temperature, and decomposes to hydrofluoric acid on contact with moisture. It is the most powerful oxidiser known. It reacts with virtually all inorganic and organic substances. Fluorine ignites on contact with ammonia, ceramic materials, phosphorus, sulphur, copper wire, acetone, and many other organic and inorganic compounds.

Containers of fluorine may explode in fire, and poisonous gases are produced in fire. Teflon is the preferred gasket material when working with fluorine gas. Workers should keep the equipment scrupulously dry and store and use only in vented gas storage cabinets or fume hoods. The reaction between metals and fluorine is relatively slow at room temperature but becomes vigorous and self-sustaining if the temperature is elevated. Process valves should be opened and closed with remote-controlled extensions passing through a suitable barricade for additional protection. Workers should use double-valve tools to facilitate the reduction in pressure from high-pressure sources of fluorine. When fluorine gas is leaked at the workplace, (i) immediately evacuate persons not wearing protective equipment from the area of leak, (ii) clean up the area completely, and (iii) ventilate the area of leak to disperse the gas. HF is very corrosive and rapidly damages tissue, workplace, and according to regulations (OSHA) an eyewash and shower should be made nearby accessible by the workplace management.

Fluoroacetamide (CAS No. 640-19-7)

Molecular formula: C_2H_4FNO
Synonyms and trade names: Acetamide, 2-fluoro-; alpha-fluoroacetamide; FAA; Fluorakil 100; 2-Fluoroacetamide; Fluoroacetic acid amide; Flutritex 1; Fussol; Megatox; Compound 1081; Monofluoroacetamide; Navron; Rodex; Yanock
Chemical name: Fluoroacetamide

Fluoroacetamide is an odourless, tasteless, white, crystalline water soluble. It is a fluorinated amide and reacts with azo and diazo compounds to generate toxic gases. Flammable gases are formed by the reaction of organic amides/imides with strong reducing agents. Fluoroacetamide is used as a rodenticide and insecticide to control rabbits but is dangerous for other species of pests, farm livestock, and humans. Fluoroacetamide is a noncombustible substance itself and does not burn but may decompose upon heating to produce irritating, corrosive, and/or toxic fumes. On decomposition, fluoroacetamide releases nitrogen oxides, carbon monoxide, irritating and toxic fumes and gases, carbon dioxide, HF gas, and nitrogen gas.

Safe Handling and Precautions

Exposure to fluoroacetamide is harmful and causes poisoning. The potential health effects include, but are not limited to, eye irritation, skin irritation, gastrointestinal irritation with nausea, respiratory tract irritation, vomiting and diarrhoea, alopecia (loss of hair), convulsions, coma, and fatal injury on accidental ingestion. Human fatalities have been reported from acute poisoning. The signs of poisoning in dogs include extreme excitation, hyperirritability, crazy running, and tonic–clonic convulsions. In horses and ruminants, there are no signs of nervous excitation, with death occurring as a result of cardiac failure manifested by tachycardia and cardiac arrhythmia. Fluoroacetamide is super toxic, and the probable oral lethal dose in humans is less than 5 mg/kg or a taste (less than 7 drops) for a 150 lb. person. Limited evidence suggests that repeated or long-term occupational exposure to fluoroacetamide produces cumulative health effects involving organs or biochemical systems. Long-term exposure to high dust concentrations causes changes in lung function, that is, pneumoconiosis caused by particles less than 0.5 μm penetrating and remaining in the lung. Chronic experience of fluoroacetamide is thought to resemble that of workers exposed to fluoroacetates (the metabolite of fluoroacetamide). Fluoroacetamide inhibits oxygen metabolism by cells with critical damage occurring to the heart, brain, and lungs resulting in heart failure, respiratory arrest, convulsions, and death.

Occupational workers should be very careful during use and handling of fluoroacetamide. Workers should avoid direct contact of fluoroacetamide with strong oxidising chemical substances/agents and strong acids, and the workplace should have proper exhaust ventilation facilities to avoid the formation or spread of chemical dust in the air. Workers should handle the chemical substance only used in fume hood, and fluoroacetamide should be kept in a cool, dry, well-ventilated area away from incompatible substances and the poison room locked. During use and handling of fluoroacetamide, workers should wear appropriate and standard protective eyeglasses or chemical safety goggles/eye and face protection equipment, wear appropriate protective gloves to prevent skin exposure, and follow standard respiratory protection requirements. Proper chemical management of fluoroacetamide is necessary to protect the health and safety of the workers and the surrounding communities and the environment. Occupational workers should avoid run-off of fluoroacetamide wastes into storm sewers and ditches, which lead to waterways. Spills of fluoroacetamide should be cleaned up immediately, observing precautions in the Protective Equipment section, swept up, then placed into a suitable container for disposal.

Fluorobenzene (CAS No. 462-06-6)

Molecular formula: C_6H_5F
Synonyms and trade names: Monofluorobenzene; Phenyl fluoride

Fluorobenzene is a colourless, highly flammable liquid, stable and incompatible with oxidising agents. It is not compatible with oxidising agents such as perchlorates, peroxides, permanganates, chlorates, nitrates, chlorine, bromine and fluorine, ammonium nitrate, chromic acid, halogens, and nitric acid. It is used as an insecticide and as a reagent for plastic and resin polymers.

Safe Handling and Precautions

Exposure to fluorobenzene is known to cause adverse health effects to students and occupational workers. The symptoms of toxicity and poisoning include, but are not limited to, nausea, headache, vomiting, CNS depression, and damage to the liver and kidneys. Repeated exposure to fluorobenzene is known to damage the lungs and affect the nervous system. In fact, literature on the chemical, physical, and toxicological properties of fluorobenzene is limited and has not been investigated completely. Students and occupational worker during use and handling of fluorobenzene should avoid breathing vapours, mist, or gas; should have adequate workplace ventilation; and should use personal protective equipment (PPE). During handling of fluorobenzene and in storage, workers should avoid all sources of ignition because the chemical vapours accumulate in low areas and form explosive concentrations. Also, workers much before use and handling of fluorobenzene should have knowledge of the chemical substance and its proper handling and storage and training. Workers should keep stored fluorobenzene in tightly closed containers in a cool, well-ventilated area and away from sources of ignition, smoking, and open flames. Metal containers involving the transfer of more of fluorobenzene are properly handled and should be grounded and bonded. Drums must be equipped with self-closing valves, pressure vacuum bungs, and flame arresters. Workers when opening and closing containers of fluorobenzene should use only nonsparking tools and equipment.

Fluoboric Acid (CAS No. 16872-11-0)

Molecular formula: HBF_4
Synonyms and trade names: Tetrafluoroboric acid; Tetrafluorohydrogen borate; Borofluoric acid; Fluoboric acid

Fluoboric acid is a colourless liquid, completely solvable in water. Fluoboric acid is a stable chemical substance and extremely reactive or incompatible with strong bases, cyanides. It is very corrosive in the presence of steel, aluminium, zinc, and copper and highly corrosive in the presence of glass and stainless steel. Fluoboric acid is used as a catalyst for alkylations and polymerisations.

Safe Handling and Precautions

Students and occupational workers whenever exposed to fluoboric acid with negligence suffer with adverse health effects. Fluoboric acid is corrosive and extremely hazardous, and in case of skin contact, eye contact (irritant), ingestion, and inhalation, it causes severe irritation or burns. Liquid or spray mist of fluoboric acid is known to produce tissue damage particularly on mucous membranes of eyes, mouth, and respiratory tract. Skin contact with fluoboric acid produces burns. Inhalation of the spray mist of fluoboric acid produces severe irritation of respiratory tract, characterised by coughing, choking, or shortness of breath. Inflammation of the eye is characterised by redness, watering, and itching and skin inflammation with symptoms of itching, scaling, reddening, and/or blistering. Exposures to fluoboric acid are toxic to lungs and mucous membranes. Repeated or prolonged exposure to spray/mist of fluoboric acid has been reported to produce respiratory tract irritation leading to frequent attacks of bronchial infection and chronic respiratory irritation.

Workers should be careful and should completely avoid any kind of negligence during use and handling of fluoboric acid. The spilled chemical is very corrosive and should not be touched without proper workplace safety. Workers should not be negligent at the workplace and use full workplace dress/suit with gloves, face shield, splash goggles, boots, and approved/certified respirator/self-contained breathing apparatus to avoid inhalation of the toxic chemical substance. Occupational workers and workplace management should control the spillage/leak of fluoboric acid with dry earth, sand, or other non-combustible material/absorption. Workers, should completely avoid water and prevent the possible entry of chemical waste into sewers, basements, or confined areas. Fluoboric acid is corrosive and corrodes, glass, and metallic surfaces. Workers should store fluoboric acid in a metallic or coated fibreboard drum using a strong polyethylene inner package and appropriate container. Fluoboric acid should be kept in a separate safety storage cabinet/room.

Fluorosulphuric Acid (CAS No. 7789-21-1)

Molecular formula: **FHSO$_3$**
Synonyms and trade names: Fluorosulphuric acid; F fluosulfonic acid

Fluorosulphuric acid (FSO$_3$H) is a free-flowing colourless fuming liquid. It is soluble in polar organic solvents such as acetic acid, ethyl acetate, and nitrobenzene but poorly soluble in nonpolar solvents such as alkanes. FSO$_3$H is corrosive to metals and body tissues, corrodes glass, and is incompatible with many metals and bases. With its strong acidity, FSO$_3$H dissolves almost all organic compounds that are even weak proton acceptors. FSO$_3$H hydrolyses slowly to HF and sulphuric acid (H$_2$SO$_4$). The related triflic acid (CF$_3$SO$_3$H) retains the high acidity of FSO$_3$H but is more hydrolytically stable. FSO$_3$H is one of the strongest known simple Bronsted acids. FSO$_3$H is useful for regenerating mixtures of HF and H$_2$SO$_4$ for etching lead glass. FSO$_3$H isomerises alkanes and the alkylation of hydrocarbons with alkenes, although it is unclear if such applications are of commercial importance. FSO$_3$H is used as a catalyst in organic synthesis and in electroplating and as a fluorinating agent and as a laboratory fluorinating agent. FSO$_3$H has been reported as a highly toxic and corrosive chemical substance. FSO$_3$H is non-combustible but enhances combustion of other substances. It gives off irritating or toxic fumes (or gases) in a fire. FSO$_3$H hydrolyses to release HF. Addition of water to FSO$_3$H causes very violent reactions similar to the addition of water to H$_2$SO$_4$. Addition of FSO$_3$H to water is a much more violent process than addition of H$_2$SO$_4$. The combination of FSO$_3$H and others is categorised as 'magic acids and super acids'. Very short contact and the fumes of FSO$_3$H cause severe painful burns.

Safe Handling and Precautions

FSO$_3$H is considered to be highly toxic and corrosive to the eyes, skin, and the respiratory tract. On contact and exposures, FSO$_3$H causes poisoning with symptoms such as cough, burning sensation, sore throat, shortness of breath, nausea, vomiting, abdominal cramps, breathing distress, shock, and/or collapse. Inhalation of vapour or aerosols of FSO_3H is known to cause severe lung oedema.

Alert!! *Students and workers during use and handling of fluorosulphuric acid should have the advice of experts and safety managers and use complete protective clothing including self-contained breathing apparatus.* In case of accidental workplace chemical spill, workers *SHOULD NOT ABSORB the chemical waste with sawdust or other combustible absorbents* and *workers should NEVER direct water jet on liquid towards the chemical leakage* but collect leaking liquid in sealable plastic containers, absorb the remaining waste liquid in sand/inert absorbent, and remove to a safe place.

Fluvalinate (CAS No. 102851-06-9)

Molecular formula: $C_{26}H_{22}CIF_3N_2O_3$
Synonyms and trade names: Apistan; Klartan; Mavrik; Mavrik aqua flow; Spur; Taufluvalinate; Yardex
IUPAC name: (RS)-alpha-cyano-3-phenoxybenzyl N-(2-chloro-a,a,a-trifluoro-p-tolyl)-D-valinate

Fluvalinate is a viscous, yellow oil in appearance. It is very soluble in organic solvents and aromatic hydrocarbons and insoluble in water. It is a synthetic pyrethroid. It is used as a broad-spectrum insecticide against moths, beetles, fleas, turf and ornamental insects, and other insect pests on cotton, cereal, grape, potato, fruit tree, vegetable, and plantation crops. It is available in emulsifiable concentrates, suspensions, and flowable formulations. Fluvalinate is a moderately toxic compound in the U.S. EPA toxicity class II. Some formulations may have the capacity to cause corrosion of the eyes. Pesticides containing fluvalinate must bear the signal word DANGER on the product label. Fluvalinate is classified as an RUP because of its high toxicity to fish and aquatic invertebrates.

Safe Handling and Precautions

Exposures to fluvalinate cause coughing, sneezing, throat irritation, itching or burning sensations on the arms or face with or without a rash, headache, and nausea. Inhalation of fluvalinate causes severe irritation to the respiratory tract. Symptoms of overexposure include headache, dizziness, nausea, shortness of breath, coughing, insomnia, diarrhoea, gastrointestinal disturbances, and back pain with urinary frequency. Reports have indicated that prolonged and severe exposure to fluvalinate is known to cause liver and kidney damage and may be fatal to exposed workers.

Fluvalinate is slightly toxic to birds. Exposure of laboratory animals exposed to fluvalinate showed no tumours. No tumours were observed in mice given doses of up to 20 mg/kg/day, nor in rats given doses as high as 2.5 mg/kg/day for over 2 years. Exposure limits: For fluvalinate, no exposure limits such as acceptable daily intake (ADI), permissible exposure limits (PEL), or TLV are available in literature. Occupational workers should keep stored fluvalinate only in the original container at a temperature not exceeding 40°C and away from food, drink, and animal feeding stuffs. Occupational workers should avoid contact of fluvalinate with skin and eyes and avoid to breathe contaminated fumes. Wear suitable protective clothing, nitrile rubber gloves, safety glasses, or face shield. Fluvalinate is an RUP and should be used and handled and/or purchased only by certified applicators. Occupational workers should keep fluvalinate out of reach of children and avoid skin and eye contact and breathing in vapour, mists, and aerosols. Fluvalinate should be stored in the closed, original container in a dry, cool, well-ventilated area out of direct sunlight, away from foodstuffs, and the chemical containers should be kept closed when not in use with regular checks to avoid leakage of the chemical substance.

Folpet (CAS No. 133-07-3)

Molecular formula: $C_9H_4Cl_3NO_2S$
Chemical name: N-[(trichloromethyl)thio]phthalimide
Synonyms and trade names: Cosan T; Faltan; Folnit; Folpan; Folpel; Folpex; Ftalan; Fungitrol 11; Intercide TMP; Orthoraltan 50; Orthophaltan; Phaltan; Phthaltan; Sanfol; Spolacid; Trifol; Vinicoil
Mixed formulations: (Folpet +) aluminium phosethyl; formulation types include wettable powders and dusts

Folpet is practically insoluble in water. It is a protective leaf fungicide. Its mode of action inhibits normal cell division of a broad spectrum of microorganisms. It is used to control cherry leaf spot, rose mildew, rose black spot, and apple scab. It is used on berries, flowers, ornamentals, fruits, and vegetables and for seed- and plant-bed treatment. It is also used as a fungicide in paints and plastics and for treatment of internal and external structural surfaces of buildings (1). It is incompatible with strongly alkaline preparations, such as lime sulphur.

Folpet is combustible under specific conditions. Liquid formulations containing organic solvents may be flammable and give off irritating or toxic fumes (or gases) in a fire.

Safe Handling and Precautions

Folpet is a member of the N-trihalomethylthio group of compounds, which are highly reactive with biological tissues. Reports have indicated that short-term exposure to folpet causes adverse health effects such as irritation of the skin, eyes, and respiratory tract. Long-term exposure to folpet causes dermatitis, skin sensitisation, and effects on the GI tract, thyroid, lymphatic and blood forming tissues, kidneys, and muscles. Folpet has been reported to cause genetic damage and retarded development of the newborn. Subchronic studies in rats demonstrated that the critical systemic toxic effect was acanthosis and hyperkeratosis and/or ulceration/erosion of the stomach following high oral doses of folpet. Repeated or prolonged contact of folpet with skin may cause dermatitis. Repeated or prolonged contact may cause skin sensitisation. Folpet has been classified as Group B2, meaning a probable human carcinogen.

Occupational exposure to folpet residues via dermal and inhalation routes can occur during handling, mixing, loading, and applying as well as during post-application activities such as harvesting avocados. Folpet may cause skin sensitisation and mutagenic and carcinogenic effects in humans. Folpet is no longer registered in Malaysia from 1998. Workers during use of folpet should avoid skin and eye contact and inhalation of spray mist or fumes and ensure adequate ventilation. Cleaning water should be disposed of in an appropriate manner. Workers should wash hands after handling of folpet and before eating, drinking, or smoking. Also, wash contaminated clothing before reuse. Workers should keep folpet stored in the original, unopened container in a cool, dry place, and protected from excessive heat. It should be kept stored away from stockfeed or foodstuffs.

Fonophos (CAS No. 944-22-9)

Molecular formula: $C_{10}H_{15}OPS_2$
IUPAC name: O-ethyl S-phenyl (RS)-ethylphosphonodithioate
Synonyms and trade names: Capfos, Cudgel, Difonate, Dyfonate, Dyphonate, and Stauffer
Toxicity class: U.S. EPA: I or II; WHO: Ia

Fonophos is a highly toxic compound. It is sparingly soluble in water but soluble in acetone, ethanol, xylene, and kerosene. It has been grouped by the U.S. EPA as RUP and hence requires special handling by qualified, certified, and trained workers. Fonophos was used as a soil insecticide, which resulted in its direct release to the environment. It was primarily used on corn crops, sugar cane, peanuts, tobacco, turf, and some vegetable crops. It controls aphids, corn borer, corn rootworm, corn wireworm, cutworms, white grubs, and some maggots. Formulations of fonophos include granular, microgranular, emulsifiable concentrate, suspension concentrate, microcapsule suspension, and for seed treatment.

Safe Handling and Precautions

Fonophos is highly toxic like many other organophosphate pesticides to humans and animals. Exposure to fonophos induces clinical signs of toxicity with typical symptoms of poisoning and cholinesterase inhibition. Accidental ingestion of fonophos by occupational workers results in signs and symptoms of acute intoxication, including muscarinic, nicotinic, and CNS manifestations. The symptoms of fonophos poisoning occur within a few minutes to 12 h after exposure. Early symptoms of poisoning include, but are not limited to, blurred vision, pinpoint pupils, headache, dizziness, depression, tremors, salivation, diarrhoea, and laboured breathing. Skin absorption of fonophos causes sweating and muscle twitching, while eye contact leads to severe tearing, pain, and blurred vision. Prolonged exposures to high concentrations of fonophos lead to respiratory failure and death. There are no reports indicating that fonophos is mutagenic, teratogenic, or carcinogenic in animals and humans. Fonophos registration was cancelled in 1999, and, therefore, it is not considered as an environmental contaminant of concern at the present time.

Formaldehyde (CAS No. 50-0-0)

Molecular formula: CH_2O
Synonyms and trade names: FA; Fannoform; Formalith; Formalin; Formalin 40; Formic aldehyde; Formol; Fyde; Hoch; Karsan; Lysoform; Methyl aldehyde; Methylene

glycol; Methylene oxide; Methanal; Morbicid; Oxomethane; Oxymethylene; Paraform; Polyoxymethylene glycols; Superlysoform

Formaldehyde, also called formic aldehyde or methyl aldehyde, has extensive application. For instance, it is used as a tissue preservative or organic chemical reagent. Thus, formaldehyde is very common to the chemical industry. In fact, formaldehyde is an important chemical used widely by industry to manufacture building materials and numerous household products. It is also a by-product of combustion and certain other natural processes. It is present in substantial concentrations both indoors and outdoors. Formaldehyde is well known as a preservative in medical laboratories, as an embalming fluid, and as a steriliser. Its primary use is in the production of resins and as a chemical intermediate. Urea–formaldehyde (uf) and phenol–formaldehyde (pf) resins are used in foam insulations, as adhesives in the production of particle board and plywood, and in the treating of textiles. Sources of formaldehyde in the home include building materials, smoking, household products, and the use of unvented, fuel-burning appliances, like gas stoves or kerosene space heaters. Formaldehyde, by itself or in combination with other chemicals, serves a number of purposes in manufactured products.

Formaldehyde itself is a colourless gas, but it is more commonly purchased and used in aqueous solution (called formalin solution), with a maximum concentration of 40%. Formalin solutions often contain some amount of methanol as well. Both formaldehyde gas and solutions have a characteristic pungent, unpleasant odour.

Safe Handling and Precautions

Formaldehyde is a colourless, pungent-smelling gas. Exposures to low levels of formaldehyde cause irritation of the eyes, nose, throat, and skin; nausea; and difficulty in breathing. Short-term exposure to formaldehyde can be fatal. Long-term exposure to low levels of formaldehyde may cause respiratory difficulty, eczema, and sensitisation. Inhaling formaldehyde fumes can cause respiratory problems and asthma-like symptoms, such as breathlessness, shortness of breath, wheezing, coughing, and/or chest tightness. Repeated exposures may cause bronchitis, with symptoms of cough and shortness of breath. Occupational workers with asthma have been found to be more sensitive to the effects of inhaled formaldehyde, and in high concentrations, formaldehyde trigger attacks in people with asthma. Low ambient concentrations of formaldehyde can cause irritation of the upper respiratory tract. At higher concentrations, the effects become more severe, with levels above 10 ppm causing coughing and chest tightness. Exposure to very high concentrations of formaldehyde is known to cause death from throat swelling and chemical burns to the lungs. Also, intake/drinking of large amounts causes severe pain, vomiting, and coma leading to death. Acute and chronic health effects of formaldehyde vary depending on the individual. The typical threshold for development of acute symptoms due to inhaled formaldehyde is 800 ppb. However, individuals and workers who are sensitive to formaldehyde have reported symptoms at formaldehyde levels around 100 ppb. Formaldehyde is common to the chemical industry. Although there is no conclusive evidence available to prove that formaldehyde is a human carcinogen, it has been shown to cause cancer in animals. Formaldehyde is therefore considered to be a probable human carcinogen, particularly as a cause of nasal and nasopharyngeal cancers as these areas are more likely to come into direct contact with formaldehyde. The 11th report on carcinogens classifies it as 'reasonably anticipated to be a human carcinogen'. According to the IARC, formaldehyde is classified as a human carcinogen and has been linked to nasal and lung cancer, with possible links to brain cancer and leukaemia. The Department of Health and Human Services (DHHS) has determined that formaldehyde may reasonably be anticipated as a carcinogen. The U.S. EPA recommends that an adult should not drink water containing more than 1 mg/L of formaldehyde for a lifetime exposure, and a child should not drink water containing more than 10 mg/L for 1 day or 5 mg/L for 10 days. The Occupational Safety and Health Administration (OSHA) set a PEL for formaldehyde at 0.75 ppm for an 8 h workday, 40 h workweek. The National Institute for Occupational Safety and Health (NIOSH) recommends an exposure limit of 0.016 ppm. Several countries around the world have regulated the sale of wood products made with urea–formaldehyde, and some common building materials have specific product standards that limit the release of the amount of formaldehyde.

Occupational workers during use and handling of formaldehyde must be familiar with the hazards of formaldehyde. Workers should always keep a Material Safety Data Sheet (MSDS) in the work area. The work area should always have adequate ventilation, a fume hood, to minimise inhalation of vapour. Occupational workers should always use chemical goggles or a face shield during handling of formaldehyde to minimise the risk of chemical splash or vapour and exposure to the corneas. Workers should wear medium or heavyweight nitrile, neoprene, natural rubber, or PVC gloves during handling concentrated formaldehyde, wear a laboratory coat, and avoid to wear shorts or open-toed shoes when handling formaldehyde and must remember to discard contaminated dress in the regular trash. Occupational workers require at least annual training and awareness of specific hazards of formaldehyde in their workplace and of the control measures required.

Appendix

Precautions: Management of Formaldehyde and Simple Safety Precautions

- Students, users, and occupational workers during handling of formaldehyde must be aware of the health hazards of formaldehyde and should completely avoid negligence. Workers should know that formaldehyde exposure has been associated with cancers of the lung, nasopharynx, oropharynx, and nasal passages in humans.

- Formaldehyde as has been stated earlier is very corrosive, and the eyes are especially very vulnerable to the vapour and fumes.

- Workers should always keep in the work area the MSDS for formaldehyde. The MSDS and this fact sheet are excellent tools for training semi-skilled workers on the hazards of formaldehyde.

- Workers should use and handle formaldehyde in work area with adequate ventilation.

- Workers should always and strictly use suitable PPE, chemical splash goggles, impervious gloves and face shield, respirators, eye protection, gloves, and other safety apparel during handling formaldehyde.

- Students and workers should always handle, mix, and/or add formaldehyde only to the specimen containers with utmost caution.

- Workers should completely avoid to store formaldehyde bottles in workplace where chemical waste and/or leak spillage may flow to a drain, and workers must immediately clean up any kind of chemical spillages.

Formamide (CAS No. 75-12-7)

Molecular formula: **HCONH₂**
Synonyms and trade names: Carbamaldehyde; Methanamide

Formamide is a synthetic reagent, solvent, and softener in gums and glues. It is a colourless, hygroscopic, viscous liquid. Formamide is a stable chemical substance. It is incompatible with strong acids, strong bases, strong oxidising agents, brass, copper, bronze, and iodine. On contact with strong oxidisers, formamide is known to cause fire or explosion. Formamide, on combustion, forms toxic gases (nitrogen oxides). It undergoes decomposition on heating at 180°C producing ammonia, water, carbon monoxide, and hydrogen cyanide. It reacts with oxidants. It attacks metals such as aluminium, iron, copper, and natural rubber.

Safe Handling and Precautions

Workers on exposure to formamide develop irritation of the eyes and skin. Formamide is also known to cause effects on the CNS. It is a teratogen – may cause reproductive abnormalities. This chemical has caused foetal abnormalities and embryonic death in laboratory animals.

Alert!! Danger! Users and occupational workers should note that formamide has been reported as a possible carcinogen, teratogen, corrosive, and irritant chemical substance. Formamide should never be handled without proper safety attire including gloves and goggles. Negligence and exposures to formamide may cause severe burns and blistering, which may not show immediate pain and related adverse health effects. Pregnant women and women of child-bearing age should minimise/avoid contact and exposure to formamide.

Occupational workers during use and handling of formamide should wear appropriate PPE and clothing including lab coat, safety glasses, gloves, and approved respirator. Workplace should have adequate ventilation. Occupational workers should avoid contact of material with skin or eyes, and the chemical should be stored at –20°C away from incompatible material. Workers should have easy and quick access to a safety shower and eyewash facility.

Formic Acid (CAS No. 64-18-6)

Molecular formula: **CH₂O₂/HCOOH**
Synonyms and trade names: Formylic acid; Hydrogen carboxylic acid; Methanoic acid; Aminic acid

Formic acid is a clear, colourless liquid with pungent odour. It is a stable corrosive, combustible, and hygroscopic chemical substance. It is incompatible with H_2SO_4, strong caustics, furfuryl alcohol, hydrogen peroxide, strong oxidisers, and bases and reacts with strong explosion on contact with oxidising agents. Formic acid is an important intermediate in chemical synthesis and naturally present most notably in the venom of bee and ant stings. The name formica is derived from Latin for ant, and formica refers to its early isolation by the distillation process of ant bodies. Ester salts and the anion derived from formic acid are referred to as formates.

Safe Handling and Precautions

Inhalation of vapours of formic acid produces severe irritation of nose, throat, and upper respiratory tract. Inhalation of higher concentrations causes CNS effects and lung damage. Accidental ingestion of formic acid causes poisoning, and the symptoms include, but are not limited to, serious burns and tissue damage to the mouth, throat, and oesophagus with severe pain, destruction of the mucous membranes and upper respiratory tract, difficulty to normal swallowing process, abdominal pain, nausea, diarrhoea, and vomiting, leading to shortness of breath and death. Severe poisoning of formic acid is also known to cause shock and kidney damage.

Students and occupational workers during use and handling of formic acid should strictly observe set regulations of safety and avoid use of strong bases, strong oxidising agents, powdered metals, and furfuryl alcohol since formic acid is combustible and hygroscopic and caused workplace accidents and chemical disasters. Also, students and workers should be well aware of the fact that pressure may build up in tightly closed bottles, requiring proper methods of chemical management and periodic checking.

Formic acid is harmful, dangerous, and a corrosive chemical substance. It has not been listed by the ACGIH, the IARC, the NIOSH, the NTP, and the U.S. OSHA as a human carcinogen.

Liquid and mist forms of formic acid have been extensively documented to cause severe burns to all body tissues of occupational workers. Negligence during handling of formic acid and swallowing and inhalation cause lung damage; vapour causes irritation to eyes and respiratory tract.

Students and workers should keep formic acid in a tightly closed container; store in a cool, dry, ventilated area away from sources of heat or ignition; and protect against physical damage. Formic acid should be stored separately and away from reactive or combustible materials and out of direct sunlight, and qualified workers should handle stainless steel, glass, ceramic, or similar corrosion-resistant materials. Containers of formic acid pose hazards when empty because empty containers are known to retain product residues (vapours, liquid). On any case of chemical leak at the workplace, students and workers should immediately remove all sources of ignition, ventilate area of leak or spill, isolate hazard area, keep away all unnecessary and unprotected personnel from entering the accident area, neutralise with alkaline material (soda ash, lime), and avoid to flush the chemical waste to sewer system. A worker who handles formic acid should thoroughly wash hands, forearms, and face with soap and water before eating, using tobacco products, using toilet facilities, applying cosmetics, or taking medication. Workers should not eat, drink, use tobacco products, apply cosmetics, or take medication in areas where formic acid or a solution containing formic acid is handled, processed, or stored.

Alert! Occupational workers and students should be aware of the fact that any kind of negligence during handling and management of formic acid is known to cause severe chemical burns, and eye exposure can result in permanent eye damage. Inhalation of vapours of formic acid causes severe irritation of upper respiratory tract with coughing, burns, breathing difficulty, and possible coma and may cause pulmonary oedema and severe respiratory disturbances.

Formonitrile (CAS No. 74-90-8)

Molecular formula: **HCN**
Synonyms and trade names: Prussic acid; Hydrocyanic acid
For more details on formonitrile, the reader may refer to Hydrocyanic acid.

Furan (CAS No. 110-00-9)

Molecular formula: C_4H_4O
Synonyms and trade names: Axole; 1,4-Epoxy-1,3-butadiene; Furfuran; Oxole; Tetrole; Divinylene oxide; Divinyl oxide; Furane; Furfuran; Oxacyclopentadiene

Furan is a stable, colourless chemical liquid. It is highly flammable and forms explosive mixtures with air. It is incompatible with many other chemical substances, for example, strong oxidising agents, acids, peroxides, and oxygen. Furan is produced commercially by decarbonylation of furfural. It is used mainly in the production of tetrahydrofuran, thiophene, and pyrrole. It also occurs naturally in certain woods and during the combustion of coal and is found in engine exhausts, wood smoke, and tobacco smoke.

Exposure to furan causes harmful health effects and poisoning. Furan is very toxic. Whenever users and occupational workers come in contact with this chemical substance and at different workplaces, severe health disorders occur. Furan enters the system through all the major routes of exposure, namely, respiratory (inhalation), oral (ingestion swallow), and skin (through skin absorption). Exposure of workers to furan causes fatal injury. Repeated and prolonged administration of furan to experimental laboratory mice and rats caused liver necrosis, liver-cell proliferation, and bile-duct hyperplasia; in rats, prominent *cholangiofibrosis* develops. While there is inadequate evidence for the carcinogenicity of furan in humans, sufficient evidence has been reported in experimental animals.

Safe Handling and Precautions

Occupational workers during handling of furan should observe strict precautions and must not be negligent. Workers should use personal protective equipment (PPE), workplace with good ventilation, safety glasses and gloves. *Furan should be used and handled by workers as a potential carcinogen.*

During handling of furan, occupational workers should use workplace with proper ventilation, explosion-proof electrical equipment, and proper lighting. Workers should avoid the use of compressed air for filling, discharging, or handling and sparking hand tools. Any kind of accidental waste chemical spill/leaking liquid at the workplace should be collected in sealable containers. The waste chemical spill should be absorbed in sand or inert absorbent and removed to a safe place and should not be washed away into sewer systems.

Furfural (CAS No. 98-01-1)

Molecular formula: $C_5H_4O_2$
Synonyms and trade names: 2-Formylfuran; 2-Furanaldehyde; 2-Furancarbonal; Furfurol; Furfurole; Cyclic aldehyde; 2-Furaldehyde; 2-Furancarboxaldehyde;2-Furaldehído; 2-Furaldéhyde; alpha-Furole; Artificial ant oil; Fural; Furole; α-Furole

Furfural is a colourless to amber-like oily liquid with an almond-like odour. On exposure to light and air, it turns reddish brown. Furfural is used in making chemicals, as a solvent in petroleum refining, a fungicide, and a weed killer. It is incompatible with strong acids, oxidisers, and strong alkalis. It undergoes polymerisation on contact with strong acids or strong alkalis. Furfural is produced commercially by the acid hydrolysis of pentosan polysaccharides from non-food residues of food crops and wood wastes. It is used widely as a solvent in petroleum refining, in the production of phenolic resins, and in a variety of other applications. Human exposure to furfural occurs during its production and use, as a result of its natural occurrence in many foods and from the combustion of coal and wood.

Safe Handling and Precautions

Exposure to furfural gets absorbed by all the routes, inhalation, skin absorption, ingestion, and skin and/or eye contact, and causes adverse health effects with symptoms of irritation of the eyes, skin, upper respiratory system; headache; dermatitis; laboured breathing; shortness of breath; sore throat; abdominal pain; diarrhoea; and vomiting. Repeated oral administration to rats causes liver necrosis and cirrhosis. Reports have shown that furfural is extensively absorbed and rapidly eliminated after inhalation by humans and rats. Reports have indicated limited evidence and inadequate evidence about the carcinogenicity of furfural in animals and humans. The IARC classified furfural as Group 3, meaning not classifiable as a human carcinogen.

Occupational workers should be alert during use and handling of furfural and must use appropriate PPE, gloves, sleeves, and encapsulating suits based on the extent of the worker's potential exposure to furfural to avoid accidental workplace exposures. Workers should be aware of the chemical hazards since furfural on contact with acids and/or bases polymerises and causes explosion hazard. It reacts violently with oxidants and attacks many plastics. Workers should keep stored furfural in a cool, dry, well-ventilated area in tightly sealed containers that are labelled in accordance with regulations and hazard communication standards. The containers of furfural should be kept protected from physical damage and should be stored separately from strong acids, oxidising materials, strong mineral acids, alkalis, or sodium hydrogen carbonate.

Bibliography

Agency for Toxic Substances and Disease Registry (ATSDR). 1999. *Toxicological Profile for Formaldehyde*. Department of Health and Human Services, Public Health Service, Atlanta, GA (updated 2008).

Agency for Toxic Substances and Disease Registry (ATSDR). 2003. *Toxicological Profile for Fluorides, Hydrogen Fluoride, and Fluorine*. U.S. Department of Health and Human Services, Public Health Service, Atlanta, GA (updated 2011).

Aments, P.W. et al. 1994. Famotidine. *Ann. Pharmacother.* 28(1):40–42.

Anonymous. 1999. *Farm Chemicals Handbook*. Meister Publishing Co., Willoughby, OH.

Anton, R. D. (ed.). 2010. *A Guide to Safe Material and Chemical Handling*. Wiley-Scrivener. Hoboken, NJ.

Calabrese, E. J. and E. M. Kenyon. 1991. *Air Toxics and Risk Assessment*. Lewis Publishers, Chelsea, MI.

Clayton, G. D. and F. E. Clayton (eds.). 1981. *Patty's Industrial Hygiene and Toxicology*, 3rd edn. John Wiley & Sons, New York.

Clayton, G. D. and F. E. Clayton. (eds.). 1993–1994. *Patty's Industrial Hygiene and Toxicology*, Vols. 2A, 2B, 2C, 2D, 2E, 2F: Toxicology. 4th edn. John Wiley & Sons, New York.

Dikshith, T. S. S. (ed.). 2011. *Handbook of Chemicals and Safety*. CRC Press, Boca Raton, FL.

Dikshith, T. S. S. and D. Prakash. (eds.). 2003. *Industrial Guide to Chemical and Drug Safety*. John Wiley & Sons, Hoboken, NJ.

Echobichon, D. J. 1996. Toxic effects of pesticides. In: *Casarett & Doull's Toxicology: The Basic Science of Poisons*, Klaassen, C. D. (ed.), 5th edn. McGraw-Hill, New York.

Elliot, M. (ed.). 1977. *Pyrethroids*. American Chemical Society, Philadelphia, PA.

Extension Toxicology Network (EXTOXNET). 1996. Fonofos. Extension Toxicology Network, Pesticide Information Profiles (PIP). University of California, Davis, CA, Oregon State University, Corvallis, OR.

Forsberg, K. and S. Z. Mansdorf. 1993. *Quick Selection Guide to ˈChemical Protective Clothing*. Van Nostrand Reinhold, New York.

Francis, J. I. and J. M. Branes. 1963. Studies on the mammalian toxicity of fenthion. *Bull. World Health. Org.* 29:205.

Gallo, M. A. and N. J. Lawryk. 1991. Organic phosphorus pesticides. In: *Handbook of Pesticide Toxicology*, Hayes, W. J., Jr. and E. R. Jr. Laws (eds.). Academic Press, New York.

Garforth, B. and R. A. Johnson. 1987. Performance and safety of the new anticoagulant rodenticide flocoumafen. *Proceedings of the Symposium on Stored Product Pest Control*, Reading University, U.K., pp. 25–27.

Gill, J. E. 1992. Laboratory evaluation of the toxicity of flocoumafen as a single-feed rodenticide to seven rodent species. *Int. Biodeter. Biodegr.* (1):65–76.

Gulmaraens, D. et al. 1994. Famotidine. *Contact Dermatitis.* 31(4):259–260.

Gunasekara, A. S., T. Tresca, K. S. Goh et al. 2006. Environmental fate and toxicology of fipronil. *J. Pest. Sci.* 32:189–199.

Hathaway, G. J., N. H. Proctor, J. P. Hughes, and M. L. Fischman. 1991. *Proctor and Hughes' Chemical Hazards of the Workplace*, 3rd edn. Van Nostrand Reinhold, New York.

Hazardous Substances Data Bank (HSDB). 2004. MTBE. Division of Specialized Information Services, Hazardous National Library of Medicine, Bethesda, MD.

Institute of Food and Agricultural Sciences (IFAS). 2006. Fenamiphos use facts and phaseout. University of Florida, IFAS, Gainesville, FL.

International Agency for Research on Cancer (IARC). 1991. Fenvalerate. In: *Occupational Exposures in Insecticide Application, and some Pesticides*. Lyon Chemical name: Ferric dimethyldithiocarbamate.

International Agency for Research on Cancer (IARC). 1995. *Furan. Summaries & Evaluations*, Vol. 63, 393. IARC, Geneva, Switzerland.

International Programme on Chemical Safety (IPCS). 2005. Fluorosulfonic acid. International Programme on Chemical Safety (IPCS) and the Commission of the European Communities (CEC), IPCS–CEC, ICSC Card No. 0996. Geneva, Switzerland.

International Program on Chemical Safety (IPCS). 2010. Flocoumafen. Material Safety Data Sheet (MSDS). ICSC Card No. 1267. IPCS (CIS). WHO, Geneva, Switzerland.

International Programme on Chemical Safety and the Commission of the European Communities (IPCS–CEC). 1997. Ferbam. International Chemical Safety Card. ICSC Card No. 072. IPCS–CEC, Geneva, Switzerland.

International Program on Chemical Safety and the Commission of the European Communities (IPCS–CEC). 1998. Furfural. ICSC Card No. 0764. IPCS–CEC, Geneva, Switzerland.

International Programme on Chemical Safety and the Commission of the European Communities (IPCS–CEC). 1999. Fenamiphos. ICSC Card No. 0483, IPCS-CEC, Geneva, Switzerland.

International Programme on Chemical Safety and the Commission of the European Communities (IPCS–CEC). 2001. Fenitrothion. ICSC No. 0622. IPCS–CEC, Geneva, Switzerland.

International Programme on Chemical Safety and the Commission of the European Communities (IPCS–CEC). 2005a. Furan. ICSC No: 1257. IPCS–CEC, Geneva, Switzerland.

International Programme on Chemical Safety and the Commission of the European Communities (IPCS–CEC). 2005b. Fenvalerate technical product, ICSC Card No. 0273. IPCS–CEC, Geneva, Switzerland.

Jennings, J. A., T. D. Canerdy, R. J. Keller et al. 2002. Human exposure to fipronil from dogs treated with Frontline. *Vet. Human Toxicol.* 44:301–303.

Kidd, H. and D. R. James (eds.). 1991. *The Agrochemicals Handbook*, 3rd edn. Royal Society of Chemistry Information Services, Cambridge, U.K.

Lewis, R. J. and R. J. Lewis, Sr. (eds.). 2008. *Hazardous Chemicals Desk Reference*. John Wiley & Sons, Hoboken, NJ.

Lund, M. 1988a. Flocoumafen—A new anticoagulant rodenticide. *Proceedings of the Thirteenth Vertebrate Pest Conference,* University of Nebraska, Lincoln, NE.

Lund, M. 1988b. Flocoumafen—A new anticoagulant rodenticide. *Proceedings of the Vertebrate Pest Conference*. University of California, Davis, CA, Vol. 13, pp. 53–58.

Material Safety Data Sheet (MSDS). Occupational safety and health guideline for formic acid.

Material Safety Data Sheet (MSDS). 1993a. Fenoxycarb. Extension Toxicology Network (EXTOXNET). USDA/Extension Service. Michigan State University, Oregon State University, Corvallis, OR.

Material Safety Data Sheet (MSDS). 1993b. Folpet. OHS Database. MDL Information Systems Inc., San Leandro, CA.

Material Safety Data Sheet (MSDS). 1994. Fenamiphos. MSDS. Extension Toxicology Network (EXTOXNET). Pesticide Information Profiles. USDA/Extension Service/Oregon State University, Corvallis, OR.

Material Safety Data Sheet (MSDS). 1995. Fenitrothion. MSDS. Extension Toxicology Network (EXTOXNET). Pesticide information profiles, USDA/Extension Service/Oregon State University, Corvallis, OR.

Material Safety Data Sheet (MSDS). 1996. Formamide. MSDS No. F5770. Mallinckrodt Baker, Inc., Phillipsburg, NJ.

Material Safety Data Sheet (MSDS). 2002. Fluoroacetamide. International Chemical Safety Cards. No ICSC. 1434. International Programme on Chemical Safety and the Commission of the European Communities (C) IPCS CEC. NIOSH, WHO. Geneva, Switzerland.

Material Safety Data Sheet (MSDS). 2003a. Safety data for fluorobenzene. Department of Physical Chemistry, Oxford University, Oxford, U.K.

Material Safety Data Sheet (MSDS). 2003b. Safety data for furan. Department of Physical and Theoretical Chemistry, Oxford University, Oxford, U.K.

Material Safety Data Sheet (MSDS). 2005a. Ferrocene. CAMEO Chemicals. National Oceanic and Atmospheric Administration (NOAA). Silver Spring, MD.

Material Safety Data Sheet (MSDS). 2005b. Safety data for ferric dimethyldithiocarbamate. Department of Physical Chemistry, Oxford University, Oxford, U.K.

Material Safety Data Sheet (MSDS). 2005c. Safety data for ferrocene. Department of Physical Chemistry, Oxford University, Oxford, U.K.

Material Safety Data Sheet (MSDS). 2005d. Safety data for fluorosulfonic acid. Department of Physical Chemistry, Oxford University, Oxford, U.K.

Material Safety Data Sheet (MSDS). 2006. Award fire ant bait. Syngenta Crop Protection, Inc., Greensboro, NC.

Material Safety Data Sheet (MSDS). 2007a. Safety data for fipronil. Department of Physical Chemistry, Oxford University, Oxford, U.K.

Material Safety Data Sheet (MSDS). 2007b. Safety data for fluoroboric acid. Department of Physical Chemistry, Oxford University, Oxford, U.K.

Material Safety Data Sheet (MSDS). 2008a. Safety data for formaldehyde, 37% solution. Department of Physical Chemistry, Oxford University, Oxford, U.K.

Material Safety Data Sheet (MSDS). 2008b. Safety data for formic acid. Department of Physical Chemistry, Oxford University, Oxford, U.K.

Material Safety Data Sheet (MSDS). 2010a. Safety data for fluoroacetamide. Department of Physical Chemistry, Oxford University, Oxford, U.K.

Material Safety Data Sheet (MSDS). 2010b. Famotidine. Sciencelab.com, Inc., Houston, TX.

Material Safety Data Sheet (MSDS). 2010c. Fenitrothion technical grade. Sumitomo Chemical Australia, New Zealand.

Material Safety Data Sheet (MSDS). 2010d. Fensulfothion. *NIOSH Pocket Guide to Chemical Hazards.* Centers for Disease Control and Prevention (CDCP), National Institute for Occupational Safety and Health (NIOSH), Atlanta, GA.

Material Safety Data Sheet (MSDS). 2010e. Formamide. Centers for Disease Control and Prevention (CDCP), National Institute for Occupational Safety and Health (NIOSH). Atlanta, GA.

Material Safety Data Sheet (MSDS). 2010f. Safety data for fluorine (and gas mixtures containing significant proportions of fluorine). Department of Physical Chemistry, Oxford University, Oxford, U.K.

Material Safety Data Sheet (MSDS). 2010g. Safety data for formamide. Department of Physical Chemistry, Oxford University, Oxford, U.K.

Material Safety Data Sheet (MSDS). 2010h. Safety data for hydrogen cyanide. Department of Physical Chemistry, Oxford University, Oxford, U.K.

Material Safety Data Sheet (MSDS). 2011i. Department of Health & Human Services (DHHS). Lincoln, NE.

Material Safety Data Sheet (MSDS). 1996a. Fluvalinate. USDA/Extension Service, Pesticide Information Profiles (PIP). Oregon State University, Corvallis, OR.

Material Safety Data Sheet (MSDS). 1996b. Fenthion. MSDS. Extension Toxicology Network, Pesticide Information Profiles. USDA/Extension Service/Oregon State University, Corvallis, OR.

Meister, R. T. 1995. *Farm Chemicals Handbook '95.* Meister Publishing Company, Willoughby, OH.

Meister, R. T. (ed.). 2000. *Farm Chemicals Handbook 2000.* Meister Publishing Co., Willoughby, OH.

Meister, R. T., G. I. Berg, C. Sine, S. Meister, and J. Poplyk. 1983. *Farm Chemicals Handbook. Pesticide Dictionary.* Meister Publishing Co., Willoughby, OH.

Morgan, D. P. 1982. *Recognition and Management of Pesticide Poisonings*, 3rd edn. U.S. Environmental Protection Agency, Washington, DC.

National Institute for Occupational Safety and Health (NIOSH). 2010a. Ferbam. *NIOSH Pocket Guide to Chemical Hazards.* Centers for Disease Control and Prevention, Atlanta, GA.

National Institute for Occupational Safety and Health (NIOSH). 2010b. Furfural. *NIOSH Pocket Guide to Chemical Hazards.* NIOSH. Centers for Disease Control and Prevention. Atlanta, GA.

National Library of Medicine (NLM). 1992a. Formic acid. Hazardous Substances Data Bank, NLM, Bethesda, MD.

National Library of Medicine (NLM). 1992b. Furfural. Hazardous Substances Data Bank, NLM, Bethesda, MD.

National Institute for Occupational Safety and Health (NIOSH). 2010. Fluorine. *NIOSH Pocket Guide to Chemical Hazards.* Centers for Disease Control and Prevention (CDCP), Atlanta, GA.

Occupational Safety & Health Administration (OSHA). 1996. Occupational safety and health guideline for furfural. OSHA, Washington, DC.

Occupational Safety & Health Administration (OSHA). 2011. Formaldehyde. OSHA, Washington, DC.

PAN Pesticide Database, Pesticide Action Network, 2010. Fenoxycarb. San Francisco, CA.

Patnaik, P. (ed.). 1992. *A Comprehensive Guide to the Hazardous Properties of Chemical Substances.* Van Nostrand Reinhold, New York.

Perry, D. L. and S. L. Phillips. 1995. *Handbook of Inorganic Compounds.* CRC Press, Boca Raton, FL.

Robert, I. K. (ed.). 2002. *Hayes' Handbook of Pesticide Toxicology*, 3rd edn. Academic Press, New York.

Safety Data for Hydrogen Cyanide. (For details of Formonitrile refer to Hydrogen cyanide.)

Schaefer, C. H., W. H. Wilder, F. S. Mulligan, and E. E. Dupras. 1987. Efficacy of fenoxycarb against mosquitoes (Diptera: Culicidae) and its persistence in the laboratory and field. *J. Econ. Entomol.* 80(1):126–130.

Singh, M., R. Vijayaraghavan, S. Pant, C. Sugendran, K. Kumar, P. Purnanand, and R. Singh. 2000. Acute inhalation toxicity study of 2-fluoroacetamide in rats. *Biomed. Environ. Sci.* 13(2):90–96.

Sittig, M. (ed.). 1985. *Handbook of Toxic and Hazardous Chemicals and Carcinogens.* 2nd edn. Noyes Publications, Park Ridge, NJ.

Thomson, W. T. 1990. *Agricultural Chemicals. Book IV: Fungicides.* Thomson Publications, Fresno, CA.

Timothy, T. M. and B. Bryan. (eds.). 2004. *Pesticide Toxicology and International Regulation.* Wiley-Interscience, Hoboken, NJ.

Tingle, C. C., J. A. Rother, C. F. Dewhurst et al. 2003. Fipronil: Environmental fate, ecotoxicology and human health concerns. *Rev. Environ. Contam. Toxicol.* 176:1–66.

Tomlin, C. D. S. (ed.). 1997. *The Pesticide Manual—A World Compendium*, 11th edn. British Crop Protection Council, Farnham, Surrey, U.K.

Tomlin, C. D. S. (ed.). 2006. *The Pesticide Manual: A World Compendium*, 14th edn. The British Crop Protection Council, Farnham, Surrey, U.K.

U.S. Department of Health and Human Services. 1993. Registry of Toxic Effects of Chemical Substances (RTECS, online database). National Toxicology Information Program, National Library of Medicine, Bethesda, MD.

U.S. Department of Transportation (DOT). 1993. *Emergency Response Guide Book*, Guide No. 60. Washington, DC.

U.S. Environmental Protection Agency (U.S. EPA). 1984. Fonofos. Pesticide Fact Sheet 36. Office of Pesticides and Toxic Substances, Washington, DC.

U.S. Environmental Protection Agency (U.S. EPA). 1986. Fluvalinate. Pesticide Fact Sheet 86: Office of Pesticides and Toxic Substances, Washington, DC.

U.S. Environmental Protection Agency (U.S. EPA). 1987a. Fonofos. Draft Health Advisory Summary: Office of Drinking Water, Washington, DC.

U.S. Environmental Protection Agency. June, 1987b. Folpet. Pesticide Fact Sheet 215.

U.S. Environmental Protection Agency (U.S. EPA). 1988a. Health and environmental effects profile for formaldehyde. EPA/600/x-85/362. Environmental Criteria and Assessment Office, Office of Health and Environmental Assessment, Office of Research and Development, Cincinnati, OH.

U.S. Environmental Protection Agency (U.S. EPA). 1988b. Health advisories for 50 pesticides. Fenamiphos. Office of Drinking Water. U.S. EPA, Atlanta, GA.

U.S. Environmental Protection Agency (U.S. EPA). 1989. Cypermethrin. Pesticide Fact Sheet 199. Office of Pesticides and Toxic Substances, Washington, DC.

U.S. Environmental Protection Agency (U.S. EPA). 2001. Chemical ingredients database on fluoro-acetamide. OPP. U.S. EPA, Atlanta, GA.

U.S. Environmental Agency (U.S. EPA). 2002. Pesticides: Reregistration, Fenamiphos facts. EPA738-F-02-003. U.S. EPA (updated 2007).

U.S. Environmental Protection Agency (U.S. EPA). 2006. Fonofos. Health Effects Support EPA-822-R-06-009 August, 2006, U.S. EPA, Atlanta, GA.

U.S. Environmental Protection Agency (U.S. EPA). 2011. Fluoroacetamide, CAMEO Chemicals. National Oceanic and Atmospheric Administration (NOAA), Washington, DC.

U.S. Environmental Protection Agency (U.S. EPA). 2007. Formaldehyde. U.S. EPA. Washington, DC.

U.S. Public Health Service. 1995. Fenthion. Hazardous Substance Data Bank, Washington, DC.

World Health Organization (WHO). 1967. Safe use of pesticides in public health. (WHO Technical Report Series No 356). Geneva, Switzerland.

World Health Organization (WHO). 1989a. Allethrins: Allethrin, D-allethrin, bioallethrin, S-bioallethrin. Environmental Health Criteria 87. International Programme on Chemical Safety, WHO. Geneva, Switzerland.

World Health Organization (WHO). 1989b. *Environmental Health Criteria for Formaldehyde*, Vol. 89. WHO, Geneva, Switzerland.

World Health Organization (WHO). 1990. Permethrin Environmental Health Criteria No. 94. International Programme on Chemical Safety, Geneva, Switzerland.

World Health Organization (WHO). 1991. Fenitrothion, International Programme on Chemical Safety (IPCS). Health and Safety Guide (HSG) No. 65. WHO, Geneva, Switzerland.

World Health Organization (WHO). 1992. Fenitrothion. International Programme on Chemical Safety (IPCS). Environmental Health Criteria no. 133. WHO, Geneva, Switzerland.

World Health Organization (WHO). 1996. Fenvalerate. WHO/FAO Data Sheets on Pesticides. No. 90. International Programme on Chemical Safety. WHO, Geneva, Switzerland.

World Health Organization (WHO). 2001. Fensulfothion. ICSC Card No. 1406. Centers for Disease Control and Prevention (CDCP). National Institute for Occupational Safety and Health (NIOSH), Atlanta, GA.

9

Hazardous Chemical Substances: G

Gadolinium
Gadolinium Fluoride
Gallium (III) Nitrate
Glass Wool Fibres (Inhalable)
Glycol Ethers
    2-Methoxyethanol
    1-Methoxy-2-Propanol
    2-Methoxy-l-Propanol
    2-Butoxyethanol
    2-Ethoxyethanol
Guthion

Gadolinium (CAS No. 7440-54-2)

Molecular formula: **Gd**
Synonyms and trade names: Gadolinia

Gadolinium is a rare earth metal. Gadolinium appears as metal foil, chunks, or powder and is insoluble in water. The powder of gadolinium is highly flammable; incompatible with strong oxidising agents, halogens, acids; and reacts with water or moisture. Gadolinium is also found in alloys and special minerals known as yttrium garnets. An alloy is made by melting and mixing two or more metals. The mixture has properties different from those of the individual metals. Gadolinium alloys are easier to work with than alloys without gadolinium. Gadolinium yttrium garnets are used in microwave ovens to produce the microwaves. Gadolinium metal is not especially reactive. It dissolves in acids and reacts slowly with cold water. It also reacts with oxygen at high temperatures. Gadolinium is used in control rods in nuclear power plants. Energy produced during nuclear fission is used to generate electricity. Nuclear fission is the process in which large atoms (usually uranium or plutonium) break apart, releasing energy. The smaller atoms produced are called fission products and are radioactive. Gadolinium compounds are used as phosphors in television tubes.

Gadolinium oxide (Gd_2O_3) is a white powder with multiple uses in medicine, chemical processes, electronics, and glass making. Gadolinium oxide is used in the creation of the phosphors used in television tubes as well as the creation of gadolinium yttrium garnets used in microwaves and materials used to absorb atomic reactions. Gadolinium has a shiny metallic lustre with a slight yellowish tint. It is both ductile and malleable.

Safe Handling and Precautions

Gadolinium also has medical uses. It is used to locate the presence of tumours in the inner ear. Gadolinium is injected into the bloodstream. It then goes to any tumour that happens to be present in the ear. The tumour appears darker when seen with x-rays. Gadolinium is used to locate tumours in the inner ear. These metals are moderately to highly toxic. The symptoms of toxicity of the rare earth elements include writhing, ataxia, laboured respiration, walking on the toes with arched back, and sedation. The rare earth elements exhibit low toxicity by ingestion exposure. However, the intraperitoneal route is highly toxic while the subcutaneous route is poisonous to moderately toxic. Information on the chemical, physical, and toxicological properties of gadolinium fluoride have been reported to be limited and require more studies. To the best of our knowledge, the chemical, physical, and toxicological properties of gadolinium fluoride have not been thoroughly investigated and recorded.

Gadolinium powder, on accidental ingestion at workplaces, causes harmful effects while any kind of contact has not been reported as hazardous. Gadolinium agents have proved useful for patients with renal impairment. However, its use in patients with severe renal failure and on dialysis has been reported to cause health risks and has been associated with nephrogenic systemic fibrosis. Users and workers, during handling the gadolinium powder, should observe safety precautions and keep the material well away from sources of ignition. Avoid contact with the chemical substance and prevent such exposures. While handling gadolinium oxide, users should wear appropriate safety glasses and avoid generating dust.

Gadolinium Fluoride (CAS No. 13765-26-9)

Molecular formula: GdF_3

Gadolinium fluoride, on heating or on contact with acids/acid fumes, decomposes and emits toxic fumes of fluorine, hydrogen fluoride vapours, and oxides of gadolinium.

Workers exposed to gadolinium fluoride suffer with harmful effects. The exposure causes redness, burning, itching, and watering of the eyes; the skin show effects of burning, redness, itching, irritation, rashes, and skin granulomas, while on repeated and prolonged exposure, skin demonstrates effects of dermatitis, sensitivity to heat, itching, and skin lesions. Users and workers, after repeated and prolonged exposure to gadolinium fluoride, demonstrate loss of weight, fibrosis, sclerosis of the bones, calcification of ligaments, mottled teeth, anorexia, anaemia, wasting, caccia and dental defects, osteosclerosis, and osteomalacia. Workers who accidentally ingest gadolinium fluoride at workplaces suffer with health disorders and poisoning. The symptoms of poisoning include, but are not limited to, nausea, vomiting, diarrhoea, abdominal burning, and cramp-like pain.

Repeated and prolonged inhalation of dusts of gadolinium fluoride causes severe poisoning and damage of the respiratory tract and mucous membrane, asthma attacks, lung damage, lung granulomas and pulmonary oedema, fluorosis, pulmonary fibrosis, writhing, loss of muscle coordination, laboured respiration, sedation, hypotension, dyspnoea, hyperaemia, liver oedema and necrosis, portal congestion, acute chemical pneumonitis, subacute bronchitis and focal hypertopic emphysema, pleural effusion and granulomatous peritonitis with serous and hemorrhagic ascites, and respiratory and cardiac failure.

Users and workers, during handling of gadolinium fluoride, must be alert and strictly observe regulations, avoid generating more workplace dust, and wear appropriate and specified workplace dress and respiratory protective equipments. The workplace must have proper ventilation. Workers should isolate spill area, vacuum up spill area using a high-efficiency particulate absolute (HEPA) air filter, and place spilled material in a closed container for proper disposal.

Gallium (III) Nitrate (CAS No. 69365-72-6)

Molecular formula: GaN_3O_9
Synonyms and trade names: Gallium (III) nitrate; Nitric acid gallium salt

Gallium (III) nitrate is a white crystal or powder. It is stable, an oxidiser, and a combustible material. It is incompatible with reducing agents.
  Alert!! Gallium (III) nitrate, on contact with combustible material, causes fire.

Glass Wool Fibres (Inhalable)

Fibreglass is an immensely versatile material which combines its lightweight with an inherent strength to provide a weather-resistant finish, with a variety of surface textures.

Glass wool fibres are a sub-category of synthetic vitreous fibres and of inorganic fibrous materials. Glass wools fibres contain aluminium or calcium silicates and a variety of materials, including rock, clay, slag, or glass. The chemical composition of glass wool products varies depending on the manufacturing requirement and end use, but almost all contain silicon dioxide as the single largest oxide ingredient for the production of glass. Commercial glasses also include additional oxides of aluminium, titanium, zinc, magnesium, lithium, barium, calcium, sodium, and potassium. Cytotoxic potencies of the glass wool fibres have been shown to be closely associated with their transforming potencies.

Glass wool fibres caused tumours in two rodent species, at several different tissue sites and by several different routes of exposure. Individual types of glass wool fibres were studied in chronic carcinogenicity bioassays in rats and/or hamsters exposed by a number of routes. Inhalation exposure studies showed that tumour incidence or lesion severity increased with the concentration of fibres in the lung. The cumulative lung burden of fibres is related to their deposition and their bio-persistence, which is the ability of fibres to remain in the lung. Also the fibre aerodynamic diameter determines whether a fibre will be deposited in the lungs or the upper airways, and the thinner the fibres, the deeper is their deposition in the lungs. It is important to consider both inhalable and respirable fibres because most human lung cancer occurs within the first five generations of the trachea-bronchial tree.

Uses of Glass Wool

Glass fibres can generally be classified into two categories based on usage: (i) low-cost, general-purpose fibres typically used for insulation applications and (ii) premium special-purpose fibres used in limited specialised applications. The primary use of glass wool is for thermal and sound insulation. The largest use of glass wool is for home and building insulation in the form of loose wool; Batts insulation in the form of a blanket, rather than a loose filling; blankets or rolls; or rigid boards for acoustic insulation. Glass wool is also used for industrial, equipment, and appliance insulation. Special-purpose glass fibres are used for a variety of applications that require either a specialised glass formulation or a particular diameter. The largest market for special-purpose glass fibres is for battery separator media; the glass wool fibres physically separate the negative and positive plates in a battery while allowing the acid electrolyte to pass through. Another important use is in high-efficiency particulate air filters for settings where high-purity air is required. Special-purpose glass fibres are also used for aircraft, spacecraft, and acoustical insulation.

Safe Handling and Precautions

Exposure to glass fibres by inhalation has been of major concern to users and occupational workers. Occupational exposure to glass wool fibres occurs during the manufacture of glass wool products and end uses associated with installation, removal, fabrication of glass wool outside the manufacturing environment.

Fibre Properties Related to Carcinogenicity

Glass fibres can be classified into two categories based on end use: insulation and special purpose (see Use section). The physicochemical properties within each category also vary, and there is some overlap of properties between the two use categories. Moreover, a specific glass wool product often contains fibres with a wide range of diameters as a result of the manufacturing process (see Properties section for a discussion of nominal diameter). For cancer hazard identification, it is important that fibres be classified according to their biological activity. For the purpose of this profile, 'inhalable' fibres include all fibres that can enter the respiratory tract.

Glass wool fibres have the potential to cause genetic damage. In vitro studies in cultured mammalian cells showed that glass wool fibres caused DNA damage, micronucleus formation, chromosomal aberrations, and DNA–DNA inter-strand cross links, and longer fibres have been reported as more potent in inducing these geno-toxic effects. Fibres of various dimensions caused DNA damage in mammalian cells. Also, intra-tracheal instillation of insulation glass wool caused DNA strand breaks in rat alveolar macrophages and lung epithelial cells. Cytotoxicity studies indicated that longer fibres of glass wool caused more toxic effects than shorter fibres to rat alveolar macrophages. Exposure to glass wool fibres also caused cytotoxicity and morphological transformation in Syrian hamster embryo cell cultures. The safety regulation limits for glass wool fibres set by the American Conference of Governmental Industrial Hygienists (ACGIH) and the National Institute for Occupational Safety and Health (NIOSH) are as follows:

1. The ACGIH: Threshold limit value (TLV) and time-weighted average (TWA) limit: 1 fibre/cm^3 for respirable fibres.
2. The NIOSH recommended exposure limit (REL) for the fibrous glass dust is 3 fibres/cm^3 (TWA) (refer to literature for details).

Glycol Ethers

2-Methoxyethanol (CAS No. 109-86-4)

Molecular formula: $C_3H_8O_2$
Synonyms and trade names: Methyl Cellosolve; Ethylene glycol monomethyl ether

1-Methoxy-2-Propanol (CAS No. 107-98-2)

Molecular formula: $C_4H_{10}O_2$
Synonyms and trade names: Dowanol® PM; Propylene glycol methyl ether; Propyleneglycol monomethyl ether

1-Methoxy-2-propanol is a colourless liquid. It is highly flammable and incompatible with strong oxidising agents, acid chlorides, acid anhydrides, and water. 1-Methoxy-2-propanol is moisture-sensitive and its vapours are heavier than air. Contact of 1-methoxy-2-propanol irritates skin, eyes, and mucous membranes. Prolonged exposure to vapours may cause coughing, shortness of breath, dizziness, and intoxication. 1-Methoxy-2-propanol is used as a solvent and as an anti-freeze agent.

Safe Handling and Precautions

1-Methoxy-2-propanol is a clear, colourless liquid with a mild ether odour and is flammable. Prolonged or repeated excessive exposure of skin to 1-methoxy-2-propanol is known to cause skin irritation, drowsiness, liver and kidney effects, and anaesthetic or narcotic effects. Inhalation at higher levels produces irritation of the eyes, nose, and throat. During handling of 1-methoxy-2-propanol, occupational workers must use personal protective equipment and eye/face protection.

2-Methoxy-l-Propanol (CAS No. 1589-47-5)

Molecular formula: $C_4H_{10}O_2$
Synonyms and trade names: PM; PGME; PM-EL; GMME; Closol; Dowanol PM; Lcinol PM; Dowanol; Solvent PM

Safe Handling and Precautions

Reports have indicated that 2-methoxy-l-propanol causes adverse health effects in laboratory animals. The principal symptoms were effects on the central nervous system, sedation, and narcosis. High doses of 2-methoxy-1-propanol affect the central nervous system, are mildly irritating in the eyes and respiratory tract, and are teratogenic in rabbits. The toxicological profile of the substance is practically identical with that of 2-methoxypropyl-1-acetate, from which the acetyl group is rapidly saponified in the organism. 2-Methoxy-1-propanol was shown to be teratogenic in the rat and rabbit. The teratogenic potential is of the same order of magnitude as that of 2-ethoxyethanol. Therefore, like 2-ethoxyethanol, 2-methoxy-1-propanol is classified in pregnancy group B.

2-Butoxyethanol (CAS No. 111-76-2)

Molecular formula: $C_6H_{14}O_2$

Synonyms and trade names: Butyl Cellosolve®; 2-*n*-Butoxyethanol; 2-Butoxy-1-ethanol; o-Butyl ethylene glycol; Butyl glycol; Butyl monoether glycol; EGBE; Ethylene glycol butyl ether; Ethylene glycol *n*-butyl ether; Ektasolve; Ethylene glycol monobutyl ether; Glycol butyl ether; Monobutyl glycol ether; 3-Oxa-1-heptano; Jeffersol EB

2-Butoxyethanol is a clear, colourless liquid with ether-like smell.2-Butoxyethanol is usually produced by a reaction of ethylene oxide with butyl alcohol, but it may also be made by the reaction of ethylene glycol with dibutyl sulphate. 2-Butoxyethanol is widely used as a solvent in protective surface coatings such as spray lacquers, quick-dry lacquers, enamels, varnishes, and latex paints. It is also used as an ingredient in paint thinners and strippers, varnish removers, agricultural chemicals, herbicides, silicon caulks, cutting oils, and hydraulic fluids and as metal cleaners, fabric dyes and inks, industrial and household cleaners (as a degreaser), and dry-cleaning compounds. It is also used in liquid soaps and in cosmetics.

2-Butoxyethanol acetate has been reported to be present in air, water, and soil as a contaminant and exposure to it occurs during its manufacture and use as an intermediate in the chemical industry, and during the formulation and use of its products in multiple industrial activities. The acetate form of 2-butoxyethanol is 2-butoxyethanol acetate and also known as ethylene glycol monobutyl ether acetate.

Safe Handling and Precautions

Occupational workers, on exposure to 2-butoxyethanol, develop health effects with symptoms such as cough; dizziness; drowsiness; headache; nausea; weakness; irritation of the eyes, skin, nose, and throat; haemolysis; abdominal pain; haematuria (blood in the urine); vomiting; and central nervous system depression. The vulnerable and target organs of 2-butoxyethanol-induced toxicity include eyes, skin, respiratory system, central nervous system, hematopoietic system, blood, kidneys, liver, and lymphoid system. Information on human health effects associated with exposure to 2-butoxyethanol is still limited, and the epidemiological studies have not identified the details. The principal human health effects attributed to 2-butoxyethanol exposure have involved the central nervous system, the blood, and the kidneys.

2-Butoxyethanol has been reported to cause moderate acute toxicity following inhalation, ingestion, or dermal exposure. It is an eye and skin irritant, but it is not a skin sensitiser. Reports have indicated that following inhalation, ingestion, or dermal exposure to 2-Butoxyethanol result in moderate acute toxicity. Heavy exposure via respiratory, dermal, or oral routes can lead to hypotension, metabolic acidosis, haemolysis, pulmonary oedema, and coma. The current TLV set by the ACGIH for worker exposure is 20 ppm in the industrial atmosphere, which is well above the odour threshold of 0.4 ppm. Blood or urine concentrations of 2-butoxyethanol or its major toxic metabolite, 2-butoxyacetic acid, may be measured using chromatographic techniques to monitor worker exposure or to confirm a diagnosis of poisoning in hospitalised patients. A biological exposure index of 200 mg 2-butoxyacetic acid per/g creatinine has been established in an end-of-shift urine specimen for exposed U.S. employees.

2-Butoxyethanol has come under scrutiny in Canada, and the Environment and Health Canada recommended that it be added to Schedule 1 of the Canadian Environmental Protection Act (CEPA). The use of some common household cleaning products containing 2-butoxyethanol could expose people to levels 12 times greater than California's 1 h guideline, especially when indoor use is considered.

2-Ethoxyethanol (CAS No. 110-80-5)

Molecular formula: $C_4H_{10}O_2$

Safe Handling and Precautions

The general public may be exposed to the glycol ethers through the use of consumer products such as cleaning compounds, liquid soaps, and cosmetics. Occupational exposure to the glycol ethers may occur for workers in the chemical industry. Acute exposure to high levels of the glycol ethers in humans results in narcosis, pulmonary oedema, and severe liver and kidney damage. Acute exposure to lower levels of the glycol ethers in humans causes conjunctivitis, upper respiratory tract irritation, headache, nausea, and temporary corneal clouding. (1) Animal studies have reported adverse effects on weight gain, peripheral blood counts, bone marrow, and lymphoid tissues from acute, inhalation exposure to 2-methoxyethanol. Chronic exposure to the glycol ethers in humans results in fatigue, lethargy, nausea, anorexia, tremor, and anaemia. Animal studies have reported anaemia, reduced body weight gain, and irritation of the eyes and nose from inhalation exposure. Anaemia and effects to the thymus, spleen, bone marrow, liver, and kidneys were reported in animals following oral exposure to the glycol ethers. No information is available on the carcinogenic effects of the glycol ethers in humans. Acute (short-term) exposure to high levels of the glycol ethers in humans results in narcosis, pulmonary oedema, and severe liver and kidney damage. Chronic (long-term) exposure to the glycol ethers in humans may result in neurological and blood effects, including fatigue, nausea, tremor, and anaemia. No information is available on the reproductive, developmental, or carcinogenic effects of the glycol ethers in humans. Animal studies have reported reproductive and developmental effects from inhalation and oral exposure to the glycol ethers. The U.S. Environmental Protection Agency (U.S. EPA) has not classified the glycol ethers for carcinogenicity.

Appendix

Safety Management of Ethylene Glycol Butyl Ether

Ethylene glycol butyl ether (EGBE) is a type of glycol ether and it remains the single most widely produced glycol ether. Primarily, ethylene glycol has applications as a solvent. EGBE evaporates quickly and is completely soluble in water. EGBE is marked under the trade name Butyl CELLOSOLVE™ solvent. Although some glycol ethers have been shown to cause adverse reproductive effects and birth defects in laboratory animals, EGBE does not show the same pattern of toxicity as these other glycol ethers. Human experience and animal studies have shown that EGBE is unlikely to cause adverse health effects when products are used as directed.

EGBE, when used improperly, with workplace negligence, has been reported to cause adverse effects to the eyes, respiratory tract, and skin with irritation and burns. Inhalation may cause headaches and haemolysis (red blood cell breakage). Ingesting products that contain EGBE can cause irritation and toxic effects.

Occupational and consumer exposure to EGBE is possible because of its extensive application in a wide variety of industrial and consumer products like cleaning products, paints, brake fluids, and inks. EGBE is unlikely to cause adverse environmental impact because it is not persistent, does not bio-accumulate, and has low toxicity to aquatic organisms.

It may cause moderate corneal injury, and the eye may be slow to heal. Repeated skin exposure to EGBE may cause irritation and even a burn. EGBE should not be ingested.

Intentional ingestion of EGBE-containing products can be toxic to humans. Because EGBE is a combustible liquid, containers, even those that have been emptied, should be kept and disposed of away from heat, sparks, and flame. Store EGBE in carbon steel, stainless steel, or Teflon containers. Do not store in aluminium, copper, galvanised iron, or galvanised steel. Do not use Viton, neoprene, nitrile, or natural rubber gaskets or seals. Avoid contact with strong acids, strong bases, and strong oxidisers. EGBE can oxidise at elevated temperatures. EGBE is thermally stable at typical use temperatures, but can oxidise at elevated temperatures. It should not be distilled to dryness because it can form peroxides. Decomposition can cause gas generation and pressure in closed systems. Thermal decomposition products can include and are not limited to aldehydes, ketones, and organic acids. Spills of EGBE on hot, fibrous insulations may result in spontaneous combustion by lowering the auto-ignition temperatures.

After extensive review of EGBE toxicity and exposure data, the EPA removed it from its list of Hazardous Air Pollutants (HAPs) in November 2004.

Guthion (CAS No. 86-50-0)

Molecular formula: $C_{10}PN_3H_{12}S_2O_3$

Guthion, also called azinphos methyl, is an organophosphorous pesticide that was used on many crops, especially apples, pears, cherries, peaches, almonds, and cotton. Many of its former uses have been cancelled by the EPA, and its few remaining uses are currently in the process of being phased out. Guthion is a synthetic substance; it does not occur naturally. Pure guthion is a colourless to white odourless crystalline solid that melts at about 72°C–74°C (162°F–165°F). Technical-grade guthion is a cream to yellow-brown granular solid. Guthion is poorly soluble in water. Pure guthion is a colourless to white odourless crystalline solid. Technical-grade guthion is a cream to yellow-brown granular solid.

Safe Handling and Precautions

Exposure to guthion by the general population occurs primarily to workers who ingest contaminated foods treated with farm product, chemical sprayers, and people who work in factories that make guthion. Exposure to guthion is most likely to occur by skin contact and inhalation. Individuals may also be exposed by going into fields too soon after spraying operations and if a family member works with guthion and residues of the chemical remain on his or her hands, clothing, or vehicle. Exposure to guthion may occur primarily by ingesting food (mostly fruits) treated with this pesticide. Exposure to high amounts of guthion may cause difficulty in breathing, chest tightness, vomiting, cramps, diarrhoea, blurred vision, sweating, headaches, dizziness, loss of consciousness, and death. Guthion has been found in at least 5 of the 1699 National Priority List (NPL) sites identified by the EPA. Most of the guthion that you may ingest will enter the bloodstream, but much less will enter if there is contact with the skin. Guthion interferes with the normal way that the nerves and brain function. If persons who are exposed to high amounts of guthion are rapidly given appropriate treatment, there may be no long-term harmful effects. If people are exposed to levels of guthion below those that affect nerve function, few or no health problems seem to occur. It is not well known how guthion affects the ability of humans to reproduce, while reports indicate that exposure to guthion did not affect fertility in animal studies. It has not been known whether guthion causes cancer in humans. Guthion was not carcinogenic in male or female mice or in female rats that were fed for more than 1 year. Some tumours were observed in male rats, but it could not be conclusively shown that guthion had caused the tumours. The Department of Health and Human Services (DHHS), International Agency for Research on Cancer (IARC), and U.S. EPA have not listed or classified guthion as carcinogenic.

Users and occupational workers, during handling guthion, should stay away from agricultural areas that have been treated with the pesticide. After the spraying operations of guthion, workers should remain indoors or leave the area for a short time. Agricultural workers who come in contact with guthion should remove contaminated clothing and wash before coming in contact with family members. Always wash fruits and vegetables before consuming them. If you pick your own fruit in an orchard, wash your hands when you get home because guthion can be absorbed through the skin. Children should avoid playing in soils near uncontrolled hazardous waste sites where guthion may have been discarded. Guthion, like other organophosphorous pesticides, interferes in the human body with an enzyme called acetylcholinesterase. A blood test that measures this enzyme in the plasma or red blood cells may be useful for detecting exposures to potentially harmful levels of a variety of pesticides, including guthion.

For more information refer to Azinphos methyl.

Bibliography

Agency for Toxic Substances and Disease Registry (ATSDR). 1998. *Toxicological Profile for 2-Butoxyethanol and 2-Butoxyethanol Acetate*. U.S. Public Health Service, U.S. Department of Health and Human Services, Atlanta, GA.

Agency for Toxic Substances and Disease Registry (ATSDR). 2004. *Toxicological Profile for Synthetic Vitreous Fibers*. ATSDR., U.S. Department of Health and Human Services. Public Health Service. Atlanta, GA.

Dement, J. M. 1975. Environmental aspects of fibrous glass production and utilization. *Environ. Res.* 9:295–312.

Gao, H. G., W. Z. Whong, W. G. Jones, W. E. Wallace, and T. Ong. 1995. Morphological transformation induced by glass fibers in BALB/c-3T3 cells. *Teratog. Carcinog. Mutagen.* 15(2):63–71.

Gosselin, R. E., R. P. Smith, H. C. Hodge, and J. E. Braddock. 1984. *Clinical Toxicology of Commercial Products*, 5th edn. Williams & Wilkins, Baltimore, MD, p. II-204.

Hesterberg, T. W. and G. A. Hart. 2001. Synthetic vitreous fibers: A review of toxicology research and its impact on hazard classification. *Crit. Rev. Toxicol.* 31(1):1–53.

Husain, A. N. 2010. The lung. In: *Robbins and Cotran Pathologic Basis of Disease*, 8th edn. Kumar, V., Abbas, A. K., Fausto, N., and Aster, J. (eds.). Elsevier Health Sciences., Philadelphia, PA, p. 723.

International Agency for Research on Cancer (IARC). 2002. Man-made vitreous fibres, *IARC Monographs on the Evaluation of Carcinogenic Risk of Chemicals to Humans*, Vol. 81, IARC, Lyon, France.

Material Safety Data Sheet (MSDS). 2003. Safety data for gadolinium. Department of Physical and Theoretical Chemistry, Oxford University, Oxford, U.K.

Material Safety Data Sheet (MSDS). 2005. Safety data for gallium (III) nitrate hydrate. Department of Physical and Theoretical Chemistry, Oxford University, Oxford, U.K.

Moorman, W. J., R. T. Mitchell, A. T. Mosberg, and D. J. Donofrio. 1988. Chronic inhalation toxicology of fibrous glass in rats and monkeys. *Ann. Occup. Hyg.* 32(Suppl 1):757–767.

National Institute for Occupational Safety and Health (NIOSH). 1997. *Pocket Guide to Chemical Hazards*. U.S. Department of Health and Human Services, Public Health Service, Centers for Disease Control and Prevention. Cincinnati, OH.

National Institute for Occupational Safety and Health (NIOSH). Agency for Toxic Substances and Disease Registry (ATSDR). 2011. Guthion. NIOSH. ATSDR. Centers for Disease Control and Prevention. Atlanta, GA.

National Toxicology Program (NTP). 2009. Report on carcinogens final background document for glass wool fibers. NTP, Research Triangle Park, NC.

Nguea, H. D., A. de Reydellet, A. Le Faou, M. Zaiou, and B. Rihn. 2008. Macrophage culture as a suitable paradigm for evaluation of synthetic vitreous fibers. *Crit. Rev. Toxicol.* 38(8):675–695.

Pradyot, P. (ed.). 2002. *Handbook of Inorganic Chemicals*. McGraw-Hill, New York.

Sax, N. I., and Lewis, R. J. Jr. (eds.). 1989. *Dangerous Properties of Industrial Materials*, 7th edn. 3 volumes. Van Nostrand Reinhold, New York.

Shannon, H., A. Muir, T. Haines, and D. Verma. 2005. Mortality and cancer incidence in Ontario glass fiber workers. *Occup. Med.* 55(7):528–534.

Sittig, M. (ed.). 1985. *Handbook of Toxic and Hazardous Chemicals and Carcinogens*, 2nd edn. Noyes Publications, Park Ridge, NJ.

Sittig, M. (ed.). 1991. *Handbook of Toxic and Hazardous Chemicals*, 3rd edn. Noyes Publications. Park Ridge, NJ.

U.S. Environmental Protection Agency (U.S. EPA). 1999. Integrated Risk Information System (IRIS). 2-Metoxyethanol. National Center for Environmental Assessment, Office of Research and Development, Washington, DC.

U.S. Environmental Protection Agency (U.S. EPA). 2001. Combined chronic toxicity, carcinogenicity testing of respirable fibrous particles. Health effects test guidelines: OPPTS 870.8355. U.S. EPA., Washington, DC.

10

Hazardous Chemical Substances: H

Hafnium Oxide
Heptachlor
Heptachlor Epoxide
Hexachlorobenzene
Hexachlorobutadiene
Hexachlorocyclopentadiene
Hydrazine
Hydrocyanic Acid

Hafnium Oxide (CAS No. 12055-23-1)

Molecular formula: HfO_2
Synonyms: Hafnis; Hafnium dioxide

Hafnium is a shiny, silvery, ductile metal and resistant to corrosion. The physical properties of hafnium metal samples are markedly affected by zirconium impurities, especially the nuclear properties, as these two elements are among the most difficult to separate because of their chemical similarity.

Hafnia is used in optical coatings and as a high-k dielectric in dynamic random-access memory (DRAM) capacitors. Hafnium (IV) oxide is a colourless, inert solid and has been reported as one of the most common and stable compounds of hafnium. It is an electrical insulator. Hafnium dioxide is an intermediate in some processes that give hafnium metal. It reacts with strong acids and strong bases. It dissolves slowly in hydrofluoric acid. At high temperatures, it reacts with chlorine in the presence of graphite or carbon tetrachloride and forms the hafnium tetrachloride. Hafnium-based oxides are currently important materials to replace silicon oxide as a gate insulator because of its high dielectric constant.

Hafnium (Hf) is found in association with zirconium ores, production based on zircon ($ZrSiO_4$) concentrates which contain 0.5%–2% hafnium. Hafnium has extensive applications in industries especially because of its resistance to corrosion. Different compounds of hafnium used in ceramics industry are hafnium boride, hafnium carbide, hafnium nitride, hafnium oxide, hafnium silicate, and hafnium titanate. Hafnium-based oxides are currently leading candidates to replace silicon oxide as a gate insulator in field-effect transistors. The compound appears to have been chosen by both IBM and Intel as a substrate for future integrated circuits, where it may help in the continuing effort to increase logic density and clock speeds or to lower power consumption, in computer processors. Because of its very high melting point, hafnia is also used as a refractory material in the insulation of devices such as thermocouples, where it can operate at temperatures up to 2500°C.

Safe Handling and Precautions

Exposure to hafnium is harmful. Hafnium is poorly soluble in water, and the pattern of its absorption is not well documented. Many hafnium compounds have been reported to cause poisoning by unspecified routes. Occupational workers on exposure to hafnium develop adverse health effects such as inflammation and irritation of the eyes, coughing, sneezing, and pulmonary irritation. Prolonged period of exposure to hafnium and its dust is known to cause liver damage and a benign pneumoconiosis. The accumulation of dust causes ventilatory effect, and without special predisposition to tuberculosis and/or lung cancer as encountered in *silicosis* and *asbestosis*. There is no published data indicating that hafnium oxide is a carcinogen and causes cancer among humans. In brief, information regarding the chemical, physical, and toxicological aspects of hafnium and hafnium oxide is sketchy in literature.

Regulations:

- The American Conference of Governmental Industrial Hygienists (ACGIH) has set the threshold limit value (TLV) for hafnium oxide at 0.5 mg/m^3.
- The U.S. Occupational Safety and Health Administration (U.S. OSHA) has set the permissible exposure limit (PEL) for hafnium oxide at 0.5 mg(Hf)/m^3.

Heptachlor (CAS No. 76-44-8)

Molecular formula: $C_{10}H_5Cl_7$
Synonyms and trade names: Arabinex 30tn; Dicyclopentadiene, 3,4,5,6,7,8,8a-heptachloro-; Drinox; Drinox h-34; E 3314; Heptachlorane; Heptagran; Heptamul; Heptox; Rhodiachlor; Velsicol 104; Velsicol heptachlor

Heptachlor is a soft, white to light tan, waxy, non-combustible, crystalline solid with a camphor-like odour. Heptachlor is a member of the cyclodiene group of chlorinated insecticides (aldrin, dieldrin, endrin, chlordane, heptachlor, and endosulfan) and has a long history following World War II. It was registered as a commercial pesticide in 1952 for foliar, soil, and structure applications and for malarial control programmes; after 1960, it was used primarily in soil applications against agricultural pests and to a lesser extent against termites.

Heptachlor is available commercially as a dust, a dust concentrate, an emulsifiable concentrate, a wettable powder, or in oil solutions. It is corrosive to metals and reacts with iron and rust to form hydrogen chloride gas. Heptachlor is incompatible with many amines, nitrides, azo/diazo compounds, alkali metals, and epoxides but is stable under normal temperatures and pressures. It may burn, but does not ignite readily. Heptachlor at high heat and temperature produces highly toxic, corrosive fumes of hydrogen chlorine gas and toxic oxides of carbon. An important metabolite of heptachlor is heptachlor epoxide which is an oxidation product formed from heptachlor by many plant and animal species. Heptachlor is almost insoluble in water but soluble in ether, acetone, benzene, and many other organic solvents.

Safe Handling and Precautions

Exposure to heptachlor occurs at workplaces through inhalation, ingestion, eye or skin contact, and absorption through the skin and is harmful. Heptachlor is likely to cause tremors, convulsions, and other central nervous system (CNS) effects in humans on acute exposure. Reports indicate that persons who lived in homes treated with heptachlor for the control of termites suffered poisoning with symptoms of blood dyscrasias and leukaemias. Chronic oral exposure to heptachlor/heptachlor epoxide caused an increase of liver carcinomas in experimental rats. The ACGIH lists heptachlor as an animal carcinogen. The U.S. Environmental Protection Agency (U.S. EPA) classified heptachlor and heptachlor epoxide as probable human carcinogens. The International Agency for Research on Cancer (IARC) observed that evidences are inadequate to list heptachlor as a carcinogen to humans.

Regulations:

- The ACGIH has set a TLV for heptachlor at 0.5 mg/m^3 (10 h – TWA).
- The OSHA has set the PEL for heptachlor at 0.5 mg/m^3 (of air 8 h – TWA).
- The National Institute for Occupational Safety and Health (NIOSH) has set the recommended exposure limit (REL) for heptachlor at 0.5 mg/m^3 (10 h – TWA).

It is important to know that use of heptachlor is restricted or no longer permitted in some countries.

Occupational workers during use and handling of heptachlor should use appropriate and adequate personal protective clothing and equipment (PPE) to prevent skin contact with heptachlor. The selection of the appropriate PPE such as gloves, sleeves, and encapsulating suits should be based on the extent of the occupational worker's potential exposure to heptachlor. Workers should store heptachlor in a cool, dry, well-ventilated area in tightly sealed containers with proper label.

Heptachlor Epoxide (CAS No. 1024-57-3)

Molecular formula: $C_{10}H_5Cl_7O$

Heptachlor epoxide is also a white powder. Bacteria and animals break down heptachlor to form heptachlor epoxide. The epoxide is more likely to be found in the environment than heptachlor. Heptachlor epoxide is a degradation product of heptachlor that occurs in soil and in or on crops when treatments with heptachlor, an insecticide, have been made. It forms readily upon exposing heptachlor to air. The U.S. EPA lists heptachlor epoxide as a possible human carcinogen.

Hexachlorobenzene (CAS No. 118-74-1)

Molecular formula: C_6Cl_6
Synonyms: No bunt; HCB

Hexachlorobenzene is a white crystalline solid. This compound does not occur naturally. It is formed as a by-product during the manufacture of chemicals used as solvents (substances used to dissolve other substances), other chlorine-containing compounds, and pesticides. Small amounts of hexachlorobenzene can also be produced during combustion processes such as burning of city wastes. It may also be produced as a by-product in waste streams of chlor-alkali and wood-preserving plants. Hexachlorobenzene was widely used as a pesticide until 1965. It was also used to make fireworks, ammunition, and synthetic rubber. Currently, the substance is not used commercially in the United States.

Hexachlorobutadiene (CAS No. 87-68-3)

Molecular formula: C_4Cl_6
Synonyms and trade names: HCBD; 1,3-Hexachlorobutadiene; Perchlorobutadiene

Hexachlorobutadiene is a colourless liquid that smells like turpentine. Hexachlorobutadiene is formed as a by-product during the manufacture of carbon tetrachloride and tetrachloroethylene. It is a chlorinated aliphatic diene with several applications and commonly used as a solvent for other chlorine-containing compounds. Applications of hexachlorobutadiene in industries are extensive, used as a solvent for rubber and other polymers, in heat transfer fluids, as a transformer liquid, as a hydraulic fluid, as a solvent and to make lubricants, and as a washing liquor for removing hydrocarbons from gas streams. The uses also include as a seed dressing and fungicide and in manufacturing processes such as production of aluminium and graphite rods. Thus, the major source of hexachlorobutadiene has been reported because of inadvertent production as a waste by-product of the manufacture of certain chlorinated hydrocarbons such as tetrachloroethylene, trichloroethylene, and carbon tetrachloride.

Safe Handling and Precautions

Occupational exposure to hexachlorobutadiene is harmful and has been observed to produce systemic toxicity and poisoning. Laboratory studies indicated that experimental rats and mice on exposure to hexachlorobutadiene through inhalation route showed an increase in the number of damaged cortical renal tubules (in mice) and kidney enlargement (in rats). Laboratory animals on exposure through oral, inhalation, and dermal routes developed health disorders such as fatty liver degeneration, epithelial necrotising nephritis, CNS depression, and cyanosis. There is no available data on the genetic and related effects of hexachlorobutadiene in humans or in rodents in vivo.

The U.S. EPA classified hexachlorobutadiene as a Group C, meaning a possible human carcinogen. The ACGIH has classified hexachlorobutadiene as an A 3, meaning a confirmed animal carcinogen with unknown relevance to humans. Reports of IARC have shown that evidences on the carcinogenicity of hexachlorobutadiene experimental animals are limited, while evidences are *inadequate* in humans. Hence, hexachlorobutadiene is listed as Group 3, meaning *not classifiable as a carcinogen to humans.*

Regulations:

- The ACGIH has set the TLV for hexachlorobutadiene at 0.02 ppm (0.21 mg/m^3 – TWA skin).
- The OSHA has set no PEL for hexachlorobutadiene.
- The NIOSH has set the REL for hexachlorobutadiene at 0.02 ppm (0.24 mg/m^3 – Ca TWA skin).

Hexachlorocyclopentadiene (CAS No. 77-47-4)

Molecular formula: C_5Cl_6
Synonyms and trade names: HCCPD; Hexachloro-1,3-cyclopentadiene; 1,2,3,4,5, 5-Hexachloro-1,3-cyclopentadiene; Perchlorocyclopentadiene

Hexachlorocyclopentadiene is a pale-yellow/lemon-yellow liquid with a characteristic musty or pungent odour (odour threshold – 0.03 ppm). Hexachlorocyclopentadiene does not occur naturally but is a manufactured chemical. It easily evaporates into the air. Hexachlorocyclopentadiene is the key intermediate in the manufacture of some pesticides, including heptachlor, chlordane, aldrin, dieldrin, and endrin. Hexachlorocyclopentadiene is also used in the manufacture of flame retardants and some resins, shock-proof plastics, fluorocarbons, and dyes. Hexachlorocyclopentadiene quickly breaks down by sunlight and reacts with other chemicals in the air.

Safe Handling and Precautions

Occupational exposure to hexachlorocyclopentadiene is harmful and very toxic to workers. Hexachlorocyclopentadiene causes poisoning with symptoms of severe lacrimation (discharge of tears), sneezing, cough, dyspnoea/breathing difficulty, salivation, pulmonary oedema, nausea, vomiting, and diarrhoea. Direct skin contact is known to cause blisters and burns; accidental inhalation of hexachlorocyclopentadiene in high concentrations has been reported to cause poisoning among workers with symptoms such as sore throat, shortness of breath, cough, chest discomfort, headache, difficulty in breathing, nervousness, and abdominal cramps. Hexachlorocyclopentadiene caused in the laboratory animals poisoning with injuries to the liver and kidneys. The U.S. EPA classified hexachlorocyclopentadiene as a Group D, meaning not classifiable as to human carcinogenicity. The IARC has not classified hexachlorocyclopentadiene for carcinogenicity.

Occupational workers during use and handling of hexachlorocyclopentadiene should strictly use workplace safety dress, splash goggles, and adequate and appropriate personal protective suit (PPE).

Regulations:

- The ACGIH has set the TLV of hexachlorocyclopentadiene at 0.01 ppm.
- The U.S. NIOSH has set the REL for hexachlorocyclopentadiene at 0.01 ppm (0.1 mg/m^3).
- There is information available on PEL for hexachlorocyclopentadiene. The OSHA is of the opinion to set a PEL for hexachlorocyclopentadiene at 0.01 ppm for an 8 h workday in workplace air which suggests substantial reduction in chemical risks.

Danger! Alert! Improper use of hexachlorocyclopentadiene and inhalation cause fatal injury and corrosion to eyes and skin. Liquid or spray mist may produce tissue damage particularly on mucous membranes of eyes, mouth, and respiratory tract. Reported to cause allergic skin reactions in susceptible individuals.

Hydrogen Iodide Anhydrous (CAS No. 10034-85-2)

Molecular formula: **HI**
Synonyms and trade names: Anhydrous hydriodic acid; Hydroiodic acid

Hydrogen iodide is a colourless to yellow/brown with an acrid odour non-flammable gas. Hydrogen iodide is incompatible with water and other halides. Hydrogen iodide, upon contact with moisture in air, releases dense vapours. Hydrogen iodide reacts with water to form corrosive acids and reacts violently with alkalis. Most metals corrode rapidly on contact with wet hydrogen iodide, and prolonged exposure of hydrogen iodide to fire or intense heat has been reported to cause the container to rupture and rocket.

Safe Handling and Precautions

Occupational exposure to hydrogen iodide causes harmful effects and poisoning. Occupational workers on prolonged exposure to hydrogen iodide develop poisoning and respiratory system damage by inhalation, and the symptoms include respiratory tract burns, skin burns, eye burns, and mucous membrane burns. It strongly irritates skin, eyes, and mucous membranes. Long-term inhalation of low concentrations or short-term inhalation of high concentrations may result in adverse health effects. The IARC, the National Toxicology Program (NTP), and the U.S. OSHA have not listed hydrogen iodide as a carcinogen to humans.

Occupational workers during handling of hydrogen iodide should strictly follow workplace safety regulations and avoid negligence to wear workplace safety dress and full self-contained breathing apparatus (SCBA) for protection against possible exposure.

Danger! Alert! Improper handling of hydrogen iodide causes severe health disorders such as, skin burns, eye damage, injury to the respiratory system, pulmonary oedema, chemical pneumonitis and fatal injury.

Hydrazine (CAS No. 302-01-2)

Molecular formula: N_2H_4
Synonyms and trade names: Diamine; Diazane

Hydrazine is a colourless liquid. It is a strong reducing agent and a flammable liquid and vapour. Hydrazine is a useful building block in organic synthesis of pharmaceuticals and pesticides.

Hydrazines are clear, colourless liquids with an ammonia-like odour. There are many kinds of hydrazine compounds, including hydrazine, 1,1-dimethylhydrazine, and 1,2-dimethylhydrazine. Small amounts of hydrazine occur naturally in plants. Most hydrazines are manufactured for use as rocket propellants and fuels, boiler water treatments, chemical reactants, medicines, and in cancer research. Hydrazines are highly reactive and easily catch fire.

Humans and occupational workers whenever exposed to hydrazines through contaminated soil or drinking or swimming at hazardous waste sites and military bases suffer of poisoning. The symptoms of acute and chronic hydrazine poisoning include, but not limited to, eye and skin burns, visual impairmeant, temporary blindness, digestive tract burns, damage to mucous membranes of the eyes, nose, mouth, and respiratory tract, chronic bronchitis, pulmonary edema, skin sensitization, allergic reaction, liver and kidney damage, anemia, blood abnormalities, cardiovascular disorders, nervous system disorders convulsions, possibly coma and severe over-exposure to hydrazines result in death.

Safe Handling and Precautions

Exposure to hydrazine is known to cause severe respiratory tract irritation. Symptoms may include coughing, shortness of breath, dizziness, nausea, and vomiting. Higher concentrations can cause trembling and convulsions. Hydrazine has been listed by the ACGIH as Group A3, meaning a confirmed animal carcinogen with unknown relevance to humans. The IARC has listed hydrazine as Group B2 carcinogen and the NTP as a suspected carcinogen. Hydrazine has shown a high tumour-generating potential in multiple studies. Hydrazine has been classified as carcinogenic in many rodent studies following long-term administration. The major target tissues include liver, lungs, and respiratory tract.

Alert! Danger!! Users and occupational workers should remember that hydrazine is incompatible with many chemical substances. Hydrazine is a highly reactive reducing agent. It is incompatible with oxidising agents (including air), acids, and some metal oxides and metals. Substance may spontaneously ignite in air when in contact with porous materials. It ignites on contact with dinitrogen oxide and tetroxide, hydrogen peroxide, tetryl, and nitric acid. It explodes on contact with dicyanofurazan, *n*-halomides, potassium, silver compounds, sodium hydroxide, titanium compounds, and trioxygen difluoride. Explosive compounds may result from contact with air, chloromethylnitrobenzene, lithium perchlorate, metal salts, methanol + nitromethane, sodium, and sodium perchlorate. Also incompatible with barium oxide or calcium oxide, benzeneseleninic acid or anhydride, calcium, carbon dioxide + stainless steel, 1-chloro-2,4-dinitrobenzene, cotton waste + heavy metals, (difluoroamino)difluoroacetonitrile, iodine pentoxide, rust, ruthenium (III) oxide, and thiocarbonyl azide thiocyanate. Workers should avoid light, ignition sources, moisture, and temperatures above 150°C. Hydrogen should be kept in a tightly closed container. Store it in a cool, dry, ventilated area away from sources of heat or ignition. Protect against physical damage. Store separately from reactive or combustible materials and out of direct sunlight.

Hydrocyanic Acid (CAS No. 74-90-8)

Molecular formula: **HCN**
Synonyms and trade names: Hydrocyanic acid; Prussic acid; Formonitrile; Carbon hydride nitride; Hydrocyanic acid, liquefied; Hydrogen cyanide; UN 1051 Formic anammonide; Methanenitrile; Prussic acid; Zyklon B
Chemical family: Inorganic gas
IUPAC name: Formonitrile; Hydridonitridocarbon

Hydrocyanic acid is a colourless liquid with almond odour. Hydrogen cyanide is a colourless or pale-blue liquid at room temperature. It is very volatile, readily producing flammable and toxic concentrations at room temperature. Hydrogen cyanide gas mixes well with air, and explosive mixtures are easily formed. Hydrogen cyanide has a distinctive bitter almond odour, but some individuals cannot detect it, and consequently, it may not provide adequate warning of hazardous concentrations. Hydrogen cyanide is incompatible and reacts with amines, oxidisers, acids, sodium hydroxide, calcium hydroxide, sodium carbonate, caustic substances, and ammonia. Hydrogen cyanide may polymerise.

It has extensive applications in industries and is produced on an industrial scale and is a highly valuable precursor to many chemical compounds ranging from polymers to pharmaceuticals.

HCN is the precursor to sodium cyanide and potassium cyanide, which are used mainly in gold and silver mining and for the electroplating of those metals, via the intermediacy of cyanohydrins; a variety of useful organic compounds are prepared from HCN including the monomer methyl methacrylate, from acetone, the amino acid methionine, via the Strecker synthesis, and the chelating agents EDTA and NTA. Via the hydrocyanation process, HCN is added to butadiene to give adiponitrile, a precursor to Nylon 66. In short, hydrogen cyanide is used in fumigating, electroplating, mining, and in producing synthetic fibres, plastics, dyes, and pesticides. It also is used as an intermediate in chemical syntheses.

Safe Handling and Precautions

Hydrogen cyanide is readily absorbed from the lungs; symptoms of poisoning begin within seconds to minutes. Exposure to hydrocyanic acid is known to cause skin irritation, rash, nausea, chest pain, irregular heartbeat, headache, blindness, bluish skin colour, suffocation, and lung congestion. Workplace negligence and inhalation and ingestion/swallowing of hydrogen cyanide cause respiratory tract irritation, paralysis, convulsions, coma, and death. Occupational workers should strictly avoid any kind of contact of the spilled material and avoid heat, flames, sparks, and other sources of ignition at the workplace. Workplace management should deny entry to unauthorised/unnecessary people.

Bibliography

Agency for Toxic Substances and Disease Registry (ATSDR). 1995. *Toxicological Profile for Hexachlorobutadiene*. U.S. Department of Health and Human Services, Atlanta, GA.

Agency for Toxic Substances and Disease Registry (ATSDR). 1999a. *Toxicological Profile for Hexachlorocyclopentadiene*. U.S. Department of Health and Human Services, Public Health Service, Atlanta, GA (updated 2011).

Agency for Toxic Substances and Disease Registry (ATSDR). 1999b. *Toxicological Profile for Hydrazines*. U.S. Department of Health and Human Services, Public Health Service. Atlanta, GA.

Agency for Toxic Substances and Disease Registry (ATSDR). 2002. *Toxicological Profile for Hexachlorobenzene*. U.S. Department of Health and Human Services, Public Health Service, Atlanta, GA (updated 2011).

Agency for Toxic Substances and Disease Registry (ATSDR). 2007. *Toxicological Profile for Heptachlor and Heptachlor Epoxide*. U.S. Department of Health and Human Services, Public Health Service, Atlanta, GA (updated 2011).

Amita Rani, B. E. and M. K. Krishnakumari. 1995. Prenatal toxicity of heptachlor in albino rats. *Pharmacol. Toxicol.* 76(2):112–114.

Bhide, S. V., R. A. D'Souza, M. M. Sawai et al. 1976. Lung tumour incidence in mice treated with hydrazine sulphate. *Int. J. Cancer* 18(4):530–535.

Burgess, W. A. 1995. *Recognition of Health Hazards in Industry: A Review of Materials and Processes*, 2nd edn. John Wiley & Sons, New York.

Dikshith, T. S. S. (ed.). 1991. *Toxicology of Pesticides in Animals*, CRC Press, Boca Raton, FL.

Dikshith, T. S. S. (ed.). 2011. *Handbook of Chemicals and Safety*. CRC Press, Boca Raton, FL.

Gosselin, R. E., R. P. Smith, and H. C. Hodge (eds.). 1984. *Clinical Toxicology of Commercial Products*. Williams & Wilkins, Baltimore, MD.

Hayes, W. J. and E. R. Laws (eds.). 1990. *Handbook of Pesticide Toxicology*, Vol. 3, Classes of Pesticides. Academic Press, New York.

International Programme on Chemical Safety and the Commission of the European Communities. (IPCS–CEC). 1998. Hydrogen iodide. ICSC Card No. 1326. IPCS-CEC, Luxembourg, Belgium.

International Programme on Chemical Safety and the Commission of the European Communities (IPCS CEC). 2003. Heptachlor. ICSC Card No. 07 43. IPCS-CEC, Luxembourg, Belgium (updated 2000).

Kidd, H. and D. R. James (eds.). 1991. *The Agrochemicals Handbook*, 3rd edn. Royal Society of Chemistry Information Services, Cambridge, U.K.

Lewis, R. J. (ed.). 2001. *Hawley's Condensed Chemical Dictionary*, 14th edn. John Wiley & Sons, New York.

Lewis R. J., Sr. (ed.). 2004. *Sax's Dangerous Properties of Industrial Materials*. Van Nostrand Reinhold, New York.

Lewis, R. J., Sr. 2007. *Hawley's Condensed Chemical Dictionary*. Wiley-Interscience, John Wiley & Sons, New York.

Lewis, R. J. and R. J. Lewis, Sr. (eds.). 2008. *Hazardous Chemicals Desk Reference*. John Wiley & Sons, Hoboken, NJ.

Manfred Rossberg, M. et al. 2006. Chlorinated hydrocarbons. In: *Ullmann's Encyclopedia of Industrial Chemistry*. Wiley-VCH Verlag GmbH & Co, Germany.

Martel, B. and K. Cassidy. 2004. *Chemical Risk Analysis: A Practical Handbook*. Butterworth-Heinemann, London, U.K.

Material Safety Data Sheet (MSDS). 1997a. Hexachlorobutadiene. ICSC Card No. 0896. International Programme on Chemical Safety (IPCS) Geneva, Switzerland.

Material Safety Data Sheet (MSDS). 1997b. Hydrazine sulfate. MSDS No. H3633.

Material Safety Data Sheet (MSDS). 2003. Safety data for hexachlorobenzene. Physical Chemistry, Oxford University, Oxford, U.K.

Material Safety Data Sheet (MSDS). 2005a. Hexachlorocyclopentadiene. ICSC Card No. 1096. International Programme on Chemical Safety (IPCS) Geneva, Switzerland.

Material Safety Data Sheet (MSDS). 2005b. Safety data for heptachlor. Department of the Physical and Theoretical Laboratory, Oxford University, Oxford, U.K.

National Institute for Occupational Safety and Health (NIOSH). Hexachlorocyclopentadiene. 1988. OSHA PEL project documentation. Centers for Disease Control and Prevention. Atlanta, GA (updated 2011).

National Library of Medicine (NLM). 1992. Heptachlor. Hazardous Substances Data Bank (HSDB). NLM. Bethesda, MD.

National Toxicology Program (NTP). 2002. Hydrazine and hydrazine sulfate. *Rep. Carcinog.* 10:138–139.

Patnaik, P. (ed.). 1992. *A Comprehensive Guide to the Hazardous Properties of Chemical Substances,* 1st edn. Van Nostrand Reinhold, New York.

Patnaik, P. (ed.). 1999. *A Comprehensive Guide to the Hazardous Properties of Chemical Substances,* 2nd edn. McGraw Hill, New York.

Sax, N. I. and R. J. Lewis, Jr. 1989. *Dangerous Properties of Industrial Materials,* 7th edn., 3 vols. Van Nostrand Reinhold, New York.

Sittig, M. (ed.). 1985. *Handbook of Toxic and Hazardous Chemicals and Carcinogens,* 2nd edn. Noyes Publications, Park Ridge, NJ.

Sittig, M. (ed.). 1991. *Handbook of Toxic and Hazardous Chemicals,* 3rd edn. Noyes Publications, Park Ridge, NJ.

Sullivan, M. D., Jr., B. John, and M. D. Gary R Krieger. 1992. *Hazardous Materials Toxicology: Clinical Principles of Environmental Health.* William & Wilkins, Baltimore, MD.

Tomlin, C. D. S. (ed.). 2006. *The Pesticide Manual—A World Compendium,* 14th edn. British Crop Protection Council (BCPC), Farnham, Surrey, U.K.

Toth, B. 1969. Lung tumor induction and inhibition of breast adenocarcinomas by hydrazine sulfate in mice. *J. Natl. Cancer Inst.* 42(3):469–475.

U.S. Environmental Protection Agency (U.S. EPA). 1984. Health assessment document for hexachlorocyclopentadiene. Environmental Criteria and Assessment Office, Office of Health and Environmental Assessment, Office of Research and Development, Cincinnati, OH.

U.S. Environmental Protection Agency (U.S. EPA). 1991. Hexachlorobutadiene. Integrated Risk Information System. U.S. EPA, Washington, DC.

U.S. Environmental Protection Agency (U.S. EPA). 1999. Integrated Risk Information System (IRIS). On hexachlorocyclopentadiene. National Center for Environmental Assessment, Office of Research and Development, Washington, DC.

World Health Organization (WHO). International Programme on Chemical Safety. 1991. Hexachlorocyclopentadiene. Environmental Health Criteria. No. 120. Geneva, Switzerland.

World Health Organization (WHO). 1993. Health and safety guide for hexachlorobutadiene. Health and Safety Guide No. 84. WHO, Geneva, Switzerland.

11

Hazardous Chemical Substances:

I

Iodine
Iodine Cyanide (Cyanogen Iodide)
Iodoethane
Iodoform
Iodomethane

Iodine (CAS No. 7553-56-2)

Molecular formula: I_2
Synonyms and trade names: Eranol, Iodin (French), Iodine Colloidal, Iodine Crystals, Iodine Sublimed, Iodine-127, Iodio (Italian), Jod (German, Polish), Jood (Dutch), Iode, Iodum, Jodum, Yodo

Iodine is a non-metallic solid element and also found in small quantities in sea water, in some seaweeds, and in various mineral and medicinal springs. Deep-sea weeds as a rule contain more iodine than those which are found in shallow waters. Iodine is a naturally occurring element that is essential for the good health of people and animals. Iodine is found in small amounts in sea water and in certain rocks and sediments. Iodine occurs in many different forms and colors such as blue, brown, greyish-black shining solid. Iodine somewhat resembles graphite. Iodine is incompatible with ammonia, acetylene, acetaldehyde, powdered aluminium, and active metals. Iodine possesses a characteristic penetrating smell, not so pungent, however, as that of chlorine or bromine. It is only very sparingly soluble in water but dissolves readily in solutions of the alkaline iodides and in alcohol, ether, chloroform, and many liquid hydrocarbons. Its solutions in the alkaline iodides and in alcohol and ether are brown in colour and appear violet in colour with chloroform and carbon bisulphide solutions. Iodine is not combustible but enhances combustion of other substances. Upon heating, iodine releases toxic fumes. The substance is a strong oxidant and reacts with combustible and reducing materials and reacts violently with many chemical substances. Iodine is an oxidiser, and its contact should be avoided with strong reducing agents, powdered metals, ammonia, ammonium salts, acetylene, acetaldehyde, combustible materials, aluminium, chemically active metals, carbides, turpentine oils, azides, carbides, ammonium hydroxide, and sodium thiosulphate, and the possible workplace fire and explosion hazard should be avoided. The soluble iodides, on the addition of silver nitrate to their nitric acid solution, give a yellow precipitate of silver iodide, which is insoluble in ammonia solution. Hydriodic acid and iodides may be estimated by conversion into silver iodide.

There are both radioactive and non-radioactive isotopes of iodine. The isotopes of iodine include iodine-123, iodine-124, iodine-125, iodine-129, and iodine-131. Iodine-129 and iodine-131 are the most important radioactive isotopes in the environment. Some isotopes of iodine, such as I-123 and I-124, are used in medical imaging and treatment but are generally not a problem in the environment because they have very short half-lives. Iodine is a trace element necessary to life in very small quantities but deadly at higher concentrations and thus required by all humans in small amounts for healthy growth and development. Radioactive iodine is used in medical tests and to treat certain diseases, such as overactivity or cancer of the thyroid gland.

Iodine is critical to normal good health of animals as well as to that of humans. Iodine is necessary for proper production of thyroid hormone, and an adult body contains about 30 mg of iodine mostly concentrated in the thyroid gland where it is needed to synthesise thyroid hormones. Iodine deficiency is also the leading cause of preventable mental retardation, a result which occurs primarily when babies or small children are rendered hypothyroidic by a lack of the element. The addition of iodine to table salt has largely eliminated the problem, while iodine deficiency remained a serious public health problem in certain countries around the world. For medicinal purposes, iodine is frequently applied externally as a counterirritant, having powerful antiseptic properties. In the form of certain salts, iodine is very widely used for internal administration in medicine and in the treatment of many conditions usually classed as surgical, bone manifestations. The most commonly used salt is the iodide of potassium; the iodides of sodium and ammonium are almost as frequently employed.

Iodine and its compounds are primarily used in pharmaceuticals, antiseptics, medicine, food supplements, dyes, catalysts, halogen lights, photography, water purifying, and starch detection. The iodine test is used to test for the presence of starch. Iodine solution – iodine dissolved in an aqueous solution of potassium iodide – reacts with starch producing a deep blue-black colour. A tincture consisting of a solution of iodine in ethyl alcohol is applied topically to wounds as an antiseptic. Tincture of iodine is 3% elemental iodine in an ethanol base. It is an essential component of any emergency survival kit, used both to disinfect wounds and to sanitise surface water for drinking.

Safe Handling and Precautions

Exposure to iodine causes adverse health effects and poisoning. The symptoms of iodine poisoning include, but are not limited to, severe irritating effects to the eyes, skin, and nose; lacrimation (discharge of tears); headache; abdominal pain; diarrhoea; nausea; vomiting; chest tightness; skin burns; rash; and cutaneous hypersensitivity. Inhalation of the vapour of iodine causes asthma-like reactions (RADS). Inhalation of the vapour may cause lung oedema. The effects may be delayed. Medical observation is indicated. Repeated and/or prolonged exposures to iodine cause skin sensitisation, asthma-like syndrome (RADS), and effects on the thyroid and are very destructive of mucous membranes and upper respiratory tract. Children are more sensitive to the harmful toxic effects of iodine and radioactive iodine than adults because their thyroid glands are still growing, and thyroid gland tissues are more easily harmed by radioactive iodine, and children need a healthy thyroid gland for normal growth. Also children need iodine to form thyroid hormones, which are important for growth and health. If infants and children do not have enough iodine in their bodies, their thyroid glands will not produce enough thyroid hormone, and they will not grow normally. If they have too much iodine in their bodies, they may develop an enlarged thyroid gland (goitre), which may not produce enough thyroid hormone for normal growth.

Human exposure to radioactive iodine will cause thyroid uptake, as with all iodine, leading to elevated chances of thyroid cancer. In brief, exposure to iodine causes deleterious effects to the eyes, skin, respiratory system, central nervous system, and cardiovascular system of workers who neglect to observe safety precautions and neglect to use safety glasses, nitrile gloves, and effective ventilation facilities.

Iodine Cyanide (Cyanogen Iodide) (CAS No. 506-78-5)

Molecular formula: **CIN**
Synonyms and trade names: Cyanogen iodide; Cyanogen monoiodide; Iodine cyanide; Iodine monocyanide; Iodocyanide; Jodcyan
IUPAC name: Carbononitridic iodide

Iodine cyanide (cyanogen iodide) appears as light pink crystalline or brown-coloured powder and is soluble in water. It is stable but sensitive to light. It is incompatible with strong acids, strong bases, and strong oxidising agents. Cyanogen iodide decomposes on contact with acids, bases, and ammonia alcohols and on heating producing toxic gases including hydrogen cyanide. It reacts with carbon dioxide or slowly with water to produce hydrogen cyanide.

Safe Handling and Precautions

Iodine cyanide (cyanogen iodide) is very readily absorbed through the skin. Iodine cyanide is corrosive, thus can cause burns and harmful effects. On exposures through inhalation, ingestion/swallowing, and/or absorption through skin contact, iodine cyanide causes toxicological health disorders such as irritation to the skin, eyes, and respiratory tract. Exposures to iodine cyanide have been reported to cause adverse effects on the intracellular oxygen metabolism and lead to severe symptoms of poisoning. The signs and symptoms of iodine cyanide include, but are not limited to, headache, vertigo (dizziness), irritation effects to mucous membranes, burning sensation of the mouth and throat, excessive salivation, nausea and vomiting, giddiness, sore throat, confusion, weakness, shortness of breath, tachycardia (rapid heart rate), hypotension (low blood pressure), bradycardia (slow heart rate), cardiac arrhythmias, respiratory depression, convulsions, paralysis, coma, and fatal injury.

During handling of cyanogen iodide, workers should use safety glasses and gloves and wear a positive-pressure, pressure-demand, full facepiece self-contained breathing apparatus (SCBA) and a fully encapsulating, chemical-resistant suit for good ventilation, and the workers also should handle the chemical substance only in a fume cupboard.

Iodoethane (CAS No. 75-03-6)

Molecular formula: C_2H_5I
Synonyms and trade names: Ethyl iodide; Monoiodoethane; Hydriodic ether

Iodoethane is a stable chemical substance. It is incompatible with strong bases, magnesium, and strong oxidising agents and gets discoloured when exposed to light, moisture, and air.

Safe Handling and Precautions

Iodoethane is a vesicant, harmful chemical substance by inhalation, ingestion, and through skin contact. Exposure to iodoethane is known to cause adverse health effects such as sensitisation and narcotic effects to workers. Iodoethane is a possible teratogen and known to cause health risk and harm to the unborn child. Users and occupational workers during handling of iodoethane should use appropriate personal protective equipment (PPE)/workplaces safety dress, gloves and safety glasses, and should have good ventilation. Iodoethane should be kept protected from light, moisture, and air.

Iodoform (CAS No. 75-47-8)

Molecular formula: CHI_3
Synonyms: Triiodomethane; Methane triiodo-

Iodoform is a stable, pale yellow crystalline solid. It is volatile with a characteristic pungent, unpleasant and penetrating odour, but with a sweetish taste. It is incompatible with strong oxidising agents, reducing agents, lithium, and metallic salts such as mercuric oxide, silver nitrate, strong bases, calomel, and tannin. Reports indicate that earlier iodoform was in use as a disinfectant and as an antiseptic or dressing and healing of wounds and sores for pets. Iodoform is the active ingredient in many antiseptic or dressing powders used for dogs and cats to prevent infection.

Safe Handling and Precautions

Exposure to iodoform causes harmful health effects to users and occupational workers. Accidental and workplace ingestion, inhalation, and/or skin contact of iodoform cause poisoning, and the symptoms include irritation of the eyes and skin; lassitude (weakness, exhaustion); dizziness; nausea; incoordination; burns of gastrointestinal tract; central nervous system depression; dyspnoea (breathing difficulty); liver, kidney, and heart damage; and visual disturbance.

Alert! Warning!

Iodoethane is harmful. Any kind of (accidental/negligence) ingestion, swallowing/inhalation and or skin absorption at workplace iodoethane causes irritation to skin, eyes and respiratory tract, affects cardiovascular system, central nervous system, liver and kidneys. The American Conference of Governmental Industrial Hygienists (ACGIH) has set the threshold limit value (TLV – airborne exposure limits) of iodoethane as 0.6 ppm (TWA). Also iodoethane is incompatible with mercuric oxide, silver nitrate, tannin, calomel, strong oxidizers, lithium, metallic salts, strong bases, acetone suggesting caution during handling of the chemical. Also workers should keep stored iodoethane in a tightly closed container, in a cool, dry, ventilated area away from sources of heat, moisture and incompatibilities.

Workers should wear impervious protective clothing (PPE), boots, gloves, lab coat, apron to prevent skin contact, chemical safety goggles.

Danger/Precautions

Users and worker should remember that iodoform is very harmful. Workplace exposures through swallowing, inhalation, and/or skin absorption cause health effects, such as damage to respiratory tract, cardiovascular system, central nervous system, liver, and kidneys.

Iodomethane (CAS No. 74-88-4)

Molecular formula: CH_3I
Synonyms and trade names: Methyl iodide; Monoiodomethane; Halon 10001; Ioguard; RCA waste number U138

Iodomethane is also commonly called methyl iodide. It is a volatile liquid related to methane by replacement of hydrogen. Iodomethane is miscible with common organic solvents. It is colourless, although upon exposure to light, samples develop a purplish tinge. Iodomethane (methyl iodide) is widely used in organic synthesis to deliver a methyl group, via the transformation called methylation. It is naturally emitted by rice plantations in small amounts. It is used for the methylation of phenols or carboxylic acids. Methyl iodide is formed during nuclear accidents by the reaction of organic matter with the 'fission iodine'. It is also used as a fungicide, herbicide, insecticide, or nematocide; as a fire extinguisher; and as a soil disinfectant, replacing bromomethane (which has been banned under the Montreal Protocol). Also, the U.S. Environmental Protection Agency approved its use as a soil fumigant in some cases, although it cannot yet be used in California (a major potential market) due to lack of state approval.

Safe Handling and Precautions

Breathing iodomethane fumes can cause lung, liver, kidney, and central nervous system damage. It causes nausea, dizziness, coughing, and vomiting. Prolonged contact with skin causes burns. Massive inhalation causes pulmonary oedema.

Iodomethane is a possible carcinogen based on its IARC, ACGIH, NTP, or EPA classification. According to the IARC, it is classified as a Group 3 substance (Group 3: the agent is not classifiable as to its carcinogenicity to humans). It causes nausea, dizziness, coughing, and vomiting. Prolonged contact with skin causes burns. Massive inhalation causes pulmonary oedema.

Bibliography

Agency for Toxic Substances and Disease Registry (ATSDR). 2004. *Toxicological Profile for Iodine*. U.S. Department of Health and Human Services, Public Health Service, Atlanta, GA (updated 2011).
Bolt, H. M. and B. Gansewendt. 1993. Mechanisms of carcinogenicity of methyl halides. *Crit. Rev. Toxicol.* 23(3):237–253.

Burgess, W. A. 1995. *Recognition of Health Hazards in Industry. A Review of Materials and Processes*, 2nd edn. John Wiley & Sons, New York.

Dikshith, T. S. S. (ed.). 2011. *Handbook of Chemicals and Safety*. CRC Press, Boca Raton, FL.

International Agency for Research on Cancer (IARC). 1999. Metabolism of iodomethane in the rat. IARC Summaries & Evaluations, Vol. 71. Geneva, Switzerland.

International Programme on Chemical Safety and the Commission of the European Communities (IPCS-CEC). 2004. Iodine. ICSC Card No. 0167. International Chemical Safety Cards. IPCS, CEC, Geneva, Switzerland.

International Programme on Chemical Safety and the Commission of the European Communities (IPCS-CEC). 2005. Iodine cyanide. ICSC No. 0662. IPCS-CEC, Geneva, Switzerland.

Material Safety Data Sheet (MSDS). 1996. Iodoform. MSDS No. 13480. Environmental Health & Safety, Mallinckrodt Baker, Inc., Phillipsburg, NJ.

Material Safety Data Sheet (MSDS). 2003. Safety data for iodoform. Department of Physical and Theoretical Chemistry, Oxford University, Oxford, U.K.

Material Safety Data Sheet (MSDS). 2005. Safety data for iodoethane. No. ICSC. 0479.

Material Safety Data Sheet (MSDS). 2009. Safety data for iodomethane. Department of Physical and Theoretical Chemistry, Oxford University, Oxford, U.K.

Material Safety Data Sheet (MSDS). 2010a. Iodoform. *NIOSH Pocket Guide to Chemical Hazards*. Centers for Disease Control and Prevention (CDCP), Atlanta, GA.

Material Safety Data Sheet (MSDS). 2010b. Safety data for cyanogen iodide. Department of Physical and Theoretical Chemistry, Oxford University, Oxford, U.K.

Material Safety Data Sheet (MSDS). 2011. Safety data for iodine. Department of Physical and Theoretical Chemistry, Oxford University, Oxford, U.K.

Patnaik, P. (ed.). 1999. *A Comprehensive Guide to the Hazardous Properties of Chemical Substances*. McGraw Hill, New York.

Sax, N. I. and R. J. Lewis, Jr. 1989. *Dangerous Properties of Industrial Materials*, 7th edn., 3 volumes. Van Nostrand Reinhold, New York.

World Health Organization (WHO). 2004. Iodoethane. *Substances*, 2nd edn. The International Programme on Chemical Safety & the Commission of the European Communities.

12

Hazardous Chemical Substances: J and K

Janus Green B
Jasmine Oil
Jemmer Stain
Jet Fuel
   Jet Fuel 8
Jojoba Oil
Kepone

Janus Green B (CAS No. 2869-83-2)

Molecular formula: $C_{30}H_{31}ClN_6$
Synonym: 3-(Diethylamino)-7-[[4-(dimethylamino)phenyl]azo]-5-phenylohenazinium chloride

Safe Handling and Precautions

Janus Green B is hazardous in case of ingestion, hazardous in case of direct skin contact, eye contact and or inhalation. No published information is available regarding the carcinogenic, mutagenic, developmental toxicity and the teratogenic effects of Janus Green B.

Students and occupational workers should be careful during handling of Janus Green B and use Personal Protection dress/full suit to avoid large spill, use splash goggles, dust respirator, boots, gloves, approved/certified respirator to avoid inhalation of the product. Workers should keep Janus Green B away from heat, away from sources of ignition.

Jemmer Stain (CAS No. 62851-42-7)

Jemmer Stain is a dark green crystalline powder. The Jenner stain solution is a mixture of several thiazin dyes in a methanol solvent. This stain is for *"In Vitro"* use only.

Safe Handling and Precautions

Exposures to Jemmer stain has been reported to cause eye and skin irritation, respiratory and digestive tract irritation. There is no detailed toxicological data of Jemmer stain.

Jet Fuel

Handling and Precautions

Keep these chemical substances away from heat, away from sources of ignition. Empty containers pose a fire risk, evaporate the residue under a fume hood. Do not breathe dust.

Storage: Keep the chemical container dry, the container tightly closed, and keep it in a cool, well-ventilated place. All combustible materials should be stored away from extreme heat and away from strong oxidising agents. Published literature has been reported as very limited. Readers should refer literature for more information.

Jet Fuel 8 (CAS No. 8008-20-6)

Trade names: JP8; JP-8; AVTUR; MIL-DTL-83133; NATO F-34

JP8 is a kerosene-based fuel used in military jets tanks and other fighting vehicles, and portable heaters. It contains benzene, toluene, xylenes, and naphthalene, as well as additives such as diethylene glycol monomethyl ether or ethylene glycol monomethyl ether.

Chemical description and physical properties: Jet fuel is a mixture of aliphatic, aromatic, and substituted naphthalene hydrocarbon compounds.

On contact, jet fuel causes skin irritation such as itching, burning, redness, rash; dermatitis, headache, fatigue, anorexia, dizziness, difficulty and poor coordination, vomiting, diarrhoea, cramps, drowsiness, restlessness, pneumonitis, irritability, loss of consciousness; and death.

The affected organs include the skin, respiratory system, and the CNS.

Alert!! Dangers of jet fuel exposure.

Jet fuel (JP 5 and JP 8) is highly flammable. Working daily with jet fuel puts people at greater risk of suffering chemical and physical burns. Jet fuel also gives off many chemicals that can damage the lungs and brain. Large and consistent exposures are needed to produce the worst health outcomes, but all people should be careful around the substance that most often comes in formulations labelled JP 5 and JP 8.

Jojoba Oil (CAS No. 61789-91-1)

Molecular formula: Not available
Synonyms: Jojoba bean oil; Jojoba liquid wax; Jojoba oil

Jojoba oil is Oily liquid in nature and has a characteristic fatty odor. Jojoba oil is golden-colored liquid wax, which are produced by the seeds of the jojoba plant Jojoba oil is used for dry and oily skin. It regulates the sebum produced on the skin, as the oil can control the greasy texture of the skin. On dry skin type, the oil acts as a moisturizer. Mix a little quantity of jojoba oil with your regular moisturizer and this works wonders on your skin. Jojoba oil also acts as a lip-balm for dry and chapped lips. Jojoba oil has many benefits, and is best used for the hair and skin.

Jojoba oil has been reported combustible at high temperature. Workers should keep Jojoba oil away from heat and away from sources of ignition. Jojoba oil contain no component at levels greater than or equal to 0.1% is identified as probable, possible or confirmed human carcinogen by International Cancer Research: Cancer (IARC). The toxicological properties of Jojoba oil have not been fully studied and not much published information is available on the safety and hazards of Jojoba oil.

Kepone (CAS No. 143-50-0)

Molecular formula: $C_{10}Cl_{10}O$
Synonyms and trade names: Chlordecone; Dechlorane; Kepone; Ferriamicide; Perchlorodihomocubane
IUPAC name: Dodecachloropentacyclodecane

Chemically, kepone is a chlorinated polycyclic insecticide and fungicide. Chlordecone was first introduced as a pesticide in 1958 and was used until 1978, when its use in the United States was discontinued (NCI 1976, IARC 1979, HSDB 2009). Chlordecone was used as an insecticide for leaf-eating insects, ants, and cockroaches; as a larvicide for flies; and for control of insects that attack structures. Chlordecone was also used on bananas, non-bearing citrus trees, tobacco, ornamental shrubs, lawns, turf, and flowers. The dry powder is readily absorbed through the skin and respiratory tract. Occupational workers handling kepone without appropriate workplace safety dress and PPE suffered chemical poisoning. The symptoms included, tremors, jerky eye movements, memory loss, headaches, slurred speech, unsteadiness, lack of coordination, loss of weight, rash, enlarged liver, decreased libido, sterility, chest pain, and arthralgia (sharp pain, extending along a nerve or group of nerves, experienced in a joint and/or joints). Kepone, also known as chlordecone, is a toxic, non-biodegradable insecticide that a chemical plant in Hopewell, Virginia, dumped into the James River from 1966 until 1975. The chemical's negative effect on the environment was documented and eventually publicised, leading authorities to shut down the Allied Chemical Corporation plant that produced kepone and to order fishing bans and advisories. The environmental and medical scandal was one of the first of its kind to play out nationally, and while it eventually led to the destruction of the Virginia fishing industry, it also led to improved environmental awareness.

Safe Handling and Precautions

Dietary administration of chlordecone caused liver cancer (hepatocellular carcinoma) in rats and mice of both sexes. In addition, the time to detection of the first hepatocellular carcinoma observed at death was shorter in male mice exposed to chordecone than in unexposed controls and appeared to be inversely related to exposure level in mice and rats of both sexes. Kepone (chlordecone) is *reasonably anticipated to be a human carcinogen* based on sufficient evidence of carcinogenicity from studies in experimental animals.

Bibliography

Agency for Toxic Substances and Disease Registry (ATSDR). 1995. *Toxicological Profile for Mirex and Chlordecone*. ATSDR, Atlanta, GA.

Armour, M.-A., 1996. *Hazardous Laboratory Chemical Disposal Guide*. Lewis Publishers, New York.

Bocquene, G. and A. Franco. 2005. Pesticide contamination of the coastline of Martinique. *Mar. Pollut. Bull.* 51(5–7):612–619.

Bretherick, I. 1990. *Handbook of Reactive Chemical Hazards*, 4th edn. CRC Press, Boca Raton, FL.

Burgess, W.A. 1995. *Recognition of Health Hazards in Industry: A Review of Materials and Processes*, 2nd edn. John Wiley & Sons, New York.

Carlton, G.N. and L.B. Smith. 2000. Exposures to jet fuel and benzene during aircraft fuel tank repair in the U.S. Air Force. *App. Occup. Environ. Hyg.* 15(6):485–491.

Carver, R. A. and Griffith, F. D. Jr. 1979. Determination of kepone dechlorination products in finfish, oysters, and crustaceans. *J. Agric. Food Chem.* 27(5):1035–1037.

Cheremisinoff, N.P. (ed.) 2003. *Industrial Solvents Handbook*, 2nd edn. Marcel Dekker, New York.

International Agency for Research on Cancer (IARC). 1979. Chlordecone. In: *Some Halogenated Hydrocarbons. IARC Monographs on the Evaluation of Carcinogenic Risk of Chemicals to Humans*, Vol. 20, IARC, Lyon, France, pp. 67–81.

Larson, P. S., J. L. Egle, Jr., G. R. Hennigar. 1979. Acute, subchronic, and chronic toxicity of chlordecone. *Toxicol. Appl. Pharmacol.* 48:29–41.

Patnaik, P. (ed.) 1999. *A Comprehensive Guide to the Hazardous Properties of Chemical Substances*, 2nd edn. McGraw Hill, New York.

Risher, J., P.M. Bittner, and S. Rhodes. 1998. *Toxicological Profile for Jet Fuels (JP-5 and JP-8)*. Agency for Toxic Substances and Disease Registry, Centers for Disease Control, Atlanta, GA, 167pp.

Ritchie, G.D., K.R. Still, J. Rossi, III, M.Y.-V. Bekkedal, A.J. Bobb, and Arfsten, D.P. 2003. Biological and health effects of exposure to kerosene-based jet fuels and performance additives. *J. Toxicol. Environ. Health B Crit. Rev.* 6(4):357–451.

Sax, N.I. and R.J. Lewis, Jr. 1989. *Dangerous Properties of Industrial Materials*, 7th edn., 3 volumes. Van Nostrand Reinhold, New York.

U.S. Environmental Protection Agency (U.S. EPA). 2009. *Toxicological Review of Chlordecone (Kepone)*. Integrated Risk Information System (IRIS), National Center for Environmental Assessment, Washington, DC.

13

Hazardous Chemical Substances: L

Lead
Lead Acetate
Lead Acetate, Trihydrate
Lead Arsenate
Lead Chloride
Lead Fluoride
Lead Iodide
Lead Nitrate
Lead (II) Phosphate
Lead Stearate
Lead Styphnate
Tetraethyl Lead
Tetramethyl Lead

Lead (CAS No. 7439-92-1)

Molecular formula: **Pb**
Synonyms and trade names: Lead metal, granular; Lead metal, foil; Lead metal, sheet; Lead metal, shot

Lead (Pb) reacts vigorously with oxidising materials and is incompatible with sodium carbide, chlorine trifluoride, trioxane + hydrogen peroxide, ammonium nitrate, sodium azide, disodium acetylide, sodium acetylide, hot concentrated nitric acid, hot concentrated hydrochloric acid, hot concentrated sulphuric acid, and zirconium.

Lead is a toxic metal that was used for many years in paint and other products found in and around our homes. Lead also can be emitted into the air from industrial sources and leaded aviation gasoline, and subsequently lead enters the sources of drinking water from plumbing materials. Exposure to lead occurs in a majority of occupations, including primary and secondary lead smelting, lead storage battery manufacturing, lead pigment manufacturing and use, solder manufacturing and use, ship building and ship repairing, auto manufacturing, and printing, through inhalation of lead-contaminated dust and lead-contaminated residential soil. Exposures to lead occur into the system through all the three important routes of absorption, namely, by inhalation (breathing), by accidental workplace ingestion, and/or through intake/eating of contaminated food and through skin absorption at different occupational workplaces.

Reports have shown that over the last two centuries, an increase in worldwide lead exposure levels have increased because of widespread use of the metal and lead compounds Manner and kinds of human exposure to lead and lead compounds is very complex. Routes of exposure to lead include are diverse. Occupational exposure is a common cause of lead poisoning among humans. Reports have indicated that more than three million workers are potentially exposed to lead and lead compoundsindifferent workplaces in United States. Exposure to lead to include, but not limited to, contaminated air, water, soil, food, and consumer products. Occupational exposure is a common cause of lead poisoning amonghumans. Reports have indicated that more than three million workers are potentially exposed to lead and lead compounds indifferent workplaces in United States. One of the largest threats to children is lead paint in many homes, especially older ones. Exposure to lead include industrial uses such as process lead-acid batteries, produce lead wire/pipes, and metal recycling and foundries, welding, manufacture of rubber, printing and smelters. Exposures to lead also occur from contact with lead in air, household dust, soil, water, and commercial products and in many more workplace conditions.

Pure lead is a heavy metal at room temperature and pressure and is a basic chemical element. It can combine with various other substances to form numerous lead compounds.

Lead poisoning was among the first known and most widely studied work and environmental hazards. One of the first metals to be smelted and used, lead is thought to have been discovered and first mined in Anatolia around 6500 BCE. Its density, workability, and corrosion resistance were among the metal's attraction. In the second century BCE, the Greek botanist Nicander described the colic and paralysis seen in lead-poisoned people. Dioscorides, a Greek physician who lived in the first century CE, wrote that lead makes the mind 'give way'.

Lead was used extensively in Roman aqueducts from about 500 BCE to 300 CE. Julius Caesar's engineer, Vitruvius, reported, 'water is much more wholesome from earthenware pipes than from lead pipes. For it seems to be made injurious by lead, because white lead is produced by it, and this is said to be harmful to the human body'. Gout, prevalent in affluent Rome, is thought to be the result of lead or leaded eating and drinking vessels. Sugar of lead (lead (II) acetate) was used to sweeten wine, and the gout that resulted from this was known as 'saturnine' gout. It is even hypothesised that lead poisoning may have contributed to the decline of the Roman Empire, a hypothesis thoroughly disputed. With the Industrial Revolution in the nineteenth century, lead poisoning became common in the work setting. The introduction of lead paint for residential use in the nineteenth century increased childhood exposure to lead; for millennia before this, most lead exposure had been occupational. An important step in the understanding of childhood lead poisoning occurred when toxicity in children from lead paint was recognised in Australia in 1897. France, Belgium, and Austria banned white lead interior paints in 1909; the League of Nations followed suit in 1922. However, in the United States, laws banning lead house paint were not passed until 1971, and it was phased out and not fully banned until 1978.

In adults, occupational exposure is the main cause of lead poisoning. People can be exposed when working in facilities that produce a variety of lead-containing products; these include radiation shields, ammunition, certain surgical equipment, foetal monitors, plumbing, circuit boards, jet engines, and ceramic glazes. In addition, lead miners and smelters, plumbers and fitters, auto mechanics, glass manufacturers, construction workers, battery manufacturers and recyclers, firing range instructors, and plastic manufacturers are at risk for lead exposure. Other occupations that present lead exposure risks include welding, manufacture of rubber, printing, zinc and copper smelting, processing of ore, combustion of solid waste, and production of paints and pigments. Parents who are exposed to lead in the workplace can bring lead dust home on clothes or skin and expose their children.

Exposure occurs through inhalation, ingestion, or occasionally skin contact. Lead may be taken in through direct contact with mouth, nose, and eyes (mucous membranes) and through breaks in the skin. Tetraethyl lead, which was a gasoline additive and is still used in fuels such as aviation fuel, passes through the skin; however, inorganic lead found in paint, food, and most lead-containing consumer products is only minimally absorbed through the skin. The main sources of absorption of inorganic lead are from ingestion and inhalation. In adults, about 35%–40% of inhaled lead dust is deposited in the lungs, and about 95% of that goes into the bloodstream. Of ingested inorganic lead, about 15% is absorbed, but this percentage is higher in children, pregnant women, and people with deficiencies of calcium, zinc, or iron. Children and infants may absorb about 50% of ingested lead, but little is known about absorption rates in children.

The main body compartments that store lead are the blood, soft tissues, and bone; the half-life of lead in these tissues is measured in weeks for blood, months for soft tissues, and years for bone. Lead in the bones, teeth, hair, and nails is bound tightly and not available to other tissues and is generally thought not to be harmful. In adults, 94% of absorbed lead is deposited in the bones and teeth, but children only store 70% in this manner, a fact which may partially account for the more serious health effects on children. The estimated half-life of lead in bone is 20–30 years, and bone can introduce lead into the bloodstream long after the initial exposure is gone. The half-life of lead in the blood in men is about 40 days, but it may be longer in children and pregnant women, whose bones are undergoing remodelling, which allows the lead to be continuously reintroduced into the bloodstream. Also, if lead exposure takes place over years, clearance is much slower, partly due to the rerelease of lead from bone. Many other tissues store lead, but those with the highest concentrations (other than blood, bone, and teeth) are the brain, spleen, kidneys, liver, and lungs. It is removed from the body very slowly, mainly through urine. Smaller amounts of lead are also eliminated through the faeces and very small amounts in hair, nails, and sweat.

Safe Handling and Precautions

Exposure to lead is known to cause a series of adverse health effects. Lead affects multiple organs including the neurological, haematological, gastrointestinal, and renal systems. Lead poisoning is also known as plumbism and saturnism and is characterised by gastrointestinal, neurological, haematological, hepatic, and renal health disorders. Lead has no known physiologically relevant role in the body, and its harmful effects are myriad. Lead and other heavy metals create reactive radicals, which damage cell structures and cell membranes. Exposure to lead in different industrial settings and occupations to users and workers is known to occur mainly from inhalation of dust or fumes. Lead dust or fumes cause effects of irritation to the upper respiratory tract, nose, throat, bronchi, and lungs, and lead dust gets absorbed through the respiratory system. The symptoms of lead poisoning include, but are not limited to, metallic taste, chest pain, decreased physical fitness, fatigue, sleep disturbance, headache, irritability, reduced memory, mood and personality changes, aching bones and muscles, constipation, abdominal pains/cramps (lead colic), spasms, decreasing appetite, nausea, vomiting, muscle weakness, hallucinations, distorted perceptions, 'lead line' on the gums, insomnia, dizziness, high lead levels in the blood and urine, and other symptoms similar to that of inhalation.

Inhalation of large amounts of lead leads to ataxia, delirium, convulsions/seizures, coma, and fatal injury. Exposure to lead causes cell membranes of red blood cells to become more fragile as a result of damage to their membranes and leads to anaemia. Lead interferes with metabolism of bones and teeth and alters the permeability of blood vessels and collagen synthesis. Lead may also be harmful to the developing immune system, causing production of excessive inflammatory proteins; this mechanism may mean that lead exposure is a risk factor for asthma in children. Lead exposure has also been associated with a decrease in activity of immune cells such as polymorphonuclear leucocytes. Lead also interferes with the normal metabolism of calcium in cells and causes it to build up within them. Repeated and prolonged period of exposure to lead causes health disorders.

Children, on repeated exposures in multiple ways to lead and lead compounds and with high blood lead level, have been associated with the development of learning disabilities such as short-term memory, cognitive abilities, decrease in intelligence, lack of skills, emotional disorders, and non-verbal reasoning. In short, children affected by lead exposure develop neurological and cognitive sequelae, including reduced cognition and behaviour scores, changes in attention, visual–motor and reasoning skills, social behaviour, and reading ability.

Studies have indicated that children exposed to high levels of lead demonstrate reduced academic performance. Similarly, adults exposed to high blood lead levels at different occupations and workplaces are closely associated with health disorders such as depression, anxiety, decreased cognitive performance or neurocognitive disorder, and impairment of central nervous system (CNS) function.

There are no reports and conclusive evidence indicating that lead causes cancer in humans. Experimental rats and mice exposed to very large doses of lead developed tumours in kidney. The Department of Health and Human Services (DHHS) observed that lead and lead compounds could reasonably be anticipated to be human carcinogens, and the U.S. Environmental Protection Agency (U.S. EPA) has determined that lead is a probable human carcinogen. Lead has been listed by the American Conference of Governmental Industrial Hygienists (ACGIH) and classified as Group A3, meaning a proven carcinogen for animals, and the International Agency for Research on Cancer (IARC) classified lead as Group 2B, meaning a possible for human carcinogen. Based on limited evidence in humans and sufficient evidence in laboratory animals, the IARC classified that inorganic lead compounds are probably a human carcinogen and that there is insufficient information to determine whether organic lead compounds will cause cancer in humans. Based on inadequate evidence, the organic lead compounds have been listed as 'not classifiable' as to their carcinogenicity in humans.

Recent studies have indicated that lead is hypothesised to be a cocarcinogen, meaning lead allows or augments the genotoxic effects of several other agents.

Lead and Global Regulations

Lead has long been recognised as a hazard to consumers. Lead is a highly toxic substance, exposure to which can produce a wide range of adverse health effects. These effects include neurological damage, delayed mental and physical development, attention and learning deficiencies, and hearing problems. Because lead accumulates in the body, even exposure to small amounts of lead can contribute to the overall bioaccumulative level of lead in the blood and to the subsequent risk of adverse health effects:

1. The U.S. Agency for Toxic Substances and Disease Registry (ATSDR) has made it mandatory to perform specific functions concerning the effect on public health of hazardous substances in the environment, including lead. These functions include public health assessments of waste sites, health consultations concerning specific hazardous substances, health surveillance and registries, response to emergency releases of hazardous substances, applied research in support of public health assessments, information development and dissemination, and education and training concerning hazardous substances.

2. The National Lead Laboratory Accreditation Program has strict protocols, criteria, and minimum performance standards for laboratory analysis of lead in paint, dust, and soil. (For details, refer to TSCA Section 405 (b).)

3. The hazardous effects of lead have resulted in its being highly regulated in a wide range of consumer products and by both federal and state agencies. The Consumer Product Safety Commission (CPSC) is the primary federal agency that regulates lead in consumer products.

4. The U.S. EPA helps regulate lead in the environment, that is, air, water, soil, emissions, and disposal.

5. The U.S. Food and Drug Administration (U.S. FDA) also has responsibility for lead in products like food, food contact articles, drugs, and cosmetics. Lead has also been the focus of state regulations such as the California Safe Drinking Water and Toxic Enforcement Act of 1986 (commonly known as Prop 65) and is enforced by the Attorney General's Office and bounty hunters. The Illinois Lead Poisoning Prevention Act is enforced by the Attorney General's Office.

6. The European Union: 2005/311/EEC regulates lead in ceramic articles.

7. Australia: Australian Statutory Rules 1977, Number 373, Food and Drug Standards (amendment no. 17), Regulations 1977 regulates materials that come into contact with foodstuff.

8. Brazil: Decree No. 27, 18 March 1996, regulates lead added to items to come in contact with food.

9. Mexico: Mexican Official Standard NOM-010-SSA1-1993, Environmental health. Glazed ceramic articles. Limits for soluble lead and cadmium; Mexican Official Standard NOM-011-SSA1-1993, Environmental health. Limits for soluble lead and cadmium in glazed pottery articles.

10. Reports indicate that the Chinese government has ordered the suspension of lead battery factories in some of the industrial provinces. The poisoning claims have spurred government into action, with the Ministry of Environmental Protection ordering the provincial governments to tighten administration of battery-producing facilities.

Lead and Renal Disorders

Exposure to high levels of lead causes kidney disorders. Evidence suggests that lower levels can damage kidneys as well. The toxic effect of lead causes neuropathy and may cause Fanconi syndrome, in which the proximal tubular function of the kidney is impaired. Long-term exposure at levels lower than those that cause lead nephropathy has also been reported as neurotoxic in patients from developed countries that had chronic kidney disease. Occupational workers and general public on repeated exposures to higher concentrations of lead have been reported to be at higher risks of health disorders – cardiac autonomic dysfunction. Lead affects the peripheral nervous system, especially motor nerves, and the CNS. Peripheral nervous system effects are more prominent in adults, and CNS effects are more prominent in children. Lead causes the axons of nerve cells to degenerate and lose their myelin coats. The brain is the organ most sensitive to lead exposure, and lead poisoning interferes with the normal development of a child's brain and nervous system. Children, therefore, are at greater risk of lead neurotoxicity than adults are. In a child's developing brain, lead interferes with synapse formation in the cerebral cortex, neurochemical development, along with neurotransmitters, and organisation of ion channels. It causes loss of neurons' myelin sheaths, reduces numbers of neurons, interferes with neurotransmission, and decreases neuronal growth.

Lead Acetate (CAS No. 301-04-2)

Molecular formula: $Pb(C_2H_3O_2)_2$
Synonyms and trade names: Lead acetate; Lead (II) acetate, trihydrate; Acetic acid lead (II) salt, trihydrate; Lead diacetate, trihydrate; Salt of Saturn

Lead acetate appears as white crystalline granules with mild acetic acid odour. Lead acetate is stable under ordinary conditions of use and storage. Lead acetate is incompatible with bromates, phenol, chloral hydrate, sulphides, hydrogen peroxide, resorcinol, salicylic acid, sulphites, vegetable infusions, alkalis, tannin, phosphates, citrates, chlorides, carbonates, tartrates, and acids. Lead (II) acetate, as well as white lead, has been used in cosmetics throughout history, though this practice has ceased in Western countries. It is still used in men's hair colouring. Lead (II) acetate paper is used to detect the poisonous gas hydrogen sulphide. The gas reacts with lead (II) acetate on the moistened test paper to form a grey precipitate of lead (II) sulphide.

Safe Handling and Precautions

The symptoms of lead acetate poisoning include abdominal pain and spasms, nausea, vomiting, and headache. Acute poisoning can lead to muscle weakness, 'lead line' on the gums, metallic taste, definite loss of appetite, insomnia, dizziness, and high lead levels in blood and urine with shock, coma, and death in extreme cases. The ACGIH classified the lead acetate as A 3, meaning a proven animal carcinogen; by the National Toxicology Program (NTP) as 2, meaning reasonably anticipated to be human carcinogens; and the IARC as Class 3, meaning not classifiable as a human carcinogen.

Users and occupational workers during use and handling must strictly observe set safety regulations. Workers should keep the container of lead acetate tightly closed and stored in a cool, dry, ventilated area. Also, lead acetate should be kept protected against physical damage, and isolate it from incompatible substances.

Occupational workers during use and handling must wear appropriate, impervious protective clothing, boots, gloves, laboratory workplace coat/apron, or coveralls to prevent contact of toxic chemical. The work area during handling of lead acetate should be limited to authorised workers/persons. Occupational workers should strictly observe disposal of spilled chemical wastes, solutions, and any by-products in accordance with and as applicable to the regional, national, and local laws and regulations.

Poison! Improper use and handling of lead acetate is hazardous.

Accidental swallowing, ingestion, or inhalation and/or skin absorption of lead acetate is known to result in fatal injury. It is a neurotoxin that causes irritation to skin, eyes, and respiratory tract and affects the gum tissue, CNS, kidneys, blood, and reproductive system.

Regulations:

- Lead acetate is classified as hazardous according to the New Zealand Hazardous Substances.

- The state of California, United States, has listed lead acetate as a known chemical substance that causes cancer and birth defects or other reproductive harm. Also, lead acetate has been considered hazardous by the OSHA Hazard Communication Standard (29 CFR 1910.1200).

- Workplace Hazardous Materials Information System (WHMIS), Canada, classified lead acetate as Class D-1B and D-2A, meaning causes immediate and other serious toxic effects.

Lead Acetate Trihydrate (CAS No. 6080-56-4)

Molecular formula: $C_4H_{12}O_7Pb$
Synonyms and trade names: Lead (II) acetate trihydrate; Lead diacetate trihydrate

Lead acetate trihydrate appears as white crystals. Lead acetate trihydrate is water soluble and absorbs carbon dioxide from the air. Lead acetate trihydrate is used in the manufacturing of other lead salts, for dyeing and printing textiles in the textile industry, as a drying agent in the paint and varnish industry, and as a purification agent in chemical synthesis. Lead acetate trihydrate is stable under normal conditions. However, it can become unstable in the presence of excess heat, flame, ignition sources, or incompatible substances, for example, bromates, phenol, chloral hydrate, sulphides, hydrogen peroxide, resorcinol, salicylic acid, sulphites, vegetable infusions, alkalis, tannin, phosphates, citrates, chlorides, carbonates, tartrates, and acids.

Safe Handling and Precautions

Exposure to lead acetate trihydrate by accidental ingestion at workplaces causes irritation to gastrointestinal tract, diarrhoea, and bloody stools, and exposure to large doses of the chemical substance by inhalation causes metallic taste, burning sensation in mouth and throat, and irritation to respiratory tract. Repeated and prolonged period of exposure to lead acetate trihydrate has been reported to cause irritability, tiredness, sterility, poor dental hygiene, birth defects, and mental retardation.

Breathing lead acetate trihydrate dust can irritate the nose, throat, and lungs. Short-term exposure by ingestion, skin absorption, or inhalation may cause toxic effects on the blood and CNS, resulting in haemolytic anaemia, nervous disorders, and kidney impairment. Typical clinical manifestations of lead poisoning include weakness, irritability, nausea, abdominal pain with constipation, and anaemia. Long-term or repeated exposure may cause cumulative toxic effects on the blood, bone marrow, cardiovascular system, kidneys, and nervous system, resulting in anaemia, an increase of blood pressure, paralysis, kidney impairment, and behavioural effects and may cause irritation by all exposure routes: kidneys, CNS, and blood-forming organs. Prolonged exposure leads to accumulation of lead in body tissues and adverse effects on heart, blood, and kidneys. Lead acetate trihydrate has not been listed as a carcinogen by the ACGIH, IARC, NTP, National Institute for Occupational Safety and Health (NIOSH), and NTP. Also, lead acetate trihydrate is not included in the Canada Ingredient Disclosure List. During use and handling of lead acetate trihydrate, workers should strictly observe safety regulations and use appropriate workplace protective dress, impervious protective clothing, boots, gloves, laboratory workplace coat/apron or coveralls, and splash-proof chemical safety goggles to prevent contact of toxic chemical.

Lead Arsenate (CAS No. 7784-40-9)

Molecular formula: **PbHAsO$_4$**
Chemical name: Acid orthoarsenate
Synonyms and trade name: Arsenic acid, lead salt; Acid lead arsenate; Dibasic lead arsenate; Gypsine, security, talbot

Lead arsenate appears as odourless white heavy powder and practically is insoluble in water. It is not combustible and emits irritating or toxic fumes (or gases) in a fire. Lead arsenate is currently used as a growth regulator on grape fruit crop. A large bulk of lead arsenate is also used annually to control cockroaches, silverfish, crickets, earthworms, and other soil-inhabiting insects on golf greens and lawns and on airport turf adjoining runways to reduce bird hazard related to earthworm. Lead arsenate is quite stable and readily accumulates in soils either as a result of use in specific soil treatments or foliar application.

Lead arsenate insecticide was used in many countries, including Australia, Canada, New Zealand, and the United States. It was used for insect pests on apples and other fruit tree, garden crops and turf grasses, on rubber and coffee trees, and for mosquito abatement in cattle dips. The use of lead arsenate was terminated in the early 1950s in Massachusetts, in the mid-1960s in New York and other states, and in 1984 in Washington State. All insecticidal uses of lead arsenate in the United States were officially banned on August 1, 1988 (U.S. EPA, 1988), with a comment that all registrations for insecticidal use had lapsed before that date. In Australia, use of lead arsenate decreased after the introduction of DDT in 1950, and it has not been used on exported crops since 1983.

Safe Handling and Precautions

Lead arsenate is a poisonous white solid used as an insecticide. Exposure of occupational workers at workplaces to lead arsenate by accidental ingestion and/or inhalation causes adverse health effects such as cough, sore throat, abdominal pain, diarrhoea, drowsiness, headache, nausea, vomiting, muscular cramp, constipation, excitation, and disorientation. Reports have indicated that long-term or repeated exposure to lead arsenate causes irritating effects to the respiratory tract, gastrointestinal tract, nervous system, kidneys, and liver and general vascular collapse leading to shock, coma, and death. Muscular cramps, facial oedema, and cardiovascular reactions are also known to occur following oral exposure to arsenic. Lead arsenate is a human carcinogen.

Lead Chloride (CAS No. 7758-95-4)

Molecular formula: $PbCl_2$
Synonyms and trade names: Plumbous chloride; Cotunnite

Lead (II) chloride is also known as lead chloride, lead dichloride, and plumbous chloride. Lead chloride is one of the most important lead-based reagents. It occurs naturally in the form of the mineral cotunnite. The solubility of lead chloride in water is low. Lead (II) chloride is the main precursor for organometallic derivatives of lead. Lead chloride has extensive applications in industries. Lead chloride is an intermediate in refining bismuth (Bi) ore. The ore containing Bi, Pb, and Zn is first treated with molten caustic soda to remove traces of acidic elements such as arsenic and tellurium.

The molten lead chloride is used in the synthesis of lead titanate ($PbTiO_3$) and barium $PbTiO_3$. It is used in organometallic synthesis to make metallocenes, known as plumbocenes. Lead chloride is used in production of infrared transmitting glass and in production of ornamental glass called aurene glass. This stained glass has an iridescent surface formed by spraying with lead chloride and reheating under controlled conditions. Stannous chloride ($SnCl_2$) is used for the same purpose.

Safe Handling and Precautions

Lead chloride like other lead-containing compounds, upon workplace exposure, is known to cause lead poisoning. Exposure to lead chloride is very hazardous and on inhalation and skin contact causes effects of irritation and corrosion to the eyes, corneal damage and/or blindness, skin inflammation, blistering and tissue damage, irritation to gastrointestinal or respiratory tract, burning sensation, sneezing, and coughing. Occupational workers overexposed to lead chloride have been reported to suffer with lung damage, choking, unconsciousness, or death. The IARC classified lead chloride as 2B, meaning a possible human carcinogen.

WHMIS, Canada, classified lead chloride as Class D-1B, meaning causes immediate and serious toxic effects, and Class D-2A, meaning causes other toxic effects. Occupational workers during use and handling of lead chloride must be very careful and must avoid breathing the chemical dust. Lead chloride container should be kept tightly closed in a cool, well-ventilated place and stored in a secure area suitable for toxic material. Workers should wear suitable, appropriate workplace protective clothing, splash goggles, vapour and dust respiratory equipment, boots, and gloves.

Lead Fluoride (CAS No. 7783-46-2)

Molecular formula: PbF_2
Synonyms and trade names: PbF_2; Plombfluorure; Lead fluoride; Plomb fluorure; Lead difluoride; Plumbousfluoride; Lead (+2) fluoride; Lead (II) fluoride

Lead fluoride appears as white crystalline powder or pieces and is odourless. It is slightly soluble in water and noncombustible. On heating, lead fluoride decomposes and releases toxic and corrosive fumes. Also, on contact with metals, it produces flammable hydrogen gas. The containers of lead fluoride explode when heated. Lead fluoride is incompatible with acids, hydrogen peroxides, strong oxidising agents, calcium carbide, and fluorine. It is stable at room temperature in closed containers under normal storage and handling conditions.

Safe Handling and Precautions

Chronic workplace exposure to lead fluoride through accidental inhalation and/or ingestion is known to cause systemic toxicity and health disorders. The symptoms of toxicity and poisoning include, but are not limited to, fluorosis and effects on blood, CNS, kidneys, liver, brain, and bone structure with skeletal abnormalities. Chronic exposures to lead result in plumbism as characterised by lead line in gum. Symptoms of lead fluoride include, but are not limited to, nausea, vomiting, diarrhoea, loss of appetite, gastrointestinal irritation, weakness, headache, insomnia, dizziness, digestive tract burns, abdominal pain, constipation, anorexia, muscle weakness, and toxic effects on the heart, liver, and kidneys. Fluoride is also known to deplete the levels of calcium and result in fatal hypocalcaemia, effects on reproduction and embryonic and foetal development, and postnatal developmental disorders.

During use and handling of lead fluoride, users and occupational workers must observe safety regulation practices. Workers and users must wear appropriate protective gloves and clothing to prevent skin exposure, self-contained breathing apparatus, and appropriate protective eyeglasses/standard chemical safety goggles. Workers and users during handling of lead fluoride should keep from contact with oxidising materials. Workers must keep the containers of lead fluoride tightly closed and stored in a cool, dry, well-ventilated area away from incompatible substances.

Lead Iodide (CAS No. 10101-63-0)

Molecular formula: PbI_2

Lead iodide appears as yellow or orange powder or solid. Lead iodide has several uses; most of its use is in colour and other pigment. Lead iodide is made with two things: first is lead and second is iodide. It is used in bronzing, printing, photography, and mosaic gold.

Safe Handling and Precautions

Users and occupational workers must be careful during handling of lead iodide because it is hazardous on contact with skin and eyes. Lead iodide is a toxic compound and should be handled with care. Exposure is known to cause poisoning with symptoms that include, but are not limited to, gastrointestinal irritation, nausea, vomiting, anorexia, muscle weakness, diarrhoea, blood abnormalities, abdominal discomfort, colic, constipation, anaemia, liver and kidney damage, cardiac disturbances, and mental changes.

Workers should know that lead iodide is poisonous, and when mixed with water, it might cause instant deaths. Workers should keep stored lead iodide in a cool place in the original container tightly closed and away from sunlight and contact with oxidising materials.

Warning! DANGER: Users and occupational workers must know that lead iodide is harmful to the eyes and to the skin, causes digestive and respiratory tract irritation and effects on CNS, causes blood abnormalities, and causes liver and kidney damage.

Regulations:

- The state of California listed lead iodide as a chemical known to cause cancer.
- The U.S. Occupational Safety and Health Administration (U.S. OSHA) defines lead iodide as hazardous and lists it in Communication Standard (29 CFR 1910.1200).
- The WHMIS, Canada, lists it as Class D-2A, meaning lead iodide causes toxic effects and is very toxic.
- During use and handling of lead iodide, occupational workers must strictly observe workplace safety regulations. Workers must wear workplace protective dress, splash goggles, gloves, and approved/certified appropriate respirator, and the workplace must have adequate ventilation.

Lead Nitrate (CAS No. 10099-74-8)

Molecular formula: $Pb(NO_3)_2$
Synonyms and trade names: Lead (II) nitrate; Lead dinitrate; Nitric acid, lead (2+); Nitric acid, lead (ii) salt; Lead (2+) bis(nitrate); Lead (2+) nitrate; Lead (ii) nitrate; Plumbous nitrate

Lead nitrate is incompatible with ammonium thiocyanate, powdered carbon, lead hypophosphite, hydrogen peroxide, combustibles, and organic materials, and workers should avoid any kind of contact with heat, flames, ignition sources, and incompatibles.

Safe Handling and Precautions

Exposure to lead nitrate is harmful and hazardous. On skin contact and inhalation, lead nitrate has been known to cause adverse health effects with symptoms that include, but are not limited to, effects of irritation of the eyes and skin, skin burns, and ulcerations. Overexposure by inhalation leads to respiratory irritation. Repeated and prolonged exposure to lead nitrate causes damage to target organs involving kidneys, reproductive system, peripheral nervous system, CNS, and blood. The IARC classified lead nitrate as 2B, meaning possibly a human carcinogen, while the ACGIH classified the chemical as A3, meaning a proven animal carcinogen.

Occupational workers must keep lead nitrate in a tightly closed container and stored in a cool, dry, ventilated area away from sources of heat and ignition and/or against physical damage and away from sunlight and separately from reactive or combustible materials. Also, the workplace access should be limited only to authorised persons.

Users and occupational workers handling lead nitrate must be careful. Lead nitrate is a neurotoxin and strong oxidising agent, and its contact with other materials may cause fire hazard; Improper use and handling of lead nitrate affects the gum tissue, central nervous system, kidneys, blood, and reproductive system. U.S. EPA has classified lead nitrate as Group B2 – probable human carcinogen.

Lead (II) Phosphate (CAS No. 7446-27-7)

Molecular formula: $Pb_3(PO_4)_2$
Synonyms and trade names: Lead (II) nitrate (1:1); Lead dinitrate; nitric acid, Lead (2+)

Lead phosphate is white powder soluble in acids and alkalis. Lead (II) phosphate is insoluble in water and alcohol but soluble in nitric acid and has fixed alkali hydroxides. On heating, it decomposes and emits very toxic fumes containing Pb and POx.

Reports have indicated that on the basis of sufficient evidence of carcinogenicity in experimental animals and inadequate evidence for the carcinogenicity in humans, lead acetate and lead phosphate are reasonably anticipated to be human carcinogens. Information on the toxicology profile of lead phosphate is limited.

Lead Stearate (CAS No. 7428-48-0)

Molecular formula: $C_{18}H_{38}O_2Pb$
Synonyms and trade names: Aid, Lead salt; Octadecanoic acid, lead salt; Lead (II) *n*-octadecanoate; Lead (II) stearate

Lead stearate appears as white fine powder. Lead stearate possesses good heat-stabilising properties, particularly when used in conjunction with other lead stabilisers. It has been used extensively in industries as a flatting agent in paint, as a component in the rubber vulcanising process, and as a high-temperature lubricant in PVC compounds.

Safe Handling and Precautions

Exposure to lead stearate by inhalation of dust causes skin and respiratory irritations. Information on the toxicological profile of lead stearate is limited. Workers during use and handling of lead stearate should be careful and use workplace protective dress, hand gloves, and dust masks.

Lead Styphnate (CAS No. 15245-44-0)

Molecular formula: $C_6HN_3O_8Pb$
Synonyms and trade names: Lead hydroxide styphnate; Lead hydroxide 2,4,6 trinitroresorcinate
Chemical name: Lead 2,4,6-trinitroresorcinate

Lead styphnate varies in colour from yellow to brown. There are two forms of lead styphnate. The longer and narrower the crystals, the more susceptible lead styphnate is to static electricity. Lead styphnate is particularly sensitive to fire and the discharge of static electricity. Lead styphnate does not react with metals and is less sensitive to shock and friction. Lead styphnate is only slightly soluble in water and methyl alcohol. It is stable in storage, even at elevated temperatures. Lead styphnate is derived from styphnic acid and is used in explosive mixtures.

Safe Handling and Precautions

Exposure to lead styphnate causes harmful health effects, and workers during use and handling of the chemical should be alert and observe set safety regulations. On accidental workplace inhalation, swallowing/ingestion, and/or skin absorption, lead styphnate has been reported to cause effects of irritation of the eyes, skin, and respiratory tract. Repeated exposure to lead styphnate causes severe effects to the kidneys and blood, possibly lung damage, and reproductive system, nervous system, and developmental effects.

Occupational workers during use and handling of lead styphnate should be very careful because when dry, lead styphnate gets readily detonated by static discharges from the human body and should avoid the dispersion of dust in air. Workers should keep stored lead styphnate in a cool, dry, well-ventilated place away from all sources of ignition.

The U.S. Toxic Substances Control Act (U.S. TSCA) has listed the component of lead styphnate on the inventory. And similarly, the Canadian WHMIS has regulated lead styphnate as Explosive Class 3.2.

Danger! ALERT! Lead styphnate is explosive. Workers should be alert and should not expose lead styphnate to mechanical shock or heat. Lead styphnate is harmful on accidental workplace inhalation, swallowing/ingestion, and/or skin absorption.

Tetraethyl Lead (CAS No. 78-00-2)

Molecular formula: $(CH_3CH_2)_4 Pb$
Synonyms and trade names: Lead tetraethyl; NCI–C54988; Plumbane, tetraethyl; Tetraethylplumbane; TEL
IUPAC name: Tetraethylplumbane

Tetraethyl lead is a colourless liquid with a characteristic odour. Pure tetraethyl lead is a colourless liquid with a characteristic odour and is a viscous liquid that is highly lipophilic and soluble in fats, oils, and lipids as well as gasoline and other nonpolar hydrocarbons. Tetraethyl lead is commonly abbreviated as TEL. Tetraethyl lead is an organolead compound, and it improves the efficiency and performance of internal combustion engines and is an inexpensive additive to gasoline (petrol) since 1920 and allows octane ratings and engine compression to be boosted significantly, increasing power and fuel economy. The use of tetraethyl lead was largely discontinued because of the toxicity of lead and its deleterious effect on catalytic converters. On exposures to elevated temperature and in fire, tetraethyl lead has been reported to cause explosion. It decomposes slowly at room temperature and more rapidly at elevated temperatures. Also, tetraethyl lead decomposes under UV light. It reacts with fats, reacts violently with oxidising agents, causing fire and explosion hazards, and attacks rubber.

Safe Handling and Precautions

Exposure to tetraethyl lead is extremely poisonous and is known to be fatal if and whenever accidentally inhaled, swallowed, or absorbed from the skin at workplaces. Contact may cause burns to skin and eyes. Most symptoms of poisoning are due to the effects of tetraethyl lead on the nervous system. Exposure to tetraethyl lead causes increased urinary output of lead.

The signs and symptoms of tetraethyl lead exposure and poisoning include, but are not limited to, anxiety, irritability, insomnia, violent/frightening dreams, headache, disorientation, hyperexcitability, delusions, hallucinations, muscular weakness, ataxia, tremors, convulsions, cerebral oedema, and possibly coma. Tetraethyl lead poisoning also causes metallic taste, sneezing, bronchitis, pneumonia, bradycardia (slow heart rate), hypotension (low blood pressure), hypothermia, and pallor. The gastrointestinal symptoms of poisoning include vomiting and diarrhoea. If a large degree of absorption of tetraethyl lead occurs from workplace inhalation or skin contact, the symptoms of poisoning include insomnia, excitability, delirium, coma, and death. Do not confuse it with inorganic lead. It has been documented that tetraethyl lead intervenes with the development of the nervous system and is therefore particularly toxic to children, thereby causing potentially permanent learning and behaviour disorders, and the symptoms of poisoning include, but are not limited to, abdominal pain, confusion, headache, and anaemia which in severe cases, it causes seizures, coma, and death.

During handling of tetraethyl lead, occupational workers must be very careful. Workers must completely avoid workplaces that are prone to sparks, flames, and other sources of ignition, and workers must keep away the chemical, keep tetraethyl lead out of water sources and sewers. On accidental chemical spill, workers should be absorbed with fly ash, cement powder, or commercial sorbents and must use/apply activated carbon on the spilled material.

Occupational workers must not be negligent during handling of tetraethyl lead and must strictly use personal protective equipment (PPE), to avoid organic vapour, canister face mask, neoprene-coated, liquid-proof gloves; protective goggles or face shield; white or light-coloured clothing; rubber shoes or boots.

Tetramethyl Lead (CAS No. 75-74-1)

Molecular formula: $C_4H_{12}Pb$
Synonyms and trade names: Tetramethylplumbane; TML

Tetramethyl lead is a flammable, colourless liquid with a slightly sweet odour. Tetramethyl lead is insoluble in water but is soluble in most organic solvents, alcohol, benzene, and petroleum ether. Tetramethyl lead is incompatible with tetrachloro trifluoromethyl phosphorus, strong oxidisers, sulphuryl chloride, and potassium permanganate. Tetramethyl lead on hazardous decomposition releases toxic gases and particulates, such as lead fumes and carbon monoxide. It may cause fires and explosions. The commercial product is often dyed red, orange, or blue. Tetramethyl lead is frequently mixed with smaller amounts of ethylene dibromide, ethylene dichloride, dyes, kerosene, stabilisers, and inert substances.

Safe Handling and Precautions

Exposure to tetramethyl lead causes health hazards. Improper use and negligence of occupational workers at workplaces allows the entry of tetramethyl through all the routes, namely, via inhalation, ingestion, and/or through dermal/skin absorption, eye, and eye contact. The signs and symptoms of acute poisoning of tetramethyl lead include, but are not limited to, fatigue, restlessness, bad dreams, irritability, weakness, tremors, effects on the CNS, and insomnia. There are no published reports regarding chronic toxicity and poisoning of tetramethyl lead among occupational workers. Workers should use and handle tetramethyl lead with care and it should be stored in a cool, dry, well-ventilated area in tightly sealed containers with proper and legible label. Workers should handle the containers of tetramethyl lead with care and should protect it from physical damage and store it separately from oxidisers, chemically active metals, heat, sparks, and open flame. Workers must use only nonsparking tools during use and handling tetramethyl lead.

Users and occupational workers during use and handling of tetramethyl lead must use and wear adequate standard quality protective clothing, safety glasses, goggles, and face shield and avoid the possibility of skin contact. The workplace should have proper/working eyewash fountains within the immediate reach of the workplace.

Bibliography

Burgess, W. A. 1995. *Recognition of Health Hazards in Industry: A Review of Materials and Processes,* 2nd edn. John Wiley & Sons, New York.

Environmental Health Organization (WHO). 1977. Lead. Environmental Health Criteria No. 3. International Programme on Chemical Safety (IPCS). WHO, Geneva, Switzerland.

Gosselin, R. E., R. P. Smith, and H. C. Hodge (eds.). 1984. *Clinical Toxicology of Commercial Products,* 5th edn. Williams & Wilkins, Baltimore, MD.

Hathaway, G. J., N. H. Proctor, J. P. Hughes, and M. L. Fischman (eds.). 1991. *Proctor and Hughes' Chemical Hazards of the Workplace,* 3rd edn. Van Nostrand Reinhold, New York.

Hazardous Substances Data Bank (HSDB). 1989. Tetramethyl lead. National Library of Medicine, Bethesda, MD.

International Agency for Research on Cancer (IARC). 1998. Summaries and evaluations. Lead and lead compounds. Lead and lead compounds: Lead and inorganic lead compounds. (Group 2B) Organolead compounds (Group 3). IARC, Lyon, France.

International Programme on Chemical Safety and the Commission of the European Communities (IPCS–CEC). 1997. Lead arsenate. ICSC Card No. 0911. IPCS–CEC, Geneva, Switzerland (updated 2005).

International Programme on Chemical Safety and the Commission of the European Communities (IPCS–CEC). 2002. Lead. ICSC Card No. 0052. IPCS–CEC, Geneva, Switzerland.

Jerome, O. N. 1983. *Lead and Lead Poisoning in Antiquity.* John Wiley & Sons, New York.

Lewis, R. J., Sr. 1996. *Sax's Dangerous Properties of Industrial Materials,* 9th edn. Wiley & Sons, New York.

Lewis, R. J., Sr. (ed.). 2004. *Sax's Dangerous Properties of Industrial Materials,* 11th edn. Wiley-Interscience, John Wiley & Sons, Hoboken, NJ.

Lewis, R. J. and R. J. Lewis, Sr. (eds.). 2008. *Hazardous Chemicals Desk Reference,* 6th edn. John Wiley & Sons, Inc., Hoboken, NJ.

Material Safety Data Sheet (MSDS) Tetraethyl lead. NOAA.

Material Safety Data Sheet (MSDS). 1999. Lead acetate. MSDS No. L 2434. Mallinckrodt Baker, Inc., Phillipsburg, NJ.

Material Safety Data Sheet (MSDS). 2000. Lead (II) acetate trihydrate. ACC No. 12530. Fisher Scientific, Fair Lawn, NJ.

Material Safety Data Sheet (MSDS). 2007. Lead (II) chloride. MSDS No. 12570. Fisher Scientific, Fair Lawn, NJ.

Material Safety Data Sheet (MSDS). 2008a. Lead nitrate. MSDS No. L3130. Mallinckrodt Baker, Inc., Phillipsburg, NJ.

Material Safety Data Sheet (MSDS). 2008b. Lead nitrate. MSDS. NOAA's Ocean Service. National Oceanic and Atmospheric Administration. Washington, DC.

National Institute for Occupational Safety and Health (NIOSH). 2010. Lead. *NIOSH Pocket Guide to Chemical Hazards.* U.S. Department of Health and Human Services, Public Health Service, Centers for Disease Control and Prevention. Cincinnati, OH (updated 2011).

National Toxicology Program (NTP). 2005. Eleventh report on carcinogens: Lead, and lead compounds. NTP. Department of Health & Human Services, Centers for Disease Prevention & Control. Washington, DC.

O'Neil, M. J. (ed.). 2006. *The Merck Index—An Encyclopedia of Chemicals, Drugs, and Biologicals.* Merck and Co., Inc., Whitehouse Station, NJ.

Patnaik, P. (ed.). 1999. *A Comprehensive Guide to the Hazardous Properties of Chemical Substances,* 2nd edn. McGraw Hill, New York.

Patnaik, P. (ed.). 2003. *Handbook of Inorganic Chemicals.* McGraw Hill, New York.

Registry of Toxic Effects of Chemical Substances (RTECS). 1989. Tetramethyl lead. RTECS. National Library of Medicine, Bethesda, MD.

Richard P. P. (ed.). 2008. *Sittig's Handbook of Toxic and Hazardous Chemicals and Carcinogens,* 5th edn. Elsevier, New York.

Sax, N. I. and R. J. Lewis, Jr. 1989. *Dangerous Properties of Industrial Materials,* 7th edn., Vol. 3. Van Nostrand Reinhold, New York.

Silbergeld, E. 2003. Facilitative mechanisms of lead as a carcinogen. *Mutat. Res.* 533(1–2):121–133.

Silbergeld, E. K., M. Waalkes, and J. M. Rice. 2000. Lead as a carcinogen: Experimental evidence and mechanisms of action. *Am. J. Ind. Med.* 38(3):316–323.

Sittig, M. (ed.). 1985. *Handbook of Toxic and Hazardous Chemicals,* 2nd edn. Noyes Publications, Park Ridge, NJ.

Sittig, M. (ed.). 2002. *Handbook of Toxic and Hazardous Chemicals and Carcinogens,* 4th edn. Norwich, New York.

Staudinger, K. C. and V. S. Roth. 1998. Occupational lead poisoning. *Am. Fam. Physician* 57:719–726.

Sun, L., J. Hu, Z. Zhao, L. Li, and H. Cheng. 2003. Influence of exposure to environmental lead on serum immunoglobulin in preschool children. *Environ. Res.* 92(2):124–128.

Susan, B. (ed.). 1995. *The Merck Index,* 12th edn. Merck & Company, Inc., Rahway, NJ.

Tabuku, A. and E. Panariti. 1996. Lead intoxication in rural Albania. *Vet. Hum. Toxicol.* 38:434–435.

U.S. Environmental Protection Agency (U.S. EPA). 1986. Lead arsenate. U.S. EPA Pesticide Fact Sheet No. 112 (12/86). PMEP, Cornell University, Ithaca, NY.

U.S. Environmental Protection Agency (U.S. EPA). 2004. Lead and compounds (inorganic). Integrated Risk Information System (IRIS), National Center for Environmental Assessment, Office of Research and Development, Washington, DC.

U.S. Environmental Protection Agency (U.S. EPA). 2011. Lead in paint, dust and soil. U.S. EPA, Washington, DC.

Vig, E. K. and H. Hu. 2000. Lead toxicity in older adults. *J. Am. Geriatr. Soc.* 48(11):1501–1506.

Williamson, A. M. and R. K. C. Teo. 1986. Neurobehavioural effects of occupational exposure to lead. *Br. J. Ind. Med.* 43(6):374–380.

World Health Organization (WHO). 1995. Lead. Environmental Health Criteria No. 165. International programme on chemical safety (IPCS). WHO, Geneva, Switzerland.

14

Hazardous Chemical Substances: M

Magnesium Phosphide
Malathion
Mercury and Its Compounds
   Mercury
   Mercuric Acetate
   Mercuric Chloride
   Mercuric (II) Cyanide
   Mercuric Nitrate
   Mercuric Oxide
   Mercuric Sulphate
   Mercuric Thiocyanate
   Mercurous Chloride
   Mercurous Nitrate
Methane
Methamidophos
Methidathion
Methoxychlor
Methyl Alcohol
Methylamine
Methyl Bromide
Methylene Chloride
Methyl Ethyl Ketone
Methyl Parathion
Methyl Vinyl Ketone
Mevinphos
Mirex and Chlordecone
Molybdenum
Mustard Gas

Magnesium Phosphide (CAS No. 12057-74-8)

Molecular formula: Mg_3P_2
Synonyms and trade names: Magnesium phosphide; Magtoxin (R); Fumi-Cel (R); Fumi-Strip (R)

Magnesium phosphide and Magtoxin have not been reported as flammable chemical substances. However, they react readily with water to produce hydrogen phosphide (phosphine, PH_3) gas, which may ignite spontaneously in air. Magnesium phosphide is stable to most chemical reactions, except for hydrolysis. Magtoxin, Fumi-Cel, and Fumi-Strip will react with moist air, liquid water, acids, and some other liquids to produce toxic and flammable hydrogen phosphide gas. Magnesium phosphide is more reactive than aluminium phosphide and will liberate hydrogen phosphide more rapidly and more completely at lower temperatures and humidities. Magnesium phosphide is not combustible but forms flammable gas on contact with water or damp air. Magnesium phosphide gives off irritating or toxic fumes (or gases) in a fire and on contact with water or damp air may cause explosion hazard.

Safe Handling and Precautions

Magnesium phosphide is a highly toxic chemical substance. Workers on mild inhalation exposure demonstrate symptoms of sickness, fatigue, nausea, chest pain, and ringing in the ears. Workers with mild poisoning of magnesium phosphide show symptoms of vomiting, pain just about the stomach, chest pain, diarrhoea, and dyspnoea (difficulty in breathing), while on severe poisoning, magnesium phosphide causes pulmonary oedema (fluid in lungs), dizziness, cyanosis, unconsciousness, and fatal injury.

Magnesium phosphide and phosphine have not been reported to cause chronic poisoning in humans. The dermal toxicity of magnesium phosphide is very low. The LD_{50} via the dermal route is estimated to be greater than 5000 mg/kg for a 1 h exposure. Primary routes of exposure are inhalation and ingestion. The inhalation LC_{50} of hydrogen phosphide gas is about 190 ppm (1 h inhalation). The acute oral LD_{50} of the Magtoxin formulation has been reported as 9.1 mg/kg of body weight. The International Agency for Research on Cancer (IARC), the National Toxicology Program (NTP), and the Occupational Safety and Health Administration (OSHA) have not listed magnesium phosphide and phosphine as human carcinogens.

During use and handling of magnesium phosphide, workers should strictly observe set safety regulations: (i) workers should strictly avoid to breathe the chemical dust and must wear full workplace suitable protective clothing/suit, suitable respiratory equipment, splash goggles, boots, and gloves; (ii) workers should avoid the contact of magnesium phosphide with water or other liquids; (iii) workers should avoid to pile up large quantities of magnesium phosphide products during fumigation or disposal; (iv) workers should open containers of Magtoxin, Fumi-Cel, or Fumi-Strip only in open air; (v) workers should strictly avoid to open containers in a flammable atmosphere. (vi) Magtoxin, Fumi-Cel, and Fumi-Strip are classified as restricted use pesticides (RUPs) due to acute inhalation toxicity of highly toxic hydrogen phosphide (phosphine, PH_3) gas; (vii) workers should be alert during waste disposal of spilled Magtoxin, Fumi-Cel, or Fumi-Strip by use according to label instruct. The chemical waste, spilled magnesium phosphide fumigants, Magtoxin, Fumi-Cel, and Fumi-Strip, must be carefully deactivated with water. Workers should not use detergent for the deactivation of these products; (viii) workers should know that the unreacted or incompletely exposed magnesium phosphide fumigants are highly toxic and are hazardous wastes; (ix) any sort of handling, sale, use, and waste disposal of Magtoxin, Fumi-Cel, and Fumi-Strip should be done ONLY by certified and trained applicators, and the workers should be under the direct supervision of the workplace supervisor.

Precautions and Storage of Magnesium Phosphide

During use and handling of magnesium phosphide, students and occupational workers should strictly follow the set workplace precautions. Magnesium phosphide decomposes on heating, producing toxic fumes including phosphorus oxides and phosphine, which increases fire hazard. Magnesium phosphide reacts violently with water, air moisture, and acids and produces phosphine causing fire and toxic hazard. Workers should avoid areas of open flames and sparks and avoid smoking and any kind of contact with water. Workers should control or prevent flames with sand, carbon dioxide, or dry extinguishing chemicals. Occupational workers should store Magtoxin, Fumi-Cel, and Fumi-Strip products under lock-and-key security cabinets in a dry, well-ventilated area away from heat. Workers should not store magnesium phosphide in buildings with human inhabitants or domestic animals. Work area where magnesium phosphide is in use should be kept under restriction, and entry of unauthorised workers/persons should be completely stopped.

DANGER! Workplace supervisor/manager should label magnesium phosphide store area.

Malathion (CAS No. 121-75-5)

Molecular formula: $C_{10}H_{19}O_6PS_2$
Synonyms and trade names: Celthion; Cythion; Dielathion; Emmaton; Exathios; Fyfanon and Hilthion; Karbofos and Maltox
IUPAC name: Diethyl(dimethoxythiophosphorylthio)succinate
Toxicity class: U.S. EPA: III; WHO: III

Malathion, a clear amber liquid, is sparingly soluble in water but soluble in a majority of organic solvents. The U.S. Environmental Protection Agency (U.S. EPA) grouped malathion under GUP, meaning general use pesticide. It is used as an insecticide as well as an acaricide for the control of pests. Malathion is used for the control of sucking insects and chewing insects on fruits and vegetables. Malathion is an effective insecticide for the control of several household pests such as houseflies, cockroaches, mosquitoes, aphids, animal ectoparasites, and human head and body lice. Malathion is also found in formulations with many other pesticides.

Safe Handling and Precautions

Acute and prolonged period of exposures to high concentrations of malathion causes poisoning in animals and humans. The symptoms of poisoning include, but are not limited to, numbness, tingling sensations, incoordination, headache, dizziness, tremor, nausea, abdominal cramps, sweating, blurred vision, difficulty in breathing or respiratory depression, and slow heartbeat. Very high doses may result in unconsciousness, incontinence, and convulsions or fatality. Malathion did not indicate any kind of delayed neurotoxicity in experimental studies with hens. Reports have indicated that ignorance and accidental exposures through severe skin absorption of malathion caused poisoning and fatalities among workers associated with the malaria control operations in Pakistan. In certain cases, development of pulmonary fibrosis following the poisoning has also been observed.

Reports have indicated that malathion is neither mutagenic nor teratogenic to animals and humans. In animals, malathion induced liver carcinogenicity at doses that were considered excessive. However, available information is not adequate to confirm the carcinogenicity of malathion to animals and humans. The IARC reported that malathion is not classifiable as a carcinogen to humans and the available data do not provide evidence that malathion or its metabolite malaoxon is carcinogenic to experimental animals, and hence malathion is listed as Group 3, meaning not classifiable as a carcinogen to humans.

During use and handling of malathion workers should wear protective clothing to prevent chemical contact, personal protective equipment (PPE), to avoid chemical dust, vapour, face mask, self-contained breathing apparatus (SCBA), liquid-proof gloves; protective goggles or face shield; white or light-coloured clothing; rubber shoes or boots.

Mercury and Its Compounds

Mercury is a naturally occurring element in the Earth's crust. Exploitation of the great majority of mercury deposits is by underground mining methods. During the process of its extracting, mercury liberates vapour, and condensation of the vapour occurs. Mercury and its derivatives, methyl, ethyl, *n*-butyl, are all known for severe toxicity as virulent neurotoxins. Mercury compounds penetrate epithelial and blood–brain barrier and persist in the intact body system. Reports have shown that mercury enters through different industrial emission sources. For example,

- Portland cement manufacturing
- Lime manufacturing
- Carbon black production
- By-product coke production
- Primary lead smelting
- Primary copper smelting
- Petroleum refining
- Municipal solid waste landfills
- Geothermal power plants
- Pulp and paper production

The uses of mercury include, but are not limited to, batteries, chlor-alkali industry, electrical equipment and measurement equipment, paint, tooth fillings, thermometers, and laboratory equipments. The significant intake of mercury of the general population has been traced and reported as from food and from dental amalgams. Fish and fish products are the dominant source of human exposure (and intake) because of their high retention of methylmercury. However, a significant intake of elemental mercury vapour with a high retention of mercury can occur from dental amalgams, dependent on the number of fillings. Exposure to inorganic mercury compounds is mainly from non-fish food with a low retention of mercury.

In brief, it may be stated that occupational exposure to mercury and compounds through acute inhalation in the form of mercury vapour causes chest pains, dyspnoea, coughing, haemoptysis, and sometimes interstitial pneumonitis leading to death. The ingestion of mercuric compounds, in particular mercuric chloride, has caused ulcerative gastroenteritis and acute tubular necrosis causing death from anuria where dialysis was not available. The central nervous system (CNS) is the critical organ for mercury vapour exposure. The subacute exposure has been known to cause psychotic reactions characterised by delirium, hallucinations, and suicidal tendency. With continuing exposure, a fine tremor develops, initially involving the hands. In the milder cases, erethism and tremor regress slowly over a period of years following removal from exposure. Decreased nerve conduction velocity has been demonstrated in mercury-exposed workers.

Global agencies have made efforts to regulate the use of mercury and its compounds and to contain the possible health hazards. All the same, it is important for workers and workplace supervisors to be very alert and to remember the importance of safety precautions and good work practices to be strictly observed during use and handling of mercury and compounds of mercury. In fact, workplace manager should permit only trained workers to use and handle compounds of mercury.

The exposure limits (TLV, PEL, and REL) of mercury compounds are very low (as reported by regulatory agencies) and require no negligence of workers during different stages of chemical management. The exposure limits are for air levels only. For details, readers should refer to literature for values for mercury and inorganic compounds and the legal airborne permissible exposure limit (PEL), the recommended airborne exposure limits.

Mercury (CAS No. 7439-97-6)

Molecular formula: **Hg**
Synonyms and trade names: Quick silver; Colloidal mercury; Metallic mercury; Liquid silver; Hydrargyrum

Mercury is a non-specific toxin, attacking many of the body's systems. At low levels of exposure, symptoms are mainly related to nerve and brain function and include memory loss, mood instability, tremor, and other stress-like symptoms: poor coordination, headache, and visual and hearing problems. Recently, reproductive health has been shown to be affected, with abnormalities in menstrual cycle, poor outcome of pregnancy, and subfertility in both men and women. The immune system is also damaged by mercury exposure.

Safe Handling and Precautions

Exposure to mercury is very hazardous. Any kind of workplace exposure through direct contact, inhalation, and/or accidental workplace ingestion causes adverse health effects and poisoning to workers. Mercury in case of contact is known to cause irritation effects to the skin and eyes, contact (irritant), of ingestion, of inhalation. Mercury is hazardous and on skin contact as liquid or spray mist is known to produce tissue damage particularly on mucous membranes of eyes, mouth, and respiratory tract. Skin contact may produce burns. Inhalation of the spray mist may produce severe irritation of respiratory tract, characterised by coughing, choking, or shortness of breath. Severe overexposure can result in death. Inflammation of the eye is characterised by redness, watering, and itching. Skin inflammation is characterised by itching, scaling, reddening, or, occasionally, blistering.

Mercuric Acetate (CAS No. 1600-27-7)

Molecular formula: **Hg(OAc)$_2$**

Synonyms and trade names: Acetic acid; Mercury (2+) salt; Bis(acetyloxy)mercury; Diacetoxymercury

ACIGH: threshold limit value (TLV): 0.1 mg/m^3 as a time-weighted average (TWA); NIOSH: mercury vapour of 0.05 mg/m^3 as a TWA

Mercuric acetate appears as white powder/crystals with characteristic odour. On exposure to heat, mercuric acetate produces toxic fumes of mercury/mercuric oxide. Mercuric acetate is incompatible with chromic acid, chromic anhydride, nitric acid, perchloric acid, permanganates, sodium peroxide, potassium hydroxide, hydrogen peroxides, acid anhydrides, and strong oxidising agents.

Safe Handling and Precautions

Workplace use and improper handling of mercuric acetate cause harmful effects such as cough, headache, respiratory distress/laboured breathing, shortness of breath, and sore throat. The exposed eyes suffer from pain, blurred vision, and severe deep burns. Accidental ingestion of mercuric acetate causes abdominal pain, burning sensation, diarrhoea, vomiting, and metallic taste. Repeated and prolonged period of exposures to mercuric acetate at workplaces through ingestion and/or inhalation cause adverse health effects. The symptoms include, but not limited to, corrosive effects to the eyes, the skin, dermatitis, respiratory tract, frothy sputum, cyanosis, lung oedema, and kidney disorders, nephrotic syndrome, leading to fatal injury. The American Conference of Governmental Industrial Hygienists (ACGIH) lists mercuric acetate as A4, meaning not classifiable as a human carcinogen, and similarly the IARC lists it as Group 3, meaning not classifiable as a human carcinogen. Improper use and handling of mercuric acetate is highly toxic, causes respiratory tract irritation, causes eye and skin irritation, and causes digestive tract irritation with nausea, vomiting, diarrhoea and severe exposure to fatal injury.

Mercuric Chloride (CAS No. 7487-94-7)

Molecular formula: **HgCl$_2$**

Synonyms and trade names: Calochlor; Corrosive mercury chloride; Corrosive sublimate; Mercury bichloride; Mercury perchloride; Mercury (II) chloride; Mercuric chloride; Mercury dichloride; Mercury bichloride; Mercury (II) chloride anhydrous; Merc; Dichloromercury; Abivit B; Fungchex; Bichloride of mercury; Sulem

ACIGH: TLV: 0.025 mg/m^3 as a TWA; NIOSH: mercury vapour of 0.05 mg/m^3 as a TWA; OSHA: PEL: 0.1 mg/m^3

Mercuric chloride is a corrosive solid. It is used for embalming chemical, disinfectant, photographic intensifier, fungicide, insecticide, and steel and iron etchant. Mercuric chloride is stable but moisture sensitive and light sensitive – decomposes in sunlight. It is incompatible with strong acids, ammonia, carbonates, metallic salts, alkalis, phosphites, phosphates, sulphites, sulphates, arsenic, antimony, and bromides.

Safe Handling and Precautions

Mercuric chloride is a corrosive and very harmful and poisonous chemical substance. It is extremely hazardous, and any kind of improper use and handling causes fatal injury.

Students and occupational workers in case of accidental ingestion and/or inhalation suffer with health disorders. Contact of mercuric chloride is corrosive, and any kind of contact with the skin and eyes causes tissue damage. It is very hazardous to the skin and causes blisters. Eye contact can result in corneal damage or blindness. Inhalation of dust will produce irritation to gastrointestinal or respiratory tract, characterised by burning, sneezing, and coughing. Severe overexposure can produce lung damage, choking, unconsciousness, or death. Inflammation of the eye is characterised by redness, watering, and itching. Skin inflammation is characterised by itching, scaling, reddening and/or blistering, local skin destruction, and/or dermatitis. Repeated and prolonged period of exposure to mercuric chloride causes toxic effects to the brain, peripheral nervous system, skin, CNS, eye, lens, or cornea. Also, mercuric chloride is toxic to the reproductive system, and repeated or prolonged exposure produces target organ damage; respiratory irritation; chronic respiratory irritation and lung damage; CNS effects including vertigo, anxiety, depression, muscle incoordination, and emotional instability; and gastrointestinal effects including gum and mouth inflammation, jaw necrosis, and loosening of the teeth. Mixture of mercuric chloride with potassium and metallic halides produces strong explosion on impact. Mixture of mercuric chloride with sodium and halide compounds produces a strong explosion in impact.

Students and occupational workers during use and handling of mercuric chloride should be alert and must use process enclosures and local exhaust ventilation to keep airborne levels below recommended exposure limits. Workers must minimise to generate dust, fume, or mist during handling and must use appropriate ventilation at the workplace. During use and handling of mercuric chloride, students and occupational workers must use gloves, splash goggles, personal protection dress synthetic apron, approved/certified vapour and dust respirator, splash goggles, and boots. An SCBA should be used to avoid inhalation of the product.

Mercuric (II) Cyanide (CAS No. 592-04-1)

Molecular formula: $Hg (CN)_2$
Synonyms and trade names: Cianurina; Dicyanomercury; Mercury cyanide; Mercurydicyanide; Mercury dicyanide; Cyanuredemercure; Mercury (+2) cyanide; Dicyanomercury (ii); Mercury (II) cyanide
ACGIH: 0.025 mg/m^3 TWA (as Hg); NIOSH: 0.05 mg/m^3 TWA; OSHA: 5 mg/m^3 TWA.

Mercuric cyanide is a white or colourless crystalline solid. It is used in medicine, germicidal soaps, photography, and in making cyanogen gas.

Safe Handling and Precautions

Occupations and exposure to mercury (II) cyanide (mercuric cyanide), improper use, and negligence in handling at workplaces cause harmful effects, skin irritation, and skin burns. Repeated overexposure by accidental ingestion and inhalation of mercury (II) cyanide causes adverse health effects such as grey skin colour; skin allergy; gastrointestinal irritation; nausea; vomiting; diarrhoea; respiratory tract irritation; headache; dizziness; weakness; irritability; sore gums; 'cyanide' rash; skin itching; macular, papular, and vesicular eruptions; memory loss; kidney and brain damage; collapse; unconsciousness; and possible death. Reports indicate that the ACGIH, the IARC, and the NTP have no sufficient data to list mercury (II) cyanide (mercuric cyanide) as a human carcinogen.

Mercuric Nitrate (CAS No. 7783-34-8)

Molecular formula: $Hg(NO_3)_2$
Synonyms and trade names: Mercury(II) nitrate

Mercuric nitrate appears as white or yellow-like deliquescent powder, with a mild nitric acid odour. It is stable and hygroscopic. Mercuric nitrate is incompatible with mercuric nitrate in contact with organic materials, powdered metals, petroleum hydrocarbons, hypophosphoric acid, unsaturates, and aromatics, which react violently.

Safe Handling and Precautions

Exposure to mercuric nitrate is harmful and causes adverse health effects. Skin exposures to mercuric nitrate and contact are known to cause irritation with itching, redness, and scaling. Mercuric nitrate gets readily absorbed through the skin and causes systemic poisoning; allergic skin reaction; skin burns; sore throat; headache; muscle weakness; coughing; pain; paleness; anorexia; ulceration of skin; gastrointestinal disturbance; breathing difficulties; respiratory distress; severe respiratory tract damage; bronchitis and pneumonitis; tightness in chest; paleness; ringing in the ears; liver changes; fever; kidney damage; exhaustion; formation of methaemoglobin, which decreases the ability of the blood to carry oxygen resulting in cyanosis; possible convulsions; tremors; coma; and death. Chronic exposure to mercuric nitrate through accidental workplace skin absorption or ingestion/inhalation has been reported to cause damage the CNS, brain, liver, and kidneys. Workers with pre-existing nervous, kidney, or respiratory disorders have been found to be more susceptible to the effects of mercuric nitrate–induced poisoning.

Mercuric Oxide (CAS No. 21908-53-2)

Molecular formula: HgO
Synonyms and trade names: Mercury (II) oxide, Yellow

Mercuric oxide is a solid/orange-red powder and very slightly soluble in cold water. Mercuric oxide is incompatible with strong reducing agents, strong oxidising agents, combustible materials, and organic materials.

Safe Handling and Precautions

Exposure to mercuric oxide is harmful and causes poisoning. Accidental exposures through ingestion and/or inhalation to mercuric oxide at workplaces have been reported to cause toxicity to workers with symptoms of poisoning such as irritation and burning sensations of mouth and throat; nausea; coughing; gastrointestinal discomfort; vomiting; wheezing; irritation to respiratory tract; shortness of breath or respiratory distress; abdominal pain; damage to liver, kidneys, lungs, nervous system, and reproductive system; and mucous membrane irritation with itching, redness, and scaling. Mercuric nitrate gets readily absorbed through the skin and causes systemic poisoning. Repeated or prolonged exposure to mercuric oxide produces damage to target organs and general deterioration of health by an accumulation in one or many human organs.

Keep container tightly closed. Keep it in a cool, well-ventilated place. Highly toxic or infectious materials should be stored in a separate locked safety storage cabinet or room.

Occupational workers during use and handling of mercuric oxide must strictly observe safety regulatory practices and must avoid workplace negligence. Workers should use with adequate ventilation and should not breathe dust or vapour; should avoid contact with skin, eyes, or clothing; and should wash hands thoroughly after handling. Workers during use and handling of mercuric oxide should wear Personal protective equipment (PPE) to prevent Chemical contact, to avoid organic vapor. Also workers should wear canister face mask, self-contained breathing apparatus, liquid proof gloves, protective goggles, rubber shoes or boots.

DANGER! Mercuric oxide is highly toxic, causes severe poisoning and irreversible effects, and causes deleterious effects to body tissues, CNS, and kidneys.

Regulations:

- Mercuric oxide has been listed by the WHMIS Canada as D1A: toxic material causing immediate/serious effects.
- California Proposition 65 has also listed mercuric oxide as hazardous chemical causing developmental and reproductive hazards.

Mercuric Sulphate (CAS No. 7783-35-9)

Molecular formula: $HgSO_4$
Synonyms and trade names: Mercury (II) sulphate (1:1); Mercury bisulphate; Sulphuric acid; Mercury (2+) salt (1:1)

Mercuric sulphate appears as white granules or crystalline powder and is odourless. Mercuric sulphate decomposes on contact with water into yellow insoluble basic sulphate and sulphuric acid. The products of decomposition of mercuric sulphate include oxides of sulphur and oxides of mercury. Mercuric sulphate is incompatible with acetylene, ammonia, and strong acids and corrosive to iron, magnesium, aluminium, zinc, lead, and copper.

Safe Handling and Precautions

Exposure to mercuric sulphate is very harmful and causes severe poisoning. The symptoms of poisoning include, but are not restricted to, sore throat, metallic taste, loss of the teeth, digestive disorders, skin rashes, skin allergies, skin to turn grey in colour, burning effects of the mucous membrane of the nose and throat, coughing, pain, tightness in chest, breathing difficulties, shortness of breath, respiratory tract damage, CNS damage, muscle tremors, memory loss, brain damage, and kidney damage. Occupational workers and individuals with pre-existing health disorders such as nervous disorders/ impaired kidney/respiratory function/a history of allergies and/or a known sensitisation to mercury have been found reported more susceptible to the poisoning effects of mercuric sulphate.

Occupational workers should remember to keep mercuric sulphate in tightly closed container and store in a cool, dry, ventilated area, free from sources of heat, ignition, and physical damage. Also, workers during use and handling of mercuric sulphate should strictly observe safety regulatory practices and avoid workplace negligence. Workers should wear approved protective clothing/workplace dress to prevent chemical contact, PPE, SCBA to avoid hazardous chemical vapour, chemical-proof gloves, protective goggles or face shield, and rubber shoes or boots.

Regulations: Mercuric sulphate contains a chemical known to the state of California to cause birth defects or other reproductive harm. Similarly, mercuric sulphate has been included in Chemical Inventory Status – Part 1 by Australia, Japan, and Korea.

DANGER! Workers should be alert: Exposure to mercuric sulphate causes fatal injuries and affects the kidneys and CNS.

Mercuric Thiocyanate (CAS No. 592-85-8)

Molecular formula: $Hg(SCN)_2$

Synonyms and trade names: Mercury (II) thiocyanate; Mercuric sulfocyanide; Mercury sulfocyanide; Mercurydithiocyanate; Mercury sulfocyanate; Mercuric sulfocyanate

Airborne exposure limits; OSHA acceptable ceiling concentration:

Mercury and mercury compounds: 0.1 mg/m^3 (TWA) skin

ACGIH threshold limit value (TLV)

Inorganic and metallic mercury, as Hg: 0.025 mg/m^3 (TWA) skin

Mercuric thiocyanate is an inorganic chemical substance. It is a stable solid at room temperature, and depending upon the purity, it appears as odourless white crystalline powder or grey. It is insoluble in water and denser than water and sinks in water. On decomposition, mercuric thiocyanate releases hazardous substances such as cyanide vapours, vapours of mercury, oxides of nitrogen (NO, NO_2), and oxides of sulphur (SO_2, SO_3). Mercury thiocyanate has limited uses in chemical synthesis.

Danger! Mercuric thiocyanate is flammable – fatal if swallowed and fatal if absorbed through the skin.

Mercuric thiocyanate causes severe eye and skin irritation with possible burns and causes digestive and respiratory tract irritation with possible burns. It may impair fertility, may cause harm to the unborn child, is harmful if inhaled, may cause allergic skin reaction, may cause kidney damage, may cause CNS effects, is light sensitive, and is a severe marine pollutant. Contact with acids liberates very toxic gas. The target organs include kidneys, CNS, reproductive system, eyes, and skin.

Safe Handling and Precautions

Exposure to mercuric thiocyanate is hazardous in case of eye contact (irritant). Inflammation of the eye is characterised by redness, watering, and itching. It is extremely hazardous in case of skin contact (permeator) and fatal if absorbed. It is hazardous in case of skin contact (irritant). Skin inflammation is characterised by itching, scaling, reddening, or, occasionally, blistering. It is extremely hazardous in case of inhalation and may be fatal if inhaled. It is hazardous in case of inhalation (lung irritant). It is extremely hazardous in case of ingestion and may be fatal if swallowed.

Occupational workers should handle mercuric thiocyanate with care and keep the material in a tightly closed light-resistant container. It should be stored in a cool, dry, ventilated area and should be protected against physical damage and direct sunlight.

Danger! Improper handling and exposure to mercuric thiocyanate cause fatal injuries. It causes damage and dangers to the eyes (lens or cornea), skin, respiratory tract, kidneys, and CNS. Mercuric thiocyanate has not been reported to cause cancer in animal and humans.

Workers should keep mercuric thiocyanate in a tightly closed light-resistant container; store in a cool, dry, ventilated area; protect against physical damage and from direct sunlight; and follow strict hygiene practices. Containers of mercuric thiocyanate material may be hazardous when empty since they retain product residues (dust, solids); workers should observe all warnings and precautions listed for the product.

Regulations:

- The U.S. regulations (CERCLA) require reporting spills and releases to soil, water, and air in excess of reportable quantities.
- SARA 302/304/311/312 regulates mercuric thiocyanate as hazardous chemicals.
- SARA 311/312 regulates mercuric thiocyanate MSDS distribution, chemical inventory, hazard, and Immediate (Acute) Health Hazard.
- WHMIS (Canada) Class D–1A: material causing immediate and serious toxic effects (VERY TOXIC).
- Class D–1B: material causing immediate and serious toxic effects (TOXIC).
- Class D–2B: material causing other toxic effects (TOXIC).
- CEPA DSL: mercuric thiocyanate.
- Pennsylvania state regulations RTK: mercuric thiocyanate (environmental hazard, generic environmental hazard).
- Massachusetts RTK: mercuric thiocyanate; mercuric thiocyanate, New Jersey; mercuric thiocyanate, California prop. 65. This product contains the following ingredients for which the state of California has found to cause cancer, birth defects, or other reproductive harm, which would require a warning under the statute.
- California Prop. 65: This product contains the following ingredients for which the state of California has found to cause birth defects, which would require a warning under the statute.
- Australian Hazchem Code: 2Z – Poison Schedule: S7.

Mercurous Chloride (CAS No. 0112-91-1)

Molecular formula: Hg_2Cl_2
Synonyms and trade names: Calomel; Mercury (I) chloride; Mercurous chloride; Mercury chloride; Mercury monochloride

Mercury (I) chloride is a dense white powder and insoluble in water and may be light sensitive. It is incompatible with strong bases, carbonates, sulphides, cyanides, alkalis, sulphites, sulphates, hydrogen peroxide, ammonia, iodine, and hydrogen bromide.

Mercurous chloride is widely used in electrochemistry for the preparation of calomel electrodes.

Safe Handling and Precautions

Users, students, and occupational workers during handling of mercurous chloride should be very careful because the chemical substance is a poison. Improper use, handling, and accidental swallowing and/or inhalation cause fatal injury to workers. Chronic exposure to mercurous chloride causes systemic toxicity with symptoms of poisoning that include build-up of mercury in the brain, liver, and kidneys; memory loss; tremors; and other serious health effects.

Mercurous Nitrate (CAS No. 10415-75-5)

Molecular formula: $Hg_2(NO_3)_2 \cdot 2H_2O$

Synonyms and trade names: Mercurous nitrate, dihydrate; Nitric acid; Mercury (1+) salt, dihydrate; Mercury protonitrate; Mercury (I) nitrate, dihydrate

Mercurous nitrate appears as colourless crystals, is odourless or has slight nitric acid odour, and is soluble in water. It is stable under ordinary conditions of use and storage. Mercurous nitrate is incompatible with ammonia, phosphorus, most common metals, combustible materials, and strong reducing agents, and mercurous nitrate solution is known to corrode metals.

Safe Handling and Precautions

Exposure to mercurous nitrate by accidental workplace inhalation, ingestion, and skin contact is harmful. It causes irritation to the respiratory tract, and the symptoms of poisoning include, but are not limited to, burning of the mouth and pharynx, abdominal pain, vomiting, corrosive ulceration, bloody diarrhoea, sore throat, coughing, pain, tightness in chest, breathing difficulties, shortness of breath, rapid and weak pulse, shallow breathing, paleness, exhaustion, pneumonitis, muscle tremors, personality and behaviour changes, memory loss, metallic taste, loosening of the teeth, digestive disorders, skin rashes, brain damage, kidney damage, collapse, and fatal injury from renal failure.

The U.S. OSHA has listed mercurous nitrate as a A4, meaning not a human carcinogen, and also the IARC has not listed mercurous nitrate as a human carcinogen.

Users, students, and occupational workers during handling of mercurous nitrate should keep it in a tightly closed container, stored in a cool, dry, ventilated area; protect it against physical damage and moisture; isolate it from any source of heat or ignition; avoid storage on wood floors; separate it from incompatibles, combustibles, organic, or other readily oxidisable materials; and follow strict hygiene practices. Containers of this material may be hazardous when empty since they retain product residues (dust, solids); observe all warnings and precautions listed for the product.

Aggravation of Pre-Existing Conditions

Persons with nervous disorders or impaired kidney or respiratory function or a history of allergies or a known sensitisation to mercury may be more susceptible to the effects of the substance.

Alert! Danger!! Exposure to mercurous nitrate is harmful, causes allergic skin reaction, and *affects the kidneys and CNS.* It is FATAL if swallowed.

Methane (CAS No. 74-82-8)

Molecular formula: CH_4
Chemical family: Hydrocarbons, gas

Methane is a natural, colourless, odourless, and tasteless gas. Methane reacts violently with chlorine dioxide and liquid oxygen and is incompatible with powerful oxidisers such as bromine pentafluoride, chlorine trifluoride, chlorine, fluorine, iodine heptafluoride, dioxygenyl tetrafluoroborate, dioxygen difluoride, trioxygen difluoride, and liquid oxygen. It is incompatible with halogens or interhalogens. It will react with bromine in light (explosively in direct sunlight).

Methane is used primarily as a fuel to make heat and light. It is also used to manufacture organic chemicals. Methane can be formed by the decay of natural materials and is common in landfills, marshes, septic systems, and sewers. It is soluble in alcohol, ether, benzene, and organic solvents. Methane is incompatible with halogens, oxidising materials, and combustible materials. Methane evaporates quickly. Methane gas is present in coal mines, marsh gas, and sludge degradations. Methane can also be found in coal gas. Pockets of methane exist naturally underground. In homes, methane may be used to fuel a water heater, stove, and clothes dryer. Also, incomplete combustion of gas also produces carbon monoxide. Methane gas is flammable and may cause flash fire. Methane forms an explosive mixture in air at levels as low as 5%. Electrostatic charges may be generated by flow and agitation.

Safe Handling and Precautions

Methane is relatively potent. It is the simplest alkane and the principal component of natural gas. Exposures to methane gas cause toxicity and adverse health effects. The signs and symptoms of toxicity include, but are not limited to, nausea, vomiting, difficulty in breathing, irregular heartbeat, headache, drowsiness, fatigue, dizziness, disorientation, mood swings, tingling sensation, loss of coordination, suffocation, convulsions, unconsciousness, and coma. While at low concentrations, methane causes no toxicity, and high doses lead to asphyxiation in animals and humans. Displacement of air by the methane gas is known to cause shortness of breath, unconsciousness, and death from hypoxemia. Methane gas does not pass readily through intact skin. Methane in its extremely cold liquefied form can, however, cause burns to the skin and eyes. No long-term health effects are currently associated with exposure to methane. There are no published reports indicating that methane is a cancer-causing agent or a human carcinogen. The IARC listed methane as Group 3, meaning not classifiable as a carcinogen to humans. No occupational exposure limits have been established for methane gas. However, the ACGIH set a TWA of 1000 ppm.

Occupational workers should store methane gas containers away from incompatible substances and handle them in accordance to standard set regulations and grounding and bonding if required. Occupational workers should be careful during handling and management of methane gas because of its severe fire and explosion hazard particularly with pressurised containers. The containers may rupture or explode if exposed to sufficient heat. Workers should avoid heat, flames, sparks, and other sources of ignition; stop leak if possible without personal risk. Workers should wear appropriate chemical-resistant gloves. Also, reduce vapours with water spray and keep away unnecessary workers/people from the place of chemical hazard. Workers should well ventilate the closed spaces before entering. Methane is not toxic; however, it is highly flammable and may form explosive mixtures with air. Methane is violently reactive with oxidisers, halogens, and some halogen-containing compounds. Methane is also an asphyxiant and, in enclosed areas, displaces oxygen. Septic tanks, cesspools, and drywells present serious hazards including septic cave-ins or collapses, methane gas explosion hazards, and asphyxiation hazards. Occupational workers/work area supervisor should note the indications of methane gas poisoning: Soon after exposure to oxygen levels of less than 15% in air, if the workers feel symptoms of dizziness, headache, and tiredness, medical advice should be provided.

Workers should remember that it is important to have a safe storage for methane gas. Containers of methane gas have to be in well-ventilated areas and away from all heat sources, including direct sunlight. If a fire does occur, it is vital to cut off the methane gas supply and keep other containers cool by soaking them in water.

Danger – Cautions – Methane gas is one of the worst greenhouse gases. Methane forms an EXPLOSIVE mixture in air at levels as low as 5%.

The leakage of methane gas is known only when commercial gas utility companies add a chemical smell to it or when it mixes naturally with hydrogen sulphide, causing a 'rotten egg' smell. Workers should be alert and aware that methane gas can asphyxiate. The area should be well ventilated or have breathing equipment. Methane gas easily displaces the oxygen in the bloodstream, and the result is serious injury and/or fatal injury.

Safety tips while working with or near methane

- Signs must be posted on doors to indicate whether they must be kept open or closed.
- Occupational workers especially miners must wear respirators in work areas where levels of methane are high.
- Never enter worksites after blasting until gases have reached safe levels.
- All methane monitors must be placed as close as possible to the working place.
- Methane tests must be conducted before and during welding operations.
- Miners must make sure that all self-rescuers are properly maintained by inspecting them regularly.
  - All tunnels and mines must have multiple escape ways, evacuation plans, and emergency action plans.
- Workers should strictly avoid to smoke in areas where methane is present.

Methamidophos (CAS No. 10265-92-6)

Molecular formula: $C_2H_8NO_2PS$
Synonyms and trade names: Monitor; Tamaron; Nitofol; Tamaron; Swipe; Nuratron; Vetaron; Filitox; Patrole; Tamanox
IUPAC name: O,S-dimethylphosramidothioate
Toxicity class: U.S. EPA: I; WHO: Ib

Methamidophos is a colourless crystalline solid with a pungent odour. It is readily soluble in water, alcohols, ketones, and aliphatic chlorinated hydrocarbons but sparingly soluble in ether and practically insoluble in petroleum ether. It is highly toxic and systemic with properties of an insecticide, acaricide, and avicide. The U.S. EPA has grouped methamidophos as an RUP, meaning use and handling of this chemical substance requires qualified, certified, and trained workers. Methamidophos is effective against chewing and sucking insects and is used to control aphids, flea beetles, worms, whiteflies, thrips, cabbage loopers, Colorado potato beetles, potato tube worms, army worms, mites, leaf-hoppers, and many others. Crop uses include broccoli, Brussel sprouts, cauliflower, grapes, celery, sugar beets, cotton, tobacco, and potatoes. It is used abroad for many vegetables, hops, corn, peaches, and other crops. Methamidophos is also a breakdown product of another OP, namely, acephate. Methamidophos is slightly corrosive to mild steel and copper alloys.

Safe Handling and Precautions

Methamidophos is highly toxic to mammals. Inhalation of methamidophos causes weakness, tightness in the chest, wheezing, headache, blurred vision, and pinpoint pupils; tearing and runny nose are common early symptoms. On accidental ingestion and with severe poisoning, methamidophos causes nausea, vomiting, diarrhoea, cramps, sweating and twitching, weakness, shakiness, blurred vision, pinpoint pupils, dyspnea (shortness of breath), tightness in the chest, sweating, confusion, changes in heart rate, convulsions, coma, respiratory failure, and death. People with health disorders such as high blood pressure and problems of gastrointestinal system, heart, liver, lung, or nervous system have been reported as more sensitive to methamidophos-induced toxicity. Reports have indicated that occupational workers exposed to methamidophos developed poisoning with symptoms such as pain (needle type) in the feet, legs, and hands; high blood pressure; gastrointestinal disorders; and heart, liver, lung, or nervous system problems and may be more sensitive to malathion. There are no evidence/published reports about the carcinogenicity of methamidophos in experimental rats and/or mice. Methamidophos has not been reported as a human carcinogen.

Regulations: Indonesia: Registration no longer permitted since 1996; Kuwait: Banned from further use since 1980; Sri Lanka: Use of methamidophos severely restricted since 1995.

Methidathion (CAS No. 950-37-8)

Molecular formula: $C_6H_{11}N_2O_4PS_3$
Synonyms and trade names: Somonic; Somonil; Supracide; Suprathion; Ultracide
IUPAC name: O,O-dimethyl S-(2,3-dihydro-5-methoxy-2-oxo-1,3,4-thiadiazol-3-methyl) phosphorodithioate
Toxicity class: U.S. EPA: I; WHO: Ib

Methidathion is a colourless crystalline pesticide at room temperature. It is sparingly soluble in water but very soluble in octanol, ethanol, xylene, acetone, and cyclohexane. Methidathion is a non-systemic organophosphorus (OP) insecticide and acaricide with stomach and contact action. It is used to control a variety of insects and mites on crops and fruit plants. Methidathion is highly toxic to animals and humans. The U.S. EPA grouped methidathion as a class I toxic substance and as an RUP.

Safe Handling and Precautions

Acute and prolonged exposures to methidathion cause poisoning in animals and humans. The symptoms include, but are not limited to, nausea, vomiting, cramps, diarrhoea, salivation, headache, dizziness, muscle twitching, difficulty in breathing, blurred vision, tightness in the chest, pulmonary oedema, respiratory depression, and respiratory paralysis. Acute exposures to high concentrations of methidathion cause intense breathing problems, including paralysis of the respiratory muscles. Methidathion did not induce any genetic changes in experimental animals. Information on the teratogenicity and carcinogenicity of methidathion is inadequate in animals and humans. The recommended acceptable daily intake (ADI) for methidathion has been set at 0.001 mg/kg/day. Occupational workers should be very careful during use and chemical management of methidathion. Because this chemical substance is a highly toxic pesticide, it has been grouped by the U.S. EPA as toxicity class I. The containers and labels of the products should bear the signal word DANGER. Methidathion is an RUP, except for use in nurseries and on safflower and sunflowers.

Methoxychlor (CAS No. 72-43-5)

Molecular formula: $C_{16}H_{15}C_{13}O_2$
Synonyms and trade names: Methoxy-DDT; Chemform; Maralate; Methoxo; Methoxcide; Metox; Moxie
IUPAC name: 1,1,1-trichloro-2,2-bis(4-methoxyphenyl)ethane
Toxicity class: U.S. EPA: IV; WHO: III

Methoxychlor is a colourless or pale yellow, crystalline synthetic organochlorine insecticide with mild fruity odour. It is available in the form of powder or crystals. Methoxychlor is moderately soluble in water and is soluble in a variety of organic solvents. Methoxychlor has been used as an insecticide against a wide range of pests, such as houseflies, mosquitoes, cockroaches, chiggers, and various arthropods commonly found on field crops, vegetables, fruits, stored grain, livestock, and domestic pets. Methoxychlor is more unstable than DDT and has less residual effect; it has been used extensively in Canada for the control of biting flies and is also effective against mosquitoes and houseflies.

Safe Handling and Precautions

Methoxychlor is an organochlorine insecticide of low toxicity. Exposure to high concentrations of methoxychlor causes poisoning with symptoms such as headache, nausea, vomiting, depression of the CNS, weakness, trembling, convulsions, diarrhoea, and adverse effects to liver, kidney, and heart. Methoxychlor in high doses causes adverse effects on liver, kidney, and the CNS. The exposed individual demonstrates neurological disturbances such as nervousness, increased salivation, decreased locomotor activity, tremors, convulsions, and death. Studies on laboratory animals have shown methoxychlor and metabolites of methoxychlor that are estrogenic can be transferred from a nursing mother to her newborn babies through breast milk. Methoxychlor and its metabolites can probably cross the placenta and have been detected in human breast milk. Methoxychlor causes adverse effects on the male and female reproductive system as well as the balance of time-sensitive hormone levels (pro-estrogenic) during foetal and postnatal development.

Methoxychlor and Cancer

Laboratory animal studies have suggested that methoxychlor is negative for cancer. In view of this, the IARC has classified methoxychlor as a Group 3 carcinogen, meaning not classifiable as to its carcinogenicity to humans. Similarly, the U.S. EPA has classified methoxychlor as a Group D, meaning not classifiable as to human carcinogen.

The ADI for methoxychlor has been set at 0.1 mg/kg/day.

Methyl Alcohol (CAS No. 67-56-1)

Molecular formula: CH_3OH

Synonyms and trade names: Methanol; Carbinol; Colonial spirit; Columbian spirit; Methylol; Methyl hydrate; Wood alcohol; Wood naphtha; Wood spirit; Methyl hydroxide; Pyroxylic spirit; RCRA waste number U154; Meths

Methyl alcohol is the simplest alcohol. It is light, volatile, colorless and highly flammable with a distinctive odour very similar to, but slightly sweeter than ethyl alcohol. Methyl alcohol reacts violently with acids, acid chlorides, acid anhydrides, oxidising agents, reducing agents, and alkali metals. Methanol is produced naturally in the anaerobic metabolism of many varieties of bacteria and is ubiquitous in the environment. Methyl alcohol at room temperature is a polar liquid used as a solvent, fuel, anti-freeze, and denaturant for ethanol. Methyl alcohol has extensive and versatile industrial applications in the production of biodiesel fuel through transesterification reactions. Methanol is converted to formaldehyde, in the production of many kinds of plastics, press crease permanent textiles, paints, explosives, etc. Methanol is used as a denaturing agent in polyacrylamide gel electrophoresis. Methanol reacts violently with oxidising materials such as perchlorates, chromium trioxide, bromine, sodium hypochlorite, chlorine, and hydrogen peroxide resulting in fire and explosive mixtures.

Safe Handling and Precautions

Methyl alcohol (methanol) is a very toxic chemical substance through all the routes of exposure to humans, namely, by ingestion, inhalation, and absorption. Repeated and prolonged period of exposure to methyl alcohol causes cough; dizziness; headache; drowsiness; nausea; weakness; vertigo; occasional convulsions; dilated pupils, with sluggish or absent light reflex; visual disturbances; and coma and results in fatal injury. Methyl alcohol is irritating to the eyes, the skin, and the respiratory tract. It causes effects on the CNS, resulting in loss of consciousness, and severe exposure leads to blindness and death.

The early signs of methyl alcohol toxicity are observed by the slight inebriation and drowsiness, and the delayed signs appear after a lapse of 8–36 h. Vision becomes blurred or dimmed and may be sufficient to impair perception of light or cause complete blindness. There is impaired pupillary response to light and contraction of visual fields, scotoma, and flashing lights. Visual disturbances can be permanent. Hyperaemia of the optic disc is common in the acute stage.

Methylamine (CAS No. 74-89-5)

Molecular formula: CH_3NH_2

Synonyms and trade names: Aminomethane; Methanamine; Monomethylamine

Methylamine is a colourless, fish-like smelling gas at room temperature. It is used in a variety of industries such as manufacture of dyestuffs, treatment of cellulose, acetate rayon, as fuel additive, rocket propellant, and leather tanning processes.

Safe Handling and Precautions

Exposures to methylamine are known to cause adverse health effects among occupational workers. The workers demonstrate symptoms of toxicity that include, but are not limited to, irritation to eyes, nose, and throat. Studies have indicated that the compound causes injury to eyes through corneal opacities and oedema haemorrhages in the conjunctiva and injury to liver. Studies of Guest and Varma indicated no significant deleterious effects on internal organs or skeletal deformities in experimental mice.

Methyl Bromide (CAS No. 74-83-9)

Molecular formula: CH_3Br
Synonyms and trade names: Brom-o-gas; Bromomethane; Curafume; Dowfume MC-2 soil fumigant; Embafume; Haltox; MBX; Metafume; Methane, bromo-; Monobromomethane; ProFume; Rotox; Terabol; Zytox

Methyl bromide (bromomethane) is an odourless, sweetish, colourless gas, incompatible with oxidising agents and strong acids. This is an ozone-depleting chemical, and its use is restricted in many countries of the world. Methyl bromide has its use as a soil fumigant and structural fumigant to control pests across a wide range of agricultural sectors. Methyl bromide is soluble in ethanol, benzene, and carbon disulphide and sparingly in water. Methyl bromide, during the 1920s, was used as an industrial fire extinguishing agent in Europe. The insecticidal value of methyl bromide was first reported by Le Goupil (1932) in France. During the 1930s, it was widely adopted for plant quarantine purposes because many plants, vegetables, and some fruits were found to be tolerant to concentrations effective against the insects concerned. During recent years, application of methyl bromide became more extensive as an industrial fumigant for the control of pests of stored products, mills, warehouses, ships, and as a sterilising agent. Now, the use of methyl bromide has been largely replaced with hydrogen cyanide. The current uses of methyl bromide include the fumigation of homes and other structures for the control of termites and other pests. Because methyl bromide depletes the stratospheric ozone layer, the amount of methyl bromide produced and imported in the United States was reduced and phased out in January 1, 2005 (based on the Act of Montreal Protocol on Substances that Deplete the Ozone Layer (Protocol) and the Clean Air Act (CAA)). Methyl bromide is not as toxic to most insect species as are some other commonly used fumigants, such as HCN, acrylonitrile, and ethylene dibromide, but other properties of methyl bromide such as its ability to penetrate quickly and deeply into sorptive materials at normal atmospheric pressure make it an important and versatile fumigant. Also, at the end of a treatment, the vapours dissipate rapidly and make possible the safe handling of bulk commodities. Another important property is the fact that many living plants are tolerant to this gas in insecticidal treatments. Methyl bromide is non-flammable and non-explosive under ordinary circumstances and may be used without special precautions against fire.

Safe Handling and Precautions

Methyl bromide is toxic; becomes fatal by ingestion, inhalation, or if absorbed through the skin; may cause CNS damage and reproductive defects, and is very destructive to mucous membranes. Exposures to methyl bromide by breathing cause injury to the brain, nerves, lungs, and throat. At high doses, breathing of methyl bromide causes injury to the kidneys and liver. The symptoms of toxicity and poisoning of methyl bromide include, but are not limited to, dizziness, headache, abdominal pain, vomiting, weakness, hallucinations, loss of speech, incoordination, laboured breathing, and convulsions. On contact with the skin and eyes, the chemical substance can lead to irritation and burns. Occupational workers after a serious exposure to methyl bromide suffer from lung and/or nervous system-related problems and permanent brain/nerve damage. Laboratory study with species of animals indicated that bromomethane does not cause birth defects and does not interfere with normal reproduction except at high exposure levels. The IARC and the U.S. EPA found inadequate evidence in humans for the carcinogenicity of methyl bromide and limited evidence in experimental animals and hence reported methyl bromide (bromomethane) as Group 3, meaning not classifiable as a human carcinogen. Occupational workers should be careful during use and management of methyl bromide, which is very reactive and produces toxic and corrosive fumes including hydrogen bromide, bromine, and carbon oxybromide. Methyl bromide reacts with strong oxidants; attacks many metals in the presence of water; attacks aluminium, zinc, and magnesium with formation of pyrophoric compounds; and causes fire and explosion hazard. The OSHA sets the limits of methyl bromide (bromomethane) in workplace air at 5.0 ppm and recommends that the exposures to this chemical substance be reduced to the lowest level feasible. The TLV of methyl bromide is set at 1.0 ppm as TWA (skin). Methyl bromide (bromomethane) should be stored in sealed containers to keep it from evaporating.

Methyl bromide (bromomethane), which was earlier in use as a soil fumigant, is phased out because of its threat to the ozone layer.

Methylene Chloride (CAS No. 75-09-2)

Molecular formula: CH_2Cl_2
Synonyms and trade names: Dichloromethane

Methylene chloride is a colourless liquid with a mild, sweet odour. It does not occur naturally in the environment. It is made from methane gas or wood alcohol. Industrial uses of methylene chloride is very extensive as a solvent in paint strippers, as a propellant in aerosols, and as a process solvent in the manufacturing of drugs. Methylene chloride is also used as a metal cleaning and finishing solvent. Methylene chloride is approved as an extraction solvent for spices and hops. Occupational workers get exposed to methylene chloride at workplaces by breathing air, breathing fumes from paint strippers that contain it (check the label), breathing fumes from aerosol cans that use it (check the label), and breathing contaminated air near waste sites.

Safe Handling and Precautions

Exposures to methylene chloride cause adverse health effects and poisoning to users. Methylene chloride harms the human CNS. The symptoms of poisoning include, but are not limited to, dizziness, nausea, tingling, and numbness in the fingers and toes. Laboratory animals exposed to very high levels of methylene chloride suffer from unconsciousness and fatal injury/death. Occupational workers who get exposed to direct skin contact with methylene chloride indicate symptoms of intense burning and mild redness of the skin and damage of the eyes and cornea. The Department of Health and Human Services (DHHS) has reported that methylene chloride may reasonably be anticipated to be a carcinogen. Breathing high concentrations of it for long periods did increase the incidence of cancer in mice. However, methylene chloride has not been shown to cause cancer in humans exposed to vapours in the workplace. The U.S. EPA recommends that water containing more than 13.3 ppm of methylene chloride for longer than 1 day or with more than 1.5 ppm for longer than 10 days is not suitable, and the National Institute for Occupational Safety and Health (NIOSH) recommends the permissible limit of 75 ppm of methylene chloride in the air over a 10 h workday.

Methyl Ethyl Ketone (CAS No. 78-93-3)

(2-butanone, 3-butanone, methyl acetone, ethyl methyl ketone)
Molecular formula: C_4H_8O
Synonyms and trade names: Ethyl methyl ketone; 2-Butanone; Methyl acetone; Methyl ethyl ketone; MEK; MEETCO

Methyl ethyl ketone (MEK) is a colourless liquid with a sweet and sharp odour. It is soluble in alcohol, ether, acetone benzene, and water. It is a solvent often found in mixtures with acetone, ethyl acetate, n-hexane, toluene, or alcohols. It has applications in the surface coating industry and in the de-waxing of lubricating oils. MEK is used in the manufacture of colourless synthetic resins, artificial leather, rubbers, lacquers, varnishes, and glues.

Safe Handling and Precautions

Symptoms of acute MEK exposure include irritation of the eyes, nose, and throat.

In human case studies, inhalation of MEK for its euphoric effect has also resulted in slight excitement, followed by somnolence or unconsciousness at higher concentrations.

Humans occupationally exposed to MEK have also complained of mild neurologic effects. Determination of acute reference exposure levels for airborne toxicants users and occupational workers on exposure to MEK shows the symptoms of toxicity and poisoning such as headaches, dizziness, skin rashes, and nausea.

Human volunteers exposed to pure MEK for a prolonged period of time showed symptoms of dermatitis, headache, fainting, mild vertigo and blurred vision, loss of appetite, weight loss, gastrointestinal disorders, dizziness, muscular hypotrophy, numbness of fingers and arms, neuropathy (not clinically recognisable), convulsions, twitching, and reduced motor nerve conduction velocity.

Methyl Parathion (CAS No. 298-00-0)

Molecular formula: $C_8H_{10}NO_5PS$
Synonyms and trade names: Folidol-M; Metacide; Dhanuman; Sweeper; Metaphos; Metron
IUPAC name: O,O-dimethyl-O-4-nitrophenylphosphorothioate
Toxicity class: U.S.E PA: I; WHO: I or Ia

Pure methyl parathion exists as white crystals, while the technical product is light to dark tan in colour. Impure methyl parathion is a brownish liquid that smells like rotten eggs. It is sparingly soluble in water but soluble in dichloromethane, 2-propanol, toluene, and in many organic solvents. Methyl parathion is used to kill insects on farm crops. Methyl parathion is a contact insecticide and acaricide used for the control of boll weevils and many biting or sucking insect pests of agricultural crops. It kills insects by contact or by stomach and respiratory action. The U.S. EPA now restricts how methyl parathion can be used and applied; only trained people are allowed to spray it. Methyl parathion can no longer be used on food crops commonly consumed by children.

Exposures to methyl parathion occur to farm workers and chemical sprayers, and people who work in factories that make methyl parathion are most likely to be exposed. People who live near farms where methyl parathion is used or near landfills where methyl parathion has been dumped may be exposed. Individuals may also be exposed by going into fields too soon after spraying.

Safe Handling and Precautions

Methyl parathion is highly toxic by inhalation and ingestion and moderately toxic by dermal adsorption. Exposures to very high concentrations of methyl parathion for a short period cause poisoning and death. Repeated exposures to methyl parathion cause poisoning with symptoms that include, but are not limited to, dullness, loss of consciousness, dizziness, confusion, headaches, respiratory distress, chest tightness, wheezing, vomiting, diarrhoea, cramps, tremors, and blurred vision. Methyl parathion at very high doses is known to cause unconsciousness, incontinence, and convulsions and/or fatal injury/death. There are no reports indicating that methyl parathion is teratogenic or mutagenic or carcinogenic to animals and humans. The U.S. EPA and IARC have observed that methyl parathion is not classifiable as a human carcinogen. The recommended ADI for methyl parathion has been set at 0.02 mg/kg/day.

Methyl Vinyl Ketone (CAS No. 78-94-4)

Molecular formula: C_4H_6O
Synonyms and trade names: Butenone; Methyl vinyl ketone; MVK; Methylene acetone; 3-Butene-2-one; Vinyl methyl ketone; Methylene acetone; Acetyl ethylene; delta-oxo-alpha-butylene; 1-Buten-3-one; 3-Oxobutene; UN 1251

Methyl vinyl ketone (MVK) is a stable, highly flammable, heat- and light-sensitive, and colourless liquid. It is incompatible with strong bases, strong oxidising agents, and strong reducing agents and undergoes autopolymerisation. MVK is a reactive organic compound classified as an enone. It is a colourless to yellow, flammable, highly toxic liquid with a pungent odour. It is easily soluble in water, methanol, ethanol, and acetic acid. It can act as an alkylating agent. Its alkylating ability is both the source of its high toxicity and the feature that makes it a useful intermediate in organic synthesis. It polymerises spontaneously and is used in the manufacture of plastic polymers. It is also an intermediate in the synthesis of steroids and vitamin A.

Safe Handling and Precautions

MEK is a chemical used mainly in fibreglass and plastic manufacture. Ingestion and other exposures to the chemical can cause various symptoms. The type and severity of symptoms vary depending on the amount of chemical involved and the nature of the exposure.

MVK is extremely hazardous upon inhalation causing coughing, wheezing, and shortness of breath even at low concentrations. It will also readily cause irritation of the skin, eyes, and mucous membranes. Occupational workers upon exposure(s) to MVK show signs and symptoms of poisoning such as severe irritation, nausea, vomiting, difficulty in breathing, headache, symptoms of drunkenness, fainting, skin colour turning bluish, lung congestion, coma, and fatal injury. MVK is very toxic by inhalation, ingestion, or skin absorption. It is a lachrymator and corrosive, causes burns, and damages the mucous membranes.

Any kind of accidental exposure at workplaces by ingestion, inhalation, or skin contact causes fatal injury. Users and occupational workers during handling of MVK should be very careful because the chemical vapour is known to cause flash fire and quickly polymerises, and the containers get ruptured and cause explosion. Workers during handling of MVK should avoid contact pints of heat, flames, sparks, and other sources of ignition and should completely avoid to touch the spilled material. Workers should keep stored MVK completely separated from incompatible chemical substances and avoid exposure to low temperatures or freezing.

Mevinphos (CAS No. 7786-34-7)

Molecular formula: $C_7H_{13}O_6P$
IUPAC name: 2-Methoxycarbonyl-1-methylvinyl dimethyl phosphate
Synonyms and trade names: Apavinphos; CMDP; ENT 22374; Fosdrin; Gesfid; Meniphos; Menite; Mevinox; Mevinphos; OS-2046; PD5; Phosdrin; Phosgene
Toxicity class: U.S. EPA: I; WHO: Ia

Pure mevinphos is a colourless liquid, while technical grade mevinphos is a pale yellow liquid with a very mild odour. It is soluble in water but very soluble in alcohols, ketones, chlorinated hydrocarbons, aromatic hydrocarbons, and many organic solvents. It is both an insecticide and acaricide. The U.S. EPA grouped mevinphos as an RUP, and hence it should be purchased and used only by certified and trained pesticide applicators. It is not registered for use in many countries including the United States. Mevinphos is used to control a broad spectrum of insects, including aphids, grasshoppers, leaf-hoppers, cutworms, caterpillars, and many other insects on a wide range of field, forage, vegetable, and fruit crops. It is also an acaricide that kills or controls mites and ticks. It acts quickly both as a contact insecticide, acting through direct contact with target pests, and as a systemic insecticide, which becomes absorbed by plants on which insects feed.

Safe Handling and Precautions

Mevinphos is highly toxic and causes poisoning in animals and humans. The symptoms of poisoning include, but are not restricted to, nausea, dizziness, blurred vision, confusion, and at very high exposures (like accidental spills), numbness, tingling sensations, incoordination, headache, dizziness, tremor, abdominal cramps, respiratory paralysis, respiratory depression, slow heartbeat, and death. Very high doses may result in unconsciousness, incontinence, convulsions, and death. Prolonged exposures to methyl parathion also cause acute pulmonary oedema.

Mevinphos and carcinogenicity: There are no reports indicating that mevinphos produces mutagenic or teratogenic or carcinogenic effects to animals and humans. There are no evidences to indicate that mevinphos is a human carcinogen. The ADI for mevinphos has been set at 0.0015 mg/kg/day, and the TLV of mevinphos at 0.092 mg/m^3 (8 h). Mevinphos is a highly toxic OP pesticide by its acute oral and dermal toxicity and residue effects to species of animals. The U.S. EPA indicated it as a toxicity class I chemical substance. All emulsifiable and liquid concentrates of mevinphos are classified as RUPs by the U.S. EPA. Therefore, mevinphos and other similar RUPs should be purchased and used only by certified and trained applicators. Products containing mevinphos must bear the signal words DANGER – POISON.

Mirex (CAS No. 2385-85-5) and Chlordecone (CAS No. 143-50-5)

Molecular formula: $C_{10}Cl_{12}$
Synonyms and trade names: 1,3,4-Metheno-1H-cyclobuta[cd]pentalene, 1,1a,2,2,3,3a,4,5,5,5a,5b, 6,-dodecachlorooctahydro-; Dechlorane

Mirex and chlordecone are two separate, but chemically similar, manufactured insecticides that do not occur naturally in the environment. Mirex is a white crystalline solid, and chlordecone is a tan-white crystalline solid. Both chemicals are odourless. Mirex and chlordecone have not been manufactured or used in the United States since 1978. Mirex was used to control fire ants and as a flame retardant in plastics, rubber, paint, paper, and electrical goods from 1959 to 1972. Chlordecone was used as an insecticide on tobacco, ornamental shrubs, bananas, and citrus trees and in ant and roach traps. Mirex was sold as a flame retardant under the trade name Dechlorane, and chlordecone was also known as Kepone. Mirex and chlordecone break down slowly in the environment, and they may stay for years in soil and water. Mirex is not readily decomposed chemically or biologically and is relatively persistent. Human exposures to mirex and chlordecone occur in multiple ways, for example by touching mirex contaminated soil near hazardous waste sites, by accidental ingestion contaminated fish or other animals living near hazardous waste sites; to nursing infants of mothers living near hazardous waste sites may be exposed to mirex through their mothers' milk; and by drinking water or breathing air is not likely to cause exposure because these compounds do not easily dissolve in water or evaporate. Mirex is not readily decomposed chemically or biologically and is relatively persistent.

Safe Handling and Precautions

Exposures to high levels of mirex and chlordecone over a long period cause harmful health effects. The symptoms of poisoning involve the nervous system, skin, liver, and male reproductive system. Laboratory animals exposed to high concentrations of mirex and chlordecone demonstrated tissue damages in the stomach, intestine, liver, kidneys, eyes, thyroid, and nervous and reproductive systems. There are no studies available on whether mirex and chlordecone are carcinogenic in people. However, studies in mice and rats have shown that ingesting mirex and chlordecone can cause liver, adrenal gland, and kidney tumours. The DHHS has determined that mirex and chlordecone may reasonably be anticipated to be carcinogens. The U.S. EPA sets a limit of 1.0 ppt of mirex in surface water (1 ppt) to protect fish and other aquatic life from harmful effects. The NIOSH recommends that average workplace air levels not exceed 1 mg/m^3 of chlordecone over a 10 h period. The U.S. Food and Drug Administration (U.S. FDA) observed that eating fish and other foods with concentrations below 100 ppt of mirex, or concentrations of chlordecone below 400 ppt, has not caused any adverse health effects in people. During use and handling of mirex, occupational workers must strictly observe set workplace regulations.

Mirex is not readily decomposed chemically or biologically and is relatively persistent; waste material should be burned in a proper incinerator designed for organochlorine waste disposal; if this is not possible, bury in an approved dump or landfill where there is no risk of contamination of surface or ground water; comply with any local legislation regarding disposal of toxic wastes. Industrial discharges from manufacturing, formulation, and technical applications should not be allowed to pollute the environment and should be treated properly. Surplus product, contaminated absorbents, and containers should be disposed of in an appropriate way and should comply with any local legislation regarding disposal of toxic wastes. The employer should ensure that workers are equipped with appropriate PPE including safety boots, masks, chaps, gloves, and coveralls, suitable for both men and women as the task indicates.

Molybdenum (CAS No. 7439-98-7)

(As Mo), Insoluble compounds (total dust)
Molecular formula: **Mo**

Molybdenum is an essential trace mineral associated with several enzyme systems required for the normal body functions. However, in a prolonged period of exposure to high concentrations, molybdenum dust is known to cause poisoning among occupational workers. Mine workers have developed symptoms of molybdenosis. The significantly expanding or softening property of the material makes it useful in applications that involve intense heat, including the manufacture of aircraft parts, electrical contacts, industrial motors, and filaments.

Safe Handling and Precautions

Chronic exposures to molybdenum in high concentrations cause health disorders, which include, but are not limited to, irritation to eyes, nose, and throat; weight loss; dizziness; listlessness; anorexia; diarrhoea; weakness; fatigue; headache; lack of appetite; epigastric pain; and joint and muscle pain. It also causes redness and moist skin, tremors of the hands, sweating, liver and kidney damage, blood disorders, dyspnoea, anaemia, gout-like symptoms such as arthralgia, articular deformities, erythema, and oedema. Reports are also available that high concentrations of molybdenum cause knock-knees (genu valgum), gastrointestinal irritation with diarrhoea, coma, and death from cardiac failure in exposed individuals.

Mustard Gas (CAS No. 505-60-2)

Molecular formula: $C_4H_8C_{12}S$

Synonyms and trade names: Sulphur mustard; 1,1'-Thiobis (2-chloroethane); Bis-(2-chloroethyl) sulphide; beta,beta'-Dichloroethyl sulphide; 2,2'-Dichloroethyl sulphide; Agent HD; H; HS; Mustard gas; Bis(beta-chloroethyl) sulphide; Bis(2-chloroethyl)sulphide; 1-Chloro-2 (beta-chlorodiethylthio) ethane; Yellow cross liquid; Kampfstoff 'Lost'; Schwefel-Lost; Senfgas; Yellow cross liquid; Yperite

Chemical Class: Alkylating Agent

Mustard gas/sulphur mustard is an organic chemical substance synthesised by treating sulphur dichloride with ethylene. Mustard gas is a chemical substance closely related to chemical warfare class agents. It is a cytotoxic and vesicant chemical substance, and exposures are known to cause blisters on the exposed skin. Pure mustard gas/sulphur mustards are colourless, viscous liquids at room temperature. However, when used in impure form, such as warfare agents, they appear as yellow brown in colour. As the name indicates, mustard gas has an odour resembling the garlic, horseradish, or mustard plants.

Mustard gas is the common name given to 1,1-thiobis(2-chloroethane), a chemical warfare agent that is believed to have first been used near Ypres in Flanders on 12 July 1917. Mustard gas is a thick liquid at ambient temperature. It is heavier than water as a liquid and heavier than air as a vapour. It does not occur naturally in the environment. Pure liquid mustard gas is colourless and odourless. It is stable, combustible, and incompatible with strong oxidising agents. Mustard gas on mixing with other chemical substances appears brown in colour and gives off a garlic-like smell. When heated, it decomposes and emits highly toxic, corrosive fumes and fumes of oxides of sulphur and chlorine-containing compounds. It is soluble in fats and oils, gasoline, kerosene, acetone, carbon tetrachloride, alcohol, tetrachloroethane, ethylbenzoate, and ether, and solubility in water is negligible. It is miscible with the OP nerve agents. During earlier years, mustard gas was in use as an important chemical warfare agent. In fact, it was used in large amounts during World Wars I and II. Mustard gas was first used by the German army in 1917. It was one of the most lethal of all the poisonous chemicals used during the war. It was reportedly used in the Iran–Iraq war in 1980–1988. It is not presently used in the United States, except for research purposes.

Safe Handling and Precautions

Mustard gas is a vesicant- or blister-causing, alkylating chemical substance. Exposures to mustard gas cause adverse health effects. The symptoms of poisoning include, but are not limited to, dyspnoea, cough, fever, headache, severe irritation, skin burns, skin blisters, swelling of eyelids, coughing, bronchitis, photophobia (sensitivity to light), lacrimation (tearing), blindness, irritation or ulceration of the respiratory tract, and respiratory disorders eventually leading to death. The skin disorders become very severe in hot and humid climatic conditions. It causes necrosis of the skin, eyes, and respiratory tract. Exposures to mustard gas cause vomiting, internal and external bleeding, damage to mucous membrane, damage to the bronchial tubes, and severe respiratory impairment. It produces cytotoxic action on the haematopoietic (blood-forming) tissues, which are especially sensitive. Mustard gas has

TABLE 14.1

Exposures to Mustard Gas and Symptoms of Poisoning

Ocular toxicity

Appearance of mild ocular toxicity and health effects after 4–12 h. The symptoms include tearing, itching, gritty feeling, and burning sensation

Appearance of moderate ocular toxicity and health effects after 3–6 h. The symptoms include tearing, itching, gritty feeling, burning sensation, redness, eyelid oedema, and moderate pain

Appearance of severe ocular toxicity and health effects after 1–2 h after exposure. The symptoms include marked lid oedema, corneal damage, severe pain, rhinorrhoea, and sneezing

Respiratory toxicity

The symptoms of mild respiratory effects include hacking cough, hoarseness, and epistaxis

Dermal toxicity

The symptoms of mild dermal toxicity include erythema after 2–24 h

The symptoms of severe dermal toxicity include vesication (skin blisters) after 2–24 h

extremely powerful vesicant effects on its victims. In addition, it is strongly mutagenic and carcinogenic because of its alkylating properties.

The target organs of mustard gas include lung, larynx, pharynx, oral cavity, bone marrow, and sexual organs. Users and occupational workers exposed to mustard gas (by accidental inhalation or ingestion) may suffer of poisoning. The symptoms include, but not limited to, nausea, vomiting lacrimation (tearing), and blindness, irritation or ulceration of the respiratory tract, dizziness, malaise (body discomfort), anorexia, and lethargy, CNS excitation, CNS depression, convulsions followed by fatal injury (Table 14.1). The DHHS and the IARC have classified mustard gas as a human carcinogen. Reports indicate that workers exposed to mustard gas during the production process as well as during war developed respiratory cancer.

Mustard Gas and Global Wars

The Germans first used mustard gas at Ypres, Belgium, on 12 July 1917. Mustard was particularly dangerous to troops because several hours must elapse before the individual detects a disabling or lethal dose. After the introduction of mustard gas, the warfare environment changed drastically. Since World War I, a number of isolated incidents have occurred in which mustard agent reportedly was used. In 1935, Italy probably used mustard against Ethiopia. Japan allegedly used mustard against the Chinese from 1937 to 1944. Egypt apparently used mustard in the 1960s against Yemen. More recently, the Iraqis used it against the Iranians and the Kurdish population in northern Iraq. Mustard agent remains an important chemical warfare agent today.

Exposures to mustard gas in the war fields caused a rapid visual damage and incapacitation of large numbers of persons. The great majority of persons with eye injury were visually made disabled for approximately 10 days with conjunctivitis, photophobia, and minimal corneal swelling. The use of mustard gas chemical substance in the past two decades and the continuation of the manufacture of chemical weapons have increased the risk that soldiers in particular and civilian population around the world may be exposed to these highly toxic chemical substances. The federal government recommends a maximum concentration for long-term exposure to sulphur mustard by the general population of 0.00002 mg/m^3 in air.

Mustard gas should be kept stored in containers made of glass for research, development, test, and evaluation (RDTE) quantities or 1 ton steel containers for large quantities. Agent containers will be stored in a single-containment system within a laboratory hood or in a double-containment system. Mustard gas is forbidden for transport other than via military (Technical Escort Unit) transport according to 49 CFR 172.

Mustard gas is an extremely hazardous chemical substance. On inhalation, mustard gas causes fatal injury. Mustard gas reacts violently with oxidising materials and reacts with water or steam to produce toxic and corrosive fumes. Workers should avoid high heat and contact with acid or acid fumes. Users should have protection equipments; should wear a positive pressure, pressure-demand, full facepiece SCBA or pressure-demand supplied air respirator with escape SCBA and a fully encapsulating, chemical-resistant suit, safety glasses, and gloves; and should have good ventilation. It should be handled as a carcinogen. This material must not be used unless a full risk assessment has been prepared in advance. It should not be used if safer alternatives are available. Disposal of waste must be carried out in a safe fashion – anything other than trivial amounts must not be vented through a fume cupboard. The use of this chemical for research purposes may be legally prohibited or restricted. Mustard agents are regulated under the 1993 Chemical Weapons Convention (CWC).

Bibliography

Agency for Toxic Substances and Disease Registry (ATSDR). 1990. *Case Studies in Environmental Medicine: Methylene Chloride Toxicity*. U.S. Department of Health and Human Services, Public Health Service, Atlanta, GA.

Agency for Toxic Substances and Disease Registry (ATSDR). 1992. *Toxicological Profile for Bromomethane*. U.S. Department of Health and Human Services, Public Health Service, Atlanta, GA (updated 2008).

Agency for Toxic Substances and Disease Registry (ATSDR). 1993. *Toxicological Profile for Methylene Chloride*. U.S. Department of Health and Human Services, Public Health Service, Atlanta, GA.

Agency for Toxic Substances and Disease Registry (ATSDR). 1996. *Mirex and Chlordecone*. ATSDR, Atlanta, GA (updated 2007).

Agency for Toxic Substances and Disease Registry (ATSDR). 1996. *Mirex and Chlordecone*. ToxFAQ (updated 2011).

Agency for Toxic Substances and Disease Registry (ATSDR). 2001. *Toxicological Profile for Methyl Parathion*. U.S. Department of Health and Human Services, Public Health Service, Atlanta, GA (updated 2007).

Agency for Toxic Substances and Disease Registry (ATSDR). 2002. *Toxicological Profile for Methoxychlor*. U.S. Department of Health and Human Services, Public Health Service, Atlanta, GA (updated 2007).

Agency for Toxic Substances and Disease Registry (ATSDR). 2003a. *Toxicological Profile for Malathion*. U.S. Department of Health and Human Services, Public Health Service, Atlanta, GA (updated 2006).

Agency for Toxic Substances and Disease Registry (ATSDR). 2003b. *Toxicological Profile for Sulfur Mustard*. U.S. Department of Health and Human Services, Public Health Service, Atlanta, GA (updated 2007).

Alexeeff, G. V. and W. W. Kilgore. 1983. Methyl bromide. *Residue Rev.* 88:102–153.

Anger, W. K., J. V. Setzer, J. M. Russo, W. S. Brightwell, R. G. Wait, and B. L. Johnson. 1981. Neurobehavioral effects of methyl bromide inhalation exposures. *Scand. J. Work Environ. Health* 7(Suppl. 4):40–47.

Baker, E. L. 1956. Epidemic malathion poisoning in Pakistan malaria workers. *Lancet* 1:31.

Bronstein, A. C. and P. L. Currance (eds.). 1994. *Emergency Care for Hazardous Materials Exposure*, 2nd edn. Mosby Lifeline, St. Louis, MO.

Burgess, W. A. 1995. *Recognition of Health Hazards in Industry. A Review of Materials and Processes*, 2nd edn. John Wiley & Sons, New York.

Dikshith, T. S. S. (ed.). 2011. *Handbook of Chemicals and Safety*. CRC Press, Boca Raton, FL.

Dive, A., P. Mahieu, R. Van Binst et al. 1994. Unusual manifestations after malathion poisoning. *Hum. Exp. Toxicol.* 13:271–274.

Gallo, M. A. and N. J. Lawryk. 1991. Organic phosphorus pesticides. In: *Handbook of Pesticide Toxicology*, Hayes, W. J., Jr. and E. R. Laws, Jr. (eds.). Academic Press, New York.

Gosselin, R. E., R. P. Smith, and H. C. Hodge. 1984. *Clinical Toxicology of Commercial Products*, 5th edn. Williams & Wilkins, Baltimore, MD, p. III-267.

Hathaway, G. J., N. H. Proctor, J. P. Hughes, and M. L. Fischman. 1991. *Proctor and Hughes' Chemical Hazards of the Workplace*, 3rd edn. Van Nostrand Reinhold, New York.

International Group of National Associations of Agrochemical Manufacturers (GIFAP) (Groupement International des Associations Nationales des Fabricants de Produits Agrochimiques). 1987. *Guidelines for the Safe Transport of Pesticides*. GIFAP, Brussels, Belgium.

International Labour Organization (ILO). 2001. *Guidelines on Occupational Safety and Health Management Systems*. ILO–OSH, Geneva, Switzerland.

International Labour Organization (ILO). 2010. Code of practice on safety and health in agriculture. Sectoral Activities Programme. Geneva, Switzerland.

International Programme on Chemical Safety (IPCS). Methanol. Poisons Information Monograph 335. Chemical.

International Programme on Chemical Safety and the Commission of the European Communities (IPCS–CEC). 1993a. Methamidophos. International Chemical Safety Cards. ICSC Card No. 0176. IPCS–CEC, Luxembourg, Belgium.

International Programme on Chemical Safety and the Commission of the European Communities (IPCS–CEC). 1993b. Mercuric acetate. ICSC Card No. 0978. IPCS–CEC, Luxembourg, Belgium.

International Program on Chemical Safety and the Commission of the European Communities (IPCS–CEC). 2004. International Chemical Safety, Magnesium Phosphide. ICSC No. 0744. IPCS–CEC, Luxembourg, Belgium.

International Programme on Chemical Safety and the Commission of the European Communities IPCS–CEC. 2004. ICSC No. 1306, Cincinnati, OH.

International Program on Chemical Safety and the Commission of the European Communities (IPCS–CEC). 2005. International Chemical Safety, Magnesium phosphide. ICSC No. 0172. IPCS–CEC, Luxembourg, Belgium.

Kidd, H. and D. R. James (eds.). 1991. *The Agrochemicals Handbook*, 3rd edn. Royal Society of Chemistry Information Services, Cambridge, U.K.

Kidd, H. and D. James (eds.). 1994. *Agrochemicals Handbook*, 3rd edn. Royal Society of Chemistry, Cambridge, England.

Lewis, R. J. (ed.). 1993. *Hawley's Condensed Chemical Dictionary*, 12th edn. Van Nostrand Reinhold Company, New York.

Lewis, R. J. (ed.). 1996. *Sax's Dangerous Properties of Industrial Materials*, 9th edn., Vols. 1–3. Van Nostrand Reinhold, New York.

Lu, F. C. 1995. A review of the acceptable daily intakes of pesticides assessed by the World Health Organization. *Regul. Toxicol. Pharmacol.* 21:351–364.

Material Safety Data Sheet (MSDS). 1995. Methamidophos. Extension Toxicology Network (EXTOXNET), Pesticide Information Profiles (PIP). USDA/Extension Service/Oregon State University, Corvallis, OR.

Material Safety Data Sheet (MSDS). 1996a. Malathion. Extension Toxicology Network (EXTOXNET), Pesticide Information Profiles (PIP). USDA/Extension Service/Oregon State University, Corvallis, OR.

Material Safety Data Sheet (MSDS). 1996b. Methidathion. Pesticide Information Profiles, Extension Toxicology Network (EXTOXNET). USDA/Extension Service/Oregon State University, Corvallis, OR.

Material Safety Data Sheet (MSDS). 1996c. Methyl parathion. Extension Toxicology Network (EXTOXNET), Pesticide Information Profiles (PIP). USDA/Extension Service/Oregon State University, Corvallis, OR.

Material Safety Data Sheet (MSDS). 1996d. Mevinphos. Extension Toxicology Network (EXTOXNET), Pesticide Information Profiles (PIP). USDA/Extension Service/Oregon State University, Corvallis, OR.

Material Safety Data Sheet (MSDS). 2000. Mercurous nitrate. ICSC Card No. 0980. International Programme on Chemical Safety and the Commission of the European Communities (IPCS–CEC), Geneva, Switzerland (updated 2001).

Material Safety Data Sheet (MSDS). 2004a. Mercuric thiocyanate. Environmental Health & Safety. MSDS Number No. M1560. Mallinckrodt Baker, Inc., Phillipsburg, NJ.

Material Safety Data Sheet (MSDS). 2004b. Methyl ethyl ketone (MEK). ICSC: 0179. International Programme on Chemical Safety and the Commission of the European Communities (IPCS–CEC), Geneva, Switzerland.

Material Safety Data Sheet (MSDS). 2005a. *Safety Data for Mercury (I) Chloride*. Department of Physical and Theoretical Chemistry, Oxford University, Oxford, U.K.

Material Safety Data Sheet (MSDS). 2005b. Safety data for methyl alcohol. Physical Chemistry at Oxford University, Oxford, U.K.

Material Safety Data Sheet (MSDS). 2005c. Safety data for methyl bromide. Physical Chemistry at Oxford University, Oxford, U.K.

Material Safety Data Sheet (MSDS). 2005d. Safety data for methyl vinyl ketone. Physical Chemistry at Oxford University, Oxford, U.K.

Material Safety Data Sheet (MSDS). 2005e. Safety data for mustard gas. Physical Chemistry at Oxford University, Oxford, U.K.

Material Safety Data Sheet (MSDS). 2006f. Safety data for mercury (II) chloride. Department of Physical and Theoretical Chemistry, Oxford University, Oxford, U.K.

Material Safety Data Sheet (MSDS). 2008. Safety data for methamidophos. Department of Physical Chemistry, Oxford University, Oxford, U.K.

Material Safety Data Sheet (MSDS). 2010. Safety data for mustard gas. Oxford University, Oxford, U.K.

Material Safety Data Sheet (MSDS). 2010. Safety data for 2-butanone. Physical Chemistry at Oxford University, Oxford, U.K.

Meister, R. T. (ed.). 1989. *Farm Chemical Handbook '89*, 75th edn. Meister Publishing, Willoughby, OH.

Meister, R. T. 1995. *Farm Chemicals Handbook '95*. Meister Publishing Company, Willoughby, OH.

National Cancer Institute (NCI). 1979. *Bioassay of Methyl Parathion for Possible Carcinogenicity*. DHEW Pub. No. (NIH) 79-1713. National Institute of Health, Bethesda, MD.

National Library of Medicine. 2006. Hazardous Substance Database. Bethesda, MD.

National Toxicology Program (NTP). 1990. Inhalation developmental toxicology studies: Teratology study of methyl ethyl ketone in mice. NTP study: TER88046. NTP, Research Triangle Park, NC.

NIOSH. 2010. Methyl bromide. *Pocket Guide to Chemical Hazards*. Centers for Disease Control and Prevention, Atlanta, GA.

Patnaik, P. (ed.). 1999. *A Comprehensive Guide to the Hazardous Properties of Chemical Substances*, 2nd edn. McGraw-Hill, New York.

Patnaik, P. (ed.). 2007. *A Comprehensive Guide to the Hazardous Properties of Chemical Substances*. John Wiley & Sons, Hoboken, NJ.

Patty, F. A. (ed.). 1963. *Industrial Hygiene and Toxicology*, 2nd revised edn., Vol. II, Toxicology. Interscience Publishers, Inc., New York.

Sax, N. I. (ed.). 1984. *Dangerous Properties of Industrial Materials*, 6th edn. Van Nostrand Reinhold, New York.

Sax, N. I. and R. J. Lewis (eds.). 1987. *Hawley's Condensed Chemical Dictionary*. Van Nostrand Reinhold Company, New York.

Sax, N. I. and R. J. Lewis, Jr. 1989. *Dangerous Properties of Industrial Materials*, 7th edn., 3 volumes. Van Nostrand Reinhold, New York.

Sittig, M. (ed.). 1985. *Handbook of Toxic and Hazardous Chemicals and Carcinogens*, 2nd edn. Noyes Publications, Park Ridge, NJ.

Sittig, M. (ed.). 2002. *Handbook of Toxic and Hazardous Chemicals and Carcinogens*, 4th edn. Noyes Publications, Norwich, NY.

Smith, A. G. 1991. Chlorinated hydrocarbon insecticides. In: *Handbook of Pesticide Toxicology*, Hayes, W. J., Jr. and E. R. Laws, Jr. (eds.). Academic Press Inc., New York.

Tomlin, C. D. S. (ed.). 2006. *The Pesticide Manual: A World Compendium*, 14th edn. British Crop Protection Council, Farnham, Surrey, U.K.

U.S. Environmental Protection Agency (U.S. EPA). 1993. *Integrated Risk Information System (IRIS) on Methoxychlor*. Environmental Criteria and Assessment Office, Office of Health and Environmental Assessment, Office of Research and Development, International Programme on Chemical Safety and the Commission of the European Communities (IPCS–CEC), Cincinnati, OH (2004).

U.S. Environmental Protection Agency (U.S. EPA). 1999. Integrated Risk Information System (IRIS) database. Reference concentration (RfC) for methyl bromide. U.S. EPA, Atlanta, GA.

U.S. Environmental Protection Agency (US EPA). 2000. Mevinphos facts United States prevention, U.S. EPA, Washington, DC.

U.S. Environmental Protection Agency. 2003a. Toxicological review of mirex. Integrated Risk Information System (IRIS). U.S. Environmental Protection Agency, Cincinnati, OH.

U.S. Environmental Agency (U.S. EPA). 2003b. *Toxicological Review of Methyl Ethyl Ketone in Support of Summary Information on the Integrated Risk Information System (IRIS)*. National Center for Environmental Assessment, Washington, DC.

U.S. Environmental Protection Agency (U.S. EPA). 2006a. Consumer factsheet on methoxychlor.

U.S. Environmental Protection Agency (U.S. EPA). 2006b. Methamidophos. Emergency first aid treatment guide. U.S. EPA, Atlanta, GA.

U.S. Environmental Protection Agency (U.S. EPA). 2006c. Methidathion. Emergency first aid treatment guide. US EPA, Atlanta, GA.

U.S. Environmental Agency (U.S. EPA). IRIS. 2011. *Methyl Ethyl Ketone (MEK)*. U.S. EPA, IRIS, Washington, DC.

Weiss, G. (ed.). 1986. *Hazardous Chemicals Handbook*. Noyes Data Corporation, Park Ridge, NJ.

Welch, L., H. Kirshner, A. Heath et al. 1991. Chronic neuropsychological and neurological impairment following acute exposure to a solvent mixture of toluene and methyl ethyl ketone (MEK). *J. Toxicol. Clin. Toxicol.* 29(4):435–445.

World Health Organization (WHO). Methoxychlor data sheets on pesticides. Data Sheets No. 28 (VBC/DS/77.28). WHO, Geneva, Switzerland.

World Health Organization (WHO). 1987. Malathion. Data Sheet on Pesticides No. 29. WHO, Geneva, Switzerland.

World Health Organization (WHO). International Programme on Chemical Safety (IPCS). 1990. Mirex Health and Safety Guide No. 39. IPCS, WHO, Geneva, Switzerland.

World Health Organization (WHO). International Programme on Chemical Safety (IPCS). 1993. Methamidophos. Health and Safety Guide No. 79. WHO, Geneva, Switzerland.

World Health Organization (WHO). 1995. Methyl bromide. International Programme on Chemical Safety (IPCS). Environmental Health Criteria No. 166. WHO, IPCS, Geneva, Switzerland.

World Health Organization (WHO). 1997a. Methanol. International Programme on Chemical Safety (IPCS). Health and Safety Guide No. 105. WHO, Geneva, Switzerland.

World Health Organization (WHO). 1997b. Methanol. Environmental Health Criteria 196. WHO, Geneva, Switzerland.

15

Hazardous Chemical Substances: N

Naphthalene
Nickel, Nickel Compounds, and Precautions
   Nickel (Ni) (Metallic)
   Nickel Carbonyl
   Nickel (II) Nitrate Hexahydrate
   Nickel Sulphate
   Nickel (II) Sulphate Hexahydrate
Nitrogen Mustards: Blister Agents HN-1, HN-2, HN-3
Nitroglycerin
2-Nitropropane
o-Nitrotoluene

Naphthalene (CAS No. 91-20-3)

Molecular formula: $C_{10}H_8$
Synonyms and trade names: Albocarbon; Camphor tar; Dezodorator; Mothballs; Moth flakes; Naftalen; Naphthaline; Tar camphor; White tar

Naphthalene occurs as transparent prismatic plates also available as white scales, powder balls, or cakes with a characteristic mothball or strong coal tar and aromatic odour. It is sparingly soluble in water but soluble in methanol/ethanol and very soluble in ether. Naphthalene is a commercially important aromatic hydrocarbon. Naphthalene occurs as a white solid or powder. Naphthalene occurs in coal tar in large quantities and is easily isolated from this source in pure condition. It volatilises and sublimes at room temperature above the melting point. The primary use for naphthalene is in the production of phthalic anhydride, also of carbamate insecticides, surface active agents and resins, as a dye intermediate, as a synthetic tanning agent, as a moth repellent, and in miscellaneous organic chemicals. Naphthalene is used in the production of phthalic anhydride; it is also used in mothballs. Naphthalene is also used in the manufacture of phthalic and anthranilic acids to make indigo, indanthrene, and triphenyl methane dyes, for synthetic resins, lubricant, celluloid, lampblack, smokeless powder, and hydronaphthalenes. Naphthalene is also used in dusting powders, lavatory deodorant discs, wood preservatives, fungicide, and as an insecticide. It has been used as an intestinal antiseptic and vermicide and in the treatment of pediculosis and scabies. Occupational exposure to naphthalene occurs in the dye industry and other chemical synthetic industries.

Safe Handling and Precautions

Exposures to naphthalene are very harmful and cause poisoning. Any kind of accidental exposure to repeated and high doses of naphthalene by inhalation, ingestion, and repeated skin contact is known to cause poisoning. The sign and symptoms of naphthalene poisoning include, but are not limited to, headache, nausea, weakness, sweating, confusion, vomiting, diarrhoea, malaise, anaemia, jaundice, dark urine, haematuria and an acute haemolytic reaction, convulsions, coma, and fatal injury. Also, severe exposures to naphthalene cause health disorders in gastroenteric system, the liver (hepatic necrosis may occur), the urinary system, the brain, and the eyes – cataract. Haemolysis is more likely to occur in individuals with a hereditary abnormality and deficiency of glucose-6-phosphate dehydrogenase, sickle cell anaemia, and sickle cell trait. Cataracts and optic neuritis have been reported in humans acutely exposed to naphthalene by inhalation and ingestion. Repeated and prolonged period of improper handling of naphthalene by individuals/and or occupational workers has been reported to cause retinal haemorrhage, inflammation of the lung, nasal membrane, hyperplasia of the respiratory epithelium in the nose, metaplasia of the olfactory epithelium, vomiting, and severe perspiration, besides severe poisoning. Reports indicate that naphthalene and coal tar exposures have been associated with laryngeal and intestinal carcinoma. Also, information in experimental animals indicates adequate evidence that naphthalene is carcinogenic to animals, while data in humans report that the evidences are inadequate. The International Agency for Research on Cancer (IARC) listed naphthalene as Group 2B, meaning possibly carcinogenic to humans, and U.S. Environmental Protection Agency (U.S. EPA) classified naphthalene as a Group C, meaning a possible human carcinogen.

Nickel, Nickel Compounds, and Precautions

Nickel is unique in its properties, and its industrial applications are many and extensive. Nickel imparts properties such as corrosion resistance, heat resistance, hardness, and strength and thus contributes a bulky share for industries that use nickel in alloys. The main uses of nickel are in the production of stainless steel, copper–nickel alloys, and other corrosion-resistant alloys. Pure nickel metal is used in electroplating, as a chemical catalyst, and in the manufacture of alkaline batteries, coins, welding products, magnets, electrical contacts and electrodes, spark plugs, machinery parts, and surgical and dental prostheses. Occupational exposure to nickel and nickel compounds is closely associated among individuals and workers involved in mining, smelting, welding, casting, spray painting and grinding, electroplating, production and use of nickel catalysts, polishing of nickel-containing alloys, and other jobs where nickel and nickel compounds are produced or used. The compounds include nickel acetate, nickel carbonate, nickel carbonyl, nickel chloride, nickel hydroxide, nickel monoxide, nickel oxide, nickel subsulphide, and nickel sulphate. Nickel carbonyl, nickel cyanide, and nickel tetracarbonyl are considered marine pollutants, and special requirements have been set for marking, labelling, and transporting these materials. Nickel (Ni) compounds are many, and users and occupational workers should be careful during use and handling to avoid health hazards. These include nickel oxide, nickel hydroxide, nickel subsulphide, nickel sulphate, nickel chloride, and nickel carbonyl.

Nickel (Ni) (Metallic) (CAS No. 7440-02-0)

Molecular formula: **Ni**

Synonyms and trade names: Alcan 756; Carbonyl nickel powder; Fibrex; Fibrex P; NI 270; Nickel 2170; Nickel sponge; Nickel catalyst; NI 0901-s; Raney alloy; Raney nickel

Nickel is a hard, silvery white, malleable metal chunk or grey powder. Nickel powder is pyrophoric – can ignite spontaneously. It may react violently with titanium, ammonium nitrate, potassium perchlorate, and hydrazoic acid. It is incompatible with acids, oxidising agents, and sulphur. The industrially important nickel compounds are nickel oxide (NiO), nickel acetate ($Ni(C_2H_3O_2)$), nickel carbonate ($NiCO_3$), nickel carbonyl ($Ni(CO)_4$), nickel subsulphide (NiS_2), nickelocene ($C_5H_5)_2Ni$, and nickel sulphate hexahydrate ($NiSO_4 \cdot 6H_2O$). Nickel compounds have been well established as human carcinogens. Investigations into the molecular mechanisms of nickel carcinogenesis have revealed that not all nickel compounds are equally carcinogenic: certain water-insoluble nickel compounds exhibit potent carcinogenic activity, whereas highly water-soluble nickel compounds exhibit less potency. The reason for the high carcinogenic activity of certain water-insoluble nickel compounds relates to their bioavailability and the ability of the nickel ions to enter cells and reach chromatin. The water-insoluble nickel compounds enter cells quite efficiently via phagocytic processes and subsequent intracellular dissolution. Nickel is classified as a borderline metal ion because it has both soft and hard metal properties and it can bind to sulphur, nitrogen, and oxygen groups. Nickel ions are very similar in structure and coordination properties to magnesium.

Safe Handling and Precautions

Exposures to compounds of nickel cause toxicity and adverse health effects: Acute poisoning causes headache, dizziness, nausea and vomiting, chest pain, tightness of the chest, dry cough with shortness of breath, rapid respiration, cyanosis, and extreme weakness. The lesions resulting from acute exposure are mainly in the lungs and brain. Soluble nickel compounds cause dermatitis or 'nickel itch' particularly among electroplaters. Nickel carbonyl is highly toxic. Nickel poisoning has induced oedema and haemorrhages in lungs, brain injury, and human fatalities. Workers with high exposures to soluble nickel compounds may develop renal tubular dysfunction, evidenced by increased renal excretion. Nickel compounds are known human carcinogens.

Nickel has a carcinogenic property and may be associated with hypersensitivity reactions. Studies have indicated that carcinogenic nickel compounds induce chromosomal aberrations, including those that are specific to heterochromatic chromosome regions found to increase the extent of DNA methylation. Although the mechanisms of DNA hypermethylation by nickel are not well documented, nickel may trigger a de novo methylation of the genome region. In addition to producing mutations, chromosome aberrations, and gene silencing by DNA methylation, exposure of cells to nickel also induces a variety of gene expression changes that yield cells with a spectra of expressed genes similar to cancer cells. Many compounds of nickel have been shown to be associated with occupational cancers in humans, and these nickel compounds also cause other health effects and disorders. As early as 1907 and 1934, reports have shown that occupational exposure to nickel in copper–nickel refineries caused respiratory tract cancer in workers. Nickel has been found to induce the hypoxia-inducible factor (HIF), which is responsible for increasing the expression of glycolytic enzymes and other genes that allow cells to survive under low oxygen tension. As suggested by the growing literature on nickel carcinogenesis, the initial events in environmentally induced cancers may be a combination of gene induction and gene silencing by epigenetic DNA methylation that leads to cancer cell selection. Determination of the extent to which this concept applies to other environmental carcinogens is the challenge of the future.

Nickel Carbonyl (CAS No. 13463-39-3)

Molecular formula: **Ni(CO)**
Synonyms and trade names: Nickel tetracarbonyl; Tetracarbonyl nickel; (T-4); Nickel carbonyl; Carbonilo de níquel (dot spanish); Nickel carbonyl (Ni(CO)$_4$); Nickel carbonyl (Ni(CO)$_4$, (t-4)-; Nickel catalyst; Nickel tetracarbonyl; Nickel-tétracarbonyle; Tetracarbonyl nickel; Tetracarbonylnickel; Tetracarbonylnickel(O)

Nickel carbonyl is a clear colourless to yellow volatile liquid, is flammable, and burns with a yellow flame. It is denser than water and insoluble in water but soluble in alcohol, benzene, chloroform, acetone, ethanol, carbon tetrachloride, and nitric acid. The vapours are heavier than air. In industries, nickel carbonyl is used in nickel coat steel and other metals and to make very pure nickel. Nickel carbonyl gets peroxidised by air as a solid deposit and decomposes to ignite.

Safe Handling and Precautions

Nickel carbonyl is very harmful and poisonous. Improper use and handling and resulting exposures to vapours of nickel carbonyl at workplaces among occupational workers has been reported to cause poisoning. The signs and symptoms of poisoning include, but are not limited to, irritation, congestion, and oedema of lungs; dermatitis, recurrent asthmatic attacks, and increased number of white blood cells (eosinophils) in respiratory tract are acute health hazards. Accidental workplace inhalation, swallowing, and/or skin absorption of nickel carbonyl usually produces mild, non-specific, immediate symptoms, including nausea, vertigo, headache, dyspnoea, and chest pain. These initial symptoms usually disappear within a few hours. In cases of accidental poisoning, autopsy examination showed congestion, collapse, and tissue destruction, as well as haemorrhage in the brain. Delayed symptoms of nickel carbonyl poisoning (after 12–36 h or more) occur and include severe cough, dyspnoea, tachycardia, cyanosis, profound weakness, and gastrointestinal and pulmonary disorders. In cases of nickel carbonyl poisoning, dizziness and headache are very common following mild or moderate nickel carbonyl exposure. In more severe cases, there may be dysphoria, somnolence, blurred vision, and rarely delirium and convulsions. Nickel carbonyl causes fatal injury from diffuse interstitial pneumonitis, cerebral haemorrhage, or cerebral oedema. Nickel carbonyl is carcinogenic and teratogenic in experimental rats.

Occupational workers should handle nickel carbonyl with extreme care because of its flammability and tendency to explode. Nickel carbonyl should be kept in tightly closed containers in a cool, well-ventilated area and away from heat flames, sparks, other sources of ignition, and oxidisers such as nitric acid and chlorine. Workers must store and handle nickel carbonyl in accordance with all current regulations and standards and protect it from physical damage. During use and handling of nickel carbonyl, workers should strictly avoid flames and sources of ignition, and nickel carbonyl should be transported in steel cylinders with proper management. Workers must remember that the odour threshold of nickel carbonyl is 0.3–3 ppm and the exposure limits are very low: Occupational Safety and Health Administration (OSHA), 0.001 ppm (Ni) (0.007 mg/m^3) time-weighted average (TWA); American Conference of Industrial Hygienists (ACGIH), 0.05 ppm (Ni) TWA; and National Institute for Occupational Safety and Health (NIOSH), 0.001 ppm (0.007 mg/m^3) TWA.

Regulations:

The U.S. regulations: The Superfund Amendments and Reauthorization Act (SARA) included nickel carbonyl with Title III and Section 304, meaning extremely hazardous substances (40 CFR 355 Subpart).

Canadian regulations: Nickel carbonyl is listed by the Workplace Hazardous Materials Information System (WHMIS) as B2, D1A, D2A, meaning complete alertness by workers during use and handling of nickel carbonyl (B2, flammable liquids with flashpoints less than 38°C; D1A, poisonous product that causes serious and immediate effects; D2A, very toxic product that causes other toxic effects).

Danger: Nickel carbonyl is an extremely flammable liquid/vapour, highly reactive, and potentially hazardous and can spontaneously explode. The vapour is heavier than air. Vapours or gases ignite from sources of distant ignition and disasters.

Nickel (II) Nitrate Hexahydrate (CAS No. 13478-00-7)

Molecular formula: $Ni(NO_3)_2\ 6H_2O$
Synonyms and trade names: Nickelous nitrate; Nitric acid nickel (+2) salt; Nickel dinitrate hexahydrate

Nickel (II) nitrate hexahydrate appears as green crystals and substantially soluble in water. It is stable, a strong oxidiser, and incompatible with reducing agents. It has wide industrial applications in nickel plating as catalysts and is indirectly associated with other industries of nickel alloy manufacturers, nickel miners, smelters and refiners, nickel platers, and exposure to alloys, for example, in coinage.

Safe Handling and Precautions

Nickel (II) nitrate hexahydrate is poisonous. Like other nitrates, nickel nitrate is oxidising. The symptoms of poisoning of nickel (II) nitrate hexahydrate include, but are not limited to, irritating effects to the eyes and skin and sensitisation reactions. Exposures to nickel (II) nitrate hexahydrate dust by inhalation causes skin allergy, dermatitis, possible destruction, ulceration, methaemoglobinemia, cyanosis (bluish discoloration of skin due to deficient oxygenation of the blood), convulsions, tachycardia, dyspnoea (laboured breathing), and inhalation of the respiratory tract. Like other nitrates, nickel nitrate hexahydrate is an oxidiser, and workers should observe caution during its use and handling with other chemical substances. The epidemiological data also indicated increased incidence of lung and nasal cavity cancers that has been noted among women in nickel smelters and refineries. Nickel (II) nitrate hexahydrate is known to cause cancer. The IARC has listed nickel (II) nitrate hexahydrate as Group 1 carcinogen, and the National Toxicology Program (NTP) classifies it as a known carcinogen, while the ACGIH has not listed it as a carcinogen. Users and occupational workers, therefore, should handle the chemical substance with care and must use adequate and standard quality personal protective equipment (PPE), safety glasses, and gloves and avoid exposure to dust. Workers should treat the chemical as a possible carcinogen. The workplace should have good ventilation.

Nickel (II) nitrate hexahydrate is a strong oxidiser; any contact with other materials may cause a fire, and its exposure causes skin allergy and it causes respiratory tract irritation on inhalation of its dust. Exposures during pregnancy cause harm to the unborn child.

Nickel Sulphate (CAS No. 7786-81-4)

Molecular formula: $NiSO_4$
Synonyms and trade names: Nickelous sulphate; Nickel (II) sulphate; Nickel monosulphate

Nickel sulphate appears as blue to blue-green transparent crystals and is an odourless soluble nickel salt. Nickel sulphate is incompatible with strong acids. Nickel sulphate has extensive industrial applications in nickel patch testing, in nickel plating, as a raw material for the production of catalysts, in dyeing and printing fabrics as a mordant, and for blackening zinc and brass and in jewellery manufacture.

Safe Handling and Precautions

Occupational workers get exposures to nickel sulphate at workplaces whenever workers use and handle the chemical with negligence and avoid proper protection dress. Nickel sulphate is known to get absorbed by inhalation or ingestion and through percutaneous absorption. On repeated inhalation, nickel sulphate is known to cause occupational asthma in metal platers.

Reports have indicated that nickel has transplacental migration property and passes to the child in maternal milk.

Nickel (II) Sulphate Hexahydrate (CAS No. 10101-97-0)

Molecular formula: $NiSO_4$
Synonyms and trade names: Nickel (II) sulphate hexahydrate (1:1:6); Sulphuric acid; Nickel (2+) salt; Hexahydrate

Nickel sulphate hexahydrate is available as odourless, blue-green crystals. It is stable at room temperature in closed containers under normal storage and handling conditions and incompatible with strong acids and strong oxidising agents. On decomposition, nickel (II) sulphate hydroxylate releases oxides of sulphur, irritating and toxic fumes and gases, and nickel oxide.

Nickel sulphate is used in the laboratory. Accidental workplace exposures to nickel sulphate hexahydrate by ingestion/inhalation and/or skin absorption cause toxicity and adverse health effects. It causes irritation to the respiratory tract. The symptoms include, but are not limited to, coughing, sore throat, shortness of breath, respiratory distress, lung allergy with asthma-type symptoms, abdominal pain, diarrhoea, nausea, vomiting, giddiness, capillary damage, myocardial weakness, central nervous system depression, and kidney and liver damage. Workers/persons with pre-existing skin disorders, impaired respiratory or pulmonary function, and history of asthma, allergies, or sensitisation to nickel compounds may be at an increased risk upon exposure to nickel sulphate hexahydrate and nickel compounds. Reports of ACGIH list nickel sulphate hexahydrate as Group A4, meaning not classifiable as a human carcinogen; the NIOSH list it as an occupational carcinogen (listed as Nickel); the NTP lists it as a suspect carcinogen (listed as Nickel); the OSHA classifies it as a possible human carcinogen (listed as Nickel); and the IARC classifies nickel sulphate hexahydrate as Group 2B carcinogen (listed as Nickel), meaning possibly carcinogenic to humans. Reports are available suggesting sufficient evidences in humans for the carcinogenicity of nickel compounds and their combinations encountered in the nickel refining industry.

Occupational workers during use and handling of nickel sulphate should strictly observe safety regulatory practices, must avoid workplace negligence, and should avoid handling of the chemical at sources of ignition, excess heat, and strong acids. Workers should wear protective clothing to prevent chemical contact, PPE to avoid organic vapour, canister face mask, self-contained breathing apparatus, liquid-proof gloves, protective goggles or face shield, white- or light-coloured clothing, and rubber shoes or boots. Workers should keep stored nickel sulphate in a tightly closed container, in a cool, dry, ventilated area and must keep it protected against any kind of physical damage. Nickel sulphate should be away/isolated from incompatible chemical substances. The work area handling nickel compounds should be properly identified with labels/appropriate signs, and the entry should be limited only to authorised workers/persons.

WARNING! Workers should remember: Nickel sulphate is harmful, causes skin allergy and respiratory reaction, and is a cancer hazard.

Nitrogen Mustards: Blister Agents HN–1 (CAS No. HN-1 538-07-8); HN–2 (CAS No. HN-2 51-75-2); HN–3 (CAS No. HN-3 555-77-1)

Nitrogen mustards are very potential chemical substances of yesteryears and produced during the 1920s and 1930s as chemical warfare weapons. They are vesicants (or blister agents) similar to the sulphur mustards. They smell fishy, musty, soapy, or fruity and are either in the form of an oily textured liquid or a vapour (the gaseous form of a liquid) or a solid. It is in liquid form at normal room temperature (70°F) with a clear, pale amber, or yellow colour. HN-1, HN-2, and HN-3 are the military designations of nitrogen mustard (for more data, refer to Muatars gas).

Nitrogen mustards (HN-1, HN-2, HN-3) are colourless to yellow, oily liquids that evaporate very slowly. HN-1 has a faint fishy or musty odour. HN-2 has a soapy odour at low concentrations and a fruity odour at higher concentrations. HN-3 may smell like butter almond. Use of nitrogen mustards is very much restricted other than for chemical warfare. In fact, presently, its use has no records. HN-1 has been used to remove warts in the past, and HN-2 has been used sparingly in chemotherapy.

Safe Handling and Precautions

Exposures to nitrogen mustards cause adverse health effects and poisoning to humans. The symptoms of poisoning caused after accidental exposures to nitrogen mustard vapour by breathing include, but are not limited to, nasal and sinus pain or discomfort, pharyngitis, laryngitis, cough, abdominal pain, diarrhoea, nausea, vomiting, and shortness of breath. Cells lining the nasal airways suffer immediate damage. Exposure to high levels of nitrogen mustards leads to fatal injury. Nitrogen mustard vapours or liquid on skin contact causes swelling and rash, followed by blistering, and the chemical substance in high concentrations causes second- and third-degree skin burns. Contact of vapour of nitrogen mustard to the eyes causes inflammation, pain, swelling, corneal damage, burns, and loss of vision. Accidental ingestion of nitrogen mustards causes burning of the mouth, oesophagus, and stomach and may further lead to damage the immune system and bone marrow. Occupational workers on exposures to very high concentrations of nitrogen mustard vapour suffer with tremors, incoordination, and seizures. Prolonged or repeated period of exposures to nitrogen mustard has caused cancer in animals and leukaemia in humans. The IARC has classified nitrogen mustard HN-2 as probably carcinogenic to humans, based on evidence that it causes leukaemia in humans and cancers of the lung, liver, uterus, and large intestine in animals.

Nitroglycerin (CAS No. 55-63-0)

Molecular formula: $C_3H_5(ONO_2)_3$
Synonyms and trade names: Anginine; Basting gelatin; Blasting oil; Glyceryl nitrate; GTN; Nitroglyn; Nitrol; NG; NTG; Nitroglycerine; Nitroglycerol; Trinitroglycerol; Trinitrin

Nitroglycerin or glyceryl trinitrate is a pale yellow oily liquid and also available in the form of rhombic crystals. It is highly explosive. It is used in combination with ethylene glycol dinitrite in the manufacture of dynamites. It is slightly soluble in water and miscible with acetone, ether, benzene, and other organic solvents. Nitroglycerin is incompatible with strong acids, such as hydrochloric acid, sulphuric acid, and nitric acid, and also with ozone and causes violent reactions. Nitroglycerin is a powerful explosive in its pure form and very sensitive to mechanical shock, heat, or UV radiation. It is a severe explosion hazard when shocked or exposed to O_3. This product is hygroscopic.

Exposures to nitroglycerin occur to workers associated with occupations/operations such as the manufacture and transportation of nitroglycerin; the manufacture of gun powder, dynamite, smokeless powders, guncotton, and other explosives; and use of the chemical substance in rocket propellants and in medicines.

Safe Handling and Precautions

Exposures to nitroglycerin cause severe health effects to users and occupational workers. Nitroglycerin is highly toxic to humans and causes rapid vasodilatation and severe throbbing. Nitroglycerin has been reported to get rapidly absorbed both through the exposed skin surface and respiratory route. Repeated exposures and high concentrations of nitroglycerin cause toxicity and poisoning with symptoms that include, but are not limited to, vomiting, nausea, cyanosis, coma, and death. Exposures to nitroglycerin lead to an increase in heart rate, myocardial contraction, and angina pectoris. Occupational exposures to nitroglycerin have also been traced to work adaptation to the chemical and the consequences of withdrawal syndrome. Nitroglycerin on skin contact causes violent headache. Therefore, whenever the spirit of glyceryl trinitrate is spilled, a solution of sodium hydroxide should be added immediately to check the chemical reaction of high explosion. The worker must remember that on evaporation of alcohol, the glyceryl trinitrate results in severe explosion. This needs care and caution in the proper management of workplaces and health of workers. The OSHA has set the permissible exposure limit (PEL) for nitroglycerin at 0.2 ppm of air; occupational workers at no time should exceed this exposure limit. The NIOSH set a short-term exposure limit (STEL) of 0.1 mg/m^3 for periods not to exceed 15 min at workplace, and the recommended exposure limit (REL) for nitroglycerin is not established. For nitroglycerin, the ACGIH has set a threshold limit value (TLV) of 0.05 ppm (TWA) for a normal 8 h workday and a 40 h workweek. Nitroglycerin should be stored in a cool, dry, well-ventilated area in tightly sealed containers with proper labels. Containers of nitroglycerin should be protected from physical damage, ignition sources, shocks (or jolts), and ultraviolet radiation and should be stored away from strong acids.

Precautions: Dangers

Occupational workers during handling of nitroglycerin should use appropriate PPE and clothing and avoid skin contact with nitroglycerin. The selection of the appropriate PPE includes gloves, sleeves, and suits (encapsulating) that are acceptable according to the regulatory standards and the worker's potential exposure to nitroglycerin. On skin contact with nitroglycerin, occupational workers should immediately flood the affected areas of the skin with water. Workers should remove and isolate all contaminated clothing and dress material. Also, workers should gently wash all affected skin areas thoroughly with soap and water and should seek immediate medical attention for health safety.

2-Nitropropane (CAS No. 000 79-46-9)

Molecular formula: $CH_3CH(NO_2)CH_3$
Synonyms and trade names: Dimethylnitromethane; Isonitropropane; Nitroisopropane; Ni-Par S-20TM (a commercial grade 2-NP); NiPar S-30TM (mixtures of 1-nitropropane and 2-NP)

2-Nitropropane (2-NP) is a clear colourless liquid with a pleasant odour. 2-NP is soluble in many organic solvents including chloroform. Its vapours may form an explosive mixture with air. 2-NP is primarily used as a solvent for organic compounds, coatings, inks, dyes, adhesives, and vinyl resins. Application of 2-NP improves drying time, insures more complete solvent release, provides better flow characteristics, and insures greater pigment dispersion (1,2,3,5). 2-NP has a minor use as an additive in explosives, propellants, and fuels (in racing cars). It also has limited use as a paint and varnish remover. 2-NP serves as an intermediate in the synthesis of some pharmaceuticals, dyes, insecticides, and textile chemicals. Occupational workers get exposures to 2-NP in many industries, construction work, maintenance, printing, highway maintenance, specifically traffic markings, shipbuilding and marine coatings, and furniture and plastic products (1–3). Solvent products containing 2-NP have been required by different industries and are produced extensively around the world. It is used for coatings as with vinyl, epoxy paints, nitrocellulose, and chlorinated rubber, in printing inks, adhesives, printing as flexographic inks, maintenance with traffic markings on roads and highways, shipbuilding and maintenance, in furniture, and plastic industries. 2-NP is also used as a solvent in food processing industries for fractionation of a partially saturated vegetable oil. Because of its large-scale use pattern, human exposure to 2-NP has become a health concern.

Safe Handling and Precautions

In laboratory studies, rats exposed to 2-NP in high concentrations of (207 ppm) developed adverse liver changes like hepatocellular hypertrophy, hyperplasia, necrosis, and liver carcinoma. It has been reported that prolonged exposure to concentrations of 20–45 ppm of 2-NP caused nausea, vomiting, diarrhoea, anorexia, and severe headaches among workers. Occupational workers handling 2-NP for the application of epoxy resins to the walls of a nuclear power plant showed the development of toxic hepatitis.

2-Nitropropane and cancer: The U.S. EPA has classified 2-NP as a Group 2B, meaning probable human carcinogen.

Care and Precautions

In view of is potential toxicity, 2-NP should be handled in the workplace as a potential human carcinogen. Strict chemical management should be observed at all levels. Occupational worker should be provided with approved personal respiratory protective devices and full-body clothing for protection against splashes.

Exposure Limits

The OSHA has set a PEL for 2-NP at 25 ppm 8 h (TWA).

---

o-Nitrotoluene (CAS No. 88-72-2)

Molecular formula: $C_7H_7NO_2$
Synonyms and trade names: Nitrotolul; o-Methylnitrobenzene; 2-Methylnitrobenzene; ortho-Nitrotoluene; 2-Nitrotoluene; ONT; 2-NT; Toluene, o-nitro

o-Nitrotoluene is a yellow-coloured liquid. It is used for the synthesis of a variety of industrial products. These include to synthesise agricultural and rubber chemicals, azo and sulphur dyes, and dyes for cotton, wool, silk, leather, and paper. O-nitrotoluene decomposes on contact with strong oxidants, reducing agents, acids, or bases producing toxic fumes, nitrogen oxides, and carbon monoxide.

Safe Handling and Precautions

Exposures to o-nitrotoluene are harmful and cause poisoning. On exposure to o-nitrotoluene, occupational workers develop symptoms that include, but are not limited to, headache, flushing of face, irritation of the skin and mucous membrane, anoxia, lassitude, weakness, exhaustion, dizziness, ataxia, dyspnoea, respiratory distress/breathing difficulty, tachycardia, vomiting, hypoxia, anaemia, depression, convulsions, and cyanosis. Users and occupational workers must remember that the REL and PEL levels of o-nitrotoluene are very low (OSHA PEL: 5 ppm and NIOSH REL: 2 ppm TWA) and must avoid workplace negligence during chemical management.

Danger: Occupational workers should be alert during handling of o-nitrotoluene. It decomposes on contact with strong oxidants, reducing agents, acids, or bases producing toxic fumes and causes fire and explosion hazard. o-nitrotoluene attacks some forms of plastic, rubber, and coatings. On combustion, it forms nitrogen oxides and carbon monoxide.

Bibliography

Agency for Toxic Substances and Disease Registry (ATSDR). 1995. *Toxicological Profile for Naphthalene.* Public Health Service, U.S. Department of Health and Human Services, Atlanta, GA (updated 2007).

Agency for Toxic Substances and Disease Registry (ATSDR). 1998. *Toxicological Profile for Nickel.* U.S. Department of Health and Human Services, Public Health Service, Atlanta, GA (update).

Agency for Toxic Substances and Disease Registry (ATSDR). 2002. *Managing Hazardous Materials Incidents. Volume III—Medical Management Guidelines for Acute Chemical Exposures: Blister Agents: HN-1, HN-2, HN-3 (Nitrogen Mustards)*. U.S. Department of Health and Human Services, Public Health Service, Atlanta, GA.

Browning, E. 1965. *Toxicity and Metabolism of Industrial Solvents*. Elsevier Publishing Co., New York, pp. 285–288.

Budavari, S. (ed.). 1989. *The Merck Index. An Encyclopedia of Chemicals, Drugs, and Biologicals*, 11th edn. Merck and Co. Inc., Rahway, NJ.

Clayton, G. and F. Clayton. (eds.). 1993. *Patty's Industrial Hygiene and Toxicology*, 4th edn. Vol. I, Part A and Part B. General Principles. John Wiley & Sons, New York.

Gosselin, R. E., R. P. Smith, and H. C. Hodge (eds.). 1984. *Clinical Toxicology of Commercial Products*. Williams & Wilkins, Baltimore, MD.

Hathaway, G. J., N. H. Proctor, J. P. Hughes, and M. L. Fischman. 1991. *Proctor and Hughes' Chemical Hazards of the Workplace*, 3rd edn. Van Nostrand Reinhold, New York.

Hazardous Substances Data Bank (HSDB). 2009. National Library of Medicine, Bethesda, MD.

International Agency for Research on Cancer (IARC). 1990. Nickel and nickel compounds. In: *Chromium, Nickel and Welding. IARC Monographs on the Evaluation of Carcinogenic Risk of Chemicals to Humans*, Vol. 49. IARC, Lyon, France.

International Programme on Chemical Safety and the Commission of the European Communities (IPCS–CEC). 1999. Naphthalene. ICSC No. 0667. IPCS–CEC, Luxembourg, Belgium (updated 2005).

International Program on Chemical Safety and the Commission of the European Communities (IPCS–CEC). 2004. o-Nitrotoluene. ICSC No. 0931. IPCS-CEC, Luxembourg, Belgium.

Kuck, P. H. 2010. Nickel. *Mineral Commodity Summaries*. U.S. Geological Survey. Denver, CO.

Kurta, D. L., B. S. Dean, and E. P. Krenzelok. 1993. Acute nickel carbonyl poisoning. *Am. J. Emerg. Med.* 11:64–66.

Kurz, J. M. 1987. Naphthalene poisoning: Critical care nursing techniques. *Dimensions Crit. Care Nurs.* 6:264–270.

Lewis, R. J., Sr. 2007. *Hawley's Condensed Chemical Dictionary*. Wiley-Interscience, John Wiley & Sons, New York.

Material Safety Data Sheet (MSDS). 2001. Nickel carbonyl. MSDS No. 0064. International Programme on Chemical Safety and the Commission of the European Communities (IPCS–CEC). IPCS, WHO, Geneva, Switzerland.

Material Safety Data Sheet (MSDS). 2003. Safety data for nickel carbonyl. Physical Chemistry, Oxford University, Oxford, U.K.

Material Safety Data Sheet (MSDS). 2004. Safety data for nickel. Department of Physical and Theoretical Chemistry, Oxford University, Oxford, U.K.

Material Safety Data Sheet (MSDS). 2005. Safety data for nickel (II) nitrate hexahydrate. Department of Physical and Theoretical Chemistry, Oxford University, Oxford, U.K.

Material Safety Data Sheet (MSDS). 2009. Nickel sulfate. MSDS No. N3122, Environmental Health & Safety. Mallinckrodt Baker Inc., Phillipsburg, NJ.

Morgan, L. G. and V. Usher. 1994. Health problems associated with nickel refining and use. *Ann. Occup. Hyg.* 38:189–198.

National Center for Environmental Assessment (NCEA). 1998. *Toxicological Review of Naphthalene*. NCEA, Cincinnati, OH.

National Institute for Occupational Safety and Health (NIOSH). 1980. Health hazard alert—2-Nitropropane (2-NP), NIOSH ALERT: DHHS (NIOSH) Publication No. 80-142. Occupational Safety and Health Administration.

National Library of Medicine (NLM). 1992. Hazardous Substances Data Bank: Nitroglycerin. NLM, Bethesda, MD.

Patnaik, P. (ed.). 1992. *A Comprehensive Guide to the Hazardous Properties of Chemical Substances*, Van Nostrand Reinhold, New York.

Sax, N. I. (ed.). 1984. *Dangerous Properties of Industrial Materials*, 6th edn. Van Nostrand Reinhold, New York.

Shi, Z. 1992. Long-term effects of exposure to low concentrations of nickel carbonyl on workers' health. *Nickel and Human Health—Current Perspectives*, Nieboer, E. and J. O. Nriagu. (eds.), Vol. 25. John Wiley & Sons, New York, pp. 273–279.

Sittig, M. (ed.). 1985. *Handbook of Toxic and Hazardous Chemicals and Carcinogens*, 2nd edn. Noyes Publications, Park Ridge, NJ.

Sittig, M. (ed.). 1991. *Handbook of Toxic and Hazardous Chemicals*, 3rd edn. Noyes Publications, Park Ridge, NJ.

Sunderman, F. W., Jr. 1988. Nickel. *Handbook on Toxicity of Inorganic Compounds*, H. G. Seiler, H. G., H. Sigel, and A. Sigel (eds.). Marcel Dekker, New York, pp. 453–468.

Sunderman, F. W. Jr., A. Aitio, L. D. Morgan, and T. Norseth. 1986. Biological monitoring of nickel. *Toxicol. Ind. Health* 2:17–78.

Sunderman, F. W. and J. F. Kincaid. 1954. Nickel poisoning—II. Studies on patients suffering from acute exposure to vapors of nickel carbonyl. *JAMA* 155:889–894.

U.S. Environmental Protection Agency (U.S. EPA). 1985. Health and environmental effects profile for 2-nitropropane. EPA/600/x-85/112. Environmental Criteria and Assessment Office, Office of Health and Environmental Assessment, Office of Research and Development, Cincinnati, OH.

U.S. Environmental Protection Agency (U.S. EPA). 1997. Health effects assessment summary tables, FY 1997 update. Office of Research and Development, Office of Emergency and Remedial Response, Washington, DC, EPA/540/R-97–036.

U.S. Environmental Protection Agency (U.S. EPA). 1999. Integrated Risk Information System (IRIS) on 2-nitropropane. National Center for Environmental Assessment, Office of Research and Development, Washington, DC.

U.S. Environmental Protection Agency (U.S. EPA). 2000. Technology transfer network air toxics web site, 2-nitropropane, Hazard summary-created in April 1992; revised in January 2000 (updated 2007).

World Health Organization (WHO). 1991. Environmental Health Criteria: 108. Nickel. WHO, Geneva, Switzerland.

World Health Organization (WHO). International Programme on Chemical Safety (IPCS). 1991. Nickel. Health and Safety Guide No. 62. WHO, Geneva, Switzerland.

World Health Organization (WHO). International Programme on Chemical Safety (IPCS). 2000. *Poisons Information Monograph No. 363*. WHO, IPCS. Geneva, Switzerland.

16

Hazardous Chemical Substances:

O

Octabenzone
Osmium Tetraoxide
Oxamyl
Oxirane
Oxydisulphoton
Oxymetholone
Ozone

Octabenzone (CAS No. 1843-05-6)

Molecular formula: $C_{21}H_{26}O_3$

Octabenzone appears as light yellow/yellowish crystals/fine powder and is odourless and freely soluble in benzene, n-hexane, and acetone; slightly soluble in ethanol; and very slightly soluble in ethane dichloride. Octabenzone is a light stabiliser with good performance, capable of absorbing the UV radiation of 240–340 nm wavelength with the characteristics of light colour, non-toxicity, good compatibility, small mobility, easy processing, etc. Octabenzone protects the polymer to its maximum extent and helps to reduce its colour. It can also delay the yellowing and impede the loss of its physical function. Industrial applications of octabenzone is very extensive and applied to PE, PVC, PP, PS, PC, organic glass, polypropylene fibre, ethylene-vinyl acetate, etc. Also, octabenzone has very good light-stability effect on drying phenol aldehyde, varnish of alcohol and acname, polyurethane, acrylate, expoxnamee, etc. The major uses include UV stabiliser, for example, polyethylene and polypropylene fibres.

Safe Handling and Precautions

Octabenzone is a skin sensitiser and has been reported to cause mechanical irritation to eyes, skin, and respiratory system. The American Conference of Governmental Industrial Hygienists (ACGIH), the National Toxicology Program (NTP), the Occupational Safety and Health Administration (OSHA), and the International Agency for Research on Cancer (IARC) have not listed octabenzone as a human carcinogen.

Occupational workers during use and handling of octabenzone must strictly observe safety regulation practices and must not be negligent. Workers should wear protective clothing to prevent chemical contact, personal protective equipment (PPE) to avoid organic vapour, canister face mask, self-contained breathing apparatus, liquid-proof gloves, protective goggles or face shield, white- or light-coloured clothing, and rubber shoes or boots. Breathing must be protected when large quantities are decanted without local exhaust ventilation. Closed containers should only be opened in well-ventilated areas. Handle it in accordance with good industrial hygiene and safety practice.

Protection against Fire and Explosion

Workers should store octabenzone in a tightly closed container and dry, cool place and away from all sources of ignition: heat, sparks, and open flame. The chemical dust can form an explosive mixture with air. During use and handling of octabenzone, workers should strictly avoid contact with strong oxidising agents, acids, and bases. Workers must handle the chemical waste or incinerate it in accordance with local regulations. The used containers and packaging must not be reused. Waste packages are recommended to be crushed, punctured, etc., to prevent unauthorised use and/or reuse purposes.

Osmium Tetroxide (CAS No. 20816-12-0)

Molecular formula: OsO_4

Osmium and compounds: Osmium is an extremely dense, blue-grey, hard but brittle metal that remains lustrous at high temperature as well. Osmium possesses quite remarkable chemical and physical properties. It has the highest melting point and the lowest vapour pressure in the platinum family. Osmium is highly volatile and extremely toxic and is rarely used in its pure state and often alloyed with other metals. Those alloys are utilised in high-wear applications. Osmium alloys such as osmiridium are very hard and, along with other platinum group metals, are used in the tips of fountain pens, instrument pivots, and electrical contacts, as they can resist wear from frequent operation.

Osmium tetroxide is a colourless to pale yellow crystalline solid with an odour that has been described as pungent or chlorine-like. Osmium tetroxide is used as a tissue fixative for electron microscopy. Osmium tetroxide has been used in fingerprint detection and for staining fatty tissues for optical and electron microscopy studies. Students and workers must take precautions in the use of osmium tetroxide. Human poisoning and symptoms of osmium tetroxide are modulated by the route of exposure. Exposures to osmium tetroxide trigger and cause adverse health effects very rapidly at very low concentrations.

Occupational exposures to osmium tetroxide cause toxicity and adverse health effect to users and workers. The symptoms of toxicity and poisoning include, but are not limited to, the development of corrosive effects to the area of contact; severe chemical burns to the skin, eyes, and respiratory tract; blisters; discoloration; pain; burning sensation; tearing; cough; headache; wheezing; shortness of breath; pulmonary oedema; and fatal injury/death. Exposures to osmium tetroxide cause skin redness or rash, visual disturbances, severe conjunctivitis, and possible permanent loss of vision. Accidental ingestion of osmium tetroxide causes abdominal cramps, burning sensation, vomiting, and collapse.

Prolonged period of inhalation exposure to osmium tetroxide causes insomnia, digestive disturbance, and distress to the pharynx and larynx.

Oxydisulphoton (CAS No. 2497-07-6)

Molecular formula: $C_8H_{19}O_3PS_3$
Synonyms and trade names: Disulfoton disulphide; Disulfoton sulfoxide; Disyston sulfoxide; O,O-Diethyl S-(2-ethylthionylethyl) phosphorodithioate; Depd; Ethylthiometon sulfoxide; Phosphorodithioic acid; O,O-diethyl S-((ethylsulfinyl)ethyl) ester; O,O-Diethyl S-(2-(ethylsulfinyl)ethyl) phosphorothioate
IUPAC name: O,O-diethyl S-2-ethylsulfinylethyl phosphorodithioate

On heating, oxydisulphoton gets decomposed and releases toxic gases and vapours (SOx, POx).

Safe Handling and Precautions

Oxydisulphoton is an organophosphate which is readily absorbed through inhalation, ingestion, and skin contact and causes inhibition of blood cholinesterase. The initial symptoms are giddiness, headache, and nausea. Subsequently, cold sweating, vomiting, abdominal cramp, diarrhoea, blurred vision, and twitching in the eyelids may occur. In severe cases, laboured breathing, wheezing, excessive sweating, salivation, mental confusion, convulsion, progressive cardiac and respiratory failure, and coma are manifested.

Signs and Symptoms of Oxydisulphoton Exposure

Exposure to oxydisulphoton may produce the following signs and symptoms: sweating, pinpoint pupils, blurred vision, headache, dizziness, profound weakness, muscle spasms, seizures, and coma. Mental confusion and psychosis may occur. Excessive salivation, nausea, vomiting, anorexia, diarrhoea, and abdominal pain may also occur. The heart rate may decrease following oral exposure or increase following dermal exposure. Chest pain may be noted. Hypotension (low blood pressure) may be observed, although hypertension (high blood pressure) is not uncommon. Respiratory signs include dyspnoea (shortness of breath), pulmonary oedema, respiratory depression, and respiratory paralysis.

Emergency Life Support Procedures

Acute exposure to oxydisulphoton may require decontamination and life support for the victims. Emergency personnel should wear protective clothing appropriate to the type and degree of contamination. Air-purifying or supplied-air respiratory equipment should also be worn, as necessary. Rescue vehicles should carry supplies such as plastic sheeting and disposable plastic bags to assist in preventing spread of contamination. All chemicals should be considered hazardous. Avoid direct physical contact. Use appropriate approved safety equipment. Untrained individuals should not handle this chemical or its container. Handling should occur in a chemical fume hood. Store in a cool, dry, well-ventilated place and keep containers closed. Protect it against physical damage.

Bibliography

Dikshith, T. S. S. 1996. *Safety Evaluation of Environmental Chemicals*. New Age International Publishers, New Delhi, India.

Dikshith, T. S. S. 2009. *Safe Use of Chemicals: A Practical Guide*. CRC Press, Boca Raton, FL.

Dikshith, T. S. S. 2011. *Handbook of Chemicals and Safety*. CRC Press, Boca Raton, FL.

Dikshith, T. S. S. and P. V. Diwan. (eds.). 2003. *Industrial Guide to Chemical and Drug Safety*, John Wiley & Sons, Hoboken, NJ.

Echobichon, D. J. 1996. Toxic effects of pesticides. *Casarett & Doull's Toxicology: The Basic Science of Poisons*, Klaassen, C. D. (ed.), 5th edn. McGraw-Hill, New York.

Gosselin, R. E., R. P. Smith, and H. C. Hodge. (eds.). 1984. *Clinical Toxicology of Commercial Products*. Williams & Wilkins, Baltimore, MD.

Lewis, R. J., Sr. 2007. *Hawley's Condensed Chemical Dictionary*. Wiley-Interscience, John Wiley & Sons, New York.

Material Safety Data Sheets (MSDS). 2009. Octabenzone. BASF Canada Inc. Mississauga, ON. Canada.

Meister, R. T. (ed.). 2000. *Farm Chemicals Handbook 2000*. Meister Publishing Co., Willoughby, OH.

Sittig, M. (ed.). 1985. *Handbook of Toxic and Hazardous Chemicals and Carcinogens*, 2nd edn. Noyes Publications, Park Ridge, NJ.

Tomlin, C. D. S. (ed.). 2006. *The Pesticide Manual: A World Compendium*, 14th edn. The British Crop Protection Council, Farnham, Surrey, U.K.

17

Hazardous Chemical Substances:
P

Palladium Chloride
Paraffin
Paraquat Dichloride
Parathion (Parathion Ethyl)
Pentaboron Nonahydride
2,2',4,4',5-Pentabromodiphenyl Ether
Pentachloroethane
Pentachlorophenol
n-Pentane
Pentanol
Perchloroethylene
Petroleum Ether
Phenol
Phenosulphonic Acid
Phorate
Phosacetim
Phosdrin
Phosfolan
Phosgene
Phosgene Oxime
Phosmet
Phosphamidon
Phosphine (Hydrogen Phosphide)
Phosphoric Acid
Phosphorus–White Phosphorus
Phosphorus Pentachloride
Phosphorus Pentoxide
Phosphorus Pentasulphide
Phthalic Anhydride
Picloram
Picric Acid
Picrotoxin
Pirimiphos Methyl
Polybrominated Diphenyl Ethers
Polybrominated Biphenyl
Polychlorinated Biphenyls
Polycyclic Aromatic Hydrocarbons

Propane
Propetamphos
Propoxur
Propylene Imine
Propylene Oxide
Pyrethrins and Pyrethroids
Pyrethrum

Palladium Chloride (CAS No. 7647-10-1)

Molecular formula: **PdCl$_2$**
Synonyms and trade names: Palladium dichloride; Palladium (II) chloride; Dichloropalladium; Enplate activator 440; Niklad 262; Palladous chloride
Chemical name: Palladium (II) chloride

Palladium chloride is a dark brown powder, hygroscopic (absorbs moisture from the air). It is incompatible with acids, aluminium, ammonia, magnesium, nitrates, zinc, heat, thiocyanates, and organic solvents. Thermal decomposition of palladium chloride may release chlorine, hydrogen chloride, and oxides of palladium. It is used as a catalyst, photographic, and electroplating reagent.

Palladium and its alloys are used as catalysts in the (petro)chemical and, above all, in the automotive industries. Applications of palladium compounds for electronics and electrical technology include use in metallisation processes (thick film paste), electrical contacts and switching systems, in the synthesis of semiconducting metal-containing polymers in which the polypyrrole backbone has a conformational energy minimum and is nearly planar. Palladium chloride is a stable chemical substance and is incompatible with strong oxidising agents.

Safe Handling and Precautions

Exposure to palladium chloride at workplaces and accidental ingestion/swallowing causes adverse health effects. It is a skin sensitiser causing allergic reaction. Inhalation of palladium chloride causes coughing, wheezing, laryngitis, shortness of breath, headache, muscular pain, nausea, irritation, and vomiting, and prolonged period of inhalation exposure causes chemical burns to the respiratory tract, inflammation, oedema of the larynx and bronchi, laryngitis, chemical pneumonitis, pulmonary oedema, and may result in fatal injury.

Palladium chloride must be kept stored in a tightly closed container, in a cool, dry, well-ventilated area away from incompatible substances and moisture. Users and occupational workers during handling of must observe precautions and use proper personal protective equipment (PPE), safety glasses, gloves, and respiratory protection. The workplace must have facilities of adequate ventilation. Workers must avoid formation/generation of dust at the workplace. They should also avoid breathing vapours, mist, or gas and must quickly evacuate personnel to safe areas.

Paraffin (CAS No. 8002-74-2)

(Paraffin wax, Hard paraffin)
Molecular formula: $C_{25}H_{52}$ (C_nH_{2n+2})

Paraffin wax, also commonly called 'paraffin', is a colourless or white, tasteless, odourless, translucent waxy solid. Paraffin wax has a typical melting point between about 46°C and 68°C. Pure paraffin wax is a combustible substance and insoluble in water but soluble in petroleum solvents and stable under normal conditions of use. Paraffin has been identified as an excellent electrical insulator. It is also used in the manufacturing of paraffin papers, candles, food packaging materials, varnishes, floor polishes, to extract perfumes from flowers, in lubricants, and cosmetics. It is also used in water-proofing wood, and cork.

Safe Handling and Precautions

Paraffin is considered non-toxic in its solid state, but the fume generated when it is in the molten state may cause discomfort and nausea. There are no reports implicating paraffin wax as a health or environmental hazard under current legislation. No significant health hazard is associated with cold paraffin wax while any kind of contact with hot material is known to cause thermal burns. Inhalation of hot wax fumes if present in sufficiently high concentration may cause eye and respiratory irritation. In short, solid material is not expected to be an irritant, but it may cause mild skin and eye irritation. Vapours from molten wax are known to cause eye irritation and, similarly, fumes of paraffin wax cause irritating effects to the eyes, nose, and throat and produce nausea. Contact with molten wax can cause serious burns. Direct ocular contact of the melted product with the eyes can cause thermal burns and ocular lesions. Fumes and vapours coming from the thermal decomposition of paraffin wax can cause irritation of the eyes. Paraffin wax is primarily inert and non-poisonous. However, one must handle it carefully and the molten materials will produce thermal burns. Reports have shown that prolonged period of inhalation of aerosols (if heated) of paraffin wax cause cough. Paraffin wax has not been classed as dangerous substance under the classification, packaging, and labelling of dangerous substances regulations. During handling molten form of paraffin wax, it is important that the workplace should have appropriate ventilation, local exhaust, or breathing protection.

Workers should use quality and approved gloves and goggles and keep away from heat and sources of ignition. Protective clothing against splashes, thermal gloves, and ocular protection must be worn to prevent wounds. Waste disposal should be through contractor to an approved waste disposal site, observing all local and national regulations.

Paraquat Dichloride (CAS No. 1910-42-5)

Molecular formula: $C_{12}H_{14}C_{12}N_2$
Synonyms and trade names: Paraquat dichloride; Methyl viologen dichloride; Crisquat; Dexuron; Esgram; Gramuron; Ortho Paraquat CL; Paraquat dichloride;

N,N'-dimethyl-gamma,gamma'-bipyridylium dichloride; Methyl viologen dichloride, 1,1'-dimethyl-4,4'-bipyridinium dichloride, Crisquat, Dexuron, Esgram, Gramuron; Paracol, Pillarxone; Tota-col; Toxer total; Cyclone; Dichloride salt of paraquat; Dimethyl-4,4'-bipyridinium dichloride; Dimethyl-4,4'-bipyridylium dichloride; Gramixel; Gramoxone; Gramoxone dichloride; Gramoxone S; Gramoxone W; Methyl viologen hydrate; Methyl viologen dichloride hydrate; AH 501

U.S. EPA toxicity: Class I. Highly toxic

Paraquat dichloride is a dark blue liquid, non-combustible, stable herbicide chemical. It is incompatible with strong oxidising agents. On contact with fire paraquat decomposes gives off irritating or toxic fumes (or gases), nitrogen oxides, and hydrogen chloride. Paraquat dichloride contact and storage destroys metal. Paraquat dichloride hydrolyses in alkaline media and reacts with aluminium to produce hydrogen gas.

Paraquat (*N,N'*-dimethyl-4,4'-bipyridinium dichloride) is one of the most widely used herbicides in the world. Paraquat dichloride is a herbicide currently registered to control weeds and grasses in many agricultural and non-agricultural areas. It is used preplant or pre-emergence on vegetables, grains, cotton, grasses, sugar cane, peanuts, potatoes, and tree plantation areas; post-emergence around fruit crops, vegetables, trees, vines, grains, soybeans, and sugar cane. It is also used on non-crop areas such as public airports, electric transformer stations, and around commercial buildings to control weeds. It has been reported that about seven pesticide products are registered which contain the active ingredient paraquat dichloride and classified as restricted use pesticides (RUPs). Paraquat dichloride and the products are primarily intended for 'Occupational Use'. In the United States, Paraquat is classified as 'restricted commercial use', and people must obtain a license to use the product.

Safe Handling and Precautions

Exposure to paraquat causes harmful effects. On accidental workplace, inhalation, ingestion, and/or skin absorption is known to cause adverse health effects with symptoms of cough, sore throat, laboured breathing, headache, nosebleeds, abdominal pain, nausea, vomiting, and diarrhoea. Paraquat has been reported to cause destructive effects of mucous membranes, burns and eye damage, and acute respiratory distress syndrome (ARDS), kidneys, liver, gastrointestinal tract, cardiovascular system and lungs, resulting in impaired functions, tissue lesions including haemorrhage and lung fibrosis. In high concentrations and after a prolonged period of exposure, paraquate cause tissue damage in the lungs, pneumonitis, asphyxiation, liver damage, kidney failure, heart failure, and fatal injury. Paraquat has been reported to cause mutagenic effects and listed as a possible carcinogen and Parkinson's disease in farm workers. Based on reports of inhalation and oral toxicity studies, paraquat has moderate to high acute toxicity and severe eye to workers. Reports have indicated that paraquat produced squamous cell carcinoma, an uncommon tumour, in the head region in both sexes of Fischer 344 rats and hence listed as a possible human carcinogen of category C.

Occupational workers should be careful when handling paraquat dichloride. Use and handling should be done only by licensed pest-control operators and primary producers. The herbicide must be kept away from food, drink, and animal feedstuff, out of REACH OF CHILDREN. Workers should wear suitable PPE, PVC gloves and apron, face shield when handling and spraying the chemical. Paraquat dichloride should be kept stored in the original container in a dry, cool, ventilated, LOCKED area, but should not be stored in prolonged sunlight. Also, paraquat dichloride should not be stored with explosives, oxidising agents, corrosive substances, or nitromethane.

Workers must handle the herbicide with the PPE; workers who handle the herbicide as herbicide mixers and loaders must wear long-sleeved shirt and long pants, chemical-resistant gloves, shoes plus socks, chemical-resistant apron, face shield, safety glasses, PVC gloves, good ventilation, and avoid prolonged or repeated exposure. In cases of accidental workplace spillage of paraquat dichloride, workers should control the spill at its source, contain the spill to prevent from spreading or contaminating soil or from entering sewage and drainage systems or any body of water. Workers should clean up the chemical waste spills immediately with absorbing material and place into compatible disposal container and seal the container with label for disposition.

Products containing paraquat must be labelled with the Signal Words DANGER—POISON. Paraquat is an RUP. RUPs may be purchased and used only by certified applicators.

Parathion (Parathion Ethyl) (CAS No. 56-38-2)

Molecular formula: $C_{10}H_{14}NO_5PS$
Synonyms and trade names: Folidol; Fostox; Ethyl parathion; Etilon; Niran; Orthophos; Panthion; Paramar; Paraphos; Parathene
IUPAC name: O,O-diethyl O-4-nitrophenyl phosphorothioate
Toxicity class: U.S. EPA: I; WHO: Ia

Pure parathion is a pale yellow liquid with a faint odour of garlic, while the technical parathion is a deep brown to yellow liquid. It is sparingly soluble in water but soluble in alcohols, aromatic hydrocarbons, esters, ethers, *n*-hexane, dichloromethane, 2-propanol, toluene, and ketones.

Parathion is one of the most acutely toxic chemical substance pesticide. The U.S. Environmental Protection Agency (U.S. EPA) classified parathion as an RUP, meaning that it should be handled by qualified trained and certified workers. In January 1992, the U.S. EPA announced the cancellation of parathion for all users on fruit, nut, and vegetable crops.

Parathion was used for the control of pests of fruits, nuts, and vegetable crops. The only uses retained are those on alfalfa, barley, corn, cotton, sorghum, soybeans, sunflowers, and wheat. Further, to reduce exposure of agricultural workers, parathion may be applied to these crops only by commercially certified aerial applicators and treated crops may not be harvested by hand. Parathion is a broad spectrum, organophosphate pesticide used to control many insects and mites.

Safe Handling and Precautions

Parathion is highly toxic by all routes of exposure. Parathion, like all organophosphate pesticides, inhibits acetylcholinesterase and alters cholinergic synaptic transmission at neuro-effector junctions (muscarinic effects), at skeletal myoneural junctions and autonomic ganglia (nicotinic effects), and in the central nervous system (CNS). Exposures to parathion cause poisoning and the symptoms include, but not limited to, abdominal cramps, vomiting, diarrhoea, pinpoint pupils, blurred vision, excessive sweating, salivation and lacrimation, wheezing, excessive tracheobronchial secretions, agitation, seizures, bradycardia or tachycardia, muscle twitching and weakness, and urinary bladder and faecal incontinence. Seizures are much more common in children than in adults. Severe exposures cause loss of consciousness, coma, excessive bronchial secretions, respiratory depression, and cardiac irregularity, eventually leading to death. Occupational workers and general public with health disorders and abnormalities such as cardiovascular, liver or kidney diseases, glaucoma, or CNS suffer with an increased risk of parathion poisoning. Further, high environmental temperatures enhance the severity of parathion poisoning.

Human fatalities associated with parathion due to accidental ingestion, spillage of sprays, suicidal and improper use are many. Reports have shown that many school-going children suffered fatalities when exposed to parathion-contaminated air and soil (earlier sprayed with parathion). Also in Kerala (India), during the shipment, food was contaminated with parathion and the consumption of the same food eventually resulted in large human fatalities. Similarly, consumption of contaminated wheat floor in Jamaica caused poisoning and fatalities. There are more reported cases of poisoning with parathion than with any other pesticide currently in use. There have been a number of cases where intoxication and death have resulted from ingestion of foodstuffs that have been grossly contaminated with parathion. In one Asian country, there were 828 cases of poisoning with 106 deaths caused by flour, sugar, and other foodstuffs becoming contaminated because parathion was transported in the same ship's hold as the food. In another Asian country, barley became contaminated with parathion. There were 38 cases of poisoning with 9 deaths. Similarly, in a country in the Americas, there were 559 cases of poisoning and 16 deaths when sacks of sugar, and possibly flour, absorbed parathion from the floor of a truck. In another country in the Americas, there were 165 known, and 445 more suspected, cases of poisoning with 63 deaths, when parathion from broken containers contaminated sacks of flour during transportation in a truck. In a European country, parathion caused 26 cases of poisoning due to transportation of food material in parathion-contaminated wagon. There is no evidence suggesting that parathion is mutagenic. Parathion produced embryocidal effects and foetal growth retardation, no malformations, in mice and rats at doses that were generally below the level that was toxic for the mother. Evidences are inadequate to evaluate the carcinogenicity of parathion in experimental animals, and the available data are insufficient to evaluate the carcinogenicity of parathion for humans. The acceptable daily intake (ADI) recommended for parathion has been reported as 0.004 mg/kg body weight.

Pentaboron Nonahydride (CAS No. 19624-22-7)

Molecular formula: B_5H_9
Synonyms and trade names: Pentaborane; Pentaboron nonahydride; 9-Pentaboron nonahydride; Dihydropentaborane

Pentaboron nonahydride is a colourless liquid, with pungent odour, and flammable. It is corrosive to natural rubber, some synthetic rubber, some greases, and some lubricants and gives off irritating or toxic fumes (or gases) in a fire. It reacts violently with fire. Pentaboron nonahydride incompatible with strong oxidants such as chromium anhydride, chlorate and potassium permanganate, and other contacts.

Safe Handling and Precautions

Pentaboron nonahydride get absorbed into the body by inhalation of its vapour, through the skin and by ingestion. Exposure to pentaboron nonahydride causes nausea, drowsiness, headache, dizziness, weakness, sleepiness, ophthalmoplegia, convulsions, and unconsciousness. Repaired and prolonged period of exposures to pentaboron nonahydride at workplaces cause severe irritation of the eyes, skin, and the respiratory tract, ataxia, muscle cramps, convulsions, CNS depression and excitation, hallucinations, incoordination, convulsions, acidosis, liver and kidney damage. In much higher concentrations, exposure to pentaboron nonahydride causes fatal injury. Pentaborane has not been listed as a carcinogen by National Toxicology Program (NTP).

During use and handling of pentaboron nonahydride, occupational workers should use chemical-resistant suit, face shield, or eye protection in combination with breathing protection. Pentaboron nonahydride should be kept stored in cool, dry, dark but well-ventilated area, in a tightly sealed container, away from any kind of ignition sources, open flames, and direct sunlight, avoid open flames, sparks, smoking and/or contact with halogens, halogenated compounds, and oxidisers. Pentaboron nonahydride should be used and handled in a chemical fume hood by trained workers and under supervision.

Danger – Label warnings – Pentaboron nonahydride is Extremely Very Flammable – Combustion Imminent.

2,2′,4,4′,5-Pentabromodiphenyl Ether (CAS No. 32534-81-9)

Molecular formula: $C_{12}H_5Br_5O$
Synonyms and trade names: BDE-99; Pentabromodiphenyl ether; PBDE-99

Pentabromodiphenyl ether (PBDE-99) is an amber coloured solid, and insoluble in water. It is a halogenated organic compound and is very unreactive. Reactivity generally decreases with increased degree of substitution of halogen for hydrogen atoms. Materials in this group are incompatible with strong oxidising and reducing agents. Also, they are incompatible with many amines, nitrides, azo/diazo compounds, alkali metals, and epoxides. 2,2′,4,4′,5-Pentabromodiphenyl ether (PBDE-99) was in use as a component of many commonly used flame retardants, including DE-71, Bromkal G1, and other popular mixtures. DE-71 was sold as a flame retardant for unsaturated polyester, rigid and flexible polyurethane foams, epoxies, laminates, adhesives, and coatings. The PBDEs are used in paint, high-impact plastic, foam, and textiles, as well as in electronic, building, automotive, furniture, and household plastic products. The U.S. manufacturers and the EU have banned the use of PBDE since 2004.

Safe Handling and Precautions

Human exposure to PBDEs may occur by inhalation as an indoor or occupational air pollutant, by dermal absorption as an occupational hazard, by contact with products containing PBDEs, or by oral ingestion in foods.

Exposure to 2,2′,4,4′,5-Pentabromodiphenyl ether (BDE-99) has been reported to cause harmful effects. It is classified as a persistent organic pollutant, and known to bioaccumulate through the food web, and pose a risk of causing adverse effects to human health and the environment.

The Persistent Organic Pollutants Review Committee identified and reported that there is sufficient evidence to show that PBDE meets the criterion on adverse effects.

Pentachloroethane (CAS No. 76-01-7)

Molecular formula: C_2HCl_5
Synonyms and trade names: Ethane pentachloride; Pentalin

Pentachloroethane is incompatible and very reactive in contact with sodium potassium (alloy + bromoform), alkalis, metals, and water. On hydrolysis, pentachloroethane produces dichloroacetic acid and the reaction with alkalis and metals produces chloroacetylenes, which becomes spontaneously explosive. Pentachloroethane is a colourless stable liquid with a camphor-like smell and non-flammable. Pentachloroethane is incompatible with strong oxidising agents and reacts violently with alkalis or metals.

Safe Handling and Precautions

Exposure to pentachloroethane causes effects of irritation to the eyes and skin. On exposure to pentachloroethane, experimental laboratory animals exhibit symptoms of toxicity such as weakness, exhaustion, restlessness, respiratory distress, muscle incoordination, and cause pathological changes/damage of the liver, kidney, lung changes. Occupational workers after an accidental workplace inhalation or ingestion and/or skin absorption show symptoms of poisoning that include, but not limited to, severe irritation of the eyes, skin, and the respiratory tract, confusion, cough, dizziness, headache, nausea, sore throat, and vomiting.

Warning – Danger: Users and occupational workers during work and handling must remember that pentachloroethane causes risk of irreversible effects and long-term harm. Pentachloroethane is banned for diffusive applications like surface or fabric cleaning. Pentachloroethane has been listed in 1999 – Schedule 2 of the Control of Substances Hazardous to Health Regulations (COSHH).

Pentachlorophenol (CAS No. 87-86-5)

Molecular formula: C_6HCl_5O
Synonyms and trade names: PCP, 2,3,4,5,6-pentachlorophenol; Pentachlorophenate; Penta; numerous trade names

Pentachlorophenol (PCP) is an odourless, white or light brown powder or crystal in appearance. It is used as herbicide and fungicide. Pentachlorophenol is incompatible with strong oxidising agents. Pentachlorophenol has a very sharp characteristic phenolic smell when hot but very little odour at room temperature. Pentachlorophenol is a synthetic substance made from other chemicals and does not occur naturally in the environment. Initially pentachlorophenol was widely used as a wood preservative. It is now used industrially as a wood preservative for power line poles, cross arms, fence post, etc. The general populations can be exposed to very low levels of pentachlorophenol in contaminated indoor and outdoor air, food, drinking water, and soil.

Safe Handling and Precautions

Pentachlorophenol (PCP) was, and still is, one of the most frequently used fungicides and pesticides. Its toxicity is due to interference with oxidative phosphorylation.

Acute and chronic poisoning may occur by dermal absorption, inhalation, or ingestion. Pentachlorophenol is readily absorbed through the skin, and is very toxic by inhalation and ingestion. It is a severe eye irritant and accidental workplace inhalation is known to cause poisoning of the circulatory system with possible heart failure and fatal injury. Children also might get exposed to pentachlorophenol by eating fish, drinking contaminated water, and other foods.

Studies in workers have shown that exposure to high levels of pentachlorophenol causes the cells in the body to produce excess heat. Prolonged period of exposure to pentachlorophenol has been shown to cause effects on liver and damage the immune system. Also, damage to the thyroid and reproductive system has been observed in laboratory animals exposed to high doses of pentachlorophenol. Damage to the thyroid and reproductive system has been observed in laboratory animals exposed to high doses of pentachlorophenol. Pentachlorophenol is very toxic to aquatic organisms. In brief, pentachlorophenol is an RUP and is used industrially as a wood preservative for utility poles, railroad ties, and wharf pilings. Pentachlorophenol was widely used as wood preservative until 1987 when its use was restricted to certified applicators. The wood preservatives containing pentachlorophenol are undergoing a six-phase RED process. Pentachlorophenol is considered a probable human carcinogen and exposure to high levels can also have other health risks. The effects and risks during use and handling of pentachlorophenol are irreversible and workers, therefore, must handle pentachlorophenol in accordance with good laboratory practices. They must wear appropriate protective clothing, safety goggles, chemical-resistant gloves, laboratory coat, and the workplace must have good ventilation.

Purchase and use of pentachlorophenol has been restricted to certified applicators since 1984.

Pentachlorophenol was one of the most widely used biocides in the United States prior to regulatory actions to cancel and restrict certain non-wood preservative uses of pentachlorophenol in 1987. It now has no registered residential uses.

---

n-Pentane (CAS No. 109-66-0)

Molecular formula: C_5H_{12}
Synonyms and trade names: Amyl hydride *n*-Pentane; Skellysolve; Pentane (normal)

n-Pentane is incompatible with oxidising agents, halogens. *n*-Pentane is a highly flammable liquid and readily forms explosive mixtures with air. Pentane is a colourless, volatile liquid obtained during the distillation of crude oil. The blends include

1. Normal Pentane/*n*-Pentane/Pentane (CAS No. 109-66-0)
2. Iso-Pentane/2-Methyl butane/Pentane blends (CAS No. 78-78-4)
3. Cyclopentane/Amyl hydride (CAS No. 287-92-3)

n-Pentane has extensive applications in a variety of industries. It is used as an aerosol propellant and as an important component of engine fuel, as a foaming (blowing) agent, used to produce foamed polystyrene and foamed polyurethane foams, as a solvent for the production of polymers. Pentanes are also used as solvents in aerosol cosmetics, aerosol paints, and car care products. In short, pentanes are used in the production of industrial and consumer products. Users and workers get exposed to pentane vapour during the period of handling and filling gas tanks of vehicles. Occupational pentane exposure does occur in a pentane distillation facility and/or in other industrial facilities or wherever pentane or pentane-containing products are used.

Safe Handling and Precautions

Studies have indicated that *n*-Pentane causes mild health hazards to users and does not cause adverse health or environmental effects. Mild skin or eye irritation may occur following contact with liquid pentane. Repeated contact with the skin cause dry skin absorption through the skin and is reported to be low because of its rapid evaporation rate. Repeated inhalation of concentrated pentane vapours is known to cause dizziness, drowsiness, unconsciousness, nausea, headache, dizziness, vomiting, and cyanosis. Repeated or prolonged exposure of skin causes irritation and dermatitis due to degreasing properties of the product.

Therefore, tissue accumulation is expected to be low. Studies have shown that prolonged period of inhalation of pentane in rats did not cause damage to the kidneys or nervous system while ingestion of pentane caused lung damage due to aspiration. *n*-Pentane is a CNS depressant. Studies with dogs have indicated that it induces cardiac sensitisation. In high concentrations, it causes incoordination and inhibition of the righting reflexes. The National Institute for Occupational Safety and Health (NIOSH) recommended limits of *n*-pentane for working areas.

Pentanol (CAS No. 71-41-0)

Molecular formula: $C_5H_{12}O$
Synonyms and trade names: Pentyl alcohol; Pentanol; Pentan-1-ol, *n*-Pentan-1-ol; Amyl alcohol; Amylol; Primary amyl alcohol; Normal amyl alcohol; *n*-Amyl alcohol; *n*-Pentyl alcohol; *n*-Pentanol; Pentasol
Chemical family: Saturated primary aliphatic alcohol

Pentanol is a flammable, colourless, combustible liquid and vapour. Vapour is heavier than air and can travel considerable distance to source of ignition and flash back. The vapour may form explosive mixtures with air at or above 49°C. Pentanol is incompatible with strong oxidising agents, alkali metals, alkaline earth metals, halogens, hydrogen trisulphide, acids, acetaldehyde, acid anhydrides, acid chlorides, lithium aluminium hydride, isocyanates, dialkylmagnesiums, *n*-halomides, ethylene oxide, hypochlorous acid, hydrogen peroxide and sulphuric acid, nitrogen tetraoxide, nitryl hypochlorite, permonosulphuric acid, and tri-isobutyl aluminium.

Exposure to pentanol is harmful if and when accidentally swallowed, inhaled, or absorbed through the skin at different workplaces. Pentanol is a severe skin and eye irritant. Exposures to pentanol cause cough, headache, nausea, dizziness, and drowsiness.

Workers during handling of 1-pentanol should wear chemical safety protective equipment of approved quality: goggles and/or face shield, skin protection of butyl or nitrile rubber or Viton gloves. Other protective impervious clothing, sleeves, coveralls, and boots sufficient to prevent contact should be worn. Other PPE, an eyewash, and safety shower should be nearby and ready for use.

Petroleum Ether (CAS No. 8032-32-4)

Molecular formula: Not applicable
Synonyms and trade names: Benzin; Ligroine; Ligroin petroleum benzene; Petroleum naphtha; Naphtha ASTM; Petroleum spirits; VM & P naphtha

Petroleum ether is a flammable liquid and is used as a universal solvent and extractant during the processing of different chemicals like fats, waxes, paints, varnishes, furniture polish thinning, as detergent, and as fuel. The majority components include paraffins, olefins, naphthenes, aromatics, and about 10%–40% ethyl alcohol. Occupational workers after an acute and prolonged exposure to petroleum ether in workplaces suffer and demonstrate a variety of health disorders in occupational workers. Symptoms of toxicity include, but not limited to, erythema, oedema, skin peeling, loss of appetite, muscle weakness, paresthesia, CNS depression, peripheral nerve disorders, skin and respiratory irritation, and chemical pneumonia. Accidental aspiration of small amounts of liquid *n*-pentane at workplaces causes bronchopneumonia or pulmonary oedema. In low concentrations it may cause narcotic effects. Symptoms may include dizziness, headache, nausea, and loss of coordination. Children exposed to petroleum spirits as thinner, show symptoms of poisoning such as skin irritation, respiratory problems, and haematologic effects. The International Agency for Research in Cancer (IARC), the NTP, and the American Conference of Industrial Hygienists (ACGIH) have not listed petroleum ether as a human carcinogen.

Occupational workers during handling of pentane and products of petroleum should strictly wear self-contained breathing apparatus (SCBA) and PPE to protect eyes, face, and skin from liquid splashes. Workplace should have adequate air ventilation, away from sources of ignition, smoke, and static discharges. Also, petroleum and products should be kept segregated and stored away from oxidant gases and other oxidants. Individuals not related with the work should be prevented from entering the workplace. Pentane is very flammable and presents a serious fire risk. Workers should strictly avoid pouring down the chemical into the sinks and sewers. Keep away from food, drink, and animal feeding stuffs. Keep working clothes separately. Chemical waste should be kept stored in a suitable container with label for disposal/waste disposal contractor. During handling of the chemical, workers should be alert and must use and wear impervious protective clothing/PPE, including boots, gloves, apron or coveralls, as appropriate, to prevent skin contact. Wear protective gloves and clean body-covering clothing. Workers should avoid inhalation of vapour and should be away from heat and sources of ignition.

Precautions – Danger!

Petroleum ether is extremely flammable. Liquid and vapour may cause flash fire, harmful and/or fatal if swallowed. Harmful if inhaled, affect CNS, cause irritation to skin, eyes, and respiratory tract.

Perchloroethylene (CAS No. 127-18-4)

Molecular formula: C_2Cl_4
Synonyms and trade names: Perk; Perclean; Tetrachloroethylene; 1,1,2,2-Tetrachloroethylene

Perchloroethylene (Tetrachloroethylene) is a colourless liquid with a slightly ethereal odour. It is marginally soluble in water and soluble in most organic solvents. It is not bioaccumulative and has been known as less or non-persistent in the environment. Perchloroethylene has a limited number of uses and applications. It is used as intermediate, as dry cleaning agent in the industrial and professional sector, as surface cleaning agent in industrial settings, as heat transfer medium in industrial settings, and in film cleaning and copying by professionals. It is also used as a chemical intermediate in the production of fluorinated compounds and in industrial surface cleaning metal degreasing.

Occupational exposure to perchloroethylene is possible in the manufacturing facilities or the industrial facilities where it is used as an intermediate. However, such type of activities occurs mostly in closed systems; exposure to perchloroethylene has been found to be fairly low while exposure to higher concentrations is possible in industrial or professional dry cleaning and surface cleaning activities.

Safe Handling and Precautions

Exposure to perchloroethylene causes effects of irritation of the skin, eyes, nose, throat, and allergic skin reaction. Repeated and long-term exposures have been reported to cause nausea, flush face, neck drowsiness or dizziness, in coordination, headache, skin erythema (skin redness), dermatitis, cracked skin, respiratory failure, damage the CNS, memory loss, confusion, kidney and liver damage, sleepiness, difficulty speaking and walking, and lightheadedness. Short-term exposure to high levels of perchloroethylene can affect the CNS, and cause unconsciousness and death. The most likely route of human exposure (workers and consumers) to perchloroethylene (tetachloroethylene) is through inhalation and/or to a less extent through dermal absorption.

Perchloroethylene is listed as 'reasonably anticipated to be a human carcinogen' in The 12th Report on Carcinogens published by the NTP because long-term exposure to perchloroethylene can cause leukaemia and cancer of the skin, colon, lung, larynx, bladder, and urogenital tract. Reports have indicated that perchloroethylene has sufficient evidence in experimental animals as a carcinogen, while the evidences are limited in humans as a human carcinogen. Perchloroethylene has therefore been listed as Group 2A, meaning a probable human carcinogen.

Release of perchloroethylene to the environment occurs mostly to the air compartment with minor emissions to the water. Because perchloroethylene is not released into the natural aquatic environment, the possible risk to the environment is low. Workers should use and handle perchloroethylene at all stages of manufacture and use with a minimal impact on the aquatic environment. Due to its unique combination of properties, perchloroethylene is a beneficial solvent for a variety of applications, stretching from industry to professionals. The properties and hazards of perchloroethylene are well known based on numerous animal and human studies as well as by decades of practical use in large volumes around the globe. Its use has been shown to be safe when appropriate technical or personal protection measures are taken, meaning PPE, protective gloves, personal respirator masks, good work practices in reducing vapour exposure and workplace hygiene.

Regulations: In 2010, perchloroethylene has been registered under the European Union REACH Regulation EC/1907/2006 and the substance was found to be safe for the uses identified. Consumer use has not been assessed for REACH. Any consumer use IS NOT supported by ECSA members. The U.S. EPA has set a maximum contaminant level (MCL) in drinking water of perchloroethylene (PCE) at 0.005 mg PCE/L water.

Alert!! WARNING! Occupational workers should note that perchloroethylene is harmful if swallowed, inhaled, or absorbed through skin, causes irritation to skin, eyes, and respiratory tract, affects CNS, liver, and kidneys. A suspect cancer hazard may cause cancer. Risk of cancer depends on the level and duration of exposure.

Phenol (CAS No. 108-95-2)

Molecular formula: C_6H_5OH
Synonyms and trade names: Carbolic acid; Phenic acid; Hydroxybenzene; Phenyl hydrate

Phenol is a stable chemical substance and appear as colourless/white crystals with a characteristic, distinct aromatic/acrid odour. It is reactive and incompatible with strong oxidising agents, strong bases, strong acids, alkalis, and calcium hypochlorite. Phenol is flammable and may discolour in light.

Phenol is used in the manufacture or production of explosives, fertiliser, coke, illuminating gas, lampblack, paints, paint removers, rubber, perfumes, asbestos goods, wood preservatives, synthetic resins, textiles, drugs, and pharmaceutical preparations. It is also extensively used as a disinfectant in the petroleum, leather, paper, soap, toy, tanning, dye, and agricultural industries.

Safe Handling and Precautions

Exposures to phenol cause adverse health effects and poisoning. Phenol is absorbed very rapidly through surfaces of skin, lungs, and stomach. The symptoms of prolonged exposures and poisoning include, but not limited to, vomiting, difficulty in swallowing, diarrhoea, lack of appetite, headache, fainting, dizziness, mental disturbances, and skin rash. Direct contact with phenol causes burning of mouth, irritation to eyes, nose, and dermatitis, discolouration of the skin damage to liver and kidney. Exposure to phenol in different concentrations is known to cause mental disturbances, depression of the CNS, and coma.

Accidental workplace ingestion or inhalation and/or skin contact of phenol is known to cause corrosive effects and causes burns and severe injuries. Occupational workers during use and handling of phenol should be very careful and avoid negligence. Workers should use adequate workplace PPE, apron, rubber boots, and goggles to protect eyes from vapours and spillage. Phenol is a systemic poison and results in serious health hazard. Students, occupational workers, and the workplace manager must be fully aware of the negligence while handling phenol. The typical maximum exposure limit (MEL) of phenol is set at 2 ppm and the typical occupational exposure limit (OEL) is set at 1.0 ppm.

Compounds of Phenol

Phenols are the simplest group of compounds. Phenols have a wide application in industries in the manufacturing of pharmaceuticals. Phenols are, in principle, a group of chemicals with severe irritation to the body system. For example, they cause severe irritation to eyes, skin, nose, respiratory tract, and mucous membranes. They are highly corrosive to skin and tissues. Creosate, a mixture of phenolic and aromatic substances, is well known as a carcinogenic agent to the skin. Phenols also cause adverse effects to CNS, cardiovascular, renal and hepatic systems of animals and humans. In view of this proper use, disposal and management is a must for all occupational workers.

Phenosulphonic Acid (CAS No. 1333-39-7)

Molecular formula: $C_6H_6O_4S/HOC_6H_4SO_3H$
Synonyms and trade names: Benzenesulphonic acid; Hydroxybenzenesulphonic acid; Phenolsulphonic acid; Sulphocarbolic acid

Phenolsulphonic acid is a mixture of ortho and para isomers. It is a yellow-coloured liquid and on exposure to air turns brown in colour. Phenolsulphonic acid is soluble in alcohol and is used as a laboratory reagent, in water analysis and in the manufacture of pharmaceuticals. Phenolsulphonic acid is a combustible material, but does not ignite readily. On heating, phenolsulphonic acid emits vapours, which may form explosive mixtures with air and cause explosion hazards. Also, on contact with metals, phenolsulphonic acid emits flammable hydrogen gas and the containers cause explosion. Phenolsulphonic acid reacts exothermically with chemical bases, for example, amines and inorganic hydroxides to form salts.

Safe Handling and Precautions

Exposure to phenolsulphonic acid causes effects of irritation to mucous membranes, skin, and eyes and is moderately toxic by ingestion. It is toxic by inhalation, ingestion, or skin contact with material and may cause severe injury or death. Accidental contact with molten substance at workplaces is known to cause severe burns to skin and eyes. Phenolsulphonic acid is hazardous in case of skin contact, permeator, and the liquid or spray mist has been reported to produce itching, scaling, reddening, or, occasionally, blistering, tissue damage on mucous membranes of eyes, mouth, and respiratory tract. Inhalation of the spray mist may produce severe irritation of respiratory tract, characterised by coughing, choking, or shortness of breath and toxic to lungs. There are no reports indicating that phenolsulphonic acid is carcinogenic or mutagenic effects and/or teratogenic to humans.

During use and handling of phenolsulphonic acid, occupational workers should completely avoid breathing vapours or dusts. Occupational workers should wear appropriate chemical protective personal protection equipment (PPE) such as splash goggles, workplace suit, boots, vapour respirator (of approved/certified respirator), gloves to avoid chemical spills, etc. The chemical container should be kept away from heat, away from sources of ignition and stored in a cool, dry place, ground all equipment containing the chemical substance. Workers should avoid accidental and workplace exposures, ingestion, to breathe the gas/fumes/vapour/chemical spray. The workplace should have sufficient ventilation, and the workers should wear suitable respiratory equipment.

Phorate (CAS No. 298-02-2)

Molecular formula: $C_7H_{17}O_2PS_3$
Synonyms and trade names: Agrimet; Geomet; Granutox; Phorate 10G; Rampart; Tarathion; Thimenox; Thimet; Timet; Vegfru; Foratox
Chemical name: O,O-Diethyl S-(ethylthio)methylphosphorodithioate; O,O-Diethyl S-ethylthiomethylthiothionophosphate
IUPAC name: O,O-diethyl S-ethylthiomethyl phosphorodithioate

Phorate is a stable, clear, pale yellow mobile liquid. Phorate is soluble and miscible with carbon tetrachloride, dioxane, vegetable oils, xylene, alcohols, ethers, and esters. Phorate get hydrolysed in the presence of moisture and by alkalis, toxic oxides of carbon, sulphur, and phosphorus.

Phorate may be used in foliar or soil treatments on alfalfa, barley, beans, brassicas, coffee, corn, cotton, grapes, hops, lettuce, oats, peanuts, potatoes, rice, sorghum, soybeans, sugarcane, sugar beets, tomatoes, watermelon, wheat, and on ornamentals and pine nursery stock. Phorate has applications for the control of a wide variety of crop pests such as mites, aphids, green-bugs, thrips, leafhoppers, sorghum shoot fly, leaf miners, corn rootworms, psyllids, cutworms, Hessian fly, foliar nematodes, wireworms, flea beetles, whiteflies, pine tip moth, and others.

Safe Handling and Precautions

Exposure to phorate causes health hazards with symptoms of toxicity. Phorate gets absorbed through the intact skin, by inhalation of spray mists or fine dust and from the gastrointestinal tract. The symptoms of poisoning include, but not limited to, effects of irritation of the eyes, skin, respiratory system, miosis, discharge of nasal mucus, headache, chest tightness, wheezing, laryngeal spasm, salivation, cyanosis, anorexia, nausea, vomiting, abdominal cramps, diarrhoea, sweating, muscle fasciculation, weakness, exhaustion, paralysis, dizziness, confusion, ataxia, convulsions, and coma. Short time and repeated exposures at workplaces to phorate affect respiratory system, CNS, cardiovascular system, and blood cholinesterase.

During use and handling of phorate and its formulations, workers must wear splash-proof or dust-resistant safety goggles and a face shield and fume mask to prevent contact with this substance. Also, within the immediate work area, emergency wash facilities for an eye wash fountain and quick drench shower should be available. Occupational workers should not touch chemical waste and the spilled material but use sand or other absorbent material for cleaning small spills and place into containers for later disposal. Workplace supervisor must keep unnecessary people away from the workplace isolate hazard area and completely ventilate closed spaces before entering.

Phosacetim (CAS No. 4104-14-7)

Molecular formula: $C_{14}H_{13}Cl_2N_2O_2PS$
IUPAC name: N'-bis (4-chlorophenoxy)phosphinothioylethanimidamide

Phosacetim is a white crystalline powder in appearance. It is a toxic organophosphate compound used as a rodenticide and acts as an inhibitor of acetylcholinesterase enzyme. On heating, phosacetim undergoes decomposition to emit toxic fumes, primarily chlorides and oxides of nitrogen, sulphur, and phosphorus. Phosacetim is susceptible to formation of highly toxic and flammable phosphine gas in the presence of strong reducing agents such as hydrides. Partial oxidation by oxidising agents may result in the release of toxic phosphorus oxides.

Safe Handling and Precautions

Phosacetim is a cholinesterase inhibitor. Exposure to the vapour or to particulate material has been reported to cause poisoning and the symptoms include, but not limited to, sweating, pinpoint pupils, blurred vision, headache, dizziness, weakness, muscle spasms, dyspnoea, seizures, and coma. Workers with phosacetim poisoning have been reported to exhibit mental confusion and psychosis, excessive salivation, vomiting, anorexia, diarrhoea, and abdominal pain, chest pain and decrease in heart rate, pulmonary oedema, respiratory depression, and respiratory paralysis.

Workers during use and handling of phosacetim should strictly wear positive pressure-demand, full face piece, SCBA or pressure-demand supplied air respirator with escape SCBA and a fully encapsulating, chemical-resistant suit.

Phosdrin (CAS No. 7786-34-7)

Molecular formula: $C_7H_{13}O_6P$
Synonyms and trade names: 3-((Dimethoxyphosphinyl)oxy)-2-butenoic acid methyl ester; Apavinphos; CMDP; Fosdrin; Gesfid; Meniphos; Mevinphos; W 10, PD5

Phosdrin is a stable, colourless, spontaneously flammable pyrophoric liquid with weak odour. It is combustible and incompatible with strong oxidising agents, corrosive to cast iron, some stainless steels, and brass.

Safe Handling and Precautions

Phosdrin is very toxic and its exposure is known to produce acute cholinesterase depression. Any kind of its accidental contact at workplaces with the skin, swallowing or ingestion and/or inhalation is known to cause poisoning. The symptoms of poisoning include, but not limited to, headache, irritation of the eyes, skin, headache, anorexia, nausea, vomiting, abdominal cramps, diarrhoea, chest tightness, wheezing, laryngeal spasm, salivation, cyanosis, wheezing, tiredness, giddiness, faintness, blurred vision, pupillary constriction and muscle twitching, ataxia, convulsions, and paralysis. In cases of severe poisoning, workers suffer from vomiting, abdominal pain, diarrhoea, sweating, salivation, confusion, ataxia, slurred speech, loss of reflexes, cardiac irregularities, CNS effects, unconsciousness, convulsions, severe respiratory depression, and fatal injury. The International Agency for Research on Cancer (IARC), the National Toxicology Program (NTP), and the Occupational Safety and Health Administration (OSHA) have not listed phosdrin as a carcinogen.

Occupational workers should be careful while handling phosdrin; it should be kept stored in its original container tightly closed in a cool, dry place. Phosdrin must be kept stored away form incompatible substances and naked flames and other sources of ignition. Also, the store place should be isolated, dry, cool, and well ventilated and secure with lock and key.

During use and handling of phosdrin, workers should use chemical-resistant PPE – protective impervious clothing, chemical-resistant gloves and apron (preferably PVC), safety glasses for eye protection, and approved respiratory protection/breathing apparatus. Workers should wear waterproof, knee-high, unlined boots when handling formulated products of phosdrin during chemical mixing and spraying.

Phosfolan (CAS No. 947-02-4)

Molecular formula: $C_7H_{14}NO_3PS_2$
Synonyms and trade names: Phosfolan; Phospholan (ISO); Cyolan; Cyalane; Cyolane insecticide; Phosphoramidic acid

Phosfolan is a solid organophosphorus insecticide in yellow colour or colourless appearance. Phosfolan is susceptible to formation of highly toxic and flammable phosphine gas in the presence of strong reducing agents such as hydrides. On heating, phosfolan gets decomposed and release/emits toxic gases and vapours, CO, NOx, SOx, and Pox. It does not undergo hazardous polymerisation reaction. Partial oxidation by oxidising agents may result in the release of toxic phosphorus oxides. Phosfolan is susceptible to formation of highly toxic and flammable phosphine gas in the presence of strong reducing agents such as hydrides. Partial oxidation by oxidising agents may result in the release of toxic phosphorus oxides. Phosfolan is a cholinesterase inhibitor. Phosfolan gets hydrolysed by alkalis. Presently, phosfolan is not available for use in the United States.

Safe Handling and Precautions

Phosfolan is an anthropogenic compound that was considered as a systemic insecticide. Exposure to phosfolan causes poisoning. The symptoms include, but not limited to, nausea, vomiting, abdominal cramps, diarrhoea, excessive salivation, headache, giddiness, vertigo and weakness, sensation of tightness in chest, blurring or dimness of vision, miosis, tearing, ciliary muscle spasm, respiratory distress, bradycardia or tachycardia, loss of muscle coordination, slurring of speech, fasciculations and twitching of muscles of tongue and eyelids, mental confusion, disorientation, cyanosis, convulsions, and coma.

During use and handling of phosfolan, users and occupational workers must wear approved, suitable protective, impervious clothing, face shields, Neoprene-coated gloves, dust-proof goggles, and dust mask equipment to avoid skin contact. The workplace must have adequate ventilation of the working area. Workers must keep phosfolan in tightly closed original container in a dry, cool, and well-ventilated place.

Danger and Precautions

Workplace supervisor/manager and the management should note that phosfolan should not be allowed to use or handle by untrained worker and the container of phosfolan must be handled only in chemical fume hood. Phosfolan burns but does not ignite readily and the container may explode in heat of fire and result in workplace disaster.

Phosgene (CAS No. 75-44-5)

Molecular formula: CCl_2O

Synonyms and trade names: CG; Carbon dichloride oxide; Carbon oxychloride; Chloroformyl chloride; Dichloroformaldehyde; Dichloromethanone

Phosgene is a colourless, reactive, non-flammable gas that is heavier than air with a musty hay odour. Phosgene is commonly stored under high pressure as a liquid. Phosgene reacts with water to form corrosive acids, reacts with most metals in the presence of moisture, liberating hydrogen, an extremely flammable gas, and reacts violently with alkalis. As an industrial and commercially important chemical, phosgene is a precursor material/chemical intermediate, and has extensive application in the manufacture of a wide range of products such as polymers – polyurethanes and polycarbonates – pesticides, medicines, dyestuffs, some insecticides, pharmaceuticals, and in metallurgy.

Safe Handling and Precautions

Exposure to phosgene causes severe poisoning and is very toxic by inhalation. On skin contact causes skin damage corrosion. Severe ocular irritation and dermal burns may occur following eye or skin exposure in workers accidentally exposed at workplaces. Acute and prolonged exposure to high concentrations phosgene of causes irreversible pulmonary changes of emphysema and fibrosis, bronchoconstriction, hyaline membrane formation, pulmonary oedema, and damage of red blood cells (haemolytic poison). There are no adequate data in experimental animals, suggesting that phosgene is a carcinogen or not. The U.S. EPA has classified phosgene as a Group D compound, meaning phosgene is not classifiable as a human carcinogen.

During use and handling of phosgene, occupational workers must be alert and ensure that the workplace has adequate ventilation facilities. Workers should use suitable chemically resistant protective clothing to protect the skin from liquid splashes, SCBA, goggles to protect eyes and face. Workers should avoid food, drinks, and smoking while handling phosgene. Workers should be alert during the use of phosgene and chemical waste/spill should not be discharged into places of its accumulation and result in accidents. Gas should be scrubbed in alkaline solution under controlled conditions and under supervision to avoid violent reaction.

Workers should return the used and empty phosgene cylinders or phosgene cylinders no longer used back to the compressed gas distributor.

Phosgene Oxime (CAS No. 1794-86-1)

Molecular formula: $CHCl_2NO$

Synonyms: Dichloroformoxime; CX

IUPAC name: Dichloroformaldoxime

Phosgene oxime is a colourless solid or yellowish-brown liquid with a disagreeable penetrating odour. Pure phosgene oxime is a colourless, crystalline solid; the munitions grade compound is a yellowish-brown liquid. Phosgene oxime is soluble in water and organic solvents, but hydrolyses rapidly, and especially in the presence of alkali. Chemically similar to but more reactive than an amide. Incompatible with strong acids and bases, and especially incompatible with strong reducing agents such as hydrides. It is also incompatible with strongly oxidising acids, peroxides, and hydroperoxides. Phosgene oxime is a very severe blistering agent. Both the liquid and the solid can give off vapours at ambient temperatures. Phosgene oxime was developed as a potential chemical warfare agent but has never been known to be used on the battlefield. Phosgene oxime (CX) is an urticant or nettle agent causing instant intolerable pain, erythema, wheals, and urticaria. It is very corrosive, capable of causing extensive tissue damage. Phosgene oxime was first produced by the Germans in 1929 as a possible warfare agent. The mechanism of action is not fully understood but the lesions produced in the skin are similar to those caused by a strong acid. Phosgene oxime will penetrate ordinary clothing and surgical gear.

Safe Handling and Precautions

Phosgene oxime is readily absorbed by the skin causing an immediate corrosive lesion. Ocular and pulmonary exposure has been reported to cause incapacitating inflammation. Exposure and direct contact with phosgene oxime results in immediate pain, irritation, and tissue necrosis. Improper handling and negligence and any kind of contact with the eyes is known to cause severe pain, conjunctivitis, and keratitis. Also, accidental exposure through inhalation causes irritation to the upper respiratory tract, pulmonary oedema, necrotising bronchiolitis, and pulmonary thrombosis. Phosgene oxime is known to cause more severe tissue damage than vesicants and other urticants but it has not been well studied and the mechanism of action is unknown. The effects of phosgene oxime poisoning are very quick and immediate.

Breathing phosgene oxime vapours can cause severe bronchitis and accumulation of fluid in the lungs. Skin contact with phosgene oxime will cause swelling and itching hives that can also result in immediate and painful skin damage. Eye contact may result in severe pain and conjunctivitis. Phosgene oxime is absorbed through the skin and eye; this can also result in pulmonary oedema. Inhaling or directly contacting significant amounts of phosgene oxime can result in death. There are no confirmed reports regarding the carcinogenicity of phosgene oxime either in laboratory animals or humans. Also, the Department of Health and Human Services (DHHS), the IARC, and the U.S. EPA have not classified phosgene oxime for carcinogenicity. Phosgene oxime is one of the least studied chemical warfare agents, and detailed and specific toxicological information is limited.

Phosmet (CAS No. 732-11-6)

Molecular formula: $C_{11}H_{12}NO_4PS_2$
Synonyms and trade names: Decemthion; FOSDAN; Fosmet; Imidan; Imidathion; INOVAT; KEMOLATE; Percolate; PMP; Phthalophos; Safidon; Simidan; Smidan; R-1504
IUPAC name: 2-(Dimethoxyphosphinothioylthiomethyl)isoindoline-1,3-dione

Phosmet is an off-white crystalline solid with an offensive odour. It is used as an insecticide and acaricide. It is very soluble in acetone, xylene, methanol, benzene, toluene, and methyl isobutyl ketone; in kerosene. Phosmet is combustible and susceptible in the presence of strong reducing agents and hydrides. It forms highly toxic and flammable phosphine gas. On partial oxidation by oxidising agents, phosmet releases toxic phosphorus oxides. On heating and/or burning, phosmet decomposes and produces toxic fumes including nitrogen oxides, phosphorous oxides, and sulphur oxides.

Phosmet is a phthalimide-derived non-systemic, organophosphate insecticide. Phosmet has extensive applications on plants for the control of pests on apple trees for control of coddling moth, though it is also used on a wide range of fruit crops, ornamentals, and vines for the control of aphids, suckers. It is also widely used for the control of lepidopterous larvae, aphids, psyllids, fruit flies, and spider mites on pome fruit, stone fruit, citrus fruit, and vines; Colorado beetles on potatoes; boll weevils on cotton; olive moths and olive thrips on olives; blossom beetles on oilseed rape; leaf beetles and weevils on lucerne; European corn borers on maize and sorghum; sweet potato weevils on sweet potatoes in storage. Non-systemic acaricide and insecticide, used on top fruit citrus, grapes, potatoes, and in forestry at rates (0.5–1.0 kg a.i./ha) such that it is safe for a range of predators of mites and therefore useful in integrated control programs. It is also used to control mites and warble fly of cattle.

Safe Handling and Precautions

Phosmet is a cholinesterase inhibitor. Exposure to phosmet at workplaces by accidental ingestion and/or inhalation causes severe poisoning. The symptoms of poisoning include, but not limited to, weakness, pupillary constriction, muscle cramp, excessive salivation, sweating, nausea, dizziness, respiratory distress/laboured breathing, abdominal cramps, vomiting, diarrhoea, effects on the nervous system convulsions, and unconsciousness.

Occupational workers during use and handling of phosmet should strictly observe safety regulations. Workers should use safety goggles or eye protection in combination with breathing protection if powder and avoid working in open flames. Workers should avoid eating, drinking, and smoking during work. Workers should wash hands before eating.

Phosphamidon

Mixture of E and Z Isomers (CAS No. 13171-21-6)

trans-Isomer/E (CAS No. 297-99-4)

cis-Isomer/Z (CAS No. 23783-98-4)

Molecular formula: $C_{10}H_{19}ClNO_5P$
Synonyms and trade names: Aimphon; Dimecron; Kinadon; Phosron; Rilan; Rimdon
IUPAC name: 2-Chloro-2-diethyl-carbamoyl-1-methylvinyl-dimethylphosphate
Toxicity class: USEPA: I; WHO: Ia.

Types of phosphamidon formulations include soluble liquid, suspension concentrate and emulsifiable concentrate, ULV liquid, and 10% granules

Phosphamidon is a pale yellow to colourless oily liquid with a faint odour. It is miscible with water and is soluble in aromatic hydrocarbons. Technical phosphamidon is a pale yellow to colourless oily liquid with a faint odour. It consists of a mixture of (Z)-isomer and (E)-isomer in the approximate proportion of 70:30. It decomposes on heating and releases highly toxic fumes such as phosphorus oxides, hydrogen chloride, and nitrogen oxides. Phosphamidon reacts and gets rapidly hydrolysed by alkalis and decomposes on heating or on burning, producing highly toxic fumes. It attacks metals such as iron, tin, and aluminium. It should be handled by trained personnel wearing protective clothing. It is used as a broad-spectrum insecticide and acaricide for the control of pests and vectors on crops like sugarcane, rice, citrus orchards, and cotton. Occupational exposures to phosphamidon occur among factory workers involved in synthesising formulation, dispensing spray operations. Human exposures also occur among crop harvesters and in vector control operations.

Safe Handling and Precautions

Exposure to phosphamidon has been reported to cause adverse health effects. It is readily absorbed from the gastrointestinal tract, through the intact skin and by inhalation of spray mists and dusts. Prolonged exposures to phosphamidon cause adverse effects and impairment on the respiratory, myocardial, and neuromuscular transmission in animals and humans. The symptoms of poisoning include, but not limited to, localised sweating and involuntary muscle contractions. Eye contact will cause pain, bleeding, tears, pupil constriction and blurred vision, nausea, vomiting, diarrhoea, abdominal cramps, headache, dizziness, eye pain, blurred vision, constriction or dilation of the pupils, tears, salivation, sweating, and confusion. Prolonged period of exposures to phosphamidon causes incoordination, slurred speech, loss of reflexes, weakness, fatigue, involuntary muscle contractions, twitching, tremors of the tongue or eyelids, and eventually paralysis of the body extremities and the respiratory muscles, involuntary defecation or urination, psychosis, irregular heart beat, unconsciousness, convulsions, coma, respiratory failure or cardiac arrest leading to death. Phosphamidon has caused clastogenic effects in bone marrow cells of rats and mice. Exposure to phosphamidon, the organophosphorus pesticide, is known to cause and can have local effects on the smooth muscles of the eyes causing early miosis and blurred vision due to spasm of accommodation, and also conjunctivitis and keratitis. The secretory glands of the respiratory tract, as well as the smooth muscles of the eyes, may be affected by minimal inhalation exposure to the organophosphates leading to watery nasal discharge and hyperemia. Acute rhinitis and pharyngitis can also occur. Exposure to phosphamidon is reported to cause similar clinical effects with respect to muscarinic, nicotinic, and CNS effects like several other organophosphorus compounds. While reports are available on the acute toxicity of phosphamidon, the chronic toxicity and the developmental toxicity of the insecticide is unknown. The IARC, and the (U.S. NTP), have not listed phosphamidon as a human carcinogen while the U.S. EPA has listed the chemical as a possible carcinogen and listed as Group C. However, the studies are found to be inadequate to arrive at meaningful conclusions about phosphamidon as a human carcinogen and no data are available. The World Health Organization (WHO) recommends that for the health and welfare of users and occupational workers and the general population, handling and application of phosphamidon should be entrusted only to competently supervised and well-trained applicators, who must follow adequate safety measures and use the chemical according to good application practices. Regularly exposed workers should receive appropriate monitoring and health evaluation. Also, use and handling of phosphamidon is restricted in the United States but is permitted for use only by certified applicators. The ADI for phosphamidon has been reported as 0.0005 mg/kg body weight. (ADI is an estimate of the amount of a pesticide, expressed on a body weight basis, that can be ingested daily over a lifetime without appreciable health risk.)

Phosphine (Hydrogen Phosphide) (CAS No. 7803-51-2)

Molecular formula: PH_3

Synonyms and trade names: Hydrogen phosphide; Phosphamine; Phosphene; Phosphorus hydride; Phosphorus trihydride; Phosphorated hydrogen; Celphos; Detia gas ex-B; PH_3

Phosphine is a pyrophoric chemical and spontaneously flammable in air. It is incompatible with strong oxidising agents, halogens, nitric acid. It has the odour of garlic or decaying fish. It is slightly soluble in water. It is flammable and is an explosive gas at ambient temperature. Phosphine decomposes on heating or on burning producing toxic fumes including phosphorus oxides. It reacts violently with air, oxygen, oxidants such as chlorine and nitrogen oxides, metal nitrates, halogens, and other toxic substances, and causes fire and explosion hazard.

Safe Handling and Precautions

Exposure to phosphine at workplaces and by accidents causes toxicity and poisoning. The early symptoms of acute phosphine poisoning include, but not limited to, pain in the diaphragm, cough, nausea, burning sensation, diarrhoea and in higher concentrations cause abdominal pain, headache, dizziness, ataxia, chest pain, tremors, breath, vomiting, anaemia, irritation of the respiratory tract, bronchitis, pulmonary oedema, convulsions, and fatal injury.

Repeated exposure of phosphine causes severe spontaneous fractures of bones, effects on the CNS, cardiovascular system, heart, gastrointestinal tract, speech and motor disturbances, liver and kidneys and related disorders leading to unconsciousness or death. Workers should handle phosphine as a pyrophoric material. During handling of phosphine, workers should avoid open flames, sparks, smoking, and/or any kind of contact with hot surfaces. Users and occupational workers during handling of phosphine should be careful and must use PPE, safety goggles for eye protection and breathing protection.

User and occupational workers during handling of phosphine should be aware of the health effects and regulations set by OSHA for phosphine. The OSHA have set a limit of 0.3 parts of phosphine per million parts of workroom air (0.3 ppm) for an 8 h work shift, 40 h work week (Table 17.1).

TABLE 17.1

Health Effects of Phosphine

Concentration (ppm)	Symptoms of Poisoning
2000	Lethal after 1–3 min
500	Fatal
35	Diarrhoea, nausea, respiratory distress
1	OSHA short-term exposure limit (STEL)
0.3	OSHA permissible exposure limit (PEL)

Source: American Industrial Hygiene Association, Falls Church, VA.

Phosphoric Acid (CAS No. 7664-38-2)

Molecular formula: H_3PO_4
Synonyms and trade names: Orthophosphoric acid

Phosphoric acid is a colourless, odourless chemical substance.

Phosphoric acid on combustion forms toxic fumes (phosphorous oxides) and decomposes on contact with alcohols, aldehydes, cyanides, ketones, phenols, esters, sulphides, and halogenated organics producing toxic fumes. It violently polymerises under the influence of azo compounds and epoxides attacks many metals forming flammable/explosive gas. Phosphoric acid violently polymerises under the influence of azo compounds and epoxides, reacting violently with bases. It has wide-scale application in industries, in the manufacture of superphosphates for fertilisers, phosphate salts, polyphosphates, and detergents, as a catalyst in ethylene manufacture and hydrogen peroxide purification as a flavour, acidulant, synergistic anti-oxidant, and sequestrant in food. In dental cements, in process engraving, metal rustproofing, latex coagulation, analytical reagent, and as a veterinary product in the treatment of lead poisoning. Food-grade phosphoric acid is used to acidify foods and beverages. It provides a tangy or sour taste and, being a mass-produced chemical, is available cheaply and in large quantities. Phosphoric acid, used in many soft drinks, has been linked to lower bone density in epidemiological studies. In brief, phosphoric acid is a strong acid and common industrial chemical used in the manufacture of a wide number of products, notably porcelain and metal cleaners, detergents, and fertilisers. It is also used as a food additive and is a major constituent of many soft drinks. Low phosphate concentrations are found in drinking water to which it is added in some areas in order to reduce lead solubility.

Safe Handling and Precautions

Exposure to phosphoric acid by accidental workplace ingestion causes abdominal pain, burning sensation, shock, or collapse. Similarly, inhalation causes burning sensation, cough, shortness of breath, and sore throat. Accidental workplace contact causes skin redness, pain, burns, and skin blisters. Workers during handling of phosphoric acid and chemical wastes should strictly observe safety regulations. Workers should completely avoid pouring water on the waste chemical spills. Workers must use protection suit/clothing combination with breathing protection and gloves during handling of phosphoric acid. Workers should sweep spilled chemical waste into covered containers, carefully collect, label the waste material, and remove to safe place.

Phosphorus: White Phosphorus (CAS No. 7723-14-0)

Molecular formula: P

White or yellow white phosphorus is a yellow waxy or colourless, transparent, volatile crystalline solid, waxy appearance with a garlic-like odour. On exposure to light, it darkens and ignites in air. It is also called yellow phosphorus colour because of impurities. White phosphorus does not occur naturally but is manufactured from phosphate rocks. It is insoluble in water, slightly soluble in benzene, ethanol, and chloroform, and is soluble in carbon disulphide. White phosphorus reacts rapidly with oxygen, easily catching fire at temperatures 10°C–15°C above room temperature. White phosphorus is used by the military in various types of ammunition and to produce smoke for concealing troop movements and identifying targets. It is also used by industry to produce phosphoric acid and other chemicals for use in fertilisers, food additives, and cleaning compounds. Small amounts of white phosphorus were used in the past in pesticides and fireworks. Phosphorus is generally stored under water. White phosphorus is used mainly for producing phosphoric acid and other chemicals. These chemicals are used to make fertilisers, additives in foods and drinks, cleaning compounds, and other products. In the military, white phosphorus is used in ammunitions such as mortar and artillery shells, and grenades. When ammunitions containing white phosphorus are fired in the field, they burn and produce smoke. White phosphorus enters the environment when industries make it or use it to make other chemicals and when the military uses it as ammunition. It also enters the environment from spills during storage.

Safe Handling and Precautions

Exposure to white phosphorus causes harmful effects and health disorders. Breathing vapours of white phosphorus have been known to cause cough or result in a health condition known as phossy jaw and is associated with poor wound healing in the mouth and deleterious effect to the jaw bone. Accidental direct contact of the material at workplaces causes burns and irritation, adverse effects to liver, kidney, heart, lung, or bone damage, and fatal injury. Exposure to fumes of white phosphorus has been reported to cause severe ocular irritation with blepharospasm, photophobia, and lacrimation. The particles and fume are caustic and seriously damage on contact with tissues. The cornea turns opaque associated with changes and development of interstitial vascularisation and episcleritis. Also, improper use, handling, and intentional ingestion of white phosphorus caused severe renal effects among individuals with symptoms of proteinuria, albuminuria, acetonuria, increased urobilinogen, oliguria, increased blood levels of urea and/or nitrogen, and increased levels of blood creatinine. There are no published data on the carcinogenic effects of white phosphorus in humans or animals. The U.S. EPA has listed white phosphorus as Group D, meaning not classifiable as a human carcinogen. The NIOSH, the OSHA, and (other recommendations) the ACGIH have all set the inhalation exposure limit for white phosphorus in the workplace during an 8 h workday at 0.1 mg of white phosphorus per cubic meter of air ($0.1 \ mg/m^3$).

Regulations: The U.S. EPA has listed white phosphorus as a hazardous air pollutant.

Phosphorus Pentachloride (CAS No. 10026-13-8)

Molecular formula: PCl_5
Synonyms and trade names: Pentachloro-; Phosphorus perchloride; Phosphoric chloride

Phosphorus pentachloride is a white to pale yellow, crystalline solid with a pungent odour. It is used in the manufacture of other chemicals, in aluminium metallurgy, and in the pharmaceutical industry. On heating it undergoes decomposition. Phosphorus pentachloride emits highly toxic fumes of chlorides and chlorine. It will react with water or steam to produce heat and toxic and corrosive fumes. It reacts violently with moisture, chlorine trioxide, fluorine hydroxylamine, magnesium oxides, diphosphorus trioxide, sodium and potassium. It is decomposed by water to form hydrogen chloride, phosphoric acids, corrosive materials, and heat. It is incompatible with chemically active metals such as sodium and potassium, alkalis, aluminium, chlorine dioxide, chlorine, diphosphorus trioxide, fluorine, hydroxylamine, magnesium oxide, 3'-methyl-2-nitrobenzanilide, nitrobenzene, sodium, urea, water.

Safe Handling and Precautions

Exposures to phosphorus pentachloride result in adverse health effects with symptoms of inflammation of the eye, redness, watering/tears, itching, skin inflammation, itching, scaling, reddening, and blisters. Inhalation of dust of phosphorus pentachloride produces irritation to gastrointestinal and the respiratory tract, burning sensation, sneezing and coughing, sore throat, cough, burning sensation, abdominal pain, shortness of breath, respiratory distress/laboured breathing. Accidental workplace ingestion and inhalation of phosphorus pentachloride has been reported to cause lung oedema, shock and/or collapse, and death due to pulmonary oedema or by circulatory shock. Severe over exposure to phosphorus pentachloride also produces lung damage, choking, unconsciousness, and fatal injury to workers.

Phosphorus Pentoxide (CAS No. 1314-56-3)

Molecular formula: P_2O_5
Synonyms and trade names: Diphosphorus pentoxide; Phosphoric anhydride; Phosphorus pentaoxide

Phosphorus pentaoxide appears as white, very deliquescent crystal or powder with pungent, sharp, irritating odour. It is not combustible but enhances combustion of other substances. Phosphorus pentaoxide in granular/powder form reacts violently with water. The products of combustion include compounds of hydrogen, phosphorus, and oxygen and toxic fumes. Phosphorus pentaoxide is stable under ordinary conditions of use and storage but reacts violently on contact with water. It is incompatible with ammonia, calcium oxide, chlorine trifluoride, hydrogen fluoride, oxygen difluoride, perchloric acid, perchloric acid and chloroform, potassium, propargyl alcohol, sodium, sodium carbonate, sodium hydroxide, peroxides, magnesium, alcohols, metals, oxidising agents, water, and a mixture of water and organic material. It also forms an explosive mixture with methyl hydroperoxide. Phosphorus pentaoxide reacts with moisture on body tissue surfaces to form phosphoric acid, which approximates sulphuric acid and hydrochloric acids in corrosive intensity. It is a usual material and reagent in chemical industry; this product is widely used in the industries of medicine, coating auxiliaries, printing and dyeing auxiliaries, anti-static additive, titanate coupling agent, phosphorus oxychloride. It is used as a strong dehydrating agent, capable even of dehydrating concentrated sulphuric acid into sulphur trioxide.

Safe Handling and Precautions

Exposures through inhalation, phosphorus pentaoxide produces damaging effects on the mucous membranes and upper respiratory tract. Symptoms of phosphorus pentaoxide poisoning include, but not limited to, irritation of the nose and throat, and laboured breathing. Phosphorus pentaoxide as airborne dust particles are harmful and corrosive to respiratory tract and its contact with skin causes severe irritation, chemical burns, painful burns to eyes, and corrosive effects to the eyes. When inhaled accidentally at workplaces, it causes severe damage to the tissue of the mucous membranes and upper respiratory tract. Breathing airborne particles or dust of phosphorus pentaoxide during mixing, spraying, sanding, grinding, etc., may cause severe irritation to respiratory tract. Accidental ingestion at workplaces causes immediate burning pain in the mouth, throat, abdominal pain, severe swelling of the larynx, disturb the ability to breathe, chronic bronchitis, eye injuries, circulatory shock, and convulsions. Phosphorus pentaoxide has not been listed as a human carcinogen by the NIOSH, ACGIH, and OSHA. Phosphorus pentaoxide is classified as hazardous under Federal OSHA regulation.

Occupational workers should keep phosphorus pentaoxide in closed containers and protect it from extreme temperatures in a cool, dry area. Workers should keep stored phosphorus pentaoxide with a label – CAUTION – Material is corrosive. Workers should keep phosphorus pentaoxide and other chemical substances out of reach of children. During use and handling of phosphorus pentaoxide, occupational workers should wear appropriate standard quality and approved chemical-resistant protective clothing and protective gloves and approved SCBA/respirator, cover all workplace dress/apron to prevent contact/minimise skin contact. Workplace should have safety eyewash station also nearby for immediate use. Workers should dispose of the chemical waste in accordance with local, state, and federal regulations. Workers should avoid breathing dust, exposures of the face/eyes to the chemical dust, and contact with water (while using the chemical). The workplace should have proper and adequate ventilation. Workers should not be negligent to wash thoroughly after handling the chemical substance. Users and occupational workers should strictly observe set regulations in handling and disposal of chemical wastes. Workers should first dilute phosphorus pentaoxide waste in a large excess of water, carefully neutralise with soda ash, and subsequently wash drain with copious plenty of water.

Phosphorus pentaoxide DANGER! CORROSIVE. Causes burns to any area of skin contact, respiratory tract burns, may cause blindness. Harmful if swallowed or inhaled. Fumes cause irritation to eyes and respiratory tract, water reactive, and reacts violently with water to generate heat and phosphoric acid.

Phosphorus Pentasulphide (CAS No. 1314-80-3)

Molecular formula: P_2S_5
Synonyms and trade names: Diphosphorus pentasulphide; Phosphorus sulphide; Phosphorus pentasulphide; Phosphorus persulphide; Thiophosphoric anhydride

Phosphorus pentasulphide appears as grey to yellow-green crystalline/solid with an odour of rotten eggs. It is a highly flammable solid and a water-reactive chemical substance. It is used for making lube oil additives, insecticides, floatation agents, safety matches, blown asphalt, and other products and chemicals. Phosphorus pentasulphide gets spontaneously heated and ignite in presence of moisture. It reaction with water and emits toxic hydrogen sulphide gas and phosphoric acid. Phosphorus pentasulphide is very dangerous when wet and reacts vigorously with strong oxidants.

Safe Handling and Precautions

Exposures to phosphorus pentoxide with workplace negligence have been reported to cause adverse health effects. The symptoms of health disorders and poisoning include, but not limited to, headaches, fatigue, irritability, eye irritation, pain, conjunctivitis, kerato-conjunctivitis, corneal vesiculation, lacrimation, sweating, nausea, vomiting, respiratory system irritation, respiratory distress, dizziness/suffocation, and burning feeling of the mucous surfaces. Exposure to higher concentrations causes pulmonary oedema, seizures, convulsions, and death.

Phthalic Anhydride (CAS No. 85-44-9)

Molecular formula: $C_8H_4O_3$
Synonyms and trade names: 1,2-Benzenedicarboxylic acid anhydride; 1,3-Dihydro-1,3-dioxoisobenzofurandione; 1,3-Phthalandion; Phthalandione; 2-Benzofuran-1,3-dione

Phthalic anhydride is a white lustrous needle-like solid, phthalic anhydride is slightly soluble in water.

Phthalic anhydride is an important chemical intermediate in the plastics industry. It has extensive industrial applications in the production of phthalic plasticisers, alkydic resins, polyesters resins, and synthetic resins. It is also used in dyes phenolphthalein, PVC stabilisers, drying agents for paints and aro. Phthalic anhydride itself is used as a monomer for synthetic resins such as glyptal, the alkyd resins, and the polyester resins. It is also used to make unsaturated polyesters that are used to manufacture fibreglass-reinforced plastics, halogenated anhydrides used as fire retardants; polyester polyols for urethanes, phthalocyanine pigments; dyes, perfumes, pharmaceuticals, tanning and curing agents, as solvents, as insect repellents, and various chemical intermedi-ates. Phthalic anhydride is released to the environment from chemical plants, mainly those that manufacture the chemical or use it in the production of plastics and resins. Phthalic anhydride is used in the synthesis of primary amines, the agricultural fungi-cide phaltan, and thalidomide. Workers also get exposed to phthalic anhydride particu-larly during the manufacture of phthalate-derived products as and when they handle the chemicals with negligence. Also, exposure to phthalic anhydride may occur from the use of plastics from which phthalate plasticisers are leached, specifically certain medical plastics such as blood bags, plastic syringes, plastic tubing, and as environ-mental pollutants.

Safe Handling and Precautions

On exposure to phthalic anhydride, occupational workers suffer with adverse health effects. The symptoms of adverse effects and poisoning include burns to skin, eyes, and gastrointestinal tract.

Acute exposure to phthalic anhydride causes irritation of the eyes, skin, and respiratory tract and lung sensitisation, allergic rhinitis, and asthma in humans. Reports have indi-cated that phthalic anhydride did not cause any effect on the dry skin, but caused burns on wet skin. Occupational workers exposed to mixtures of phthalic anhydride and phthalic acid developed conjunctivitis, bloody nasal discharge/nasal ulcer bleeding, rhinitis, bronchitis, atrophy of the nasal mucosa, hoarseness, cough, occasional bloody sputum, emphysema, and signs of CNS excitation. There are no published reports or information on the carcinogenicity of phthalic anhydride in humans. There is no published information on the carcinogenic effects of phthalic anhydride in humans and the U.S. EPA has not classified phthalic anhydride as a human carcinogen.

Picloram (CAS No. 1918-02-1)

Molecular formula: $C_6H_3Cl_3N_2O_2$
Synonyms and trade names: Access; Grazon; Pathway; Tordon
Chemical name: 4-Amino-3,5,6-trichloropyridine-2-carboxylic acid

Picloram is a colourless crystal. It is very soluble in acetone, ethanol, benzene, and dichloromethane. It is a systemic herbicide used for general woody plant control, sold under the trade names Tordon and Grazon. It also controls a wide range of broad-leaved weeds, but most grasses are resistant. It is used in formulations with other herbicides such as bromoxynil, diuron, 2,4-D, MCPA, triclorpyr, and atrazine. It is also compatible with fertilisers. Picloram, in the pyridine family of compounds, is a systemic herbicide used for control of woody plants and a wide range of broad-leaved weeds. Most grasses are resistant to picloram, so it is used in range management programs. Picloram is formulated either as an acid (technical product), a potassium or triisopropanolamine salt, or an isooctyl ester, and is available as either soluble concentrates, pellets, or granular formulations. The materials in this document refer to the technical acid form unless otherwise indicated. Picloram is stable under acidic, neutral and basic conditions. Picloram is formulated either as an acid (technical product), a potassium or triisopropanolamine salt, or an isooctyl ester, and is available as either soluble concentrates, pellets, or granular formulations and related manufacturing impurities.

Safe Handling and Precautions

Picloram is of moderate toxicity to the eyes and only mildly toxic on the skin. Absorption of picloram through the gastrointestinal tract is rapid and almost complete. Laboratory animals exposed to picloram indicated that the principal target organ is the liver; effects on the kidney occur at higher doses. In short-term and subchronic feeding studies on rats, dose-related elevations in liver/body weight ratios and histopathological changes (slight centrilobular hypertrophy) were observed. No-observed-adverse-effect levels (NOAELs) for effects on the liver were 200 and 50 mg/kg bw/day in 14 and 90 day studies, respectively. Based on studies, picloram has been classified as Group E, meaning no evidence as a human carcinogen.

Regulations: Picloram is a slightly toxic compound in US EPA toxicity class III. Products containing it must bear the Signal Word CAUTION on the label. All products except for Tordon RTU and Pathway are RUPs. RUPs may be purchased and used only by certified applicators. Further, all picloram products are classified as RUPs based on hazard to non-target plants, and may be applied only by or under the direct supervision of certified applicators.

Picric Acid (CAS No. 88-89-1)

Molecular formula: $C_6H_2(NO_2)_3OH$
Synonyms and trade names: Picronitric acid; 2,4,6-Trinitrophenol; Trinitrophenol

Picric acid is a white to yellowish crystalline substance, soluble in most organic solvents and highly flammable. Picric acid is a derivative of phenol. It reacts with metals to form metal picrates, which like picric acid itself are highly sensitive. It is often used for tissue fixative (Bouin solution) for histology specimens, as a booster to detonate another, less sensitive explosive, such as trinitotoluene (TNT). It is used in the manufacture of fireworks, matches, electric batteries, coloured glass, dyes, antiseptics, explosives, disinfectants, leather industries, pharmaceutical, and textile. Picric acid is also used as a yellow dye, as an antiseptic, and in the synthesis of chloropicrin, or nitro-trichloromethane. Picric acids are highly sensitive to heat, shock, or friction and because of the explosive nature it is among the most hazardous substances found in the laboratory.

Safe Handling and Precautions

Picric acid is a poisonous chemical substance. Exposures to picric acid cause different adverse effects on the skin of animals and humans like allergies, dermatitis, irritation, and sensitisation. Picric acid is harmful by accidental workplace ingestion, swallowing, if inhaled, or direct skin contact and skin absorption. It produces respiratory tract irritation, causes damage of the liver, kidneys, and blood. It reacts with proteins in the skin to give a dark brown colour that may last as long as a month. Workers on exposures to picric acid suffer from symptoms of headache, dizziness, nausea, vomiting, abdominal pain and diarrhoea, coughing, shortness of breath, irritation to the respiratory tract, and more severe and heavy exposures cause destruction of red blood cells and bloody urine, liver and kidney damage, convulsions, weakness, muscle pain, coma, and death. Prolonged or repeated eye contact with picric acid is known to cause conjunctivitis and dermatitis and repeated inhalation of picric acid at workplaces causes yellowing of the teeth and skin. Accidental workplace contact of picric acid with the eyes results in corneal damage or blindness and skin contact produces inflammation and blistering. Inhalation of picric acid dust is known to produce irritation to gastrointestinal or respiratory tract, characterised by burning, sneezing, and coughing, and severe over exposure produces lung damage. Prolonged or repeated exposures can cause liver, kidney, and blood effects. The colour of the conjunctiva of the eye, hair, and skin becomes yellow. Occupational workers with pre-existing health disorders become more susceptible to the exposure of picric acid and suffer from aggravation of skin, blood, liver, and kidney disorders. Severe and heavy exposures to picric acid cause destruction of red blood cells and bloody urine, liver, and kidney damage, convulsions, weakness, muscle pain, coma, and death. Picric acid has not been listed by IARC, NIOSH, ACGIH, NTP, or by the OSHA as a human carcinogen.

Students and workers must be very alert during use and handling of picric acid. Workers should be careful in handling and management of picric acid and protect the chemical against physical damage. Workers must store picric acid in a cool, dry well-ventilated location, away from incompatibles, metals, copper, lead, and zinc aluminium, ammonia, concrete, plaster, salts, oxidisers, gelatin, and alkaloids. Any kind of contact of picric acid with salts becomes more explosive sensitive than picric acid itself. All work with picric acid should be done in a chemical fume hood to minimise inhalation exposure. Workers must wear proper and adequate PPE (workplace dress, coat with sleeves) eye protection, splash goggles, and neoprene gloves.

Workers should be strictly alert and must not touch dry picric acid or picrate salts or moved under any circumstances. Industrially, picric acid is especially hazardous because it is volatile and slowly sublimes even at room temperature. Over time, the buildup of picrates on exposed metal surfaces can constitute a grave hazard.

Workers should keep stored container of picric acid in a cool, well-ventilated area away from incompatibles such as oxidising agents, reducing agents, metals, alkalis, away from heat. Keep away from sources of ignition, away from direct sunlight. Picric acid should not be allowed to dry out and must be kept soared wetted with a minimum of 30% water.

In brief, users and occupational workers who handle picric acid must be trained in its hazards and in procedures for use and disposal. Dangerous explosion hazard when dry becomes increasingly shock, heat, and friction sensitive as it loses moisture and workers should not touch dry picric acid, avoid to use picric acid when working alone or after working hours, and should not use or store in metal containers. Workers should avoid using metal spatulas to handle solids. Use and store picric acid in containers made of polyethylene, polypropylene, Teflon, or glass. Wet wipe screw closures before sealing to prevent solids formation in threads. Workers should clean very small spills and absorb with wet paper towels, keep wet and collect for disposal. Collect all picric acid-containing wastes in plastic or glass bottles for disposal by the department.

Caution – Danger!! Workers and Workplace Manager should remember ...

Workers should store the least amount of picric acid at the workplace and workers must remember that picric acid should never be allowed to dry out in containers, especially on metal or concrete surfaces. Keep picric acid wet – becomes explosive if picric acid becomes dry. Containers should have LEGIBLE BOLD LABEL AS PICRIC ACID.

- Not to purchase large quantities of picric acid, but only purchase the minimum amount of picric acid for your work.
- When possible, purchase picric acid in solution, not as a dry solid.
- Label all picric acid containers with date received.
- Store solid picric acid or picrate salts in distilled water.
- Should not store picric acid solution or solid in containers with metal caps, and do not use metal spatulas with picric acid solids.
- Students and occupational workers should use and perform all work of picric acid only in a chemical fume hood to minimise inhalation exposure.
- Workers should periodically check the hydration of picric acid every 2 months and add distilled water as necessary.
- All picric acid waste should be properly packaged and clearly labelled for disposal.
- Workers should date all bottles of picric acid with 'Date Received' and 'Date Opened'.
- Workers should keep an accurate and up-to-date chemical inventory of all laboratory spaces to reduce the potential for old and overlooked picric acid.
- During handling of picric acid, students and workers must wear eye protection and/or face shield to prevent possible eye contact. With the use of picric acid, workplace dress/clothing must be changed daily.

Picrotoxin (CAS No. 124-87-8)

Molecular formula: $C_{30}OH_{34}O_{13}$
Synonyms and trade names: Cocculin

Picrotoxin, also known as cocculin, appears as white to light beige crystalline material. It is isolated from *Cocculus indicus* (*Fructus cocculi*), fishberries, or Indian berries. Picrotoxin is a colourless, flexible, shining, prismatic crystals, or a micro-crystalline powder. It is odourless, has a very bitter taste, and is permanent in the air. Picrotoxin is soluble in hot water, readily soluble in strong ammonia water, and in aqueous solutions of sodium hydroxide, soluble in dilute acids, and alkalis as well as in glacial acetic acid, sparingly soluble in chloroform and very slightly soluble in cold water and alcohol. Picrotoxin is stable under normal temperatures and pressures and decompose when exposed to light. Picrotoxin is incompatible with strong oxidising agents, strong acids, strong bases, light.

Safe Handling and Precautions

Picrotoxin is used as a CNS stimulant, antidote, convulsant, and gamma aminobutyric acid (GABA) antagonist. It is a non-competitive antagonist at GABAA receptors and thus a convulsant. It blocks the research tool, and has been used as a CNS stimulant and an antidote in poisoning by CNS depressants, especially barbiturates also used for the treatment of respiratory distress. Exposure to picrotoxin causes eye and skin irritation, respiratory and digestive tract irritation and CNS effects. Laboratory studies have indicated that *C. indicus* when administered to experimental animals caused severe poisoning and CNS disorders. Workers on accidental ingestion/swallowing have been reported to show severe poisoning with symptoms that include, but not limited to, respiratory tract irritation, gastric irritation, giddiness, respiratory tract irritation, irritation of the gastrointestinal tract, behavioural disorders such as excitement, somnolence, tremors, rigidity, convulsions, effects on the CNS, cardiovascular system, peripheral nervous system – peripheral nerve and sensation and coma. The ACGIH, the NIOSH, the IARC, the OSHA, and the NTP have not listed picrotoxin as a human carcinogen.

Regulations: OSHA has included picrotoxin as hazardous under the definition of Hazard Communication Standard (29 CFR 1910.1200).

Pirimiphos Methyl (CAS No. 2 9232-93-7)

Molecular formula: $C_{11}H_{20}N_3O_3PS$
Synonyms and trade names: Actellic; Actellifog; Blex; Pyridimine phosphate; Plant protection; Silosan; Sybol
IUPAC name: O,2-Diethylamino-6-methylpirimidin-4-yl O,O-dimethyl phosphorothioate
WHO classification: Pirimiphos-methyl is class III–Slightly hazardous

Pirimiphos methyl together with related manufacturing impurities and shall be a clear or faintly turbid, mobile, red-brown liquid at temperatures above 18°C. Pirimiphos methyl is slightly volatile, has low solubility in water, and is readily soluble in organic solvents. It is a broad-spectrum organophosphorous insecticide and acaricide, with contact and fumigant action. In plants, It penetrates leaf tissue and exhibits translaminar action but is of short persistence. It is used for controlling a wide range of chewing, sucking, and boring insects and mites in warehouses, stored grain, animal houses, domestic and industrial premises. Pirimiphos methyl is a post-harvest insecticide used on stored corn and sorghum grain and seed, incorporated into cattle ear tags, and used for the fogging treatment of iris bulbs. It is used to control various insects such as mealy bugs and mites (on iris bulbs), horn and face flies (on cattle), and cigarette beetle, confused flour beetle; corn sap beetle; flat grain beetle; hairy fungus beetle, red flour beetle, granary weevil, maize weevil, merchant grain beetle, rice weevil, lesser grain borer, and angoumois grain moth, Indian Meal moth and almond moth on corn and sorghum grain and seed. Organophosphorous insecticides represent one group of pesticides that is widely used and has been known to have toxic effects in human and animals. Reports have indicated that pirimiphos methyl is rapidly absorbed, metabolised, and excreted in species of laboratory animals.

Safe Handling and Precautions

Pirimiphos methyl is used primarily for grain, corn, wheat storage, and occupational workers as sprayers get exposure to the pesticide predominantly through skin absorption or dermal route. Exposure to pirimiphos methyl is harmful and hazardous. Human exposure to low doses of pirimiphos methyl has been reported to cause mild types of health effects. At high doses, pirimiphos methyl, like other organophosphorous insecticides, causes severe inhibition of the activity of the enzyme acetylcholinesterase in the nervous system. The symptoms of poisoning include, but not limited to, nausea, vomiting, abdominal cramps, diarrhoea, dizziness, sweating, extreme weakness, ataxia, blurred vision, twitching, tremor, slow heartbeat, respiratory distress/difficulty breathing, tightness in chest, pulmonary oedema, seizures, respiratory paralysis, and fatal injury. Reports have shown that although pirimiphos methyl is highly toxic to birds and fish, these risks are not of concern based on the use pattern of pirimiphos methyl. Further, studies on occupationally exposed workers and human volunteers did not indicate any kind of poisoning attributed to pirimiphos methyl.

Polybrominated Diphenyl Ethers (PBDEs)

Polybrominated diphenyl ethers (PBDEs) are flame-retardant chemicals, meaning they resist catching fire. PBDEs are added to plastics and foam to make them hard to burn. These chemicals can leave these products and enter the environment. There are different kinds of PBDEs. Some PBDE mixtures, called decaBDEs, are made in many places around the world. Exposure to polybrominated diphenyl ethers has been reported to occur through indoor dust—considered to be one of the primary sources of human exposure.

Breathing contaminated air containing PBDEs has been found to be one of the possible reasons of human health disorders attributable to PBDEs. The general population is exposed to very low levels as PBDEs in the air. Indoor concentrations can lead to a slightly higher exposure, especially in areas with computer equipment, television, and other electronic equipment. The concentrations that people are exposed to during daily life are typically well below what may cause health problems. Nursing babies can be exposed to PBDEs through breast milk. PBDEs can be found in water but this is unusual.

Safe Handling and Precautions

PBDE is a common flame retardant used to reduce the risk of fire in a wide variety of products, such as children's pajamas and your computer. PBDEs are excellent flame retardants, but the chemicals have been accumulating in the environment and in human bodies. Relatively recent reports have indicated that exposure to low concentrations of these chemicals may result in irreparable damage to the nervous and reproductive systems.

There is no definite information on health effects of PBDEs in people. Studies were done on rats and mice. These animals ate food with moderate amounts of PBDEs for a few days. Effects occurred in the thyroid gland. There is limited evidence of adverse effects of PBDEs in humans. Studies conducted in New York and the Netherlands have measured PBDEs in the bodies of pregnant mothers and/or in the umbilical cord blood at birth and then followed the children as they matured. Higher PBDEs levels in mothers have been associated with lower measures of intelligence, attention, and fine motor skills in their children. Higher PBDEs in mothers were also associated with longer time to become pregnant and lower thyroid hormones during pregnancy. Based on the safety data of experimental animals, the U.S. EPA classified decabromodiphenyl ether as a possible human carcinogen. More studies are needed to determine the sources and pathways of PBDE exposures and whether these exposures have adverse effects on human health.

Regulations: Deca-BDE is prohibited in televisions, computers, and residential upholstered furniture. Before this prohibition could take effect, the law required the Department of Ecology and the Department of Health to identify a safer and technically feasible alternative that met fire safety standards. This was accomplished and published in the January 2009 report.

Report: Environmental Health, Safety and Toxicology. Washington State Department of Health, Olympia, WA.

Polybrominated Biphenyls

There are no known natural sources of polybrominated biphenyls (PBBs) in the environment. PBB are solid and colourless to off-white chemical substances. They are a class of structurally similar brominated hydrocarbons in which 2–10 bromine atoms are attached to the biphenyl molecule. PBBs are chemicals that were added to plastics used in a variety of consumer products, such as computer monitors, televisions, textiles, and plastic foams, to make them difficult to burn. Because PBBs were mixed into plastics rather than bound to them, they were able to leave the plastic and find their way into the environment. PBBs are no longer manufactured in North America, but very small amounts of PBBs may be released into the environment from poorly maintained hazardous waste sites and improper incineration of plastics that contain PBBs. Occupational exposure to PBBs may be primarily by breathing air that contains PBBs.

Because PBBs were mixed into plastics rather than bound to them, they were able to leave the plastic and find their way into the environment. Commercial production of PBBs began in the 1970s. Manufacture of PBBs was discontinued in the United States in 1976.

Concern regarding PBBs is mainly related to exposures resulting from an agriculture contamination episode that occurred in Michigan over a 10 month period during 1973–1974.

There are no known natural sources of PBBs in the environment. PBBs are solids and are colourless to off-white. PBBs enter the environment as mixtures containing a variety of individual brominated biphenyl (for PBBs) components, known as congeners. Some commercial PBB mixtures are known in the United States under the industrial trade name, FireMaster®. However, other flame-retardant chemicals also may be identified by this name. PBBs are no longer used in North America because the agriculture contamination episode that occurred in Michigan in 1973–1974 led to the cessation of its production.

Safe Handling and Precautions

Studies in experimental animals indicated that exposures to large amounts of PBBs for a short period or to smaller amounts over a longer period caused severe health effects and poisoning. The symptoms included weight loss, skin disorders, and effects on the CNS system, immune system, liver, kidney, and thyroid glands. Reports have been published indicating that exposure of laboratory animals to polybrominated diphenyl ethers (PBDEs) in the womb and through nursing has caused thyroid effects and neurobehavioural alterations in newborn animals, but did not cause any kind of birth defects in animals. However, there are no reports to confirm whether or not exposure to PBDEs causes birth defect in children.

The IARC has classified PBBs as the possible carcinogens.

Polychlorinated Biphenyls (CAS No. 1336-36-3)

Aroclor 1232 (CAS No. 11141-16-5); Aroclor 1242 (CAS No. 53469-21-9)

Aroclor 1248 (CAS No. 12672-29-6); Aroclor 1254 (CAS No. 11097-69-1)

Aroclor 1260 (CAS No. 11096-82-5); Aroclor 1262 (CAS No. 37324-23-5)

Aroclor 1268 (CAS No. 11100-14-4)

Synonyms and trade names: Aroclor; Aroclor 1221; Aroclor 1242; Aroclor 1254; Biphenyl, polychloro-; Chlorinated biphenyl; Chlorinated diphenylene; Clophen; PCBs; Polychlorinated biphenyls; Pyranol

Polychlorinated biphenyls (PCBs) describe a group of synthetic chlorinated organic chemicals.

PCB are a group of 209 different chemicals that share a common structure but vary in the number of attached chlorine atoms. They are used in capacitors and transformers because they combine dielectric properties with chemical stability and fire resistance. Approximately twice as many pounds of PCBs are used in the manufacture of capacitors as in the manufacture of transformers. Much before the global environmental concern regarding the PCBs persistence and ubiquitousness, these chemical compounds were of extensive industrial use as fluids for heat transfer systems, hydraulic systems, gas turbines, and vacuum pumps, as fire retardants, and as plasticisers in adhesives, textiles, surface coatings, sealants, printing, and carbonless copy paper.

In fact, much earlier and around 1970s application of PCBs got restricted both for its domestic and industrial applications in capacitors and transformers. PCBs are a group of chemicals that have extremely high boiling points and are practically non-flammable. Polychlorinated biphenyls are mixtures of up to 209 individual chlorinated compounds (known as congeners). There are no known natural sources of PCBs. PCBs are either oily liquids or solids that are colorless to light yellow. Some PCBs can exist as a vapor in air. PCBs have no known smell or taste. Many commercial PCB mixtures are known in the U.S. by the trade name Aroclor.

PCBs are well known as persistent organic pollutants and as such have entered the environment because of improper management of chemical wastes over the years. In recent years, global concern has emerged regarding hazardous chemical substances and human health. The environmental transport of PCBs is complex and is nearly global in scale.

PCBs get released into the general environment from poorly maintained toxic waste sites; by illegal or improper dumping of polychlorinated biphenyls wastes, such as transformer fluids; through leaks or fugitive emissions from electrical transformers containing polychlorinated biphenyls. PCBs were in use earlier as coolants and as insulating fluids (transformer oil) and for capacitors used in old fluorescent light ballasts. Earlier, PCBs were also used in paints and cements as plasticisers and as stabilising additives in flexible PVC coatings of electrical wiring and electronic components. They were also used as flame retardants, lubricating oils, hydraulic fluids, in building construction works, wood floor finishes, and in surgical implants.

Safe Handling and Precautions

It is now well known that the major and potential routes of human exposure to polychlorinated biphenyls are the three normal routes – ingestion, inhalation, and through skin absorption and/or dermal contact. The release of PCBs from prior industrial uses and their persistence in the environment have resulted in widespread contamination of water and soil. A major source of human exposure to PCBs is dietary. Because polychlorinated biphenyls are soluble in fats and oils, the major U.S. commodities in which PCBs have been found are fish, cheese, eggs, and animal feed. Residues of PCBs have been detected in human milk and fat samples collected from the general U.S. population. PCBs undergo poor metabolic degradation and tend to accumulate in animal tissues, including humans. The accumulation particularly in tissues and organs rich in lipids appears to be higher in the case of penta and more highly chlorinated biphenyls.

During a study conducted in 1978, the average daily human intake of PCBs through food had been estimated as 0.027 µg/kg, which by 1991 was shown to decline to less than 0.001 µg/kg. The PCBs have been frequently identified at relatively high concentrations in the blood, fat, and milk of native populations living in Arctic regions, whose diet is high in fish and marine animals. For example, the mean concentration of PCBs in fat tissue collected from a native population in Greenland was 5719 µg/kg of lipid, and the concentrations were highest in older individuals. PCBs also found accumulated in the breast milk of women in this population.

People exposed directly to high levels of PCBs, either via the skin, by consumption, or in the air, have experienced irritation of the nose and lungs, skin irritations such as severe acne (chloracne) and rashes, and eye problems. Reports have indicated that the known adverse health effects among occupational workers and common public exposed to PCBs include skin lesions, chloracne, brown pigmentation of the skin and nails, distinctive hair follicles, increased eye discharge (lacrimation), swelling of eyelids, transient visual disturbance, and systemic gastrointestinal symptoms, jaundice and distinctive hair follicles. Also, studies/surveys have indicated that women exposed to PCBs before and/or during pregnancy are proned to give birth to children with significant neurological and motor control problems, including lowered IQ and poor short-term memory.

Studies of PCBs in humans have found increased rates of melanomas, liver cancer, gall bladder cancer, biliary tract cancer, gastrointestinal tract cancer, and brain cancer, and may be linked to breast cancer. PCBs are known to cause a variety of types of cancer in rats, mice, and other study animals. Reports have indicated that potential health effects after long-term exposure to polychlorinated biphenyls above the MCL cause health disorders such as, but not limited to, skin changes, thymus gland problems, immune deficiencies, reproductive or nervous system difficulties, and increased risk of cancer. The IARC concluded that polychlorinated biphenyls (PCB 126 (3,3′,4,4′,5-penta chlorobiphenyl)) was a complete carcinogen in experimental animals. Based on extensive evidence reports indicate that polybrominated biphenyls (PBBs) acts through the same aryl-hydrocarbonreceptor-mediated mechanism as 2,3,7,8-tetra-chlorodibenzo-p-dioxin (TCDD, or dioxin). The IARC classified PCBs as carcinogenic in humans. Oral exposure to PCBs caused liver tumours in mice and rats. The most commonly observed health effects in people exposed to large amounts of PCBs are skin conditions such as acne and rashes. Studies in exposed workers have shown changes in blood and urine that may indicate liver damage. PCBs are reasonably anticipated to be human carcinogens based on sufficient evidence of carcinogenicity from studies in experimental animals. Reports have also indicated that not all PCB mixtures cause tumours in experimental animals.

Precautions: Workers should keep stored polybrominated biphenyls in tightly closed containers in a cool, well-ventilated area away from strong oxidisers, such as chlorine, bromine, and fluorine. A regulated marked area with label should be selected for safe storage. Store in a cool, dry, well-ventilated location.

Polycyclic Aromatic Hydrocarbons

Polycyclic aromatic hydrocarbons (PAHs) are a group of over 100 different chemicals that are formed during the incomplete burning of coal, oil and gas, garbage, or other organic substances like tobacco or charbroiled meat. PAHs are usually found as a mixture containing two or more of these compounds, such as soot. Some PAHs are manufactured. These pure PAHs usually exist as colourless, white, or pale yellow-green solids. PAHs are found in coal tar, crude oil, creosote, and roofing tar, but a few are used in medicines or to make dyes, plastics, and pesticides. Reports have indicated that human exposures to PAHs mostly occur through inhalation/breathing of polluted air, smoke, auto emissions, or industrial exhausts since these exhausts do contain many different and large number of PAH compounds. Occupational workers with the highest exposures to smoke, who work as roofers, road builders, and who live near major highways or industrial sources, have been shown to be the common victims to PAH-related health disorders. PAHs do present everywhere and are the by-products of combustion, from sources as varied as coal and coke burners, diesel-fuelled engines, grilled meats, and cigarettes. PAH residues are often associated with suspended particulate matter in the air, and thus inhalation is a major source of PAH exposure. Also, PAHs enter the living environment and air mostly as releases from volcanoes, forest fires, burning coal, and automobile exhaust. PAHs have been found to occur in air attached to dust particles. Some PAH particles can readily evaporate into the air from soil or surface waters. PAHs do break down by reacting with sunlight and other chemicals in the air, over a period of days to weeks. PAHs enter water through discharges from industrial and wastewater treatment plants. Most PAHs do not dissolve easily in water. They stick to solid particles and settle to the bottoms of lakes or rivers. Microorganisms can break down PAHs in soil or water after a period of weeks to months. In soils, PAHs are most likely to stick tightly to particles; certain PAHs move through soil to contaminate underground water. PAH contents of plants and animals may be much higher than PAH contents of soil or water in which they live. Breathing air containing PAHs from cigarette smoke, wood smoke, vehicle hausts, asphalt roads, or agricultural burn smoke. Coming in contact with air, water, or soil near hazardous waste sites. Eating grilled or charred meats; contaminated cereals, flour, bread, vegetables, fruits, meats; and processed or pickled foods. Drinking contaminated water or cow's milk. Nursing infants of mothers living near hazardous waste sites may be exposed to PAHs through their mother's milk.

Safe Handling and Precautions

The toxicity of PAHs is structurally dependent, with isomers.

Mice that were fed high levels of one PAH during pregnancy had difficulty reproducing and so did their offspring. It is not known whether these effects occur in people. Animal studies have also shown that PAHs can cause harmful effects on the skin, body fluids, and ability to fight disease after both short- and long-term exposure. But these effects have not been seen in people. The DHHS reported that some polycyclic aromatic hydrocarbons are reasonably expected to be carcinogens. Workers who were exposed and/or breath or touch mixtures of PAHs and other hazardous chemical substances repeatedly and for prolonged periods of time have developed cancer. Also, studies with laboratory animals indicated that PAHs caused lung cancer when animals breathed PAH-contaminated air, developed stomach cancer when ingested PAH-contaminated food, and developed skin cancer when PAH was applied on the skin of experimental animals. In general, polycyclic aromatic hydrocarbons affect the same organ systems in all people who are exposed. However, the seriousness of the effects may vary from person to person. Studies have shown that PAHs are lipophilic and are stored in the fat tissue of the breast and produce increased risk for breast cancer through a variety of mechanisms. The most common PAHs are weakly estrogenic (oestrogen mimicking) due to interactions with the cellular oestrogen receptor.

The following PAHs have been reported as very hazardous and use and handling is restriction:

1. Benzo[a]pyrene (BaP); Benzo[e]pyrene (BeP)
2. Benzo[a]anthracene (BaA)
3. Chrysen (CHR)
4. Benzo[b]fluoroanthene (BbFA)

5. Benzo[j]fluoroanthene (BjFA)

6. Benzo[k]fluoroanthene (BkFA)

7. Dibenzo[a,h]anthracene (DBAhA)

Regulations:

- The U.S. Government agencies have established standards that are relevant to PAHs exposures in the workplace and the environment. The Regulatory Agency has determined the MCL for benzo (a)pyrene, one of the polycyclic aromatic hydrocarbons (PAH), as 0.2 ppb.
- Air.

The OSHA has determined the PEL for PAHs in the workplace as 0.2 mg/m^3. Standards and Regulations for Polycyclic Aromatic Hydrocarbons (PAHs)
Workplace air levels by different agencies:

- The NIOSH recommends that the average workplace air levels for coal tar products not exceed 0.1 mg/m^3 for a 10 h workday, within a 40 h workweek
- 0.2 mg/m^3 for benzene-soluble coal tar pitch fraction: ACGIH and the OSHA
- mg/m^3 for coal tar pitch volatile agents: NIOSH
- 0.0001 mg/L: U.S. EPA
- Drinking water
- The MCL goal for benzo(a)pyrene in drinking water is 0.2 ppb: U.S. EPA

Propane (CAS No. 74-98-6)

Molecular formula: $\mathbf{C_3H_8}$
Synonyms and trade names: *n*-Propane, propan; Dimethylmethane; Propyldihydride; Propyl hydride; Propane liquefied; Petroleum gas, liquefied; Bottled gas

Propane is colourless and odourless, with a mercaptan odour. Like all fossil fuels, propane is a non-renewable energy source. Propane is a gas derived from natural gas and petroleum. It is found mixed with natural gas and petroleum deposits. Propane is called a 'fossil fuel' because it was formed millions of years ago from the remains of tiny sea animals and plants. Propane is a clean-burning, versatile fuel. It is used by nearly everyone, in homes, on farms, by business, and in industry mostly for producing heat and operating equipment. Propane is one of the many fossil fuels included in the liquefied petroleum gas (LPG) family. Because propane is the type of LPG most commonly used in the United States, propane and LPG are often used synonymously. Butane is another LPG often used in lighters.

Propane is released to the living environment from automobile exhausts, burning furnaces, natural gas sources, and during combustion of polyethylene and phenolic resins. Propane is both highly inflammable and explosive and needs proper care and management of workplaces. Its use in industry includes source for fuel and propellant for aerosols.

Occupational workers exposed to liquefied propane have demonstrated skin burns and frost bite. Propane also causes depression effects on the CNS. Exposures to propane cause dizziness, confusion, excitation, asphyxia, and liquid frostbite. Repeated exposures cause adverse effects on the CNS.

Propetamphos (CAS No. 31218-83-4)

Molecular formula: $C_{10}H_{20}NO_4PS$
Synonyms and trade names: Blotic, Safrotin, and Seraphos
IUPAC name: (E)-O-2-isopropoxycarbonyl-1-methylvinyl O-methyl ethyl-phosphoramido-thioate
Toxicity class: U.S. EPA: II; WHO: Ib

Propetamphos technical is a yellowish, oily liquid at room temperature and a moderately toxic acaricide and insecticide. It is sparingly soluble in water, but very soluble in acetone, chloroform, ethanol, and hexane. U.S. EPA has classified propetamphos both as a GUP and RUP, indicating that its handling should be done by certified, qualified, and applicators. Propetamphos is used to control cockroaches, flies, ants, ticks, moths, fleas, and mosquitoes in households and where vector eradication is necessary to protect public health. It is also used in veterinary applications to combat parasites such as ticks, lice, and mites in livestock. The formulations of propetamphos include aerosols, emulsified concentrates, liquids, and powders. In veterinary applications, the insecticide is used for the control of ticks, lice, and mites in livestock. Commercial products include aerosols, emulsified concentrates, liquids, and powders.

Safe Handling and Precautions

Propetamphos is a moderately toxic organophosphate insecticide. It inhibits the cholinesterase enzyme in animals and humans leading to the overstimulation of the CNS. Prolonged period of exposures to high concentrations of propetamphos causes poisoning with symptoms that include, but not limited to, nausea, headache, dizziness, confusion, numbness, tingling sensations, in-coordination, tremor, abdominal cramps, sweating, blurred vision, breathing difficulty, unconsciousness, convulsions, slow heartbeat, respiratory paralysis, and death. There is no evidence to suggest that propetamphos causes mutagenic, teratogenic, or carcinogenic effects in animals or humans. Propetamphos has not been listed as a human carcinogen.

Propoxur (CAS No. 114-26-1)

Molecular formula: $C_{11}H_{15}NO_3$
Synonyms and trade names: Baygon; Bifex; Blattanex; Brifur; Bolfo; BO Q 5812315
IUPAC name: 2-Isopropoxyphenyl methylcarbamate

Toxicity class: U.S. EPA: II; WHO: II

ENT 25671; Invisi-Gard; OMS 33; PHC; Pillargon; Prentox; Propogon; Proprotox; Propyon; Rhoden; Sendran; Suncide; Tendex; Tugen; Unden; Undene.

Technical propoxur is a white-to-cream-coloured crystalline solid. Various formulations of propoxur has been classified by the U.S. EPA under different categories. It is a GUP, although some formulations may be for professional use only.

Propoxur is a non-systemic insecticide, which was introduced in 1959. It is compatible with most insecticides and fungicides except alkalines, and may be found in combination with azinphosmethyl, chlorpyrifos, cyfluthrin, dichlorvos, disulfoton, or methiocarb. It is used on a variety of insect pests such as chewing and sucking insects, ants, cockroaches, crickets, flies, and mosquitoes, and may be used for control of these in agricultural or (as Baygon) in non-agricultural, for example, in private or public facilities and ground applications. Agricultural applications include cane, cocoa, fruit, grapes, maize, rice, sugar, vegetables, cotton, lucerne, forestry, and ornamentals. It has contact and stomach action that is long-acting when it is in direct contact with the target pest. Propoxur is available in several types of formulations and products, including emulsifiable concentrates, wettable powders, baits, aerosols, fumigants, granules, and oil sprays.

Safe Handling and Precautions

Propoxur is highly toxic via the oral route. The acute oral LD_{50} is 50 mg/kg in rats and mice, 40 mg/kg for guinea pigs. Propoxur is only slightly toxic via the dermal route, with acute dermal LD_{50} of more than 5000 mg/kg in rats. The acute LC_{50} for rats (4 h) is more than 0.5 mg/L. Studies have shown that propoxur does not cause skin or eye irritation in rabbits. Like other carbamates, propoxur can inhibit the action of cholinesterase and disrupt nervous system function. Depending on the severity of exposure, this effect may be short term and reversible. The signs of propoxur intoxication include nausea, vomiting, abdominal cramps, sweating, diarrhoea, excessive, salivation, weakness, imbalance, blurring of vision, breathing difficulty, increased blood pressure, incontinence, or death. In rats, propoxur poisoning resulted in brain pattern and learning ability changes at lower concentrations than those, which caused cholinesterase-inhibition and/or organ, weight changes. During wide-scale spraying of propoxur in malarial control activities conducted by the WHO, only mild cases of poisoning were noted. Applicators who used propoxur regularly showed a pronounced daily fall in whole blood cholinesterase activity and a distinct recovery after exposure stopped. No adverse cumulative effects on cholinesterase activity were demonstrated.

Human adults have ingested single doses of 50 mg of propoxur without apparent symptoms. Prolonged or repeated exposure to propoxur may cause symptoms similar to acute effects. Propoxur is very efficiently detoxified (transformed into less toxic or practically non-toxic forms), thus making it possible for rats to tolerate daily doses approximately equal to the LD_{50} of the insecticide for long periods, provided that the dose is spread out over the entire day, rather than ingested all at once. When high dietary doses of approximately 18 mg/kg/day of propoxur were given to female rats, as a part of a three-generation reproduction study, reduced parental food consumption, growth, lactation, and litter size were observed. At 25 mg/kg/day administered to pregnant rats, there was a decrease in the number of offspring.

Dietary doses (2.25 mg/kg/day) of propoxur did not affect fertility, litter size, or lactation, but parental food intake and growth were depressed in the exposed group.

This evidence suggests that reproductive effects in humans are unlikely at expected exposure levels. Offspring of female rats fed 5 mg/kg/day of propoxur during gestation and weaning exhibited reduced birth weight, retarded development of some reflexes, and evidence of CNS impairment. In another rat study, growth reduction was observed in the offspring of pregnant rats given doses of 3, 9, and 30 mg/kg/day, but no other physiological or anatomical abnormalities were observed. The evidence suggests that teratogenic effects will only occur at high doses. Propoxur did not cause mutations in six different types of bacteria. The evidence indicates that propoxur is not mutagenic. Reports have shown that animal tests and data from human autopsies in poisoned individuals change in the nervous system and liver because of propoxur. During wide-scale spraying of propoxur in malarial control activities conducted by the WHO, only mild cases of poisoning were noted. Applicators who used propoxur regularly showed a pronounced daily fall in whole blood cholinesterase activity and a distinct recovery after exposure stopped. No adverse cumulative effects on cholinesterase activity were demonstrated. Human adults have ingested single doses of 50 mg of propoxur. In a 2 year feeding study of propoxur on mice, the dietary concentrations of 500, 2000, or 8000 ppm were given to male mice and at concentrations of 2000 ppm and higher caused an increased incidence of benign liver adenomas in male mice. Laboratory rats exposed to propoxur for 2 years in a single type of diet showed development of urinary bladder neoplasias. No carcinogenic effects of propoxur on humans have been reported. Propoxur is not listed by NTP, IARC, or by OSHA as a human carcinogen.

Occupational worker's handling and store of propoxur should be careful. Workers should store propoxur in a cool, dry place and prevent cross-contamination with other pesticides, fertilisers, or food and feed. Propoxur must be kept away from sunlight, radiators, stoves and other heat material, and in a locked storage area. Handle an open container in a manner as to prevent spillage.

Propylene Imine (CAS No. 75-55-8)

Molecular formula: C_3H_7N
Synonyms and trade names: 2-Methylaziridine; 2-Methylethyleneimine; Propyleneimine; Propylene imine (inhibited)

Propylene imine is a colourless oily fuming liquid, with pungent odour. The vapour is heavier than air and may travel along the ground; distant ignition possible. It polymerises under the influence of acids with fire or explosion hazard. The substance decomposes on heating producing toxic fumes including nitrogen oxides. Attacks some forms of plastic, coatings, and rubber.

Propylene imine is incompatible with acids, strong oxidisers, water, carbonyl compounds, quinones, and sulphonyl halides. It undergoes violent polymerisation on contact with acids and hydrolyses in water to form methylethanolamine. Propylene imine is highly flammable and gives off irritating or toxic fumes (or gases) in a fire. The vapour/air mixtures are explosive and also of risk of fire and explosion on contact with acids, oxidants.

Propylene imine is used as a chemical intermediate in the modification of latex surface coating resins, polymers in textile and paper industries, dyes, photography, gelatins, oil additives, and organic synthesis. It has been in use for polymers with methacrylic acid and esters.

Safe Handling and Precautions

Exposures to propylene imine on inhalation, skin absorption, and ingestion cause adverse health effects. The symptoms include cough, sore throat, burning sensation, headache, shortness of breath, breathing distress, nausea, vomiting, and dizziness. Propylene imine on contact with the skin causes severe irritation effects, and the eyes develop redness, pain, blurred vision; on inhalation the worker develops symptoms of respiratory tract irritation and lung oedema. Propylene imine (2-Methylaziridine) has been listed as Group 2B, meaning a possible human carcinogen.

Propylene Oxide (CAS No. 75-56-9)

Molecular formula: C_3H_6O
Synonyms and trade names: Epoxypropane; 1,2-Epoxy-propane; Ethylene oxide, Methyl oxirane; Methyl ethylene oxide; Oxirane, methyl-; NCI-C50099; Oxyde de propylene; 1-2-Propylene oxide; Propene oxide; Propylene epoxide

Propylene oxide is a colourless liquid. It is soluble in water, miscible in acetone, benzene carbon tetrachloride, methanol and ether. It is used in the production of polyethers (the primary component of polyurethane foams) and propylene glycol. Propylene oxide is used in the fumigation/sterilisation of packaged food stuffs and plastic medical instruments and in the manufacture of dipropylene glycol and glycol ethers, as herbicides, as solvents, and in the preparation of lubricants, surfactants, and oil demulsifiers. Also it has extensive industrial application as a chemical intermediate in the production of propylene glycol, glycol ethers, and as a solvent and in the preparation of surfactants and oil demulsifiers. Occupational exposure of workers to propylene oxide is known to occur at workplaces – production, storage, transport, and use – by the inhalation and dermal routes. Also, reports have indicated that propylene oxide has been detected in fumigated food products; consumption of contaminated food is another possible route of exposure.

Safe Handling and Precautions

Acute exposure of humans and animals to propylene oxide has caused eye and respiratory tract irritation. As a respiratory irritant, coughing, dyspnoea (difficulty in breathing), and pulmonary oedema may result from inhalation exposure and possibly lead to pneumonia. Dermal contact, even with dilute solutions, has caused skin irritation and necrosis. Propylene oxide is a mild CNS depressant and exposures to high concentrations have been reported to poisoning with symptoms such as headache, motor weakness, incoordination, ataxia, and coma in humans. Chronic inhalation exposure to propylene oxide may cause some neuropathological changes in experimental rats and monkeys. Propylene oxide has been observed to cause tumours in rodents, causing fore-stomach tumours. Exposure to propylene oxide via gavage and after inhalation also caused nasal tumours. There is no literature regarding health effects of propylene oxide after chronic exposure in humans. The U.S. EPA classified propylene oxide as Group B2, meaning a probable human carcinogen.

Pyrethrins and Pyrethroids (CAS No. 8003-34-7)

Molecular formula: $C_{21}H_{20}Cl_2O_3$
Permethrin is a synthetic pyrethrin.

Pyrethrum (CAS No. 8003-34-7)

Molecular formula: $C_{43}H_{56}O_8$
Synonyms and trade names: Pyrethrin-based insecticide; Buhach; Pyrethrum extract; Pyrethrum oleoresin

Pyrethrins/Pyrethrum are natural insecticides produced by certain Chrysanthemum species of plants. In contrast, Permethrin ('per-meth-rin') is a synthetic, man-made insecticide, whose chemical structure is based on natural pyrethrum. Pyrethrum was first recognised as having insecticidal properties around 1800 in Asia and was used to kill ticks and various insects such as fleas and mosquitoes. Six individual chemicals have active insecticidal properties in the pyrethrum extract, and these compounds are called pyrethrins. Pyrethrum are viscous, tan-coloured brown resins, liquids, or solids which inactivate readily in air. Pyrethrum are soluble in organic solvents such as alcohol, kerosene, nitromethane, petroleum ether, carbon tetrachloride, and ethylene dichloride. A quick acting liquid insecticide especially suited for control of insect pests of vegetables, ornamentals, and exotic crops. The piperonyl butoxide ingredient is used to enhance the activity of the permethrin (synergist).

Pyrethrum has been used effectively to control insects for decades and is non-persistent, decomposing rapidly in the environment. This rapid degradation of pyrethrum has resulted in little known cases of insect resistance, making it an excellent choice for the control of agricultural pests, control of insects on pets and/or livestock. Pyrethrum powder is toxic to ants, roaches, silverfish, bed bugs, fleas, wasps, spiders, crickets, mosquitoes, and just about every other category of unwanted house or garden pest. Because it decomposes rapidly in the environment, pyrethrum has been approved for a wide range of indoor and outdoor uses, including homes, restaurants, broad-scale spraying operations, and organic farms.

Pyrethrins break down quickly in the environment, especially when exposed to natural sunlight. Pyrethroids are manufactured chemicals that are very similar in structure to the pyrethrins, but are often more toxic to insects, as well as to mammals, and last longer in the environment than pyrethrins. Pyrethrins and pyrethroids are often combined commercially with other chemicals called synergists, which enhance the insecticidal activity of the pyrethrins and pyrethroids. The synergists prevent some enzymes from breaking down the pyrethrins and pyrethroids, thus increasing their toxicity. Technical-grade (concentrated) pyrethrins and pyrethroids are usually mixed with carriers or solvents to produce a commercial-grade formulated product.

Safe Handling and Precautions

Pyrethrum has been extensively studied. It is low in acute toxicity to man and other vertebrate animals, is non-carcinogenic, causes no adverse reproductive affects and is non-mutagenic. Exposure to pyrethroids at high concentrations repeatedly for prolonged periods causes health disorders. After an acute inhalation exposure, occupational workers suffer with symptoms of cough, wheezing, shortness of breath, runny or stuffy nose, chest pain, or difficulty breathing, on skin contact workers develop skin rash, itching and/or skin blisters. Long-term effects cause gastrointestinal effects such as nausea, vomiting, and diarrhoea. Inhalation of high vapour or mist concentrations or ingestion of large quantities can result in nervous system effects such as dizziness, headache, loss of coordination, tremors, loss of consciousness, and coma. Symptoms usually regress with no long-lasting effects. Occupational workers on exposure to large amounts of pyrethrins or pyrethroids and on dermal absorption or accidental workplace ingestion develop symptoms of toxicity that include feelings of numbness, itching, burning, stinging, tingling, dizziness, headache, nausea, and muscle twitching. Workers on exposure to very high doses of pyrethrins or pyrethroids become unconscious and suffer with convulsions. Pyrethrins and pyrethroids interfere with the way that the nerves and brain function.

The natural pyrethrins are contact poisons that quickly penetrate the nerve system of the insect and very quickly interfere with the locomotor system of the insect. A few minutes after application, the insect cannot move or fly away, and the 'knockdown dose' does not mean a killing dose. The natural pyrethrins swiftly get detoxified by enzymes in the insect. Thus, some pests will recover. Semi-synthetic derivatives of the chrysanthemumic acids have been developed as insecticides. These are called pyrethroids and tend to be more effective than natural pyrethrins while they are less toxic to mammals. The common synthetic pyrethroid is Allethrin (for details, refer literature).

The U.S. EPA classified pyrethrin-I as an RUP and according to regulations, occupational workers and certified applicators are permitted to handle the RUP. This product is not listed as a human carcinogen by the NTP, the IARC, the OSHA, and by the ACGIH.

Bibliography

Agency for Toxic Substances and Disease Registry (ATSDR). 1995. *Toxicological Profile for Polycyclic Aromatic Hydrocarbons (PAHs)*. U.S. Department of Health and Human Services, Public Health Service, Atlanta, GA (updated 2011).

Agency for Toxic Substances and Disease Registry (ATSDR). 1997. *Toxicological Profile for White Phosphorus*. U.S. Department of Health and Human Services, Public Health Service, Atlanta, GA (updated 2011).

Agency for Toxic Substances and Disease Registry (ATSDR). 2000. *Toxicological Profile for Polychlorinated Biphenyls (PCBs)*. U.S. Department of Health and Human Services, Public Health Service, Atlanta, GA (updated 2011).

Agency for Toxic Substances and Disease Registry (ATSDR). 2001. *Toxicological Profile for Pentachlorophenol*. U.S. Department of Health and Human Services, Public Health Service, Atlanta, GA.

Agency for Toxic Substances and Disease Registry (ATSDR). 2002a. *Managing Hazardous Materials Incidents. Volume III—Medical Management Guidelines for Acute Chemical Exposures: Phosgene Oxime*. U.S. Department of Health and Human Services, Public Health Service, Atlanta, GA.

Agency for Toxic Substances and Disease Registry (ATSDR). 2002b. *Managing Hazardous Materials Incidents. Volume III—Medical Management Guidelines for Acute Chemical Exposures: Phosphine*. U.S. Department of Health and Human Services, Public Health Service, Atlanta, GA (updated 2011).

Agency for Toxic Substances and Disease Registry (ATSDR). 2003. *Toxicological Profile for Pyrethrins and Pyrethroids*. U.S. Department of Health and Human Services, Public Health Service, Atlanta, GA.

Agency for Toxic Substances and Disease Registry (ATSDR). 2004a. *Toxicological Profile for PBBs and PBDEs*. U.S. Department of Health and Human Services, Atlanta, GA.

Agency for Toxic Substances and Disease Registry (ATSDR). 2004b. *Toxicological Profile for Polybrominated Biphenyls and Polybrominated Diphenyl Ethers (PBBs and PBDEs)*. U.S. Department of Health and Human Services, Public Health Service, Atlanta, GA (updated 2011).

Agency for Toxic Substances and Disease Registry (ATSDR). 2006. *Medical Management Guidelines for Parathion*. Division of Toxicology and Environmental Medicine, Atlanta, GA.

Ahrens, W. H. 1994. *Herbicide Handbook*, 7th edn. Weed Science Society of America, Champaign, IL.

Aldridge, W. N. 1990. An assessment of the toxicological properties of pyrethroids and their neurotoxicity. *Toxicology* 21(2):89–104.

American Cyanamid. 1992. *Toxicological Summary for Thimet*. American Cyanamid, Wayne, NJ.

Arcaro, K. F., P. W. O'Keefe, Y. Yang et al. 1999. Antiestrogenicity of environmental polycyclic aromatic hydrocarbons in human breast cancer cells. *Toxicology* 133:115–127.

Baan, R., Y. Grosse, K. Straif et al. 2009. A review of human carcinogens—Part F: Chemical agents and related occupations. *Lancet Oncol.* 10(12):1143–1144.

Baron, R. L. 1991. Carbamate insecticides. *Handbook of Pesticide Toxicology*, Hayes, W. J., Jr. and E. R. Laws, Jr. (eds.). Academic Press, New York.

Beyer, A. and M. Biziuk. 2009. Environmental fate and global distribution of polychlorinated biphenyls. *Rev. Environ. Contam. Toxicol.* 201:137–158.

Blair, D. M. 1961. Dangers in using and handling sodium pentachlorophenate as a molluscicide. *Bull. World Health Organ.* 25:597–601.

Bonner, M. R., D. Han, J. Nie et al. 2005. Breast cancer risk and exposure in early life to polycyclic aromatic hydrocarbons using total suspended particulates as a proxy measure. *Cancer Epidemiol. Biomarkers Prev.* 14:53–60.

Bradman, A., L. Fenster, A. Sjodi, R. S. Jones, D. G. Patterson, and B. Eskenazi. 2007. Polybrominated diphenyl ether levels in the blood of pregnant women living in an agricultural community in California. *Environ. Health Perspect.* 115(1):71–74.

Brown, F. R., J. Winkler, P. Visita, J. Dhaliwal, and M. Petreas. 2006. Levels of PBDEs. PCDDs, PCDFs, and coplanar PCBs in edible fish from California coastal waters. *Chemosphere* 64:276–286.

Budavari, S. (ed.). 1989. *The Merck Index. An Encyclopedia of Chemicals, Drugs, and Biologicals*, 11th edn. Merck & Co. Inc., Rahway, NJ.

Budavari, S. (ed.). 1997. *The Merck Index. An Encyclopedia of Chemicals, Drugs, and Biologicals*, 12th edn. Merck & Co. Inc., Rahway, NJ.

Carpenter, D. O. 1998. Polychlorinated biphenyls and human health. *Int. J. Occup. Med. Environ. Health* 11(4):291–303.

Carpenter, D. O. 2006. Polychlorinated biphenyls (PCBs): Routes of exposure and effects on human health. *Rev. Environ. Health* 21(1):1–23.

Carpenter, C. P., C. S. Weil, U. C. Pozzani, and H. F. Smyth, Jr. 1950. Comparative acute and subacute toxicities of allethrin and pyrethrins. *AMA Arch. Ind. Hyg. Occup. Med.* 2:420–432.

Carson, P. A. and C. J. Mumford (eds.). 2002. *Hazardous Chemicals Handbook*. Butterworth-Heinemann, Woburn, MA.

CDC (Centers for Disease Control). 2003. Phosgene oxime. Department of Health and Human Services, Centers for Disease Control and Prevention, Atlanta, GA.

Centers for Disease Control and Prevention (CDC). 2005. Pirimiphos-methyl. Third National Report on Human Exposure to Environmental Chemicals. Atlanta, GA.

Cheremisinoff, N. P. (ed.). 1999. *Handbook of Hazardous Chemical Properties*. Butterworth, Heinemann, Boston, MA.

Clayton, G. D. and F. E. Clayton (eds.). 1981. *Patty's Industrial Hygiene and Toxicology*, 3rd revised edn., Vol. IIA. John Wiley & Sons, New York.

Diggory, H. J. P., P. J. Landrigan, K. P. Latimer et al. 1977. Fatal parathion poisoning caused by contamination of flour in international commerce. *Am. J. Epidemiol.* 106:145–153.

Dikshith, T. S. S. (ed.). 1991. *Toxicology of Pesticides in Animals.* CRC Press, Boca Raton, FL.

Dikshith, T. S. S. (ed.). 2009. *Safe Use of Chemicals: A Practical Guide and Handbook of Chemicals and Safety.* CRC Press, Boca Raton, FL.

Dikshith, T. S. S. (ed.). 2011. *Handbook of Chemicals and Safety.* CRC Press, Boca Raton, FL.

Dikshith, T. S. S. and P. V. Diwan (eds.). 2003. *Industrial Guide to Chemical and Drug Safety.* John Wiley & Sons, Inc., Hoboken, NJ.

Dikshith, T. S. S., W. Rockwood, R. Abraham, and F. Coulston. 1975. Effects of polychlorinated biphenyl (Aroclor 1254) on rat testis. *Exp. Mol. Pathol.* 22(3):376–385.

Downey, D. 1989. Contact mucositis due to palladium. *Contact Dermatitis* 21:54.

Echobichon, D. J. 1996. Toxic effects of pesticides. *Casarett & Doull's Toxicology: The Basic Science of Poisons,* 3rd edn., Klaassen, C. D. (ed.). McGraw-Hill, New York.

Ecobichon, D. J. 2007. Toxic effects of pesticides. *Casarett and Doull's Toxicology, The Basic Science of Poisons,* 5th edn., Klaassen, C. D. McGraw-Hill, New York, NY.

Edwards, I. R., D. G. Ferry, and W. A. Temple. 1991. Fungicides and related compounds. *Handbook of Pesticide Toxicology,* Hayes, W. J. and E. R. Laws (eds.). Academic Press, New York.

Eldridge, S. R., M. S. Bogdanffy, M. P. Jokinen, and L. S. Andrews. 1995. Effects of propylene oxide on nasal epithelial cell proliferation in F344 rats. *Fundam. Appl. Toxicol.* 27(1):25–32.

Elliot, M., N. F. Janes, E. C. Kimmel, and J. E. Casida. 1972. Metabolic fate of pyrethrin I, pyrethrin II, and allethrin administered orally to rats. *J. Agric. Food Chem.* 20:300–312.

European Commission Joint Research Centre. 2003. European Union risk assessment report *n*-pentane. European Chemicals Bureau, Oslo, Norway.

Extension Toxicology Network (EXTOXNET). 1994. Pyrithrins. Oregon State University, USDA/Extension Service, Corvallis, OR.

Extension Toxicology Network (EXTOXNET). 1996a. Paraquat. Pesticide Information Profiles (PIP). USDA, Extension Service, Oregon State University, Corvallis, OR.

Extension Toxicology Network (EXTOXNET). 1996b. Picloram. Pesticide Information Profiles (PIP). USDA/Extension Service, Oregon State University, Corvallis, OR.

Gallo, M. A. and N. J. Lawryk. 1991. Organic phosphorus pesticides. *Handbook of Pesticide Toxicology,* Vol. 2, Classes of Pesticides, Hayes, W. J., Jr. and E. R. Laws, Jr. (eds.). Academic Press, San Diego, CA.

García-Ortega, S., P. J. Holliman, and D. L. Jones. 2006. Toxicology and fate of pestanal and commercial propetamphos formulations in river and estuarine sediment. *Sci. Total Environ.* 366(2–3):826–836.

Gosselin, R. E., H. P. Smith, and H. C. Hodge. 1984. *Clinical Toxicology of Commercial Products,* 5th edn., Section III, Therapeutics index. Williams & Wilkins, Baltimore, MD, pp. 352–355.

Gray, L. E. et al. 1995. Functional developmental toxicity of low doses of 2,3,7,8-tetrachlorodibenzo-*p*-dioxin and a dioxin-like PCB (169) in Long Evans rats and Syrian hamsters: Reproductive, behavioral and thermoregulatory alterations. *Organohal. Comp.* 25:33.

Guidelines for safe use and management of picric acid. 2010. Laboratory Fact Sheet. Texas A&M University, Environmental Health & Safety Department.

Hayes, W. J., Jr. 1982. *Pesticides Studied in Man.* Williams & Wilkins, Baltimore, MD.

Hayes, W., Jr. (ed.). 1992. *Pesticides Studied in Man.* Williams & Wilkins. Baltimore, MD.

Hayes, J. R., L. W. Condie, and J. F. Borzelleca. 1996. Acute, 14-day repeated dosing, and 90-day subchronic toxicity studies of potassium picloram. *Fundam. Appl. Toxicol.* 7:464.

Hayes, W. J. and E. R. Laws (ed.). 1991. *Handbook of Pesticide Toxicology.* Academic Press, New York.

Hazardous Substances Data Bank (HSDB). TOXNET. 1992. Cyolane. HSDB, Bethesda, MD.

Hoenig, S. L. (ed.). 2007. *Compendium of Chemical Warfare Agents.* Springer-Verlag, New York.

International Agency for Research on Cancer (IARC). 1978. Polychlorinated biphenyls. *Polychlorinated Biphenyls and Polybrominated Biphenyls. IARC Monographs on the Evaluation of Carcinogenic Risk of Chemicals to Humans,* Vol. 18. IARC, Lyon, France.

International Agency for Research on Cancer (IARC). 1987. Polychlorinated biphenyls. *Overall Evaluations of Carcinogenicity. IARC Monographs on the Evaluation of Carcinogenic Risk of Chemicals to Humans*, Suppl. 7. IARC, Lyon, France.

International Chemical Safety Cards. 2002. Pentachloroethane. ICSC Card No. 1394. Geneva, Switzerland.

International Programme on Chemical Safety and the Commission of the European Communities (IPCS–CEC). 1990. Parathion. ICSC Card No. 0006. Commission of the European, Luxembourg (updated 1999).

International Programme on Chemical Safety (IPCS). 1992a. Parathion. Health and Safety Guide No. 74. World Health Organization, Geneva, Switzerland.

International Programme on Chemical Safety (IPCS). 1992b. Polychlorinated biphenyls. Environmental Health Criteria No. 140. IPCS, WHO, Geneva, Switzerland.

International Programme on Chemical Safety & the Commission of the European Communities (IPCS–CEC). 1993a. Mevinphos (isomer mixture). ICSC No. 0924. IPCS–CEC, Geneva, Switzerland.

International Programme on Chemical Safety and the Commission of the European Communities (IPCS–CEC). 1993b. Phosmet. ICSC Card No. 0543. IPCS–CEC, Luxembourg.

International Programme on Chemical Safety and the Commission of the European Communities (IPCS–CEC). 1994. Tetrachloroethylene. ICSC Card No. 0076. International Programme on Chemical Safety. Geneva, Switzerland (updated 2000).

International Programme on Chemical Safety and the Commission of the European Communities (IPCS–CEC). 1997a. Phosphorus pentachloride. ICSC Card No. 0544. IPCS–CEC, Geneva, Switzerland.

International Programme on Chemical Safety and the Commission of the European Communities (IPCS-CEC). 1997b. Phosphorus pentoxide. ICSC Card No. 0545. IPCS–CEC, Geneva, Switzerland.

International Programme on Chemical Safety and the Commission of the European Communities (IPCS–CEC). 1998. Pentaboron nonahydride. ICSC No. 0819. IPCS-CEC, Geneva, Switzerland.

International Programme on Chemical Safety and the Commission of the European Communities (IPCS–CEC). 2000. Phosphoric acid. ICSC Card No. 1008. Commission of the European, Luxembourg (updated 2005).

International Program on Chemical Safety and the Commission of the European Communities (IPCS–CEC). 2003a. Propane. ICSC No. 0319. IPCS–CEC, Luxembourg.

International Programme on Chemical Safety (IPCS) and the Commission of the European Communities (CEC). 2003b. Parffin wax. ICSC No. 1457. IPCS–CEC, Geneva, Switzerland.

International Programme on Chemical Safety and the Commission of the European Communities (IPCS–CEC). 2003c. Phthalic anhydride. ICSC Card No. 0315. IPCS–CEC, Geneva, Switzerland.

International Programme on Chemical Safety and the Commission of the European Communities (IPCS–CEC). 2004. Phosphorus pentasulfide. ICSC Card No. 1407. IPCS–CEC, Geneva, Switzerland.

International Programme on Chemical Safety and the Commission of the European Communities (IPCS–CEC). 2005a. Phosphine. ICSC Card No. 0694. IPCS–CEC, Geneva, Switzerland.

International Programme on Chemical Safety and the Commission of the European Communities (IPCS–CEC). 2005b. Phosphamidon. ICSC Card No. 0189. Commission of the European Communities, Luxembourg.

International Program on Chemical Safety and the Commission of the European Communities (IPCS–CEC). 2010. Propylene imine. ICSC No. 0322. IPCS–CEC, Luxembourg.

Jacobson, J. L. and S. W. Jacobson. 1996. Intellectual impairment in children exposed to polychlorinated biphenyls in utero. *N. Engl. J. Med.* 335(11):783–789.

Johnson, B. L., H. E. Hicks, W. Cibulas et al. 1999. *Public Health Implications of Exposure to Polychlorinated Biphenyls (PCBs)*. Agency for Toxic Substances and Disease Registry, Atlanta, GA.

Jokanovic, M., M. Maksimovic, and R. M. Stepanovic. 1995. Interaction of phosphamidon with neuropathy target esterase and acetylcholinesterase of hen brain. *Arch. Toxicol.* 69(6):425–428.

Kidd, H. and D. R. James (eds.). 1991. *The Agrochemicals Handbook*, 3rd edn. Royal Society of Chemistry Information Services, Cambridge, U.K.

Kimbrough, R. D. 1987. Human health effects of polychlorinated biphenyls (PCBs) and polybrominated biphenyls (PBBs). *Annu. Rev. Pharmacol. Toxicol.* 27:87–111.

Korrick, S. A. and L. Altshul. 1998. High breast milk levels of polychlorinated biphenyls (PCBs) among four women living adjacent to a PCB-contaminated waste site. *Environ. Health Perspect.* 106(8):513.

Kosswig, K. 2002. Surfactants. In: *Ullmann's Encyclopedia of Industrial Chemistry*. Wiley-VCH, Weinheim, Germany.

Krieger, R. (ed.). 2001. *Handbook of Pesticide Toxicology*. Academic Press, New York.

Kuper, C. F., G. J. Reuzel, and V. J. Feron. 1988. Chronic inhalation toxicity and carcinogenicity study of propylene oxide in Wistar rats. *Food Chem. Toxicol.* 26(2):159–167.

Lewis, R. J., Sr. (ed.). 1992. *Sax's Dangerous Properties of Industrial Materials*, 8th edn. Van Nostrand Reinhold, New York.

Lewis, R. J. (ed.). 1999. *Sax's Dangerous Properties of Industrial Materials*, 10th edn., Vols. 1–3. John Wiley & Sons, New York.

Lewis, R. J., Sr. 2001. *Hawley's Condensed Chemical Dictionary*, 14th edn. John Wiley & Sons, New York.

Lewis, R. J., Sr. (ed.). 2002. *Hazardous Chemicals Desk Reference*, 5th edn. John Wiley & Sons, New York.

Lewis, R. J. and R. J. Lewis, Sr. 2008. *Hazardous Chemicals Desk Reference*. John Wiley & Sons, New York.

Material Safety Data Sheet (MSDS). Picrotoxin. MSDS ACC No. 45574. Acros Organics, Fair Lawn, NJ.

Material Safety Data Sheet (MSDS). 1993. Parathion. Extension Toxicology Network (EXTOXNET), Pesticide Information Profiles (PIP). USDA/Extension Service, Oregon State University, Corvallis, OR.

Material Safety Data Sheet (MSDS). 1996. Propetamphos. Extension Toxicology Network (EXTOXNET). Pesticide Information Profiles (PIP). USDA/Extension Service/Oregon State University, Corvallis, OR.

Material Safety Data Sheet (MSDS). 2003. Safety data for pentachlorophenol. Department of Physical and Theoretical Chemistry, Oxford University, Oxford, U.K.

Material Safety Data Sheet (MSDS). 2005a. Chemical safety data: 1-Pentanol. Department of Physical Chemistry at Oxford University, Oxford, U.K.

Material Safety Data Sheet (MSDS). 2005b. Safety data for pentachloroethane. Department of Physical and Theoretical Chemistry, Oxford University, Oxford, U.K.

Material Safety Data Sheet (MSDS). 2005c. Safety data for phosphine. Department of Physical Chemistry, Oxford University, Oxford, U.K.

Material Safety Data Sheet (MSDS). 2005d. Safety data for phosphoric acid. Department of Physical Chemistry, Oxford University, Oxford, U.K.

Material Safety Data Sheet (MSDS). 2005e. Tetrachloroethylene. MSDS No. T0767. Mallinckrodt Baker, Inc., Phillipsburg, NJ.

Material Safety Data Sheet (MSDS). 2007a. Phosphorus pentoxide. MSDS No. P4116. Environmental Health & Safety. Mallinckrodt Baker, Inc., Phillipsburg, NJ.

Material Safety Data Sheet (MSDS). 2007b. Safety data for paraquat. Department of Physical and Theoretical Chemistry, Oxford University, Oxford, U.K.

Material Safety Data Sheet (MSDS). 2008a. Picric acid. Saturated Environmental Health and Safety, UCSD Health System and VA Medical Center, San Diego, CA.

Material Safety Data Sheet (MSDS). 2008b. Safety data for phosdrin. Department of Physical and Theoretical Chemistry, Oxford University, Oxford, U.K.

Material Safety Data Sheet (MSDS). 2009. Safety data for pentane. Department of Physical Chemistry at Oxford University, Oxford, U.K.

Material Safety Data Sheet (MSDS). 2010. Safety data for palladium chloride. Department of Physical and Theoretical Chemistry, Oxford University, Oxford, U.K.

Meister, R. T. (ed.). 1995. *Farm Chemicals Handbook '95*. Meister Publishing Company, Willoughby, OH.

Mendola, P., G. M. Buck, L. E. Sever et al. 1997. Consumption of PCB-contaminated freshwater fish and shortened menstrual cycle length. *Am. J. Epidemiol.* 145(11):955.

Menon, J. A. 1958. Tropical hazards associated with the use of pentachlorophenol. *Br. Med. J.* 1:1156–1158.

Morgan, D. P. 1982. *Recognition and Management of Pesticide Poisonings*, 3rd edn. U.S. Environmental Protection Agency, Washington, DC.

National Cancer Institute (NCI). 1979a. Bioassay of phthalic anhydride for possible carcinogenicity. Technical Report 159. Public Health Service, Bethesda, MD.

National Cancer Institute (NCI). 1979b. *Bioassay of Phthalic Anhydride for Possible Carcinogenicity*. NCI, U.S. DHEW, Bethesda, MD.

National Institute for Occupational Safety and Health (NIOSH). 1997. *Pocket Guide to Chemical Hazards*. U.S. Department of Health and Human Services, Public Health Service, Centers for Disease Control and Prevention, Cincinnati, OH.

National Institute for Occupational Safety and Health (NIOSH). 2003. *Pocket Guide to Chemical Hazards*. U.S. Department of Health and Human Services, Atlanta, GA.

National Institute for Occupational Safety and Health (NIOSH). 2010a. Phosphoric acid. *NIOSH Pocket Guide to Chemical Hazards*. NIOSH, Centers for Disease Control and Prevention, Atlanta, GA.

National Institute for Occupational Safety and Health (NIOSH). 2010b. Phosphine. *NIOSH Pocket Guide to Chemical Hazards*. NIOSH, Centers for Disease Control and Prevention, Atlanta, GA.

National Institute for Occupational Safety and Health (NIOSH). 2010c. Phthalic anhydride. *NIOSH Pocket Guide to Chemical Hazards*. U.S. Department of Health and Human Services, Public Health Service, Centers for Disease Control and Prevention, Cincinnati, OH (updated 2011).

National Institute for Occupational Safety and Health (NIOSH). 2010d. Propylene imine. *NIOSH Pocket Guide to Chemical Hazards*. NIOSH Centers for Disease Control and Prevention, Atlanta, GA (updated 2011).

National Institute for Occupational Safety and Health (NIOSH). 2010e. Pyrethrum. *NIOSH Pocket Guide to Chemical Hazards*. Centers for Disease Control and Prevention, Atlanta, GA (updated 2011).

National Toxicology Program. 2005. Tetrachloroethylene, perchloroethylene. U.S. Department of Health and Human Services (U.S. EPA 2006).

Ngoula, F., P. Watcho, M.-C. Dongmo, A. Kenfack, P. Kamtchouing, and J. Tchoumboué. 2007. Effects of pirimiphos-methyl (an organophosphate insecticide) on the fertility of adult male rats. *Afr. Health Sci.* 7(1):3–9.

O'Neil, M. J. (ed.). 2001. *The Merck Index—An Encyclopedia of Chemicals, Drugs, and Biologicals*, 13th edn. Merck & Co., Inc., Whitehouse Station, NJ.

Ohnishi, A., T. Yamamoto, Y. Murai, Y. Hayashida, and I. Hori Tanaka. 1988. Propylene oxide causes central-peripheral distal axonopathy in rats. *Arch. Environ. Health* 43(5):353–356.

Patnaik, P. (ed.). 1992. *A Comprehensive Guide to the Hazardous Properties of Chemical Substances*. Van Nostrand Reinhold, New York.

Patty, F. A. (ed.). 1963. *Industrial Hygiene and Toxicology*, 2nd rev. edn., Vol. II, Toxicology. Interscience Publishers, Inc., New York.

Pillai, C. K. S. and U. S. Nandi. 1977. Interaction of palladium (II) with DNA. *Biochim. Biophys. Acta* 474:11–16.

Pohanish, R. P. (ed.). 2008. *Sittig's Handbook of Toxic and Hazardous Chemical Carcinogens*, 5th edn., Vol. 1: A-H, Vol. 2: I-Z. William Andrew, Norwich, NY.

Pohanish, R. P. (ed.). 2011. *Sittig's Handbook of Toxic and Hazardous Chemicals and Carcinogens*, 6th edn. William Andrew, Norwich, NY.

Pohanish, R. P. and S. A. Greene (eds.). 2009. *Wiley Guide to Chemical Incompatibilities*. John Wiley & Sons, New York.

Proctor, N. H. and G. J. Hathaway. 2004. *Proctor and Hughes' Chemical Hazards of the Workplace*, 5th edn. Wiley-Interscience, Hoboken, NJ.

Product Safety Information. 2012. Perchloroethylene. Brussels, Belgium.

Reynolds, J. E. F. and A. B. Prasad (eds.). 1982. *Martindale: The Extra Pharmacopoeia*, 28th edn. The Pharmaceutical Press, London, U.K.; *Farm Chemicals Handbook 87*. 1987. Meister Publishing Co., Willoughby, OH.

Safe, S. H. 1994. Polychlorinated biphenyls (PCBs): Environmental impact, biochemical and toxic responses, and implications for risk assessment. *Crit. Rev. Toxicol.* 24(2):87–149.

Sagar, D. B., W. Shih-Schoeder, and D. Girard. 1987. Effect of early postnatal exposure to polychlorinated biphenyls (PCBs) on fertility of male rats. *Bull. Environ. Contam. Toxicol.* 38(6):946–953.

Sagiv, S. K., M. M. Gaudet, S. M. Eng et al. 2009. Polycyclic aromatic hydrocarbon-DNA adducts and survival among women with breast cancer. *Environ. Res.* 109:287–291.

Sax, N. I. (ed.). 1984. *Dangerous Properties of Industrial Materials*, 6th edn. Van Nostrand Reinhold Company, New York.

Sax, N. I. (ed.). 1995. *Dangerous Properties of Industrial Materials*. Van Nostrand Reinhold, New York.

Sax, N. I. and R. J. Lewis (eds.). 1987. *Hawley's Condensed Chemical Dictionary*. Van Nostrand Reinhold Company, New York.

Sax, N. I. and R. J. Lewis, Jr. (eds.). 1989. *Dangerous Properties of Industrial Materials*, 7th edn., 3 volumes. Van Nostrand Reinhold, New York.

Sax, N. I. and R. J. Lewis (eds.). 2004. *Dangerous Properties of Industrial Materials*, 11th edn. Van Nostrand Reinhold Company, New York.

Schell, L. M., L. Hubicki, A. DiCaprio et al. 2000. Polychlorinated biphenyls and thyroid function in adolescents of the Mohawk Nation at Akwesasne. *Proceedings of the Ninth International Conference*. Turin, Italy.

Silberhorn, E. M., H. P. Glauert, and L. W. Robertson. 1990. Carcinogenicity of polyhalogenated biphenyls: PCBs and PBBs. *Crit. Rev. Toxicol.* 20(6):440–496.

Sittig, M. (ed.). 1985. *Handbook of Toxic and Hazardous Chemicals and Carcinogens*, 2nd edn. Noyes Publications, Park Ridge, NJ.

Spencer, E. Y. 1982. *Guide to the Chemicals Used in Crop Protection*, 7th edn. Publication No. 1093. Research Institute, Agriculture Canada, Information Canada, Ottawa, Ontario, Canada.

Stewart, P., J. Reihman, E. Lonky, T. Darvill, and J. Pagano. 2000. Prenatal PCB exposure and neonatal behavioral assessment scale (NBAS) performance. *Neurotoxicol. Teratol.* 22:21–29.

The International Programme on Chemical Safety (IPCS). 2002. Palladium. Environmental Health Criteria No. 226. World Health Organization, Geneva, Switzerland.

Tomlin, C. D. S. (ed.). 1997. *The Pesticide Manual: A World Compendium*, 11th edn. British Crop Protection Council, Farnham, Surrey, U.K.

Tomlin, C. D. S. (ed.). 2006. *The Pesticide Manual: A World Compendium*, 14th edn. The British Crop Protection Council, Farnham, Surrey, U.K.

Toutonghi, G., D. Echeverria, M. Morgan et al. 1994. Characterization of exposure to perchloroethylene, with biological monitoring, in commercial dry cleaning workers. *Scand. J. Work Indust. Health*.

U.S. Department of Agriculture (USDA). 2000. Picloram. Herbicide Information Profile. Forest Service Pacific Northwest Region. Forest Service, Portland, OR.

U.S. Department of Labor. Occupational Safety & Health Administration (OSHA). 1999. *Phosdrin*. OSHA, Washington, DC.

U.S. Environmental Protection Agency (U.S. EPA). 1985. Phorate. Pesticide Fact Sheet No. 34.1. OPTS, Washington, DC.

U.S. Environmental Protection Agency (U.S. EPA). 1986. Health and environmental effects profile on phthalic anhydride. Prepared by the Office of Health and Environmental Assessment, Environmental Criteria and Assessment Office, Cincinnati, OH.

U.S. Environmental Protection Agency (U.S. EPA). 1987a. Pentachlorophenol. Integrated Risk Information System (IRIS). U.S. EPA, Washington, DC (updated 2011).

U.S. Environmental Protection Agency (U.S. EPA). 1987b. *Summary Review of the Health Effects Associated with Propylene Oxide*. Environmental Criteria and Assessment Office, Office of Health and Environmental Assessment, Office of Research and Development, Research Triangle Park, NC.

U.S. Environmental Protection Agency (U.S. EPA). 1988. Phosphamidon. Pesticide Fact Sheet No. 154. US EPA, Washington, DC.

U.S. Environmental Protection Agency (U..S EPA). 1992. *Ethyl Parathion—Correction to the Amended Cancellation Order*. OPP, U.S. EPA, Washington, DC.

U.S. Environmental Protection Agency (U.S. EPA). 1997. Paraquate dichloride—RED FACTS. (EPA-738-F-96–018). U.S. EPA, Atlanta, GA.

U.S. Environmental Protection Agency (U.S. EPA). 1998. Cancer assessment document: Evaluation of the carcinogenic potential of propetamphos. Final Report. 13 October 1998. Cancer Assessment Review Committee, Health Effects Division, Office of Pesticide Programs, US EPA, Washington, DC.

U.S. Environmental Protection Agency (U.S. EPA). 1999a. Integrated Risk Information System (IRIS) on White phosphorus. National Center for Environmental Assessment, Office of Research and Development, Washington, DC.

U.S. Environmental Protection Agency (U.S. EPA). 1999b. Integrated Risk Information System (IRIS) on Phthalic anhydride. National Center for Environmental Assessment, Office of Research and Development, Washington, DC.

U.S. Environmental Protection Agency (U.S. EPA). Integrated Risk Information System (IRIS). 1999c. Propylene oxide. National Center for Environmental Assessment, Office of Research and Development, Washington, DC.

U.S. Environmental Protection Agency (U.S. EPA). 2000a. Hudson River PCBs Reassessment RI/FS Phase 3 Report: Feasibility Study. U.S. Environmental Protection Agency and U.S. Army Corps of Engineers.

U.S. Environmental Protection Agency (U.S. EPA). 2000b. *Propetamphos Facts. United States Prevention, Pesticides, Environmental Protection and Toxic Substances*. U.S. EPA, Atlanta, GA.

U.S. Environmental Protection Agency (U.S. EPA). 2000c. Phosgene. Technology Transfer Network. U.S. EPA, Research Triangle Park, NC (updated 2007).

U.S. Environmental Protection Agency (U.S. EPA). 2002. Phosmet. Integrated Risk Information System (IRIS) (updated 2011).

U.S. Environmental Protection Agency (U.S. EPA). 2006a. Finalization of interim registration eligibility decision for pirimiphos-methyl. Case No. 2535. U.S. EPA, Atlanta, GA (updated 2010).

U.S. Environmental Protection Agency (U.S. EPA). 2006b. Pirimiphos-methyl IRED facts. U.S. EPA, Atlanta, GA (updated 2011).

U.S. Environmental Protection Agency (U.S. EPA). 2010. An exposure assessment of polybrominated diphenyl ethers (PBDEs) (Final). (EPA/600/R-08/086F, 2010). U.S. EPA, Washington, DC.

U.S. Environmental Protection Agency (U.S. EPA). 2011a. Phthalic anhydride. Integrated Risk Information System (IRIS). US EPA, IRIS, U.S. EPA, Research Triangle Park, NC.

U.S. Environmental Protection Agency (U.S. EPA). 2011b. Paraquat. Integrated Risk Information System (IRIS). U.S. EPA, IRIS, Washington, DC.

U.S. Public Health Service. 1996. Propetamphos. Hazardous Substance Data Bank, Washington, DC.

United States Department of Agriculture (USDA). 2000. Picloram. Herbicide information profile. Forest Service Pacific Northwest Region. Forest Service, ortland, Oregon.

Vettorazzi, G. 1979. Phorate. *International Regulatory Aspects for Pesticide Chemicals*, Vol. 1. CRC Press, Boca Raton, FL.

Weed Science Society of America. 1994. *Herbicide Handbook*, 7th edn. Weed Science Society of America, Champaign, IL.

Weisglas-Kuperus, N., S. Patandin, G. A. Berbers et al. 2000. Immunologic effects of background exposure to polychlorinated biphenyls and dioxins in Dutch preschool children. *Environ. Health Perspect*. 108(12):1203.

World Health Organization (WHO). 1994. Polybrominated biphenyls. Environmental Health Criteria No. 152. WHO, International Programme on Chemical Safety (IPCS), Geneva, Switzerland (updated 2008).

World Health Organization (WHO). 1996. Pirimiphos-methyl. Data sheets on pesticides. No. 49. WHO, Geneva, Switzerland.

World Health Organization (WHO). 1997. *Dry Cleaning, Some Chlorinated Solvents and Other Industrial Chemicals. IARC Monographs on the Evaluation of Carcinogenic Risks to Humans*, Vol. 63. International Agency for Research on Cancer (IARC), Lyon, France.

World Health Organization (WHO). 2001a. Phosphamidon. Data Sheets on Pesticides No. 74, (WHO/VBC/DS/87.74). WHO, Geneva, Switzerland.

World Health Organization (WHO). 2001b. International Programme on Chemical Safety (IPCS). *Phosphamidon*. Poisons Information Monograph No. 454. WHO, Geneva, Switzerland.

World Health Organization (WHO) International Programme on Chemical Safety and the European Commission (IPCS-EC). 2004. Paraquat dichloride. MSDS No. ICSC. 0005. WHO. Geneva, Switzerland.

Worthing, C. R. and S. B. Walker (eds.). 1987. *The Pesticide Manual—A World Compendium*, 8th edn. The British Crop Protection Council, Thornton Heath, U.K.

Yobs, A. R. 1972. Polychlorinated biphenyls in adipose tissue of the general population of the nation. *Environ. Health Perspect.* 1:79–80.

18

Hazardous Chemical Substances:
Q

Quinalphos
Quinone (*p*-Benzoquinone)
Quinoline
Quintozene

Quinalphos (CAS No. 82-68-8)

Molecular formula: $C_6Cl_5NO_2$
WHO classification: Class II – Moderately hazardous
Synonyms and trade names: Nitropentachlorobenzene; PCNB; Avicol; Batrilex; Botrilex; Brassicol; Chinozan; Folosan; Fomac 2; Fungichlor; Kobu; Kobutol; Marisan forte; Olpisan; Pentagen; Quintocene; Quintozene; Terrachlor; Tilcarex; Tri-pcnb; Tritisan

Quinalphos is an off-white powder. It is stable and incompatible with strong bases and strong oxidising agents. Quinalphos is used as a fungicide and herbicide. Quinalphos has different formulations such as emulsifiable concentrate 25% w/w, granules 5%, and dust 1.5%. Quinalphos is compatible with most insecticides and fungicides.

Quinalphos has been found to control effectively pests like caterpillars on fruit trees, cotton, vegetables, and peanuts and scale insect on fruit trees and pest complex on rice. Quinalphos also controls aphids, bollworms, borers, leafhoppers, mites, thrips, etc., on vines, ornamentals, potatoes, soya beans, tea, coffee, cocoa, and other crops.

Safe Handling and Precautions

Exposures to quinalphos cause toxicity and adverse health effects. On contact with skin, quinalphos causes effects of sensitisation. The acute oral LD_{50} of technical quinalphos was found to be 19.95 mg/kg in male rats and 13.78 mg/kg in female rats. Administration by oral route of 0.75, 1.50, or 3.0 mg/kg/day technical quinalphos for a period of 90 days to experimental laboratory rats produced poisoning and death. Pattern of mortality, enzyme profiles, and cholinesterase inhibition of subchronic toxicity studies suggested that male rats were found to be more susceptible to quinalphos than female rats. Quinalphos is an experimental carcinogen and neoplastigen.

Quinone (*p*-Benzoquinone) (CAS No. 106-51-4)

Molecular formula: $C_6H_4O_2$
Synonyms and trade names: 1,4-Benzoquinone; Benzoquinone; Chinone; *p*-Benzoquinone.

Quinone (*p*-benzoquinone) exists as a large yellow, monoclinic prism with an irritating odour resembling that of chlorine. Quinone is extensively used as a chemical intermediate, a polymerisation inhibitor, an oxidising agent, a photographic chemical, a tanning agent, and a chemical reagent. Quinone (*p*-benzoquinone) was first produced commercially in 1919 and has since been manufactured in several European countries. Its major use is in hydroquinone production, but it is also used as a polymerisation inhibitor and as an intermediate in the production of a variety of substances, including rubber accelerators and oxidising agents. It is used in the dye, textile, chemical, tanning, and cosmetic industries. In chemical synthesis for hydroquinone and other chemicals, quinone is used as an intermediate. It is also used in the manufacturing industries and chemical laboratory associated with protein fibre, photographic film, hydrogen peroxide, and gelatin making. Occupational exposure to quinone may occur in the dye, textile, chemical, tanning, and cosmetic industries. Inhalation exposure to quinone may occur from tobacco smoke.

Safe Handling and Precautions

Acute exposure to quinone (*p*-benzoquinone) is known to cause local skin changes including discolouration, erythema, and the appearance of papules; necrosis can occur. Exposure to vapours induces serious vision disturbances; injury extends through the entire conjunctiva and cornea. High levels of quinone, via inhalation in humans, are highly irritating to the eyes, resulting in discolouration of the conjunctiva and cornea, while dermal exposure causes dermatitis with skin discolouration and erythema. Animal studies have reported effects on the kidneys from exposure to quinone. Exposures to vapour of quinone are highly irritating to the eyes and may be followed by corneal opacities and structural changes in cornea and loss of visual activity. Solid quinone may produce discolouration, severe irritation, swelling, and formation of papules and vesicles. Chronic dermal contact to quinone in humans may result in skin ulceration, while chronic inhalation exposure may result in visual disturbances. There is no published information available in literature on the carcinogenic effects of quinone in humans.

The International Agency for Research on Cancer (IARC) observed inadequate or insufficient evidence in experimental animals and humans to conclude quinone (*p*-benzoquinone) as a human carcinogen. Quinone (*p*-benzoquinone) has, therefore, been listed by IARC as Group 3, meaning not classifiable as a human carcinogen or to its carcinogenicity to humans, while the U.S. Environmental Protection Agency (U.S. EPA) has not classified quinone for carcinogenicity.

Quinoline (CAS No. 91-22-5)

Molecular formula: C_9H_7N
Synonyms and trade names: 1-Azanaphthalene; 1-Benzazine; 2,3-Benzo(b)pyridine; Chinoleine; Chinoline; Leucoline; Leucol

Quinoline is a colourless hygroscopic liquid with characteristic odour. On exposure to light, it turns brown in colour. Quinoline decomposes on heating, and on burning produces toxic fumes including nitrogen oxides. Quinoline reacts with strong oxidants, acids, and anhydrides. Quinoline is only slightly soluble in cold water but dissolves readily in hot water and most organic solvents. Quinoline is combustible. It gives off irritating or toxic fumes (or gases) in a fire. Quinoline is incompatible with strong acids, oxidisers, dinitrogen tetroxide, linseed oil, thionyl chloride, maleic anhydride, and perchromates and reacts violently with most incompatibles. Quinoline is used extensively in the manufacturing of dyes, preparation of hydroxyquinoline sulphate and niacin, as a solvent for resins and terpenes, and as an intermediate in the manufacture of other products.

Quinoline is used mainly as an intermediate in the manufacture of other several products, as a catalyst, as a corrosion inhibitor, in metallurgical processes, in the manufacture of dyes, as a preservative for anatomical specimens, in polymers and agricultural chemicals, and as a solvent for resins and terpenes. Quinoline is also used as an anti-malarial medicine. Because of its solubility in water, quinoline has significant potential for mobility in the environment, which may promote water contamination. Potential exposure to quinoline also occurs from the inhalation of cigarette smoke. Quinoline breaks down quickly in the atmosphere and water.

Safe Handling and Precautions

Studies have shown that occupationally workers in certain industries get exposed to quinoline by inhalation, ingestion of particulates, or dermal contact. Also, workers get exposed to quinoline by consumption of contaminated water. However, quinoline breaks down quickly in water. An increased incidence of liver vascular tumours has been observed in rats and mice orally exposed to quinoline. Workers with negligence and lack of proper workplace protective dress and after short-term/acute exposure to quinoline have been reported to develop poisoning with symptoms of headache, dizziness, nausea, lethargy, irritation, sore throat, nose bleeding, hoarseness, cough, chest tightness, difficulty in breathing, nausea, vomiting, dizziness, and fever. Quinoline has been reported to be toxic to the retina or optic nerve. Inhalation of quinoline vapours causes irritation of the eyes, skin, throat, and respiratory tract and respiratory muscle paralysis. Extreme exposures could result in a build-up of fluid in the lungs (pulmonary oedema) that might be fatal in severe cases.

The target organs of quinoline-induced toxicity include liver, respiratory system, eyes, skin, and optic nerve. There are no published information on the chronic (long-term) toxicity and reproductive, developmental, or carcinogenic effects of quinoline in humans. Liver damage has been observed in rats chronically exposed to quinoline by ingestion. Exposures to quinoline at high concentrations cause coma in humans. While the American Conference of Governmental Industrial Hygienists (ACGIH), National Institute for Occupational Safety and Health (NIOSH), and Occupational Safety and Health Administration (OSHA) have not listed quinoline as a human carcinogen, the U.S. EPA has provisionally classified quinoline as a Group C, meaning possible human carcinogen.

Occupational workers should handle quinolone with care and should keep quinoline stored away from heat, light, flame, ignition sources, light, moisture, and incompatibles. And quinoline should be kept in a tightly closed container, in a cool, dry, ventilated area and protected against physical damage. Users and occupational workers should wear impervious protective clothing, including safety goggles and full-face shield to avoid chemical splashing and boots, gloves, lab coat, apron or coveralls, as appropriate, to prevent skin contact.

Warning! Quinoline is harmful if swallowed, inhaled, or absorbed through skin. Improper handling causes poisoning: It causes irritation to skin, eyes, and respiratory tract. It affects the central nervous system. It is a combustible liquid and vapour.
    Regulations:

- *United States*: 21 CFR (Part 74) lists quinoline in the approved colour as D&C Yellow No. 10. The specifications are listed in 21 CFR (Section 74.1710) which requires that the dye contain not less than 75% of the monosulphonated component and not more than 15% of the disulphonated component. Presently, D&C Yellow No. 10 is approved for use in drugs and cosmetics and not approved for food uses. However, quinoline is not acceptable for use in foods or drugs in Europe due to a difference in the specifications of the monosulphonated and disulphonated components of the dye.

- *European Union*: EC Directive 94/36/EC lists quinoline in the approved colour – Yellow (E104). The specific purity requirements are listed in EC Directive 95/45/EC, which requires that the material contain not less than 80% of the disulphonated component and not more than 15% of the monosulphonated component. Quinoline Yellow (E104) is approved in Europe for use in food and drugs. Therefore, by definition, a material that meets both the United States and the European specifications for use in drugs is available. Presently, D&C Yellow No. 10 and Quinoline Yellow (E104) are NOT approved for use in foods or drugs.

- *Canada*: The Canadian Workplace Hazardous Materials Information System (WHMIS) classifications D2B, D1B, and D2A classify D&C Yellow No. 10 and Quinoline Yellow (E104) in accordance with the hazard criteria of the Controlled Products Regulations, and the Material Safety Data Sheet (MSDS) contains all of the information required by those regulations.

Quintozene (CAS No. 82-68-8)

Molecular formula: $C_6Cl_5NO_2$
Synonyms: PCNB; Pentachloronitrobenzene
Chemical class: organochlorine
IUPAC: Pentachloronitrobenzene

Quintozene appears as a light brown-coloured granule with a mild chemical odour. Quintozene is soluble in all organic solvents such as acetonitrile, cyclohexane, ethanol, ethyl acetate, heptane, methanol, and toluene. On hazardous decomposition, quintozene releases phosgene, hydrogen chloride, oxides of nitrogen, and chlorine-containing compounds, and other unknown materials may be released in a fire situation. Incomplete combustion may lead to the formation of oxides of carbon.

    As a fungicide, quintozene is registered for use in Australia as a seed dressing; as a seedling drench; a pre-plant soil-applied fungicide for vegetables, cotton, and ornamentals; and as a pre-emergence fungicide for cotton. Quintozene is a soil-applied fungicide that controls a wide range of soil-borne plant diseases. It has registrations covering vegetable crop, seed beds, turf, and ornamental crops. It is also used to control fungal diseases on bowling greens and golf greens and for a small number of post-emergence uses on lettuce, peanuts, apples, and ornamentals. Available information suggests that the majority of its use is as a seed dressing and on high-value turf such as bowling and golf greens. The product is not believed to be supplied or used in the home garden or in commercial turf production.

Safe Handling and Precautions

Occupational exposure to quintozene is known to cause adverse health effects to workers. The symptoms of poisoning include, but are not limited to, vomiting, hyperirritability, convulsions, and the possible development of pulmonary oedema.

A human health risk assessment carried out by the Office of Chemical Safety and Environmental Health in the Department of Health and Ageing for the Australian Pesticides and Veterinary Medicines Authority (APVMA) indicated that dioxin levels in quintozene products could be a risk to the health of occupational workers during use in some situations. There is limited information on the effects of quintozene in the general environment. It has been shown to be toxic for earthworms in laboratory tests. Data on other organisms suggest that quintozene does not pose a problem in the general environment.

Occupational workers should keep quintozene out of reach of children and should store the product in its original container, tightly closed, in a dry, ventilated area. Occupational workers should wear clean rubber gloves while handling the product. Workers should wash thoroughly after handling the chemical and before eating, drinking, or smoking and take care to avoid contamination of feed or foodstuffs. Also, during waste disposal, workers should not contaminate ponds, lakes, streams, or other bodies of water since this product is toxic to fish. Dispose of the container in accordance with provincial requirements.

CAUTION! Occupational workers should avoid breathing dusts and prolonged or repeated skin contact of quintozene – it causes skin irritation and harmful if swallowed or ingested.

Regulations: Quintozene containing more than 1 g/kg of HCB or more than 10 g/kg pentachlorobenzene is prohibited in the European Union in use as plant protection products.

Formulations of quintozene include emulsifiable concentrate (EC), dustable powder (DP), flowable concentrate for seed treatment (FS), granule (GR), solution for seed treatment (LS), for seed coated with a pesticide (PS), as suspension concentrate (flowable concentrate, SC), and wettable powder (WP).

- *Canadian regulations:* Quintozene 75WP is registered under the Pest Control Product Act of Canada. It is a violation of Canadian Law to use this product in any manner inconsistent with its labelling. Read and follow all label directions. This product has been classified according to the hazard criteria of the CPR, and the MSDS contains all the information required by the CPR.

- *European Union regulations:* Quintozene 75WP is not currently registered in the European Union. Quintozene is a fungicide registered for use in Australia as a seed dressing; as a seedling drench; a pre-plant soil-applied fungicide for vegetables, cotton, and ornamentals; and as a pre-emergence fungicide for cotton. It is also used to control fungal diseases on bowling greens and golf greens and for a small number of post-emergence uses on lettuce, peanuts, apples, and ornamentals. Information to date suggests that the majority of its use is as a seed dressing and on high-value turf such as bowling and golf greens. The product is not believed to be supplied or used in the home garden or in commercial turf production. Ten products are registered in Australia. These products are relatively expensive to use on an area basis, and use volumes are therefore low. Testing undertaken by the APVMA on particular batches of the quintozene active constituent and products containing quintozene indicated that those batches contained undeclared high levels of dioxins. A human health risk assessment carried out by the Office of Chemical Safety and Environmental Health in the Department of Health and Ageing for the APVMA has shown that dioxin levels in quintozene products from those batches could be a risk to the health of workers applying the products in some situations. The contaminated active constituent from which the Australia products are made is imported from overseas. Other countries, such as the United States and Canada, use the same active constituent source.

Bibliography

Anderson, B. and F. Oglesby. 1958. Corneal changes from quinone-hydroquinone exposure. *Arch. Ophthalmol.* 59:495.

Asakura, S., S. Sawada, T. Sugihara et al. 1997. Quinoline-induced chromosome aberrations and sister chromatid exchanges in rat liver. *Environ. Mol. Mutagen.* 30(4):459–467.

Clayton, G. D. and F. E. Clayton (eds.). 1981. *Patty's Industrial Hygiene and Toxicology*, Vol. IIA, 3rd revised edn. John Wiley & Sons, New York.

Dikshith, T. S. S. (ed.). 1991. *Toxicology of Pesticides in Animals.* CRC Press, Boca Raton, FL.

Dikshith, T. S. S. 2011. *Handbook of Chemicals and Safety*, CRC Press, Boca Raton, FL.

International Programme on Chemical Safety and the Commission of the European Communities (IPCS-CEC). 2005. Quinoline. International Chemical Safety Card. ICSC Card No. 0071. IPCS-CEC, Geneva, Switzerland (updated 2008).

Material Safety Data Sheet (MSDS). 2005. Quinalphos. Department of Physical Chemistry. Oxford University, Oxford, U.K.

NIOSH Pocket Guide to Chemical Hazards. 2010. Quinone. Centers for Disease Control and Prevention. Atlanta, GA.

Raizada, R. B., M. K. Srivastva, R. P. Singh, R. P. Kaushal, K. P. Gupta, and T. S. Dikshith. 1993. Acute and subchronic oral toxicity of technical quinalphos in rats. *Vet. Hum. Toxicol.* 35(3):223–225.

Seutter, E. and A. H. M. Sutorius. 1972. Quantitative analysis of hydroquinone in urine. *Clin. Chim. Acta.* 38:231.

Sittig, M. (ed.). 1985. *Handbook of Toxic and Hazardous Chemicals and Carcinogens*, 2nd edn. Noyes Publications, Park Ridge, NJ.

U.S. Department of Health and Human Services. 1993. Quinone. Hazardous Substances Data Bank (HSDB). National Toxicology Information Program, National Library of Medicine, Bethesda, MD (updated 2007).

U.S. Environmental Protection Agency (U.S. EPA). 1985. Health and environmental effects profile for quinoline. (EPA/600/x-85/355). Environmental Criteria and Assessment Office, Office of Health and Environmental Assessment, Office of Research and Development, Cincinnati, OH.

U.S. Environmental Protection Agency (U.S. EPA). 2001. *Toxicological Review of Quinoline in Support of Summary Information on the Integrated Risk Information System (IRIS).* National Center for Environmental Assessment, Washington, DC.

U.S. Environmental Protection Agency (U.S. EPA). 2007. Quinoline. Integrated Risk Information System (IRIS), National Center for Environmental Assessment, Washington, DC (updated 2011).

19

Hazardous Chemical Substances: R

Reserpine
Resorcinol
Rhodium Chloride, Trihydrate
Rhodium Sulphate

Reserpine (CAS No. 50-55-5)

Synonyms and trade names: 17-alpha-dimethoxy-, methyl ester; 20-alphaYohimbane-16-beta-carboxylate; Abesta; Abicol; Adelfan; Adelphane; Adelphin; Alkarau; Alkserp; Apoplon; Ascoserp; Ascoserpina; Banasil; Bioserpine; Crystoserpine; Deserpine; Ebserpine; Elserpine; Gammaserpine; Helfoserpin; Hexaplin; Orthoserpina; Raulolycin; Rauloydin; Raumorine; Rauserpen-Alk; Rauserpin, alkaloid

Reserpine is an odourless, tasteless, fine white or yellow crystalline powder and is insoluble in water but soluble in alcohol, acetone, chloroform, and acetic acid. Reserpine on exposure to light turns to dark colour. Reserpine is an herbal extract and an alkaloid. Reserpine is produced by several members of the genus *Rauwolfia*, a climbing shrub indigenous to Southern and Southeast Asia. Extracts of *Rauwolfia serpentina* have been used medicinally in India for centuries. They were used in traditional Hindu medicine for a variety of conditions, including snakebite, hypertension, insomnia, and insanity. Reserpine has also been used as a tranquiliser and sedative in animal feeds. Reports also indicate that reserpine also has been used as a radioprotective agent and experimentally as a contraceptive.

Reserpine is used to treat high blood pressure, to treat severe agitation in patients with mental disorders, and to reduce the heart rate and as a tranquiliser and sedative in humans. The use of reserpine as a drug may result in its release to the environment in various waste streams. Occupational exposure to reserpine has been reported to occur at workplaces through inhalation and/or dermal contact.

Safe Handling and Precautions

Exposure to reserpine is known to cause side effects with symptoms that include, but are not limited to, dizziness, loss of appetite, diarrhoea, upset stomach, vomiting, stuffy nose, headache, and dry mouth. The severe adverse health effects of reserpine are depression, nightmares, fainting, slow heartbeat, chest pain, and swollen ankles or feet. Exposure to reserpine works by slowing the activity of the nervous system, causing the heartbeat to slow and the blood vessels to relax.

Exposure to high doses of reserpine caused fibroadenoma of the breast in experimental animals.

Accidental ingestion of reserpine causes poisoning with symptoms such as lethargy, sedation, depression, hypothermia, facial flushing, nausea, vomiting, abdominal cramping, cardiovascular toxicity, hypotension, bradycardia, and coma.

The International Agency for Research on Cancer (IARC) has reported that evidences on the carcinogenic potentials of reserpine in animals are limited and inadequate in humans. Hence, reserpine has been listed as Group 3, meaning not classifiable as a human carcinogen.

Resorcinol (CAS No. 108-46-3)

Molecular formula: $C_6H_6O_2$
Synonyms and trade names: Resorcin; 1,3-Benzenediol; *m*-Dihydroxybenzene

Resorcinol is a white solid/powder and flammable in the presence of open flames and sparks and heat. Resorcinol is a combustible chemical substance. It is soluble in cold water and diethyl ether. Resorcinol is incompatible with acetanilide, albumin, alkalies, antipyrine, camphor, ferri salts, menthol, spirit of nitrous ether, urethane, and periodate. When heated to decomposition, it emits acrid smoke and irritating fumes. Resorcinol has been reported to cause potentially explosive reaction with nitric acid.

Resorcinol is an important chemical intermediate in specialty chemicals manufacturing, such as light screening agents used to protect plastics from exposure to ultraviolet light. Wide are the industrial applications of resorcinol, for instance, manufacture of dyestuffs, pharmaceuticals, flame retardants, agricultural chemicals, fungicidal creams and lotions, explosive primers, anti-oxidants, a chain extender for urethane elastomers, a treatment to improve mechanical and chemical resistance of paper machine fabrics, and as an important component of an adhesive system used in the tyre manufacturing process and other fibre-reinforced rubber mechanical goods. The adhesives formulated from resorcinol–formaldehyde resins or phenol-modified resorcinol–formaldehyde resins are the criteria for wood bonding applications demanding room temperature cure, structural integrity, and waterproof characteristics. Resorcinol is a specific inhibitor of polyphenol oxidase, and, therefore, it can act as an anti-browning agent.

Safe Handling and Precautions

Laboratory animals exposed to high doses of resorcinol showed clinical signs of toxicity such as hyperexcitability, tachypnoea, ataxia, prostration, and tremors, which were observed in mice and rats in subacute, subchronic, and chronic toxicity studies. On contact at workplaces, resorcinol gets absorbed through the skin, by inhalation, and by accidental ingestion. Users and occupational workers exposed to resorcinol for a prolonged period of time and repeatedly have been reported to develop adverse health effects and target organ damage and injury to the skin, eyes, blood, liver, endocrine, and cardiovascular systems.

The American Conference of Governmental Industrial Hygienists (ACGIH), the IARC, and the National Toxicology Program (NTP) have all reported that there was no evidence to classify resorcinol as a human carcinogen.

During use, handling, and chemical waste disposal of resorcinol, workers should be alert and should use workplace dress and approved/certified personal protective equipment (PPE) such as splash goggles, gloves, and dust respirator. Workers should strictly avoid breathing dust or vapour and contact of dust with skin, eyes, or clothing.

Rhodium Chloride, Trihydrate (CAS No. 13569-65-8)

Molecular formula: $RhCl_3 3H_2O$
Synonyms and trade names: Rhodium (III) chloride, anhydrous

Rhodium trichloride is an odourless, red-brown to black crystalline powder. It is corrosive to steel and aluminium. It is hygroscopic and absorbs moisture or water from the air. It is incompatible with strong oxidising agents. On decomposition, rhodium trichloride emits hydrogen chloride, carbon monoxide, carbon dioxide, and rhodium/rhodium oxides.

Safe Handling and Precautions

Exposure to rhodium trichloride causes severe eye irritation and possible injury.

Warning! It may be harmful if swallowed. Target organs: respiratory system, eyes, and skin.

Potential health effects:

- *Eyes*: Causes eye irritation, risk of serious damage to eyes.
- *Skin*: May cause skin irritation. May be harmful if absorbed through the skin.
- *Ingestion*: May cause gastrointestinal irritation with nausea, vomiting, and diarrhoea. May be harmful if swallowed.
- *Inhalation*: May cause respiratory tract irritation. May be harmful if inhaled.
- *Chronic*: Adverse reproductive effects have been reported in animals. Animal studies have reported the development of tumours.

Reports indicate that rhodium trichloride has not been listed as a carcinogen by the following: ACGIH, IARC, and NTP.

Rhodium chloride and trihydrate decomposes on heating producing toxic and corrosive fumes including hydrogen chloride (see ICSC 0163). It reacts violently with iron pentacarbonyl and zinc causing explosion hazard. Rhodium (III) chloride is not listed under Annex I of Directive 67/548/EEC, but is usually classified as harmful if swallowed. Some Rh compounds have been investigated as anti-cancer drugs. It is listed in the inventory of the Toxic Substances Control Act (TSCA).

During use of rhodium (III) chloride trihydrate, workers should use PPE and must avoid dust formation and breathing dust and should take care of adequate ventilation at the workplace. Rhodium (III) chloride trihydrate should be kept separated from incompatible materials.

Rhodium Sulphate (CAS No. 10489-46-0)

Rhodium plating solution: Rhodium plating solution is a yellow-orange liquid and is stable.

When heated to decomposition, this product can emit acid mists and toxic gases (including oxides of sulphur and rhodium compounds). This product is not compatible with bases, halides, water, cyclopentadiene, cyclopentanone, oxime, nitroaryl amines, haxalithium disilicide, phosphorus (III) oxide, chlorine bromine pentafluoride, trifluoride, and oxygen difluoride (OF_2). Avoid contact with metals. This product can react with water to generate heat.

Safe Handling and Precautions

Rhodium plating solution is a corrosive solution/liquid. Occupational workers exposed to solutions/liquids and vapours of rhodium plating solution get rapid harmful effects as demonstrated by tissue damage.

Rhodium plating solution on repeated contact is also known to cause chemical burns and severe irritation of the contaminated tissues – the skin, eyes, and mucous membranes. Inhalation of vapours or liquid may cause lung injury, the effects of which may not be apparent for up to 48 h. Rhodium plating solution, if ingested or inhaled or swallowed, is known to cause fatal injury. The prolonged or repeated inhalation overexposures can cause burns and ulcers to the nose and throat, dental erosion, bronchitis, and stomach pain. Prolonged or repeated skin exposure can cause dermatitis. Target organs of rhodium-induced toxicity include skin, eyes, and respiratory system. The severity of tissue damage depends on the concentration of the solution and the duration of tissue contact. The symptoms of toxicity and hazard of rhodium plating solution include mists or vapours of this product can cause nasal irritation, sore throat, choking, coughing, and breathing difficulties.

Alert! Precautions

Though unlikely to cause severe injuries in small volume, it is important to note that inhalation of mists of rhodium plating solution (even for a few minutes) can cause severe lung damage with potentially life-threatening pulmonary oedema (accumulation of fluid in the lungs). Symptoms of pulmonary oedema include shortness of breath and chest pains; symptoms can be delayed for up to 48 h after exposure. Prolonged or repeated overexposures to this solution can cause burns and ulcers to the nose and throat, dental erosion, bronchitis, and stomach pain.

Repeated contact of rhodium plating solution and its vapours with skin or eyes can cause severe burns; dermatitis; red, cracked, and irritated skin; and ulceration, depending on the concentration and duration of exposure. Contact with eyes may result in permanent scarring and/or blindness. Skin absorption is not expected to be a significant route of occupational exposure for any component of this product. The only anticipated route of skin absorption is through breaks in the skin. This corrosive solution will cause burns of contaminated areas (see 'Contact with skin and eyes').

Although not anticipated to be a significant route of occupational overexposures, ingestion of this product may be fatal. Swallowing this material may cause burns in the mouth, throat, oesophagus, and other tissues. Symptoms can include difficulty in swallowing, intense thirst, nausea, vomiting, diarrhoea, and, in severe cases, collapse and death. Small amounts of acid can be aspirated during vomiting and may cause serious lung injury.

Safe Handling

Occupational workers and users who handle the chemical and hazardous materials should be trained to handle it safely. Workers should always use the chemical product in well-ventilated area and must ensure containers of this product are properly labelled. Open containers carefully, on a stable surface. Close containers tightly after use. When diluting this solution, slowly add the product to the water to prevent splattering. Workers should keep the containers in a cool, dry location, away from direct sunlight and sources of intense heat and away from incompatible materials and in secondary containment. Inspect all incoming containers before storage to ensure containers are properly labelled and not damaged. Users and workers should periodically inspect the containers of rhodium plating solution for leaks or damage. Empty containers may contain corrosive liquids or vapours and should be handled with care.

Workers should avoid exposure to incompatible materials or to extreme heat, as product can decompose, producing acid mists and toxic gases – sulphur oxides and rhodium compounds.

Alert – Rhodium Plating Solution – For industrial use only. DANGER! CAUSES SEVERE BURNS – Contains Sulphuric Acid – Keep away from children.

Bibliography

Anonymous. 1976. Severe depression caused by reserpine. *Med. Lett. Drugs Ther.* 18(4):19–20.

Darwin, R. L. and W. M. O'Fallon. 1980. Reserpine and breast cancer. A community-based longitudinal study of 2,000 hypertensive women. *JAMA* 243:2304–2310.

Dikshith, T. S. S. (ed.). 2011. *Handbook of Chemicals and Safety.* CRC Press, Boca Raton, FL.

Dikshith, T. S. S. and P. V. Diwan. (eds.). 2003. *Industrial Guide to Chemical and Drug Safety.* John Wiley & Sons, Hoboken, NJ.

Ellenhorn, M. J., S. Schonwald, G. Ordog, and J. Wasserberger. 1997. *Ellenhorn's Medical Toxicology: Diagnosis and Treatment of Human Poisoning,* 2nd edn. Williams & Wilkins, Baltimore, MD.

International Agency for Research on Cancer (IARC). 1982. Reserpine. *Chemicals, Industrial Processes and Industries Associated with Cancer in Humans. IARC Monographs on the Evaluation of Carcinogenic Risk of Chemicals to Humans,* Supplement 4, Lyon, France.

International Agency for Research on Cancer (IARC) – Summaries and evaluations. 1987. Reserpine. Supplement No. 7: 330.

International Programme on Chemical Safety (IPCS). 2006. Concise International Chemical Assessment document; 71. Resorcinol. World Health Organization, Geneva, Switzerland.

International Programme on Chemical Safety and the Commission of the European Communities (IPCS–CEC). 2005. Rhodium (III) chloride trihydrate. Material Safety Data Sheet (MSDS).

Katin, M. J., B. P. Teehan, M. H. Sigler, C. R. Schleifer, and C. S. Gilgore. 1977. Resorcinol-induced hypothyroidism in a patient on chronic hemodialysis. *Ann. Int. Med.* 86:447–449.

Loggie, J. M. H., H. Saito, I. Kahn, A. Fenner, and T. E. Gaffney. 1967. Accidental reserpine poisoning: Clinical and metabolic effects. *Clin. Pharmacol. Ther.* 8:692–695.

Lynch, B. S., E. S. Delzell, and D. H. Bechtel. 2002. Toxicology review and risk assessment of resorcinol: Thyroid effects. *Regul. Toxicol. Pharmacol.* 36:198–210.

National Toxicology Program (NTP). 2011. Reserpine. *Report on Carcinogens*, 12th edn. NTP, Department of Health and Human Services (DHHS). Research Triangle Park, NC.

NIOSH Pocket Guide to Chemical Hazards. 2010. Rhodium (soluble compounds, as Rh). Centers for Disease Control and Prevention, Atlanta, GA (updated 2011).

Samuels, A. H. and A. J. Taylor. 1989. Reserpine withdrawal psychosis. *Aust. NZ J. Psychol.* 23:129–130.

Segal, M. S. 1969. Bronchospasm after reserpine. *N Engl. J. Med.* 281(25):1426–1427.

Technical Report. 1987. Technical guidance for hazards analysis emergency planning for extremely hazardous substances. U.S. Environmental Protection Agency (U.S. EPA), Federal Emergency Management Agency (FEMA), U.S. Department of Transportation (DOT), Washington, DC.

20

Hazardous Chemical Substances:
S

Samarium
Samarium Compounds
   Samarium (III) Fluoride
   Samarium (III) Nitrate Hexahydrate
   Samarium Oxide
Sarin
Selenium and Its Compounds
   Selenious Acid
   Selenium Chloride
   Selenium (IV) Oxide
   Selenium Sulphide
Silicon Sulphide
Styrene
Sulphur Dioxide

Samarium (CAS No. 7440-19-9)

Molecular formula: **Sm**

Samarium occurs as silver-coloured solid/foils or grey powder and is an odourless, flammable, and water-reactive solid. All forms of samarium are known to react with dilute acids emitting flammable/explosive hydrogen gas. Samarium on contact with water reacts and liberates extremely flammable gases. Samarium is incompatible with strong acids, strong oxidising agents, and halogens. The major commercial application of samarium is in samarium–cobalt magnets. These magnets possess permanent magnetisation property. Samarium compounds have been shown to withstand significantly higher temperatures, above 700°C, without losing their magnetic properties. The radioactive isotope samarium-153 is the major component of the drug samarium 153Sm lexidronam (Quadramet). These are used in the treatment of cancers of lung, prostate, and breast and osteosarcoma. Samarium is also used in the catalysis of chemical reactions, radioactive dating, and in x-ray laser. Samarium is used as a catalyst in certain organic reactions: Samarium iodide (SmI_2) is used by organic research chemists to make synthetic versions of natural products.

Samarium occurs with concentration up to 2.8% in several minerals including cerite, gadolinite, samarskite, monazite, and bastnäsite, the last two being the most common commercial sources of the element. These minerals are mostly found in China, the United

States, Brazil, India, Sri Lanka, and Australia; China is by far the world leader in samarium mining and production. Radioactive isotope samarium-153 is the major component of the drug samarium (153Sm) lexidronam (Quadramet), which kills cancer cells in the treatment of lung cancer, prostate cancer, breast cancer, and osteosarcoma. The isotope, samarium-149, is a strong neutron absorber and is therefore added to the control rods of nuclear reactors. Samaria oxide is used for making special infrared-adsorbing glass and cores of carbon arc-lamp oxide electrodes and as a catalyst for the dehydration and dehydrogenation of ethanol. Its compound with cobalt ($SmCo_5$) is used in making a new permanent magnet material. Samarium has no biological role, but it has been noted to stimulate metabolism. Soluble samarium salts are mildly toxic by ingestion, and there are health hazards associated with these because exposure to samarium causes skin and eye irritation. One of the most important applications of samarium is in samarium–cobalt magnets, which have a nominal composition of $SmCo_5$ or Sm_2Co_{17}. They have high permanent magnetisation, which is about 10,000 times that of iron and is second only to that of neodymium magnets.

Safe Handling and Precautions

Exposures to samarium cause eye irritation, lacrimation (tearing), blurred vision, and photophobia. Occupational exposure to and accidental ingestion of samarium cause nausea, vomiting, dizziness, fatigue, abdominal pain, increased salivation, irritation of the digestive tract, cardiac disturbances, central nervous system effects, and memory difficulties. Samarium has not been listed by the International Agency for Research on Cancer (IARC) or American Conference of Governmental Industrial Hygienists (ACGIH) or National Toxicology Program (NTP) as a human carcinogen. There is no published information regarding the possible adverse health effects after chronic exposures.

Occupational workers should be careful during handling of samarium because it is known to cause eye and skin irritation, respiratory tract irritation, and air sensitivity. The toxicological properties of this material have not been fully investigated. Users and occupational workers must be alert during handling of samarium and should avoid creating fine dusts, because as a powder, this product is capable of creating a dust explosion. Workers should use personal protective equipment (PPE), safety glasses with side shields, and chemical workers' dust-proof goggles, and the workplace must maintain a sink, safety shower, and eyewash fountain in the work area. Workers should keep stored samarium in cool, dry place in tightly closed containers. Samarium is air and moisture sensitive and workers should keep it stored away from oxidisers and other incompatible materials.

Samarium Compounds

Samarium compounds are extremely stable magnetic materials, having high-energy products, high remembrance, and high coercive field strengths used in small motors, printers, quartz watches, headphones, loudspeakers, magnetic storage, and travelling-wave tubes. Samarium is a rare earth element that is extracted from natural ore materials such as mineral monazite and bastnäsite. As such, the use of this material may require a radioactive materials licence. It is recommended that you contact your state radiation protection programme for guidance in this matter. Specifically, samarium powder may be associated with sufficient concentrations of source material that can result in having a licensable radioactive quantity: (i) samarium fluoride, (ii) samarium nitrate, and (iii) samarium oxide.

Samarium (III) Fluoride (CAS No. 13765-24-7)

Molecular formula: SmF_3
Synonyms and trade names: Samarium trifluoride; Trifluorosamarium

Samarium (III) fluoride is a yellow solid or odourless powder and slightly hygroscopic. It is insoluble in water. Workers should avoid open flame, moisture, and strong acids.

Samarium (III) Nitrate Hexahydrate (CAS No. 13759-83-6)

Molecular formula: $Sm(NO_3)_3 \cdot 6H_2O$

Samarium (III) nitrate hexahydrate is a yellow to tan powder and chunks in physical appearance and is an oxidiser. It has application as a chemical reagent. It is hygroscopic – absorbs moisture from the air.

Safe Handling and Precautions

Exposure to samarium (III) nitrate hexahydrate causes harmful effects. The signs and symptoms of poisoning include, but are not limited to, nausea, vomiting, confusion, dizziness, drowsiness, headache, shortness of breath, respiratory depression, cyanosis (bluish discolouration of skin due to deficient oxygenation of the blood), rapid heart rate and chocolate-brown-coloured blood, convulsions, and death. The toxicological properties of this substance have not been fully investigated. Ingestion of nitrate-containing compounds can lead to methaemoglobinaemia.

Exposure to samarium (III) nitrate hexahydrate is known to cause severe eye, skin, and respiratory and digestive tract irritation. Samarium (III) nitrate hexahydrate causes methaemoglobinaemia. The toxicological properties of this material have not been fully investigated.

Alert! Danger! Occupational workers should be careful during use and handling of samarium (III) nitrate hexahydrate, in a well-ventilated area. Minimise dust generation and accumulation at the workplace. Workers should wear appropriate protective eyeglasses or chemical safety goggles, eye and face protection, and appropriate protective gloves to prevent skin exposure. Workers should avoid contact of the chemical with the eyes, skin, and clothing.

Workers should keep the chemical container tightly closed, in a cool, dry, well-ventilated area away from incompatible substances and away from heat, sparks, flame, and combustible materials.

Samarium Oxide (CAS No. 12060-58-1)

Molecular formula: Sm_2O_3

Samarium oxide is a light yellow, odourless powder and incompatible with strong acids, oxidising agents, and water/moisture.

Samarium is considered a rare earth metal. These metals are moderately to highly toxic. The symptoms of toxicity of the rare earth elements include writhing, ataxia, laboured respiration, walking on the toes with arched back, and sedation. Rare earth oxides are much less toxic than the chlorides or citrates, known to cause skin and lung granulomas, writhing, ataxia, laboured respiration, and lung granulomas. Repeated exposures affect the blood and lungs. The chemical, physical, and toxicological properties of samarium oxide have not been thoroughly investigated and recorded. Workers should use good housekeeping and sanitation practices and maintain eyewash capable of sustained flushing, safety drench shower, and facilities for washing very close to workplace.

Sarin (CAS No. 107-44-8)

Molecular formula: $C_4H_{10}FO_2P$
Chemical name: O-isopropyl methylphosphonofluoridate
Synonyms and trade names: Phosphonofluoridic acid, methyl-, isopropyl ester; Phosphonofluoridic acid, methyl-, 1-methylethyl ester; Methylfluorophosphonic acid, isopropyl ester; Isopropoxymethylphosphonyl fluoride; Isopropyl methylfluorophosphate; Isopropoxy-methylphosphoryl fluoride; Zarin

Sarin, also known as 'nerve agent Gas B' or GB, is an organophosphorus compound, a colourless and odourless liquid, and a potent inhibitor of the cholinesterase enzyme. Sarin reacts with steam or water to produce toxic and corrosive gases. Sarin is incompatible with tin, magnesium, cadmium-plated steel, and some aluminium and reacts with copper, brass, and lead. Sarin is the most volatile of the nerve agents, which means that it can easily and quickly evaporate from a liquid into a vapour and spread into the environment. People can be exposed to the vapour even if they do not come in contact with the liquid form of sarin.

Safe Handling and Precautions

Sarin is a highly toxic nerve agent produced for chemical warfare. Sarin is an extremely potent acetylcholinesterase (AchE) inhibitor with high specificity and affinity for the enzyme. Exposure to sarin causes poisoning within seconds of exposure with symptoms that include, but are not limited to, runny nose; tearing (watery eyes); small, constricted, and pinpoint pupils; eye pain; blurred vision; excessive sweating; cough; chest tightness; muscle twitching; jerking; staggering; rapid breathing; diarrhoea; increased urination; confusion; drowsiness; weakness; headache; nausea; vomiting; abdominal pain; cardiac disorders; convulsions; respiratory failure; and fatal injury. The main clinical symptoms of acute toxicity of sarin are seizures, tremors, and hypothermia. Recent studies showed that long-term exposure to low levels of sarin caused neurophysiological and behavioural alterations. Toxicity from sarin significantly increased following concurrent exposure to other chemicals such as pyridostigmine bromide. Miosis and copious secretions from the respiratory and gastrointestinal tracts (muscarinic effects) were common in severely to slightly affected victims. Weakness and twitches of muscles (nicotinic effects) appeared in severely affected victims.

Neuropathy and ataxia were observed in a small number of victims for a brief period. Leukocytosis and high serum CK levels were common. To state in brief, sarin is a human-made chemical warfare nerve agent. Nerve agents are the most toxic and rapidly acting of the known chemical warfare agents, which are similar but much more potent to certain kinds of organophosphates pesticides. Regarding exposure severity of sarin, there is only a slight difference between a fatal dose and a dose that produces more mild health effects. Occupational workers must wear PPE, appropriate protective eyeglasses or chemical safety goggles, and suitable workplace dress to avoid exposure to chemical dust and fumes.

Selenium (CAS No. 7782-49-2) and Its Compounds

Molecular formula: **Se**

Selenium is a naturally occurring mineral element that is distributed widely in nature in most rocks and soils. In its pure form, it exists as metallic grey to black hexagonal crystals, but in nature it is usually combined with sulphide or with silver, copper, lead, and nickel minerals. Most processed selenium is used in the electronics industry, but it is also used as a nutritional supplement; in the glass industry; as a component of pigments in plastics, paints, enamels, inks, and rubber; in the preparation of pharmaceuticals; as a nutritional feed additive for poultry and livestock; in pesticide formulations; in rubber production; as an ingredient in anti-dandruff shampoos; and as a constituent of fungicides. Radioactive selenium is used in diagnostic medicine.

The general population is exposed to very low levels of selenium in air, food, and water. The majority of the daily intake comes from food. People working in or living near industries where selenium is produced, processed, or converted into commercial products may be exposed to higher levels of selenium in the air. People living in the vicinity of hazardous waste sites or coal-burning plants may also be exposed to higher levels of selenium.

Safe Handling and Precautions

Selenium has both beneficial and harmful effects. Low doses of selenium are needed to maintain good health. However, exposure to high levels can cause adverse health effects. Short-term oral exposure to high concentrations of selenium may cause nausea, vomiting, and diarrhoea. Chronic oral exposure to high concentrations of selenium compounds can produce a disease called selenosis. The major signs of *selenosis* are hair loss, nail brittleness, and neurological abnormalities (such as numbness and other odd sensations in the extremities). Brief exposures to high levels of elemental selenium or selenium dioxide in air can result in respiratory tract irritation, bronchitis, difficulty in breathing, and stomach pains. Prolonged period of exposure to either of these airborne forms of selenium has been reported to cause respiratory irritation, bronchial spasms, and coughing. However, levels of these forms of selenium that would be necessary to produce such health effects are normally not seen outside of the workplace. Reports have indicated that inhalation of hydrogen selenide causes pulmonary oedema. Selenium dusts produce respiratory tract irritation. Exposure to selenium dioxide fumes causes signs and symptoms of metal fume fever. Dermal exposure to selenium dioxide or selenium oxychloride may produce skin burns.

Chronic selenosis is extremely rare but has been reported in an occupational setting. It may be associated with nausea and vomiting, muscle tenderness, tremor, emotional instability, garlicky breath, bitter metallic taste in the mouth, brittle hair and nails, and skin lesions.

The IARC evaluated the literature relating selenium to carcinogenesis in both humans and animals. The IARC has observed that the available data provide no indication suggesting that selenium is carcinogenic in humans. Hence, the IARC has not listed selenium and selenium compounds as human carcinogens and classified as Group 3, meaning not classifiable as a human carcinogens. The U.S. Environmental Protection Agency (U.S. EPA) listed selenium sulphide as a probable human carcinogen, and based on animal studies data, the U.S. NTP grouped selenium sulphide as a suspected carcinogen.

Selenious Acid (CAS No. 7783-00-8)

Molecular formula: H_2SeO_3

Selenious acid is a colourless, deliquescent crystal in appearance. It decomposes on heating, producing water and toxic fumes of selenium oxides.

It is incompatible with strong reducing agents, organic materials, and finely powdered metals and reacts on contact with acids producing toxic gaseous hydrogen selenide. Selenious acid is a non-combustible chemical. On fire/burning, it emits irritating or toxic fumes (or gases).

The major use of selenious acid is in changing the colour of steel, especially the steel in guns from silver grey to blue grey. It is also used for the chemical darkening and patination of copper brass and bronze, producing a rich dark-brown colour that can be further enhanced with mechanical abrasion.

Safe Handling and Precautions

Exposures to selenious acid by accidental workplace ingestion or by inhalation cause burning sensation, cough, respiratory distress/laboured breathing, sore throat, abdominal pain, confusion, nausea, weakness, and low blood pressure. Selenious acid is corrosive to the eyes, to the skin, and to the respiratory tract. Selenious acid on prolonged period of exposures has been reported to cause allergic-type reaction of the eyelids *(rose-eye)*, dermatitis, nasopharyngeal irritation, gastrointestinal distress, and persistent garlic odour.

Like many selenium compounds, selenious acid is highly toxic, and ingestion of any significant quantity of selenious acid is usually fatal. Symptoms of selenium poisoning can occur several hours after exposure and may include stupor, nausea, severe hypotension, and death; the IARC, the National Institute for Occupational Safety and Health (NIOSH), the NTP, and the ACGIH have not listed selenious acid as a human carcinogen.

Occupational workers must handle selenious acid with care, and it should be stored in a cool, dry place in a tightly closed container and with the label 'POISON' in a locked room. The workplace management must provide adequate ventilation to workers during handling of chemicals. During use and handling of selenious acid, workers should be very alert to avoid any kind of contact or to breathe dust, vapour, mist, or gas of the chemical. The workers should wear full protective workplace dress (PPE) and must use approved and appropriate respiratory protection.

Selenium Chloride (CAS No. 10026-03-6)

Molecular formula: $SeCl_4$

Selenium chloride is white-yellow deliquescent powder/pieces in appearance with a pungent/garlic odour. Selenium chloride is non-flammable and on exposure to heat undergoes decomposition and emits toxic fumes of selenium and chloride.

Safe Handling and Precautions

Selenium chloride, like all other selenium compounds, is toxic, and exposures by inhalation and intravenous routes cause poisoning. Long-term exposure to selenium chloride is known to cause amyotrophic lateral sclerosis in humans, just as it may cause 'blind staggers' in cattle. Elemental selenium has low acute systemic toxicity, but dust or fumes cause serious irritation of the respiratory tract and gastrointestinal disturbances. Inorganic selenium compounds cause dermatitis.

On accidental exposures by inhalation and/or ingestion at workplaces, selenium chloride is dangerous, corrosive, and poisonous and causes severe irritation. It may cause irritation to the respiratory tract, acute selenium poisoning, nervousness, depression, and digestive disturbances. Chronic exposure of selenium chloride at high doses has been reported to result in fatal injuries.

Selenium (IV) Oxide (CAS No. 7446-08-4)

Selenium dioxide
Synonym: Molecular formula: SeO_2

Selenium oxide is a crystalline powder in appearance and white, off-white, or light beige in colour with a pungent odour. It is soluble in water and hygroscopic and absorbs moisture or water from the air. Selenium oxide is incompatible with strong oxidising agents, reducing agents, strong acids, ammonia, organics, and phosphorus trichloride.

Safe Handling and Precautions

Exposure to selenium oxide is harmful. On swallowing, inhalation, and/or through skin absorption at workplaces, it causes poisoning. Severe exposure has been reported to cause skin burns, eye irritation and corneal injury, puffy eyelids due to an allergic reaction, and skin sensitisation. Selenium oxide causes health disorders in blood, kidneys, liver, spleen, respiratory system, and gastrointestinal system. Repeated inhalation of high concentrations of selenium oxide at workplaces has been shown to cause metallic taste, pallor, garlic breath, anaemia, fever, chills, cough, weakness, nausea, headache, metal fume fever, dizziness, respiratory tract irritation, chest pain, muscle pain, increased white blood cell count, and central nervous system effects characterised by unconsciousness and coma. Reports have indicated that selenium oxide has not been listed as a human carcinogen by the IARC, the ACGIH, and the NTP.

Selenium dioxide is formed when selenium is heated in air. Direct exposure to selenium dioxide at workplaces is, therefore, primarily an occupational hazard and not likely to be a risk at hazardous waste sites. Selenium dioxide forms selenious acid on contact with water, including perspiration, and can cause severe irritation.

Acute inhalation exposure to selenium dust and selenium dioxide at workplaces has been shown to cause irritation of the mucous membranes of the nose and throat, leading to coughing, nosebleed, loss of olfaction, and, in heavily exposed workers, dyspnoea, bronchial spasms, bronchitis, and chemical pneumonia.

Acute inhalation of large quantities of selenium dioxide powder can produce pulmonary oedema as a result of the local irritant effect on alveoli. Bronchial spasms, symptoms of asphyxiation, and persistent bronchitis have been noted in workers briefly exposed to high concentrations of selenium dioxide.

Occupational workers during use and handling of selenium oxide should keep the container tightly sealed. Selenium oxide should be stored in cool, dry conditions; away from water/moisture and acids in properly sealed containers; under lock and key; and with entry permission restricted to technical experts only. Selenium oxide is hygroscopic and should be kept protected from humidity and water.

Selenium Sulphide (CAS No. 7446-34-6)

Molecular formula: SeS_2
Synonyms: Selenium (IV) disulphide (1:2); Sulphur selenide; Selenium disulphide

Selenium sulphide is a selenium salt that exists as a yellow-orange to bright-orange tablet or powder at room temperature. It is insoluble in water or ether and soluble in carbon disulphide. It is incompatible with acids, metals, strong oxidising agents, chromium trioxide, potassium bromate, silver oxide, and ammonia. Selenium sulphide is used as an active ingredient in anti-dandruff shampoos and as a constituent of fungicides.

Selenium sulphide is an anti-fungal agent as well as a cytostatic agent, slowing the growth of hyperproliferative cells in seborrhoea. Selenium sulphide is the active ingredient often used in shampoos for the treatment of dandruff, seborrheic dermatitis, and tinea capitis, a fungal infection that is primarily a disease of preadolescent children. Selenium sulphide is highly active in inhibiting the growth of *P. ovale*. It is also a proven cytostatic agent, slowing the growth of both hyperproliferative and normal cells in dandruff and seborrheic dermatitis. A 0.6% micronised form of selenium sulphide is also safe and effective for dandruff. Selenium sulphide is not present in foods and is a very different chemical from the organic and inorganic selenium compounds found in foods and in the environment.

Safe Handling and Precautions

The routes of potential human exposure to selenium sulphide are dermal contact. Accidental inhalation and occasional ingestion of selenium sulphide cause skin irritation; skin eruptions; coating of the tongue; anaemia; irritation of the mucous membrane; gastrointestinal irritation with nausea, vomiting, and diarrhoea; severe salivation; teeth decay or discolouration; partial loss of hair and nails; pallor; garlic breath; metallic taste; and liver and spleen damage. Repeated exposures to high concentrations of selenium sulphide cause central nervous system effects including nervousness, drowsiness, and convulsions; respiratory tract irritation; delayed pulmonary oedema; lumbar pain; and possible liver and spleen damage. Selenium sulphide is reasonably anticipated to be a human carcinogen based on sufficient evidence of carcinogenicity from studies in experimental animals.

Selenium compounds are considered hazardous materials; careful management and special requirements have been set for marking, labelling, and transporting these materials. Workers should be alert and strictly avoid negligence during use, storage, and handling of selenium.

- Workers should have a well-ventilated workplace and must be alert to change the contaminated clothing and wash before reuse.
- Workers should minimise dust generation and accumulation at the workplace and avoid contact of dust material with the eyes, skin, and clothing.
- Workers should avoid breathing the contaminated dust, vapour, mist, or gas.
- Workers should store the chemical in a tightly closed container; in a cool, dry, well-ventilated area away from incompatible substances; and preferably below 40°C.
- Occupational workers must avoid negligence to use PPE, must wear appropriate protective eye glasses or standard chemical safety goggles for face protection, and must wear appropriate protective clothing to prevent skin exposure and a respiratory protection.

Styrene (CAS No. 100-42-5)

Molecular formula: C_8H_8
Synonyms and trade names: Cinnamene; Cinnamol; Ethenylbenzene; Phenylethylene; Styrol; Verschueren; Vinylbenzene

Styrene is a colourless liquid that evaporates easily and has a sweet and pleasant smell, but when it contains other chemicals, it has a penetrating, sharp, and unpleasant smell. Styrene has extensive application in industries. It is widely used to make plastics and rubber; products containing styrene include insulation, plastic pipes, automobile parts, shoes, drinking cups and other food containers, and carpet backing. Styrene is used predominantly in the production of polystyrene plastics and resins; fibreglass products used for boats are also made from polyester resins dissolved in styrene as an intermediate in the synthesis of materials used for ion exchange resins and to produce copolymers such as styrene acrylonitrile (SAN) and acrylonitrile butadiene styrene (ABS).

Styrene can be found in air, water, and soil after release from the manufacture, use, and disposal of styrene-based products. Exposure to styrene occurs while breathing air contaminated with styrene vapors released from building materials, cigarette smoke, photocopy machines, automobile exhaust, Exposure to styrene also occurs because of contaminated workplace air, or skin contact of liquid styrene and resins.

Safe Handling and Precautions

Exposure to styrene in high concentrations causes health disorders with symptoms of poisoning. Studies have shown that styrene gets absorbed by inhalation and dermal transfer in both man and laboratory animals. The symptoms of poisoning include, but are not limited to, changes in colour vision, tiredness, thirst, concentration problems, balance problems, irritation of the upper respiratory tract, mucous membrane irritation, and effects on the nervous system. Changes in the lining of the nose and damage to the liver have also been observed in animals exposed to high concentrations of styrene, which must have been due to the more sensitive nature of animals than humans. Several studies have also indicated that inhalation exposure of humans to styrene causes mild or no effects on the blood. Several occupational studies have indicated styrene caused potential endocrine effects in reinforced plastics industry workers based on significant increases in serum prolactin levels in male and female workers. Reports have indicated that chronic exposures to styrene cause a variety of neurological effects in workers such as altered vestibular function, impaired hearing, decreased colour discrimination, altered performance on neurobehavioural tests, and increased clinical symptoms. The epidemiological studies of industrial workers associated with the use of styrene in the production of glass-reinforced plastic products, styrene monomer, and styrene polymerization and styrene–butadiene rubber products indicated development of more and increased malignancies of the lymphatic and haematopoietic system. The carcinogenic potential of styrene in workers at styrene manufacturing and polymerisation facilities, reinforced plastics facilities, and styrene–butadiene manufacturing facilities exposed to elevated styrene workers indicated an increased risk of leukaemia and lymphoma, which however is inconclusive because of inadequate data. The IARC has, therefore, listed styrene as a possible human carcinogen.

Sulphur Dioxide (CAS No. 7446-09-5)

Molecular formula: SO_2

Uses and Exposures

Sulphur dioxide is a colourless gas with a pungent odour. It is a liquid when under pressure, and it dissolves in water very easily. In nature, sulphur dioxide can be released to the environment from volcanic eruptions, from burning of coal and oil at power plants and or from copper smelting industries that are the major sources of sulphur dioxide in the living environment. Occupational workers and general public become exposed to sulphur dioxide while working in the manufacture of sulphuric acid, paper, food preservatives, or fertilisers and also living near heavily industrialised activities where sulphur dioxide occurs.

Safe Handling and Precautions

Prolonged exposures to sulphur dioxide causes burning sensation to the nose and throat, breathing difficulties, and severe airway obstructions. Inhaling sulphur dioxide causes increased respiratory symptoms and disease, difficulty in breathing, and premature death. Sulphur dioxide forms sulphurous acid when it contacts mucous membranes. Exposure can damage the eyes and lead to blindness, and inhalation can cause severe respiratory damage or even death. Sulphur dioxide has the potential for violent or explosive reactions with certain substances. It is very important to separate sulphur dioxide from powdered metals and strong alkalis (sodium hydroxide, fluorine).

Sulphur Dioxide and Cancer

There is inadequate evidence for the carcinogenicity in humans of sulphur dioxide, sulphites, bisulphites, and metabisulphites. There is limited evidence for the carcinogenicity in experimental animals of sulphur dioxide. There is inadequate evidence for the carcinogenicity in experimental animals of sulphites, bisulphites, and metabisulphites. Overall evaluation is as follows: sulphur dioxide, sulphites, bisulphites, and metabisulphites are not classifiable as to their carcinogenicity to humans.

The ACGIH lists sulphur dioxide as A4, meaning not classifiable as a human carcinogen. The IARC has classified sulphur dioxide as Group 3, meaning not classifiable as to human carcinogenicity. The U.S. EPA has set a limit of 0.03 ppm for sulphur dioxide for long-term exposure. The Occupational Safety and Health Administration (OSHA) has set a limit of 2 ppm over an 8 h workday (TWA) for sulphur dioxide.

Occupational workers during handling of sulphur dioxide should wear appropriate PPE such as self-contained breathing apparatus with full-face mask, gloves, hard hat, safety shoes, long-sleeved shirts, pants, and clothing, which should be free of oil or grease.

Bibliography

Abu-Qare, A. W. and M. B. Abou-Donia. 2002. Sarin: Health effects, metabolism, and methods of analysis. *Food Chem. Toxicol.* 40(10):1327–1333.

Agency for Toxic Substances and Disease Registry (ATSDR). 1998. *Toxicological Profile for Sulfur Dioxide.* U.S. Department of Health and Human Services, Public Health Service, Atlanta, GA (updated 2007).

Agency for Toxic Substances and Disease Registry (ATSDR). 2003. *Toxicological Profile for Selenium.* U.S. Department of Health and Human Services, Public Health Service. Atlanta, GA (updated 2011).

Agency for Toxic Substances and Disease Registry (ATSDR). 2010. *Toxicological Profile for Styrene.* U.S. Department of Health and Human Services, Public Health Service, Atlanta, GA.

Barceloux, D. G. 1999. Selenium. *Clin. Toxicol.* 37:145–172.

Benignus, V. A., A. M. Geller, W. K. Boyes et al. 2005. Human neurobehavioral effects of long-term exposure to styrene: A meta-analysis. *Environ. Health Perspect.* 113:532–538.

Dalton, P., P. S. Lees, M. Gould et al. 2007. Evaluation of long-term occupational exposure to styrene vapor on olfactory function. *Chem. Senses* 32(8):739–747.

Gosselin, R. E., R. P. Smith, H. C. Hodge et al. 1984. *Clinical Toxicology of Commercial Products*, 5th edn. Williams & Wilkins, Baltimore, MD.

Hazardous Substances Data Bank (HSDB). 2009. Styrene. National Library of Medicine, Washington, DC.

International Program on Chemical Safety and the Commission of the European Communities (IPCS-CEC). 2000. Selenious acid. ICSC No. 0945. IPCS-CEC, Luxembourg, Belgium.

International Programme on Chemical Safety and the Commission of the European Communities (IPCS-CEC). 2009. Selenium. ICSC Card No. 0072. IPCS-CEC, Geneva, Switzerland.

Koppel, C., H. Baudisch, K. H. Beyer, I. Kloppel, and V. Schneider. 1986. Fatal poisoning with selenium dioxide. *Clin. Toxicol.* 24:21–35.

Lewis, R. J. (ed.). 2000. *Sax's Dangerous Properties of Industrial Materials*, Vol. 3 Set, 11th edn. John Wiley & Sons, Hoboken, NJ.

Material Safety Data Sheet (MSDS). 1994. Sulfur oxide. International Programme on Chemical Safety and the Commission of the European Communities (IPCS CEC). MSDS, ICSC 0074 (updated 2006).

Material Safety Data Sheet (MSDS). 2008. Samarium. ACC No. 16097. Acros Organics N.V., Fair Lawn, NJ.

National Cancer Institute (NCI). 1980. Bioassay of selenium sulfide (dermal study) for possible carcinogenicity. Technical Report Series No. 197. U.S. Department of Health Education and Welfare, NCI, Bethesda, MD.

National Institute for Occupational Safety and Health (NIOSH). 2006. Facts about sarin. Centers for Disease Control and Prevention, Atlanta, GA.

National Institute for Occupational Safety and Health (NIOSH/OSHA). 1997. *Pocket Guide to Chemical Hazards*. DHHS (NIOSH), Publication No. 97–140, U.S. Government Printing Office, Washington, DC.

NIOSH Pocket Guide to Chemical Hazards. 2nd edn. 2005. J. J. Keller & Associates.

NIOSH Pocket Guide to Chemical Hazards. 2010. Selenium. NIOSH. Centers for Disease Control and Prevention, Atlanta, GA.

Olson, O. E. 1986. Selenium toxicity in animals with emphasis on man. *J. Am. Coll. Toxicol.* 5(1):45–70.

Patnaik, P. (ed.). 1999. *A Comprehensive Guide to the Hazardous Properties of Chemical Substances*, 2nd edn. Wiley-Interscience, New York.

Scarlato, E. A. and J. Higa. 1990. Selenium. Poison information. Monograph No. 483. INCHEM. International Programme on Chemical Safety, Geneva, Switzerland (updated 2001).

Sittig, M. (ed.). 1985. *Handbook of Toxic and Hazardous Chemicals and Carcinogens*, 2nd edn. Noyes Publications, Park Ridge, NJ.

World Health Organization (WHO). 2000. International Programme on Chemical Safety and the Commission of the European Communities (IPCS–CEC). Selenium trioxide. ICSC Card No. 0949. Geneva, Switzerland.

U.S. Environmental Protection Agency (U.S. EPA). 1985. Drinking water criteria document for styrene. Final draft. U.S. EPA. Cincinnati, OH.

Yanagisawa, N., H. Morita, and T. Kakajima. 2006. Sarin experiences in Japan: Acute toxicity and long-term effects. *J. Neurol. Sci.* 248(1):76–85.

21

Hazardous Chemical Substances:
T

Terbufos
Tetrachloroethylene
Tetrahydrofuran
Thallium (Soluble Compounds)
   Thallium Bromide
   Thallium Iodide
   Thallium Nitrate
   Thallium Oxide
Thiodicarb
Thiourea
Thiram
Thorium
Thorium Nitrate Anhydrous
Tin and Its Compounds
   Tin (IV) Chloride
   Tin Organic Compounds as Sn
   Tin Oxide
Toluene (Technical)
Tributyl Phosphate
Tributyltin Oxide
Trichloroethylene (Technical)
1,2,3-Trichloropropane
Triethanolamine
Triforine
Trinitrotoluene
Triphenyl Phosphate
Trisodium Phosphate
Trisodium Phosphate (Anhydrous)

Terbufos (CAS No. 13071-79-9)

Molecular formula: $C_9H_{21}O_2PS_3$
Synonyms and trade names: Aragran; Contraven; Counter; Plydox
Chemical name: S-*tert*-butylthiomethyl O,O-diethyl phosphorodithioate

Terbufos is an organothiophosphate insecticide–nematicide. Terbufos does hydrolyse quickly. The principal degradates of terbufos are its sulphone, sulphoxide, and oxon sulphone metabolites. Terbufos is a pale yellow/slightly brownish-yellow liquid. It is soluble in acetone, aromatic hydrocarbons, chlorinated hydrocarbons, and alcohols and negligibly soluble in water. Terbufos is usually formulated into granules for agricultural applications. It is used to control pests on corn, sugar beets, and grain sorghum and to control wireworms, seedcorn maggots, white grubs, corn rootworm larvae, and other pests. Terbufos is used to control soil insects and nematodes. Terbufos has no residential uses and, thus, is not registered for use in residential settings. Terbufos is stable for more than 2 years at room temperature. It decomposes upon prolonged heating at temperatures above 120°C. It is subject to alkaline hydrolysis in the presence of strong bases.

Safe Handling and Precautions

Terbufos has been classified as a highly toxic chemical substance and of toxicity class I. Exposure to terbufos has been known to cause cholinesterase inhibition, overstimulation of the nervous system, and poisoning in humans. Terbufos is readily absorbed by the oral, inhalation, and dermal routes. Terbufos does not accumulate in body tissues. Terbufos is highly toxic by both the oral and dermal routes. As an organophosphorus insecticide, the toxic effect of terbufos is inhibition of acetylcholinesterase. At high doses, terbufos is known to cause congestion of the liver, kidneys, and lungs with symptoms that include muscle tremors, salivation, diuresis, hyperpnoea, and tachycardia. The symptoms of poisoning also include nausea, dizziness, chest tightness, wheezing, blurred vision, fatigue, headache, slurred speech, confusion, abdominal cramps, vomiting, salivation, excessive sweating, and diarrhoea within 45 min of ingestion. Exposures at very high doses lead to respiratory paralysis, and fatal injury/death result from respiratory arrest, respiratory muscle paralysis, and/or constriction of the lungs. Occupational workers should be alert during handling of the chemical and should know that the no-observed-adverse-effect level (NOEL) is as low as 0.0025 mg/kg bw. Terbufos has been listed as a human carcinogen.

Regulations: Terbufos is classified as toxicity class I (highly toxic). Products containing 15% or more terbufos are classified as restricted use pesticides (RUPs), meaning to be purchased, used, and handled only by certified applicators and/or occupational workers.

Tetrachloroethylene (CAS No. 127-18-4)

Molecular formula: C_2Cl_4
Synonyms and trade names: Ethylene tetrachloride; Tetrachloroethene; Perchloroethylene; Carbon bichloride; Carbon dichloride

Tetrachloroethylene (perchloroethylene) is a colourless liquid chlorocarbon. It is widely used for dry-cleaning. It has a sweet odour. It is hazardous to human health and is generally made and used in closed systems by trained professionals with safety equipment. It is highly recommended that only workers with specific training be allowed to handle this substance. Perchloroethylene is well suited for recycling and constant re-use.

Safe Handling and Precautions

Exposure to perchloroethylene is known to cause irritation of the eyes, skin, nose, throat, and respiratory system; nausea; flushed face and neck; dizziness and incoordination; headache and drowsiness; skin erythema (skin redness); liver damage; and potential occupational carcinogen.

Regulations: Tetrachloroethylene (perchloroethylene) contains a chemical known to cause cancer. Based on evidences of carcinogenicity studies in experimental animals tetrachloroethylene is reasonably anticipated to be a human carcinogen. The state of California has listed it as a hazardous chemical that cause cancer. (For more information, refer to Perchloroethylene.)

Tetrahydrofuran (CAS No. 109-99-9)

Molecular formula: C_4H_8O
Synonyms: Tetramethylene oxide; THF

Tetrahydrofuran is a synthetic solvent used in the production of resins. Exposure to tetrahydrofuran has been known to occur through breathing vapours, through skin contact, through accidental ingestion at workplaces, and through ingestion of contaminated water and food.

Tetrahydrofuran is a clear, colourless liquid with ether-like odour. It is highly flammable.

Contact of tetrahydrofuran with strong oxidising agents may cause explosions. Tetrahydrofuran may polymerise in the presence of cationic initiators. Contact with lithium–aluminium hydride, with other lithium–aluminium alloys, or with sodium or potassium hydroxide can be hazardous.

Safe Handling and Precautions

Exposure to tetrahydrofuran causes irritation to the skin and eyes. Prolonged and/or repeated breathing of tetrahydrofuran in the form of gases, vapours, or mists causes severe health effects. The health disorders include narcotic effects, skin irritation, defatting, mucous membrane irritation; gastrointestinal tract, digestive tract, respiratory tract, and pulmonary/bronchial disease; breathing difficulties; unconsciousness; or respiratory arrest. The symptoms of poisoning during occupational exposures to high levels of tetrahydrofuran included irritation of mucous membranes, nausea, headache, dizziness, and possible cytolytic hepatitis. The effects on mucous membranes and the central nervous system, however, returned to normalcy within a few hours after cessation of exposure. Although sufficient animal studies data exist, the inadequate data in humans indicated that tetrahydrofuran needs varied types of listing by the U.S. Environmental Protection Agency (EPA) as a probable carcinogen, by the U.S. Department of Health and Human Services (DHHS) as a reasonably anticipated human carcinogen, and by the International Agency for Research on Cancer (IARC) as a probable human carcinogen.

Workers should handle tetrahydrofuran with care because it is normally stabilised with an anti-oxidant. Workers should keep tetrahydrofuran (with an inhibitor) stored in a cool, dry, well-ventilated area in tightly sealed metal or amber glass containers with label in accordance with set standard regulations. The containers of tetrahydrofuran should be kept, stored, and protected from physical damage and separated from oxidisers, heat, sparks, and open flame. Workers should strictly observe set regulations during handling and the management of tetrahydrofuran and should wear approved protective clothing, safety glasses, goggles, or face shields to prevent skin contact with tetrahydrofuran.

Thallium (Soluble Compounds) (CAS No. 7440-28-0)

Molecular formula: **TI**

Pure thallium is a bluish-white metal that is found in trace amounts in the Earth's crust. Thallium salts are tasteless, odourless, and colourless. In the past, thallium was obtained as a by-product from smelting other metals. Thallium is a non-volatile heavy metal and, if released to the atmosphere by anthropogenic sources, may exist as an oxide (thallium oxide), hydroxide (TlOH), sulphate (thallium sulphate), or sulphide. Thallium exists in two chemical states (thallous and thallic). The thallous state is the more common and stable form. Thallous compounds are the most likely form to which common exposures occur in the environment. Thallium is present in air, water, and soil. Thallium is used mostly in the manufacture of electronic devices, switches, and closures. It also has limited use in the manufacture of special glasses and in medical procedures that evaluate heart disease. The levels of thallium in air and water are very low. The greatest exposure occurs from food, mostly home-grown fruits and green vegetables, contaminated by thallium. Small amounts of thallium are released into the air from coal-burning power plants, cement factories, and smelting operations. This thallium falls out of the air onto nearby fruit and vegetable gardens. Thallium enters food because it is easily taken up by plants through the roots. Very little is known on how much thallium is in specific foods grown or eaten. Cigarette smoking is also a source of thallium. People who smoke have twice as much thallium in their bodies than non-smokers. Although fish take up thallium from water, we do not know whether eating fish can increase thallium levels in our body. It has been estimated that the average person eats, on a daily basis, 2 parts thallium per billion parts (ppb) of food. Even though rat poison containing thallium was banned in 1972, accidental poisonings from old rat poison still occur, especially in children.

Thallium is a heavy metallic element that exists in the environment mainly combined with other elements (primarily oxygen, sulphur, and halogens) in inorganic compounds. Thallium is quite stable in the environment, since it is neither transformed nor biodegraded.

Safe Handling and Precautions

Occupational exposure to thallium may be significant for workers in smelters, power plants, cement factories, and other industries that produce or use thallium compounds or alloys. Exposure may occur by dermal absorption from handling thallium-containing compounds, ores, limestone, or cement or by inhalation of workplace air. Reports have indicated that human exposure to thallium may occur by inhalation, ingestion, or dermal absorption. The general population is exposed most frequently by ingestion of thallium-containing foods and/or drink/water, through thallium in the air, and workplace dermal contact.

Also, occupational workers in industries producing or using thallium-containing materials have been known to have potentially high exposures to thallium compounds. When thallium is swallowed, most of it is absorbed and rapidly goes to various parts of the body, especially the kidney and liver. Thallium does not persist but slowly get eliminated from the system. Most of the thallium leaves the system in urine and to a lesser extent in faeces. Studies have indicated that thallium is found eliminated in urine within 1 h after exposure and after a lapse of 24 h, and increasing amounts of thallium have been found rejected in faeces. It can be found in urine as long as 2 months after exposure; about half of thallium that enters various parts of the body is released in a period of 3 days. The significant, likely routes of exposure near hazardous waste sites are through swallowing thallium-contaminated soil or dust, drinking contaminated water, and skin contact with contaminated soil. Populations with potentially high exposures are those that live near coal-burning power plants, metal smelters, or cement plants. The airborne particulate emissions from these industrial plants may have high thallium levels, especially on the small diameter. Human populations living in the vicinity of these plants may be exposed by inhalation or by ingestion of fruits and vegetables home grown in contaminated soils. Workers in industries producing or using thallium-containing materials also have potentially high exposures.

Reports have indicated that thallium causes poisoning with symptoms such as headache, vertigo, lethargy, nausea, diarrhoea or constipation, abdominal pain, vomiting, blurred vision, ptosis, strabismus, peripheral neuritis, tremor, paraesthesia of legs, rapid heart rate, retrosternal tightness, chest pain, myalgia, arthralgia, pulmonary oedema, sleep disturbances (insomnia), convulsions, chorea, psychosis, liver, kidney damage, blackening of hair roots, alopecia, and dystrophy of nails. Some neurological effects seem to be caused by direct action, such as ataxia and tremor by cerebellar alterations or alterations in endocrine activity through changes in the hypothalamus. The autonomic nervous system, mainly the adrenergic, may be activated by thallium. In peripheral nerves, thallium seems to interfere presynaptically with the spontaneous release of transmitter by antagonising these calcium-dependent processes. Exposure to thallium affects the nervous system, lung, heart, liver, and kidney if large amounts are eaten or drunk for short periods of time. Temporary hair loss, vomiting, and diarrhoea can also occur, and death may result after exposure to large amounts of thallium for short periods. Thallium can be fatal from a dose as low as 1 g. No information was found on health effects in humans after exposure to smaller amounts of thallium for longer periods. Birth defects observed in children of mothers exposed to small amounts of thallium did not occur more often than would be expected in the general population. The length of time and the amount of thallium eaten by the mothers are not known exactly. As in humans, animal studies indicate that exposure to large amounts of thallium for brief periods of time can damage the nervous system and heart and can cause death. Further, it has been observed that in the male experimental animals, drinking small amounts of thallium-contaminated water for 2 months caused damage of the reproductive organs – the testes. These effects have not been seen in humans. No information was found on effects in animals after exposure to small amounts of thallium for longer periods of time. Exposure to thallium during the first trimester of pregnancy has been reported to cause poisoning and health effects including skeletal deformations, alopecia, low birth weight, and premature birth. Long-term studies on the carcinogenicity of thallium have not indicated whether thallium causes cancer in experimental laboratory animals and in humans. The details of thallium toxicity are still not known and require more confirmatory data. Studies have indicated that thallium intoxication causes selective impairment in behaviour. This has been correlated with biochemical effects and cellular damage in certain regions of the brain. Not much information is available in the literature about the fate, transport, or potential for human exposure to thallium.

Thallium Bromide (CAS No. 7789-40-4)

Molecular formula: TlBr

Thallium bromide is a yellow-white solid and hygroscopic. It is similar to silver bromide. It gets darker when light shines on it. It is very toxic, like all thallium compounds. It does not dissolve in water. Thallium bromide is used in gamma ray and x-ray detectors and also in semiconductors. Prolonged period of exposure to thallium compounds has been reported to cause adverse health effects such as joint pain, severe pain in legs, loss of appetite, fatigue, and albuminuria.

Thallium Iodide (CAS No. 7790-30-9)

Synonyms: Thallium iodide (TlI), dimer; Thallium monoiodide; Thallium (1+) iodide; Thallium (I) iodide; Thallous iodide

Thallium (I) iodide is a yellow solid, nearly insoluble in water. It is used as infrared radiation transmitter (crystals mixed with thallium bromide).

Safe Handling and Precautions

Thallium (I) iodide is harmful and poisonous. Exposures to thallium (I) iodide are known to cause muscle weakness and lowered blood pressure in an oral toxic-dose study of men. It causes irritation and blindness and fatal injury by inhalation and ingestion and is of danger on cumulative effects. The target organs include kidneys, cardiovascular system, skin, liver, male reproductive system, nerves, and eyes.

Thallium Nitrate (CAS No. 10102-45-1)

Molecular formula: $TlNO_3$
Synonyms and trade names: Nitrate de thallium; Nitric acid, thallium (1+) salt; Nitric acid, thallium (I) salt; Nitric acid, thallous salt; Thallium (I) nitrate; Thallium mononitrate; Thallium (1+) nitrate; Thallous nitrate

Thallium nitrate is a colourless crystalline solid. It is toxic upon ingestion and skin absorption and is used to make other chemicals. Thallium nitrate is an oxidising agent, and contact with organic materials catches fire and contact with mixtures with alkyl esters thallium nitrate causes explosions because of the formation of alkyl nitrates. Mixing it with phosphorus, tin (II) chloride, or other reducing agents may cause an explosive reaction.

Thallium nitrate exposures and negligence at the workplace may result in occupational workers developing symptoms of thallium nitrate poisoning. These include, but are not limited to, mild irritation to the respiratory system, nausea, vomiting, dizziness, abdominal cramps, bloody diarrhoea, weakness, convulsions, and collapse. Repeated exposures to small doses of thallium nitrate lead to weakness, general depression, headache, and mental impairment. Thorium nitrate is a confirmed human carcinogen producing *anglosarcoma*, liver and kidney tumours, lymphoma, and other tumours of the blood system.

Thallium Oxide (CAS No. 1314-32-5)

Molecular formula: Tl_2O_3

Thallium oxide is a solid, very dark brown and odourless chemical substance. Thallium oxide is stable under normal temperatures and pressures.

It is insoluble in water. Thallium oxide has been used to produce glasses with a high index of refraction.

Safe Handling and Precautions

Users and occupational workers, on acute exposure to thallium oxide, develop adverse health effects and poisoning. Thallium oxide is absorbed through the skin and inhalation is known to cause hair loss and fatigue. The main symptoms of thallium poisoning are peripheral neuropathy and loss of hair. The symptoms of acute poisoning include, but are not limited to, paraesthesias, ataxia, alopecia, fever, coryza, abdominal pain, nausea, vomiting, lethargy, speech disorders, tremors, and cyanosis. Repeated and chronic exposures to thallium oxide have been reported to cause convulsions, pulmonary oedema, bronchopneumonia, and kidney, liver, and brain damage leading to fatal injury. There are no reports indicating that thallium oxide is a human carcinogen and has not been listed by the American Conference of Governmental Industrial Hygienists (ACGIH) or IARC and National Toxicology Program (NTP) as a human carcinogen.

Thiodicarb (CAS No. 59669-2600)

Molecular formula: $C_{10}H_{18}N_4O_4S_3$
Synonym and trade name: Larvin
Chemical name: Dimethyl N,N'-[thiobis[(methylimino)carbonyloxy]]bis[ethanimidothioate]

Thiodicarb is a white crystalline powder with a slight sulphurous odour. Thiodicarb is stable in light and ambient conditions and unstable in alkaline conditions. Thiodicarb is a carbamate insecticide. Thiodicarb is commonly used to protect agricultural crops from major lepidopterous insect pests and suppresses coleopterous and some hemipterous insect pests. Thiodicarb acts as an ovicide against cotton bollworms and budworms. Thiodicarb is used primarily on cotton, sweet corn, and soybeans. Thiodicarb is formulated to include several liquid products and one powdered product that must be mixed with water before field application. Thiodicarb is reclassified as an RUP. Thiodicarb degrades rapidly to methomyl, which is already a restricted use chemical.

Safe Handling and Precautions

Exposures to thiodicarb through negligence are harmful and cause poisoning by inhibiting the activity of the enzyme acetylcholinesterase. The signs and symptoms of thiodicarb toxicity include, but are not limited to, increased salivation, headache, muscle weakness, dizziness, nausea, vomiting, abdominal cramping, and profuse sweating. It causes mild skin irritation and skin sensitisation. Thiodicarb has been classified as a Group B2, meaning a probable human carcinogen.

Occupational workers should handle thiodicarb with caution and keep the original container in a cool, ventilated, dry, locked area. Also it should be kept out of reach of children and domestic animals and completely away from food, feedstuffs, fertilisers, and seed.

Regulations: Based on the reviews of the generic data for the active ingredient thiodicarb, the agency has sufficient information on the health effects of thiodicarb and on its potential for causing adverse effects in fish and wildlife and the environment. The agency has determined that thiodicarb products, labelled and used as specified in the Reregistration Eligibility Decision, will not pose unreasonable risks to humans or the environment. The agency concludes that products containing thiodicarb for all uses are eligible for reregistration.

Thiourea (CAS No. 62-56-6)

Molecular formula: CH_4N_2S

Synonyms and trade names: Isothiourea; Pseudothiourea; Thiocarbamide; 2-Thiopseudourea; b-Thiopseudourea; 2-Thiourea; THU

Thiourea appears as white crystal/powder, is combustible, and on contact with fire, gives off irritating or toxic fumes/gases. Thiourea is a reducing agent used primarily in the production of bleached recycled pulp. In addition, it is also effective in the bleaching of stone groundwood, pressurised groundwood. Thiourea undergoes decomposition on heating and produces toxic fumes of nitrogen oxides and sulphur oxides. It reacts violently with acrolein, strong acids, and strong oxidants. The main application of thiourea is in textile processing and also is commonly employed as a source of sulphide. Thiourea is a precursor to sulphide to produce metal sulphides, for example, mercury sulphide, upon reaction with the metal salt in aqueous solution. The industrial uses of thiourea include production of flame-retardant resins and vulcanisation accelerators. Thiourea is used as an auxiliary agent in diazo paper, light-sensitive photocopy paper, and almost all other types of copy paper. Thiourea is used in many industrial applications, including as a chemical intermediate or catalyst, in metal processing and plating, and in photoprocessing.

Safe Handling and Precautions

Exposure to thiourea causes adverse health effects and poisoning. It is absorbed into the body by inhalation of its aerosol and by ingestion. Repeated or prolonged contact of thiourea is known to cause skin sensitisation and diverse health effects on the thyroid. Reports have indicated that information and evidences on the carcinogenic potential of thiourea in experimental animals are limited and in humans are inadequate. Thiourea has been reported as not classifiable and is listed as Group 3, meaning not classifiable as a human carcinogen. Thiourea has not been listed as a human carcinogen by IARC, NTP, and Occupational Safety and Health Administration (OSHA).

Thiram (CAS No. 137-26-8)

Molecular formula: $C_6H_{12}N_2S_4$
Synonyms and trade names: Pomarso; Rhodiason
IUPAC name: Tetramethylthiuram disulphide
Chemical name:
Toxicity class: U.S. EPA: III; WHO: III

Thiram is a dimethyl dithiocarbamate compound and appears as a white to yellow crystalline powder with a characteristic odour.

Thiram is used to prevent crop damage in the field and to protect harvested crops from deterioration in storage or transport. Thiram is a broad-spectrum protectant fungicide for use on fruit, vegetables, ornamentals, and amenity turf. In addition, it is used as an animal repellent to protect fruit trees and ornamentals from damage by rabbits, rodents, and deer. Thiram is available as dust; flowable, wettable powder; water-dispersible granules; and water suspension formulations and in mixtures with other fungicides. Thiram has been used in the treatment of human scabies, as a sunscreen, and as a bactericide applied directly to the skin or incorporated into soap.

Safe Handling and Precautions

Thiram is slightly toxic by ingestion and inhalation, but it is moderately toxic by dermal absorption. Acute exposure in humans may cause headaches, dizziness, fatigue, nausea, diarrhoea, and other gastrointestinal complaints. Thiram is an irritant to the eyes, skin, and respiratory tract of humans. It is a skin sensitiser. Symptoms of acute inhalation exposure to thiram include itching, sensitisation by skin contact, scratchy throat, hoarseness, sneezing, coughing, inflammation of the nose or throat, bronchitis, dizziness, headache, fatigue, nausea, diarrhoea, and other gastrointestinal complaints. Persons with chronic respiratory or skin disease are at increased risk of exposure to thiram. Ingestion of thiram and alcohol together may cause stomach pains, nausea, vomiting, headache, slight fever, and possible dermatitis.

Regulations:

- Thiram and its products must be used only for the treatment of seeds; they *MUST NOT be sprayed onto growing crops.*
- Pesticide products containing thiram must bear the signal words *Caution* and *Dangerous for the environment* on the label.
- Classified as Dangerous Goods according to the Land Transport Safety Authority (LTSA), New Zealand.

Occupational workers, during use and handling of thiram, should wear appropriate and suitable workplace dress, respiratory protective equipment to avoid exposure to mist, protective gloves and water-repellent barrier cream when handling concentrate, safety goggles, and face shield.

Thorium (CAS No. 7440-29-1)

Molecular formula: **Th**

Thorium is a silvery-white metal that is air-stable and retains its lustre for several months. When contaminated with the oxide, thorium slowly tarnishes in air, becoming grey and finally black. The physical properties of thorium are greatly influenced by the degree of contamination with the oxide. Thorium oxide (ThO_2), one of thorium's compounds, has many uses. Thorium was discovered by Jöns Jacob Berzelius, a Swedish chemist, in 1828. He discovered it in a sample of a mineral that was given to him by the Reverend Has Morten Thrane Esmark, who suspected that it contained an unknown substance. Esmark's mineral is now known as thorite ($ThSiO_4$). Thorium makes up about 0.0007% of the Earth's crust and is primarily obtained from thorite, thorianite (ThO_2), and monazite (($Ce, La, Th, Nd, Y)PO_4$).

Thorium oxide (ThO$_2$), also called thoria, has one of the highest melting points of all oxides (3300°C). When heated in air, thorium metal turnings ignite and burn brilliantly with a white light. Because of these properties, thorium has found applications in lightbulb elements, lantern mantles, arc-light lamps, welding electrodes, and heat-resistant ceramics. Glass containing thorium oxide has a high refractive index and dispersion and is used in high-quality lenses for cameras and scientific instruments.

Thorium is a naturally occurring, radioactive substance. In the environment, thorium exists in combination with other minerals, such as silica. Small amounts of thorium are present in all rocks, soil, water, plants, and animals. Soil contains an average of about 6 parts of thorium per million parts of soil. More than 99% of natural thorium exists in the form of thorium-232 and later breaks down into two parts – a small part called 'alpha' radiation and a large part called the decay product. The decay product is also not stable and continues to break down through a series of decay products until a stable product is formed. During these decay processes, radioactive substances are produced. These include radium and radon. These substances give off radiation, including alpha and beta particles and gamma radiation. Rocks of certain underground mines contain thorium in a more concentrated form. After these rocks are mined, thorium is usually concentrated and changed into thorium dioxide or other chemical forms. After most of the thorium is removed, the rocks are called 'depleted' ore or tailings. Soil commonly contains an average of around 6 ppm of thorium. Thorium is more abundant than uranium and is widely distributed in nature as an easily exploitable resource in many countries and has not been exploited fully and commercially. Thorium fuels, therefore, complement uranium fuels and ensure long-term sustainability of nuclear power. Thorium fuel cycle is an attractive way to produce long-term nuclear energy with low radiotoxicity waste. In addition, the transition to thorium could be done through the incineration of weapons-grade plutonium (WPu) or civilian plutonium.

Thorium has extensive societal applications: to make ceramics, gas lantern mantles, and metals used in the aerospace industry and in nuclear reactions. In India, there has always been a strong incentive for development of thorium fuels and fuel cycles because of large thorium deposits compared to the very modest uranium reserves. Thorium oxide is also used to make glass with a high index of refraction that is used to make high-quality camera lenses. Thorium oxide is used as a catalyst in the production of sulphuric acid (H$_2$SO$_4$), in the cracking of petroleum products, and in the conversion of ammonia (NH$_3$) to nitric acid (HNO$_3$). Thorium exists in nature in a single isotopic form – Th-232 – which decays very slowly (its half-life is about three times the age of the Earth). The decay chains of natural thorium and uranium give rise to minute traces of Th-228, Th-230, and Th-234, but the presence of these in mass terms is negligible. Also, ThO$_2$ is relatively inert and does not oxidise unlike UO$_2$, which oxidises easily to U$_3$O$_8$ and UO$_3$. Hence, long-term interim storage and permanent disposal in repository of spent ThO$_2$-based fuel are simpler without the problem of oxidation.

Safe Handling and Precautions

Studies on thorium workers have shown that breathing high levels of thorium dust results in an increased chance of getting lung disease. Liver diseases and effects on the blood were found in people injected with Thorotrast, a thorium compound injected into the body as a radiographic contrast medium between the years 1928 and 1955. Animal studies have shown that breathing thorium may result in lung damage.

Studies on exposed human populations have not reported any birth defects or effects on a person's ability to have children. Workers who had high exposures to cigarette smoke, radon gas, and thorium had cancers of the lung, pancreas, and blood. People who had large amounts of thorium injected into their blood for special x-ray tests had more than the usual number of liver tumours, cancers of the blood such as leukaemia, and tumours of the bone, kidney, spleen, and pancreas; breathing thorium dust may cause an increased chance of developing lung disease and cancer of the lung or pancreas many years after being exposed. Changes in the genetic material of body cells have also been shown to occur in workers who breathed thorium dust. Liver diseases and effects on the blood have been found in people injected with thorium in order to take special x-rays. Many types of cancer have also been shown to occur in these people many years after thorium was injected into their bodies. Since thorium is radioactive and may be stored in bone for a long time, bone cancer is also a potential concern for people exposed to thorium.

Animal studies have shown that breathing in thorium may result in lung damage. Other studies in animals suggest drinking massive amounts of thorium can cause death from metal poisoning. The presence of large amounts of thorium in the environment could result in exposure to more hazardous radioactive decay products of thorium, such as radium and thoron, which is an isotope of radon. Thorium has not been reported to cause birth defects or to affect the ability to have children.

Thorium Nitrate Anhydrous (CAS No. 13823-29-5)

Molecular formula: $Th(NO_3)_4 \cdot 4H_2O$
Synonyms and trade names: Thorium nitrate; Thorium nitrate tetrahydrate; Thorium (IV) nitrate; Thorium (IV) nitrate hydrate; Thorium nitrate hydrate; Thorium tetranitrate; Nitric acid, Thorium (4+) salt; Nitric acid, thorium (4++) salt; Thorium (4+) nitrate

Thorium nitrate is a white crystalline mass. It is soluble in water and very soluble in alcohol (ethanol) and acids and is not combustible but accelerates burning of combustible materials. Thorium nitrate is highly reactive and incompatible with combustible materials, organic materials, and finely powdered metals. It may explode in a fire if in large quantities or if the combustible material is finely divided. Toxic oxides of nitrogen are produced in fires. Thorium nitrate, on contact with oxidising agent, oxidisable materials, and combustible and/or organic materials, causes fire and chemical hazards and violent explosion.

Safe Handling and Precautions

In the workplace, accidental inhalation and/or ingestion of thorium nitrate causes poisoning and hazards. The symptoms of poisoning include, but are not limited to, nausea, vomiting, and dizziness. Improper use and handling of thorium nitrate and contact with eyes cause irritation. Repeated and prolonged exposure to thorium nitrate through ingestion and/or inhalation causes skin burns, skin ulcerations and respiratory irritation, nausea, vomiting, headache, dizziness, abdominal cramps, ulceration or bleeding from the small intestine, bloody diarrhoea, weakness, changes in the blood and urine, convulsions, collapse, general depression, and mental impairment. Thorium nitrate has been reported as a suspected carcinogen. Thorium nitrate emits radiation that could cause cancer, but no evidence of cancer has yet been confirmed.

Repeated and/or prolonged exposure to thorium nitrate has been reported to affect the liver, kidneys, lungs, and bone marrow, and may reduce the ability of the bone marrow to make blood cells. Prolonged or repeated inhalation of thorium nitrate dust has been reported to cause scarring of the lungs, anglosarcoma, liver and kidney tumours, lymphoma, and other tumours of the blood system. Occupational workers must remember that accumulation of thorium in the bones, lungs, and lymph system and for long periods of time increases the risk of radiation-induced cancer in tissues where it is retained.

Occupational workers, during use and handling of thorium nitrate, must strictly observe safety and workplace regulatory practices and must avoid workplace negligence. Workers should keep thorium nitrate away from heat, sources of ignition, and combustible material. Workers should avoid ingesting and breathing dust. Workers should wear protective clothing to prevent chemical contact, personal protective equipment (PPE), to avoid organic vapour, canister face mask, self-contained breathing apparatus, liquid-proof gloves, protective goggles or face shield, white or light-coloured clothing, and rubber shoes or boots.

Tin and Its Compounds

Tin is a soft, white, silvery metal that is insoluble in water. Tin, in contact with water, does not corrode but, in contact with acids and alkalis, causes damage to the metal. Tin takes high polishing and has been in industrial use for the protective coating of metals. Tin is a metal with a melting point of 232°C. Tin has two oxidation states, the II (stannous) and IV (stannic) states. Tin compounds are divided into (i) soluble compounds ($SnBr_4$, $SnCl_2$, $SnCl_4$, SnF_2, SnI_2, SnI_4) and insoluble salts (SnO, SnO_2, SnP_2O_7, SnS, $SnSO_4$). The soluble salts have melting points between −33°C and 143°C. The insoluble compounds have much higher melting points, more than 880°C. Tin is mainly used as a protective coating for other metals due to its resistance to corrosion. Tin bonds readily to iron and is used for coating lead or zinc and steel to prevent corrosion. Tin has extensive applications, for instance, in electroplating, electronic components, integrated circuits, clips, store pharmaceutical chemical solutions, capacitor electrodes, fuse wires, and ammunitions. Alloys of tin are important, as soft solder, pewter, bronze, and phosphor bronze. Tin compounds that contain lead, barium, calcium, and copper are very important in the production of electric capacitors. Tin (II) chloride is used as a reducing agent and as a mordant, and the organic tin compounds have been in large use in agriculture as fungicides and insecticides for the control of crop pests, to treat wood, textile, and paper as preservatives. Tin metal is used to line cans of food, beverages, and aerosols. Tin is present in brass, bronze, pewter, and some soldering materials. Tin is a metal that can combine with other chemicals to form various compounds. When tin is combined with chlorine, sulphur, or oxygen, it is called an inorganic tin compound. Tin, when combined with carbon, forms organotin compounds. The organotin compounds have wide industrial applications and are used in plastics. In fact, for the stabilisation of PVC plastics, the organotin compounds have wide application. Tin has been used for food packages, plastic pipes, pesticides, paints, wood preservatives, and rodent (rats and mice) repellents. In common terms, tin is grouped as (a) tin inorganic compounds as Sn (except oxides), (b) tin organic compounds as Sn, and (c) tin oxide as Sn, total dust.

Safe Handling and Precautions

Tin is present in small amounts in all human and animal organs. A normal adult human contains about appreciable level of tin in his system. Tin metal itself, taken orally, is practically innocuous, while chronic inhalation of tin dust and/or fumes at industrial workplaces has been reported to cause benign pneumoconiosis. The inorganic salts are caustic and produce variable toxicity. Reports indicate that tin and inorganic tin compounds have poor absorption properties in the gastrointestinal tract and do not get accumulated in tissues but are rapidly excreted in the faeces. The symptoms of adverse health effect of tin on humans who drink fruit juices and beverages stored in tin containers include nausea and diarrhoea and irritation of the mucosal surfaces or acute gastric irritation. Exposures to vey high levels of tin can also cause skin or eye problems like inflammation or irritation, and very high exposure, specifically to trimethyltin and triethyltin compounds, has been reported to cause injuries and has been proven fatal.

Similarly, tin miners showed no increased incidences of lung cancer. Limited information is available concerning the effects of tin (compounds) in humans after inhalatory exposure. In humans exposed to SnO_2 dust or fumes for 3–50 years, no effects on lung function were found despite abnormal findings in radiographs and at autopsy. Aggregates of dust-containing macrophages in the lungs were found presenting a pneumoconiosis called stannosis. No fibrosis of the lungs associated with exposure was reported. Increased incidences of lung cancer in tin miners were not attributed to exposure to tin. Thus, there are no reports or evidences regarding the adverse effects of tin and cancer in humans.

The U.S. EPA, the DHHS, and the IARC have not classified inorganic tin compounds nor metallic tin as carcinogens to humans.

Tin (IV) Chloride (CAS No. 7646-78-8)

Molecular formula: $SnCl_4$
Synonyms and trade names: Stannic chloride; Tin tetrachloride

Tin (IV) chloride appears as white crystals with a strong pungent chlorine odour. On heating, tin (IV) chloride decomposition emits acrid fumes. At room temperature, it is colourless and releases fumes on contact with air, giving a stinging odour. Stannic chloride was used as a chemical weapon during World War I. It is also used in the glass container industry for making an external coating that toughens the glass. Stannic chloride is used in chemical reactions with fuming (90%) nitric acid for the selective nitration of activated aromatic rings in the presence of unactivated ones. Tin (IV) chloride reacts violently with water or moist air to produce corrosive hydrogen chloride. Tin (IV) chloride reacts with turpentine, alcohols, and amines, causing fire and explosion hazard. It attacks many metals, some forms of plastic, rubber, and coatings.

Safe Handling and Precautions

Exposure to tin (IV) chloride causes harmful effects and poisoning and adverse effects to kidneys, liver, and brain. The signs and symptoms of poisoning because of improper handling of tin (IV) chloride by ingestion or inhalation include, but are not limited to, redness, tearing, itching, and burning effects to the eyes; damage to cornea; conjunctivitis; and loss of vision, and the skin also develops redness, blistering, burning, itching, and tissue destruction. Improper handling and accidental workplace ingestion cause coughing, wheezing, headache, nausea, vomiting, burning, diarrhoea, shortness of breath, spasm, inflammation and oedema of bronchi, pneumonitis, ulceration, convulsions, and shock. Repeated/prolonged skin contact may cause thickening, blackening, or cracking. Repeated eye exposure may cause corneal erosion or loss of vision.

Occupational workers, during use and handling of tin (IV) chloride (tin tetrachloride), should strictly observe safety regulatory practices and avoid workplace negligence. Workers should wear protective workplace dress/clothing to prevent inhalation of dust or vapour and contact with the eyes and skin. Workers should wear face mask, self-contained breathing apparatus, liquid-proof gloves, protective goggles, and rubber shoes or boots. Occupational workers should store tin (IV) chloride in corrosive cabinet, in a cool, dry, well-ventilated, locked store room away from incompatible materials.

Regulations:

- The OSHA has set the permissible exposure limit (PEL) of tin chloride as 2 mg/m^3.
- The ACGIH has set the threshold limit value (TLV) as 2 mg/m^3, short term exposure limit (STEL).

Tin (IV) chloride should be kept and stored in a safe area [white storage] with other corrosive items, dedicated corrosive cabinet. Workers should store tin (IV) chloride in a cool, dry, well-ventilated, locked store room away from incompatible materials.

Tin Organic Compounds as Sn

Organotin compounds are versatile agents in a wide range of industrial applications.

Organotin includes many compounds. These include tributyltin fluoride (CAS no. 1983-10-4), dibutyltin dilaurate (CAS no. 77-58-7), tributyltin benzoate (CAS no. 4342-36-3), dibutyltin maleate (CAS no. 77-58-7), stannous-2-ethyl hexanoate (CAS no. 301-10-0), butyltin trichloride (CAS no. 1118-46-3), bis(tributyltin) oxide (CAS no. 56-35-9), triphenyltin hydroxide (CAS no. 76-87-9), methyltin mercaptide (CAS no. 57583-35-4), and tetramethyl tin (CAS no. 594-27-4).

The common organotins are extensively used in chemical synthesis as catalysts. These are hydrated monobutyltin oxide, butyl chlorotin dihydroxide, butyltin tris(2-ethyl-hexoate), dibutyltin diacetate, dibutyltin oxide, dibutyltin dilaurate, butyl stannoic acid, dioctyltin dilaurate, and dioctyltin maleate. Mono- and diorganotins are used extensively as heat stabilisers for processing polyvinyl chloride (PVC). Tin mercaptide stabilisers are some of the most effective PVC stabilisers available. The main applications for tin stabilisers are building products, such as pipe and fittings, and siding and profiles (windows, etc.), packaging, and flexible PVC.

Monoorganotins are used on glass containers in hot-end coatings (HECs). In HECs, a metallic oxide is deposited on the hot glass surface of bottles thereby preventing microfissures.

Tributyltin (TBT) is unique among the organotins in that it is used as a biocide. In the marine anti-foulant (MAF) paint market, TBT is used as a biocide in paint formulations. These paints are then used to protect the underwater surface area of a ship's hull against barnacles, algae, etc. in order to avoid increased fuel consumption and premature dry-docking. Triorganotins were introduced for this application 30–40 years ago. Originally, tributyltin oxide (TBTO) was freely dispersed in what were called free association paints (FAP). These paints had uncontrolled, rapid leaching rates of the biocide.

Triorganotin is also used as ingredients of acaracides and fungicides for the control of crop pests of citrus, top fruit, vines, vegetables, and hops. These include triphenyltin hydroxide (TPTH or fentin hydroxide), tricyclohexyltin hydroxide (TCTH or cyhexatin), tricyclohexyltin triazole (TCTT or azocyclotin), trineophenyltin oxide (TNTO or fenbutatin oxide), and triphenyltin acetate (TPTA or fentin acetate).

Major environmental implications are related to triorganotins used as agricultural and industrial biocides and to a lesser extent to the application of some diorganotin and mono-organotin compounds in rigid PVC. Triorganotin compounds are generally more toxic than organotins in other classes. Trialkyltins with linear organic groups cannot be used as agricultural biocides due to their high toxicity to plants (phytotoxicity). Diorganotins show no antifungal activity. Their anti-bacterial and toxic activity is low, except for the diphenyl derivatives.

Regulations: Global regulatory agencies have set safety limits of tin at workplaces:

- The OSHA has set the workplace exposure limits: organotin compounds, 0.1 mg/m^3 per cubic metre of air at the workplace; inorganic tin compounds, 2 mg/mic tin compounder of air.
- The National Institute for Occupational Safety and Health (NIOSH) set the immediately dangerous to life or health (IDLH) concentration limit of TCTH as 25 mg/m^3.
- The U.S. FDA regulates the use of some organic tin compounds in coatings and plastic food packaging and has set limits for the use of stannous chloride as an additive for food.

Tin Oxide (CAS No. 21651-19-4)

Molecular formula: **SnO**
Synonyms and trade names: Stannous oxide; Tin protoxide; Tin monoxide; Tin (II) oxide

Tin oxide is insoluble in water but soluble in acids and alkalis and slightly soluble in ammonium chloride. Tin oxide is incompatible with acids and/or alkalis.

Safe Handling and Precautions

Occupational exposure to tin oxide occurs through inhalation, ingestion, and eye or skin contact. Tin oxide is a mild irritant and causes pulmonary effects in humans. Exposure has resulted in mild irritation to the eyes, skin, and mucous membranes.

Chronic exposure to tin oxide has been reported to produce the development of stannosis, a benign pneumoconioses. Tin oxide at its early stages of exposure can cause injury such as chest pain and stannosis that are indistinguishable from silicosis.

Toluene (Technical) (CAS No. 108-88-3)

Molecular formula: $C_6H_5CH_3$

Toluene is a clear, colourless liquid with a sweet, benzene-like odour. Toluene occurs naturally in crude oil and in the toluene tree. It is also produced in the process of making gasoline and other fuels from crude oil and making coke from coal. Toluene is used in making paints, paint thinners, fingernail polish, lacquers, adhesives, and rubber and in some printing and leather tanning processes. Toluene is also used in the production of polymers used to make nylon, plastic soda bottles, and polyurethanes and for pharmaceuticals, dyes, cosmetic nail products, and the synthesis of organic chemicals.

Occupational exposures to toluene normally occur through breathing contaminated workplace air or automobile exhaust, at workplaces using gasoline, kerosene, heating oil, paints, and lacquers, by drinking contaminated water and at hazardous waste sites containing toluene products.

Toluene has been reported as the most commonly abused hydrocarbon solvent, primarily through 'glue sniffing'. The common possibilities of exposure to high levels of toluene include indoor air from the use of household products such as paints, paint thinners, adhesives, synthetic fragrances, and many other sources.

Safe Handling and Precautions

Exposure to toluene is harmful and causes poisoning. The signs and symptoms of poisoning include, but not limited to, irritation to eyes and nose, weakness, headache, dilated pupils, lacrimation exhaustion, confusion, euphoria, dizziness, anxiety, muscle fatigue, insomnia, and paraesthesia. Reports have indicated that occupational negligence during prolonged period of handling of toluene causes adverse developmental effects in offspring. Repeated and prolonged period of exposure to high levels of toluene is known to cause severe health effects such as dermatitis, liver and kidney damage, neurobehavioural effects, CNS depression, unconsciousness, and fatal injury. The U.S. EPA classified toluene as a Group D, meaning not classifiable as to human carcinogenicity.

Regulations:

- The ACGIH recommends a TLV of 50 ppm (8 h time-weighted average (TWA)).
- The NIOSH recommends 100 ppm (TWA).
- The OSHA sets the level of toluene at the workplace air as 200 ppm (8 h TWA).
- U.S. EPA sets the maximum contaminant level (MCL) of toluene in drinking water as 1.0 ppm (1.0 mg/L).

Tributyl Phosphate (CAS No. 126-73-8)

Molecular formula: $(C_4H_9O)_3 PO$

Synonyms and trade names: TBP; Phosphoric acid, tributyl ester; *n*-Butyl phosphate; Tributyl phosphate

Tributyl phosphate (TBP) is combustible and gives off irritating or toxic fumes (or gases) in a fire.

On decomposition, TBP releases COx, toxic fumes of phosphoric acid, phosphorus oxides, and/or phosphine. TBP is incompatible with strong oxidising agents and alkalis. The major uses of TBP in industry are as a component of aircraft hydraulic fluid and as a solvent for rare earth extraction and purification. Minor uses of TBP include use as a defoamer additive in cement casings for oil wells, an anti-air entrainment additive for coatings and floor finishes, as well as a carrier for fluorescent dyes. The major uses of TBP comprise over 80% of the volume produced. No current consumer product uses of TBP have been identified. The primary occupational exposure to TBP results from its use as an ingredient in aircraft hydraulic fluids. The potential for exposure to TBP varies with the type of maintenance activity but is almost always via a dermal pathway.

Safe Handling and Precautions

The primary and most probable route of exposure to TBP is through dermal contact in the occupational setting. The dermal toxicity of TBP has been shown to be very low. A secondary, infrequent route of exposure is to mist resulting from accidental leaks in an operating hydraulic system. Occupational workers associated with the manufacture, formulation processes, and/ or distribution of TBP may be exposed during these activities. These workers may be exposed by activities such as transfer of TBP from partial tank trucks to process storage tanks or smaller end-use containers sampling and maintenance of processing facilities. This potential worker exposure would be mainly dermal. Reports have indicated that the potential exposure to TBP varies with the type of maintenance activity but is almost always through a dermal pathway. Any exposure to mists of hydraulic fluid reported has been as a result of accidental leaks during hydraulic system operation or maintenance, or from venting of pressure prior to draining a system. Any exposure of aircraft passengers and/or crew to TBP as a constituent of aircraft hydraulic fluid is the result of a mechanical component failure during operations.

Occupational workers on exposure develop symptoms of health effects and poisoning. Accidental ingestion/swallowing and improper use and handling of TBP at workplaces cause cholinesterase inhibition. The symptoms of poisoning include, but not limited to, salivation, sweating, headache, nausea, eye and skin irritation, irritation of the lungs, muscle twitching, tremors, incoordination, blurred vision, tears, abdominal cramps, diarrhoea, and chest discomfort. Repeated exposures and accidental ingestion of TBP cause adverse effects on the bladder, resulting in tissue lesions. The ACGIH, the IARC, and the NTP have not listed TBP as a human carcinogen.

Occupational workers during use and handling of TBP must use proper and appropriate PPE, store TBP in a cool, dry place away from oxidising materials, and keep the chemical container tightly closed.

Regulation: Canada—WHMIS classified TBP as D2B, meaning a toxic material.

Tributyltin Oxide (CAS No. 56-35-9)

Molecular formula: $C_{24}H_{54}OSn_2$
Synonyms and trade names: Bis(tri-*n*-butyltin) oxide; Butinox; Hexabutylditin; Tributyltin oxide; TBT; TBTO; Bis(tributyloxide) of tin; Bis(tributylstannyl)oxide; Oxybis-(tributyltin)
Chemical class: Trialkyl organotin compound

TBTO (bis(tri-*n*-butyltin)oxide) appears as thin, colourless to pale yellow, flammable and combustible liquid. It is soluble in organic solvents. TBTO, or bis(tri-*n*-butyltin)oxide, is an organotin compound used as a biocide, fungicide, and molluscicide. Uses of tributyltin also include as an anti-fouling chemical in marine paints for boats, anti-fungal agent in textiles and industrial water systems, in cooling tower and refrigeration water systems, wood pulp preservative in paints and paper mill systems, inner surfaces of cardboard, and in the manufacturing processes of leather goods, textiles, wood, plastics, and mothproof stored garments. In fact, TBT compounds are considered the most hazardous of all tin compounds.

Safe Handling and Precautions

Exposure to TBTO is harmful and causes severe pain to skin and irritation to the skin and eyes. Accidental ingestion at workplaces causes serious health damage. TBTO is a metabolic poison and is highly toxic. It penetrates the intact skin and gets absorbed from the gastrointestinal tract and by inhalation. The signs and symptoms of TBTO poisoning include, but are not limited to, irritation of the skin and eyes, profuse lacrimation, headaches, inflammation of the skin, sore throat, coughing, nausea, dizziness, photophobia, and weakness. Workplace accidental ingestion of TBTO has been reported to cause vomiting, abdominal pain, and diarrhoea. Prolonged period of exposure to TBTO is known cause severe poisoning, irritation of the upper respiratory tract, severe dermatitis, injuries to the liver and kidneys, tremors, incoordination, depression of the immune function, convulsions, and flaccid paralysis. Reports have also shown that large number of workers in a rubber factory using TBTO in the vulcanising process suffered irritation of the upper respiratory tract.

Regulations: Regulatory agencies have set the safety limits of TBT in workplaces as follows:

- The ACGIH set the TLV ([TWA]) in air as 0.1 mg tin/m^3 and the short-term TLV at 0.2 mg tin/m^3.
- The Federal Republic of Germany recommends the limit of organotin compounds in air as 0.1 mg tin/m^3.
- The U.K. recommended the occupational exposure limit as 0.1 mg tin/m^3 and the acceptable daily intake (ADI) as 7.2.5.
- Japan adopted a tentative ADI of 1.6 er of/day.
- Occupational workers, during handling of TBTO, should remember the safety regulations and management of toxic chemical substances at the workplaces. The occupational exposure limits of TBTO are very low.
- TLV (as Sn) 0.1 mg/m^3 as TWA, 0.2 mg/m^3 as STEL, and skin A4 (not classifiable as a human carcinogen) (ACGIH 2008).
- MAK (as Sn) 0.004 ppm 0.02 mg/m^3, peak limitation category I(1), skin absorption (H), carcinogen category 4, and pregnancy risk group C (DFG 2009).

TBT is a highly toxic biocide and causes a serious problem in the aquatic environment because it is extremely toxic to non-target organisms, is linked to immunosupression and imposex (development of male characteristics in females) in snails and bivalves, and can be persistent. Occupational workers handling anti-fouling paint containing TBT should use and dispose of the chemical waste in accordance with generally acceptable practices to minimise contact with the soil, water, and aquatic life and to avoid contamination and damage to non-target organisms.

Occupational workers, during use and handling of TBTO, should strictly observe safety regulatory practices and avoid workplace negligence. Workers should collect all contaminated chemical waste and TBT paint spill and dispose of in accordance with state, local, and federal requirements. Workers should wear protective clothing, PPE, to prevent chemical contact and canister face mask, self-contained breathing apparatus, liquid-proof gloves, protective goggles or face shield, and rubber shoes or boots to avoid aerosol spray, organic mist, vapour, and chemical spillage. It is important that the concentrated material should be handled only by trained workers. Occupational workers, such as shipyard workers exposed to TBT dust and vapours while repairing a submarine, mixers, and applicators, should wear protective workplace dress and protective impermeable boots.

The International Convention on the Control of Harmful Anti-fouling Systems on Ships prohibits the use of harmful organotins in anti-fouling paints used on ships and establishes a mechanism to prevent the potential future use of other harmful substances in anti-fouling systems. Also, many countries have restricted the use of TBT anti-fouling paints as a result of effects on shellfish.

Trichloroethylene (Technical) (CAS No. 79-01-6)

Molecular formula: C_2HCl_3
Synonyms and trade names: Acetylene trichloride; Ethinyl trichloride; Ethylene trichloride; 1-Chloro-2,2-dichloroethylene; 1,1-Dichloro-2-chloroethylene; 1,1,2-Trichloroethylene; TCE; TRI; Trichlor

Trichloroethylene is a colourless liquid with a mild sweet odour. Trichloroethylene is moderately stable. On contact with air, it slowly decomposes and forms phosgene, hydrogen chloride, and dichloroacetyl chloride. Trichloroethylene in contact with water becomes corrosive and forms dichloroacetic acid and hydrochloric acid. It is soluble in methanol, diethyl ether, and acetone.

Safe Handling and Precautions

Exposure to trichloroethylene is harmful and causes poisoning. Occupational exposure and negligence during use and handling of trichloroethylene cause adverse effects and poisoning with symptoms of dizziness, headache, sleepiness, nausea, confusion, blurred vision, facial numbness, weakness, and effects on the kidneys, nervous system, liver, heart, upper respiratory tract, and immune and endocrine systems. Studies have shown that simultaneous alcohol consumption and trichloroethylene inhalation increases the toxicity of trichloroethylene in humans. There is some evidence from human epidemiological studies and strong evidence from animal studies that trichloroethylene causes cancer. The IARC lists trichloroethylene as Group 2A, meaning a probable human carcinogen. The Canadian Environmental Act (CEPA) also lists trichloroethylene as Group II, meaning probably carcinogenic to humans. The U.S. EPA currently has no consensus classification for the carcinogenicity of trichloroethylene. The agency is currently reassessing the cancer classification of trichloroethylene and its potential as a human carcinogen. The recent epidemiological studies suggest trichloroethylene exposure to be associated with several types of cancers in humans, especially kidney, liver, cervix, and lymphatic system. Animal studies have reported increases in lung, liver, kidney, and testicular tumours and lymphoma.

Occupational workers during use and handling of trichloroethylene must strictly observe safety regulatory practices and avoid excessive heat, open flames, sparks, electrical arcs, and sources of high temperature such as welding arcs, hot surfaces, and sunlight. Workers should avoid moisture. The workplace should have adequate ventilation and should be free from welding operations, cutting, soldering, drilling, or other hot work.

Regulations: Canada – WHMIS categorises trichloroethylene as Class D-1B, meaning chemical substance causes immediate and serious toxic effects, and Class D-2B, meaning it causes other toxic effects.

1,2,3-Trichloropropane (CAS No. 96-18-4)

Molecular formula: $C_3H_5Cl_3$
Synonyms and trade names: Allyl trichloride; Glycerol trichlorohydrin; Glyceryl trichloro-hydrin; Trichlorohydrin; Trichloropropane

1,2,3-Trichloropropane is a synthetic chemical that is also known as allyl trichloride, glycerol trichlorohydrin, and trichlorohydrin. It is a colourless, heavy liquid with a sweet but strong chloroform-like odour and is combustible. 1,2,3-Trichloropropane is slightly soluble in water but soluble in chloroform, diethyl ether, and ethanol. On contact with heat/ fire, 1,2,3-trichloropropane releases off irritating or toxic fumes (or gases). It evaporates very quickly and small amounts dissolve in water. It is mainly used to make other chemicals. 1,2,3-Trichloropropane was used in the past mainly as a solvent and extractive agent, including as a paint and varnish remover and as a cleaning and degreasing agent. It is now used mainly as a chemical intermediate, for example, in the production of polysulphone liquid polymers, dichloropropene and hexafluoropropylene and as a cross-linking agent in the synthesis of polysulphides.

Safe Handling and Precautions

The general population may potentially be exposed to low levels of 1,2,3-trichloropropane through ingestion of contaminated well water or inhalation of contaminated air. Exposure is more likely for individuals who live near facilities that use or produce 1,2,3-trichloropropane or near hazardous waste disposal facilities. Workplace accidental and improper handling of 1,2,3-trichloropropane causes adverse health disorders among occupational workers. The symptoms of toxicity and poisoning include, but not limited to, cough, irritation to the eyes, sore throat, headache, drowsiness, and respiratory tract irritation, and prolonged and repeated exposures to high concentrations of 1,2,3-trichloropropane cause effects on the liver and kidneys, resulting in impaired functions and unconsciousness. Reports indicate that 1,2,3-trichloropropane has sufficient evidence in experimental animals as a carcinogen, but evidences in humans are inadequate as a human carcinogen. Hence, 1,2,3-trichloropropane has been classified as Group 2A, meaning a probable human carcinogen.

Triethanolamine (CAS No. 102-71-6)

Molecular formula: $C_6H_{15}NO_3$
Synonyms and trade names: Alkanolamine 244; Daltogen; Nitrilotriethanol; Sterolamide; Sting-Kill; TEA; TEA (amino alcohol); TEOA; Triethanolamin; tris(β-Hydroxyethyl)amine; tris(2-Hydroxyethyl)amine; Thiofaco; Trolamine
IUPAC name: 2,2′,2″-Nitrilotriethanol

Triethanolamine is a viscous, colourless/pale yellow liquid with a weak ammoniacal odour. Triethanolamine is incompatible with copper, copper alloys, galvanised iron, acids, and oxidisers. Reports indicate that in India itself, as many as six companies manufacture triethanolamine and it is manufactured by many different countries around the world. Global production and industrial application of triethanolamine is very extensive.

In industries, triethanolamine is used as a corrosion inhibitor in metal-cutting fluids; a curing agent for epoxy and rubber polymers; a copper–triethanolamine; in emulsifiers, thickeners, and wetting agents in the formulation of consumer products such as cosmetics, detergents, shampoos, and other personal products; and a neutraliser-dispersing agent in agricultural herbicide formulations. In brief, triethanolamine has wide applications as a corrosion inhibitor, a surface-active agent, and an intermediate in various products including metalworking fluids, oils, fuels, paints, inks, cement, cosmetic, and personal products and formulations of algicides and herbicides.

Safe Handling and Precautions

Exposure to triethanolamine causes harmful effects such as irritation of the eyes and skin, redness, pain, and dermatitis and adverse effects such as damage to the skin and skin sensitisation and dermatitis. Prolonged and/or repeated workplace exposures of triethanolamine through inhalation and ingestion cause health disorders and poisoning with symptoms that include, but not limited to, nausea, vomiting, gastrointestinal irritation, and diarrhoea. Exposures to vapours of triethanolamine cause irritation of the respiratory tract, nasal discomfort, coughing, respiratory distress, or breathing difficulty. Exposures to high concentrations of triethanolamine cause burns in the mouth, pharynx, and oesophagus; abdominal pain; nausea; vomiting; diarrhoea; and injuries to the liver and kidney. The IARC listed triethanolamine as Group 3, meaning not classifiable as a carcinogen to humans.

Occupational workers should use and handle triethanolamine with care and keep the chemical in a tightly closed container, stored in a cool, dry, ventilated area, protected against physical damage, and isolated from any source of heat or ignition, avoiding contact with copper and copper alloys. Material is suitably handled in stainless steel equipment. Do not use aluminium for storage of aqueous solutions. Outside or detached storage is preferred. Isolate chemical from acidic materials. It may separate and freeze below 16°C (60°F). Thaw and mix before sampling or using. Do not store above 43°C (110°F).

Triforine (CAS No. 26644-46-2)

Molecular formula: $C_{10}H_{14}Cl_6N_4O_2$
Synonyms and trade names: Aprol; Basforin; Denarin; Funginex; Triforine; Triforin; cme74770; Triforine Rose Fungicide; Saprol

Triforine is a clear light yellow colour with an ethanol (alcohol) odour and a highly flammable liquid. It is used as a fungicide for the control of black spot, powdery mildew, and diseases of rust in roses, fruits, vegetables, cereals, and ornamentals. Application of triforine acts both as a preventative and a curative fungicide and is known to destroying diseases already in the plant and also preventing disease infestations. It is also used as a post-harvest control of brown rot on peaches, nectarines, apricots, cherries, and plums.

Safe Handling and Precautions

Exposure to triforine is harmful. Accidental workplace exposure and negligence during use have been reported to cause toxicity with symptoms of irritation to mouth; irritation to the eyes, with effects including tearing, pain, stinging, and blurred vision; irritation to the skin, with redness, skin inflammation, and itching; throat mucous build-up; irritation to the tongue and lips; stomach; irritation to the nose and respiratory system; nausea; dizziness; headache; vomiting; central nervous system depression; confusion; and intoxication (because of ethanol ingredient).

Occupational workers during use of triforine should maintain adequate ventilation and a local exhaust ventilation system. Workers should avoid accumulation of vapours in hollows or sumps and sources of heat, ignition, and open flames at the work area. Triforine should be kept stored in a cool place, away from direct sunlight, strong alkalis, acids, combustibles, and oxidising agents.

Regulations: Classified as a highly flammable chemical.

Danger! Triforine is a highly flammable liquid. Avoid all sources of ignition, heat, and naked flames.

Trinitrotoluene (CAS No. 118-96-7)

Molecular formula: $C_7H_5N_3O_6$
Synonyms and trade names: 2-Methyl-1,3,5-trinitrobenzene; 1-Methyl-2,4,6-trinitrobenzene; TNT trinitrotoluol; sym-Trinitrotoluene; TNT; 2,4,6-Trinitrotoluene
IUPAC name: 2-Methyl-1,3,5-trinitrobenzene

2,4,6-Trinitrotoluene (TNT) is a yellow, odourless, unstable solid. TNT does not occur naturally in the environment. TNT is an explosive used in military shells, bombs, and grenades; in industrial uses; and in underwater blasting. TNT is a high explosive that is unaffected by ordinary shocks and therefore must be set off by a detonator. TNT is often mixed with other explosives such as ammonium nitrate to form amatol. Because it is insensitive to shock and must be exploded with a detonator, it is the most favoured explosive used in munitions and construction. TNT reacts violently, is potentially explosively, reacts with heavy metals, and is a chemical with risk of explosion if heated or struck.

Safe Handling and Precautions

Exposure to TNT is harmful and fatal. It causes effects of irritation to the eyes, the skin, and respiratory tract and effects on the blood leading in haemolysis, peripheral neuritis, muscle pain, kidney damage, cardiac irregularities, cataracts, sensitisation dermatitis, anaemia, leucocytosis, formation of methaemoglobin, and fatal injury or death.

IARC reports that evidences to list TNT as a carcinogen in animal or humans are inadequate and hence listed as Group 3, meaning not classifiable as a carcinogen.

Occupational workers should be very careful with precautions and completely avoid negligence during handling of TNT.

Regulations:

1. The NIOSH has set the recommended exposure limit (REL) of TNT as 0.5 mg/m^3 TWA (skin).
2. The ACGIH has set the TLV of TNT as 0.1 mg/m^3 TWA (skin).
3. The U.S. OSHA has set the PEL on skin of TNT in general industrial workplaces as 1.5 mg/m^3.
4. The OSHA at construction industrial area has set the – PEL as 15 mg/m^3 TWA (skin).

Triphenyl Phosphate (CAS No. 115-86-6)

Molecular formula: $C_{18}H_{15}O_4P$

Safe Handling and Precautions

Hazardous in case of skin contact (irritant, permeator), of eye contact (irritant), of ingestion, of inhalation.

Storage: Keep container dry. Keep in a cool place. Ground all equipment containing material. Keep container tightly closed. Keep in a cool, well-ventilated place. Combustible materials should be stored away from extreme heat and away from strong oxidizing agents.

Personal Protection in Case of a Large Spill: Splash goggles. Full suit. Dust respirator. Boots. Gloves. A self contained breathing apparatus should be used to avoid inhalation of the product. Suggested protective clothing might not be sufficient; consult a specialist BEFORE handling this product.

Gloves. Lab coat. Dust respirator. Be sure to use an approved/certified respirator or equivalent. Wear appropriate respirator when ventilation is inadequate and splash goggles.

Trisodium Phosphate (Anhydrous) (CAS No. 7601-54-9)

Molecular formula: Na_3PO_4
Synonyms and trade names: 12-Hydrate; Phosphoric acid, trisodium salt, dodecahydrate; Sodium phosphate, tribasic; Trisodium orthophosphate

Trisodium phosphate (anhydrous) is a white, granular or crystalline solid, highly soluble in water and produces a strong alkaline solution. On exposure to heat, trisodium phosphate decomposes and produces toxic and corrosive fumes including phosphorous oxides.

The major use for trisodium phosphate is as a cleaning agent, food additive, stain remover, and degreaser. Trisodium phosphate of commercial grade is often partially hydrated and ranges from anhydrous trisodium phosphate, Na_3PO_4, to the dodecahydrate, $Na_3PO_4 \cdot 12H_2O$. Most often found in white powder form, it is also called trisodium orthophosphate or just plain sodium phosphate. Trisodium phosphate reacts violently with water and acids to liberate heat. Trisodium phosphate is corrosive and in the presence of water attacks many metals.

Trisodium phosphate is an approved flux for use in hard soldering joints in medical grade copper plumbing. The flux is applied as a concentrated water solution and dissolves copper oxides at the temperature used in copper brazing. Residues are fully water soluble and can be rinsed out of plumbing before it is put in service. Also, trisodium phosphate is still in vast use for the cleaning, degreasing, and deglossing of walls prior to painting. In fact, application of trisodium phosphate breaks the gloss of oil-based paints and opens the pores of latex-based paint providing a surface better suited for the adhesion of the subsequent layer of paint.

Safe Handling and Precautions

Exposure to trisodium phosphate causes corrosive effects to the eyes, skin, and respiratory tract. It is corrosive on ingestion. Inhalation of dust may cause lung oedema. The alkaline nature of trisodium phosphate is known to cause injuries to the oesophagus and digestive tract and irritation to the respiratory tract. Symptoms may include coughing and shortness of breath. Prolonged exposure and improper use and handling of the chemical substance result in the destruction of mucous membranes, permanent tissue damage, asthmatic bronchitis, chemical pneumonitis, or pulmonary oedema.

Occupational workers must be alert during use and handling of trisodium phosphate. Workers should not be negligent to wear appropriate workplace dress: impervious protective clothing, boots, gloves, apron or coveralls, chemical safety goggles, and/or full face shield to avoid dusting or splashing of solutions and, as appropriate, to prevent the possible chemical injuries. Workers must keep chemical in a tightly closed container, stored in a cool, dry, well-ventilated area and protected against physical damage and away from children.

Warning! Trisodium phosphate is harmful. Avoid swallowing or inhaling chemical.

Trisodium Phosphate (Anhydrous) (CAS No. 7601-54-9)

Molecular formula: Na_3PO_4
Synonyms and trade names: 12-Hydrate; Phosphoric acid, trisodium salt, dodecahydrate; Sodium phosphate, tribasic; Trisodium orthophosphate

Trisodium phosphate is a white, granular or crystalline solid, highly soluble in water and produces a strong alkaline solution. On exposure to heat, trisodium phosphate decomposes and produces toxic and corrosive fumes including phosphorous oxides.

The major use for trisodium phosphate is as a cleaning agent, food additive, stain remover, and degreaser. Trisodium phosphate of commercial grade is often partially hydrated and ranges from anhydrous trisodium phosphate, Na_3PO_4, to the dodecahydrate, $Na_3PO_4 \cdot 12H_2O$. Most often found in white powder form, it is also called trisodium orthophosphate or just plain sodium phosphate. Trisodium phosphate reacts violently with water and acids to liberate heat. Trisodium phosphate is corrosive and in the presence of water attacks many metals.

Trisodium phosphate is an approved flux for use in hard soldering joints in medical grade copper plumbing. The flux is applied as a concentrated water solution and dissolves copper oxides at the temperature used in copper brazing. Residues are fully water soluble and can be rinsed out of plumbing before it is put in service. Also, trisodium phosphate is still in vast use for the cleaning, degreasing, and deglossing of walls prior to painting. In fact, application of trisodium phosphate breaks the gloss of oil-based paints and opens the pores of latex-based paint providing a surface better suited for the adhesion of the subsequent layer of paint.

Safe Handling and Precautions

Exposure to trisodium phosphate is corrosive on ingestion and also affects the eyes, the skin, and the respiratory tract. Inhalation of trisodium phosphate dust causes lung oedema. The alkaline nature of trisodium phosphate is known to cause injuries to the oesophagus and digestive tract and irritation to the respiratory tract. Symptoms may include coughing and shortness of breath. Prolonged exposure and improper use and handling of the chemical substance result in the destruction of mucous membranes, permanent tissue damage, asthmatic bronchitis, chemical pneumonitis, or pulmonary oedema.

Occupational workers must be alert during use and handling of trisodium phosphate. Workers should not be negligent to wear appropriate workplace dress: impervious protective clothing, boots, gloves, apron or coveralls, chemical safety goggles, and/or full face shield to avoid dusting or splashing of solutions and, as appropriate, to prevent the possible chemical injuries. Workers must keep chemical in a tightly closed container, stored in a cool, dry, well-ventilated area and protected against physical damage and away from children.

Warning! Trisodium phosphate is harmful. Avoid swallowing or inhaling chemical.

Bibliography

Abel, R. 1996. European policy and regulatory action for organotin-based antifouling paints. In: *Organotin: Environmental Fate and Effects*, Champ, M. A. and P. F. Seligman (eds.). Chapman & Hall, London, U.K., Chapter 2.

Agency for Toxic Substances and Disease Registry (ATSDR). 1990. *Toxicological Profile for Thorium*. U.S. Department of Health and Human Services, Public Health Service, Atlanta, GA (updated 2011).

Agency for Toxic Substances and Disease Registry (ATSDR). 1992. *Toxicological Profile for Thallium*. U.S. Department of Health and Human Services, Public Health Service, Atlanta, GA (updated 2011).

Agency for Toxic Substances and Disease Registry (ATSDR). 1995. *Toxicological Profile for 2,4,6-Trinitotoluene (TNT)*. ATSDR, Atlanta, GA.

Agency for Toxic Substances and Disease Registry (ATSDR). 2000. *Toxicological Profile for Toluene*. U.S. Department of Health and Human Services, Public Health Service, Atlanta, GA (update 2011).

Agency for Toxic Substances and Disease Registry (ATSDR). 2003. *Trichloroethylene*. U.S. Department of Health and Human Services, Public Health Service, Atlanta, GA.

Agency for Toxic Substances and Disease Registry (ATSDR). 2005. *Toxicological Profile for Tin and Compounds*. U.S. Department of Health and Human Services, Public Health Service, Atlanta, GA (updated 2011).

Barnes, J. M. and H. B. Stoner. 1959. The toxicology of tin compounds. *Pharmacol. Rev.* 11:211–231.

Beyer, K. H., Jr., W. F. Bergfeld, W. O. Berndt, R. K. Boutwell, W. W. Carlton, D. K. Hoffmann, and A. L. Schroeder. 1983. Final report on the safety assessment of triethanolamine, diethanolamine and monoethanolamine. *J. Am. Coll. Toxicol.* 2:183–235.

Bingham, E., B. Cohrssen, and C. H. Powell. 2001. *Patty's Toxicology*, 5th edn., Vols. 1–9. John Wiley & Sons, New York.

Blum, A. and G. Lischka. 1997. Allergic contact dermatitis from mono-, di- and triethanolamine (short communication). *Contact Derm.* 36:166.

Cascieri, T., E. J. Ballester, L. R. Serman, R. F. Mcconnell, J. W. Thackara, and M. J. Fletcher. 1985. Subchronic toxicity study with tributyl phosphate in rats. *Toxicologist*, 5:97.

Cheremisinoff, N. P. (ed.). 2003. *Industrial Solvents Handbook*, 2nd edn. Marcel Dekker, Inc., New York.

Conibear, S. A. 1983. Long term health effects of thorium compounds on exposed workers: The complete blood count. *Health Phys.* 44(Suppl 1):231–237.

Downs, W. L., J. K. Scott, L. T. Steadman, and E. A. Maynard. 1960. Acute and sub-acute toxicity studies of thallium compounds. *Am. Ind. Hyg. Assoc. J.* 21:399–406.

Gallo, M. A. and N. J. Lawryk. 1991. Organic phosphorus pesticides. In: *Handbook of Pesticide Toxicology*, Hayes, W. J., Jr. and E. R. Laws, Jr. (eds.). Academic Press, New York.

Gillner, M. and I. Loeper. 1993. Health effects of selected chemicals. 2. Triethanolamine. *Nord* 29:235–260.

Gosselin, R. E., R. P. Smith, and H. C. Hodge (eds.). 1984. *Clinical Toxicology of Commercial Products.* Williams & Wilkins, Baltimore, MD.

Greene, S. A. and R. P. Pohanish (eds.). 2007. *Sittig's Handbook of Pesticides and Agricultural Chemicals.* William Andrew Publishing, Norwich, NY.

Hathaway, G. J., N. H. Proctor, J. P. Hughes, and M. L. Fischman. 1991. *Proctor and Hughes' Chemical Hazards of the Workplace,* 3rd edn. Van Nostrand Reinhold, New York.

Hayes, W. J. and E. R. Laws (eds.). 1990. *Handbook of Pesticide Toxicology,* Vol. 3, Classes of Pesticides. Academic Press, Inc., New York.

Hazardous Substances Data Bank (HSDB). 1989. Tetrahydrofuran. HDSB, National Library of Medicine, Bethesda, MD.

Inoue, K., T. Sunakawa, K. Okamoto, and Y. Tanaka. 1982. Mutagenicity tests and in vitro transformation assays on triethanolamine. *Mutat. Res.* 101:305–313.

International Programme on Chemical Safety (IPCS). Poisons Information Monograph No. 525. IPCS, Geneva, Switzerland.

International Agency for Research on Cancer (IARC). 1994. Some industrial chemicals. *IARC Monographs on the Evaluation of Carcinogenic Risks to Humans,* Vol. 60, IARC, Lyon, France, pp. 161–180.

International Programme on Chemical Safety and the Commission of the European Communities (IPCS–CEC), Material Safety Data Sheet (MSDS). 1995a. Triethanolamine. MSDS No. ICSC No. 1178. IPCS–CEC, Luxembourg (updated 2004).

International Programme on Chemical Safety and the Commission of the European Communities (IPCS–CEC). International Chemical Safety Cards (ICSC). 1995b. Trisodium phosphate (anhydrous). ICSC No. 1178. IPCS–CEC, Luxembourg (updated 2004).

International Programme on Chemical Safety and the Commission of the European Communities (IPCS–CEC). 1999. Trinitotoluene. ICSC Card No. ICSC: 0987. IPCS–CEC, Luxembourg, Germany (updated 2005).

International Chemical Safety Cards (ICSC). Material Safety Data Sheet (MSDS). 2003. Triethanolamine. MSDS No. ICSC. 1034. International Programme on Chemical Safety and the Commission of the European Communities (IPCS–CEC) (updated 2004).

International Programme on Chemical Safety and the Commission of the European Communities (IPCS, CEC). 2005a. Tributyltin oxide. ICSC Card No. ICSC: 1282. IPCS–CEC, Luxembourg (updated 2005).

International Programme on Chemical Safety and the Commission of the European Communities (IPCS-CEC). 2005b. 1,2,3-Trichloropropane. ICSC No. 0683. IPCS–CEC, Geneva, Switzerland.

Johansen, C. 1967. Tumors in rabbits after injections of various amounts of thorium dioxide. *NY Acad. Sci.* 145:724.

Kennedy, P. and J. B. Cavanagh. 1976. Spinal changes in the neuropathy of thallium poisoning. *J. Neurol. Sci.* 29:295–301.

Kidd, H. and D. R. James (eds.). 1991. *The Agrochemicals Handbook,* 3rd edn. Royal Society of Chemistry Information Services, Cambridge, U.K.

Konishi, Y., A. Denda, K. Uchida, Y. Emi, H. Ura, Y. Yokose, K. Shiraiwa, and M. Tsutsumi. 1992. Chronic toxicity carcinogenicity studies of triethanolamine in B6C3F1 mice. *Fundam. Appl. Toxicol.* 18:25–29.

Krieger, R. I. and C. K. William. 2001. *Handbook of Pesticide Toxicology,* 2nd edn. Elsevier Inc., New York.

Lewis, R. J. (ed.). 1993. *Hawley's Condensed Chemical Dictionary,* 12th edn. Van Nostrand Reinhold Company, New York.

Lewis, R. J., Sr. (ed.). 1997. *Hawley's Condensed Chemical Dictionary,* 13th edn. John Wiley & Sons, New York.

Lewis, R. J., Sr. 2007. *Hawley's Condensed Chemical Dictionary.* Wiley-Interscience, John Wiley & Sons, New York.

Lewis, R. J. and R. J. Lewis. 2008. *Hazardous Chemicals Desk Reference.* John Wiley & Sons, New York.

Marcus, R. L. 1985. Investigation of a working population exposed to thallium. *J. Soc. Occup. Med.* 35:4–9.

Material Safety Data Sheet (MSDS). 2004. Tin (IV) chloride. ICSC No. 0953. International Programme on Chemical Safety and the Commission of the European Communities (IPCS–CEC), Geneva, Switzerland.

Material Safety Data Sheet (MSDS). 2005a. Tetrachloroethylene. MSDS No. T0767. Mallinckrodt Baker, Inc., Phillipsburg, NJ.

Material Safety Data Sheet (MSDS). International Chemical Safety Cards. 2005b. Tri-*n*-butyl phosphate. ICSC No. 0584. International Programme on Chemical Safety & the Commission of the European Communities (IPCS CEC), Geneva, Switzerland.

Material Safety Data Sheet (MSDS). 2006. Tributyl phosphate. ACC No. 01643. Fisher Scientific, Fair Lawn, NJ.

Meister, R. T. (ed.). 1992. *Farm Chemicals Handbook '92*. Meister Publishing Company, Willoughby, OH.

Meister, R. T., G. L. Berg, C. Sine, S. Meister, and J. Poplyk (eds.). 1984. *Farm Chemicals Handbook*, 70th edn. Meister Publishing Co., Willoughby, OH.

Merck Index. 1983. 10th edn. Merck Co., Rahway, NJ.

Morris, H. P., A. Dubnik, and A. Dalton. 1946. Effect of prolonged ingestion of thiourea on mammary glands and the appearance of mammary tumors in adult C3H mice. *J. Natl. Cancer Inst.* 7:159.

National Institute for Occupational Safety and Health (NIOSH). 1997. Toluene. *Pocket Guide to Chemical Hazards*. U.S. Department of Health and Human Services, Public Health Service, Centers for Disease Control and Prevention, Cincinnati, OH (updated 2010).

National Toxicology Program (NTP). 1999. *Toxicology and Carcinogenesis Studies of Triethanolamine (CAS No. 102-71-6) in F344/N Rats and B6C3F1 Mice (Dermal Studies)*. (NTP TR 449; NIH Publ. No. 00-3365). Research Triangle Park, NC.

NIOSH. 2010. Tetrachloroethylene. *Pocket Guide to Chemical Hazards*. NIOSH, Centers for Disease Control and Prevention, Atlanta, GA.

O'Neil, M. J. (ed.). 2001. *The Merck Index—An Encyclopedia of Chemicals, Drugs, and Biologicals*, 13th edn. Merck & Co., Inc., Whitehouse Station, NJ.

Patnaik, P. (ed.). 1999. *A Comprehensive Guide to the Hazardous Properties of Chemical Substances*, 2nd edn. McGraw Hill, New York.

Patnaik, P. (ed.). 2007. *A Comprehensive Guide to the Hazardous Properties of Chemical Substances*. John Wiley & Sons, Hoboken, NJ.

Pesticides Safety Directorate, Ministry of Agriculture, Fisheries and Food. 2000. Thiodicarb. *Pesticide Residues in Food*. York, U.K.

Pohanish, R. P. (ed.). 2008. *Sittig's Handbook of Toxic and Hazardous Chemicals and Carcinogens*, 5th edn. William Andrew, Norwich, NY.

Sax, N. I. (ed.). 1984. *Dangerous Properties of Industrial Materials*, 6th edn. Van Nostrand Reinhold Company, Inc., New York.

Sax, N. I. and R. J. Lewis, Jr. 1989. *Dangerous Properties of Industrial Materials*, 7th edn., 3 volumes. Van Nostrand Reinhold, New York.

Scott, R. M. (ed.). 1989. *Chemical Hazards in the Workplace*. CRC Press, Boca Raton, FL.

da Silva Horta, J. 1956. Late lesions in man caused by colloidal thorium dioxide (Thorotrast). *Arch. Pathol.* 62:403–418.

Stewart, R. D., H. C. Dodd, H. H. Gay et al. 1970. Experimental human exposure to trichloroethylene. *Arch. Environ. Health* 20:64–71.

Tomlin, C. D. S. (ed.). 1994. *The Pesticide Manual: A World Compendium*, 10th edn. The British Crop Protection Council, Surrey, U.K.

Tomlin, C. D. S. (ed.). 2006. *The Pesticide Manual: A World Compendium*, 14th edn. The British Crop Protection Council, Surrey, U.K.

Tomlin, C. D. S. (ed.). 2011. *The Pesticide Manual: A World Compendium*. British Crop Protection Council, Surrey, U.K.

U.S. Department of Labor (DOL). Occupational Safety & Health Administration (OSHA). *Thallium-Soluble Compounds (as Tl). Chemical Sampling Information*. OSHA, Washington, DC.

U.S. Department of Labor (DOL). 1993. *Occupational Safety and Health Guideline for Tetrahydrofuran*. Occupational Safety & Health Administration, Washington, DC.

U.S. Environmental Protection Agency (U.S. EPA). 1988a. *Health Advisory: Terbufos.* Office of Drinking Water, Washington, DC.

U.S. Environmental Protection Agency (U.S. EPA). 1988b. Terbufos. Pesticide Fact Sheet No. 5. Office of Pesticide and Toxic Substances, Washington, DC.

U.S. Environmental Protection Agency (U.S. EPA). 1989. Pesticide tolerance for terbufos. *Fed. Reg.* 54:35896–35897.

U.S. Environmental Protection Agency (U.S. EPA). 1996. *Occupational Safety and Health Guideline for Tin Oxide.* U.S. EPA, Washington, DC.

U.S. Environmental Protection Agency (U.S. EPA). Integrated Risk Information System (IRIS). 1999. Toluene. National Center for Environmental Assessment, Office of Research and Development, Washington, DC.

U.S. Environmental Protection Agency (U.S. EPA). Integrated Risk Information System (IRIS). 2000. Trichloroethylene—Hazard summary. National Center for Environmental Assessment, Office of Research and Development, Washington, DC (updated 2007).

Urben, P. G. (ed.). 2006. *Bretherick's Handbook of Reactive Chemical Hazards,* 7th edn. Elsevier, Academic Press, Amsterdam, the Netherlands

West, R. J. and S. J. Gonsior. 1996. Biodegradation of triethanolamine. *Environ. Toxicol. Chem.* 15:472–480.

World Health Organization (WHO). 1985. Trichloroethylene. Environmental Health Criteria No. 50. International Programme on Chemical Safety (IPCS). WHO, IPCS, Geneva, Switzerland.

World Health Organization (WHO). 1991. Tributyl phosphate. Criteria No. 112. IPC Environmental Health. WHO, Geneva, Switzerland.

World Health Organization (WHO). 1996a. Thallium. Health and Safety Guide No. 102. International Programme on Chemical Safety (IPCS), WHO, Geneva, Switzerland.

World Health Organization (WHO). 1996b. Thallium. Environmental Health Criteria. No. 182. International Programme on Chemical Safety (IPCS), WHO, Geneva, Switzerland.

World Health Organization (WHO). 1996c. Tributyltin (TBT). Tributyltin compounds. Environmental Health Criteria No. 116. International Programme on Chemical Safety (IPCS), WHO, Geneva, Switzerland.

World Health Organization (WHO). 1999. Tributyltin. Concise International Chemical Assessment Document No. 14. International Programme on Chemical Safety (IPCS), WHO, Geneva, Switzerland.

World Health Organization (WHO). 2000. Thiourea. International Chemical Safety Card No. 0680. International Programme on Chemical Safety (IPCS), WHO, Geneva, Switzerland.

World Health Organization (WHO). 2003a. 1,2,3-Trichloropropane. Concise International Chemical Assessment Document No. 56. International Programme on Chemical Safety (IPCS), WHO, Geneva, Switzerland.

World Health Organization (WHO). 2003b. Thiourea. Concise International Chemical Assessment Document No. 49. WHO, Geneva, Switzerland.

22

Hazardous Chemical Substances:
U

Uracil Mustard
Uranium
Urethane (Ethyl Carbamate)

Uracil Mustard (CAS No. 66-75-1)

Molecular formula: $C_8H_{11}Cl_2N_3O_2$
Synonyms: Aminouracil mustard; Chlorethaminacil; Demethyldopan; Nordopan; Uracillost; Uramustin; Uramustine

Uracil mustard appears as creamy/off-white, odourless, crystalline powder. It is used as an anti-cancer medicine. Uracil mustard is a chemotherapy drug that belongs to the class of alkylating agents. It is used for its anti-neoplastic properties. It works by damaging deoxyribonucleic acid (DNA), primarily in cancer cells that preferentially take up the uracil due to their need to make nucleic acids during their rapid cycles of cell division. At high concentrations of the drug, cellular RNA and protein synthesis are also suppressed. The DNA damage leads to apoptosis of the affected cells. Chemically it is a derivative of nitrogen mustard and uracil. Uracil mustard is a non-combustible substance itself; it does not burn but may decompose upon heating to produce corrosive and/or toxic fumes. Some are oxidisers and may ignite combustibles – wood, paper, oil, clothing, etc. Contact with metals may evolve flammable hydrogen gas. Containers may explode when heated.

Safe Handling and Precautions

Uracil mustard is a hazardous and harmful chemical. Exposures have been known to cause effects such as nausea, vomiting, diarrhoea, dermatitis, irritability, depression, leukopaenia, thrombocytopaenia, and anaemia. Inhalation, ingestion, or skin contact with uracil mustard is known to cause severe injury or death. Contact with molten substance may cause severe burns to skin and eyes. Effects of contact or inhalation may be delayed. Fire may produce irritating, corrosive, and/or toxic gases. Run-off of uracil mustard from fire control or dilution water may be corrosive and/or toxic and causes pollution. The guanine and cytosine content correlates with the degree of uracil mustard-induced cross-linking. Uracil mustard's production and use as an anti-neoplastic agent may result in its release to the environment through various waste streams. If released to soil, uracil mustard should have high mobility. Volatilisation of uracil mustard should not be important from moist or dry soil surfaces. Available data are insufficient to determine the rate or importance of biodegradation of uracil mustard in soil or water. Uracil mustard is an alkylating agent and listed as Group 2B, meaning possibly carcinogenic to humans. Data are not available on the genetic and related effects of uracil mustard in humans.

Workers should keep uracil mustard in a cool, dry, dark location in a tightly sealed container or cylinder away from incompatible materials and ignition sources and in a secure area. Uracil mustard should be handled in a chemical fume hood, and no untrained worker/individual should handle uracil mustard or its container.

Workers should handle uracil mustard with gloves that should be inspected prior to use to avoid skin contact with the chemical. Dispose of contaminated gloves after use in accordance with applicable laws and good laboratory practices and then wash and dry hands. Workers should have adequate ventilation at the workplace, and they should wear respiratory protection to avoid breathing vapours, mist, or gas.

Uranium (CAS No. 7440-61-1)

Molecular formula: **U**

Uranium is a silver-white, lustrous, heavy, mildly radioactive metal. Its appearance will change upon exposure to air or water, as oxidation occurs. Its colour darkens through brass, from brown to charcoal grey. Powders, fines, chips, or turnings oxidise rapidly, yielding a dull or flat dark grey or brown colour. Uranium is almost as hard as steel and much denser than lead. Natural uranium is used to make fuel for nuclear power plants; depleted uranium is the leftover product. Some alloys will oxidise more slowly, retaining the silver-white and then brassy colour. No odour is found. Uranium is used as an abundant source of concentrated energy. Uranium occurs in most rocks in concentrations of 2–4 parts per million and is as common in the Earth's crust as tin, tungsten, and molybdenum. Uranium occurs in seawater and can be recovered from the oceans.

Uranium is a naturally occurring radioactive element. Natural uranium is a mixture of three isotopes: ^{234}U, ^{235}U, and ^{238}U. The most common isotope is ^{238}U; it makes up about 99% of natural uranium by mass. Depleted uranium is a mixture of the same three uranium isotopes except that it has very little ^{234}U and ^{235}U. It is less radioactive than natural uranium. The high density of uranium means that it also finds uses in the keels of yachts and as counterweights for aircraft control surfaces, as well as for radiation shielding. Uranium metal is known to react dangerously with carbon tetrachloride, chlorine, fluorine, nitric acid, nitric oxide, selenium, sulphur, and water (in finely divided form). On decomposition with fire, it produces uranium metal fume and/or oxide. Radioactive progenies (daughters), thorium-234, protactinium-234, and protactinium-234m (metastable), are produced by natural radioactive decay; they are the source of the majority of the penetrating radiation. These isotopes can be concentrated in situations where the metal is melted, condensed, or dissolved, potentially elevating the observed external dose rate. Many industries involved in mining, milling, and processing of uranium can also release it into the environment. Inactive uranium industries may continue to release uranium into the environment. Uranium exists in air as dust, and very small dustlike particles of uranium in the air settle from the air onto water, plants, and land. Rain washes uranium from the air and increases the amount of uranium that will settle to the ground. Workers involved in the mining, milling, processing, and/or production of uranium products and workers associated with phosphate fertiliser industries have also been found exposed to higher levels of uranium.

Safe Handling and Precautions

Uranium and its salts are both toxic and radioactive. Most of the uranium that is absorbed into the body is found excreted from the body through urine and faeces and is not absorbed into the system. The main target for inhaled, soluble and moderately soluble uranium compounds in humans is the kidneys. Uranium that is not absorbed remains for a long time in the lungs. Dermatitis, renal damage, acute necrotic arterial lesions, and possibly death may occur from extreme exposure. Inhalation of fine uranium particles presents increased radiation hazards; isolated uranium particles in the lungs may be a long-term cancer hazard. Uranium dusts are respiratory irritants, with coughing and shortness of breath as possible outcomes. Prolonged skin contact can cause damage to the basal cells. Radioactivity is the property of the spontaneous emissions of alpha or beta particles and gamma rays, by the disintegration of the nuclei of the atoms. Exposure to uranium is known to cause nausea, vomiting, shortness of breath, coughing, pneumoconiosis, pulmonary fibrosis, lymphoma, osteosarcoma, and lung cancer. Exposures to uranium oxide peroxide (UH_2O_4) through inhalation are known to cause stridor (noisy breathing), dyspnoea, upper airway injury, and pulmonary oedema. The National Toxicology Program, the International Agency for Research on Cancer, and the U.S. Environmental Protection Agency have classified natural uranium or depleted uranium as a human carcinogen. Reports indicate that the carcinogenicity of uranium is dependent on isotopic composition and solubility of uranium compounds, a finding consistent with the epidemiological and experimental studies. Occupational workers, during handling and storage of uranium and its compounds, should follow precautions. Workers should strictly follow all appropriate federal, state, and/or local regulations governing the disposal of radioactive waste and contaminated materials. Occupational workers, during handling of uranium and its compounds, should wear appropriate personal protective equipment, impervious gloves, boots, aprons, etc., as appropriate, to prevent prolonged or repeated skin contact. Occupational workers, during handling of uranium, should avoid contact of oxidisers and generation of dust at the workplace.

Urethane (Ethyl Carbamate) (CAS No. 51-79-6)

Ethyl carbamate is a white crystalline substance produced by the ammonia on ethyl chloroformate.

Ethyl carbamate has been produced commercially in the United States for many years. It has been used as an anti-neoplastic agent and for other medicinal purposes.

Bibliography

Agency for Toxic Substances and Disease Registry (ATSDR). 2011. *Toxicological Profile for Uranium* (Draft for Public Comment). U.S. Department of Health and Human Services, Public Health Service, Atlanta, GA.

Belles, M., M. L. Albina, V. Linares, M. Gomez, D. J. Sanchez, and J. L. Domingo. 2005. Combined action of uranium and stress in the rat. I. Behavioral effects. *Toxicol. Lett.*, 158:176–185.

Buskrirk, H. H., J. A. Crim, H. G. Petering et al. 1965. Effect of uracil mustard and several antitumor drugs on the primary antibody response in rats and mice. *J. Natl. Cancer Inst.* 34(6):747–758.

Feugier, A., S. Frelon, P. Gourmelon, and M. Claraz. 2008. Alteration of mouse oocyte quality after a subchronic exposure to depleted uranium. *Reprod. Toxicol.* 26:273–277.

Guseva, C. I., S. Jacob, E. Cardis, P. Wold et al. 2011. Uranium carcinogenicity in humans might depend on the physical and chemical nature of uranium and its isotopic composition: Results from pilot epidemiological study of French nuclear workers. *Cancer Causes Control* 22(11):563–1573.

IARC Monographs on the Evaluation of Carcinogenic Risks to Humans. 1987. Overall evaluations of carcinogenicity: An updating of IARC monographs, Vols. 1–42 (suppl. 7). IARC, Geneva, Switzerland.

International Agency for Research on Cancer (IARC)—Summaries and evaluations. Uracil mustard (Group 2B). IARC, Geneva, Switzerland.

Patnaik, P. (ed.). 1999. *A Comprehensive Guide to the Hazardous Properties of Chemical Substances,* 2nd edn. McGraw Hill, New York.

World Health Organization (WHO). Chemical teratogens, carcinogens, mutagens. International Agency for Research on Cancer (IARC), Geneva, Switzerland.

23

Hazardous Chemical Substances: V

Vanadium Pentoxide
Vinyl Acetate Monomer
Vinyl Chloride

Vanadium Pentoxide (CAS No. 1314-62-1)

Molecular formula: V_2O_5
Synonyms and trade names: Vanadium oxide dust; Vanadic anhydride dust; Divanadium pentoxide dust; Vanadium pentaoxide dust; Vanadium dust

Vanadium pentoxide is a yellow to red colour solid and is odourless. Vanadium pentoxide dust is the particulate form of a non-combustible, odourless, yellow-orange or dark grey crystalline solid.

On decomposition by heating, vanadium pentoxide produces toxic fumes. Vanadium is widely distributed in the Earth's crust in a wide range of minerals and in fossil fuels. Vanadium pentoxide, the major commercial product of vanadium, is mainly used in the production of alloys with iron and aluminium. It is also used as an oxidation catalyst in the chemical industry and in a variety of minor applications.

Safe Handling and Precautions

Exposure to vanadium pentoxide and dust causes harmful effects and poisoning. Exposure to vanadium pentoxide in the workplace occurs during the refining and processing of vanadium-rich mineral ores, during the burning of fossil fuels especially petroleum, during the handling of vanadium catalysts in the chemical manufacturing industry, and during the cleaning of oil-fired boilers and furnaces. Vanadium pentoxide gets absorbed into the body by inhalation and by ingestion. The aerosol of vanadium pentoxide is irritating to the eyes, the skin, and the respiratory tract. The symptoms of poisoning include irritation of the mucous membranes, upper respiratory tract, bronchi, lungs, eyes, and skin of exposed humans. Acute intoxication may cause systemic symptoms. Chronic and or repeated exposures to vanadium pentoxide dust have been reported to cause productive cough, increased mucus production, shortness of breath, fatigue, chronic bronchitis, chronic rhinitis, and allergic dermatitis. Reports indicate that vanadium pentoxide has sufficient evidence in experimental animals as a carcinogen but inadequate evidence in humans as a carcinogen. The International Agency for Research on Cancer (IARC) has listed vanadium as Group 2B, meaning possibly carcinogenic to humans.

Vinyl Acetate Monomer (CAS No. 108-05-4)

Molecular formula: $C_4H_6O_2$

Synonyms and trade names: Vinyl ester acetic acid; Ethenyl ester acetic acid; Vinyl acetate monomer (VAM); Ethenyl acetate; 1-Acetoxyethylene; Acetic acid ethenyl ester

Vinyl acetate monomer (VAM) is a colourless liquid, immiscible or slightly soluble in water. VAM is a flammable liquid. VAM has a sweet, fruity smell (in small quantities), with sharp, irritating odour at higher levels. VAM is an essential chemical building block used in a wide variety of industrial and consumer products. VAM is a key ingredient in emulsion polymers, resins, and intermediates used in paints, adhesives, coatings, textiles, wire and cable polyethylene compounds, laminated safety glass, packaging, automotive plastic fuel tanks, and acrylic fibres. Vinyl acetate is used to produce polyvinyl acetate emulsions and resins. Very small residual levels of vinyl acetate have been found present in products manufactured using VAM, such as moulded plastic items, adhesives, paints, food packaging containers, and hairspray.

Safe Handling and Precautions

Exposure to vinyl acetate is harmful and hazardous. Exposure causes cough, shortness of breath, drowsiness, headache, pain, sore throat, redness of the skin, blisters, and mild burns. Short-term exposure to vinyl acetate causes irritation of the eyes, the skin, and the respiratory tract and effects on the lungs, resulting in tissue lesions. Skin contact of vinyl acetate is known to cause sensitisation and an allergic skin reaction in a small proportion of human workers. Animal studies found that long-term exposure to VAM can cause a carcinogenic response. Reports indicate that vinyl acetate is possibly carcinogenic to humans (Group 2B). The American Conference of Governmental Industrial Hygienists (ACGIH) has classified VAM as A3 carcinogen, meaning a confirmed animal carcinogen with unknown relevance to humans.

Users and occupational workers should keep vinyl acetate properly stored in a cool, dry, well-ventilated area in tightly sealed containers, away from sources of ignition, heat, and flame. The containers should be labelled in accordance with regulations, should be protected from physical damage and should be stored separately from incompatible chemicals. Ground all equipment containing material. In case of insufficient ventilation, wear suitable respiratory equipment. Workers should avoid contact with vinyl acetate with skin and eyes. Vinyl acetate, if not properly managed, causes a serious fire and/or health hazard. Any kind of cross-contamination of vinyl acetate with other chemicals, especially oxidising materials or strong acids or bases, may lead to spontaneous polymerisation and fire. Wastes containing vinyl acetate (VAM) must be treated or disposed of at an authorised facility.

During use and handling of vinyl acetate (VAM), workers should wear protective personal equipment (PPE) and appropriate, approved, and adequate workplace safety dress: flame-retardant clothing, chemical splash goggles, hard hat and safety shoes, organic vapour respirators with full facepieces, and chemical-resistant gloves. Only professionally trained personnel provided with full personal protective gear should be involved to manage vinyl acetate (VAM) spill, chemical waste clean-up, etc. Only trained personnel should attend the disposal of contaminated chemical waste, dirt, or absorbent material properly and in compliance with applicable local, regional, and/or national waste regulations.

Regulations:

- The ACGIH has listed the threshold limit value (TLV) for vinyl acetate at 10 ppm (35 mg/m^3).
- The ACGIH has set a short-term exposure limit (STEL) of 15 ppm (53 mg/m^3 for periods not to exceed 15 min).
- The National Institute for Occupational Safety and Health (NIOSH) has set the recommended exposure limit (REL) of for vinyl acetate at 4 ppm.
- The Occupational Safety and Health Administration (OSHA) does not have any current regulation about vinyl acetate.

Alert!! Danger!! Occupational workers handling vinyl acetate (VAM), or in the vicinity of VAM, should be fully aware of its hazards and informed of appropriate safe handling and emergency response.

Any kind of workplace contact *should strictly be avoided* between vinyl acetate and acids – chlorosulphonic, hydrochloric, hydrofluoric, nitric, sulphuric, or phosphoric – bases, silica gel, alumina, oxidisers, azo compounds, 2-aminoethanol, ethylenediamine, ethyleneimine, oleum, peroxides, strong caustics (sodium hydroxide or potassium hydroxide), and ozone.

Vinyl Chloride (CAS No. 75-01-4)

Molecular formula C_2H_3Cl
Synonyms and trade names: Chloroethene; Chloroethylene; 1-Chloroethylene; Ethylene monochloride

Vinyl chloride is a colourless gas with a mild, sweet odour, is slightly soluble in water, and is flammable.

Safe Handling and Precautions

Acute exposures to high concentrations of vinyl chloride have caused adverse effects such as dizziness, drowsiness, headaches, giddiness, and irritation of the eyes and respiratory tract. Vinyl chloride, on exposure, has been reported to cause adverse effects to blood, kidneys, liver, mucous membranes, lymphatic system, upper respiratory tract, skin, eyes, and central nervous system. Prolonged exposures to high levels of vinyl chloride have caused loss of consciousness, effects on the lungs and kidney irritation, and inhibition of blood clotting in humans and cardiac arrhythmias in animals. Vinyl chloride exposure has also caused liver damage, peripheral neuropathy, tingling, numbness, weakness, and pain in fingers among occupational workers. The U.S. Environmental Protection Agency (U.S. EPA) classified vinyl chloride as a Group A, meaning a human carcinogen. The U.S. Department of Health and Human Services has determined that vinyl chloride is a known carcinogen. Studies in workers who have breathed vinyl chloride over many years showed an increased risk of liver, brain, and lung cancer, and some cancers of the blood have also been observed in workers.

Regulations:

- The ACGIH has set the TLV for vinyl chloride at 1 ppm (TWA–8 h).
- The OSHA has set the STEL at 5 ppm (15 min).
- The OSHA has set a limit of 1.0 ppm for vinyl chloride in the workplace air.

Occupational workers, during handling and storage of vinyl chloride, should follow precautions:

Vinyl chloride should be handled with adequate ventilation. Self-contained breathing apparatus (SCBA) with a full facepiece should be used to avoid inhalation of the product.

Bibliography

Agency for Toxic Substances and Disease Registry (ATSDR). 2006. *Toxicological Profile for Vinyl Chloride*. U.S. Department of Health and Human Services, Public Health Service. Atlanta, GA (updated 2011).

International Programme on Chemical Safety (IPCS) and the Commission of the European Communities (CEC). 2005a. Vinyl acetate. ICSC No. 0347.

International Programme on Chemical Safety and the Commission of the European Communities (IPCS–CEC). 2005b. Vanadium pentoxide. MSDS No. 0596. IPCS–CEC (updated 2006).

Kidd, H. and D. R. James. (eds.). 1991. *The Agrochemicals Handbook*, 3rd edn. Royal Society of Chemistry Information Services, Cambridge, U.K.

Lewis, R. J. (ed.). 1993. *Lewis Condensed Chemical Dictionary*, 12th edn. Van Nostrand Reinhold Company, New York.

Material Safety Data Sheet (MSDS). 2005. Vinyl acetate safe handling guide. Vinyl Acetate Council, Washington, DC.

Occupational Safety & Health Administration (OSHA). 1996. Occupational safety and health guideline for vanadium pentoxide dust. OSHA. Washington, DC.

Sittig, M. (ed.). 1985. *Handbook of Toxic and Hazardous Chemicals*, 2nd edn. Noyes Publications, Park Ridge, NJ

Tomlin, C. D. S. (ed.). 2006. *The Pesticide Manual: A World Compendium*, 14th edn. British Crop Protection Council (BCPC), Surrey, U.K.

U.S. Environmental Protection Agency (U.S. EPA). Vinyl chloride. 1992. Technology Transfer Network Air Toxics. Hazard Summary (revised 2007).

24

Hazardous Chemical Substances: W

Warfarin
Welding Fumes (NOC: Not Otherwise Classified)
Wood Dust (Certain Hardwoods as Beech and Oak)

Warfarin (CAS No. 81-81-2)

Molecular formula: $C_{19}H_{16}O_4$
Synonyms and trade names: 3-(alpha-Acetonylbenzyl)-4-hydroxycoumarin; Acetonylbenzyl-4-hydroxycoumarin; Athrombin-K; 1-(4'-Hydroxy-3'-coumarinyl)-1-phenyl-3-butanone; 2H-1-Benzopyran-2-one; 4-Hydroxy-3-(3-oxo-1-phenylbutyl)-; Coumadin; Coumarins; Coumafen; Liquatox; Rodafarin; Ro-Deth; Rat-Ola; Rat-Kill; Rat-Mix; Solfarin; Warfarin; Warficide; Warfarat
Chemical name: 3-(alpha-Acetonylbenzyl)-4-hydroxycoumarin
Chemical class: Rodenticide, anti-coagulant

Warfarin is a solid, odourless, and tasteless chemical substance. Warfarin is very harmful and hazardous. Warfarin is a general use pesticide (GUP). Warfarin was the first anti-coagulant rodenticide introduced and was first registered for use in the United States in 1952. Warfarin is used for controlling rats and house mice in and around homes, animal and agricultural premises, and commercial and industrial sites. In liquid form, warfarin rodenticides may be flammable and explosive and should be kept away from heat, sparks, and flames. In solid form, warfarin rodenticides are not combustible. Warfarin incompatibilities with strong oxidisers may cause fires and explosions.

Warfarin comes in water-soluble, ready-to-use bait, concentrate, powder, liquid concentrate, nylon pouch, coated talc, and dust formulations. The compound also comes in mixed formulations with pindone, calciferol, and sulphaquinoxaline. It is considered compatible with other rodenticides and the toxicological action is not rapid; usually about a week is required before a marked reduction in the rodent population is noticeable. Rodents do not tend to become bait-shy after once tasting warfarin; they continue to consume it until its anti-clotting properties have produced death through internal haemorrhaging. The pro-thrombin content of the blood is reduced and internal bleeding is induced. Repeated ingestion is needed to produce toxic symptoms. This rodenticide can be used year after year wherever a rodent problem exists. Mice are harder to control than rats, and complete control may take a longer period.

Safe Handling and Precautions

Exposure to warfarin is very harmful, hazardous, and toxic to kidneys and liver. Repeated or prolonged exposure to the substance can produce target organ damage. Repeated exposure to a highly toxic material may produce general deterioration of health by an accumulation in one or many human organs. The symptoms of warfarin poisoning include, but are not limited to, epistaxis (nosebleed), bleeding gums, pallor, petechial rash, haematomas around the joints or on the buttocks, back pain, bleeding lips, mucous membrane haemorrhage, abdominal pain, vomiting, blood in the urine and faeces, paralysis due to cerebral haemorrhage, and finally haemorrhagic shock. Warfarin is extremely hazardous if swallowed and on prolonged skin contact. Accumulation of warfarin in the body is possible with significant skin absorption over an extended period when the symptoms develop after several hours or days. Warfarin causes organ damage by inhibiting blood coagulation. Absorption by the lungs may result in haemorrhagic effects. Severe overexposure can result in death.

Occupational workers should be careful during use and handling of warfarin. Waste must be disposed of in accordance with federal, state, and local environmental control regulations. Workers should use workplace safety dress including splash goggles, gloves, full suit, dust respirator, boots, and a self-contained breathing apparatus.

Alert! Warning!! Warfarin reacts violently with strong oxidants causing fire and explosion hazard and decomposes on heating, producing irritating fumes.

Regulations: Global regulatory agencies have set the following exposure limits for warfarin:

- American Conference of Governmental Industrial Hygienists (ACGIH): 0.1 mg/m³ time-weighted average (TWA)
- Occupational exposure limits (OEL) for Ontario workplaces (Canada): 0.1 mg/m³ TWA and short-term exposure limit (STEL) of 0.3 mg/m³
- Alberta, Canada: 0.1 mg/m³ OEL
- The National Institute to Occupational Safety and Health (NIOSH): 0.1 mg/m³
- Occupational Safety and Health Administration (OSHA): 0.1 mg/m³ TWA

Welding Fumes (NOC: Not Otherwise Classified)

Welding fumes are the fumes that result from various welding operations. The primary components are oxides of the metals involved such as zinc, iron, chromium, aluminium, or nickel. Welding fumes typically have a metallic odour, and their specific composition varies considerably and they are released to the work area during welding and metalworking operations.

Welding fumes and gases form a complex mixture. The composition and quantity of welding fumes are dependent upon the metal being welded, the process, the procedure, and the electrodes used. Other conditions that also influence the composition and quantity of the fumes and gases to which workers may be exposed include the coatings on the metal being welded (such as paint, plating, or galvanising), the number of welders and the volume of work area, the quality and amount of ventilation, the position of the welder's head with respect to the fume plume, as well as the presence of contaminants in the atmosphere (such as chlorinated hydrocarbon vapours from cleaning and degreasing activities). Welding arc and sparks can ignite combustible and flammables. Electric arc welding may create one or more of the following health hazards: fumes and gases can be dangerous to your health. Short-term (acute) overexposure to welding fumes may result in discomfort such as dizziness, nausea, or dryness or irritation of nose, throat, or eyes. Long-term overexposure to manganese compounds may affect the central nervous system. Symptoms include muscular weakness and tremors similar to Parkinson's disease. Long-term overexposure has been reported to cause *siderosis* (iron deposits in lungs) and is believed by some investigators to affect pulmonary functions.

Workers should be aware that the composition and quantity of fumes and gases to which they may be exposed are influenced by coatings that may be present on the metal being welded (such as paint, plating, or galvanising), the number of welders in operation and the volume of the work area, the quality and amount of ventilation, the position of the welder's head with respect to the fume plume, as well as the presence of contaminants in the atmosphere (such as chlorinated hydrocarbon vapours from cleaning and degreasing procedure). When the electrode is consumed, the fumes and gas decomposition products generated are different in percent and form from the ingredients listed in Section II. The composition of these fumes and gases is the concerning matter and not the composition of the electrode itself. Decomposition products include those originating from the volatilisation, reaction, or oxidation of the ingredients shown in Section II, plus those from the base metal, the coating, and the other factors noted earlier. Gaseous reaction products may include carbon monoxide and carbon dioxide. Ozone and nitrogen oxides may be formed by the radiation from the arc. One method of determining the composition and quantity of the fumes and gases to which the workers are exposed is to take an air sample from inside the welder's helmet while worn or within the worker's breathing zone.

Safe Handling and Precautions

Although the inhalation of tungsten has the potential for causing transient or permanent lung damage, it is generally considered to exhibit a low degree of toxicity. For the reduction of accumulation of dusts, routine wet mopping or vacuuming with an explosion-proof vacuum, fitted with a high-efficiency particulate absolute (HEPA) filter, has been suggested. This would include wearing welder's gloves and a protective face shield and may include arm protectors, apron, hats, shoulder protection, as well as dark substantial clothing. Welders should be trained not to allow electrically live parts to come in contact with the skin or wet clothing and gloves. The welders should insulate themselves from the work and ground. Workplace should have plenty of ventilation and/or local exhaust at the arc to keep the fumes and gases below the threshold limit value within the worker's breathing zone and the general work area. Welders should be advised to keep their head out of the fumes.

Occupational workers should wear approved and appropriate workplace protective clothing equipment, safety dress, screens and flash goggles, respirable fume respirator or air-supplied respirator, and head, hand, and body protection to help prevent injury from radiation, sparks, and electrical shock. Workers should discard any product, waste residue, disposal container, or liner in an environmentally acceptable manner approved by state and local regulations.

Wood Dust (Certain Hardwoods as Beech and Oak)

Synonyms: Hardwood dust; Softwood dust; Western red cedar dust

Wood dust is a complex mixture. Its chemical composition depends on the species of tree and consists mainly of cellulose, polyoses, and lignin, with a large and variable number of substances. Wood dust is a light brown or tan fibrous powder. Wood dust has limited commercial uses. Wood is one of the world's most important renewable resources and grows in forests all over the world. Wood dust is used to prepare charcoal and as an absorbent for nitroglycerin and a filler in plastics and linoleum and paperboard (Radian 1991). Another commercial use for wood dust is in wood composts. Wood dust is common and occurs in the environment in areas where machinery or tools are used to cut or shape wood.

Safe Handling Precautions

Exposure to wood dust at workplaces has been reported to cause irritation of the eyes, epistaxis (nosebleed), dermatitis, respiratory disorders, hypersensitivity, granulomatous pneumonitis, asthma, cough, wheezing, sinusitis, and prolonged colds. Reports also indicate that negligence to use appropriate workplace dress during handling of wood dust may lead to potential occupational carcinogenicity. Wood dust is known to be a human carcinogen, based on sufficient evidence of carcinogenicity from studies in humans. It has been demonstrated through human epidemiologic studies that exposure to wood dust increases the occurrence of cancer of the nose (nasal cavities and *paranasal sinuses*). Other types of nasal cancers (squamous cell carcinoma of the nasal cavity) and cancers at other sites, including the *nasopharynx* and the *larynx*, and Hodgkin's disease have been associated with wood dust exposure in several epidemiologic studies. The International Agency for Research on Cancer (IARC) classified wood dust as Group I carcinogen, meaning carcinogenic to humans.

Regulations: The OSHA regulates wood dust under the Occupational Safety and Health Act of 1970 and the Construction Safety Act. Personal protective equipment (PPE) is required for working with any process that might produce wood dust, including the use of hand and power tools. The act states that a collecting system is required for sawmills that produce wood dust:

- The ACGIH assigned threshold limit values of 1 mg/m^3 for certain hardwoods, such as beech and oak, and 5 mg/m^3 for softwoods except western red cedar, as TWAs for a normal 8 h workday and a 40 h workweek.
- The **OSHA** established a permissible exposure limit (PEL) of 15 mg/m^3 of air for the total dust and 5 mg/m^3 for the respirable fraction of wood dust, all softwoods and hardwoods.
- The NIOSH established a recommended exposure limit of 1 mg/m^3 for dust from all wood except western red cedar, as a TWA for up to a 10 h workday and a 40 h workweek.

Occupational workers for the safety management of wood dust and possible hazards should wear appropriate and approved workplace dress and head, hand, and body protection. Occupational workers must wear workplace safety dress with spectacles with side protection; goggles or face shields are required when chipping, woodworking, sawing, drilling, chiselling, or powered fastening, to protect from flying fragments, objects, large chips, particles, sand, and dirt.

Workers using hand and power tools and exposed to the hazard of falling, flying, abrasive, and splashing objects or exposed to harmful dusts, fumes, mists, vapours, or gases shall be provided with the particular PPE necessary to protect them from the hazard.

Bibliography

Bhattacharjee, J. W., R. K. S. Dogra, M. M. Lal, and S. H. Zaidi. 1979. Wood dust toxicity: In vivo and in vitro studies. *Environ. Res.* 20:455–464.

Budavari, S. 1996. *The Merck Index: An Encyclopedia of Chemicals, Drugs and Biologicals*, 12th edn. Merck & Co. Inc., Rahway, NJ.

Hartley, D. and H. Kidd. 1983. *The Agrochemicals Handbook*. The Royal Society of Chemistry, The University, Nottingham, England.

Hazardous Substances Data Bank. 1987. Warfarin. HSDB Bank, National Library of Medicine. Bethesda, MD.

International Programme on Chemical Safety (IPCS) and the Commission of the European Communities (CEC). 2010. Warfarin. ICSC Card No: 0821. Geneva, Switzerland.

Kidd, H. and D. R. James. (eds.). 1991. *The Agrochemicals Handbook*, 3rd edn. Royal Society of Chemistry Information Services, Cambridge, U.K.

Meister, R. T. 1994. *Farm Chemicals Handbook*. Meister Publishing Co. Willoughby, OH.

National Institute for Occupational Safety and Health (NIOSH) et al. Occupational safety and health guideline for welding fumes.

National Institute for Occupational Safety and Health (NIOSH). 1995. Registry of toxic effects of chemical substances: Welding fumes. Cincinnati, OH.

Stellman, S. D. and L. Garfinkel. 1984. Cancer mortality among woodworkers. *Am. J. Ind. Med.* 5:343–357.

Tomlin, C. D. S. (ed.). 2006. *The Pesticide Manual: A World Compendium*, 15th edn. British Crop Protection Council (BCPC), Hampshire, U.K.

U.S. Department of Health and Human Services, Public Health Service. 2000. Wood dust. National Toxicology Program. Research Triangle Park, NC.

U.S. Environmental Protection Agency. June, 1991. R.E.D. Facts: Warfarin. U.S. EPA, Office of Pesticides and Toxic Substances. Washington, DC.

Vandenplas, O., F. Dargent, J. Auverdin, J. Boulanger, J. Bossiroy, D. Roosels, and R. Vande Weyer. 1995. Occupational asthma due to gas metal arc welding on mild steel. *Thorax* 50:587–589.

Vaughan, T. L. and S. Davis. 1991. Wood dust exposure and squamous cell cancers of the upper respiratory tract. *Am. J. Epidemiol.* 133:560–564.

World Health Organization (WHO). 1995. Anticoagulant rodenticides. Environmental Health Criteria No. 175. WHO, Geneva, Switzerland.

25

Hazardous Chemical Substances:
X

Xylene

Xylene (CAS No. 1330-20-7)

Molecular formula: $CH_3C_6H_4CH_3$

Synonyms and trade names: Dimethylbenzene; Methyl toluene; 1,4-dimethylbenzene; Violet 3; Xylol

Xylene is an aromatic hydrocarbon solvent commonly found in paints and laboratories. Xylene is insoluble in water and soluble in alcohol, ether, acetone, and benzene.

Xylene is used as a solvent. In this application, the mixture of isomers is often referred to as xylenes or xylol. Solvent xylene often contains a small percentage of ethylbenzene. Like the individual isomers, the mixture is colourless, sweet smelling, and highly flammable. Application of xylene is extensive and includes, but is not limited to, printing, rubber, and leather industries.

Similarly, it is used as a cleaning agent for steel and silicon wafers. In the petroleum industry, xylene is also a frequent component of paraffin solvents, used when the tubing becomes clogged with paraffin wax. Xylene is incompatible with strong oxidisers and is known to cause fires and explosions. There are three forms of xylene in which the methyl groups vary on the benzene ring: (i) *meta*-xylene, (ii) *ortho*-xylene, and (iii) *para*-xylene. These forms are referred to as isomers. Xylene is a colourless, sweet-smelling liquid. Xylene occurs naturally in petroleum and coal tar. Chemical industries produce xylene from petroleum. It is also used as a cleaning agent and a thinner for paint and in paints, in glues, in printing inks, and in varnishes. Xylene evaporates quickly from the soil and surface water into the air.

Safe Handling and Precautions

Xylene is toxic and poisonous. Exposure to xylene at workplaces is known to occur through all major routes, meaning, by inhalation, ingestion, and/or contact with skin or eye. Improper use and handling of xylene cause irritation of the eyes and mucous membranes, and at high concentrations, xylene is narcotic. The signs and symptoms of xylene poisoning include, but are not limited to, headache; fatigue; flushing; redness of the face; a sensation of increased body heat; increased salivation; irritability; lassitude; nausea; irritation of the eyes, nose, and throat; conjunctivitis; dryness of the nose; dermatitis; anorexia; flatulence;

dizziness; confusion; motor incoordination; impairment of equilibrium; tremors; and cardiac irritability. Prolonged period of exposure to high concentrations of xylene causes tissue damage of kidney and liver, memory difficulties, unconsciousness, and fatal injury. The U.S. Environmental Protection Agency observed that there is insufficient information to classify xylene as a human carcinogen. Similarly, the International Agency for Research on Cancer (IARC) observed that evidences on xylene are inadequate and list xylene as Group 3, meaning not classifiable as a human carcinogen.

Workers and workplace supervisor during use and handling of xylene should be very alert and should remember the importance of precautions. Xylene should be appropriately kept with proper label in a cool, dry, well-ventilated area in tightly sealed containers and in accordance with the local, state, and global safety regulations. Storage of xylene is preferred outside or detached storage area with explosion-proof ventilation. The containers of xylene should be protected from physical damage and away from strong oxidisers, heat, sparks, and open flame. Occupational workers during use and handling of xylene should observe safety regulatory practices and must avoid workplace negligence. Workers should wear appropriate protective dress/clothing to prevent chemical contact, personal protective equipment (PPE), to avoid organic vapour, canister face mask, self-contained breathing apparatus/respiratory protection, impervious/liquid-proof gloves; protective goggles or face shield; white or light-coloured clothing; and rubber shoes or boots.

Danger! Inhalation of xylene vapour causes depression of the central nervous system (CNS). Workers must remember that the odour threshold for xylene is 1 ppm (1 part per million parts of air).

Bibliography

Agency for Toxic Substances and Disease Registry (ATSDR). 2007. *Toxicological Profile for Xylenes.* U.S. Department of Health and Human Services, Public Health Service, Atlanta, GA.

Browning, E. (ed.). 1965. Xylene. In: *Toxicity and Metabolism of Industrial Solvents.* Elsevier, New York.

International Programme on Chemical Safety and the Commission of the European Communities (IPCS, CEC). 2002. P-Xylene. ICSC No. 0086, Material Safety Data Sheet (MSDS). (updated 2005).

National Library of Medicine (NLM). 1986. Xylenes. NLM, Hazardous Substances Data Bank. Bethesda, MD.

Sax, N. I. and R. J. Lewis. (eds.). 1989. *Dangerous Properties of Industrial Materials,* 7th edn. Van Nostrand Reinhold Company, New York.

Sittig, M. (ed.). 1985. *Handbook of Toxic and Hazardous Chemicals,* 2nd edn. Noyes Publications, Park Ridge, NJ.

26

Hazardous Chemical Substances: Y

Yttrium Barium Copper Oxide
Ytterbium (III) Chloride Hexahydrate
Yttrium Fluoride
Ytterbium (III) Oxide

Yttrium Barium Copper Oxide (CAS No. 107539-20-8)

Molecular formula: $YBa_2Cu_3O_7$

Yttrium barium copper oxide (YBCO) is a crystalline compound with important property of 'high temperature superconductor' above the boiling point (77 K) of liquid nitrogen. YBCO is sensitive to air and moisture. Exposures to YBCO have been reported to cause skin irritation, serious eye irritation, and respiratory irritation. Symptoms of systemic copper poisoning may include capillary damage, headache, cold sweat, weak pulse, kidney and liver damage, central nervous system excitation followed by depression, jaundice, convulsions, paralysis, and coma. Death may occur from shock or renal failure. Chronic copper poisoning is typified by hepatic cirrhosis, brain damage and demyelination, kidney defects, and copper deposition in the cornea as exemplified by humans with Wilson's disease. It has also been reported that copper poisoning has led to haemolytic anaemia and accelerates arteriosclerosis. There is no more published information on the chemical, physical, and toxicological properties of YBCO. Occupational workers, during handling and storage of YBCO, should be alert and follow precautions. The container should be tightly closed and stored in a cool, dry, and well-ventilated place. Workers should handle YBCO in accordance with good industrial hygiene and safety practice.

Ytterbium (III) Chloride Hexahydrate (CAS No. 10035-01-5)

Molecular formula: $YbCl_3$

Ytterbium (III) chloride hexahydrate is a stable, hygroscopic (absorbs moisture from the air), moisture-sensitive compound. It is incompatible with strong oxidising agents. Exposure to ytterbium (III) chloride hexahydrate has been reported to cause eye irritation, skin irritation, and respiratory tract irritation. Ytterbium (III) chloride hexahydrate has not been listed by the American Conference of Governmental Industrial Hygienists (ACGIH), the International Agency for Research on Cancer (IARC), the National Institute for Occupational Safety and Health (NIOSH), the Occupational Safety and Health Administration (OSHA), and the National Toxicology Program (NTP) as a human carcinogen.

Yttrium Fluoride (CAS No. 13709-49-4)

Molecular formula: YF_3

There are no published reports on the chemical, physical, and toxicological properties of yttrium fluoride, and yttrium fluoride has not been thoroughly investigated and recorded.

Ytterbium (III) Oxide (CAS No. 1314-37-0)

Molecular formula: Yb_2O_3

Ytterbium (III) oxide is white powder and pieces, has no odour, is insoluble in water, and is incompatible with strong acids, oxidising agents, and carbon dioxide.

Ytterbium and compounds are considered rare earth metal. These metals are moderately to highly toxic. The symptoms of toxicity of the rare earth elements include writhing, ataxia, laboured respiration, walking on toes with arched back, and sedation. The rare earth elements exhibit low toxicity by ingestion exposure. However, the intraperitoneal route is highly toxic, while the subcutaneous route is poisonous to moderately toxic. The production of skin and lung granulomas after exposure to them requires extensive protection to prevent such exposure.

There are no published reports on the chemical, physical, and toxicological properties of ytterbium (III) oxide that have not been thoroughly investigated and recorded.

27

Hazardous Chemical Substances:
Z

Zinc
Zinc Compounds
   Zinc Chloride Anhydrous
   Zinc Oxide
   Zinc Phosphide
Zineb
Ziram

Zinc (CAS No. 7440-66-6)

Molecular formula: **Zn**

Zinc is one of the most common elements in the Earth's crust. Metal zinc was first produced in India and China during the Middle Ages. Zinc is found in air, soil, and water and chemical waste sites and is present in all foods. Pure zinc is a bluish-white shiny metal. Zinc metal has extensive industrial and commercial applications. Zinc metal is used to coat steel for corrosion protection, galvanising, electroplating, and electrogalvanising; as an alloying element in bronze, brass, aluminium, and other metal alloys; for zinc die casting alloys; for zinc dry cell batteries; as coatings to prevent rust; mixed with other metals to make alloys like brass and bronze; for coinage applications; in the production of zinc sheet for architectural and as a reducing agent in organic chemistry; and for other chemical applications. Zinc is also used for metal coatings to prevent rust and mixed with other metals to make alloys like brass and bronze. Zinc combines with other elements to form zinc compounds. Zinc compounds are widely used in industry to make paint, rubber, dye, wood preservatives, and ointments.

Safe Handling and Precautions

Zinc is an essential element required as part of a healthy diet. Laboratory animals with zinc deficiency demonstrated reduced fertility, malformations of foetal nervous system, and growth retardation in late pregnancy. Reports have also indicated that humans with low maternal serum zinc concentrations had labour abnormalities, congenital malformations, and preterm labour even in otherwise healthy women. Numerous studies have examined pregnancy outcomes following zinc supplementation. Simmer et al. found significant intrauterine growth retardation.

Exposure to zinc and zinc compounds has been reported as relatively non-toxic. However, workplace negligence and accidental ingestion of large quantities of zinc and zinc salts cause harmful effects. The overexposure to zinc oxide fume is known to cause metal fume fever, characterised by flu-like symptoms such as chills, acute gastroenteritis, fever, nausea, and vomiting; *sideroblastic anaemia*; leucopenia; microcytic anaemia; neutropenia; bleeding gastric erosion; hepatitis; liver failure; intestinal bleeding; acute tubular necrosis; interstitial nephritis; stomach cramps; and related health disorders. Chronic exposures to zinc chloride fumes cause irritation, pulmonary oedema, pulmonary fibrosis, bronchopneumonia, and cyanosis. It also causes anaemia, pancreas damage, and lower levels of high-density lipoprotein cholesterol. Breathing large amounts of zinc (as dust or fumes) can cause a specific short-term disease called metal fume fever. Repeated skin exposure to zinc dust or powder at workplaces is known to cause dryness, irritation, and cracking (dermatitis) since zinc is astringent and may tend to draw moisture from the skin. Also, it is important to know that very low levels of zinc in the diet cause loss of appetite, decreased sense of taste and smell, slow wound healing and skin sores, or damage to the immune system. The dermal exposure to zinc or zinc compounds has not caused any kind of noticeable toxic effects. The American Conference of Governmental Industrial Hygienists (ACGIH), the National Toxicology Program (NTP), the Occupational Safety and Health Administration (OSHA), and the International Agency for Research on Cancer (IARC) have not listed zinc as a human carcinogen.

The recommended dietary allowance (RDA) for zinc is 15 mg/day for men, 12 mg/day for women, 10 mg/day for children, and 5 mg/day for infants. Reports have indicated that pregnant woman with low zinc intake have babies with growth retardation. Harmful health effects generally begin at levels from 10 to 15 times the RDA (in the 100 to 250 mg/day range). Information on the possible toxicological effects followed by prolonged period of exposures to high concentrations of zinc is not known.

Zinc Compounds

Industrially important compounds of zinc include (i) zinc chloride ($ZnCl_2$), (ii) zinc oxide (ZnO), (iii) zinc stearate ($Zn(C_{16}H_{35}O_2)_2$), and (iv) zinc sulphide (sphalerite, ZnS).

Zinc Chloride Anhydrous (CAS No. 7646-85-7)

Molecular formula: $ZnCl_2$
Synonyms and trade names: Zinc butter; Zinc dichloride

Zinc chloride is a white deliquescent salt. It forms acidic solutions in water and in polar organic solvents such as ethanol, acetone, and ether. Anhydrous zinc chloride hydrolyzes with moisture to form hydrochloric acid. It also forms complex ions with water, ammonia, and some organic solvents. Zinc chloride reacts with sulphide to minimise release of H_2S gas in waste treatment facilities. Zinc chloride 50% solution also serves as a high-quality mercerising agent for cotton. Zinc chloride is incompatible with strong oxidising agents, moisture, cyanides, sulphides, and potassium.

Safe Handling and Precautions

Exposure to zinc chloride causes harmful health effects. Zinc chloride is an acidic material. Improper workplace handling and ingestion and/or inhalation of zinc chloride have been reported to produce burns to mouth, throat, and stomach and irritation to gastrointestinal or respiratory tract, characterised by burning, sneezing, and coughing. Even brief contact of the eyes with zinc chloride in water is known to produce permanent damage. Overexposure by inhalation may cause respiratory protection. The amount of tissue damage caused by zinc depends upon the length of contact, resulting in skin inflammation and blistering. Workers with pre-existing skin disorders or eye problems have been found to be more susceptible to the toxicity of zinc chloride.

Zinc chloride has not been listed as human carcinogen by the ACGIH, National Institute for Occupational Safety and Health (NIOSH), IARC, NTP, and OSHA. Occupational workers during use and handling of zinc chloride should observe safety regulatory practices and must avoid workplace negligence. The workplace should have adequate ventilation. Workers should avoid inhaling mist, dust, and fumes of zinc chloride. Workers must keep the container of zinc chloride closed. Workers should strictly avoid to dispose of or flush the chemical wastes to the drain, but dispose of in accordance with all applicable federal, state, and local regulations. Workers during use and handling of zinc chloride must wear self-contained breathing apparatus to avoid inhalation of hazardous decomposition products. Workers should wear protective clothing to prevent chemical contact, personal protective equipment (PPE) to avoid organic vapour, canister face mask, self-contained breathing apparatus, liquid-proof gloves; protective goggles or face shield; white or light-coloured clothing; and rubber shoes or boots.

Danger! Occupational workers should remember that zinc chloride is corrosive. Any kind of contact with skin causes irritation and possible burns, especially if the skin is wet or moist. Workers should also remember that exposure to high concentrations of zinc chloride fume causes respiratory distress syndrome leading to pulmonary fibrosis and death.

Zinc Oxide (CAS No. 1314-13-2)

Molecular formula: **ZnO**
Synonyms and trade names: Zinc peroxide; Chinese white; Zinc white; Flowers of zinc; Calamine

Zinc oxide is a white to yellowish-white amorphous, odourless powder. It is insoluble in water and alcohol and soluble in dilute acids. Zinc oxide is amphoteric, that is, it reacts with both acids and alkalis. Zinc oxide is incompatible with magnesium and strong acids. It reacts violently with magnesium and linseed oil. Zinc oxide and magnesium can react explosively when heated and incompatible when mixed with chlorinated rubber. Zinc oxide is an inorganic chemical substance used mainly as a white powder. It has wide industrial applications, for instance as an additive to commercial products, plastics, ceramics, glass, cement, rubber, car tyres, lubricants, paints, ointments, adhesives, sealants, and pigments. In fact, the rubber industry uses as large as 50% of ZnO, and similarly the concrete manufacturing industries also use large quantities of zinc oxide. Zinc oxide is widely used to treat a variety of other skin conditions, in products like baby powder, diaper rash creams, antiseptic ointments, and anti-dandruff shampoos. Also, paints containing zinc oxide powder have long been utilised as anti-corrosive coatings for metals, especially effective for galvanised iron. Zinc oxide reacts with fatty acids such as stearic directly by mixing and heating the components above the acid melting point. Zinc oxide exposed to air absorbs both water vapour and carbon dioxide. This results in the formation of basic zinc carbonate.

Safe Handling and Precautions

Inhalation of zinc oxide fumes during industrial processing of zinc alloys is hazardous. Exposures to zinc oxide have been reported to cause metal fume fever with chills, muscle ache, nausea, fever, dry throat, cough, lassitude (weakness, exhaustion), metallic taste, headache, blurred vision, low back pain, vomiting, malaise (vague feeling of discomfort), chest tightness, dyspnoea (breathing difficulty), and decreased pulmonary function. Zinc oxide dust inhalation may cause coughing, choking, or irritation of the nose, throat, and upper respiratory tract. The zinc oxide fumes cause irritation to the respiratory tract, irritation of mucous membranes, coughing, respiratory distress/shortness of breath, flu-like illness, or 'metal fume fever' characterised by chills, fever, aching muscles, dryness in the mouth and throat, and headache. Inhalation of freshly formed zinc oxide can lead to metal fume fever. The dust may cause eye and skin irritation. Zinc oxide is not carcinogenic. Occupational workers should keep zinc oxide waste and handle and dispose of it in accordance with applicable regulations. Keep it in a tightly closed container, stored in a cool, dry, ventilated area. Protect it against physical damage. Isolate it from incompatible substances.

Zinc Phosphide (CAS No. 1314-84-7)

Molecular formula: Zn_3P_2
Synonyms and trade names: Blue-ox; Kilrat; Mous-con; Phosphure de zinc; Phosvin; Rumetan; Zinco(fosfuro di); Zinc phosphide; Zinc(phosphure de); Zinc-tox; Zinkfosfide; Zinkphosphid

Zinc phosphide is a rodenticide. It is stable when dry but decomposes slowly in moist air. It has a strong garlic-like odour. Zinc phosphide reacts violently with acids, with decomposition to the spontaneously inflammable phosphine gas. Reports have indicated that zinc phosphide is used to control the field menace of grasshoppers, moles, rats, and squirrels. It is federally approved as a mole poison. Zinc phosphide–laced bait has a strong garlic-like odour.

Safe Handling and Precautions

Exposures to zinc phosphide repeatedly cause kidney damage and hyaline degeneration of the myocardium. Hepatocytes demonstrate changes with cloudy swelling, hyaline degeneration, and necrosis at the centre of liver lobules. Zinc phosphide turns into a gas when it reaches fluid in the stomach. This gas destroys cells throughout the bloodstream and body and results in organ and tissue damage. It takes from 15 min to 4 h for the mole to die, according to the Department of Natural Resources and Environment (DNRE). The death is a painful one and involves abdominal pain, convulsions, nausea, vomiting, and paralysis. Exposure to zinc phosphide causes cough, headache, fatigue, nausea, photophobia, vomiting, abdominal pain, ataxia, diarrhoea, dizziness, respiratory distress, breathing disorders, irritation to the respiratory tract, lung oedema, and effects on the liver, kidneys, heart, and nervous system. Exposure to high concentrations of zinc phosphide has been reported to cause death, unconsciousness, and fatal injury.

According to the reports of DNRE, exposures to zinc phosphide cause death to dogs and cats through secondary exposure, and birds are especially affected by the mole poison. It takes several months for the bait to be degraded by weather or for the mole to degrade. Dry bait may stay toxic indefinitely. The exposure limits of zinc phosphide are very low, and the regulatory agencies demand for full precautions during different stages in handling and management of zinc phosphide.

Workers and workplace supervisor should be very alert during use and handling of zinc phosphide, which should be stored in clearly labelled, sealed impermeable containers. Zinc phosphide should be handled by trained workers and kept away from oxidising agents and acids and stored under lock-and-key safety. Zinc phosphide should not be stored in place of moisture or be allowed to become damp, which may lead to chemical hazards and disasters.

During use and handling of zinc phosphide, access to unauthorised/untrained workers should be strictly prohibited. Workplace manager and management must provide adequate washing facilities for the workers close to the workplace at all times during handling of zinc phosphide.

Danger! Zinc phosphide is a 'poison'. Label it with skull and crossbones insignia. Zinc phosphide is a poison used in bait form as a rodenticide. Workers should strictly avoid dust/fumes of zinc phosphide and keep the chemical baits out of reach of children and domestic animals and completely away from foodstuffs, animal feed, etc. Remember to keep zinc phosphide dry and away from acids. During use and handling of zinc phosphide, entry/access to workplace by unauthorised individuals should be completely stopped.

Zineb (CAS No. 12122-67-7)

Molecular formula: $C_4H_6N_2S_4Zn$
Synonyms and trade names: Amitan; Aspor; Devizeb; Dithane; Dithane Z-78; Hexathane; Phytox; Polyram-Z; Zebtox; Zimate; Zinosan
IUPAC name: Zinc ethylenebisdithiocarbamate
Toxicity class: U.S. EPA: IV; WHO: III

Zineb is a light-coloured powder/crystal and insoluble in water. Zineb is one of the non-systemic (surface-acting) fungicides of the group ethylenebisdithiocarbamates (EBDCs). Zineb is a polymer of ethylene (bis)thiocarbamate units linked with zinc and used for the control and prevention of crop damage in the field and also to protect harvested crops from deterioration during storage or transport. Zineb was used to protect fruit and vegetable crops from a wide range of foliar and other diseases. It was available in the United States as wettable powder and dust formulations. Zineb can be formed by combining nabam and zinc sulphate in the spray tank.

Safe Handling and Precautions

Exposure to zineb is harmful. It is moderately irritating to the skin, eyes, respiratory tract, and mucous membranes. It may also be a dermal sensitiser, with possible cross-sensitisation to maneb and mancozeb. This irritation may result in itching, scratchy throat, sneezing, coughing, inflammation to the nose or throat, and bronchitis. Early symptoms from exposure of humans of zineb include tiredness, dizziness, and weakness. In cases of severe poisoning after repeated exposures, workers have been reported to suffer with symptoms of headache, nausea, fatigue, slurred speech, allergic skin reaction, respiratory irritation, convulsions, and unconsciousness.

Ethylene thiourea (ETU), a potentially toxic metabolite of zineb, may be involved in thyroid effects. Occupational inhalation of zineb can lead to changes in liver enzymes, moderate anaemia and other blood changes, increased incidence of poisoning symptoms during pregnancy, and chromosomal changes in the lymphocytes. Liver functioning was affected in workers exposed to zineb. Repeated or prolonged dermal exposure may cause dermatitis or conjunctivitis. Farm workers who were repeatedly exposed to zineb, in fields sprayed with 0.5% suspension of the fungicide, reported severe and extensive contact dermatitis. ETU formation during metabolism of zineb or other EBDC pesticides may potentially result in goitre, a condition in which the thyroid gland is enlarged.

Ziram (CAS No. 137-30-4)

Molecular formula: $C_6H_{12}N_2S_4Zn$
Synonyms and trade names: Ziram; Zinc, bis(dimethyldithiocarbamato)-; Aavolex; Aazira; Accelerator L; Aceto ZDED; Aceto ZDMD; Alcobam ZM; Bis(dimethyldithiocarbamato) zinc; Carbamodithioic acid, dimethyl-, zinc salt; Carbazinc; Corona corozate; Corozate; Cuman; Cuman L; Cymate; Dimethyldithiocarbamic acid, zinc salt; Eptac 1; Fuclasin; Fuclasin-ultra; Fuklasin; Hermat ZDM; H exazir; Karbam white; Methasan; Methazate; Methyl zimate; Methyl zineb; Methyl ziram; Milbam; Molurame; Orchard brand ziram; Pomarzo
IUPAC name: Zinc bis(dimethyldithiocarbamate)
Toxicity class: U.S. EPA: III; WHO: III

Ziram is a carbamate and an agricultural fungicide. It may be applied to the foliage of plants, but it is also used as a soil and/or seed treatment. Ziram is used primarily on almonds and stone fruits. It is also used as an accelerator in rubber manufacturing, packaging materials, adhesives, and textiles. Another use of the compound is as a bird and rodent repellent. Ziram is used primarily as a rubber vulcanisation accelerator but is also used as a foliar fungicide, mainly on fruit and nuts. It has been in commercial use since the 1930s. Ziram has been formulated for use as a wettable powder, a paste, and water-dispersible granules and also in combination with other pesticides. Exposure can occur during its production, its use in the rubber industry, and its application as a fungicide, and at much lower levels, from consumption of foods containing residues.

Safe Handling and Precautions

The oral LD_{50} for ziram is 1400 mg/kg in rats and 480 and 400 mg/kg for mice and rabbits, respectively. Ziram has an LD_{50} of 100–150 mg/kg in guinea pigs. The acute dermal LD_{50} for rats is greater than 6000 mg/kg. Ziram can cause skin and mucous membrane irritation. Humans with prolonged inhalation exposure to ziram have developed nerve and visual disturbances. Ziram is corrosive to eyes and may cause irreversible eye damage.

Female rats administered relatively small amounts of ziram in their diets (2.5 mg/kg/day) for 9 months showed decreased antibody formation. Rats fed diets containing 0.25% ziram for an unknown time period exhibited poor growth and development. In a 1-year feeding study with rats, no effects were seen at the low dose of 5 mg/kg/day nor were any effects seen in weanlings receiving 5 mg/kg in their diet for 30 days. At unknown doses and duration, rats developed a peculiar hind leg grasping reaction plus other motor changes when given ziram. Dogs also showed convulsive seizures. Another study with dogs fed ziram in their diets also showed no harmful effects for 12 months at 5 mg/kg/day.

Ziram was tested adequately for carcinogenicity by oral administration in one study in mice and one study in rats. In mice, the incidence of benign lung tumours was increased in females. In rats, a dose-related increase in the incidence of C-cell thyroid carcinomas was observed in males. In single studies, ziram caused embryotoxicity and minor malformations in rats and embryolethality in chicks hatched from injected ova. An increased frequency of chromatid and chromosomal aberrations was seen in peripheral blood lymphocytes of workers who handled and packaged ziram. Ziram was clastogenic in mammalian cells in vivo and in vitro and induced mutations in cultured rodent cells and in insects and bacteria. Ziram has been listed as Group 3, meaning not classifiable as a human carcinogen.

Occupational workers and users during use and handling of ziram and ziram products should wear protective equipment (PPE), workplace dress to prevent chemical contact and to avoid organic vapour. The personal protective equipment includes, but not limited to face mask, self-contained breathing apparatus, liquid-proof gloves, protective goggles, rubber shoes or boots. Workers should disposed of chemical spill and waste in accordance with set regulations and good laboratory practices.

Bibliography

Agency for Toxic Substances and Disease Registry (ATSDR). 1989. *Toxicological Profile for Zinc*. Agency for Toxic Substances and Disease Registry, U.S. Public Health Service, Atlanta, GA.

Agency for Toxic Substances and Disease Registry (ATSDR). 1994. *Toxicological Profile for Zinc*, U.S. Public Health Service, Agency for Toxic Substances and Disease Registry, Atlanta, GA.

Bai, K. M., M. K. Krishnakumari, H. P. Ramesh, T. Shivanandappa, and S. K. Majunder. 1980. Short-term toxicity study of zinc phosphide in albino rats. *Indian J. Exp. Biol.* 18:854–857.

Bumbrah, G. S., K. Krishan, T. Kanchan, M. Sharma, and G. S. Sodhi. 2012. Phosphide poisoning: A review of literature. *Forensic Sci. Int.* 214:1–6.

Centers for Disease Control and Prevention (DCP). 2010. Zinc oxide. Atlanta, GA.

Chugh, N., H. K. Aggarwal, and S. K. Mahajan. 1998. Zinc phosphide intoxication symptoms: Analysis of 20 cases. *Int. J. Clin. Pharmacol. Ther.* 36:406–407.

Edwards, I. R., D. G. Ferry, and W. A. Temple. 1991. Fungicides and related compounds. In: *Handbook of Pesticide Toxicology*, Vol. 3, Classes of Pesticides, W. J. Hayes and E. R. Laws. (eds.). Academic Press, New York.

Gosselin, R. E., R. P. Smith, and H. C. Hodge. 1984. *Clinical Toxicology of Commercial Products*. Williams & Wilkins, Baltimore, MD.

International Agency for Research on Cancer (IARC)—Summaries & evaluations. 1991. Ziram. IARC. Vol. 53, p. 423.

International Programme on Chemical Safety and the Commission of the European Communities (IPCS, CEC). 2001. Zinc phosphide. ICSC Card No. 0602. IPCS-CEC, Luxembourg, Belgium (updated 2004).

International Programme on Chemical Safety and the Commission of the European Communities (IPCS, CEC). 2002. Zinc dichloride. ICSC Card No. 1064. IPCS-CEC, Luxembourg, Belgium.

Klaassen, C. D., M. O. Amdur, and J. Doull. (eds.). 1995. *Casarett and Doull's Toxicology. The Basic Science of Poisons*, 5th edn. McGraw-Hill, New York.

Material Safety Data Sheet (MSDS). 1991. Ziram. FMC Corporation. Philadelphia, PA.

Meister, R. T. (ed.). 1992. *Farm Chemicals Handbook*. Meister Publishing Co., Willoughby, OH.

Muthu, M., M. K. Krishnakumari, V. Muralidhara, and S. K. Majumder. 1980. A study on the acute inhalation toxicity of phosphine to albino rats. *Bull. Environ. Contam. Toxicol.* 24:404–410.

National Institute of Occupational Safety and Health (NIOSH). 1990. *NIOSH Pocket Guide to Chemical Hazards*. Department of Health and Human Services, Bethesda, MD.

National Library of Medicine. 1993. Hazardous Substances Data Bank. TOXNET, Medlars Management Section, Bethesda, MD.

Raina, A., H. C. Shrivastava, and T. D. Dogra. 2003. Validation of qualitative test for phosphine gas in human tissues. *Indian J. Exp. Biol.* 41:909–911.

Rodenberg, H. D., C. C. Chang, and W. A. Watson. 1989. Zinc phosphide ingestion: A case report and review. *Vet. Hum. Toxicol.* 31:559–562.

Simmer, K., L. Lort-Phillips, C. James et al. 1991. A double-blind trial of zinc supplementation in pregnancy. *Eur. J. Clin. Nutr.* 45:139–144.

Stephenson, J. B. P. 1967. Zinc phosphide poisoning. *Arch. Environ. Health* 15:83–88.

Tomlin, C. D. S. 2006. *The Pesticide Manual: A World Compendium*, 14th edn. British Crop Production Council, Hampshire, U.K.

World Health Organization (WHO). 1976. International Programme on Chemical Safety (IPCS). Data Sheets on Pesticides. No. 24. Zinc phosphide. WHO, Geneva, Switzerland.

World Health Organization (WHO). 1988a. International Programme on Chemical Safety (IPCS). Phosphine and selected metal phosphides. Environmental Health Criteria No. 73. WHO, Geneva, Switzerland.

World Health Organization (WHO). 1988b. Dithiocarbamate pesticides, ethylenethiourea and propylenethiourea. Environmental Health Criteria No. 78. WHO, Geneva, Switzerland.

Zinc Phosphide. 2009. Tolerance for residues. Code of Federal Regulations, Section 180.284, Title 40.

28

Hazardous Chemical Substances and Ocular Disorders (Injuries to Eyes): Perspectives and Scenarios

Praveen Murthy and T.S.S. Dikshith

Introduction

Various hazardous chemical substances have been known to cause ocular disorders because of negligence and lack of appropriate and suitable workplace dress by occupational workers and the general public. This has caused global concern. Injuries to human eyes range from very minor, mild, to very severe type, resulting in permanent loss of vision or loss of the eyes. Injuries to human eyes have been known to occur in the workplace, at home, in other accidents, or while participating in sports and wherever workers neglect to observe safety precautions and elements of workplace safety dress.

All chemicals can have toxic effects at certain dose levels and particular routes of exposure. It is therefore wise to minimise exposure to chemicals. Chemicals can have local or systemic effects. Local toxicity refers to the direct action of chemicals at the point of contact. Systemic toxicity occurs when the chemical agent is absorbed into the bloodstream and distributed throughout the body, affecting one or more organs. Health effects can be acute or chronic. Acute effects last for a relatively short time and then disappear. Chronic effects are not reversible.

Do not confuse acute and chronic exposure with acute and chronic effects. Acute exposures to chemicals are for short periods. Chronic health effects can develop from acute exposures depending on the properties and amount of the chemical. Acute or chronic adverse health effects can also occur with chronic (repeated) exposure to chemicals, even at low concentrations.

It is therefore wise to minimise exposure to chemicals. Chemicals can have local or systemic effects. Local toxicity refers to the direct action of chemicals at the point of contact. Systemic toxicity occurs when the chemical agent is absorbed into the bloodstream and distributed throughout the body, affecting one or more organs. Health effects can be acute or chronic. Acute effects last for a relatively short time and then disappear. Chronic effects are not reversible.

Hazardous chemical substances have the potential to inflict serious to very serious visual disorders, impairment, injuries, and complete loss of sight. The type and degree of injury and/or injuries depend upon the specific site, nature of the chemical, and the period of exposure. It is important that workers should be aware of the possible chemical hazards and the kinds of ocular injuries and disorders.

Ocular exposure: The eyes are of particular concern, due to their sensitivity to irritants. Ocular exposure can occur via splash or rubbing eyes with contaminated hands. Few substances are innocuous with eye contact, and several can cause burns and loss of vision. The eyes have many blood vessels and rapidly absorb many chemicals.

Studies have indicated that the important function of the ocular surface of the eye is to provide a clear and undisturbed vision to the healthy individual. The eye is extremely sensitive and a very delicate vital organ. Eyes are very susceptible to infection and to all kinds of injuries, especially at workplaces. Occupational workers and workplace management should strictly observe and use protective dress to prevent permanent damage to the eyes because of burns from chemicals, such as acids and caustic soda, and stronger chemicals that cause irritation and stinging, such as ammonia, vinegar, alcohol, or household bleach, and burns caused by the flash and arc welder.

Ocular disorder causes different signs and symptoms, which include, but are not limited to, pain and irritation, sensitivity to light, severe watering of eyes, reddened eyeballs, swollen eyelids, arc welder's burns, blowout fracture of the orbit, subconjunctival haemorrhage or bleeding, corneal abrasions – a scratch or a traumatic defect in the surface of the cornea – chemical conjunctivitis, corneal abrasion, eye contusion from sports or altercations, contusion injury, and retinal detachment. Physical or chemical injuries to the eyes are serious health problems among occupational workers and children. The most common and obvious symptom of the eye injury is redness and pain of the affected eyes that require appropriate and timely care. Direct exposure to chemical substances in the form of dust, fibres, fume, vapour, and solid and liquid particles is known to cause injuries. The most common symptoms are pain or intense burning. Eyes soon begin to tear profusely and become red, the eyelids become swollen, and the vision becomes blurred and develops pain (hyphema).

Blurred vision: Lack of sharpness of vision with, as a result, the inability to see fine detail. Blurred vision can occur when a person who wears corrective lens is without them. Blurred vision can also be an important clue to eye disease.

Glaucoma: A common eye condition in which the fluid pressure inside the eyes rises because of slowed fluid drainage from the eye. If untreated, it may damage the optic nerve and other parts of the eye, causing the loss of vision or even blindness.

Iritis: Inflammation of the iris. The iris is the circular, coloured curtain in the front of the visible part of the eye. (The opening of the iris forms the pupil.)

Keratitis: Inflammation of the cornea (the transparent structure in the front of the eye).

Retina: The retina is the nerve layer that lines the back of the eye, senses light, and creates impulses that travel through the optic nerve to the brain. There is a small area, called the macula, in the retina that contains special light-sensitive cells. The macula allows us to see fine details clearly.

Retinal: Pertaining to the retina, the extraordinary layer of neurons (nerve cells) that line the back of the eye, which can sense light and create impulses capable of voyaging through the optic nerve to the brain where the impulses are recognised as an image.

Retinopathy: Any disease of the retina, the light-sensitive membrane at the back of the eye. The type of retinopathy is often specified. Arteriosclerotic retinopathy is retinal disease due to arteriosclerosis ('hardening of the arteries'). Diabetic retinopathy is retinal disease associated with diabetes. Hypertensive retinopathy is retinal disease due to high blood pressure.

Sclera: The tough white outer coat over the eyeball that covers approximately the posterior five-sixths of its surface. The sclera is continuous in the front of the eye with the cornea and in the back of the eye with the external sheath of the optic nerve.

Tear: A drop of the salty secretion of the lacrimal glands, which serves to moisten the conjunctiva and cornea.

Very important:

Before working with chemical substances, occupational workers/individuals should be trained in its proper handling and storage. Workers should also know how to use proper and adequate personal protective workplace dress and equipment.

Occupational workers should have decontamination of work surfaces: Protect work surfaces from contamination by using 'bench paper' (disposable plastic-backed absorbent paper) or stainless steel trays. Place the plastic side down and the absorbent side facing up. Change worn or contaminated bench paper and dispose of properly. Decontaminate other items and equipment with appropriate solvents when contaminated during experiments.

Types of Eye Injuries

There are many types of eye injuries. The most common eye injuries include (a) *corneal abrasions*: damage to the cornea (the protective layer of transparent tissue at the front of the eye) caused by scratching or grazing; (b) *iritis (uveitis)*: inflammation (swelling) of the iris (the coloured part of the eye that controls the amount of light that enters); it can be caused by a trauma to the eye (traumatic iritis) or by another condition (non-traumatic iritis); and (c) *foreign bodies*: material that accidentally gets into the eye, such as metal, wood, plastic, or dust. These ocular injuries/disorders are preventable provided adequate precautionary measures are taken by worker. To maintain safety and quality standards in terms of human health, safety, and overall progress in industries, factories, and different commercial workplaces, observance of an element of safety regulations has been found very essential. The great increase and diversification of mechanisation in industry, with associated increased demand on our social lives, result in increased risk and occurrence of all types of trauma especially ocular trauma. The eyes are exposed to a variety of injurious agents depending on the type of industry. Ocular injury is commonly due to occupational hazards. A study carried out at Wolverhampton, a highly industrialised area of the United Kingdom, showed that 73.8% of all ocular traumas seen over a 10 year period occurred in industries. A similarly high figure of 71% was reported as early as 1923. However, a much lower figure of 15.4% was reported from a much less industrialised area of Northern Ireland. There are other sources of industrial ocular hazard, which include exposure to dangerous rays. Acute exposure to ultraviolet (UV) radiation results in photokeratitis, which is characterised by pain and grittiness. This may result in decreased corneal sensitivity and damage to the corneal endothelium. Long-term exposure may be partly responsible for conditions such as pterygia, pingueculae, band-shaped keratopathy, and climatic droplet keratopathy.

Repeated exposure to radiant energy on glassblowers, steel workers, blast furnace attendants, and blacksmiths can result in glassblower cataract. The damage to the lens is a consequence of absorption of direct and indirect infrared radiation. There are several chemicals used in the refining process in the petroleum industry. Exposure to these chemicals may cause adverse ocular side effects, which may manifest mainly in the technical workers. The aim of this study is to identify the pattern of ocular disorders among occupational workers.

Hazardous Chemicals

Alkaline Compounds

Sodium hydroxide is extremely corrosive. The severity of injury increases with the concentration of the solution, the duration of exposure, and the speed of penetration into the eye. Ocular damage has been reported to range from severe irritation and mild scarring to blistering, disintegration, ulceration, severe scarring, and clouding. Conditions which affect vision, such as glaucoma and cataracts, are possible late developments. In severe cases, there is progressive ulceration and clouding of eye tissue, which may lead to permanent blindness.

Asbestos

Asbestos is a group of minerals found in nature. It has wide applications and is used regularly for building structures because of its strong, heat-resistant properties and is also relatively less expensive. The automotive industry uses asbestos in vehicle brake shoes and clutch pads. Asbestos has also been used in ceiling and floor tiles, paints, coatings, and adhesives and in vermiculite-containing garden products and some talc-containing crayons.

Asbestos has been classified as a known human carcinogen.

Asbestos is not a single substance, but is the generic name for a family of six related polysilicate fibrous minerals of which one (chrysotile) belongs to the serpentine family and five (actinolite, amosite, anthophyllite, crocidolite, and tremolite) belong to the amphibole family. These minerals differ from each other in physical and chemical properties, and each mineral can exist in a wide range of fiber sizes. Exposure to asbestos is common during home construction, remodeling of buildings, or renovating or razing of old buildings and structures. Improper use and handling of asbestos by workers is known to cause health hazards and dangers. Any negligence by workers during handling of asbestos causes serious health hazards. Asbestos enters the body through the eyes or lungs-primary targets. The complications and eye-related injuries caused by asbestos include damage to corneal and conjunctival epithelium, goblet cells, stromal keratocytes, corneal extracellular matrix, blood vessels, ciliary body, and trabecular meshwork. Inhalation of asbestos fibers may lead to fibrotic lung disease (asbestosis), pleural plaques and thickening, and cancer of the lung, the pleura, and the peritoneum. Workers should wear appropriate safety dress and safety equipment at all workplaces to avoid health hazard and injuries. Asbestos has been known and classified as a potential human carcinogen.

The complications and eye-related injuries include damage to corneal and conjunctival epithelium, goblet cells, stromal keratocytes, corneal extracellular matrix, blood vessels, ciliary body, and trabecular meshwork.

1. Clinical features
   a. Immediate rise in the pH following alkaline solution exposure to eye
   b. Symptoms: ocular pain, lacrimation, blepharospasm
   c. Signs
      i. In mild cases: epithelial erosion, mild corneal haze, and conjunctival injection
      ii. In moderate cases: cornea may opacify with slight ischemia of limbus
      iii. In severe cases: significant ischemia of the sclera, avascularity of the limbus, blanching of conjunctiva, and severe corneal haze
2. Complications

3. Eyelid scarring

4. Corneal opacification, severe dry eye, corneal ulcer, perforation with potential secondary intraocular infection

5. Conjunctival scarring, symblepharon, or ankyloblepharon

6. Aqueous dynamic changes with increased or decreased intraocular pressure

7. Cataract and phthisis bulbi

Hazardous Chemicals and Eye Burn

Chemical exposure to any part of the eye or eyelid may result in a chemical eye burn. Chemical burns represent 7%–10% of eye injuries. About 15%–20% of burns to the face involve at least one eye. Although many burns result in only minor discomfort, every chemical exposure or burn should be taken seriously. Permanent damage is possible and can be blinding and life altering.

The severity of a burn depends on what substance caused it, how long the substance had contact with the eye, and how the injury is treated. Damage is usually limited to the front segment of the eye, including the cornea (the clear front surface of the eye responsible for good vision, which is most frequently affected), the conjunctiva (the layer covering the white part of the eye), and occasionally the internal structures of the eye, including the lens. Burns that penetrate deeper than the cornea are the most severe, often causing cataracts and glaucoma.

Causes of Eye Pain Include Ocular Pain and Orbital Pain

Ocular pain is eye pain coming from the outer structures of the surface of the eye. Conjunctivitis is one of the most common eye problems. Conjunctivitis (pinkeye is a non-medical term) can be an allergic, bacterial, chemical, or viral inflammation of the conjunctiva – the delicate membrane lining the eyelid and covering the eyeball.

Chemical Burns

Chemical burns and flash burns are significant causes of eye pain. Chemical burns come from eye exposure to acid or alkaline substances, such as laboratory chemicals, household cleaners, and/or bleach. Flash burns have been reported to occur because of direct exposure to sources of very bright or intense light and without workplace-appropriate protective dress and/or goggles. Even an intense sunny day can cause a flash burn.

Corneal Abrasion

Exposure of the unprotected eye to UV light from sun lamps or welding arcs can cause changes in the corneal surface resembling corneal abrasions.

A corneal injury may occur when something gets into your eye, for example, when the wind blows a dried leaf particle into your eye or when paint chips fall into your eye while you are scraping off old paint. This material may scratch the cornea.

In addition to causing corneal injury, high-speed particles may penetrate your eye and injure deeper structures. An example of this would be a small metal fragment flying into the eye when a person is using a grinding wheel without protective eyewear. This may cause a serious injury that demands immediate medical attention to guard against permanent loss of vision.

The symptoms of cornea injury include, in brief, (i) tearing of the eyes, (ii) eye redness, (iii) blurred vision or distortion of vision, and (iv) spasm of the muscles surrounding the eye causing to squint.

Good eyesight is important in order to effectively do your job and to perform other tasks such as driving and reading.

- However, you may injure your eyes if you do not properly protect them.
- Eyes exposed to airborne dust and debris can become itchy, irritated, and uncomfortable.
- Exposure to toxic chemicals, pokes from plants and tree branches, harmful gases and vapours, and flying objects from equipment can cause serious eye injuries that may require medical treatment and lead to permanent damage.
- Excessive exposure to sunlight may lead to cataracts, which can impair your vision.
- You can save your eyes from serious injury by using the right equipment.
- Keep your eye protection clean and ready to use.
- Most safety glasses have side shields or come in a 'wrap-around' style that can protect your eyes from dust, particles, sharp branches, and flying objects. The lenses are made to protect against impact. The lenses are also designed to block out damaging rays from the sun.
- However, when applying powdered or liquid pesticides, safety goggles provide better protection than safety glasses.
- When using safety goggles, make sure they fit tightly on your face.
- In high-chemical-exposure situations, it is important to wear a face shield over your safety glasses or goggles for skin protection.
- Do not use sunglasses or reading glasses for impact protection. They will not adequately protect your eyes from flying particles.
- Do not wear contact lenses when handling chemicals that can splash you in the face. This can be a risk, even if protective eyewear is worn over them. We recommend that you consult with a health-care professional if you wear contact lenses and will be handling chemicals.
- Replace broken or unsafe eye protection.
- If wood chips, dust, or other particles get into your eyes, look down and flush out your eyes at the nearest eyewash station. If there is no eyewash station available, carry a squeeze bottle of water or an eyewash dispenser for use in case of an emergency. See your supervisor if you do not have one.
- If a pesticide gets into your eyes, immediately flush them, preferably with cool to lukewarm water for at least 15 min and seek medical help. Continue to flush your eyes during the drive to the hospital. Chemical burns to the eyes require immediate medical attention.

- Carefully review the material safety data sheet (MSDS) and the pesticide label before you start the job for additional first aid instructions.

- Apply cold packs to eyes that are hit by flying objects from equipment. If the injury becomes discoloured, seek medical attention.

- When working outdoors, try to position yourself according to wind direction so your eyes are not exposed to blowing dirt, dust, and debris.

- Protect your eyes from the sun, even when you are not working with chemicals. Do this by wearing sunglasses (or safety glasses if needed) that filter at least 90% of the sun's UV rays. Be sure the label indicates that they filter both 'UVA' and 'UVB' rays. Also, wear a hat that can keep the sun away from your eyes.

- Take care of your eyes at home as well as at work. Give your eyes a break when reading or watching television for long periods of time so you do not develop eyestrain.

- Be aware that household cleaners and other commonly used household products often contain chemicals that can be dangerous to your eyes.

- Ask for eye protection that will fit over your prescription eyewear.

- Remember that eye injuries can be very costly – both in terms of medical costs and in terms of permanent loss of vision.

Chemical Substances and Eye Protection

Acids

For solutions of hydrochloric acid with a pH less than or equal to 3.0, persons should wear, at a minimum, an 8 in. face shield. Splash-proof goggles are also recommended where mists of hydrochloric acid solution could contact the eyes.

Occupational workers and general public should follow basic safety practices during use and handling of hazardous chemicals to minimise health hazards and to prevent accidents.

Standard Operating Procedures

Standard operating procedures (SOPs) are intended to provide general guidance on how to safely work with a specific class of chemical or hazard. This SOP is generic in nature. It addresses the use and handling of substances by hazard class only. In some instances, multiple SOPs may be applicable for a specific chemical (i.e., both the SOPs for flammable liquids and carcinogens would apply to benzene). If you have questions concerning the applicability of any items listed in this procedure, contact the principal investigator of your laboratory. Specific written procedures are the responsibility of the principal investigator.

All locations within the laboratory where acutely toxic chemicals are handled when work involves chemicals or hazardous materials should be demarcated with designated area caution tape and/or posted with designated area caution signs.

Eye protection in the form of safety glasses must be worn at all times when handling acutely toxic chemicals. Ordinary (street) prescription glasses do not provide adequate protection. (Contrary to popular opinion, these glasses cannot pass the rigorous test for industrial safety glasses.) Adequate safety glasses must meet the requirements of the regulations and must be equipped with side shields. Safety glasses with side shields do not provide adequate protection from splashes; therefore, when the potential for splash hazard exists, other eye protection and/or face protection must be worn.

Where the eyes or body of any person may be exposed to acutely toxic chemicals, suitable facilities for quick drenching or flushing of the eyes and body shall be provided within the work area for immediate emergency use. Bottle-type eyewash stations are not acceptable.

Manipulation of acutely toxic chemicals should be carried out in a fume hood. If the use of a fume hood proves impractical, refer to the section on special ventilation. Workplace areas where acutely toxic chemicals are stored or manipulated must be labelled as a designated area.

Gloves should be worn when handling acutely toxic chemicals. Disposable latex or nitrile gloves provide adequate protection against accidental hand contact with small quantities of most laboratory chemicals. Laboratory workers should ask their supervisor for advice on chemical-resistant glove selection when direct or prolonged contact with hazardous chemicals is anticipated.

Workplace safety dress, closed toed shoes, and long-sleeved clothing should be worn by workers when handling acutely toxic chemicals. Additional protective clothing should be worn if the possibility of skin contact is likely. Safety shielding is required any time there is a risk of explosion, splash hazard, or a highly exothermic reaction. All manipulations of acutely toxic chemicals that pose this risk, should occur in a fume hood with the sash in the lowest feasible position. Portable shields, which provide protection to all laboratory occupants, are acceptable.

Manipulation of acutely toxic chemicals outside of a fume hood may require special ventilation controls in order to minimise exposure to the material. Fume hoods provide the best protection against exposure to acutely toxic chemicals in the laboratory and are the preferred ventilation control device. Where possible, handle acutely toxic chemicals in a fume hood. If the use of a fume hood proves impractical, attempt to work in a glove box or in an isolated area on the laboratory bench top. If available, consider using a biological safety cabinet. The biological safety cabinet is designed to remove the acutely toxic chemicals before the air is discharged into the environment. Acutely toxic chemicals that are volatile must not be used in a biological safety cabinet unless the cabinet is vented to the outdoors. If your research does not permit the handling of acutely toxic chemicals in a fume hood, biological safety cabinet, or glove box, all areas where acutely toxic chemicals are stored or manipulated must be labelled as a designated area.

All materials contaminated with acutely toxic chemicals should be disposed of as hazardous waste. Wherever possible, attempt to design research in a manner that reduces the quantity of waste generated.

In brief, occupational workers and general public should follow basic safety practices at workplaces.

Workers should wear safety glasses or goggles to prevent eye injuries:

- Flush eyes with water for at least 15 min if chemicals are splashed into them.
- Wear a face shield over safety glasses or goggles for skin protection in high-chemical-exposure situations.

Common public can protect the eyes from the sun by wearing tinted safety glasses and a hat when working outdoors:

- Always report any kind of eye injury to the workplace supervisor no matter how minor the injury is.

Workers should

- Wear safety glasses for chemical splash protection. Use safety goggles instead.
- Wear sunglasses or reading glasses for eye protection from impact or particles.
- Care for eyes at home and elsewhere off the job.
- Seek immediate medical attention if eyes are injured. Remember that eye injuries can result in permanent loss of vision.

Using eye protection can save your eyes from serious injury from flying particles, chemicals, and pokes from branches.
Wear safety goggles, instead of safety glasses, when handling hazardous chemicals.

- Immediately flush your eyes with water if they are exposed to chemicals.

Different Kinds of Eye Protection Equipment

- Glass.
- *Plastic and polycarbonate*: Protect against welding splatter.
- *Shielded safety glasses*: Prescription glasses (with or without side shields) are *not* an acceptable substitution for safety glasses. Prescription safety glasses are available. Safety glasses do *not* provide complete protection against splash or spray because they do not fit tightly to your face. Safety glasses must meet quality standards. Safety glasses must be worn *anytime* chemicals or chemical products are handled.
- *Goggles*: Goggles provide greater protection from splashes, liquids, and dusts than shielded safety glasses and are the best protection tool against liquid pesticides and other toxic chemicals.
- *Splash goggles*: Splash goggle must be worn *anytime* there is the chance of a chemical splash or spray. Safety glasses are *not* an acceptable substitution for goggles and do not provide complete protection against splash or spray because they do not fit tightly to your face. Operations requiring goggles include but are *not* limited to pouring, scrubbing, rinsing, spraying (aerosols), washing, and dispensing. Splash goggles must also meet quality standards.
- *Full-face shields*: Face shields protect the eyes, face, and neck from chemical splashes and spray as well as flying particles. Face shields should *not* be worn independently. In other words, safety glasses or goggles must be worn underneath face shields for complete protection. Face shields are necessary *anytime* there is a severe risk of splash or spray or if the material in use is highly hazardous, for example, highly corrosive alkaline material.
- *Glasses*: Glass lenses provide good scratch resistance. They can withstand chemical exposure.
- Wear goggles or a face shield around flying chips or particles; electrical arcing or sparks; chemical gases or vapours; harmful light liquid chemicals, acids, or caustics; molten metal; dusts; or swinging objects like ropes or chains. Remove protective eyewear only after turning off the tool.
- Replace cracked, pitted, or damaged goggles or glasses. Be certain that protective eyewear is approved for the hazardous environment you are in. Keep sharp or pointed objects away from the face and eyes.

- If you get dust, a wood chip, or another small particle in your eye, look down and flush it out with eyewash solution. Use water if eyewash solution is not available. If a pesticide gets into your eyes, immediately use a portable eye flush dispenser or call for help if needed to get to an eyewash station. Flush your eye with eyewash solution for 15 min. Have someone call for medical attention while you are flushing.

GENERAL CHEMICAL SAFETY GUIDELINES

Maintain an organised and orderly facility:
- Work area:
  - Keep the work area clean and uncluttered.
  - Never play practical jokes or engage in horseplay.
  - Always use adequate safety measures and never leave the following unattended:
  - Ongoing chemical reactions in laboratories
  - Exposed sharps (needles, razor blades, etc.)
  - Energised electrical, mechanical, or heating equipment
- Chemical storage and inventory:
  - Follow chemical storage and compatibility guidelines.
  - Maintain lean, well-managed chemical inventories to avoid fire code violations and subsequent inventory reduction measures.
- Corridors:
- Keep corridors free of hazardous materials at all times, without exception:

Communicate hazards to everyone entering the facility:
- Post
  - Warning signs near any dangerous equipment, reactions, or conditions,
  - A list of chemical abbreviations (PDF) (Word file) used on chemical container labels (including hazardous waste) near the lab entrance
  - Personal protective equipment requirements for entering the facility, if applicable
- Label all containers.
- Keep containers closed except when in use, including hazardous waste containers:

Follow safe handling practices:
- Evaluate the hazards.
  - Read the MSDS before beginning work with a chemical.
  - Follow SOPs for extremely hazardous materials.
  - Pay particular attention to control measures for chemicals that are known to be extremely hazardous or chemical carcinogens.
- Do not underestimate risk:
  - Never pipette by mouth.
  - Never smell chemicals to identify them.
  - Assume that
- Any mixture will be more hazardous than its most toxic component.
- All substances of unknown toxicity are highly toxic.

GENERAL CHEMICAL SAFETY GUIDELINES (continued)

- Engineering controls.
  - Use chemical fume hoods and other engineering controls as needed.
  - Building vacuum system (Figure 28.1)
  - Never pull liquids, solids, or hazardous gases into the vacuum system.
  - Turn off when not in use.
  - Building vacuum system alert.
- Wear appropriate personal protective equipment.
- Be aware of electrical hazards:
  - Keep electrical panels clearly visible and unobstructed.
  - Know how your circuits are labelled so equipment can be de-energised quickly in an emergency.
  - Never use extension cords as permanent wiring. Unplug them at the end of the workday.
  - Mount multi-plug adaptors a few inches off the floor to avoid possible water damage.
  - Never use multi-plug adaptors in series.
  - Replace any damaged or frayed electrical cords immediately.
  - Do not eat, drink, store food, smoke, or apply cosmetics in areas where chemicals are in use (except in clearly marked clean areas). Wash hands frequently and before eating.

FIGURE 28.1
Building vacuum system alert.

(continued)

GENERAL CHEMICAL SAFETY GUIDELINES (continued)

Follow safe handling practices for specific chemicals:
- See specific chemical guidelines.

Prepare for accidents and emergencies:
- Do not work alone.
- Prepare for spills:
  - Clean up only very small quantities and only if you have been properly trained. All other spills should be cleaned up by specially trained personnel.
  - Keep a fully stocked chemical spill kit easily accessible.
  - Train personnel on how to use the spill kit and when it is safe to do so.
- Know the locations and how to use emergency equipment:
  - Telephones.
  - Emergency guide.
  - First aid kit.
- Fire extinguishers and fire alarm pull stations:
  - Check your fire extinguisher monthly to ensure it is charged and accessible.
  - Eyewash and emergency showers.
  - Dispose of chemical waste promptly and according to guidelines.
  - Disposal of hazardous waste using sinks, intentional evaporation, or as regular trash is against the law. Laboratories must abide by strict state and federal waste disposal requirements. You may be held liable for violations of applicable laws.

Source: University of California, San Diego. General chemical safety guidelines, 2012. http://blink.ucsd.edu/safety/research-lab/chemical/general/index.html#Follow-safe-handling-practices.

EYE PROTECTION IN THE WORKPLACE

What Contributes to Eye Injuries at Work?
- Not wearing eye protection. BLS reports that nearly three out of every five workers injured were not wearing eye protection at the time of the accident.
- Wearing the wrong kind of eye protection for the job. About 40 of the injured workers were wearing some form of eye protection when the accident occurred. These workers were most likely to be wearing eyeglasses with no side shields, though injuries among employees wearing full-cup or flat-fold side shields occurred as well.

EYE PROTECTION IN THE WORKPLACE (continued)

What Causes Eye Injuries?

- *Flying particles*: BLS found that almost 70% of the accidents studied resulted from flying or falling objects or sparks striking the eye. Injured workers estimated that nearly three-fifths of the objects were smaller than a pinhead. Most of the particles were said to be travelling faster than a hand-thrown object when the accident occurred.
- Contact with chemicals caused one-fifth of the injuries. Other accidents were caused by objects swinging from a fixed or attached position, like tree limbs, ropes, chains, or tools, which were pulled into the eye while the worker was using them.

Where Do Accidents Occur Most Often?

Craft work and industrial equipment operation. Potential eye hazards can be found in nearly every industry, but BLS reported that more than 40% of injuries studied occurred among craft workers, like mechanics, repairers, carpenters, and plumbers. Over a third of the injured workers were operatives, such as assemblers, sanders, and grinding machine operators. Labourers suffered about one-fifth of the eye injuries. Almost half the injured workers were employed in manufacturing; slightly more than 20% were in construction.

How Can Eye Injuries Be Prevented?

Always wear effective eye protection: Occupational Safety and Health Administration (OSHA) standards require that employers provide workers with suitable eye protection. To be effective, the eyewear must be of the appropriate type for the hazard encountered and properly fitted. For example, the BLS survey showed that 94% of the injuries to workers wearing eye protection resulted from objects or chemicals going around or under the protector. Eye protective devices should allow for air to circulate between the eye and the lens. Only 13 workers injured while wearing eye protection reported breakage.

Nearly one-fifth of the injured workers with eye protection wore face shields or welding helmets. However, only 6% of the workers injured while wearing eye protection wore goggles, which generally offer better protection for the eyes. Best protection is afforded when goggles are worn with face shields.

Better training and education: BLS reported that most workers were hurt while doing their regular jobs. Workers injured while not wearing protective eyewear most often said they believed it was not required by the situation. Even though the vast majority of employers furnished eye protection at no cost to employees, *about 40% of the workers received no eye safety training on where and what kind of eyewear should be used.*

Maintenance: Eye protection devices must be properly maintained. Scratched and dirty devices reduce vision, cause glare, and may contribute to accidents.

(continued)

EYE PROTECTION IN THE WORKPLACE (continued)

Where Can I Get More Information?

- The OSHA website or your nearest OSHA area office. Safety and health experts are available to explain mandatory requirements for effective eye protection and answer questions. They can also refer you to an on-site consultation service available in nearly every state through which you can get free, penalty-free advice for eliminating possible eye hazards, designing a training programme, or other safety and health matters.
- Don't know where the nearest federal or state office is? Call an OSHA Regional Office at the U.S. Department of Labor in Boston, New York, Philadelphia, Atlanta, Chicago, Dallas, Kansas City, Denver, San Francisco, or Seattle.
- The National Society to Prevent Blindness. This voluntary health organisation is dedicated to preserving sight and has developed excellent information and training materials for preventing eye injuries at work. Its 26 affiliates nationwide may also provide consultation in developing effective eye safety programmes. For more information and a publications catalogue, write the National Society to Prevent Blindness, 79 Madison Ave., New York, NY 10016-7896.

Eye Protection Works!

BLS reported that more than 50% of workers injured while wearing eye protection thought the eyewear had minimised their injuries. But nearly half the workers also felt that another type of protection could have better prevented or reduced the injuries they suffered.

It is estimated that 90% of eye injuries can be prevented through the use of proper protective eyewear.

Source: U.S Department of Labor, OSHA Fact Sheet 92-03.

Bibliography

American Thoracic Society. 1990. Health effects of tremolite. *Am Rev Respir Dis* 142(6):1453–1458.

Bateman ED, Benatar SR. 1987. Asbestos-induced diseases: Clinical perspectives. *Q J Med* 62:183–194.

U.S. Department of Health and Human Services (U.S. DHHS). 2001. Agency for Toxic Substances and Disease Registry (ATSDR). 2001. Toxicological profile for asbestos. U.S. DHHS, ATSDR. Atlanta, GA (updated 2011).

29

Hazardous Chemical Substances: Global Regulations

Introduction

Chemical substances have become an indispensable part of our life, sustaining many of our activities, preventing and controlling many diseases, and increasing agricultural productivity. The benefits are incalculable, but on the other hand, chemicals may endanger our health and poison our environment. The net consumption of chemicals varies among countries, depending on factors such as national economy, industries, and agriculture. New chemicals are synthesised every year and evaluated for their potential advantages over their predecessors and their commercial viabilty. There are about 100,000 chemical substances in present commercial use, and about 2,000 new ones enter the market annually. The chemical scene is in constant flux as old substances are superseded by new ones, and as their effectiveness and demand determine quantities produced. Wherever and whenever chemicals are in use, safety is an issue. In the event of an accident, the correct information needs to be available on the spot. A chemical inventory system should provide details about exactly what chemicals are in stock and where they are. Safety information about those chemicals should be readily available, whether as a material safety data sheet (MSDS) or as customised handling instructions.

There are risks from exposure during production, storage, handling, transport, use, and waste disposal of chemicals because of accidental leakage or illegal dumping, released inappropriately into the living environment. In the event of an accident, safety information (MSDS or customised handling instructions) needs to be readily available. A chemical inventory system should provide details about the chemicals in stock and where they are stored. Chemicals of many kinds and of many categories at different workplaces have become a common situation of societal life and are a global phenomenon. The amount of the chemical may appear very minute, but some chemicals accumulate in the body over long periods. Some chemicals cause harm many years after the exposure. Although exposure duration may be short, exposure may occur frequently and to excessive concentrations. Children, the aged, pregnant women, and those weakened by disease may be more susceptible than the healthy adult.

Chemical Substances, Global Regulatory Systems, and Regulations

Global advances in scientific researches, technology and development, and identification and formulations of newer chemical products for societal development and for the improvement of the quality of life along with economic development have created newer avenues

445

and opportunities. This has necessitated an essential requirement of science-based regulatory system for the management of chemical substances and to arrive at decisions to meet the global regulatory challenges. In this direction, international collaborations and partnerships enhanced newer scientific innovations and greater mobility of a variety of products across the countries. This also has provided increased speed to communicate and share the benefits of technology between governments. The data requirements for the registration and distribution of chemical substances such as pesticides, drugs and pharmaceuticals, cosmetics, food products, and many more have been made uniform.

Global Regulatory Agencies

Global regulatory agencies specially related to chemicals, drugs, and petrochemicals are many and in brief include the following (besides many more):

- U.S. Food and Drug Administration (U.S. FDA)
- U.S. Environmental Protection Agency (U.S. EPA)
- Insecticides Act, 1968. Ministry of Agriculture and Cooperation, Government of India, and the Environment Protection Act, 1986, India
- Pest Management Regulatory Agency (PMRA), Health Canada
- PMRA, U.S. EPA
- Federal Insecticide, Fungicide and Rodenticide Act (FIFRA), United States
- National Toxicology Program (NTP), United States
- National Administration of Drugs, Food, and Medical Technology, Argentina
- The European Agency for the Evaluation of Medicinal Products
- Federal Institute for Drugs and Medical Devices, Germany
- Medicines and Healthcare Products Regulatory Agency (MHRA), United Kingdom
- World Health Organization (WHO), Geneva, Switzerland
- International Conference on Harmonisation (ICH), Geneva, Switzerland

Global Chemical Production

Many other countries manufacture and trade in chemicals, but not on the same scale, or with the broad range of precursor chemicals, as the countries in this section, which is mostly reproduced from the U.S. Department of State 2012 International Narcotics Control Strategy Report.

Argentina

Argentina is one of South America's largest producers of precursor chemicals and remains a source of potassium permanganate. The Government of Argentina (GOA) has banned imports or exports of ephedrine. The GOA has enhanced its precursor chemical regulatory framework and port and border controls and related criminal investigations in combating the traffic in precursor chemicals. Argentina is a party to the 1988 United Nations (UN) Drug Convention and has laws meeting the convention's requirements to track chemicals. Argentina restricted the importation and exportation of ephedrine, both as a raw material and as an elaborated product, in 2008, resulting in a substantial decrease in legal ephedrine imports in both 2009 and 2010. In addition, the GOA has taken steps to implement Commission on Narcotic

Drugs (CND) resolution 49/3. In August 2010, Argentina implemented the International Narcotics Control Board INCB's online Public Education Network (PEN) system.

Brazil

Brazil's chemical industry continues to grow as expanding exports of manufactured products and growing domestic markets have increased the demand for chemicals. One of the world's 10 largest chemical producers and the leader in Latin America, Brazil is also the only country that borders all three Andean cocaine-producing countries. Brazil is a party to the 1988 UN Convention and passed its first chemical control law in 2001 with an updated 2003 decree imposing strict controls on 146 chemicals that could be used to produce narcotic substances. The Brazilian Federal Police (DPF) established regulatory guidelines for all chemical handlers in the country. It (DPF) implemented the National Computerized System of Chemical Control in 2008, which is used to monitor all movements of chemicals in the country, including imports/exports, and licensing. This system requires all companies to use an on-line system for registration and to report all activity being conducted, including the submission of mandatory monthly reports of all chemical related movements. Based on analysis of activity reports, the DPF conducts inspections of suspect companies. Strict restrictions on ether and acetone shipments have caused traffickers to use substitutes for cocaine processing, such as cement and lime. These two materials, as well as kerosene and gasoline, are controlled by Brazil for exports to Bolivia, Colombia, and Peru as essential substances for cocaine production, but are not controlled domestically in Brazil. Brazil uses the PEN to report legitimate exports. Brazil currently controls both potassium permanganate and acetic anhydride for quantities in excess of 1 kg/L and is in the process of drafting new legislation to implement stricter controls. The DPF Chemical Diversion Investigations unit works closely with the United States Government (USG) and with its neighbours to target diversion.

Canada

Canada continues to be a destination and transit country for the precursor chemicals used to produce synthetic drugs, particularly methamphetamine and ecstasy. According to the 2010 annual report of the Criminal Intelligence Service of Canada (CISC), Canadian-sourced pseudoephedrine has been found in raids of clandestine U.S. methamphetamine laboratories. Though methamphetamine use in Canada has stabilised, according to CISC, production has continued to increase to supply export markets. CISC asserts that criminals export significant quantities of methamphetamine to the United States, Japan, Australia, and New Zealand. Canadian officials find that smugglers move ecstasy precursor chemicals into Canada from source countries to include China and India. The United States works closely with Canada to target precursor chemicals and to identify and dismantle methamphetamine laboratories. Canada participates actively in a large annual conference, the National Methamphetamine and Pharmaceuticals Initiative, that brings together law enforcement officials, regulators, scientists and health professionals, prosecutors, and policymakers to focus on the diversion and illicit production and trafficking of precursor chemicals (as well as controlled prescription drugs). Canadian officials confirm, however, that domestic production of methamphetamine and ecstasy continues to increase. U.S. officials continue to work closely with Canadian partners to identify and dismantle ecstasy and methamphetamine laboratories. Canada is a party to the 1988 UN Convention and complies with its record-keeping requirements. Canada participates in Project Prism, targeting synthetic drug chemicals, and is a member of the North American working group. It also supports Project Cohesion.

Chile

Chile has a large petrochemical industry engaged in the manufacturing, importation, and exportation of chemical products. Although it has been a source of ephedrine for methamphetamine processing in Mexico, no ephedrine has been seized by Chilean counterparts since 2009. Chile is also a potential source of precursor chemicals used in coca processing in Peru and Bolivia. Despite Chile's chemical control laws, monitoring of diversion and smuggling is limited by the bureaucratic structure, lack of efficient registration system, and lack of sufficient personnel. The regulatory function for chemicals belongs to the Special Register of Controlled Chemical Handlers (REUSQC) under the Ministry of Interior. Chilean law enforcement entities have specialised chemical diversion units and dedicated personnel assigned with the responsibility for investigating chemical and pharmaceutical diversion cases. Customs, which is not a traditional law enforcement agency, has a risk analysis unit that profiles suspicious imports and exports, which may include chemical precursors.

Companies that import, export, or manufacture chemical precursors must register with REUSQC and maintain customer records and are subject to inspections. Through 2011, approximately 200 importers and exporters had registered with the Government of Chile, but there are potentially many more companies who should be registered, but are not, due to inefficiency within the registration system. There is legislation pending in the Chilean Congress to expand the list of companies subject to inspection, and Chilean authorities continue to work with the United States. The majority of chemical imports originate in India and China and the diversion of such chemicals is primarily directed to Bolivia, Peru, and Mexico. Chemicals destined for Peru and Bolivia are transported by land, while chemicals sent to Mexico are transported via air cargo and maritime shipments.

Mexico

Significant methamphetamine production continues in Mexico and importations of precursor chemicals are on the rise. During 2011, the quantity of precursor chemicals seized as reported by the Government of Mexico (GOM) has totalled over 527,077 kg or more than 527 metric tons. A strong bilateral working relationship between U.S. and Mexican authorities continues, involving information exchange and operational cooperation, through participation in the National Methamphetamine and Pharmaceuticals Initiative conference. The two governments also cooperate to convey best practices to Central American countries that have become affected by the trafficking of precursor chemicals. Mexico is a party to the 1988 UN Drug Convention and has laws and regulations that meet the convention's chemical control requirements. GOM outlawed imports of pseudoephedrine, except for liquid pseudoephedrine for hospital use, in 2008. In November 2009, Mexico enhanced its regulatory laws pertaining to the import of precursor chemicals, which tightened the regulations for imports of phenyl acetic acid, its salts and derivatives, methylamine, hydriodic acid, and red phosphorous. In June of 2010, the GOM again strengthened its regulatory laws. However, potassium permanganate and acetic anhydride are not regulated in Mexico.

Mexico has several major chemical manufacturing and trade industries that produce, import, or export most of the chemicals required for illicit drug production, including potassium permanganate (for cocaine) and acetic anhydride (for heroin). While Mexico is a major supplier of methamphetamine, the country currently has no facilities or chemical plants that can synthesise or manufacture pseudoephedrine or ephedrine powder. Imports of both precursor and essential chemicals are limited by the GOM to specific ports of entry. Mexico has a total of 49 ports of entry, of which only 17 are authorised for the importing of essential chemicals. Mexican authorities have also detected shipments entering Mexico from the United States. The import, export, and trade of PAA are regulated according to an agreement issued by Mexico's Health Secretariat in 2009. In May 2010, officials seized 88 tons of ethyl phenylacetate, a pre-precursor chemical used to make phenyl-2-propanone (P2P), a precursor to methamphetamines, at the Port of Manzanillo, representing the largest single seizure of the chemical. The chemical was found in five shipping containers sent from China.

United States

The United States manufactures and/or trades in all 23 chemicals listed in Tables I and II of the 1988 UN Drug Convention. It is a party to the 1988 UN Convention and has laws and regulations meeting its chemical control provisions. The basic U.S. chemical control law is the Chemical Diversion and Trafficking Act of 1988. This law and subsequent chemical control amendments were all designed as amendments to U.S. controlled substances laws, rather than stand-alone legislation. The Drug Enforcement Administration (DEA) is responsible for administering and enforcing them. The Department of Justice, primarily through its U.S. Attorneys' Offices, handles criminal prosecutions and cases seeking civil penalties for regulatory violations. In addition to registration and record-keeping requirements, the legislation requires traders to file import/export declarations at least 15 days prior to shipment of regulated chemicals. The DEA uses the 15-day period to determine if the consignee has a legitimate need for the chemical. Diversion investigators and special agents work closely with exporting and receiving country officials in this process. If legitimate end use cannot be determined, the legislation gives DEA the authority to stop shipments. One of the main goals of DEA's Diversion Control Program is to ensure that U.S. registrants' (those companies registered with DEA to handle List I chemicals) products are not diverted for illicit drug manufacture. U.S. legislation also requires chemical traders to report to DEA suspicious transactions such as those involving extraordinary quantities or unusual methods of payment. Close cooperation has developed between the U.S. chemical industry and DEA in the course of implementing the legislation.

The United States has played a leading role in the design, promotion, and implementation of cooperative multi-lateral chemical control initiatives. The United States also actively works with other concerned nations, and with the United Nations Office on Drugs and Crime (UNODC), and the INCB to develop information-sharing procedures to better control precursor chemicals, including pseudoephedrine and ephedrine, the principal precursors for methamphetamine production. U.S. officials participate in the task forces for both Project Cohesion and Project Prism. The United States has established close operational cooperation with counterparts in major chemical manufacturing and trading countries. This cooperation includes information sharing in support of chemical control programmes and in the investigation of diversion attempts.

Asian Countries

China

China's chemical industry, with an estimated 80,000 individual chemical companies in 2009, presents widespread opportunities for chemical diversion. Effective regulatory oversight of this industry remains a major challenge for China's central authorities. China produces and monitors all 23 of the chemicals on the tables included in the 1988 Drug Convention. China continues to cooperate with the United States and other concerned countries in implementing a system of PEN for dual-use precursor chemicals. China regulates the import and export of precursor chemicals covered by the 1988 UN Convention, but does not currently notify other countries of non-regulated chemicals on the INCB's surveillance list. Chinese authorities successfully investigated 234 cases of illegal trade and smuggling of precursor chemicals in 2010 and seized 869.11 tons of precursor chemicals.

India

India is one of the world's largest manufacturers of precursor chemicals and in 2010 was the top exporter of both ephedrine (65,000 kg) and pseudoephedrine (458,000 kg). India is a party to the 1988 UN Drug Convention, but it does not have controls on all the chemicals listed in the convention. The Narcotic Drug and Psychotropic Substances Act requires records on all transactions of acetic anhydride, ephedrine, and pseudoephedrine. Exports of ephedrine and pseudoephedrine require a 'No Objection' Certificate from the Indian Narcotics Commissioner, who issues a PEN to the competent authority in the importing country as well as the INCB. India continues to work closely with the INCB and with international partners. India is, nevertheless, a key source of diverted precursor chemicals for methamphetamine and heroin. Seizures in South and Central America continue to indicate that traffickers are targeting India. And several large shipments of ephedrine and pseudoephedrine tablets were seized in Mexico. Large shipments of bulk pseudoephedrine from India were formed into tablets in Bangladesh and sent to countries in Central America and the Caribbean.

To exercise all the powers vested under all act and rules pertaining to protection of environment and control of pollution, Republic of India enacted several legislations, rules, and regulations. These are implemented and enforced in all environment legislation agency activities. The selected acts and regulations include, but are not limited to, the following:

- Drugs and Cosmetics Act, 1940
- The Prevention of Food Adulteration Act, 1954
- Insecticides Act, 1968
- The Water (Prevention and Control of Pollution) Act, 1974
- Air (Prevention and Control of Pollution) Act, 1981
- Narcotic Drugs and Psychotropic Substances Act 1985
- The Environment (Protection) Act, 1986
- Manufacture, Storage and Import of Hazardous Chemicals Rules, 1989
- Hazardous Wastes (Management and Handling) Rules, 1989
- Water Prevention and Control of Pollution Act, 1974
- Hazardous Waste (Management and Handling) Rules, 1989

- Manufacture, Storage and Import of Hazardous Chemical Rules, 1989
- Rules for manufacture, use, import and storage of Hazardous Microorganisms, genetically Engineered Micro-organism or Cells, 1989
- Bio-medical Waste (Management and Handling) Rules, 1998
- The Recycled Plastic Manufacture and Usage Rules, 1999
- The Ozone Depleting Substances (Regulation and Control) Rules, 2000
- The Noise Pollution (Regulation and Control) Rules, 2000
- The Batteries (Management and Handling) Rules, 2001

Republic of Korea

In 2010, South Korea was the third largest importer of ephedrine and the second largest importer of pseudoephedrine. With one of the most developed commercial infrastructures in the region, the Republic of Korea (ROK) is an attractive location for criminals to obtain precursor chemicals. As of 2011, 30 precursor chemicals were controlled by Korean authorities. Both the Korea Customs Service (KCS) and the Korean Food and Drug Administration (KFDA) participate in the INCB's Projects Cohesion and Prism. In this role, they closely monitor imports and exports of precursor chemicals, particularly acetic anhydride. Korean law enforcement authorities also cooperate with Southeast Asian nations to verify documents and confirm the existence of importing businesses and send representatives to the region to investigate. In April 2011, the National Assembly passed a new law that requires manufacturers and exporters of precursor chemicals to register with the government and provides education to Korean businesses to prevent them from unknowingly exporting such chemicals to fraudulent importers. South Korean authorities work closely with the U.S. authorities to track suspect shipments.

Singapore

In 2010, Singapore's exports and imports of both ephedrine and pseudoephedrine decreased slightly. However, in 2010, Singapore was ranked the fifth largest importer of ephedrine and the fourth largest importer of pseudoephedrine (ranked first in 2009). Authorities indicate that the amounts not re-exported are used primarily by the domestic pharmaceutical industry and by the large number of regional pharmaceutical companies served by Singapore's largest port. Singapore is one of the largest distributors of acetic anhydride in Asia. Used in film processing and the manufacture of plastics, pharmaceuticals, and industrial chemicals, acetic anhydride is also the primary acetylating agent for heroin. Singapore participates in a multi-lateral precursor chemical control programme, including Operation Cohesion and Operation Prism, and works closely with the USG. Singapore controls precursor chemicals, including pseudoephedrine and ephedrine, in accordance with the 1988 UN Drug Convention provisions and accordingly tracks exports and works closely with industry officials.

Taiwan

In 2010, Taiwan was the fourth largest exporter of ephedrine and ranked as the third largest exporter of pseudoephedrine. Taiwan was also ranked the fourth largest importer of ephedrine in 2010. Taiwan law enforcement has long recognised that certain Taiwan-based chemical companies divert chemicals, which may be used to manufacture illicit substances in countries such as Cambodia, Thailand, Mexico, Honduras, and Belize. The Ministry of Economic Affairs' Industrial Development Bureau serves as the regulatory agency for chemicals, including those controlled under the 1988 UN Convention, and other non-regulated chemicals. In 2011, Taiwan exported 848 kg of acetic anhydride worldwide. Taiwan does not have control regulations for the trade of ephedrine/pseudoephedrine combination in over-the-counter pharmaceutical preparations. However, companies engaging in their import/ export must register their transactions with the Taiwan's Department of Health, which may elect to examine relevant shipping records. Taiwan does have control regulations for the export and import of ephedrine and pseudoephedrine. In 2011, Taiwan's Ministry of Economic Affairs added eight new precursor chemicals to the control list. Taiwan's law enforcement agencies work closely with U.S. law enforcement officials.

Sri Lanka

Control of Pesticides Act No. 33 of 1980 regulates the import, packing, labelling, storage, formulation, transportation, safety, and use of all pesticides in Sri Lanka. The provisions of the act include the following for the purpose of preventing the misuse or abuse of pesticides that are known to cause health hazard to humans and the environment: (a) appointment of licensing authority (Registrar of Pesticides); (b) appointment of pesticide technical and advisory committee; (c) appointment of authorised officers for enforcement of regulations; (d) designation of authorised analysts; (e) control of imports; (f) requirements for packaging, labelling, advertisement, marketing, and quality control; (g) requirements for storage and transport; (h) requirement on preharvest intervals and residue limits; and (i) penalties for violators of the act and the regulations; *Cosmetics, Devices and Drugs Act, No. 27 of 1980, Sri Lanka.*

Thailand

Thailand is not a chemical manufacturer or producer, but the government imports chemicals in bulk for licit domestic requirements. Thai officials are concerned by a dramatic increase in pseudoephedrine diversion, the diversion of pseudoephedrine preparation of 60 mg tablets manufactured in South Korea identified in late 2009. Thailand has taken a proactive stance in precursor chemical control and has laws to regulate and control precursor chemicals. In 1975, the Psychotropic Substances Act was passed wherein drugs and chemicals were placed into four categories or schedules similar to the drug schedule developed in the United States. Pseudoephedrine and ephedrine, for example, are listed as Type II controlled drugs under the 1975 Psychotropic Substances Act. Drugs listed under Type II can pose serious health risks and include strict controls. Thailand submits information to the INCB on the country's illicit trade and legitimate uses of and requirements for substances. During 2010, the Thai Office of Narcotics Control Board (ONCB) reported the seizure of more than 33 million pseudoephedrine tablets destined for Thailand's neighbouring countries. As of June 2011, the Thai ONCB had reported the seizure of 7.5 million pseudoephedrine tablets. As a result, Thai FDA and ONCB officials are working with parliament to adopt stricter import regulations and controls for ephedrine and pseudoephedrine pharmaceutical products.

Europe

Chemical diversion control within the European Union (EU) is regulated by EU regulations binding on all 27 member states. The regulations are updated regularly, most recently in 2005. The EU regulations meet the chemical control provisions of the 1988 UN Convention, including provisions for record-keeping on transactions in controlled chemicals, a system of permits or declarations for exports and imports of regulated chemicals, and authority for governments to suspend chemical shipments. The EU regulations are directly applicable in all member states. Only a few aspects require further implementation through national legislation, such as law enforcement powers and sanctions. The EU regulations govern the regulatory aspects of chemical diversion control and set up common risk management rules to counter diversion at the EU's borders. Member states are responsible for investigating and prosecuting violators of national laws and creating regulations necessary for implementing the EU regulations.

The U.S.–EU Chemical Control Agreement, signed 28 May 1997, is the formal basis for U.S. cooperation with the European Commission and EU member states in chemical control through enhanced regulatory cooperation and mutual assistance. The agreement calls for annual meetings of a Joint Chemical Working Group to review implementation of the agreement and to coordinate positions in other areas. The annual meeting coordinates national or joint positions on chemical control matters before larger multi-lateral fora, including the CND.

Bilateral chemical control cooperation continues between the United States and EU member states. Many states participate in voluntary initiatives such as Project Cohesion and Project Prism. In 2007, the EU established guidelines for private sector operators involved in trading in precursor chemicals, with a view to offering practical guidance on the implementation of the main provisions of EU legislation on precursor chemicals, in particular the prevention of illegal diversion.

Germany and the Netherlands, with large chemical manufacturing or trading sectors and significant trade with drug-producing areas, are considered the major European source countries and points of departure for exported precursor chemicals. Other European countries have important chemical industries, but the level of chemical trade with drug-producing areas is not as large and broad scale as these countries. Belgium and the United Kingdom are also included this year because of their large exports of ephedrine and pseudoephedrine.

Belgium

Belgium is not a major producer of illicit drugs or chemical precursors used for the production of illicit drugs. However, Belgium has a substantial pharmaceutical product sector that manufactures ephedrine and pseudoephedrine for licit products to a very limited extent. Belgium has reporting requirements for the import and export of precursor chemicals (bulk pseudoephedrine/ephedrine). Shipments of pharmaceutical preparations containing pseudoephedrine and ephedrine are only controlled on a regulatory level by the Belgian Ministry of Safety and Public Health.

Seeking to circumvent the ban on pseudoephedrine and ephedrine by Mexico and many countries in Central America, traffickers are extracting these substances from pharmaceutical preparations or cold medicine from Belgium for methamphetamine production. As a result, Belgium and other Western European countries have seen an increase in transshipments of ephedrine and other methamphetamine precursors embodied in uncontrolled pharmaceutical preparations. The United States continues to coordinate with Belgian authorities to identify and investigate both suppliers and shippers of precursor chemicals.

Germany

Germany continues to be a leading manufacturer of licit pharmaceuticals. In 2010, Germany was the second largest exporter of ephedrine and pseudoephedrine worldwide. Most of the 23 scheduled substances under international control as listed in Tables I and II of the 1988 UN Drug Convention and other chemicals, which can be misused for the illicit production of narcotic drugs, are manufactured and/or sold by the German chemical and pharmaceutical industry. Germany is a party to the 1988 UN Convention. In Germany, the National Precursor Monitoring Act complements EU regulations. Although Germany's developed chemical sector makes it susceptible to chemical diversion. National and EU regulations, law enforcement action, and voluntary industry action tightly control the movement of chemicals throughout the country. In 2010, the number of cases regarding acetic anhydride decreased from 2009. Cooperation between the chemical/pharmaceutical industry, merchants, and investigation authorities is a key element in Germany's chemical control strategy. Germany works very closely with UNODC in the field of drug control.

The United States works closely with Germany's chemical regulatory agency, the Federal Institute for Drugs and Medical Devices, on chemical control issues, including exchanging information and cooperating both bilaterally and multi-laterally, to promote transnational chemical control initiatives. Germany supports INCB precursor chemical control activities and continues to participate in significant UN projects. Germany was recognised for its contributions to Operation PAAD, a Project Prism initiative that focused on shipments of phenylacetic acid and its derivatives that are being used to illicitly produce P2P in clandestine laboratories.

The Netherlands

Drug traffickers continue to target the Netherlands, which has a large chemical industry with large chemical storage facilities, and use Rotterdam as a major chemical shipping port. However, the Netherlands has strong legislation and regulatory controls and the police force tracks domestic shipments and works closely with its international partners. The Netherlands is a party to the 1988 UN Drug Convention and 1990 EU regulations. Trade in precursor chemicals is governed by the 1995 Act to Prevent Abuse of Chemical Substances (WVMC). The law seeks to prevent the diversion of legal chemicals into the illegal sector. The National Crime Squad's synthetic drug unit and the public prosecutor's office have strengthened cooperation with countries playing an important role in precursor chemicals used in the manufacture of ecstasy. The Netherlands signed an MOU with China concerning chemical precursor investigations. The Netherlands is an active participant in the INCB/Project Prism task force. The Dutch continues to work closely with the United States on precursor chemical controls and investigations. In April 2009, the Netherlands established a separate Expertise Center on Synthetic Drugs and Precursors (ESDP) and a precursor task force. Trade and industry report suspect transactions of registered chemicals, and in 2010, there were 82 investigations of suspicious transaction reports, up from 59 in 2009. The latest ESDP report indicated two new types of precursors: PMK-glycidate and alpha-phenylacetoacetonitrile (APAAN) used in methamphetamine production. Criminal groups use these new types of precursors and preprecursors to circumvent national and international legislation.

The United Kingdom

In 2010, the United Kingdom was the fifth largest worldwide exporter of ephedrine. The United Kingdom strictly enforces national precursor chemical legislation in compliance with EU regulations and is a party to the 1988 UN Drug Convention. In 2008, the Controlled Drugs Regulations (Drug Precursors) (Intra- and External Community Trade) were implemented, bringing UK law in line with pre-existing EU regulations. Licensing and reporting obligations are requirements for those that engage in commerce of listed substances, and failure to comply with these obligations is a criminal offence. The Home Office Drug Licensing and Compliance Unit is the regulatory body for precursor chemical control in the United Kingdom. However, the Serious Organized Crime Agency (SOCA) and the police have the responsibility to investigate suspicious transactions. Revenue and Customs monitors imports and exports of listed chemicals.

Process Safety Management System

In 1980, the International Programme on Chemical Safety (IPCS) was jointly established by three cooperating organisations – WHO, ILO, and UNEP – to govern activities related to chemical safety. Under the executive authority of the WHO, the main role of the IPCS is to establish the scientific basis for chemical safety and to strengthen national capabilities and capacities for it. Any safety management system is a constituent of the overall management system of an organisation/industry/establishment. The implementation of the guidelines set out by an organisation/industry should reflect in the overall management philosophy and system to achieve the goal. In the present context, the book discusses in its own limited manner only the safety management system with special reference and emphasis on hazardous chemical substances and health of users and occupational workers, the environment, quality, and workplaces, to develop and to cover major workplace accidents because of improper use/misuse/negligence during handling of hazardous chemical substances and the possible controls and preventive steps thereof.

The very first step to achieve a meaningful success of safety management is that the workplace manger and team manager should establish and implement written procedures (SOPs) to maintain the ongoing integrity of handling of hazardous chemical substances and process equipment. Training to handle hazardous chemical substances that the workers use and work with must include the following:

- Appropriate work practices in the safe handling and use of chemical substances.
- Prevention of injury from mixing incompatible chemical substances, for instance, mixing bleach with an ammonia cleaning product.
- Proper dilution of concentrated chemical substances.
- Appropriate labelling and safe storage of chemical substances. Occupational workers must know proper methods of use of personal protective equipment (PPE) such as (1) gloves, (2) rubber apron or protective clothing, (3) safety glasses with side shields, (4) splash goggles, (5) face shield, and (6) non-slip safety shoes.
- Workplace should include qualified personnel who know how to manage any kind of chemical spills and methods of chemical cleanup.
- Workplace should have spill kits designed specifically for the workplace and the chemicals in use there. Spill kits may include (1) materials to absorb liquids such as a chemical spill powder, (2) an absorbent material such as a chemical spill pad,

(3) a neutralising agent, (4) waste containers, (5) a brush and scoop, (6) PPE, and (7) other products that are necessary to neutralise on-site chemicals.

- Hazardous material spills can endanger workers, workplace, and the environment. Spills of chemical substances contaminate and/or destroy property if not managed properly. Stocking up on hazmat socks is the responsible solution to the problem of potential hazardous spills or leaks.

- Occupational workers should follow the prescribed safety precautions during use/handling of hazardous chemical substances at their respective workplace. There are requirements for PPE, spill response, and disposal of wastes that each worker should also know.

According to OSHA, 'unexpected releases of toxic, reactive, or flammable liquids and gases in processes involving highly hazardous chemicals (HHCs) have been reported for many years in various industries that use chemicals with such properties. Regardless of the industry that uses these HHCs, there is a potential for an accidental release any time they are not properly controlled, creating the possibility of disaster.' An effective process safety management (PSM) programme can help prevent releases and prepare for emergency response in the event of a chemical release.

To help ensure safe and healthful workplaces, the Occupational Safety and Health Administration (OSHA) has issued the PSM of HHC standard that contains requirements for the management of hazards associated with processes using HHCs.

PSM is intended to prevent an incident like the 1984 Bhopal disaster. A process is any activity or combination of activities including any use, storage, manufacturing, handling, or the on-site movement of HHCs. A process includes any group of vessels that are interconnected or separate and contain HHCs that could be involved in a potential release. A PSM incident is the 'Unexpected release of toxic, reactive, or flammable liquids and gases in processes involving highly hazardous chemicals'. Incidents continue to occur in various industries that use highly hazardous chemicals, which exhibit toxic, reactive, flammable, or even explosive properties, or may exhibit a combination of these properties. Regardless of the industry that uses these highly hazardous chemicals, there is a potential for an accidental release any time they are not properly controlled by a properly designed PSM programme. This, in turn, creates the possibility of disaster.

All industrial facilities must comply with OSHA's PSM regulations as well as the quite similar U.S. EPA Risk Management Program (RMP) regulations. The Center for Chemical Process Safety Management (CCPS) of the American Institute of Chemical Engineers (AIChE) has published a widely used book that explains various methods for identifying hazards in industrial facilities and quantifying their potential severity.

Hazardous chemical releases pose a significant threat to workers. The key provision of PSM is process hazard analysis (PHA), a careful review of what could go wrong and what safeguards must be implemented to prevent releases of hazardous chemicals. The following references help begin a PHA by recognising process hazards. Responsible care management system should confirm that all participating

- Workers should know the hazardous chemicals in the work area.
- Workers should know the location of any plan to handle the hazardous chemical substances and MSDs.
- Workers should know and understand any labelling system, including the MSDS system, and how to use the appropriate hazard information.

- Workers should know how to detect the presence or release of hazardous chemical substances.
- Occupational workers should know that the containers/packages contain hazardous substances and should know the information on health hazards, the characteristics of chemical substances they may encounter in the workplace, and protective measures and precautions for the safe handling, use, and storage of each chemical substance.

It should be well understood by all, meaning students, occupational workers, and general public, *that no chemical is safe, but there are safe methods to use and handle* different chemical substances. To state again, occupational workers should always remember that there are no non-hazardous chemicals, but there are several non-hazardous methods and procedures to use and handle chemicals for betterment and societal development.

Hazardous chemicals do cause acute and chronic health disorders – headaches, rashes, skin burns, respiratory problems, lung and liver damage, reproductive damage, cancer, and fatal injury. Also, the physical hazards of chemicals include flammability, burning, *fire and explosion*, and reactivity. flammable liquids/solids, combustible liquids, compressed gases, explosive materials, unstable materials, water reactive materials.

Safety precautions to be strictly followed by occupational workers and students:

- Workers should wear the proper PPE.
- Workers should check labels prior to use and handling for hazard warnings.
- Workers should read the labels on the container/package.
- Workers should discuss with the workplace supervisor and treat unlabeled containers as dangerous.
- Workers should not remove or destroy labels – keep the labels safe.
- Workers should look for National Fire Protection Association (NFPA) labels: (a) blue is toxic, (b) red is flammable, (c) yellow is reactive, and (d) numbers 3 and 4 are very hazardous.
- Workers should use MSDS for more information: (i) physical characteristics, (ii) fire and explosion data, (iii) dangerous properties, (iv) reactivity data, (v) precautions, (vi) strictly follow and use appropriate and recommended type of PPE – gloves, aprons or full-body suits, safety glasses, goggles, face shields, respirators, dust mask, head protection, and foot protection.
- Workers *should NOT use a chemical without first reading its MSDS.*

The MSDS must contain the following:

- The identity that is used on the container label.
- The chemical and common name of all ingredients.
- The physical and chemical characteristics of the hazardous components.
- The primary routes of exposure to the chemical substance.
- Whether or not listed as a hazardous chemical by the regulatory authorities (ACGIH, IARC, OSHA, NIOSH, NTP).

- Identity of the hazardous chemicals, with appropriate hazard warnings.
- Safety precautions to be observed by workers during use, handling, and waste disposal and during accidental workplace accidents, fire, explosion hazard, chemical spill/leak, and cleanup.
- Important control measures and the emergency first-aid procedures.
- Workers should keep *readily available address of the supplier, manufacturer/importer, of the chemical substance.*
- Occupational workers should contact the facility supervisor before planning to execute the work with which he or she is not familiar with.
- Occupational workers should be aware of the compatibility between chemicals and containers, because some chemicals cannot be used with plastic beakers, while some other chemicals such as HF cannot be used with glass beakers; he/she should not cross-contaminate beakers.
- Occupational workers should always use the fume hood while working with chemicals; all kinds of acid and base work must be done in an exhausted fume hood.
- Workers should store acids in the cabinet labelled 'ACIDS' and bases in the cabinet labelled 'BASES'. And workers should *never work with acid and bases side by side* and should avoid violent reactions. And workers should *never* pour any chemical wastes into the drain.

Students and occupational workers should be very careful about the disposal of multi-hazardous chemical wastes and should consult the workplace supervisor/manager. Multi-hazardous and mixed waste must be secured or held under constant surveillance to prevent unauthorised removal or access.

Mixed and hazardous chemical waste containing radioactive material must only be stored in laboratories posted for use of radioactive material and NEVER place mixed waste in corridors (even while awaiting pickup). Workers should ensure that all waste containers are closed securely to prevent leaks, spills, or escape of vapours.

Workers and the management should strictly follow local and international regulations during the disposal of the multi-hazardous chemical wastes within 60 days of the collection start date.

Multi-hazardous chemical wastes include, but are not limited to, the following:

- Aqueous radioactive wastes containing chloroform or heavy metals
- Methanol/acetic acid solutions from electrophoresis procedures containing radioactive material
- Hazardous liquid scintillation counting fluids with radioactive content
- Radioactive trichloroacetic acid solutions
- Phenol/chloroform mixtures used to extract DNA from radiolabelled cells
- Vacuum pump oil contaminated with radioactive material
- Chemical or radioactive wastes containing infectious agents
- Used animal bedding contaminated with at least two of the earlier listed hazard types (chemical, radioactive, and infectious)
- Lead contaminated with radioactive material
- Aqueous radioactive liquids with pH

In brief, the basic elements of safety management of hazardous chemicals could be described as follows.

Workplace Monitoring and Follow-Up Actions

- The auditor and the group should review the information obtained from regular inspections and identify the immediate corrective actions needed. Inspection report should show the following:
- Priorities for corrective action
- Need for improving safe work practices
- Possible reasons about accidents occurring in particular areas
- Any kind of negligence in proper workplace supervision or training of workers in certain areas or equipment

The health and safety committee should review the progress of the recommendations, especially on the education and training of actual workers. It is also the committee's responsibility to study the information from regular inspections. This will help in identifying trends for the maintenance of an effective health and safety programme:

- Essential inputs for meaningful workplace inspection
- Information on chemicals in use
- Storage areas
- Workforce size, shifts, and supervision
- Workplace rules and regulations
- Work procedures and safe work practices
- Manufacturer's specifications on chemicals in use
- Details of PPE
- Engineering controls
- Emergency procedures – fire, first aid, and rescue
- Accident and investigation reports
- Worker complaint reports regarding particular hazards in the workplace
- Recommendations of the health and safety committee
- Previous inspections – details
- Maintenance reports, procedures, and schedules
- Regulator inspection reports or other external audits of specialist
- Monitoring reports – details of chemical, physical, or biological hazards
- Reports of unusual operating conditions
- Names of inspection team members and any technical experts assisting

For more information, readers should refer to the sources/references.

Bibliography

Centers for Disease Control and Prevention (CDCP). International Chemical Safety Cards (ICSC), Atlanta, GA.

International Agency for Research on Cancer (IARC). 2008. *IARC Monographs on the Evaluation of Carcinogenic Risks to Humans*. IARC, Lyon, France.

Moran, L. and T. Masciangioli (eds.). 2011. *Chemical Laboratory Safety and Security: A Guide to Prudent Chemical Management*. The National Academies Press, Washington, DC.

Patnaik, P. 2007. *A Comprehensive Guide to the Hazardous Properties of Chemical Substances*, 3rd edn. John Wiley & Sons, New York.

1991. *Patty's Industrial Hygiene and Toxicology*, 3rd edn., 4 vols. Wiley-Interscience, New York.

United Nations. 2007. *Globally Harmonized System of Classification and Labeling of Chemicals (GHS)*, 2nd Revised Edition. UN, New York.

U.S. Department of Labor. 1997. Process safety management of highly hazardous chemicals (§1926.64). OSHA Office of Training and Education, Washington, DC. http://www.osha.gov/doc/outreachtraining/htmlfiles/psm.html.

U.S. Environmental Protection Agency (U.S. EPA). Clean Air Act. Washington, DC.

U.S. Environmental Protection (U.S. EPA). Drug Safety Act. Washington, DC.

U.S. Environmental Protection Agency (U.S. EPA). *Advanced Chemical Safety*. San Diego, CA.

U.S. Environmental Protection Agency (U.S. EPA). Toxic Substances Control Act (TSCA).

World Health Organization (WHO). *Safety and Health in the Use of Chemicals at Work*. International Programme on Chemical Safety (IPCS). WHO, Geneva, Switzerland.

30

Hazardous Chemical Substances and Safety Management System

Introduction

Exposure to chemical substances of many kinds and at many categories of workplaces has become a common situation of societal life and is a global phenomenon. Whether it causes short-term or chronic harm to us depends on the quantity, duration, and frequency of exposure and the toxicity, as well as the sensitivity of the individual. The amount may be minute, but some chemicals accumulate in the body over long periods. Some hazardous chemicals have been known to cause delayed and deleterious effects many years after the exposure. Although the duration and period of exposure may be short, exposure may occur frequently and in excessive concentrations. Children, the aged, pregnant women, and those weakened by disease may be more susceptible than a healthy adult.

The growth of chemical industries, in developing as well as developed countries, is predicted to go on increasing for the next century. Chemical safety, meaning control, prevention, and management of chemical hazards, is essential for the healthy growth and advantage of the society and of the living environment.

At the United Nations Conference on Environment and Development (UNCED) held in Rio de Janeiro, Brazil, in June 1992, representatives of more than 150 countries adopted at the highest political level Agenda 21 – an action plan to guide national and international activities for the years to come. A specific chapter of Agenda 21 is devoted to the 'Environmentally sound management of toxic chemicals including prevention of illegal international traffic in toxic and dangerous products'. UNCED recognised that many countries lack the scientific knowledge of judging the impact of toxic chemicals on human health and the environment. As a result, and all too often, toxic chemicals are being produced, transported, used, and disposed of without taking the necessary precautions to prevent chemical contamination and grave damage to human health and the environment.

Any safety management system (SMS) is a constituent part of the overall management system of an organisation/industry/establishment. The implementation of the guidelines set out by an organisation/industry should reflect the overall management philosophy and system to achieve the goal. In the present context, the book discusses in its own limited manner only the SMS with special reference and emphasis on hazardous chemical substances and health of users and occupational workers, the environment, quality, and workplaces. This is to develop and to cover major workplace accidents because of improper use/misuse/negligence of hazardous chemical substances during handling (use, mixing, storage, and waste disposal) and the possible controls and preventive steps thereof.

Employers have many important roles in the worker's compensation system:

- Employers are responsible for maintaining a safe workplace and working in partnership with workers and the WCB to prevent workplace injuries and illnesses from happening.
- If a workplace injury occurs, the employer must understand what is required during the claim process and assist the worker to return to work safely.
- Management/employers pay premiums which are used to pay benefits to workers who are injured or become ill in the workplace. It is important that the employer maintains registration and coverage.
- Management should train the awareness of occupational workers regarding the safety and health at workplaces.
- Management should promote a safety culture of workplace injury and prevention.
- Management should develop and maintain occupational health and safety standards.

Workplace Safety

Workplace safety is a joint responsibility of all individuals to create a healthy and safe workplace:

- Hazards of chemical substances exist and occur everywhere in the workplace. Students and occupational workers should act with safety in mind all day, every day.
- Establish and clarify the safety rights and responsibilities in all workplaces.

However, it is the responsibility of management to

- Educate workers and employers in building and sustaining healthy and safe workplaces.
- Intervene when safety responsibilities are not carried out.
- Provide information and resources on a variety of safety topics and to prevent workplace injury and illness.

An SMS and its meaningfulness depend on the accurate functioning of the management system, the associated personnel, and the compliance of set regulations. The working elements, in brief, include, but are not limited, to the following:

- Organisation and personnel
- Identification and evaluation of major hazards associated with the candidate chemical substance
- Management change, if any
- Planning for emergencies
- Monitoring performance and frequency of monitoring (active monitoring should include inspections of safety critical plant, equipment, and instrumentation as well as assessment of compliance with training, instructions, and safe working practices)

Here are some important questions to consider in the formation of a monitoring programme:

1. Why are workplace inspections important?
2. What is the purpose of inspections?
3. How do you plan for inspections?
4. What types of hazards do we look for in a workplace?
5. What type of information do I need to complete an inspection report?
6. Are there other types of inspection reports that may be useful?
7. Should supervisors be on the inspection team?
8. How long should an inspection take to do?
9. How frequent should inspections be done?
10. How are inspections actually done?
11. What should the final report have in it?
12. What should I know about follow-up and monitoring?
13. Example of workplace inspection report.

An SMS also includes several other important inputs. The following could be listed:

In addition to the routine monitoring of performance, the operator should carry out periodic audits of its SMS as a normal part of its business activities. An audit should determine whether the overall performance of the SMS conforms to requirements, both external and those of the operator. The results of these audits should be used to decide what improvements should be made to the elements of the SMS for a meaningful implementation of the work practices.

The SMS requires periodical inspections of the workplace and workers to control and prevent the possible occurrences of injuries, accidents, and related problems. Through critical examination of the workplace, inspections identify and record hazards for corrective action. Regular workplace inspections are an important part of the overall occupational health and safety programme.

The health and safety programme system includes one of the important parts of workplace inspection by a qualified inspection team. This helps to know the real problems, if any, of the workers and workplace managers/supervisors. Periodical inspections help to identify jobs and tasks of the workers, the potential hazards that the workers are exposed to, and the possible reasons for occurrences of hazards. Inspections also help to know the use of personal protective equipment (PPE), engineering controls, workplace procedures, etc., to contain the possible dangers of hazardous chemical substances and necessary corrective action by the management, if any.

Safety Management Systems: Workplace Inspections and Monitoring

Purpose

Workplace inspections are an essential part of a health and safety system. Critical and meaningful inspections and monitoring of the workplace help to identify and record the possible chemical hazards. Regular workplace inspections should be an important part of the overall occupational health and safety system. Essentially, majority of chemical substances – solid, liquid, vapour, gas, dust, fume, or mist – cause workplace hazards and health disturbances more whenever workers are careless and negligent to observe set methods of precautions. As an essential part of a health and safety programme, workplaces should be inspected. Regular and qualitative inspections of the workplace are important because workplace inspections provide correct directions to improve the chemical safety system. Workplace audit/inspection should help each occupational worker and the management about the minor and major problems. Inspection should address the implementation of immediate action and permanent correction as needed. It is important that workplace inspection and audit team must include personnel with technical knowledge of health and safety aspects of chemical substances and familiarity with workplace processes and practices. Therefore, every workplace inspection plan must address the following: who, what, where, when, why, and how.

- Inspectors/auditors must listen to the concerns of workers, supervisors, and any kind of earlier complaints of hazards in the workplace.
- Inspections help to gain further understanding of jobs and tasks.
- Inspections help to identify the existing and potential hazards related with the handling of candidate chemical substance and or their formulations and to determine the causes of hazards.
- Inspections help to monitor the hazard controls: PPE, procedures, etc.
- Inspections help to suggest/recommend and report appropriate, suitable, and timely corrective actions.
- Audit/inspection report should indicate whether or not any sort of suggestions/recommendations made earlier by periodic, regular, planned inspections were implemented by the management.
- Inspections help to comply to set area regulations and responsibilities and to prepare meaningful quality checklist of the workplace.

Chemical Inventory

Determine which chemicals are used in the workplace and whether material safety data sheets are available. Find out whether actual and potential sources of chemical exposure are properly controlled. Make sure that all workers have received training in handling chemicals. Check that all chemicals are labelled with pertinent information (such as handling, storage, and waste disposal) according to Workplace Hazardous Materials Information System (WHMIS) requirements.

Workplace Exposures to Hazardous Chemical Substances

- Does the worker strictly observe workplace safety regulations?
- Does the workplace have eyewash fountains and safety showers facilities in proper areas?
- Do workers use personal protective clothing and equipment: gloves, goggles for eye protection, and respirators during handling of chemicals?
- Are flammable or toxic chemicals kept properly and safely in closed containers?

- Have standard operating procedures (SOPs) been established, and are they being followed?
- Are respirators kept in a convenient and clean location?
- Does the worker know the regulations and avoid food, drinks, smoking, etc., during use of chemicals?
- Has the management provided the required essential and quality PPE to the workers?
- Do workers maintain the exposure to chemicals within acceptable levels?
- Does the management maintain records of written SOP for selecting and using respirators where needed?
- Appropriate supervisors and managers will be contacted to ensure follow-up of corrective action.
- Is the status of the chemical store area clean and free of objectionable objects/materials?
- Is the worker aware of MSDS and its availability?
- The workers knowledge about the disposal of outdated chemicals/chemical wastes/chemical containers.
- Is the worker aware of the management of chemical wastes/chemical spills?
- Are the corrosives, flammable chemicals, incompatibles, and explosive chemicals kept labelled, separately, in proper and specified store areas?
- Is the workplace equipped with appropriate ventilation and proper entry and exit openings?

Workplace Inspection Principles

When conducting inspections, the inspector should follow these basic principles:

- The inspector should draw the attention to the presence of workplace danger etc. in the final report.
- The inspector should close and 'lock out' any kind of hazardous items/materials that cannot be brought to a safe operating standard until properly repaired.
- The inspector should not operate any newly installed equipment himself or herself alone. The details should be enquired, and ask the operator/workplace supervisor for a demonstration.
- The inspector should be alert and totally avoid negligence, be methodical and thorough during inspection, and should not spoil the workplace inspection with a 'once-over-lightly' approach.
- The inspector should clearly describe each hazard and its exact location.
- The inspector should allow 'on-the-spot' recording of all findings before they are forgotten.
- The inspector should record what you have or have not examined in case the inspection is interrupted.
- The inspector should ask questions and should avoid work disruption.
- The inspector should record and make efficient assessment of the job function.
- If a machine is shut down during inspection, the inspector should consider postponing the inspection until it is functioning again.

- The inspector should discuss as a group, about the problem, hazard, or accident generated from this situation when looking at the equipment, the process, or the environment and identify what corrections or controls are appropriately required.
- The inspector should not try to detect all hazards simply by looking at them during the inspection. It is necessary to monitor the equipment to measure the levels of exposure to chemicals, noise, radiation, biological agents, pollution, etc.

What Should the Final Inspection/Auditor's Report Have in It?

To make a report, first copy all unfinished items from the previous report on the new report. Then, write down the observed unsafe condition and recommended methods of control. Enter the department or area inspected and the date and the inspection team's names and titles on top of the page. Number each item consecutively, followed by a hazard classification of items according to the chosen scheme.

State exactly what has been detected and accurately identify its location. Instead of stating 'machine unguarded', state 'guard missing on upper pulley #6 lathe in North Building'.

Assign a priority level to the hazards observed to indicate the urgency of the corrective action required. For example,

A = Major – requires immediate action
B = Serious – requires short-term action
C = Minor – requires long-term action

Make management aware of the problems in a concise, factual way. Management should be able to understand and evaluate the problems, assign priorities, quickly reach decisions, and take immediate action as needed. When permanent correction takes time, take any temporary measures you can, such as roping off the area, tagging out equipment, or posting warning signs.

After each listed hazard, specify the recommended corrective action and establish a definite correction date. Each inspection team member should review for accuracy, clarity, and thoroughness.

Follow-Up Steps of Workplace Regular Inspections: Monitoring

Workplace manager and management and related unit should review the information obtained from regular inspections to identify where immediate corrective action is needed. Identify trends and obtain timely feedback. Analysis of inspection reports may show the following:

- Priorities for corrective actions
- Need for improvement/modifications, if any, in safe work practices
- Insight about why workplace accidents are occurring in specific/particular workplace
- Need for training in certain areas
- Areas and safety equipment that require more in-depth hazard analysis

The health and safety committee should review the progress of the recommendations, especially when they pertain to the education and training of employees. It is also the committee's responsibility to study the information from regular inspections. This will help in identifying trends for the maintenance of an effective health and safety programme.

Inspection Information Requirements

- Proper and correct information on hazardous chemical substances that are in current use
- Storage areas of chemical substance
- Work force size, shifts, and supervision staff that handle chemical substances
- Workplace rules and regulations
- Work procedures and safe work practices
- Manufacturer's specifications regarding use and handling of candidate chemical substances
- PPE in working condition – yes/no
- Engineering controls updated or not
- Emergency procedures such as fire, first aid, and rescue readily available – yes/no
- Workplace chemical accident/disaster and investigation reports
- Worker complaints/reports regarding particular hazards in the workplace
- Recommendations of the health and safety management committee
- Previous inspections – frequencies
- Maintenance of reports, procedures, and schedules
- Regulator inspection reports or any other external audits/specialist
- Monitoring reports: chemicals, physical, or biological hazards
- Identification/reports of unusual operating conditions
- Details of names of workplace inspection team members – technical expertise

Hazardous Chemical Substances: Handling and Precautions

Safety Precautions

It has been said that there are no non-hazardous chemicals, only non-hazardous ways of dealing with them. There are a variety of ways by which a user and the occupational worker get exposed and injured by chemical substances. Similarly, there are also a variety of methods and regulations, when meaningfully observed, to help the user and worker achieve workplace safety and individual protection and safeguard the living environment. The workplace management and the worker should first properly understand and plan the nature of work, purpose of work, and the chemical safety – handling – steps of precautions.

Users and occupational workers much before handling a chemical substance should read the **Workplace Safety Precautions and Regulations** and workers must completely avoid negligence.

Occupational Workers Should
- Read the all labels on the container.
- Read the materials safety data sheets (MSDS).
- Wear the proper PPE.
- Check labels prior to use for chemical hazard warnings, if any.
- Treat unlabelled containers as dangerous.
- Not remove labels: (i) Look for NFPA labels, (ii) blue is toxic, (iii) red is flammable, (iv) yellow is reactive, and (v) hazardous/very hazardous, etc.

The MSDS Should Provide
- Name of the chemical substance/material
- Physical characteristics of the candidate chemical
- Dangerous properties: carcinogenic/corrosive/explosive, flammable/mutagenic/reactive, etc.
- Precautions for safe use and handling
- Ready availability and use of PPE: aprons or full body suits, safety glasses, goggles, gloves, face shields/respirators and dust masks, head protection, and foot protection

Hazardous Chemicals: Development of Health Disorders – A MUST
for Workplace Precautions – Avoid Negligence
- Headaches, rashes, and burns
- Respiratory problems
- Lung and liver damage
- Reproductive damage
- Cancer
- Fatal injury/death

Possible Types of Physical Hazards at the Workplace
- Flammable liquids or solids
- Combustible liquids
- Compressed gases
- Explosive materials
- Unstable materials
- Water-reactive chemical/materials

Possible Manner of Exposure to the Chemical Substance without Safety and PPE
- Breathing through nose or mouth
- Contact with the skin and or eyes
- Swallowing/ingestion during handling, eating, drinking, etc.

Users and Occupational Workers Should Strictly Observe Workplace and Safety Regulations

1. Avoid eating, drinking, or smoking when working with hazardous chemicals.
2. Wash or store your PPE away from family clothing.
3. Always wash hands, arms, and face with soap and water after use.
4. Avoid negligence by checking the PPE for damage before use.
5. Observe/self-check for any kind of signs and symptoms at the workplace and during handling of hazardous chemical substances:
   a. Unpleasant/foul odour
   b. metallic taste
   c. presence of foreign particles
   d. cloudy air
6. Sudden development of headache, dizziness, blurred eye sight, watery eyes/tears, giddiness, rashes, feeling of burning, etc.

Glossary

AARC: Alternative Agriculture Research and Commercialization Corporation.

Abatement: Reducing the degree or intensity of, or eliminating, pollution.

Abdomen: The area of the body that contains the pancreas, stomach, intestines, liver, gallbladder, and other organs.

Abdominal: Relating to the abdomen, the belly, which is the part of the body that contains all of the structures between the chest and the pelvis. The abdomen is separated anatomically from the chest by the diaphragm, the powerful muscle spanning the body cavity below the lungs.

Abdominal pain (pain in the belly): Abdominal pain can come from conditions affecting a variety of organs. The abdomen is an anatomical area that is bounded by the lower margin of the ribs above, the pelvic bone (pubic ramus) below, and the flanks on each side. Although abdominal pain can arise from the tissues of the abdominal wall that surround the abdominal cavity (the skin and abdominal wall muscles), the term abdominal pain generally is used to describe pain originating from organs within the abdominal cavity (from beneath the skin and muscles). These organs include the stomach, small intestine, colon, liver, gallbladder, and pancreas.

Abiotic transformation: Any process in which a chemical in the environment is modified by non-biological mechanisms.

Abiotic, non-biological: Term describing anything that is characterised by the absence of life or incompatible with life. In toxicology and ecotoxicology, the term indicates physical (heat, sunlight) or chemical processes (hydrolysis) that modify chemical structures. Thus, biotic transformation is a process in which a chemical substance in the living environment is modified by non-biological mechanisms.

Abnormal: Not normal; different from the usual structure, position, condition, or behaviour. With reference to a tissue growth, abnormal may mean that it is cancerous or premalignant (likely to become cancer).

Abortifacient: Chemical substance that causes pregnancy to end prematurely and causes an abortion.

Abrasion: The process of wearing away by rubbing.

ABS (acrylonitrile–butadiene–styrene): A terpolymer and an amorphous resin. It is manufactured by combining three different compounds (acrylonitrile, butadiene, and styrene). ABS has the unique position of being the 'bridge' between utility and engineering thermoplastics.

Abscess: An enclosed collection of pus in tissues, organs, or confined spaces in the body. An abscess is a sign of infection and is usually swollen and inflamed.

Absence seizure: A kind of brief seizure with an accompanying loss of awareness or alertness. It is also often termed as a petit mal seizure.

Absolute lethal concentration (LC_{100}): The lowest concentration of a chemical substance in an environmental medium that kills 100% of test organisms or species under defined conditions.

Absolute lethal dose (LD_{100}): The lowest amount of a chemical substance that kills 100% of test animals under defined conditions. (It is important to note that the value of LD100 is dependent on the number of organisms used in its assessment.)

Absolute risk: The excess risk due to exposure to a hazard is referred to as the absolute risk.

Absorbance: The logarithm to the base of 10 of the reciprocal of transmittance.

Absorbate: Substance/chemical that has been retained by the process of absorption.

Absorbed dose (of a chemical substance): The amount (of a chemical substance) taken up by an organism or into organs or tissues of interest (internal dose).

Absorbent: Material/chemical substance in which absorption occurs.

Absorption: (i) The process of taking in. For a person or an animal, absorption is the process of a chemical substance getting into the body through the eyes, skin, stomach, intestines, or lungs. (ii) The penetration of a chemical substance into the body of another. (ii) The process of soaking up or taking up hazardous chemical substances to prevent enlargement of the contaminated area. (iii) The movement of chemical substances into the blood vascular system or into the tissues of the organism/animal/human.

Abuse (of drugs, chemical substances, solvents, etc.): Improper use of drugs, industrial chemical substances, toxic materials, and many other chemical substances.

Accelerator: A chemical that accelerates a chemical reaction in the production of rubber or plastics.

Acceptable daily intake (ADI): (i) An estimate of the amount of a substance in food or drinking water, expressed on a body-weight basis, that can be ingested daily over a lifetime without appreciable risk (standard human = 60 kg). The ADI is listed in units of mg per kg of body weight. (ii) The concept of the ADI has been developed principally by the World Health Organization (WHO) and Food and Agriculture Organization (FAO) and is relevant to chemical substances such as additives to foodstuffs, residues of pesticides, and veterinary drugs in foods. *See also* Tolerable Daily Intake section. The ADI is considered a safe intake level for a healthy adult of normal weight who consumes an average daily amount of the substance in question. Persons with overweight or underweight and health problems should seek medical advice.

Acceptable risk: (i) Denotes and relates to the probability of suffering disease or injury that will be tolerated by an individual, group, or society. Acceptability of risk depends on scientific data; social, economic, and political factors; and on the perceived benefits arising from a chemical or process. (ii) Probability of suffering disease or injury that is considered to be sufficiently small to be 'negligible'. The calculated risk of an increase of one case in a million people per year for cancer is usually considered to be negligible.

Accession number: An identification number that is used to assign (for cataloguing purposes) volumes of studies submitted to regulatory authorities.

Acclimatisation: The physiological and behavioural adjustments of an organism to changes in its environment.

Accreditation: A formal recognition that a laboratory is competent to carry out specific tests or specific types of tests.

Accredited laboratory: A laboratory that has been evaluated and given approval to perform a specified measurement or task, usually for a specific parameter and a specified period of time.

Accretion: A phenomenon consisting of the increase in size of particles by the process of external additions.

Accumulation: Repeated doses of a chemical substance may result in its progressive increase in concentration in an organism, organ, or tissue, and the toxic effects may become more marked with successive doses. Factors involved in accumulation include selective binding of the chemical to tissue molecules, concentration of fat-soluble chemicals in body fat, absent or slow metabolism, and slow excretion. Accumulation is a mass balance effect where input exceeds output.

Accuracy: The extent to which a given measurement/value agrees with the standard value for that measurement – a quality indicator.

Acetals (polyoxymethylene): Crystalline thermoplastics introduced in the 1960s made by the polymerisation of formaldehyde. Acetals have a unique balance of physical properties not available with metals or most other plastics.

Acetaminophen: A pain reliever and fever-reducing pharmaceutical drug (brand name: Tylenol). Acetaminophen relieves pain by elevating the pain threshold (i.e. by requiring a greater amount of pain to develop before it is felt by a person). Acetaminophen reduces fever through its action on the heat-regulating centre (the 'thermostat') of the brain.

Active: The chemical in a pesticide formulation that has an ingredient that has a direct effect on a pest.

Acetylation: The introduction of an acetyl group using a reactant such as acetyl chloride or acetic anhydride. An acetyl group is an acyl group having the formula $-C-CH_3$.

Acetylcholine: A chemical made by some types of nerve cells. It is used to send messages to other cells, including other nerve cells, muscle cells, and gland cells. It is released from the nerve ending and carries signals to cells on the other side of a synapse (space between nerve cells and other cells). Acetylcholine helps control memory and the action of certain muscles. It is a type of neurotransmitter.

ACGIH (American Conference of Governmental Industrial Hygiene): ACGIH is a professional organisation composed of personnel in governmental agencies or educational institutions engaged in occupational safety and health/industrial hygiene programmes; it develops and publishes recommended occupational exposure limits (OELs) for hundreds of chemical substances and physical agents. The Threshold Limit Value (TLV) Committee and Ventilation Committee of the ACGIH publish exposure guidelines that are used worldwide.

ACGIH carcinogen: Confirmed human carcinogen – 1; suspected human carcinogen – 2; confirmed animal carcinogen – 3; not classifiable carcinogen – 4; Not suspected carcinogen – 5.

Acid: (i) A chemical substance that dissolves in water and releases hydrogen ions (H+); acids cause irritation, bums, or more serious damage to tissue, depending on the strength of the acid, which is measured by pH. (ii) Any kind of typically water-soluble and sour compounds that in solution are capable of reacting with a base to form a salt, that redden litmus, that have a pH less than 7, and that are hydrogen-containing molecules or ions able to give up a proton to a base or are substances able to accept an unshared pair of electrons from a base.

Acid rain: Acidified particulate matter in the atmosphere that is deposited by precipitation onto a surface, often eroding the surface away.

Acidification: This process happens when compounds like nitrogen oxides and sulphur oxides are converted in a chemical reaction in the gas phase or in clouds into acidic substances. These acids are rained out or dry deposited.

Acidity: The quantitative capacity of aqueous solutions to react with hydroxyl (OH) ions. It is measured by titration with a standard solution of base to a specified end point.

Acidosis: An abnormal increase in the acidity of the body's fluids, caused either by accumulation of acids or by depletion of bicarbonates.

Acquired immunity: A resistance to reinfection of a disease after the body has recovered from the original infection (S80).

Acre-foot: The volume of water, 43,560 ft³, that will cover an area of 1 acre to a depth of 1 ft; a term used in sewage treatment in measuring the volume of material in a trickling filter.

Acromegaly: A condition in which the pituitary gland makes too much growth hormone after normal growth of the skeleton is finished. This causes the bones of the hands, feet, head, and face to grow larger than normal. Acromegaly can be caused by a pituitary gland tumour.

Acrylics: Polymethyl methacrylate offers excellent clarity and weatherability making it a very good candidate for exterior applications. Acrylic can be offered in many colours, transparent, translucent, opaque, frosted, and special effects.

ACTH: Adrenocorticotropic hormone. A hormone made in the pituitary gland. ACTH acts on the outer part of the adrenal gland to control its release of corticosteroid hormones. More ACTH is made during times of stress. Also called corticotropin.

Action level: (i) A level of the chemical substance similar to a tolerance level except it is not established through formal regulatory proceedings. It is an informal judgment by a regulatory agency on what amount of a chemical should be allowed in food products. (ii) A concentration designated by Occupational Safety and Health Administration (OSHA) for a specific chemical substance and calculated as an 8 h time-weighted average (TWA), which initiates certain required activities such as exposure monitoring and medical surveillance. (iii) The exposure concentration at which certain provisions of the National Institute for Occupational Safety and Health (NIOSH)-recommended standard must be initiated.

Activated charcoal: Charcoal that has been heated to increase its absorptive capacity. Activated charcoal is sold as an over-the-counter (OTC) product to help relieve intestinal gas. It is also used to absorb poisons (as in gas mask filters), neutralise poisons that have been swallowed, and filter and purify liquids.

Activated sludge process: A process that helps to remove organic matter from sewage by saturating it with air and microbial organisms.

Activation: Term used in the treatment of a chemical substance by heat, radiation, or activating reagent to produce a more complete or rapid chemical reaction or physical change.

Active ingredient: The part of a product that actually does what the product is designed to do. It is not necessarily the largest or most hazardous part of the product. For example, an insecticidal spray may contain less than 1% pyrethrin, the ingredient that actually kills insects. The remaining ingredients are often called inert ingredients. Active ingredients are often used to determine which products must comply with the national Workplace Hazardous Materials Information System (WHMIS). (i) Pesticides are regulated primarily on the basis of active ingredients; (ii) Chemical component of a pesticide product that can kill, repel, attract, mitigate, or control a pest or that acts as a plant growth regulator, desiccant, or nitrogen stabiliser. The remainder of a formulated pesticide product consists of one or more inert ingredients such as water, solvents, emulsifiers, surfactants, clay, and propellants that are also present for reasons other than pesticide activity.

Acuity: Acuteness of vision or perception.

Acute: Sudden or brief. Acute can be used to describe either an exposure or a health effect. An acute exposure is a short-term exposure. Short term means lasting for minutes, hours, or days. An acute health effect is an effect that develops either immediately or a short time after an exposure. Acute health effects may appear minutes, hours, or even days after an exposure. (i) Effects occurring over a short time, of abrupt onset, in reference to a disease. Acute often also connotes an illness that is of short duration, rapidly progressive, and in need of urgent care. (ii) In animal testing, it pertains to administration of an agent in a single dose, not to be confused with the clinical term for a disease having a short and relatively severe course, and (iii) in clinical medicine, it is sudden and severe, having a rapid onset.

Acute bronchitis: Inflammation of the tubes that carry air into the lungs.

Acute dermal toxicity: Adverse effects occurring within a short time of dermal application of a singular dose of a test chemical.

Acute effects: Adverse symptoms that occur immediately or shortly after an exposure to a chemical substance. Common symptoms of toxicity after an acute exposure to a chemical substance include, but not limited to headache, dizziness, or nausea.

Acute exposure: (i) A single exposure to a toxic chemical substance that may result in severe biological harm or death. Acute exposures are usually characterised as lasting no longer than a day, as compared to the continued exposures over a period of time.

Acute hepatitis: A newly acquired symptomatic hepatitis virus infection, usually less than 6 months of duration.

Acute inhalation toxicity: Adverse effects produced by a test chemical following a single uninterrupted exposure through inhalation/respiratory route over a short period of time (24 h or less).

Acute leukaemia: A rapidly progressing cancer that starts in blood-forming tissue such as the bone marrow and causes large numbers of white blood cells (WBCs) to be produced and enter the bloodstream.

Acute noncancer effect: A biochemical change, functional impairment, or pathological lesion that is produced within a short period of time following an exposure and that affects the performance of the whole organism or reduces the organism's ability to respond to additional environmental challenges.

Acute oral toxicity: Adverse effects produced within a short time of oral administration of a single dose of a test chemical substance or its multiple doses given within 24 h.

Acute-severe syndrome: The onset is acute, severe, and life-threatening. For chemical substances, the length of exposure is less than 24 h. The patient is likely to be admitted to the hospital.

Acute test: A test lasting for a short period of time – 14 days.

Acute toxicity: (i) The capacity of a chemical substance to cause adverse health effects/ poisonous effects or death as a result of a single or short-term exposure. (ii) Any poisonous effect produced within a short period of time following exposure, usually up to 24–96 h, resulting in biological harm and often death. (iii) Adverse effects occurring within a short time of administration of a single dose of a chemical, or immediately following short or continuous exposure and/or multiple doses over 24 h or less.

Average daily dose: Dose rate averaged over a pathway-specific period of exposure expressed as a daily dose on a per-unit-body-weight basis. The ADD is usually expressed in terms of mg/kg-day or other mass-time units.

Addison disease: A rare disorder in which the adrenal glands do not make enough of certain hormones. Symptoms include weight loss, loss of appetite, nausea and vomiting, diarrhoea, muscle weakness, fatigue, low blood sugar, low blood pressure, and patchy or dark skin. Most cases of the disorder are caused by immune system problems but may also be caused by infection, cancer, or other diseases. Also called adrenal insufficiency.

Additive effect: (i) A biological response to exposure to multiple chemical substances that equals the sum of responses of all the individual chemical substances. (ii) An additive effect is the overall consequence that is the result of two chemical substances acting together and that is the simple sum of the effects of the chemical substances acting independently also (*See* Antagonistic effect, Synergistic effect). (iii) The combined effect produced by the action of two or more agents, being equal to the sum of their separate effects.

Adduct: A covalent compound formed between a carcinogen or its metabolites and a protein or nucleic acid (either deoxyribonucleic acid (DNA) or ribonucleic acid (RNA)).

Adenocanthoma, ovarian: (i) Denotes an adenocarcinoma in which some or the majority of the cells exhibit squamous (scaly or platelike) differentiation. (ii) The endometrial adenocarcinomas commonly contain foci of squamous epithelium, in addition to the glandular elements. In the past, these mixed tumours were known as adenocanthomas if they were well differentiated and as adenosquamous carcinomas if they were poorly differentiated.

Adenocarcinoma: (i) A malignant neoplasm of epithelial cells with a glandular or glandlike pattern (synonym: glandular cancer/glandular carcinoma). (ii) A term applied to a malignant tumour originating in glandular tissue. (iii) Tumours of the linings of organs. (iv) Carcinoma derived from glandular tissue or in which the tumour cells form recognisable glandular structures.

Adenoma: A tumour that is not cancer. It starts in glandlike cells of the epithelial tissue (thin layer of tissue that covers organs, glands, and other structures within the body). This term refers a tumour that is usually benign, in glandular tissue. Adenoma can be precancerous in cases such as polyps in the colon.

Adenomatoid: A tissue change resembling an adenoma.

Adenomatous: Relating to an adenoma and to some types of glandular hyperplasia.

Adenosarcoma: A tumour that is a mixture of an adenoma (a tumour that starts in the glandlike cells of epithelial tissue) and a sarcoma (a tumour that starts in bone, cartilage, fat, muscle, blood vessels, or other connective or supportive tissue). An example of an adenosarcoma is Wilms tumour.

Adenopathy: Large or swollen lymph glands.

ADR (adverse drug reaction): ADR and event is the occurrence of any kind of noxious, undesired, or unintended response to a drug, which occurs at dosages used for prophylaxis, diagnosis, therapy, or modification of physiological functions.

Adrenal cortical steroids: Steroid hormones produced in the cortex of the adrenal gland.

Adrenal gland: A hormone-secreting organ located above each kidney.

Adrenergic: Secreting adrenaline (epinephrine) and (or) related substances, in particular referring to sympathetic nerve fibres.

Adsorbate: Chemical that has been retained by the process of adsorption.

Adsorbent: A solid material on the surface of which adsorption takes place.

Adsorption: (i) A physical process in which molecules of gas of dissolved chemicals or liquids adhere in an extremely thin layer to the surfaces of solid bodies with which they are in contact. (ii) the adhesion of an extremely thin layer of solid, liquid, or vapour molecules to the surface of a solid or liquid.

Adulterated food: Food and food products that are generally impure, unsafe, or unwholesome. In terms of the Federal Food, Drug, and Cosmetic Act (FFDCA); the Federal

Meat Inspection Act; the Poultry Products Inspection Act; and the Egg Products Inspection Act, the term 'adulterated food' in separate language defines in very specific (and lengthy) terms how the term 'adulterated' will be applied to the foods each of these laws regulate. Products found to be adulterated under these laws cannot enter into commerce for human food use.

Advanced waste treatment: Any treatment method or process employed following biological treatment (1) to reduce pollution load, (2) to remove substances that may be harmful to receiving waters or the environment, and (3) to produce a high-quality effluent suitable for reuse in any specific manner or for discharge under critical conditions. The term *tertiary treatment* is commonly used to denote advanced waste treatment methods.

Adverse effect: (i) An abnormal, undesirable, or harmful effect to an organism. The symptoms normally include mortality, altered food consumption, altered body and organ weights, altered enzyme levels, or visible pathological change. An effect may be classed as adverse if it causes functional or anatomical damage, causes irreversible change in the homeostasis of the organism, or increases the susceptibility of the organism to other chemical or biological stress. Adverse effects occur within a short time of administration of a single dose of a chemical substance or immediately following short or multiple doses over 24 h or less. (ii) An unexpected medical problem that happens during treatment with a drug or other therapy. Adverse effects do not have to be caused by the drug or therapy but by several hazardous chemical substances as well. The adverse effects may be mild, moderate, or severe.

Adverse health effect: (i) Abnormal or harmful effect to an organism (e.g. a person) caused by exposure to a chemical. It includes results such as death, other illnesses, altered body and organ weights, and altered enzyme levels. (ii) A change in body functions or cell structure that might lead to disease or health problems.

AE (adverse event): Any adverse change in health or side effect that occurs in a person who participates in a clinical trial while receiving treatment (study medication, application of the study device, etc.) or within a previously specified period of time after the treatment has been completed.

AEGLs (acute exposure guideline levels): The primary purpose of the AEGLs is to develop guideline levels for once-in-a-lifetime, short-term (not repeated chronic) exposures to airborne concentrations of acutely toxic, high-priority chemical substances. AEGLs are needed for a wide range of applications in chemical emergency planning, prevention, and response programmes. The AEGLs are intended to protect most individuals in the general population, including those that might be particularly susceptible to the deleterious effects of the chemicals.

Aerobic biological oxidation: Any waste treatment process or other processes utilising aerobic organisms, in the presence of air or oxygen, as the agent for reducing pollution load, oxygen demand, or the amount of organic substance in waste. The term is used in reference to secondary treatment of wastes.

Aerobic conditions: In the presence of oxygen.

Aerosol: (i) A collection of very small particles suspended in air. The term is also commonly used for a pressurised container (aerosol can) that is designed to release a fine spray of a material such as paint. Inhalation of aerosols is a common route of exposure to many chemicals. Also, aerosols may be fire hazards. This term is broadly applied to any suspension of solid or liquid particles in a gas. The small particles, usually in the range of 0.01–100 μm, dispersed in air, include liquid (mist) and solid particles (dust) and (ii) a fine suspension in the air of small particles (e.g. smoke or fog).

Aerosol particles: One of the components of an atmospheric air mixture comprised of minute solid particles, part of which is almost certainly water.

Aetiology: In medicine, the term refers to the science of the investigation of the cause or origin of disease.

Aflatoxins: (i) A harmful substance made by certain types of mould (*A. flavus* and *Aspergillus parasiticus*) that is often found on poorly stored grains and nuts. Consumption of foods contaminated with aflatoxin is a risk factor for primary liver cancer.

Agent: (i) A chemical, physical, mineralogical, or biological entity that may cause deleterious effects in an organism after the organism is exposed to it (U.S. EPA, 1992: GL for Exposure Assessment). (ii) Suter et al. (1994) suggested it as an alternative for the term stressor. It is considered to be more neutral than stressor and is used in Environmental Protection Agency (EPA)'s Guidelines for Exposure Assessment (U.S. EPA, 1997: Guidance on Cumulative Risk Assessment, Planning and Scoping). (iii) Any physical, chemical, or biological entity that can be harmful to an organism (synonymous with stressor) (U.S. EPA, 1997a: EPA terms of environment). (iv) Any physical, chemical, or biological entity that can induce an adverse response (synonymous with stressor) (U.S. EPA, 1998a: Guidelines for ecological risk assessment). (v) A chemical, physical, mineralogical, or biological entity that may cause deleterious effects in a target after contacting the target (Zartarian, et al., 1997: Quant. Def. of Exp. & Related Concepts).

Agent Orange: *See* 2,4-dichlorophenoxyacetic acid (2, 4-D) and 2,4,5-trichlorophenoxyacetic acid (2,4,5-T), dioxin.

Agglomeration: A process of contact and adhesion whereby the particles of a dispersion form clusters of increasing size.

Agitation: The stirring or mixing of a spray solution in a sprayer.

Agricultural pollution: Wastes, emissions, and discharges arising from farming activities. Causes include run-off and leaching of pesticides and fertilisers, pesticide drift and volatilisation, erosion and dust from cultivation, and improper disposal of animal manure and carcasses. Some agricultural pollution is point source, for example, large feedlots, which require permits under the Clean Water Act, but much is non-point source, meaning that it derives from dispersed origins, for example, blowing dust or nutrients leaching from fields. As most pollution control programmes have focused on particular categories of point sources, unregulated and non-point sources account for an increasingly large proportion of remaining pollution. Based on state surveys, EPA concludes that agricultural sources account for over one-half the pollution impairing surface water quality in the United States. The Clean Water Act mandates that states develop and implement management programmes to control non-point sources of water pollution.

Agricultural quarantine inspection (AQI): A programme, administered by departments of agriculture of different countries to inspect about animal and plant health of incoming passengers, luggage, and cargo at ports of entry of the country in order to protect U.S. agriculture from foreign animal and plant pests and diseases.

Agricultural Research Service (ARS): A U.S. Department of Agriculture agency employing federal scientists to conduct basic, applied, and developmental research in the following fields: livestock; plants; soil, water, and air quality; energy; food safety and quality; nutrition; food processing, storage, and distribution efficiency; non-food agricultural products; and international development.

AIDS: A disease caused by the human immunodeficiency virus (HIV). People with AIDS are at an increased risk for developing certain cancers and for infections that usually occur only in individuals with a weak immune system. Also called acquired immunodeficiency syndrome.

AIHA: American Industrial Hygiene Association.

AIN (acute interstitial nephritis): A form of nephritis affecting the interstitium of the kidneys surrounding the tubules. This disease can be either acute, meaning it occurs suddenly, or chronic, meaning it is ongoing and eventually ends in kidney failure. Acute interstitial nephritis (AIN) is a rapidly developing inflammation that occurs within the interstitium. It can produce a variety of clinical symptoms, depending upon the severity and extent of kidney involvement.

Air pollution: (i) Contamination of atmospheric air with substances/chemicals not considered unsuitable for health. (ii) Contamination of the atmosphere by any toxic or radioactive gases and particulate matter as a result of human activity. (iii) Presence of substances in the atmosphere resulting either from human activity or natural processes, in sufficient concentration, for a sufficient time and under circumstances such as to interfere with comfort, health, or welfare of persons or to harm the environment. (iv) Contamination of the atmosphere by substances that, directly or indirectly, adversely affect human health or welfare. Air pollution results from human activities, both deliberate releases (as from smokestacks) and fugitive emissions (as dust blown from streets or fields), and from natural sources, including sea spray, volcanic emissions, and pollen. The Clean Air Act (CAA) authorities the U.S. EPA to regulate air pollution.

Air quality standard (AQS): *See* Environmental quality standard.

Alanine transaminase: An enzyme found in the highest amounts in the liver. Injury to the liver results in release of ALT into the blood. This test is used to determine if a patient has liver damage.

ALARA Committee: Multidisciplined forum that reviews and advises management on improving progress towards minimising radiation exposure and radiological releases (U.S. DOE, 1998: *Radiological Control Manual*).

Albinism: A group of genetic conditions marked by little or none of the pigment melanin in the skin, hair, and/or eyes. People with albinism may have vision problems and white or yellow hair; reddish, violet, blue, or brown eyes; and pale skin. Albinism is inherited and is not contagious. People are born with albinism because they inherit an albinism gene or genes from their parents.

Albino: An organism exhibiting deficient pigmentation in skin, eyes, and/or hair.

Albuminuria: Presence of albumin, derived from plasma, in the urine.

Alcohol: Any class of organic compounds containing the OH group, OH. Specifically, the term is applied to ethyl alcohol (C_2H_5OH). Alcohol is an organic chemical in which one or more OH groups are attached to carbon (C) atoms in place of hydrogen (H) atoms. Alcohols are grouped into different classes depending on how the –OH group is positioned on the chain of carbon atoms: (i) In a primary (1°) alcohol, the carbon that carries the –OH group is only attached to one alkyl group; (ii) in a secondary (2°) alcohol, the carbon with the –OH group attached is joined directly to two alkyl groups; and (iii) in a tertiary (3°) alcohol, the carbon atom holding the –OH group is attached directly to three alkyl groups, which may be any combination of same or different.

The common alcohols include ethyl alcohol or ethanol (found in alcoholic beverages), methyl alcohol or methanol (can cause blindness), and propyl alcohol or propanol (used as a solvent and antiseptic). Rubbing alcohol is a mixture of acetone, methyl isobutyl ketone, and ethyl alcohol. In everyday talk, alcohol usually refers to ethanol as, for example, in wine, beer, and liquor. It can cause changes in behaviour and be addictive.

Alcohol dependence: A chronic disease in which a person craves drinks that contain alcohol and is unable to control his or her drinking. A person with this disease also needs to drink greater amounts to get the same effect and has withdrawal symptoms after stopping alcohol use. Alcohol dependence affects physical and mental health and can cause problems with family, friends, and work. Regular heavy alcohol intake increases the risk of several types of cancer. This is also called alcoholism.

Algaecide: (i) A chemical agent added to water to destroy algae. (ii) A pesticide that controls algae.

Alkaline: A material whose index is opposed to acidity.

Alkalinisation: A process that lowers the amount of acid in a solution. In medicine, an alkali, such as sodium bicarbonate, may be given to patients to lower high levels of acid in the blood or urine that can be caused by certain medicines or conditions.

Alkalinity: The capacity of water to neutralise acids, a property imparted by the water's content of carbonate, bicarbonate, hydroxide, and on occasion borate, silicate, and phosphate. It is expressed in milligrams per litre of equivalent calcium carbonate ($mg/L\ CaCO_3$).

Alkaloid: A member of a large group of chemical substances found in plants and in some fungi. Alkaloids contain nitrogen and can be made in the laboratory. Nicotine, caffeine, codeine, and vincristine are alkaloids. Some alkaloids, such as vincristine, are used to treat cancer.

Alkylating agent: (i) Any chemical substance that introduces an alkyl radical into a compound in place of a hydrogen atom. (ii) A substance that causes the incorporation of single-bonded carbon atoms into another molecule. A type of drug that is used in the treatment of cancer. It interferes with the cell's DNA and inhibits cancer cell growth.

Allele: One of the variant forms of a gene at a particular locus on a chromosome. Different alleles produce variation in inherited characteristics such as hair colour or blood type. In an individual, one form of the allele (the dominant one) may be expressed more than another form (the recessive one). 'Genes' are considered simply as segments of a nucleotide sequence and allele refers to each of the possible alternative nucleotides at a specific position in the sequence.

Allelopathy: The production of compounds by one plant that retards or inhibits the growth of another plant.

Allergen: (i) A descriptor term for a chemical substance that produces an allergy. (ii) Immunostimulant antigenic chemical substance that may or may not cause a clinically significant effect but that is capable of producing immediate hypersensitivity.

Allergy (hypersensitivity): (i) A broad term applied to disease symptoms following exposure to a previously encountered chemical substance (allergen), often one that would otherwise be classified as harmless, essentially a malfunction of the immune system. (ii) The most common forms of allergy are rhinitis, urticaria, asthma, and contact dermatitis. (iii) An altered immune response to a specific substance on reexposure. (iv) An exaggerated immune response to a foreign substance causing tissue inflammation and organ dysfunction.

Allometry: A term in biology to measure the rate of growth of a part or parts of an organism relative to the growth of the whole organism.

Alloy: A metallic material, homogeneous to the naked eye, consisting of two or more elements so combined that they cannot be readily separated by mechanical means. Alloys are considered to be mixtures for the purpose of classification under the Globally Harmonised System (GHS).

Alopecia: (i) Loss of hair. The lack or loss of hair from areas of the body where hair is usually found. Alopecia can be a side effect of some cancer treatments baldness and absence or thinning of hair from areas of skin where it is usually present.

Alpha particle: Nucleus of a helium atom emitted by certain radioisotopes upon disintegration.

Alternating current (ac): Current that reverses its direction at regular intervals, such as a common 115 V circuit.

Alum: Technically, a double sulphate of ammonium or a univalent or trivalent metal but commonly used to denote aluminium sulphate ($Al_2(SO_4)_3$.

Aluminium: A metallic element that is found combined with other elements in the Earth's crust. It is also found in small amounts in soil, water, and many foods. It is used in medicine and dentistry and in many products such as foil, cans, pots and pans, airplanes, siding, and roofs. High levels of aluminium in the body can be harmful.

Alveoli: Tiny sacs at the ends of bronchioles in the lungs where carbon dioxide and oxygen are exchanged with red blood cells (RBCs) in adjacent capillaries. (i) The term alveoli is the plural of alveolus. (ii) Tiny sacs at the ends of bronchioles in the lungs where carbon dioxide and oxygen are exchanged with RBCs in adjacent capillaries. (iii) The term alveoli usually refers to small, sac-like pouches in the portion of the lungs.

Alzheimer's dementia: A brain disorder that usually starts in late middle age or old age and gets worse over time. Symptoms include loss of memory, confusion, difficulty thinking, and changes in language, behaviour, and personality. This is also called Alzheimer's disease.

Ambient: Surrounding or encompassing, for instance, the ambient environment.

Ambient air: (i) Surrounding, as in the surrounding environment. The medium surrounding or contacting an organism (e.g. a person), such as outdoor air, indoor air, water, or soil, through which chemicals or pollutants can be carried and can reach the organism. (ii) Air surrounding on all sides.

Ambient measurement: (i) A measurement (usually of the concentration of a chemical or pollutant) taken in an ambient medium, normally with the intent of relating the measured value to the exposure of an organism that contacts that medium. (ii) A measurement of the concentration of a substance or pollutant within the immediate environs of an organism, taken to relate it to the amount of possible exposure.

Ambient standard: *See* Environmental quality standard.

Ames test: A method of an experiment performed using bacteria as a test system to determine the mutagenic potential of a substance/chemical. (ii) In vitro test for mutagenicity using mutant strains of the bacterium *Salmonella typhimurium* that cannot grow in a given histidine-deficient medium: mutagens can cause reverse mutations that enable the bacterium to grow on the medium. The test can be carried out in the presence of a given microsomal fraction (S-9) from rat liver (see Microsome) to allow metabolic transformation of mutagen precursors to active derivatives.

Amines: A class of organic compounds of nitrogen that may be considered as derived from ammonia (NH_3) by replacing one or more of the hydrogen atoms by organic radicals, such as CH_3 or C_6H_5, as in methylamine and aniline. The former is a gas at ordinary temperature and pressure, but other amines are liquids or solids. All amines are basic in nature and usually combine readily with hydrochloric or other strong acids to form salts.

Amino acids: (i) Amino acids play central roles both as building blocks of proteins and as intermediates in metabolism. There are about 20 amino acids. (ii) Amino acids are the chemical units or 'building blocks' of the body that make up proteins. Protein substances make up the muscles, tendons, organs, glands, nails, and hair. Growth, repair, and maintenance of all cells are dependent upon them. Next to water, protein makes up the greatest portion of our body weight. Amino acids that must be obtained from the diet are called 'essential amino acids'; other amino acids that the body can manufacture from other sources are called 'non-essential amino acids'.

Ammonia: A gas made of nitrogen and hydrogen. It has a strong odour and can irritate the skin, eyes, nose, throat, and lungs. Ammonia is made by bacteria and decaying plants and animals and is found in water, soil, and air. Ammonia is also made by the body when proteins break down. In the laboratory, ammonia can be changed to a liquid and is used in medicines, fertilisers, household cleaning liquids, and other products.

Amnesic shellfish poisoning (ASP): A kind of serious illness that is a consequence of consumption of bivalve shellfish (molluscs) such as mussels, oysters, and clams that have ingested, by filter feeding, large quantities of micro-algae containing poisonous acid; acute symptoms include vomiting, diarrhoea, and, in some cases, confusion, loss of memory, disorientation, and even coma.

Anabolism: Biochemical processes by which smaller molecules are joined to make larger molecules.

Anaemia: (i) An abnormal deficiency in the oxygen-carrying component of the blood – a condition suggesting lack of RBCs. (ii) Decreased haemoglobin or number of RBCs. (iii) A qualitative or quantitative deficiency of haemoglobin, a protein found inside RBCs.

Anaerobic: Living or occurring only in the absence of free oxygen. (ii) Requiring the absence of oxygen.

Anaerobic biological treatment: Any waste treatment process utilising anaerobic or facultative organisms in the absence of air to reduce the organic matter in water.

Anaerobic respiration: Living or functioning in the absence of oxygen. Cellular respiration in the absence of oxygen.

Anaerobic waste treatment (sludge processing): Waste stabilisation brought about through the action of microorganisms in the absence of air or elemental oxygen.

Anaesthesia: Temporary loss of consciousness induced by high concentrations of organic solvents.

Anaesthetic: An agent that causes loss of sensation with or without the loss of consciousness.

Analgesic: A drug that alleviates pain without causing loss of consciousness, a pain reliever; with an effective analgesic, there is an inability to feel pain while still conscious. From the Greek an-, without, and algesis, sense of pain and structural support.

Analyte: Any chemical substance measured in the laboratory.

Analytic epidemiological study: A study that evaluates the association between exposure to hazardous chemical substances and disease by testing scientific hypotheses.

Anaphylactic shock: A severe and sometimes life-threatening immune system reaction to an antigen that a person has been previously exposed to. The reaction may include itchy skin, oedema, collapsed blood vessels, fainting, difficulty in breathing, and death.

Anaphylaxis: Life-threatening type 1 hypersensitivity reaction occurring in a person or animal exposed to an antigen or hapten to which they have previously been sensitised. The consequences of the reaction may include angio-oedema, vascular collapse, shock, and respiratory distress.

Androgen suppression: Treatment designed to suppress or block the production of male hormones.

Aneuploidy: The occurrence of one or more extra or missing chromosomes leading to an unbalanced chromosome complement or any chromosome number that is not an exact multiple of the haploid number.

Angio-oedema: A reaction in the skin and underlying tissue showing swelling and red blotches.

Angstrom: A unit of length, used especially in expressing the length of light waves, equal to one ten thousandth of a micron or one hundred millionth of a centimetre ($1 \times 10E-8$ cm).

Anhydrous: Without water.

Animal and Plant Health Inspection Service (APHIS): A USDA agency established to conduct inspections and regulatory and control programmes to protect animal and plant health. It utilises border inspections to prevent international transmission of pests and disease, administers quarantine and eradication programmes, and certifies that U.S. exports meet importing countries' animal and plant health standards.

Animal model: An animal with a disease either the same as or like a disease in humans. Animal models are used to study the development and progression of diseases and to test new treatments before they are given to humans. Animals with transplanted human cancers or other tissues are called xenograft models.

Anion: Ion having a negative charge; an atom with extra electrons. Atoms of non-metals, in solution, become anions.

Annuals: Plants that live 12 months or less.

Anorexia: An abnormal loss of the appetite for food. Anorexia can be caused by cancer, **AIDS,** a mental disorder, meaning anorexia nervosa, or other diseases.

Anosmia: Loss of the sense of smell. Anosmia can be temporary or permanent and could be induced by exposure to chemical substances or injuries to brain and head. Anosmia can even be fatal.

Anoxia: The absence of oxygen in inspired/inhaled gases or in arterial blood and/or in the tissues. This is closely related to hypoxia – severe oxygen deficiency in the tissues – anoxia as the most extreme case of hypoxia.

ANSI: American National Standards Institute.

Antagonism: (i) Combined effect of two or more factors that is smaller than the solitary effect of any one of those factors. (ii) Adverse effect or risk from two or more chemicals interacting with each other is less than what it would be if each chemical was acting separately. (iii) An effect that occurs when two or more pesticides are mixed together and their combined activity is less than if they were used separately.

Antagonist: A chemical substance that acts within the body to reduce the physiological activity of another chemical substance (as an opiate), especially one that opposes the action on the nervous system of a drug or a substance occurring naturally in the body by combining with and blocking its nervous receptor. Any chemical substance that binds to a cell receptor normally responding to a naturally occurring chemical substance and that prevents a response to the natural substance.

Antagonistic: Reduction of the effect of one chemical substance by the other when interacted.

Antagonistic effect: (i) A biological response to exposure to multiple substances that is less than would be expected if the known effects of the individual substances were added together. (ii) Antagonistic effect is the consequence of one chemical substance or group of chemical substances counteracting the effects of another: in other words, the situation where exposure to two or a group of chemical substances together has less effect than the simple sum of their independent effects; such chemical substances are said to show antagonism.

Anthelmint(h): A chemical substance intended to kill or cause the expulsion of parasitic intestinal worms, such as helminths.

Anthracosis (coal miners' pneumoconiosis): A form of pneumoconiosis caused by accumulation of anthracite carbon deposits in the lungs due to inhalation of smoke or coal dust.

Anthrax: A disease of mammals and humans caused by a spore-forming bacterium called *Bacillus anthracis*. Anthrax has an almost worldwide distribution and is a zoonotic disease, meaning it may spread from animals to humans. All mammals appear to be susceptible to anthrax to some degree, but ruminants such as cattle, sheep, and goats are the most susceptible and commonly affected, followed by horses and then swine.

Anthropogenic: (i) Effects produced as a result of human activities. (ii) Effects caused by or influenced by human activities and the impact on nature.

Antibiotics: (i) A chemical substance that can destroy or inhibit the growth of microorganisms. Antibiotics are widely used in the prevention and treatment of infectious disease. (ii) Chemical substance produced by and obtained from certain living cells (especially bacteria, yeasts, and moulds) or an equivalent synthetic substance, which is biostatic or biocidal at low concentrations to some other forms of life, especially pathogenic or noxious organisms. (iii) Chemical substances produced by microorganisms or synthetically that inhibit the growth of, or destroy, bacteria. Rules guiding the use of veterinary drugs and medicated animal feeds, including tolerance levels for drug residues in meats for human consumption, are set by the Center for Veterinary Medicine of the Food and Drug Administration (FDA). The Food Safety and Inspection Service (FSIS) enforces the FDA rules through a sampling and testing programme that is part of its overall meat and poultry inspection programme.

Antibody(ies): A protein made by plasma cells (a type of WBC) in response to an antigen (a substance that causes the body to make a specific immune response). Each antibody can bind to only one specific antigen. The purpose of this binding is to help destroy the antigen. Some antibodies destroy antigens directly. Others make it easier for WBCs to destroy the antigen. (i) Specific proteins produced by the body's immune system that bind with foreign proteins (antigens). (ii) Protein (immunoglobulin) produced by the immune system in response to exposure to an antigenic molecule and characterised by its specific binding to a site on that molecule.

Anticoagulant: Chemical substance that prevents blood clotting, for example, warfarin.

Antidote: (i) A chemical substance capable of specifically counteracting or reducing the effect of a potentially toxic substance in an organism by a relatively specific chemical or pharmacological action. (ii) An agent that counteracts a poison and neutralises its effects.

Antiepileptics (antiepileptic drugs): Antiepileptic drugs are those medicines that reduce the frequency of epileptic seizures. Antiepileptic drugs have a large number of side effects and possible adverse effects.

Antigen: (i) Any substance that causes the body to make a specific immune response. (ii) A foreign substance that provokes immune response when introduced into the body and the body reacts by making antibodies. The descriptor term is applied to any chemical substance that produces a specific immune response when it enters the tissues of an animal.

Antihistamine: Chemical substance that blocks or counteracts the action of histamine.

Antimicrobic: A drug used to inhibit or kill microorganisms.

Antimuscarinic: (i) A chemical substance inhibiting or preventing the actions of muscarine and muscarine-like agents, for example, atropine, on the muscarinic acetylcholine receptors. (ii) Inhibiting or preventing the actions of muscarine and muscarine-like agents on the muscarinic acetylcholine receptors.

Antimycotic: A chemical substance used to kill a fungus or to inhibit its growth – a fungicide.

Antinicotinic: (i) A chemical substance inhibiting or preventing the actions of nicotine and nicotine-like agents, for example, suxamethonium chloride, on the nicotinic acetylcholine receptors.

Antioxidants: (i) A substance that protects cells from the damage caused by free radicals (unstable molecules made by the process of oxidation during normal metabolism). Free radicals may play a part in cancer, heart disease, stroke, and other diseases of ageing. Antioxidants include beta-carotene; lycopene; vitamins A, C, and E; and other natural and manufactured substances. (ii) Chemical substances added to food to prevent the oxygen present in the air from causing undesirable changes in flavour and colour. BHA, BHT, and tocopherols are examples of antioxidants.

Antipyretic: A chemical substance that relieves or reduces fever.

Anxiety: Feelings of fear, dread, and uneasiness that may occur as a reaction to stress. A person with anxiety may sweat, feel restless and tense, and have a rapid heartbeat. Extreme anxiety that happens often over time may be a sign of an anxiety disorder.

Anxiolytic agent: A drug used to treat symptoms of anxiety, such as feelings of fear, dread, uneasiness, and muscle tightness, that may occur as a reaction to stress. Most anxiolytic agents block the action of certain chemicals in the nervous system. This is also called antianxiety agent and anxiolytic.

AOR (authorised organisation representative): An AOR serves as an organisation's communication link and face to the outside world. Informally, such a representative can serve for a private company in a variety of capacities. More specifically, organisation representatives are the conduit between a chartered group and a community or government agency and are a key part of ensuring proper organisation.

Aphasia: Loss or impairment of the power of speech or writing or of the ability to understand written or spoken language or signs, due to a brain injury or disease.

Aphicide: A chemical substance intended to kill aphids.

Aphid: Common name for a harmful plant parasite in the family Aphididae, some species of which are vectors of plant virus diseases.

Aphotic zone: The deeper part of lakes/sea/ocean where light does not penetrate.

Aphytic zone: Part of the lake floor where vegetation is not available.

Aplasia: Lack of development of an organ or tissue or of the cellular products from an organ or tissue.

Aplastic anaemia: (i) Bone marrow failure with markedly decreased production of WBCs, RBCs, and platelets leading to increased risk of infection and bleeding. (ii) One type of anaemia caused by injury to blood-forming tissues and associated with occupational exposure to chemical substances such as TNT, benzene, and ionizing radiation.

Apnoea: Temporary absence or cessation of breathing.

Application factor: Refers to number used to estimate concentration of a substance/chemical that will not produce significant adverse effects/harm to a population during chronic exposure. The factor is based on the formula: application factor = MATP.

Application rate: The amount of pesticide applied to a site, usually expressed as the amount of liquid or dry material to apply per acre, square foot, animal, etc.

Applied dose: (i) The amount of a substance in contact with the primary absorption boundaries of an organism, for example, skin, lung, and gastrointestinal tract, and available for absorption (U.S. EPA, 1992: GL for exposure assessment) (REAP, 1995: Residential exposure assessment project).

Aquaculture: (i) Breeding and rearing of fish in captivity, also termed as pisciculture. (ii) The production of aquatic plants or animals in a controlled environment, such as ponds, raceways, tanks, or cages, for all or part of their life cycle. In the United States, baitfish, catfish, clams, crawfish, freshwater prawns, mussels, oysters, salmon, shrimp, tropical (or ornamental) fish, and trout account for most of the aquacultural production. Less widely established but growing species include alligator, hybrid striped bass, carp, eel, red fish, northern pike, sturgeon, and tilapia.

Aquatic organisms: Organisms/animals related to living water bodies.

Aqueous: Related to watery solution.

Aquifer: A subsurface geological structure that contains water.

Arboreal: Relating to, resembling, or consisting of trees/plants.

ARF (acute renal failure): Acute kidney failure is the sudden loss of your kidneys' ability to perform their main function of eliminating excess fluid and salts (electrolytes) as well as waste material from the blood. When the kidneys lose their filtering ability, dangerous levels of fluid, electrolytes, and wastes accumulate in the body.

Argyria and argyrosis: Pathological condition characterised by grey-bluish or black pigmentation of tissues (such as skin, retina, mucous membranes, internal organs) caused by the accumulation of metallic silver, due to reduction of a silver compound that has entered the organism during (prolonged) administration or exposure and/or workplace exposure.

Aromatic amines: Petrochemical compounds with a pungent odour (are known to produce cancer).

Aromatic hydrocarbons: Hydrocarbon compounds in which the carbon atoms are connected by a ring structure that is planar and joined by sigma and pie bonds between the carbon atoms. A class of synthetic compounds used as solvents and grease cutters, and these are members of the carcinogenic benzene family of chemicals.

Aromatic: (i) Technical term about a chemical compound which contains one or more benzene rings. (ii) An organic chemical (hydrocarbon) characterised by the presence of a benzene ring.

Arrhythmia: Any variation from the normal rhythm of the heartbeat.

Arteriosclerosis: Hardening and thickening of the walls of the arteries.

Arthralgia: Pain in body joint or joints.

Arthritis: Chronic inflammation of a joint, usually accompanied by pain and often by changes in structure.

Artificial respiration: Making someone breathe by mechanically forcing air into and out of the lungs.

Artificial colouring: A colouring chemical substance containing any dye or pigment manufactured by a process of synthesis or other similar artifice, or a colouring that was manufactured by extracting naturally produced dyes or pigments from a plant or other material.

Artificial flavouring: Artificial flavours are restricted to an ingredient that was manufactured by a process of synthesis or similar process. The principal components of artificial flavours usually are esters, ketones, and aldehyde groups. These ingredients are declared in the ingredient statement as 'artificial flavours' without naming the individual components.

Asbestos: A naturally occurring mineral, fibrous, with tiny fibres found in certain types of rock formations. Asbestos has been used as insulation against heat and fire in buildings. Loose asbestos fibres breathed into the lungs can cause several serious diseases, including lung cancer and malignant mesothelioma (cancer found in the lining of the lungs, chest, or abdomen). Asbestos that is swallowed may cause cancer of the gastrointestinal tract.

Asbestosis: (i) A lung disease caused by breathing in particles of asbestos (a group of minerals that take the form of tiny fibres). Symptoms include coughing, trouble breathing, and chest pain caused by scarring and permanent damage to lung tissue. (ii) A form of pneumoconiosis caused by inhalation of asbestos fibres.

Ascaricide: Chemical substance intended to kill roundworms.

Ascorbic acid: A nutrient that the body needs in small amounts to function and stay healthy. Ascorbic acid helps fight infections, heal wounds, and keep tissues healthy. It is an antioxidant that helps prevent cell damage caused by free radicals (highly reactive chemicals). Ascorbic acid is found in all fruits and vegetables, especially citrus fruits, strawberries, cantaloupe, green peppers, tomatoes, broccoli, leafy greens, and potatoes. It is water soluble (can dissolve in water) and must be taken in every day. Ascorbic acid is being studied in the prevention and treatment of some types of cancer. Also called vitamin C.

Ash: Mineral content of a product that remains after complete combustion.

Asphyxiant: (i) A substance that causes injury by decreasing the amount of oxygen available to the body. Asphyxiants may act by displacing air from an enclosed space or by interfering with the body's ability to absorb and transport oxygen. (ii) A chemical substance capable of inducing asphyxia, which is a lack of oxygen or an excess of carbon dioxide in the body, usually caused by the interruption of breathing and resulting in unconsciousness. (iii) Asphyxiants work by displacing so much oxygen from the ambient atmosphere that the haemoglobin in the blood cannot pick up enough oxygen from the lungs to fully oxygenate the tissues leading to slow suffocation of the victim. Asphyxiation is an extreme hazard when working in enclosed spaces. Be sure you are trained in confined space entry before working in sewers, storage tanks, etc., where gases such as methane are known to displace oxygen from the atmosphere.

Aspiration: Entry of a liquid or solid chemical product into the trachea and lower respiratory system directly through the oral or nasal cavity or indirectly from vomiting.

Aspirin: A good example of a trade name that entered into the language, aspirin was once the Bayer trademark for acetylsalicylic acid.

AST: Aspartate aminotransferase (also known as serum glutamic–oxaloacetic transaminase [SGOT]) is an enzyme normally present in liver and heart cells. AST is released into blood when the liver or heart is damaged. The blood AST levels are thus elevated with liver damage, for example, from viral hepatitis, or with an insult to the heart, for example, from a heart attack. Some medications can also raise AST levels.

Asthenia: Weakness; lack of energy and strength.

Asthma: (i) A chronic disease in which the bronchial airways in the lungs become narrowed and swollen, making it difficult to breathe. Symptoms include wheezing, coughing, tightness in the chest, shortness of breath, and rapid breathing. An attack may be brought on by pet hair, dust, smoke, pollen, mould, exercise, cold air, or stress.

Astringent: (i) Chemical substance causing contraction, usually locally after topical application. (ii) A chemical substance causing cells to shrink, thus causing tissue contraction or stoppage of secretions and discharges; such substances may be applied to skin to harden and protect it.

Ataxia: Unsteady or irregular manner of walking or movement caused by loss or failure of muscular coordination.

Atelectasis: (i) Atelectasis is the collapse of part or all of a lung. It is caused by a blockage of the air passages (bronchus or bronchioles) or by pressure on the lung. (ii) Failure of the lung to expand (inflate) completely because of a blocked airway, a tumour, general anaesthesia, pneumonia or other lung infections, lung disease, or long-term bed rest with shallow breathing. This is also called a collapsed lung.

Atherosclerosis: Pathological condition in which there is thickening, hardening, and loss of elasticity of the walls of blood vessels, characterised by a variable combination of changes of the innermost layer consisting of local accumulation of lipids, complex carbohydrates, blood and blood components, fibrous tissue, and calcium deposits. In addition, the outer layer becomes thickened and there is fatty degeneration of the middle layer.

Atmosphere: The sum total of all the gases surrounding the Earth, extending several hundred kilometres above the surface in a mechanical mixture of various gases in fluidlike motion.

Atmosphere, an: A unit of pressure equal to the pressure exerted by a vertical column of mercury 760 mm high at a temperature of 0°C and under standard gravity.

Atmosphere, the: The gaseous envelop surrounding a planet; the Earth's atmosphere is surrounded by a whole mass of air largely composed of oxygen (20.9%) and nitrogen (79.1%) by volume and carbon dioxide (0.03%) and traces of noble gases, water vapour, organic matter and suspended solid particles, etc.

Atmospheric dispersion: The mechanism of dilution of gaseous or smoke pollution leading to progressive decrease of pollutants.

Atom: The smallest part of a substance that cannot be broken down chemically. Each atom has a nucleus (centre) made up of protons (positive particles) and neutrons (particles with no charge). Electrons (negative particles) move around the nucleus. Atoms of different elements contain different numbers of protons, neutrons, and electrons.

Atomic absorption: Quantitative chemical method used for the analysis of elemental constituents.

Atomic mass: The mass of an atom expressed in atomic mass units (amu); the total number of protons and neutrons in the nucleus.

Atomic mass unit (amu): A unit of mass equal to 1/12 the mass of the carbon isotope with mass number 12, approximately $1.6604 \times 10E-24$ g.

Atomic number: The number of protons in the nucleus of an atom.

Atomic weight: The average weight of an atom of an element, usually expressed relative to one atom of the carbon isotope taken to have a standard weight of 12.

ATP (adenosine triphosphate): A substance present in all living cells that provides energy for many metabolic processes and is involved in making RNA. ATP made in the laboratory is being studied in patients with advanced solid tumours to see if it can decrease weight loss and improve muscle strength.

Atrophy: A wasting or decrease in size of a body organ, tissue, or part owing to disease, injury, or lack of use involving a decrease in size and (or) numbers of cells. The process is observed during the wasting of a tissue or an organ.

ATSDR: Agency for Toxic Substances and Diseases Registry.

Attention: The ability to focus selectively on a selected stimulus, sustaining that focus and shifting it at will; the ability to concentrate.

Audit: A process involving a systematic, independent, and documented evaluation for obtaining evidence to determine the extent to which defined criteria are fulfilled. The audit should be conducted by qualified, competent, and certified individuals and/or authorities internal or external to the workplace and who are independent of the activity being audited.

Authoritative scientific or regulatory organisation: Organisations that either have regulatory authority over a subject (such as control of certain chemicals in certain contexts) or are widely recognised as using the best available scientific practices and peer review processes in developing their policies and recommendations about that subject. Our lists of recognised health hazards come from lists already put together by authoritative organisations.

Auto-ignition temperature: The auto-ignition temperature is the lowest temperature at which a material begins to burn in air in the absence of a spark or flame. Many chemicals will decompose (break down) when heated. Auto-ignition temperatures for a specific material can vary by 100°C or more, depending on the test method used. Therefore, values listed in documents such as a material safety data sheet (MSDS) may be rough estimates. To avoid the risk of fire or explosion, hazardous chemical substances/materials must be stored and handled at temperatures well below the auto-ignition temperature.

Autoimmune disease: Pathological condition resulting when an organism produces antibodies or specific cells that bind to constituents of its own tissues (autoantigens) and cause tissue injury: examples of such disease may include rheumatoid arthritis, myasthenia gravis, systemic lupus erythematosus, and scleroderma. A condition in which the body recognises its own tissues as foreign and directs an immune response against them.

Autoimmunity: A health condition in which the immune responses of an animal are directed against own tissues.

Auto-oxidation: (i) Self-catalysed oxidation reaction that occurs spontaneously in an aerobic environment. (ii) Oxidation caused by the atmosphere; an oxidation reaction begun only by an inductor.

Autophagosome: The name of the membrane-bound body occurring inside a cell and containing decomposing cell organelles.

Autopsy (necropsy): Post-mortem examination of the organs and body tissue to determine cause of death or pathological condition.

Autotrophic: Related with those organisms that produce their own organic constituents from inorganic compounds utilising sunlight for energy or by oxidation process.

Ayurveda: A medical system from India that has been used for thousands of years. The goal is to cleanse the body and to restore balance to the body, mind, and spirit. It uses diet, herbal medicines, exercise, meditation, breathing, physical therapy, and other methods. It is a type of complementary and alternative medicine (CAM) therapy. Also called ayurvedic medicine.

B cell: A type of immune cell that makes proteins called antibodies, which bind to microorganisms and other foreign substances and help fight infections. A B cell is a type of WBC. Also called B lymphocyte.

Back siphon: A reverse flow that can draw a spray mixture from the tank into the water supply.

Back siphonage: The backflow of contaminated or polluted water, from a plumbing fixture or cross-connection into a water supply line, due to a lowering of the pressure in the line.

Backbone: The bones, muscles, tendons, and other tissues that reach from the base of the skull to the tailbone. The backbone encloses the spinal cord and the fluid surrounding the spinal cord. Also called spinal column, spine, and vertebral column.

Backflow prevention: A system designed to protect potable water from wastewater contamination that could occur if wastewater pressure exceeds potable water pressure over a cross-connection where one or more check valves fail.

Background levels: Two types of background levels may exist for chemical substances: (a) naturally occurring levels (ambient concentrations of substances present in the environment, without human influence) and (b) anthropogenic levels (concentrations of substances present in the environment due to human-made, non-site sources [e.g. automobiles, industries]).

Backwashing: The process of cleaning a rapid sand or mechanical filter by reversing the flow of water.

Bacteria: (i) A large group of single-cell microorganisms. Some cause infections and disease in animals and humans. The singular of bacteria is bacterium. (ii) Living single-cell organisms. Bacteria can be carried by water, wind, insects, plants, animals, and people and survive well on skin and clothes and in human hair. They also thrive in scabs, scars, mouth, nose, throat, intestines, and room-temperature foods. Often bacteria are maligned as the causes of human and animal disease, but there are certain types that are beneficial for all types of living matter. (iii) Any of numerous unicellular microorganisms of the class *Schizomycetes*, occurring in a wide variety of forms, existing either as free-living organisms or parasites, and having a wide range of biochemical, often pathogenic, properties. Some bacteria are capable of causing human, animal, or plant diseases; others are essential in pollution control because they break down organic matter in air and water.

Bacterial examination: The examination of water and wastewater to determine the presence, number, and identification of bacteria. Also called as bacterial analysis.

Bactericide: A chemical substance intended to kill bacteria.

Bagassosis: Lung disease caused by the inhalation of dust from sugar cane.

Banks, sludge: Accumulations of solid, sewage, or industrial waste deposits on the bed of a waterway.

Barbiturates: A drug used to treat insomnia, seizures, and convulsions and to relieve anxiety and tension before surgery. It belongs to the family of drugs called central nervous system (CNS) depressants.

Base: Any substance that contains OH groups and furnishes hydroxide ions in solution; a molecular or ionic substance capable of combining with a proton to form a new substance; a substance that provides a pair of electrons for a covalent bond with an acid; a solution with a pH of greater than 7.

Base pairing: A process involving the linking of the complementary pair of polynucleotide chains of DNA. The linking process is by means of hydrogen bonds between the opposite purine and pyrimidine pairs. Stable base pairs form only between adenine and thymine (A–T) and guanine and cytosine (G–C). In RNA, uracil replaces thymine and can form a base pair with adenine.

BCF: Bioconcentration factor.

BEI: Biological Exposure Index.

BEN: Balkan endemic nephropathy.

Bench to bedside: A term used to describe the process by which the results of research done in the laboratory are directly used to develop new ways to treat patients.

Benchmark dose (BMD) or concentration (BMC): A dose or concentration that produces a predetermined change in response rate of an adverse effect (called the benchmark response or BMR) compared to background.

Benchmark response (BMR): An adverse effect, used to define a BMD from which a chronic reference dose (RfD) or chronic reference concentration (Rfc) can be developed. The change in response rate over background of the BMR is usually in the range of 5%–10%, which is the limit of responses typically observed in well-conducted animal experiments.

Benign: (i) This adjective is applied to any growth that does not invade surrounding tissue (*See* malignant, tumour). (ii) Not cancerous; cannot invade neighbouring tissues or spread to other parts of the body; a condition of tissue growth that is harmless.

Benign tumour: (i) A slow-growing set of cells with abnormal look of a tumour. (This term refers to any tissue growth that does not invade surrounding tissues in the body.) (ii) A growth that is not cancer. (iii) A tumour that does not spread to a secondary localisation but may impair normal biological function through obstruction or may progress to malignancy later.

Benzene: Benzene is an aromatic hydrocarbon chemical. It is a widely used chemical formed from both natural processes and human activities. Breathing benzene can cause drowsiness, dizziness, and unconsciousness. Breathing very high levels of benzene can result in death.

Benzo(a)pyrene: A chemical that comes from certain substances when they are not burned completely. It is found in car exhaust, smoke from wood fires, tobacco, oil and gas products, charred or grilled foods, and other sources. It may also be found in water and soil. Benzo(a)pyrene can cause a skin rash, a burning feeling, skin colour changes, warts, and bronchitis. It may also cause cancer. It is a type of PAH. Also called 3,4-benzpyrene.

Beryllium disease (berylliosis): Serious and usually permanent lung damage resulting from chronic inhalation of beryllium.

Best available technology for toxics (T-BAT): The term denotes an emission standard that reflects the maximum reduction in emissions of, and risk from, a toxic air contaminant (TAC) that the district determines that can reasonably be achieved by a process or process equipment, taking into account energy, environmental, and economic impacts and other costs, and health and welfare benefits.

T-BAT includes many other inputs such as work practices, raw material substitutions, alternative processes and process design characteristics, air pollution control equipment, pollution prevention measures, and equipment maintenance measures (including leak detection and repair).

Best management practices (BMP): Procedures or controls other than effluent limitations to prevent or reduce pollution of surface water including run-off control, spill prevention, and operating procedures.

Beta burns: Beta-emitting isotopes from smoke and fallout can cause desquamation from high-dose local radiation delivered to exposed skin surfaces but only if these isotopes are in contact with the skin for longer than 1 h. Since beta radiation is not as penetrating as gamma radiation, dry desquamating skin lesions secondary to beta burns may not be as serious as web desquamating lesions.

Bias: (i) A technical term for playing favourites in choosing study subjects or in assessing exposure. (ii) Any effect at any stage of investigation or inference tending to produce results that depart systematically from the true values (to be distinguished from *random error*). The term 'bias' does not necessarily carry an imputation of prejudice or other subjective factor, such as the experimenter's desire for a particular outcome. This differs from conventional usage in which bias refers to a partisan point of view. Many varieties of bias have been described, for instance, bias in analytic research (*J. Chron. Dis.* 32:51–53 (1979)) (i) *ascertainment bias* (systematic error, arising from the kind of individuals or patients, for example, slightly ill, moderately ill, acutely ill, and seriously ill, that the individual observer is seeing. Also systematic error arising from the diagnostic process [which may be determined by the culture, customs, or individual idiosyncrasy of the person providing care for the patient]); (ii) *bias due to instrumental error* (systematic error due to faulty calibration, inaccurate measuring instruments, contaminated reagents, incorrect dilution or mixing of reagents, etc.); (iii) *bias, in assumption, or conceptual bias* (error arising from faulty logic or premises or mistaken beliefs on the part of the investigator, or false conclusions about the explanation for associations between variables. For example, having correctly deduced the mode of transmission of cholera, John Snow concluded that yellow fever was transmitted by similar means. In fact, the 'miasma' theory would better fit the facts of yellow fever transmission); (iv) *bias in autopsy series* (systematic error resulting from the fact that autopsies represent a nonrandom sample of all deaths); and (v) *bias in handling* (error arising from a failure to discard an unusual value occurring in a small sample or due to exclusion of unusual values that should be included).

Bilateral: Affecting both the right and left sides of the body.

Bile: A fluid made by the liver and stored in the gallbladder. Bile is excreted into the small intestine, where it helps digest fat.

Biliary tract: The organs and ducts that make and store bile (a fluid made by the liver that helps digest fat) and release it into the small intestine. The biliary tract includes the gallbladder and bile ducts inside and outside the liver. Biliary tract is also called biliary system.

Bilirubin: (i) Bilirubin is the yellow breakdown product of normal haem. (ii) Orange-yellow pigment, a breakdown product of haem-containing proteins (haemoglobin, myoglobin, cytochromes), which circulates in the blood plasma bound to albumin or as water-soluble glucuronide conjugates and is excreted in the bile by the liver.

Bilirubinuria (urinary bilirubin): (i) Excretion of bilirubin in the urine. (ii) The presence of bile pigments on the urine.

Bioaccumulants: Substances that increase in concentration in living organisms as they take in contaminated air, water, or food because of the very slow metabolic conversion or excretion of the substances.

Bioaccumulation: (i) Bioaccumulation describes the process by which a chemical accumulates in a living organism – either from the surrounding environment (air, water, or soil) or from other sources, for example, by consuming food containing the substance. (ii) The absorption and storage of toxic chemicals, heavy metals, and certain pesticides in plants and animals. For example, lead that is ingested by calves can bioaccumulate in their bones, interfering with calcium absorption and bone development. Stored chemicals may be released to the bloodstream at a later time, for example, during gestation or weight loss. Bioconcentration is a synonym for bioaccumulation. It is the process by which chemicals concentrate in an organism. For example, dichlorodiphenyltrichloroethane (DDT) concentrates in fish and birds that eat fish. This concentration effect is expressed as the ratio of the concentration of the chemical in an organism (like a fish) to its concentration in the surrounding medium (usually water). Bioaccumulation refers to the uptake of chemicals both from water (bioconcentration) and from ingested food and sediment.

Bioaccumulation factor: (i) The ratio of concentration of a chemical in an organism to its concentration in the food. (ii) Progressive increase in the amount of a substance in an organism or part of an organism that occurs because the rate of intake exceeds the organism's ability to remove the substance from the body.

Bioactivation: Metabolic conversion of a xenobiotic to a more toxic derivative or one that has more of an effect on living organisms.

Bioassay: (i) The quantitative measurement of the effects of a chemical substances on the organism under standard conditions. (ii) Procedure for estimating the concentration or biological activity of a substance such as vitamin, hormone, plant growth factor, antibiotic, and enzyme by measuring its effect on a living system compared to a standard system. (iii) An assay method using a change in biological activity as a qualitative or quantitative means of analysing a material response to industrial waste and other wastewater by using viable organisms or live fish as test organisms. An assay for determining the potency (or concentration) of a substance that causes a biological change in experimental animals.

Bioavailability: (i) The degree to which a substance becomes available to the target tissue after administration or exposure. Bioavailability is the measurement of the extent of a therapeutically active drug that reaches the blood and is available at the site of action (in the body). (ii) Bioavailability is a measurement of the degree to which a chemical in the environment can be taken up into a living organism. Reduced bioavailability, which can occur if the chemical is bound to something else, means a lower exposure.

Biochemical mechanism: General term for any chemical reaction or series of reactions, usually enzyme catalysed, which produces a given physiological effect in a living organism.

Biochemical oxygen demand (BOD): (i) A measure of the amount of oxygen consumed in 5 days due to natural, biological processes that break down organic matter, such as those that take place when manure or sawdust is put in water. High levels of oxygen-demanding wastes in waters deplete dissolved oxygen (DO), thereby endangering aquatic life. Chemical oxygen demand (COD) is a measure of the oxygen consumed when organic matter is broken down chemically rather than naturally. COD can be determined much more quickly than BOD and more accurately reflects the amount of organic matter in a water sample. (ii) The amount of oxygen used for biochemical oxidation by a unit volume of water at a given temperature and for a given period of time. In the measurement of degree of water pollution, BOD finds great application. (iii) The quantity of oxygen required for the biochemical oxidation of organic matter in a specified time, at a specified temperature, and under specified conditions; standard test used in assessing wastewater BOD.

Biochemicals: Chemical substances that are either naturally occurring or identical to naturally occurring substances. Examples include hormones, pheromones, and enzymes. Biochemicals function as pesticides through non-toxic, nonlethal modes of action, such as disrupting the mating pattern of insects, regulating growth, or acting as repellents. Biochemicals tend to be environmentally compatible and are thus important to integrated pest management (IPM) programmes.

Biocide: (i) A general term for any substance that kills or inhibits the growth of microorganisms (mould, slime, bacterium, fungus). (ii) Chemical agents with the capacity to kill biological life forms, for example, bactericides, insecticides, and pesticides.

Bioconcentration: The process by which a chemical can build up in a living organism to levels higher than those found in the surrounding environment. For example, a fish will have a higher concentration of a chemical in its body than is present in the water it swims in. Bioconcentration refers to the uptake of the chemical from water only – not all sources of exposure to the chemical (*See also* Bioaccumulation). (i) A process whereby living organisms acquire chemical substances from water through gills/integument and store in their bodies at concentration higher than in the environment. (ii) A process leading to a higher concentration of a substance in an organism than in environmental media to which it is exposed.

Biodegradability: The susceptibility of a substance to decomposition by microorganisms, specifically the rate at which compounds may be chemically broken down by bacteria and/or natural environmental factors.

Biodegradable: The capability of an organism/biological system to break the organic chemical substances.

Biodegradation: A measure of the ability of a chemical to be broken down into smaller units by living organisms. Biodegradation is a key process for the natural reduction of hazardous chemicals. (i) Biodegradation is the chemical dissolution of materials by bacteria or other biological means. The term is often used in relation to ecology. (ii) Biodegradation is the decomposition of organic material by microorganisms. The term biodegradation is often used in relation to sewage treatment. (iii) Transformation process of organic chemical substances into new compounds through biochemical reactions or the actions of microorganisms such as bacteria.

Bioelimination: (i) Removal, usually from the aqueous phase, of a test substance in the presence of living organisms by biological processes supplemented by physico-chemical reactions. (ii) Removal, usually from the aqueous phase, of a test substance in the presence of living organisms by biological processes supplemented by physico-chemical reactions.

Bioequivalence: Relationship between two preparations of the same drug in the same dosage form that have a similar bioavailability.

Biohazardous infectious material: A material that contains organisms that can cause disease in humans or animals. Included in this category are bacteria, viruses, fungi, and parasites. Because these organisms can live in body tissues or fluids (blood or urine), the tissues and fluids are also treated as toxic. For example, a person exposed to a blood sample from someone with hepatitis B may contract the disease.

Biological agent: A substance that is made from a living organism or its products and is used in the prevention, diagnosis, or treatment of cancer and other diseases. Biological agents include antibodies, interleukins, and vaccines. Also called biologic agent and biological drug.

Biological control: Control of pests by using living organisms.

Biological control of pests. Control, but not total eradication, of insect pests achieved by using natural enemies, either indigenous or imported, or diseases to which the pest is susceptible. It includes such non-toxic pesticides as *Bacillus thuringiensis* (Bt).

Biological cycle: Complete circulatory process through which a substance passes in the biosphere. It may involve transport through the various media (air, water, soil), followed by environmental transformation and carriage through various ecosystems.

Biological degradation: The breakdown of a pesticide in the soil due to the activities of living organisms, such as bacteria and fungi.

Biological effect monitoring (BEM): Continuous or repeated measurement of early biological effects of exposure to a substance to evaluate ambient exposure and health risk by comparison with appropriate reference values based on knowledge of the probable relationship between ambient exposure and biological effects.

Biological half-life ($t_{1/2}$): The time required for the amount of a particular substance in a biological system to be reduced to one-half of its value by biological processes when the rate of removal is approximately exponential. Substances with a long biological half-life will tend to accumulate in the body and are, therefore, particularly to be avoided. Substances with a short biological half-life may accumulate if some becomes tightly bound, even if most is cleared from the body rapidly. There is also the possibility of cumulative effects of a chemical substance that has a short residence time in the body.

Biological indicator: Species or group of species that is representative and typical for a specific status of an ecosystem, which appears frequently enough to serve for monitoring and whose population shows a sensitive response to changes, for example, the appearance of a toxicant in an ecosystem.

Biological marker (biomarker): Indicators of changes or events in human biological systems. Biological markers of exposure refer to cellular, biochemical, or molecular measures that are obtained from biological media such as human tissues, cells, or fluids and are indicative of exposure to environmental contaminants (NRC, 1991: Human Exp. for Airborne Pollutants).

Biological monitoring: (i) A procedure of periodic examination of biological specimens for the purposes of monitoring. It is usually applied to exposure monitoring but can also apply to effect monitoring. The term is also used to mean assessment of the biological status of populations and communities of organisms at risk, in order to protect them and to have early warning of possible hazards to human health. (ii) Continuous or repeated measurement of any naturally occurring or synthetic chemical, including potentially toxic substances or their metabolites or biochemical effects in tissues, secreta, excreta, expired air, or any combination of these in order to evaluate occupational or environmental exposure and health risk by comparison with appropriate reference values based on knowledge of the probable relationship between ambient exposure and resultant adverse health effects.

Biological oxygen demand (BOD): An indirect measure of the concentration of biologically degradable material present in organic wastes. It usually reflects the amount of oxygen consumed in 5 days by biological processes breaking down organic waste.

Biological pesticides: (i) Chemicals that are derived from plants, fungi, bacteria, or other non-man-made synthesis and that can be used for pest control. These agents usually do not have toxic effects on animals and people and do not leave toxic or persistent chemical residues in the environment.

Biological uptake: The transfer of substances from the environment to plants, animals, and humans.

Biological wastewater treatment: Forms of wastewater treatment in which bacterial or biochemical action is intensified to stabilise, oxidise, and nitrify the unstable organic matter present. Intermittent sand filters, contact beds, trickling filters, and activated sludge processes are examples.

Biologically based dose–response (BBDR) model: A predictive tool used to estimate potential human health risks by describing and quantifying the key steps in the cellular, tissue, and organismal responses as a result of chemical exposure.

Biologically effective dose: (i) The amount of the deposited or absorbed contaminant that reaches the cells or target site where an adverse effect occurs or where an interaction of that contaminant with a membrane surface occurs (NRC, 1991: Human Exp. for Airborne Pollutants).

Biomagnification: (i) A phenomenon where the bioaccumulated chemical substances increase in concentration as they pass upwards through two or more trophic levels. (ii) A general term applied to the sequence of processes in an ecosystem by which higher concentrations are attained in organisms of higher trophic level, meaning of higher levels in the food chain. This is a process by which xenobiotics increase in body concentration in organisms through a series of prey–predator relationships from primary producers to ultimate predators, often human beings. The tissue concentration increases as trophic level increases. (i) A parameter used to identify a toxic effect in an individual organism and can be used in extrapolation between species. (ii) A chemical, biochemical, or functional indicator as an effect of exposure to environmental chemical, physical, or biological agent. (iii) An indicator signalling an event or condition in a biological system or sample and giving a measure of exposure, effect, or susceptibility.

Biomass: (i) Total amount of biotic material, usually expressed per unit surface area or volume, in a medium such as water, and (ii) material produced by the growth of microorganisms, plants, or animals.

Biomedical testing: The testing of persons/workers to find out whether or not a change in the body function might have occurred because of exposure to hazardous chemical substances.

Biomolecule: A chemical substance that is synthesised by and occurs naturally in living organisms.

Biomonitoring: The use of living organisms to test the suitability of effluent for discharge into receiving waters and to test the quality of such waters downstream from a discharge.

Biopesticide: Biological agent with pesticidal activity, for example, the bacterium Bt when used to kill insects.

Biopolymers: Macromolecules (including proteins, nucleic acids, and polysaccharides) formed by living organisms.

Biopsy: (i) The removal of cells or tissues for examination by a pathologist. The pathologist may study the tissue under a microscope or perform other tests on the cells or tissue. There are many different types of biopsy procedures. The most common types include (ii) incisional biopsy, in which only a sample of tissue is removed; (iii) excisional biopsy, in which an entire lump or suspicious area is removed; and needle biopsy, in which a sample of tissue or fluid is removed with a needle. When a wide needle is used, the procedure is called a core biopsy. When a thin needle is used, the procedure is called a fine-needle aspiration biopsy. (iv) Excision of a small piece of living tissue for microscopic or biochemical examination; usually performed to establish a diagnosis.

Biospecimens: Samples of material, such as urine, blood, tissue, cells, DNA, RNA, and protein from humans, animals, or plants. Biospecimens are stored in a biorepository and are used for laboratory research. If the samples are from people, medical information may also be stored along with a written consent to use the samples in laboratory studies.

Biosystem: A group of molecules that interact as a whole. One type of biosystem is a biological pathway, which can consist of interacting genes, proteins, and small molecules. Another type of biosystem is a disease, which can involve components such as genes, biomarkers, and drugs.

Biota: The plants and animals in an environment. Some of these plants and animals might be sources of food, clothing, or medicines for people.

Biotechnology: (i) The use of technology, based on living systems, to develop processes and products for commercial, scientific, or other purposes. These include specific techniques of plant regeneration and gene manipulation and transfer. (ii) The application of living organisms to produce new products/substances. (iii) Agricultural biotechnology is a collection of scientific techniques, including genetic engineering, that are used to create, improve, or modify plants, animals, and microorganisms. Using conventional techniques, such as selective breeding, scientists have been working to improve plants and animals for human benefit for hundreds of years. Modern techniques now enable scientists to move genes (and therefore desirable traits) in ways they could not before – and with greater ease and precision.

Biotin: A nutrient in the vitamin B complex that the body needs in small amounts to function and stay healthy. Biotin helps some enzymes break down substances in the body for energy and helps tissues develop. It is found in yeast, whole milk, egg yolks, and organ meats. Biotin is water-soluble (can dissolve in water) and must be taken in every day. Not enough biotin can cause skin, nerve, and eye disorders. Biotin is present in larger amounts in some cancer tissue than in normal tissue. Attaching biotin to substances used to treat some types of cancer helps them find cancer cells. Also called vitamin H.

Biotransformation: This is the process by which living organisms transform chemicals in the environment. (i) In this process, a chemical substance is modified by a living organism in contrast to abiotic processes referred to earlier. The enzyme-mediated transformation of xenobiotics, frequently involving phase 1 and phase 2 reactions. (ii) A chemical conversion of a substance that is mediated by living organisms or enzyme preparations.

Black carbon: Emitted during the burning of coal, diesel fuel, natural gas, and biomass.

Bladder: The organ that stores urine.

Blights: Diseases that hurt and sometimes destroy plants. Blights will cause a plant to wither and stop growing or cause all or parts of it to die.

BLL (blood lead level): Lead in the human body can be measured in blood, urine, bones, teeth, or hair. The most frequent test is to measure the blood lead level (BLL). Measuring an individual's BLL can detect lead poisoning in adults or children. RBCs increase erythrocyte protoporphyrin (EP) when blood lead is high. Children with an EP of 35 μg/dL should be tested for a BLL. Children with a BLL of 20 μg/dL or higher should be screened for lead poisoning. Medical treatment is necessary if the BLL is higher than 45 μg/dL.

Blood: (i) The tissue with RBCs, WBCs, platelets, and other substances suspended in fluid called plasma. Blood takes oxygen and nutrients to the tissues and carries away wastes. (ii) The blood is transported throughout the body by the circulatory system. Blood functions in two directions: arterial and venous. Arterial blood is the means by which oxygen and nutrients are transported to tissues, while venous blood is the means by which carbon dioxide and metabolic by-products are transported to the lungs and kidneys, respectively, for removal from the body, bodily organs, and tissues.

Blood–placenta barrier: Physiological interface between maternal and foetal blood circulations that filters out some substances that could harm the foetus while favouring the passage of others such as nutrients: many fat-soluble substances such as alcohol are not filtered out and several types of virus can also cross this barrier. It is important to remember that the effectiveness of the interface as a barrier varies with species and different forms of placentation.

Blood pressure: The force of circulating blood on the walls of the arteries. Blood pressure is taken using two measurements: systolic (measured when the heart beats, when blood pressure is at its highest) and diastolic (measured between heartbeats, when blood pressure is at its lowest). Blood pressure is written with the systolic blood pressure first, followed by the diastolic blood pressure (e.g. 120/80).

Blood urea nitrogen (BUN): Nitrogen in the blood that comes from urea (a substance formed by the breakdown of protein in the liver). The kidneys filter urea out of the blood and into the urine. A high level of urea nitrogen in the blood may be a sign of a kidney problem. Blood–brain barrier is the physiological interface between brain tissues and circulating blood created by a mechanism that alters the permeability of brain capillaries, so that some substances are prevented from entering brain tissue, while other substances are allowed to enter freely.

Blowdown: Removal of liquids and/or solids from a process vessel or storage vessel or line by the use of pressure; often used to remove materials that, in high concentrations, could cause damage to the vessel or line or exceed limits established by best engineering practices.

Blowoff: A controlled outlet on a pipeline, tank, or conduit that is used to discharge water or accumulations of material carried by the water.

BOD: Acronym for biological oxygen demand or biochemical oxygen demand. It is used as a method of determining how much contamination has entered a water supply. It is used primarily for waters that receive pollution from sewage and industrial wastes.

BOD test: A test that measures the amount of oxygen that is consumed as microbial life.

Body burden: Also known as chemical load; the amount of harmful chemicals present in the body of a person. The biomonitoring studies estimate the exposure by measuring the chemicals or their metabolites in human specimens such as blood or urine. Results are usually expressed in mass units, such as grams and milligrams. These chemical residues – termed the 'chemical body burden' – are detected in blood, urine, and breast milk. It is important that all of us should be aware of chemical compounds in their bodies and more so occupational workers in chemical industries.

Boiling point: Temperature at which the material changes from a liquid to a gas. Below the boiling point, the liquid can evaporate to form a vapour. As the material approaches the boiling point, the change from liquid to vapour is rapid and vapour concentrations in the air can be extremely high. Airborne gases and vapours may pose fire, explosion, and health hazards. The boiling point of a mixture is given as a range of temperatures. This is because the different ingredients in a mixture can boil at different temperatures. If the material decomposes (breaks down) without boiling, the temperature at which it decomposes may be given with the abbreviation 'dec'. Some of the decomposition chemicals may be hazardous.

Bolus: (i) A single dose of a chemical substance, originally a large pill; (ii) a dose of a chemical substance administered by a single rapid intravenous injection; and (iii) a concentrated mass of food ready to be swallowed.

Bone marrow: Flexible tissue found in the hollow interior of bones such as legs, arms, and hips. Marrow produces platelets, RBCs, and WBCs, the primary agents of the body's immune system.

Botanical pesticide: A chemical substance with activity against pests that is produced naturally within a plant and may act as a defence against predators.

Botulism: (i) Acute food poisoning caused by botulinum toxin produced in food by the bacterium *Clostridium botulinum* and characterised by muscle weakness and paralysis; disturbances of vision, swallowing, and speech; and a high mortality rate. (ii) There are three main kinds of botulism, one of which is foodborne botulism caused by eating foods that contain the botulism toxin. Foodborne botulism can be especially dangerous because many people can be poisoned by eating a contaminated food. All forms of botulism can be fatal and are considered medical emergencies. Good supportive care in a hospital is the mainstay of therapy for all forms of botulism.

Bowen disease: A skin disease marked by scaly or thickened patches on the skin and often caused by prolonged exposure to arsenic. The patches often occur on sun-exposed areas of the skin and in older white men. These patches may become malignant (cancer). Also called precancerous dermatitis and precancerous dermatosis.

Bowman's capsule (glomerular capsule): A cuplike sac at the beginning of the tubular component of a nephron in the mammalian kidney. Glomerulus is enclosed in the sac. It is the outer cortex of the kidney and the main units for blood filtering.

Bradycardia: Slow heart rate, usually fewer than 60 beats/min in an adult human.

Bradypnoea: Abnormally slow breathing.

Breakdown: Failure of insulator or insulating medium to prevent discharge or current flow.

Breathing zone: (i) Space within a radius of 0.5 m from a person's face. (ii) The zone/location in the atmosphere at which individuals (animals/humans) breath.

Bremsstrahlung radiation: Secondary photon radiation (x-ray) produced by the deceleration of charged particles through matter.

Brine: (i) A strong solution of water and salt, and a sweetener such as sugar, molasses, honey, or corn syrup may be added to the solution for flavour and to improve browning. (ii) Water saturated with or containing large amounts of a salt, especially sodium chloride. (iii) Water saturated or strongly impregnated with common salt.

British thermal unit (btu): The quantity of heat necessary to raise the temperature of 1 lb of water by 1°F.

Bronchial: Having to do with the bronchi, which are the larger air passages of the lungs, including those that lead from the trachea (windpipe) to the lungs and those within the lungs.

Bronchiole: (i) One of the small airways leading to the lungs. (ii) Bronchioles – the narrowest airways that branch from the bronchi of the trachea.

Bronchiolitis: Inflammation of the bronchioles, usually caused by a viral infection.

Bronchitis: (i) The inflammation, swelling, and reddening of the bronchi – air passages of the lungs consisting of muscle tissue lined with mucous membranes. Chronic bronchitis is defined by the presence of a mucus-producing cough most days of the month, 3 months of a year for 2 successive years without other underlying disease to explain the cough. Bronchitis can be acute or chronic. Acute bronchitis is a short-term condition that usually resolves itself and is not uncommon after a cold or single exposure to a respiratory irritant. (ii) Inflammation of the bronchioles, restricting airflow to and from the lungs. Acute bronchitis is caused by a viral or bacterial infection and is aggravated by physical activity. Chronic bronchitis may be induced by smoking. (iii) Chronic or acute inflammation of the large airways.

Bronchoconstriction: Narrowing of the air passages through the bronchi of the lungs.

Bronchodilator: A drug that relaxes the smooth muscles of the airways and relieves constriction of the bronchi.

Bronchopneumonia: (i) Inflammation of small areas of the lung. (ii) An inflammatory reaction of the lungs that usually begins in the terminal brachioles. These become clogged with a mucopurulent exudate forming consolidated patches in adjacent lobules. The disease is frequently secondary in character followed by infections of the upper respiratory tract. In infants and debilitated persons, the disease may occur as a primary affection. The symptoms include bronchial pneumonia and lobular pneumonia.

Bronchopulmonary: Pertaining to the lungs and air passages.

Bronchospasm: (i) Spasmodic narrowing of the large airways. (ii) Intermittent violent contraction of the air passages of the lungs.

Buffer: (i) A solution selected or prepared to minimise changes in hydrogen ion concentration that would otherwise occur as a result of a chemical reaction. (ii) A solution that maintains constant pH by resisting changes in pH from dilution or addition of small amounts of acids and bases.

Bulking agent: A fine, solid material that is sometimes added to a wastewater stream to produce clarification or coagulation by adding bulk to the solids.

BUN (blood urea nitrogen): A measure of the urea level in blood. Urea is cleared by the kidney. The BUN determines how well the kidneys are working. It measures the amount of nitrogen in your blood. The nitrogen is present in a chemical called urea. Urea is a waste product produced as your body digests protein. Urea is carried by the blood to the kidneys, which filter the urea out of the blood and into the urine.

Burn: A type of injury caused by heat, chemical substances, cold, electricity, radiation, and friction. Burns are characterised as follows: 1st-degree redness, 2nd-degree blisters, and 3rd-degree ulcers that heal by scarring. The severity of a burn is variable and depends on the affected or damaged area with subsequent pain due to profound injury to nerves.

Burning rate: The time it takes a sample of solid material to burn a specified distance.

Byssinosis: (i) An occupational respiratory disease caused by the long-term inhalation of cotton, flax, or hemp dust and characterised by shortness of breath, coughing, and wheezing. Also called *brown lung disease*. (ii) Pneumoconiosis caused by inhalation of dust and associated microbial contaminants and observed in cotton, flax, and hemp workers.

Bystander exposure: Liability of members of the general public to come in contact with chemical substances arising from operations or processes carried out by other individuals in their vicinity. The most common examples are (a) exposures to cigarette smoke, (b) industrial fumes, and (c) automobile exhausts.

CAA: U.S. Clean Air Act. The CAA is the law that defines EPA's responsibilities for protecting and improving the nation's air quality and the stratospheric ozone layer. The *Federal Clean Air Act* (FCAA) is the federal law passed in 1970 and last amended in 1990, which forms the basis for the national air pollution control effort. Basic elements of the act include national ambient air quality standards (AQSs) for major air pollutants, hazardous air pollutant standards, state attainment plans, motor vehicle emission standards, stationary source emission standards and permits, acid rain control measures, stratospheric ozone protection, and enforcement provisions.

Cachexia: Loss of body weight and muscle mass and weakness that may occur in patients with cancer, AIDS, or other chronic diseases.

CAD (coronary artery disease): A disease in which there is a narrowing or blockage of the coronary arteries (blood vessels that carry blood and oxygen to the heart). CAD is usually caused by atherosclerosis (a build-up of fatty material and plaque inside the coronary arteries). The disease may cause chest pain, shortness of breath during exercise, and heart attacks. The risk of CAD is increased by having a family history of CAD before age 50, older age, smoking tobacco, high blood pressure, high cholesterol, diabetes, lack of exercise, and obesity. CAD is also called coronary heart disease.

Calcification: Deposits of calcium in the body tissues. Calcification in the breast can be seen on a mammogram, but cannot be detected by touch. There are two types of breast calcification, macrocalcification and microcalcification. Macrocalcifications are large deposits and are usually not related to cancer. Microcalcifications are specks of calcium that may be found in an area of rapidly dividing cells. Many microcalcifications clustered together may be a sign of cancer.

Calcination: The calcining process is used to remove some or all unwanted volatiles from a material, for example, H_2O and CO_2, and/or to convert a material into a more stable or durable state. Varying temperatures are required to calcine various materials. For example, kaolins are calcined to form molochite. Normally materials are ground after calcining. Often calcining can produce a less stable form of a material that gradually wants to revert to the former carbonated or hydrated state. For a good example of this, mix calcium carbonate with kaolin and make a bar and fire it. Out of the kiln, it will appear to be a hard ceramic, but after several days, it will absorb CO_2 from the air and completely fracture into a powder. Pour water on it and it will immediately fracture and generate an amazing amount of heat.

Calcitonin: A hormone formed by the C cells of the thyroid gland. It helps in maintaining a healthy level of calcium in the blood. When the calcium level is too high, calcitonin lowers it.

Calcium: The mineral needed for healthy teeth, bones, and other body tissues. It is the most common mineral in the body. A deposit of calcium in body tissues, such as breast tissue, may be a sign of disease.

Calcium gluconate: The mineral calcium combined with a form of the sugar glucose. It is also being studied in the treatment of bone loss and nerve damage caused by chemotherapy.

Calorie: The measurement of the energy content of food. The body needs calories as to perform its functions, such as breathing, circulating the blood, and physical activity. When a person is sick, their body may need extra calories to fight fever or other problems.

Cancer: (i) The term for diseases where abnormal cells divide without control and can invade nearby tissues. Cancer cells can also spread to other parts of the body through the blood and lymph systems. There are several main types of cancer. Carcinoma is a cancer that begins in the skin or in tissues that line or cover internal organs. Sarcoma is a cancer that begins in bone, cartilage, fat, muscle, blood vessels, or other connective or supportive tissue. Leukaemia is a cancer that starts in blood-forming tissue such as the bone marrow and causes large numbers of abnormal blood cells to be produced and enter the blood. Lymphoma and multiple myeloma are cancers that begin in the cells of the immune system. CNS cancers are cancers that begin in the tissues of the brain and spinal cord. Also called malignancy. (ii) The injurious malignant growth of potentially unlimited size of cells and tissue invading local tissues and spreading to distant areas of the body. (iii) Cancer disease occurs when a cell, or group of cells, grows in an unchecked, uncontrolled, or unregulated manner. Cancer disease can involve any tissue of the body and can have many different forms in each body area. Most cancers are named for the type of cell or the organ in which they begin, such as leukaemia or lung cancer.

Cancer effect level (CEL): The lowest dose of chemical in a study, or group of studies, that produces significant increases in the incidence of cancer/tumours between the exposed population and its appropriate control.

Cancer Genome Anatomy Project (CGAP): A project to determine the gene expression profiles of normal, precancer, and cancer cells, leading eventually to improved detection, diagnosis, and treatment for the patient.

Cancer potency estimate: An estimate of a chemical's likelihood to cause cancer, generally derived from animal studies and extrapolated to humans.

Cancer risk: A theoretical risk for getting cancer if exposed to a chemical substance every day for 70 years (a lifetime exposure). The true risk might be lower.

Candidosis: A condition in which *Candida albicans*, a type of yeast, grows out of control in moist skin areas of the body. It is usually a result of a weakened immune system but can be a side effect of chemotherapy or treatment with antibiotics. Candidosis usually affects the mouth (oral candidosis); however, rarely, it spreads throughout the entire body. Also called candidiasis and thrush.

CANUTEC: Canadian Transport Emergency Centre.

Capillaries: The tiniest blood vessels; capillary networks connect the arterioles (the smallest arteries) and the venules (the smallest veins). The wall of a capillary is thin and leaky, and capillaries are involved in the exchange of fluids and gases between tissues and the blood.

Capsule: A sac of tissue and blood vessels that surrounds an organ, joint, or tumour. A capsule is also a form of medicine that is taken by mouth. It usually has a shell made of gelatin with the medicine inside.

Carbamates: A group of insecticides belonging to class carbamate. These insecticides interrupt nerve conduction and cause the accumulation of acetylcholine at nerve endings by reversibly binding with the acetylcholinesterase enzyme.

Carbamide: A substance formed by the breakdown of protein in the liver. The kidneys filter carbamide out of the blood and into the urine. Carbamide can also be made in the laboratory. A topical form of carbamide is being studied in the treatment of hand-foot syndrome (pain, swelling, numbness, tingling, or redness of the hands or feet that may occur as a side effect of certain anticancer drugs).

Carbohydrate: A sugar molecule. Carbohydrates can be small and simple (e.g. glucose) or they can be large and complex (e.g. polysaccharides such as starch, chitin, or cellulose).

Carbon monoxide: A poisonous gas that has no colour or odour. It is given off by burning fuel (as in exhaust from cars or household heaters) and tobacco products. Carbon monoxide prevents RBCs from carrying enough oxygen for cells and tissues to live.

Carboxylhaemoglobin: Compound that is formed between carbon monoxide and haemoglobin in the blood of animals and humans and that is incapable of transporting oxygen.

Carcinogen: A substance that can cause cancer. (i) A carcinogen is any agent, chemical, physical, or biological, that can act on living tissue in such a way as to cause a malignant neoplasm. More simply, a carcinogen is any substance that causes cancer. (ii) A chemical that can increase the incidence of cancer in exposed populations. Chemicals are classified by the International Agency for Research on Cancer (IARC) as known, probable, or possible human carcinogens based on available epidemiological and toxicological evidence.

Carcinogenesis: (i) A biological process involving the transformation of a normal cell into cancer cell. Carcinogenesis is the process that leads to development of cancer. Carcinogenesis may be a matter of induction by chemical, physical, or biological agents of neoplasms that are usually not observed, an earlier induction of neoplasms that are usually observed, and/or the induction of more neoplasms than are usually found; (ii) the origin or production of a benign or malignant tumour. The carcinogenic event modifies the genome and/or other molecular control mechanisms of the target cells, giving rise to a population of altered cells.

Carcinogenic: A term applied to any chemical substance or physical agent that can cause cancer or produce cancer.

Carcinogenicity: The ability of a substance to cause cancer. (i) The process of induction of malignant neoplasms and thus cancer, by chemical, physical, or biological agents. (ii) Cancer-causing potential (of a chemical substance or agent).

   When classifying hazardous chemical substances/materials for the workplace, under the global regulations, chemical substances/materials are identified as carcinogens if they are recognised as carcinogens by the ACGIH or the IARC. Under the U.S. OSHA Hazcom Standard, chemical substances/materials are identified as carcinogens if they are listed as either carcinogens or potential carcinogens by IARC or the U.S. NTP, if they are regulated as carcinogens by OSHA, or if there is valid scientific evidence in man or animals demonstrating a cancer-causing potential. The lists of carcinogens published by the IARC, ACGIH, and NTP include known human carcinogens and some materials that cause cancer in animal experiments.

Certain chemicals may be listed as suspect or possible carcinogens if the evidence is limited or so variable that a definite conclusion cannot be made.

Carcinogenicity test: Long-term (chronic) test designed to detect any possible carcinogenic effect of a test substance.

Carcinoid: A slow-growing type of tumour usually found in the gastrointestinal system (most often in the appendix) and sometimes in the lungs or other sites. Carcinoid tumours may spread to the liver or other sites in the body, and they may secrete substances such as serotonin or prostaglandins, causing carcinoid syndrome.

Carcinoma: Malignant (cancerous) growth that arises from the epithelium (the covering of internal and external surfaces of the body, including the lining of vessels and other small cavities). This includes the skin and lining of the organs such as breast, prostate, lung, stomach, or bowel. Carcinomas tend to spread (a process called metastasis) through the blood vessels, lymph channels, or spinal fluid to other organs such as the bone, liver, lung, or brain. Carcinoma ranges to about 90% of all types of cancer. According to the American Cancer Society, at least 80% of all cancers are carcinomas.

Cardiopulmonary: Related to or involving both the heart and the lungs.

Cardiotoxic: Chemically harmful to the cells of the heart.

Carotenoid: A yellow, red, or orange substance found mostly in plants, including carrots, sweet potatoes, dark green leafy vegetables, and many fruits, grains, and oils. Some carotenoids are changed into vitamin A in the body and some are being studied in the prevention of cancer. A carotenoid is a type of antioxidant and a type of provitamin.

Cartridge: A cylinder-shaped part of a respirator, which absorbs fumes from the air before you breathe them.

CAS number: Chemical Abstracts Service Registry Number, a unique number for each chemical in the format CAS xxx-xx-x. (ii) A unique number assigned to a substance or mixture by the American Chemical Society (ACS) CAS. The CAS number provides a single unique identifier. A unique identifier is necessary because the same material can have many different names. For example, the name given to a specific chemical may vary from one language or country to another. The CAS Registry Number is similar to a telephone number and has no significance in terms of the chemical nature or hazards of the material. The CAS Registry Number can be used to locate additional information on the material, for example, when searching in books or chemical databases. Are unique, numerical identifiers for chemical substances (chemical compounds, polymers, mixtures, alloys, and biological sequences).

Case study: A medical or epidemiological evaluation of one person or a small group of people to gather information about specific health conditions and past exposures.

Case–control study: (i) An epidemiological study contrasting those with the disease of interest (cases) to those without the disease (controls). The groups are then compared with respect to exposure history, to ascertain whether they differ in the proportion exposed to the chemical(s) under investigation. (ii) A study starts with the identification of persons with the disease (or other outcome variable) of interest and a suitable control (comparison, reference) group of persons without the disease. The relationship of an attribute to the disease is examined by comparing the diseased and non-diseased with regard to how frequently the attribute is present or, if quantitative, the levels of the attribute, in the two groups.

Catabolism: Reactions involving the oxidation of organic substrates to provide chemically available energy (e.g. ATP) and to generate metabolic intermediates. Generally, the process involves the breakdown of complex molecules into simpler ones, often providing biologically available energy.

Catalase: A haem-based enzyme that catalyses the breakdown of hydrogen peroxide into oxygen and water. It is found in all living cells, especially in the peroxisomes.

Catatonia: Schizophrenia marked by excessive and sometimes violent motor activity and excitement or by generalised inhibition.

CBD: Chronic beryllium disease.

CCAP: Cancer Chromosome Aberration Project. CCAP was designed to expedite the definition and detailed characterisation of the distinct chromosomal alterations that are associated with malignant transformation.

CCF: Common cause failure.

CCOHS: Canadian Centre for Occupational Health and Safety.

CCPR: The Codex Committee on Pesticide Residues. The Codex Alimentarius Commission (CAC) was created in 1963 by FAO and WHO to develop food standards, guidelines, and related texts such as codes of practice under the Joint FAO/WHO Food Standards Programme.

CCPS: Center for Chemical Process Safety, a part of AIChE that addresses process safety issues across the chemical, petroleum, and pharmaceutical industries.

CCR: Coal combustion residue.

CCRIS: Chemical Carcinogenesis Research Information System.

CDC (Centers for Disease Control and Prevention): An agency within the U.S. Department of Health and Human Services (DHHS) that monitors and investigates foodborne disease outbreaks and compiles baseline data against which to measure the success of changes in food safety programmes.

CDDs: Chlorinated dibenzo-*p*-dioxins (*See also* CFCs).

CDFs: Chlorinated dibenzofurans (*See also* CFCs).

CEI: Chemical Exposure Index.

Ceiling: The concentration that should not be exceeded during any part of the working exposure (ACGIH). *See* Threshold limit value.

Ceiling value (CV): Maximum permissible airborne concentration of a potentially toxic substance and is a concentration that should never be exceeded in the breathing zone. (ii) The maximum permissible concentration of a material in the working environment that should never be exceeded for any duration.

Cell: The smallest structural unit of all living organisms.

Cell cycle: Regulated biochemical steps that cells go through involving DNA replication.

Cell line: A defined population of cells that has been maintained in a culture for an extended period and that has usually undergone a spontaneous process of transformation conferring an unlimited culture life span on the cells.

Cell-mediated hypersensitivity: State in which an individual reacts with allergic effects caused by the reaction of antigen-specific T lymphocytes following exposure to a certain substance (allergen) after having been exposed previously to the same substance or chemical group.

Cell-mediated immunity: Immune response mediated by antigen-specific T lymphocytes.

Cell proliferation: Rapid increase in cell number.

CEPA: Canadian Environmental Protection Act.

CEPA Toxics List: List of substances classified by the Canadian federal government under CEPA as having the potential to cause immediate or longer-term harmful effects on the environment or human health.

CERCLA (Comprehensive Environmental Response, Compensation, and Liability Act of 1980): CERCLA, also known as Superfund, is the federal law that concerns the removal or clean-up of hazardous substances in the environment and at hazardous waste sites. Agency for Toxic Substances and Disease Registry (ATSDR), which was created by CERCLA, is responsible for assessing health issues and supporting public health activities related to hazardous waste sites or other environmental releases of hazardous substances. This law was later amended by the Superfund Amendments and Reauthorization Act (SARA).

Cerebral oedema: Accumulation of fluid in and resultant swelling of the brain.

Certified applicator: A person who is authorised to apply 'restricted use pesticide' (RUP) as result of meeting requirements for certification under Federal Insecticide, Fungicide, and Rodenticide Act (FIFRA)-mandated programmes. Applicator certification programmes are conducted by states, territories, and tribes in accordance with national standards set by EPA. 'RUPs' may be used only by or under the direct supervision of specially trained and certified applicators.

Certified pesticide applicator: Any individual who is certified under Section 4 of the FIFRA as authorised to use or supervise the use of any pesticide that is classified for restricted use. Any applicator who applies registered pesticides, only to provide a service of controlling pests without delivering any additional pesticide supplies, is not deemed to be a seller or distributor of pesticides under FIFRA.

CFCs (chlorofluorocarbons): A family of inert, non-toxic, and easily liquefied chemicals used in refrigeration, air conditioning, packaging, and insulation or as solvents and aerosol propellants. Because CFCs are not destroyed in the lower atmosphere, they drift into the upper atmosphere where their chlorine components destroy ozone.

CFR: U.S. Code of Federal Regulations.

CGA: Compressed Gas Association.

Chelation: The combination of a metallic ion with heterocyclic ring structures in such a way that the ion is held by bonds from each of the rings.

Chelation therapy: Treatment with a chelating agent to enhance the elimination or reduce the toxicity of a metal ion.

Chelators: These form water-soluble complexes with heavy metal ions in the wash water and brought in by the clothes. This helps to stabilise the bleach that is rapidly decomposed by heavy metal ions. It also helps stain removal of highly coloured stains such as tea, coffee, and red wine.

Chemical asphyxiant: A poison that blocks either the transport or use of oxygen by living organisms.

Chemical element: A substance that cannot be separated into simpler substances by chemical means. There are approximately 114 known chemical elements, 83 of which are naturally occurring. Common examples include carbon, hydrogen, sodium, and iron.

Chemical family: Classification describing the general nature of the chemical. Chemicals belonging to the same family often share certain physical and chemical properties and toxic effects. However, there may also be important differences. For example, toluene and benzene both belong to the aromatic hydrocarbon family. However, benzene is a carcinogen, and toluene is not.

Chemical formula: Sometimes called the molecular formula, tells which elements (carbon, hydrogen, oxygen, and so on) make up a chemical. Chemical formula expresses the exact composition of a molecule or substance using the chemical abbreviations of the chemical elements. It also gives the number of atoms of each element in one unit or molecule of the chemical. The chemical formula can be used to confirm the identity of ingredients or to indicate the presence of a potentially hazardous element. For example, zinc yellow has the chemical formula $ZnCrO_4$, which shows that it contains not only zinc (Zn) but also chromium (Cr).

Chemical hygiene officer: A designated person who provides technical guidance in the development and implementation of the chemical hygiene plan.

Chemical hygiene plan: A written programme that outlines procedures, equipment, and work practices that protect employees from the health hazards present in the workplace.

Chemical incident: An unforeseen event involving any nonradioactive substance, resulting in a potential toxic risk to public health or leading to the exposure of two or more individuals resulting in illness or potential illness.

Chemical interaction: The process where two or more chemical substances interact with each other, resulting in either antagonistic or synergistic effects.

Chemical inventory: A list of the hazardous chemicals known to be present using an identity that is referenced on the appropriate material safety data sheet the list may be compiled for the workplace as a whole or for individual work areas.

Chemical name: (i) The scientific designation of a chemical in accordance with the nomenclature system developed by the International Union of Pure and Applied Chemistry (IUPAC) or the CAS rules of nomenclature, or a name which will clearly identify the chemical for the purpose of conducting a hazard evaluation. (ii) The chemical name is a proper scientific name for an ingredient of a product. For example, the chemical name of the herbicide 2,4-D is 2,4-dichlorophenoxyacetic acid. The chemical name can be used to obtain additional information.

Chemical oxygen demand (COD): A measure of the oxygen required to oxidise all compounds/organic and inorganic matter in a sample of water. COD is expressed as parts per million (ppm) of oxygen taken from a solution of boiling potassium dichromate for 2 h. COD test is used to assess the strength of sewage and waste.

Chemical reactivity: The ability of a material to undergo a chemical change. A chemical reaction may occur under conditions such as heating, burning, contact with other chemicals, or exposure to light. Undesirable effects such as pressure build-up, temperature increase, or formation of other hazardous chemicals may result.

Chemical register: A list of all hazardous chemical substances used in the workplace and safety data sheets (SDS) for each chemical substance.

Chemical-resistant hood: A cover/hood that goes over the head, shoulders, and upper torso of the wearer. Designed to be worn by users and occupational workers during handling of toxic and hazardous chemical substances, in conjunction with the gas mask to protect the wearer against chemical agents such as 'liquid mustard' that can cause severe burns to the head, neck, and body.

Chemical safety: Practical certainty that there will be no exposure of organisms to toxic amounts of any substance or group of substances: this implies attaining an acceptably low risk of exposure to potentially toxic substances.

Chemical Safety and Surveillance Committee (CSSC): The CSSC is advisory to the chancellor on all matters relating to the safe use of hazardous chemicals. The primary charge to the committee is to reduce risks associated with hazardous chemicals, establish policies and procedures that meet or exceed applicable norms, monitor new regulations, and implement adopted policies and procedures for hazardous chemicals. Should there be a wilful or negligent violation of University of California, San Diego (UCSD)-established chemical safety practices and procedures, the committee has the authority to impose disciplinary measures that are subject to review and/or modification by the chancellor or his or her designated representative.

Chemical suit: A full-body suit that protects the entire body from chemical burns and possible contamination of all biohazardous materials. Large enough in the back area for air tanks and/or backpacks. The Chemturion 'Ready 1' suit carried by Safety Central is the finest quality chemical protection suit in the world. It was designed and is manufactured by the same company that manufactures the Apollo, Skylab, and Shuttle spacesuits. The Saratoga suit is a military version used by the U.S. Army and Marine Corps.

Chemical treatment: The chemical substance is treated in a way that renders the material non-recoverable.

Chemicals: The term as used in these guidelines refers to all chemicals used in the clandestine manufacture of ATS drugs (Amphetamine-type stimulants), (primarily amphetamine, methamphetamine, malondialdehyde (MDA), and dimethylamine (DMA)). It also includes chemicals used to synthesise precursors and chemicals used to process opium/morphine to heroin.

CHEMNET: Chemical Industry Mutual Aid Network.

Chemosis: Any swelling around the eye caused by exposure to a chemical substance as a result of oedema of the conjunctiva.

Chemotherapy: The development and use of chemical compounds that are specific for the treatment of diseases.

Chloracne: A skin disease characterised by acne that is caused by exposure to dioxin, pentachlorophenol, PCBs, and other chlorinated hydrocarbon compounds.

CHEMTREC: Chemical Transportation Emergency Center (part of the American Chemistry Council).

Civilian mask: A gas mask that is styled for use by civilians. As effective as a military gas mask for saving the wearer's life from NBC (nuclear, biological, chemical) agents.

Chlorinated hydrocarbons: (i) Chemical substances containing only chlorine, carbon, and hydrogen. These include a class of persistent, broad-spectrum insecticides that linger in the environment and accumulate in the food chain. Among them are DDT, aldrin, dieldrin, heptachlor, chlordane, lindane, endrin, mirex, hexachloride, and toxaphene. Other examples include 1,1,1-trichloroethylene (TCE), used as an industrial solvent. (ii) Any chlorinated organic compounds including chlorinated solvents such as dichloromethane, trichloromethylene, and chloroform.

Chlorination: Adding chlorine to water or wastewater, generally for the purpose of disinfection but frequently for accomplishing other biological or chemical results. Chlorine also is used almost universally in manufacturing processes, particularly for the plastics industry.

Chlorofluorocarbons (CFCs): A family of chemicals commonly used in air conditioners and refrigerators as coolants and also as solvents and aerosol propellants. CFCs drift into the upper atmosphere where their chlorine components destroy ozone. CFCs are thought to be a major cause of the ozone hole over Antarctica.

Choline acetylase: An enzyme system that helps to generate acetylcholine, an important neurotransmitter. Acetylcholine is synthesised from choline and acetyl CoA by the enzyme choline acetylase.

Choline acetyltransferase: An enzyme that controls the production of acetylcholine; appears to be depleted in the brains of individuals with Alzheimer.

Cholinesterase: The enzyme found primarily at nerve endings that catalyses the breakdown of acetylcholine in animals. It regulates nerve impulses by the inhibition of acetylcholine; cholinesterase inhibition in animal indicates a variety of acute symptoms such as nausea, vomiting, blurred vision, stomach cramps, and rapid heart rate.

Cholinesterase and pseudocholinesterase inhibitors: Chemical substances that inhibit the enzyme cholinesterase and thus enhance and subsequently prevent transmission of nerve impulses from one nerve cell to another or to a muscle.

CHRIS: Chemical Hazards Response Information System.

Chromaticity: A method for evaluating colour in the ceramic tile industry.

Chromatid: One of the two identical copies of DNA making up a replicated chromosome that are joined at their centromere for the process of cell division.

Chromosomal aberration: Any abnormality of chromosome number or structure could be described as an aberration.

Chromosome-type aberration: The damage expressed in both sister chromatids at the same locus.

Chromosome: The structure found in the nucleus of a cell and these structures bears the DNA with genes. The genes carry the genetic code of the organism. Chromosomes form the basis of heredity and carry genetic information in DNA in the form of a sequence of nitrogenous bases.

Chronic: Long term or prolonged. It can describe either an exposure or a health effect. A chronic exposure is a long-term exposure. Long term means lasting for months or years. A chronic health effect is an adverse health effect resulting from long-term exposure or a persistent adverse health effect resulting from a short-term exposure. The Canadian CPRs describe technical criteria for identifying materials that cause chronic health effects. These regulations are part of the WHMIS.

   (i) An event or occurrence that persists over a long period of time, lasting a long time; this term in medicine comes from the Greek chronos, meaning time. (ii) In experimental toxicology, the term chronic refers to mammalian studies lasting considerably more than 90 days or to studies ranging a large part of the lifetime of an organism. (iii) Adverse effects resulting from repeated doses of, or exposures to, a chemical substance by any route for more than 3 months.

Chronic effect: An adverse effect on any living organism in which symptoms develop slowly over a long period of time or recur frequently.

Chronic exposure: The exposures to a chemical substance over a long period of time for more than 1 year. (ii) The long-term exposure or a continued exposure or exposures occurring over an extended period of time, or a significant fraction of the test species' or of the group of individuals' or of the population's lifetime.

Chronic interstitial nephritis (CIN): (i) CIN is a form of nephritis affecting the interstitium of the surrounding tubules of the kidneys. This disease can be either acute or chronic. CIN eventually leads to kidney failure. (ii) Interstitial nephritis is a kidney disorder in which the spaces between the kidney tubules become swollen (inflamed). The inflammation of the tubules and the spaces between the tubules and the glomeruli affect the kidneys' function, including their ability to filter waste. The condition may be a temporary lesion, or it may be chronic and progressive.

Chronic reference concentration (RfC): An estimate (with uncertainty spanning perhaps an order of magnitude) of a continuous inhalation exposure for a chronic duration (up to a lifetime) to the human population (including sensitive subgroups) that is likely to be without an appreciable risk of deleterious effects during a lifetime. It can be derived from a no-observed-adverse-effect level (NOAEL), lowest-observed-adverse-effect level (LOAEL), or BMC, with uncertainty factors generally applied to reflect limitations of the data used. Generally used in EPA's noncancer health assessments.

Chronic reference dose (RfD): An estimate (with uncertainty spanning perhaps an order of magnitude) of a daily oral exposure for a chronic duration (up to a lifetime) to the human population (including sensitive subgroups) that is likely to be without an appreciable risk of deleterious effects during a lifetime. It can be derived from a NOAEL, LOAEL, or BMD, with uncertainty factors generally applied to reflect limitations of the data used. Generally used in EPA's noncancer health assessments.

Chronic syndrome: The onset of symptoms is gradual over a period longer than 2 months.

Chronic toxicity: (i) The capacity of a substance to cause adverse effects/harmful effects in the organism after long-term exposure. (ii) Adverse health effects from repeated doses of a toxic chemical or other toxic substances over a relatively prolonged period of time, generally greater than 1 year. With experimental animals, this usually means a period of exposure of more than 3 months. Chronic exposure studies over 2 years using rats or mice are used to assess the carcinogenic potential of chemical substances.

Cirrhosis: An abnormal liver condition characterised by irreversible scarring of the liver. Alcohol and viral hepatitis B and C are among the many causes of cirrhosis. Cirrhosis can cause yellowing of the skin (jaundice), itching, and fatigue. Diagnosis of cirrhosis can be suggested by physical examination and blood tests and can be confirmed by liver biopsy in some patients. Complications of cirrhosis include mental confusion, coma, fluid accumulation (ascites), internal bleeding, and kidney failure. Treatment of cirrhosis is designed to limit any further damage to the liver as well as complications. Liver transplantation is becoming an important option for patients with advanced cirrhosis.

Clastogen: Any chemical substance that causes chromosomal breaks and the consequent gain, loss, or rearrangement of pieces of chromosomes.

Clastogenesis: A process resulting in chromosomal breaks and gain, loss, or rearrangement of pieces of chromosomes.

Clastogenic: A chemical substance or process that causes chromosomal breaks.

Clinical: Related to the examination and treatment of patients; applicable to patients; a laboratory test may be of clinical value (of use to patients).

Clinical research: Research conducted with human subjects. This includes (i) patient-oriented research, research conducted with human subjects (or on material of human origin such as tissues, specimens, and cognitive phenomena), (ii) mechanisms of human disease, (iii) therapeutic interventions, (iv) clinical trials, (v) development of new technologies, (vi) epidemiological and behavioural studies, and (vii) outcomes research and health services research.

Clinical toxicology: Scientific study involving research, education, prevention, and treatment of diseases caused by chemical substances such as drugs and toxins. Clinical toxicology often refers specifically to the application of toxicological principles to the treatment of human poisoning.

Clinical trial: A prospective biomedical or behavioural research study of human subjects. Clinical trial is designed to answer specific questions about biomedical or behavioural interventions such as drugs, treatments, devices, or new ways of using known drugs. Clinical trials are used to determine whether new biomedical or behavioural interventions are safe, efficacious, and effective. The biomedical clinical trials of an experimental drug proceed through four phases: *Phase I* (tests a new biomedical intervention in a small group of people, for example, 20–80 for the first time to determine efficacy and evaluate safety, namely, to determine a safe dosage range and identify side effects), *Phase II* (studies the biomedical or behavioural intervention in a larger group of people (several hundred) to determine efficacy and further evaluate safety), *Phase III* (studies to determine efficacy of the biomedical or behavioural intervention in large groups of people (from several hundred to several thousand) by comparing the intervention to other standard or experimental interventions as well as to monitor adverse effects and to collect information that will allow the interventions to be used safely), and *Phase IV* (studies conducted after the intervention has been marketed). These studies are designed to monitor the effectiveness of the approved intervention in the general population and to collect information about any adverse effects associated with widespread use.

Clostridium botulinum: The name of a group of bacteria commonly found in soil. These rod-shaped organisms grow best in low oxygen conditions. The bacteria form spores that allow them to survive in a dormant state until exposed to conditions that can support their growth. *C. botulinum* is the bacterium that produces the nerve toxin that causes botulism.

CMA: Chemical Manufacturers Association (now the American Chemistry Council).

CNS (central nervous system): The part of the nervous system that consists of the brain and the spinal cord.

CNS depression: A state of health indicating the reduced level of consciousness.

CNS solvent syndrome: Organic solvents can affect the CNS both acutely (increased reaction time and anaesthesia) and chronically (permanent brain damage).

Cocarcinogen: A chemical substance/agent that assists carcinogens to cause cancer. (ii) Chemical, physical, or biological factor that intensifies the effect of a carcinogen.

COD (chemical oxygen demand): COD is the amount of oxygen used to oxidise reactive chemicals in water. It is a measure of water quality. The term COD is used as a measure of oxygen requirement of a sample that is susceptible to oxidation by strong chemical oxidant. The dichromate reflux method is preferred over procedures using other oxidants such as potassium permanganate, because of its superior oxidising ability, applicability to a wide variety of samples, and ease of manipulation. Oxidation of most organic compounds is 95%–100% of the theoretical value.

Code of Federal Regulations (CFR): (i) A document that codifies all rules of the executive departments and agencies of the federal government. It is divided into 50 volumes, known as titles. Title 40 of the CFR (referenced as 40 CFR) lists all environmental regulations. (ii) The codification of the general and permanent rules published in the Federal Register by the executive departments and agencies of the federal government. The code is divided into 50 titles that represent broad areas subject to regulation. Most regulations directly related to agriculture are in Title 7. Each title is divided into chapters that usually bear the name of the issuing agency, followed by subdivisions into parts covering specific regulatory areas.

Codex Alimentarius Commission (CAC): (i) The CAC is the highest international body on food standards. (ii) A joint commission of the FAO and the WHO, comprised of some 146 member countries, created in 1962 to ensure consumer food safety, establish fair practices in food trade, and promote the development of international food standards. The commission drafts non-binding standards for food additives, veterinary drugs, pesticide residues, and other substances that affect consumer food safety. It publishes these standards in a listing called the *Codex Alimentarius*.

Codon: The sequence of three nucleotides in DNA or mRNA that specifies a particular amino acid during protein synthesis; also called a triplet. Of the 64 possible codons, 3 are stop codons that do not specify amino acids.

COE (coefficient of thermal expansion): A measure of the reversible volume or length change of a ceramic material with temperature. The more it expands during heating, the more it contracts while cooling down. Glazes that do not have a similar thermal expansion to the body cause problems like crazing, shivering, and weakened ware.

Cohort: A group of individuals, identified by a common characteristic, who are studied over a period of time as part of an epidemiological investigation.

Cohort study: (i) A study in which a group of people with a past exposure to chemical substances or other risk factors are followed over time and their disease experience compared to that of a group of people without the exposure. (ii) In a cohort study, the subjects who presently have a certain condition and/or receive a particular treatment are followed over time and compared with another group who are not affected by the condition under investigation. For research purposes in a cohort study, any group of individuals who are linked in some way or who have experienced the same significant life event within a given period. There are many kinds of cohorts, including birth (e.g. all those who were born between 1970 and 1975), disease, education, employment, and family formation. (iii) Any study in which there are measures of some characteristic of one or more cohorts at two or more points in time is called cohort analysis.

Cohort study advantages: The advantage of cohort analysis is that the study design does not require strict random assignment of subjects, which is, in many cases, unethical or improbable. As in the case of smoking versus nonsmoking cohort study, random assignment is not a feasible or ethical alternative (who wants to be assigned to a smoking group if he or she is a nonsmoker?). Cohort analysis is an appealing and useful technique because it is highly flexible. It provides insight into the effects of maturation and social, cultural, and political change. In addition, it can be used with either original data or secondary data. In some instances, a cohort analysis can be less expensive than experiments or surveys.

Colic: Acute abdominal pain, especially in infants.

Colloid: Material in the nanometre to micrometre size range whose characteristics and reactions are largely controlled by surface properties. Colloidal particles are so small and light that they do not settle in water. The movement of water molecules is enough to keep them in suspension. It is important to remember that colloidal particles occur in a suspension, not a solution (if a beam of light is visible through a liquid, it is likely a solution, although it could be a suspension of low specific gravity). Bentonite contains colloidal particles (which also carry an electrolytic charge). Materials can be ground to nanosized colloidal particles in a ball mill, for example.

Colorant: Chemical substances/materials that transforms a glossy or white glaze into a coloured glaze. Colorants can be raw metal oxides (e.g. iron oxide, chrome oxide) or smelted (e.g. stains). Potters and smaller companies often use raw colorants, whereas industry employs stains. Unlike stains that are prefired, the colour of a raw powder colorant likely bears no resemblance to the colour it will produce in a glaze. In ceramics, colour is a matter of chemistry. The colour produced depends on the chemistry of the host glaze and of the mix of colorants added. The same metal oxide can participate in many colour systems. Some colorants produce the same colour across a wide range of host glazes (cobalt), and others are very sensitive to the presence or absence of specific helper or hostile oxides (chrome–tin). Colours are the most vibrant in transparent glazes where there is depth. In opaque glazes, colorants tend to produce pastel shades. Some colours are potent, and 1% can produce a strong colour. Others are weak and 10% may be needed.

Coma: A state of deep unarousable unconsciousness; a state of profound loss of consciousness.

Combustible: Able to burn. A chemical substance that ignites, burns, or supports combustion. Broadly speaking, a material is combustible if it can catch fire and burn. However, in many jurisdictions, the term combustible is given a specific regulatory meaning. The terms combustible and flammable both describe the ability of a material to burn. Commonly, combustible materials are less easily ignited than flammable materials.

Combustible liquid: A liquid with a flashpoint at a temperature lower than the boiling point; according to the National Fire Protection Association (NFPA) and the U.S. Department of Transportation (DOT), it is a liquid with a flashpoint of 100°F (37.8°C) or higher. Under the Canadian CPRs, a combustible liquid has a flashpoint from 37.8°C to 93.3°C (100°F to 200°F) using a closed-cup test. The CPR is part of the national WHMIS. The U.S. OSHA Hazcom Standard uses a similar definition. This range of flashpoints is well above normal room temperature. Combustible liquids are, therefore, less of a fire hazard than flammable liquids. If there is a possibility that a combustible liquid will be heated to a temperature near its flashpoint, appropriate precautions must be taken to prevent a fire or explosion.

Combustible material: Materials that ignite above 65°C and burn relatively slowly.

Commercial applicator: A person applying pesticides as part of a business applying pesticides for hire or a person applying pesticides as part of his or her job with another (not for hire) type of business, organisation, or agency. Commercial applicators often are certified but need to be so only if they use RUP.

Common mechanism of toxicity: Two or more chemicals or other substances that cause common toxic effect(s) by the same, or essentially the same, sequence of major biochemical events (i.e. interpreted as mode of action).

Common name: Any designation or identification such as code name, code number, trade name, brand name, or generic name used to identify a chemical other than by its chemical name.

Comparative effect level (CEL): Dose by which potency of chemicals may be compared, for example, the dose causing a maximum of 15% cholinesterase inhibition.

Compatible: Able to effectively combine or mix with something.

Compatible materials: Chemical substances that do not react together to cause a fire, explosion, or violent reaction or lead to the evolution of flammable gases or otherwise lead to injury to people or danger to property.

Compatibility agents: Chemicals that enhance the effective mixing of two or more pesticide products.

Complete carcinogens: Chemical substances that will both initiate and promote cancer.

Compressed gas: A material that is a gas at normal room temperature and pressure but is packaged as a pressurised gas, pressurised liquid, or refrigerated liquid. A substance in a container with an absolute pressure greater than 276 kPa or 40 psi at 21°C, or an absolute pressure greater than 717 kPa (40 psi) at 54°C. Regardless of whether a compressed gas is packaged in an aerosol can, a pressurised cylinder, or a refrigerated container, it must be stored and handled very carefully. Puncturing or damaging the container or allowing the container to become hot may result in an explosion.

Concentration: The amount of a substance present in a certain amount of soil, water, air, food, blood, hair, urine, breath, or any other media.

Concentration–response curve: A graph produced to show the relation between the exposure concentration of a drug or xenobiotic and the degree of response it produces, as measured by the percentage of the exposed population showing a defined, often quantal, effect. If the effect determined is death, the curve may be used to estimate a median lethal concentration (LC_{50}) value.

Condensation: The process of converting a chemical in the gaseous phase to a liquid/solid state by decreasing temperature, by increasing pressure, or both.

Confined space: (i) A space in which a hazardous gas, vapour, dust, or fume may collect or in which oxygen may be used up because of the construction of the space, its location, contents, or the work activity carried out in it. (ii) Confined space has limited or restricted means for entry or exit and not designed for continuous occupancy. Confined spaces include storage tanks, bins, boilers, ventilation and exhaust ducts, pits, manholes, vats, and reactor vessels.

Congenital: A trait or condition of disorder that exists in the organism/animal from birth.

Conjugate: In chemistry, a water-soluble derivative of a chemical formed by its combination with glucuronic acid, glutathione, sulphate, acetate, glycine, etc. Usually conjugation takes place in the liver and facilitates excretion of a chemical substance that would otherwise tend to accumulate in the body because of their solubility in body fat.

Conjunctiva: Mucous membrane that covers the eyeball and the undersurface of the eyelids.

Conjunctivitis: Inflammation of the membranes of the eye (conjunctiva).

Consumer Product Safety Commission (CPSC): An independent U.S. federal regulatory agency that protects the public against unreasonable risk of injury and death associated with consumer products.

Contact dermatitis: Dermatitis caused by contact with irritating or allergenic chemical substances.

Contact sensitiser: A substance that will induce an allergic response following skin contact. The definition for 'contact sensitiser' is equivalent to 'skin sensitiser'.

Contaminant: (i) A compound, element, or physical parameter, resulting from human activity, or found at elevated concentrations, that may have a harmful effect on public health or the environment. (ii) A chemical substance that is present in an environment at levels sufficient to cause harmful or adverse health effects to organisms, animals, and humans.

Contaminated water: Water rendered unwholesome by contaminants and pollution.

Contamination: Introduction into water, air, and soil of microorganisms, chemicals, toxic substances, wastes, or wastewater in a concentration that makes the medium unfit for its next intended use. Also applies to surfaces of objects, buildings, and various household and agricultural use products.

Control limit: A regulatory value applied to the airborne concentration in the workplace of a potentially toxic substance that is judged to be 'reasonably practicable' for the whole spectrum of work activities and that must not normally be exceeded.

Controlled products: Under the Canadian Controlled Products Regulations [part of the WHMIS], a controlled product is defined as a material, product, or substance that is imported or sold in Canada and meets the criteria for one or more of the following classes:

Class A – Compressed gas
Class B – Flammable and combustible material
    Division 1 – Flammable gas
    Division 2 – Flammable liquid
    Division 3 – Combustible liquid
    Division 4 – Flammable solid
    Division 5 – Flammable aerosol
    Division 6 – Reactive flammable material
Class C – Oxidising material
Class D – Poisonous and infectious material
    Division 1 – Material causing immediate and serious toxic effects
        Subdivision A – Very toxic material
        Subdivision B – Toxic material
    Division 2 – Material causing other toxic effects
        Subdivision A – Very toxic material
        Subdivision B – Toxic material
    Division 3 – Biohazardous infectious material
Class E – Corrosive material
Class F – Dangerously reactive material

Controlled Products Regulations (CPRs): The CPRs are Canadian federal regulations developed under the Hazardous Products Act. They are part of the national WHMIS. The regulations apply to all suppliers (importers or sellers) in Canada of controlled products intended for use in Canadian workplaces. The regulations specify the criteria for identification of controlled products. They also specify what information must be included on labels and MSDSs.

Controlled substance: Drug or chemical substance whose manufacture, possession, or use is controlled by law.

Control sample: A sample with predetermined characteristics that undergoes sample processing identical to that carried out for test samples and that is used as a basis for comparison with test samples. Examples of control samples include reference materials, spiked test samples, method blanks, dilution water (as used in toxicological testing), and control cultures, namely, samples of known biological composition.

Convulsion: Abnormal and involuntary jerks and quick movements of the body.

COPD (chronic obstructive pulmonary disease): One of the most common lung diseases that causes difficulties to normal breathing process. The American Lung Association lists coal miners, grain handlers, metal moulders, and other workers exposed to dust as being at a higher risk for chronic bronchitis. Workers in such professions should use appropriate protective measures such as respirators, and they should avoid tobacco smoke.

Corrosive: (i) Any liquid or solid chemical substance that causes visible destruction/ irreversible alteration of skin tissue at the place of contact. (ii) A chemical substance capable of causing visible destruction of, and/or irreversible changes to, living tissue by chemical action at the site of contact (i.e. strong acids, strong bases, dehydrating agents, and oxidising agents).

Corrosive material: A chemical substance that attacks (corrodes) metals or human tissues such as the skin or eyes. Corrosive materials may cause metal containers or structural materials to become weak and eventually leak or collapse. Corrosive materials can burn or destroy human tissues on contact and can cause effects such as permanent scarring or blindness. The WHMIS and the U.S. OSHA Hazcom Standard specify technical criteria for identifying materials that are classified as corrosive materials for the purposes of each regulation.

Corrosive to tissue: Any chemical substance that destroys tissues on direct contact.

Corrosive to metal: A substance or a mixture, which by chemical action, will materially damage, or even destroy, metals.

Corrosivity: A waste is corrosive if it dissolves metals and other materials or burns or damages the skin or eyes on contact. (i) Is aqueous with a pH of 2 or less or 12 or higher or a powder that when added to water produces the same pH values. (ii) It is a liquid and corrodes steel at a rate greater than 6.35 mm/year.

   Examples of corrosive wastes include acids (hydrochloric acid, sulphuric acid, nitric acid, acetic acid) and bases (sodium hydroxides, ammonium hydroxide, aqueous ammonia).

Cost–benefit analysis: A quantitative evaluation and decision-making technique where comparisons are made between the costs of a proposed regulatory action on the use of a substance/chemical with the overall benefits to society of the proposed action; often converting both the estimated costs and benefits into health and monetary units.

Counts per minute (cpm): The number of counts or nuclear events detected by a radiation survey device such as a Geiger counter. Since not all events that occur are detected, cpm are always less than actual disintegrations per minute (dpm) emanating from a radioactive material.

CPC: Chemical protective clothing.

CPQRA: Chemical process quantitative risk assessment.

CPSC: Consumer Product and Safety Commission.

Crepitations: Abnormal respiratory sounds heard on auscultation of the chest, produced by passage of air through passages that contain secretion or exudate or which are constricted by spasm or a thickening of their walls.

CRBOH: Canadian Registration Board of Occupational Hygienists.

Cristobalite: A crystalline form of silica (quartz is also) available as a raw material and formed by natural processes during firing of certain bodies. During cooling, cristobalite changes from beta to alpha form around 220°C. This change is accompanied by a 3% volume contraction. Cristobalite forms spontaneously at temperatures above 1100°C from very fine quartz found in some clays, from finely ground silica, and from molecular silica liberated during the formation of mullite from kaolin.

Critical concentration: An ambient chemical concentration expressed in units of $\mu g/m^3$ and used in the operational derivation of the inhalation RfC. This concentration will be the NOAEL human equivalent concentration (HEC) adjusted from principal study data.

Critical control point: An operation (practice, procedure, process, or location) at or by which preventive or control measures can be exercised that will eliminate, prevent, or minimise one or more hazards. Critical control points are fundamental to Hazard Analysis and Critical Control Point (HACCP) systems.

Critical effect: The first adverse effect, or its known precursor, that occurs to the most sensitive species as the dose rate of an agent increases.

Critical end point: Toxic effect used by the U.S. EPA as the basis for a reference dose.

Critical organ: That part of the body that is most susceptible to action of chemical substances and/or radiation damage under the specific conditions.

Critical study/principal study: The study that contributes most significantly to the qualitative and quantitative assessment of risk.

Cross-contamination: The transfer of harmful substances or disease-causing microorganisms to food by hands, food-contact surfaces, sponges, cloth towels, and utensils that touch raw food, are not cleaned, and then touch ready-to-eat foods. Cross-contamination can also occur when raw food touches or drips onto cooked or ready-to-eat foods.

CSB: U.S. Chemical Safety and Hazard Investigation Board.

CSF, cerebrospinal fluid: CSF is a clear, plasma-like fluid that circulates around the outside of the brain, in cavities within the brain (ventricles), and in the space surrounding the spinal cord. It mainly cushions the brain to make knocks to the head have less of an effect. It also helps nourish, support, and remove metabolic wastes. Analysing CSF by a spinal tap helps to diagnose some disorders including meningitis and haemorrhage in the CNS.

CSSB: Canadian Society of Safety Engineering.

Cumulative effect: Overall change that occurs in an organism/animal after exposures to repeated doses of a chemical substance or radiation.

Cutaneous hazard: A chemical substance that causes damage to or disease in the skin or dermal layer, which is the body's largest organ.

CW: Cooling water.

CWP: Coal workers' pneumoconiosis.

Cyanosis: Bluish discoloration of the skin due to deficient oxygenation of the blood.

Cytochrome 450: (i) A haem-containing protein that takes part in the phase I reactions of xenobiotics during biotransformation processes. (ii) The iron-containing proteins are important in cell respiration as catalysts of oxidation–reduction reactions.

Cytogenetics: Part of the science of genetics that correlates the structure and number of chromosomes with heredity and genetic variability.

Cytokine: Any of a group of soluble proteins that are released by a cell causing a change in function or development of the same cell (autocrine), an adjacent cell (paracrine), or a distant cell (endocrine); cytokines are involved in reproduction, growth and development, normal homeostatic regulation, response to injury and repair, blood clotting, and host resistance (immunity and tolerance).

Cytoplasm: In cell biology, the ground substance of the cell in which the cell organelles such as the nucleus, mitochondria, endoplasmic reticulum, and ribosomes are situated.

Cytotoxic: Chemical substance causing harmful effects/damage to cell structure and function and ultimately causing cell death.

Dangerously reactive material: The Canadian CPRs (part of the WHMIS) describe technical criteria for identifying materials that are classified as dangerously reactive. A dangerously reactive chemical substance/material can react (i) vigorously with water to produce a very toxic gas, (ii) on its own by polymerisation or decomposition, and/or (iii) under conditions of shock or an increase in pressure or temperature. The ANSI defines a dangerously reactive material as one that is able to undergo a violent self-accelerating exothermic chemical reaction with common chemical substances/materials or by itself. A dangerously reactive chemical substances/material may cause a fire, explosion, or other hazardous condition. It is very important to know which conditions (such as shock, heating, or contact with water) may set off the dangerous reaction so that appropriate preventive measures can be taken.

Dechlorination: Removal of chlorine and chemical replacement with hydrogen or hydroxide ions to detoxify a substance.

Decontamination: To make safe by eliminating poisonous or otherwise harmful substances, such as noxious chemicals or radioactive material, from people, buildings, equipment, and landscape.

Defoliant: A chemical substance used for removal of leaves by its toxic action on living plants.

Degradation: The destructive effect a chemical substance may cause on a piece of chemical protective clothing (CPC). Protective clothing that has been degraded may be partially dissolved, softened, hardened, or completely destroyed. If not destroyed, the material may have reduced strength and flexibility. This may result in easy tearing or punctures, opening up a direct route to skin contact by penetration.

Dehydrogenase: An enzyme that catalyses oxidation of compounds by removing hydrogen.

Delaney Clause: FFDCA statement that no additive shall be deemed to be safe for human food if it is found to induce cancer in man or animals. It is an example of the zero tolerance concept in food safety policy. The Delaney prohibition appears in three separate parts of the FFDCA: Section 409 on food additives; Section 512, relating to animal drugs in meat and poultry; and Section 721 on colour additives. Section 409 prohibition applied to many pesticide residues until enactment of the Food Quality Protection Act of 1996 (P.L. 104–170, 3 August 1996). This legislation removed pesticide residue tolerances from Delaney Clause constraints.

Delayed effects (latent effect): Consequences of effects occurring after a latent period following the end of exposures to toxic chemical substance or other harmful environmental factors.

Dementia: (i) Marked decline in mental function. (ii) A word for a group of symptoms caused by disorders that affect the brain. (iii) *Dementia is not a specific disease. It is an overall term that describes a wide range of symptoms* associated with a decline in memory or other thinking skills severe enough to reduce a person's ability to perform everyday activities.

Demyelination: Destruction of the myelin sheath of a nerve.

Denaturant: A substance added to ethyl alcohol to prevent its being used for internal consumption.

Denaturation: Addition of methanol, acetone, or other suitable chemical(s) to alcohol to make it unfit for drinking.

Dendrite: Any of the branched extensions, or processes, of the neuron along which nerve impulses travel towards the cell body.

Denitrification: Reduction of nitrates to nitrites, nitrogen oxides, or dinitrogen (N_2) catalysed by facultative aerobic soil bacteria under anaerobic conditions.

Density: The density of a chemical substance/material is its weight for a given volume. Density is usually given in units of grams per millilitre (g/mL) or grams per cubic centimetre (g/cm³). Density is closely related to specific gravity (relative density). The volume of a material in a container can be calculated from its density and weight.

Deoxyribonucleic acid (DNA): The constituent of chromosomes that stores the hereditary information in the form of a sequence of nitrogenous bases, purine and pyrimidine. Much of the information is related with the synthesis of proteins and acts as a determinant of all physical and functional activities of the cell and consequently of the whole organism.

Department of Transportation (DOT): U.S. federal agency that regulates the labelling and transportation of hazardous materials. U.S. EPA, U.S. federal agency that develops and enforces regulations to protect human health and the natural environment.

Depilatory: Chemical substance that causes loss of hair.

Dermal: Of the skin. Dermal absorption means absorption through the skin.

Dermal absorption (dermal penetration): (i) The transfer of contaminant across the skin and subsequent incorporation into the body. (ii) Movement of a pesticide into and through the skin and includes those taken up into the systemic circulation and those retained in the skin compartment. (iii) Process by which a chemical penetrates the skin and enters the body as an internal dose.

Dermal exposure: (i) Contact with the skin by any medium containing chemicals, quantified as the amount on the skin and available for adsorption and possible absorption. (ii) Contact between a chemical substance and the skin. (iii) This term refers to a quantifiable measure of the amount of residue deposited on skin, normally expressed as a density, or mass per unit time, deposited on a defined skin surface area (e.g. mg/h hand exposure); equivalent to potential dose for the dermal route.

Dermal irritation: A localised skin reaction resulting from either single or multiple exposures to a physical or chemical agent at the same site. It is characterised by redness and swelling and may be accompanied by local cell death.

Dermal layer: The skin of the human body, the largest human organ. The dermal layers that make up human skin are the body's first defence against heat, light, injury, and infection. The skin has two main layers: the outer layer, called the *epidermis*, and the inner layer, called the *dermis*. The epidermis is the layer of skin injured in first-degree burns. It is made up of three kinds of cells: on the outermost layer are flat, scale-like cells called keratinocytes, the (i) *squamous cells*, which constitute 95% of the epidermis. Beneath these are round cells called the (ii) *basal cells*. The innermost part of the epidermis is made up of (iii) *melanocytes*, which give skin its colour. The dermis, the inner layer of skin, contains blood vessels, nerves, lymph vessels, hair follicles, and sweat glands. This layer of skin, along with the epidermis, is injured in second-degree burns.

Dermatitis: Inflammation of the skin. Occurrence of contact dermatitis is due to local exposure to chemical substances and may be caused by irritation, allergy, or infection.

Decibel (dB): A unit used to measure sound intensity and other physical quantities.

Descriptive epidemiology: Study of the occurrence of disease or other health-related characteristics in populations, including general observations concerning the relationship of disease to basic characteristics such as age, sex, race, occupation, and social class; it may also be concerned with geographic location. The major characteristics in descriptive epidemiology can be classified under the following headings: individuals, time, and place.

Detergent: A cleaning or wetting agent, classed as anionic if it has a negative charge and cationic if it has a positive charge.

Detoxification (detoxication): (i) The process, or processes, of chemical modification that makes a toxic molecule less toxic. (ii) Treatment of patients suffering from poisoning in such a way as to promote physiological processes that reduce the probability or severity of adverse effects.

Detoxify: To reduce the toxicity of a chemical substance either (i) by making it less harmful or (ii) by treating patients suffering from poisoning in such a way as to reduce the probability and/or severity of harmful effects, (iii) to reduce or eliminate the toxicity of a chemical substance or poison, and to promote the recovery of a person from an addictive drug such as alcohol or heroin.

Developmental toxicity: Adverse effects on the developing organism (including structural abnormality, altered growth, or functional deficiency or death) resulting from exposure through conception, gestation, during prenatal development, and organogenesis till the time of sexual maturation. The major manifestations of developmental toxicity include death of the developing organism, structural abnormality, altered growth, and functional deficiency.

Device: A thing made for a special purpose, such as a respiratory protective device.

Diaphoretic: Chemical substance that causes sweating.

Dioxins: A group of chemical compounds that share certain similar chemical structures and biological characteristics. Dioxins are present in the environment all over the world. Within animals, dioxins tend to accumulate in fat. About 95% of the average person's exposure to dioxins occurs through consumption of food, especially food containing animal fat. Scientists and health experts are concerned about dioxins because studies have shown that exposure may cause a number of adverse health effects.

Diploid: The state in which the chromosomes are present in homologous pairs. It is important to remember that the normal human somatic (nonreproductive) cells are diploid (they have 46 chromosomes), whereas reproductive cells, with 23 chromosomes, are haploid.

Diplopia: Appearance of temporary double vision. Diplopia, commonly known as double vision, is the simultaneous perception of two images of a single object that may be displaced horizontally, vertically, or diagonally (i.e. both vertically and horizontally) in relation to each other. It is usually the result of impaired function of the extraocular muscles (EOMs), where both eyes are still functional but they cannot converge to target the desired object.

Disease: (i) Literally, disease, lack of ease; pathological condition that presents a group of symptoms peculiar to it and that establishes the condition as an abnormal entity different from other normal or pathological body states. (ii) Sickness often characterised by typical patient problems (symptoms) and physical findings (signs).

Discharge: Any pollutant or chemical waste or combination of pollutants added to the environment from any point or area source.

Disinfectant: A chemical substance that destroys vegetative forms of harmful microorganisms but does not ordinarily kill bacterial spores.

Distribution: A general term for the dispersal of a xenobiotic and its derivatives throughout an organism or environmental system.

Diuresis: Excretion of urine, especially in excess.

Diuretic: A chemical substance or agent that increases urine production.

DNA (deoxyribonucleic acid): (i) A nucleic acid that carries the genetic information in the cell and is capable of self-replication and synthesis of RNA. (ii) DNA is the chemical inside the nucleus of a cell that carries the genetic instructions for making living organisms. DNA is composed of two antiparallel strands, each a linear polymer of nucleotides. Each nucleotide has a phosphate group linked by a phosphoester bond to a pentose (a five-carbon sugar molecule, deoxyribose), which in turn is linked to one of four organic bases, adenine, guanine, cytosine, or thymine, abbreviated A, G, C, and T, respectively. The bases are of two types: purines, which have two rings and are slightly larger (A and G), and pyrimidines, which have only one ring (C and T). Each nucleotide is joined to the next nucleotide in the chain by a covalent phosphodiester bond between the 5′ carbon of one deoxyribose group and the 3′ carbon of the next. DNA is a helical molecule with the sugar–phosphate backbone on the outside and the nucleotides extending towards the central axis. There is specific base pairing between the bases on opposite strands in such a way that A always pairs with T and G always pairs with C.

Domestic Substances List (DSL): The Domestic Substances List (DSL) was created in accordance with the CEPA by Environment Canada. The DSL defines 'existing' substances for the purposes of implementing CEPA and is the sole basis for determining whether a substance is 'existing' or 'new' to Canada. Substances that are not on the DSL may require notification and assessment before they can be manufactured or imported into Canada.

Dominant: Allele that expresses its phenotypic effect when present in either the homozygous or the heterozygous state.

Dosage (of a chemical substance): A measurable amount of exposure to a chemical substance or a hazard. Dose divided by product of mass of organism and time of dose.

Dose: (i) The total quantity of a chemical substance administered to, taken up, or absorbed by an organism, organ, or tissue. (ii) The amount of a toxic chemical substance taken into the body over a given period of time. (iii) *Absorbed dose* – the amount of a chemical substance penetrating the exchange boundaries of an organism after contact. Calculated from intake and absorption efficiency and expressed as mg/kg-day. (iv) *Administered dose* – mass of a substance given to an organism and in contact with an exchange boundary, expressed as mg/kg-day. (v) *Applied dose* – the mount of a chemical substance given to an organism, especially through dermal contact. (vi) The amount of a chemical substance to which an individual worker/user is exposed. Dose often takes body weight into account. (vii) The amount of a pollutant that is absorbed. A level of exposure that is a function of a pollutant's concentration, and the length of time a subject is exposed. (viii) In pharmacology, the term dose indicates a known quantity of drug to be administered at one time, such as a specified amount of medication. (ix) In radiobiology, the term generally suggests the quantity of radiation or energy absorbed (for special purposes, the doe must be appropriately qualified, and if the complete term is unqualified, the term refers to absorbed dose).

Unit of absorbed dose is the rad. The SI unit is the grey (1 Gy = 1 J kg^{-1} = 100 R). Source – Gr. dosis = a giving (Graham, 1998). (x) For example, to receive equivalent doses of medicine, children are given smaller amounts than adults. (xi) Total dose is the sum of doses received by a person from a contaminant in a given interval resulting from interaction with all environmental media that contain the contaminant. Units of dose and total dose (mass) are often converted to units of mass per volume of physiological fluid or mass of tissue. (xii) The amount of a pesticide systemically available. (xiii) The amount of chemical substance(s) available for interaction with metabolic processes or biologically significant receptors after crossing the outer boundary of an organism (*See also* Potential dose, Applied dose, and Internal dose).

Dose–effect curve: (i) This is a graph drawn to show the relationship between the dose of a drug or xenobiotic and the magnitude of the graded effect that it produces.

Dose–response assessment/relationship: (i) The determination of the relationship between the magnitude of an administered, applied, or internal dose and a specific biological response. Response can be expressed as measured or observed incidence or change in level of response, per cent response in groups of subjects (or populations), or the probability of occurrence or change in level of response within a population. (II) The amount of a chemical substance that an organism/ experimental animal is *exposed to is the dose,* and the *severity of the effect of that exposure is called the response.* A dose–response assessment is a scientific study to determine the relationship between dose and response.

Dose–response curve: A graph to show the relation between the dose of a drug or xenobiotic and the degree of response it produces, as measured by the percentage of the exposed population showing a defined, often quantal effect. If the effect determined is death, such a curve may be used to estimate an LD_{50} value.

Dosimetry: A system used for determining absorbed dose, consisting of dosimeters, measurement instruments, and their associated reference standards and procedures for the system's use.

DOT: U.S. Department of Transportation.

Draize test: Evaluation of materials for their potential to cause dermal or ocular irritation and corrosion following local exposure, generally using the rabbit model (almost exclusively the New Zealand White) although other animal species have been used.

Drug: Any chemical substance which when absorbed into a living organism may modify one or more of its functions. A common term generally accepted for a chemical substance taken for a therapeutic purpose but is also commonly used for substances of abuse.

DSL: Domestic Substances List.

Dysarthria: Imperfect articulation of speech due to neuromuscular damage.

Dysfunction: Abnormal, impaired, or incomplete functioning of an organism, organ, tissue, or cell.

Dysosmia: A distorted sense of smell.

Dysphagia: Difficulty in swallowing.

Dysplasia: Abnormal development of an organ or tissue identified by morphological examination.

Dyspnoea: Difficulty in breathing or shortness of breath.

Ebola: Ebola is a filovirus named after a river in Zaire, its first site of discovery. This filovirus is usually fatal. All forms of viral haemorrhagic fever begin with fever and muscle aches. Depending on the particular virus, disease can progress until the patient becomes very ill with respiratory problems, severe bleeding, kidney problems, and shock. With Ebola, persons develop fever, chills, headaches, muscle aches, and loss of appetite. As the disease progresses, vomiting, bloody diarrhoea, abdominal pain, sore throat, and chest pain can occur. The blood fails to clot and patients bleed from injection sites as well as into the gastrointestinal tract, skin, and internal organs. Basically, you just bleed from every orifice.

ECD: Electron capture detector.

ECn: A commonly used abbreviation for the exposure concentration of a toxicant causing a defined effect on n% of a test population.

Ecology: Branch of biology that studies the interactions between living organisms and all factors (including other organisms) in their environment: such interactions encompass environmental factors that determine the distributions of living organisms.

Ecosystem: (i) A dynamic complex of plant, animal, and microorganism communities and their abiotic environment interacting as a functional unit. Grouping of organisms (microorganisms, plants, animals) interacting together, with and through their physical and chemical environments, to form a functional entity within a defined environment. (ii) The interacting synergism of all living organisms in a particular environment; every plant, insect, aquatic animal, bird, or land species that forms a complex web of interdependency. An action taken at any level in the food chain: use of a pesticide, for example, has a potential domino effect on every other occupant of that system.

Ecotoxicology: (i) Ecotoxicology is the science devoted to the study of the production of harmful effects by substances entering the natural environment, especially effects on populations, communities, and ecosystems; an essential part of ecotoxicology is the assessment of movement of potentially toxic substances through environmental compartments and through food webs; (ii) Ecotoxicology is a research field that explores how exposure to a toxicant negatively affects single organisms, populations, communities, and ecosystems.

Eczema: Acute or chronic skin inflammation with erythema, papules, vesicles, pustules, scales, crusts, or scabs, alone or in combination, of varied aetiology.

Edema (oedema): Presence of abnormally large amounts of fluid in intercellular spaces of body tissues.

EDn: A commonly used abbreviation for the dose of a toxicant causing a defined effect on n% of a test population.

Effective concentration (EC): Concentration of a substance that causes a defined magnitude of response in a given system. Note that EC_{50} is the median concentration that causes 50% of maximal response.

Effective dose (ED): Dose of a chemical substance that causes a defined magnitude of response in a given system. Note that the ED_{50} is the median dose that causes 50% of maximal response.

Effluent: Fluid, solid, or gas discharged from a given source into the external environment.

EINECS: The European Inventory of Existing Commercial Substances, which originally listed all substances that were reported to be on the market in Europe between 1 January 1971, and 18 September 1981. The substances placed on the market for the first time after this target date are 'new'. There are 100, 196 different substances in the EINECS. The EINECS was created by the European Community Commission Decision 81/437/EEC.

Electroneurography (ENG): Recording and measuring the electrical signals generated by nerves by means of an electromyograph. ENG is used in testing the effects of neurotoxic substances on humans.

Electrophoresis: The migration of charged molecules in solution in response to an electric field.

Electrophysiology: Measuring and recording the electrical activity of the brain or nerve cells by means of electrodes.

Element: One of the 103 known chemical substances that cannot be broken down further.

Elimination: The process whereby a substance or other material is expelled from the body (or a defined part thereof), usually by a process of extrusion or exclusion but sometimes through metabolic transformation. The combination of chemical degradation of a xenobiotic in the body and excretion by the intestine, kidneys, lungs, and skin and in sweat, expired air, milk, semen, menstrual fluid, or secreted fluids.

Embryo: The earliest stage of development of a plant or animal. The embryo is generally contained in another structure, the seed, egg, or uterus. An embryo is an organism in the early stages of its development prior to birth. In humans, the embryo is the developing child from conception to the end of the second month of pregnancy.

Embryonic period: Period from fertilisation to the end of major organogenesis.

Embryotoxic: Adjectival term applied to any chemical substance that is harmful, in any sense, to an embryo. Embryotoxic means harmful to the embryo.

Embryotoxicity: Embryotoxicity is the ability of a substance to cause harm to the embryo. (i) Production of toxic effects by a chemical substance in progeny in the first period of pregnancy between the conception and the foetal stage. (ii) Any kind of toxic effect on the conceptus as a result of prenatal exposure to toxic substances/agents during the embryonic stages of development: these effects may include malformations and variations, malfunctions, altered growth, prenatal death, and altered postnatal function.

Emesis: Vomiting.

Emission: The release or discharge of a substance into the environment. Generally refers to the release of gases or particulates into the air.

Emission standard: This regulatory value is a quantitative limit on the emission or discharge of a potentially toxic substance from a source. The simplest form for regulatory purposes is a uniform emission standard (UES) where the same limit is placed on all emissions of a particular contaminant (*See* Limit values).

Emulsifiable: A chemical property that allows you to mix two liquids that normally do not mix, such as oil and water.

Encephalopathy: Degeneration of the brain.

Endemic: Present in a community or among a group of people; said of a disease prevailing continually in a region.

Endocrine: Pertaining to hormones or to the glands that secrete hormones directly into the bloodstream.

Endocrine system: A series of ductless glands that produce chemical messages or extracellular signalling molecules called hormones. The endocrine system is important or responsible for regulating metabolism, growth, development and puberty, and tissue function and also plays a part in determining mood.

Endocrine toxicity: Any adverse structural and/or functional changes to the endocrine system (the system that controls hormones in the body) that may result from exposure to chemical substances. Endocrine toxicity can harm human and animal reproduction and development.

Endocrine disruptor chemicals: (i) Different groups of exogenous chemical substances that act like hormones in the endocrine system and disrupt the physiological function of endogenous hormones. (ii) Exogenous chemical substances that alters function of the endocrine system and consequently causes adverse health effects in an intact organism, its progeny or (sub)populations. (iii) Chemical substances that interfere with the production, release, transport, metabolism, binding, action, or elimination of natural substances such as sex hormones or phyto-oestrogens. The substances can come from industrial, agricultural, and municipal wastes and can cause deformities and embryo mortality, impaired reproduction and development, abnormal reproduction, depressed thyroid and immune functions, and feminisation.

Endogenous: Produced within or caused by factors within an organism/animal system.

Endoplasmic reticulum: In cell biology, this is a complex pipelike system of membranes that occupies much of the cytoplasm in cells and that contains many of the enzymes that mediate biodegradation of xenobiotics. (ii) Intracellular complex of membranes in which proteins and lipids, as well as molecules for export, are synthesised and in which the biotransformation reactions of the monooxygenase enzyme systems occur. It may also be isolated as microsomes following cell fractionation procedures.

Endothelial: Pertaining to the layer of flat cells lining the inner surface of blood and lymphatic vessels and the surface lining of serous and synovial membranes.

Endothelium: Layer of flattened epithelial cells lining the heart, blood vessels, and lymphatic vessels.

End point: An observable or measurable biological event or chemical concentration (e.g. metabolite concentration in a target tissue) used as an index of an effect of a chemical exposure.

Endothermic: A term used to characterise a chemical reaction that requires absorption of heat from an external source.

Endotoxin: Toxin that forms an integral part of the cell wall of certain bacteria and is released only upon breakdown of the bacterial cell. Endotoxins do not form toxoids.

Enteritis (intestinal inflammation): Enteritis is usually caused by eating or drinking substances that are contaminated with bacteria or viruses. The germs settle in the small intestine and cause inflammation and swelling, which may lead to abdominal pain, cramping, diarrhoea, fever, and dehydration.

Environment: (i) The term environment has several meanings and depends on specific context. In general, the term environment refers to the natural environment that includes all living and nonliving things that occur naturally on Earth. Environment consists of all, or any, of the following: the air, the water (oceans, rivers, lakes, and ponds), and the land (mountains, forests, fields, and gardens). (ii) Aggregate, at a given moment, of all external conditions and influences to which a system under study is subjected.

Environmental contaminants: The list of environmental contaminants is very large and the major and selected ones could be listed as (a) persistent organic pollutants (POPs), (b) organochlorines (OCs), and (c) polycyclic aromatic hydrocarbons (PAHs). (d) Chlorine gas is an extremely reactive and poisonous substance that rarely occurs in nature but bonds quickly with organic matter to form a new class of chemicals called OCs. Over 11,000 different OCs are manufactured today, used in products ranging from pesticides such as DDT and chlordane and insecticides such as toxaphene and mirex to plastics, toothpaste, mouthwash, and solvents. Their production and use produces thousands more unwanted OC by-products. OCs also include industrial chemicals such as PCBs and industrial waste products such as dioxin, which results from the manufacture of polyvinyl chloride or PVC, as well as furans. They may be present in old electrical transformers or may be formed when burning garbage. DDT, chlordane, PCBs, and toxaphene are banned substances in Canada; (e) the PAHs are derived from the incomplete combustion of fossil fuels producing asphalt, coal tar, and creosote, a wood preservative. As a group, they are common water contaminants but are not quite as persistent and mobile as the OCs. PAHs are not thought to constitute a major human health concern in the Arctic. (f) Another important environmental contaminant is ozone depletion, which is a thinning of the ozone layer, the blanket of ozone gas that shields us from the sun's damaging ultraviolet (UV) radiation. (g) Build-up of greenhouse gases from increased carbon dioxide emissions from fossil fuels burned by cars, industry, and power plants as well as from CFCs, methane, and nitrous oxides released into the atmosphere. Warmer temperatures could dramatically alter ecosystems in all regions of Canada and bring a range of problems such as droughts, flooding, forest fires, insect infestations, and melting permafrost. (h) Acid rain is another contaminant of the environment. Acid rain occurs when pollutants such as nitrogen oxides and sulphur dioxide from power plant emissions, metal smelting, motor vehicles, and industry combine with water in the atmosphere to form droplets of very weak acid. Acid rain, an air and water pollutant, damages lakes, threatens wildlife, and affects air quality. (i) Many heavy metals are also environmental contaminants. These include lead, mercury, and cadmium, which are toxic heavy metals that are naturally present in rocks and soils. Metal mining and smelting can result in additional human-made releases of mercury and cadmium.

Environmental damage: Adverse effects to the natural environment. Major factors associated with environmental damage include hydroelectric, mining, energy, and forestry projects; chemical pollution from industry; nuclear energy and weapons testing; toxic waste mismanagement; depletion of the ozone layer; and global warming, to name a few pressing environmental problems that have created serious human health hazards.

Environmental exposure level (EEL): The level (concentration or amount or a time integral of either) of a chemical substance to which an organism or animal or other component of the environment is exposed in its natural surroundings.

Environmental fate: The environmental fate of a chemical is a statement of what tends to happen to the chemical when it is released into the environment. Fate of a hazardous chemical depends on properties of the candidate chemical substance such as solubility, vapour pressure, partition coefficient, stability, and reactivity. The hazardous chemical may preferentially end up in air, in water, or in soil and may break down quickly or slowly. The destiny of a hazardous chemical substance or biological pollutant after release into the natural environment.

Environmental hazard: Chemical or physical agent capable of causing harm to the environment.

Environmental health: Human welfare and its influence by the environment, including technical and administrative measures for improving the human environment from a health point of view.

Environmental health criteria documents: Critical publications of the IPCS containing reviews of methodologies and existing knowledge – expressed, if possible, in quantitative terms – of selected chemical substances (or groups of chemical substances) on identifiable, immediate, and long-term effects on human health and welfare.

Environmental health impact assessment (EHIA): Estimate of the adverse effects to health or risks likely to follow from a proposed or expected environmental change or development.

Environmental hygiene (environmental sanitation): Practical control measures used to improve the basic environmental conditions affecting human health, for example, clean water supply, human and animal waste disposal, protection of food from biological contamination, and housing conditions, all of which are concerned with the quality of the human environment.

Environmental impact assessment (EIA): Appraisal of the possible environmental consequences of a past, ongoing, or planned action, resulting in the production of an environmental impact statement or 'finding of no significant impact (FONSI)'.

Environmental monitoring: Continuous or repeated measurement of agents in the environment to evaluate environmental exposure and possible damage by comparison with appropriate reference values based on knowledge of the probable relationship between ambient exposure and resultant adverse effects.

Environmental protection: (i) Actions taken to prevent or minimise adverse effects to the natural environment and (ii) complex measures including monitoring of environmental pollution and development and practice of environmental protection principles (legal, technical, and hygienic), including risk assessment, risk management, and risk communication.

Environmental Protection Agency (EPA): An agency of the federal government charged with a variety of responsibilities relating to protection of the quality of the natural environment. It includes research and monitoring, promulgation of standards for air and water quality, and control of the introduction of pesticides and other hazardous materials into the environment. EPA leads the nation's environmental science, research, education, and assessment efforts. The mission of the EPA is to protect human health and the environment. Since 1970, EPA has been working for a cleaner, healthier environment for the American people. (ii) The federal agency's mission is to protect human health and to safeguard the natural environment &151; air, water, and land &151; upon which life depends. EPA is responsible for researching and setting national standards for a variety of environmental programmes, delegating to states and tribes the responsibility for issuing permits, and monitoring and enforcing compliance.

Environmental quality objective (EQO): A regulatory value defining the quality to be aimed for in a particular aspect of the environment, for example, 'the quality of water in a river such that coarse fish can maintain healthy populations'. Unlike an environmental quality standard (EQS), an EQO is not usually expressed in quantitative terms and it is not legally enforceable.

Environmental quality standard (EQS): Regulatory value that defines the maximum concentration of a potentially toxic substance that can be allowed in an environmental compartment, usually air (AQS) or water, over a defined period.

Enzyme induction: A process whereby an enzyme is synthesised in response to a specific chemical substance or to other agents such as heat or a metal species.

Enzyme: (i) A protein that acts as a selective catalyst permitting reactions to take place rapidly in living cells under physiological conditions. (ii) A macromolecule that functions as a biocatalyst by increasing the specific reaction.

Enzymic (enzymatic) process: Any chemical reaction or series of reactions catalysed by an enzyme or enzymes.

Epidemiologist: A medical scientist who studies the various factors involved in the incidence, distribution, and control of disease in a population.

Epidemiology: (i) The study of the distribution and determinants of health-related states or events in specified populations. It deals with the statistical study of categories of persons and the patterns of diseases from which they suffer, with the aim of determining the events or circumstances causing these diseases.

(ii) Study of the distribution of disease, or other health-related conditions and events in human or animal populations, in order to identify health problems and possible causes.

Epigastric: Pertaining to the upper-middle region of the abdomen.

Epigenetic changes: Any changes in an organism brought about by alterations in the action of genes are called epigenetic changes. Epigenetic transformation refers to those processes that cause normal cells to become tumour cells without any mutations having occurred (*See also* Mutation, Transformation, Tumour).

Epilepsy: A collective term for a variety of different types of seizures; all forms of epilepsy start with a random discharge of nerve impulses into the brain. Antiepileptic drugs act by either raising the seizure threshold or by limiting the spread of impulses from one nerve to another inside the brain.

Epileptiform: Reactions that occur in severe or sudden spasms, as in convulsion or epilepsy.

Epithelioma: Any tumour derived from epithelium.

Epithelium: Sheet of one or more layers of cells covering the internal and external surfaces of the body and hollow organs.

Equivocal evidence: When carcinogenicity is demonstrated by studies that are interpreted as showing a chemically related marginal increase of neoplasms.

Erosion: The wearing away of soil by wind or water, intensified by land-clearing practices related to farming, residential or industrial development, road building, or logging.

Erythema: In medicine, redness of the skin due to blood vessel distension and by congestion of the capillaries.

Eschar: In medicine, a slough or dry scab that forms, for example, on an area of skin that has been burnt or exposed to corrosive agents.

ESIS: European Chemical Substances Information System.

Estuary: A complex ecosystem between a river and near-shore ocean waters where fresh- and saltwater mix. These brackish areas include bays, mouths of rivers, salt marshes, wetlands, and lagoons and are influenced by tides and currents. Estuaries provide valuable habitat for marine animals, birds, and other wildlife.

ET_{50}: The exposure time required to produce a defined effect when a test population is exposed to a fixed concentration or specified dose of a toxicant.

Etiology (aetiology): The science dealing with the cause or origin of disease.

Evaporation rate: A measure of how quickly the chemical substance/material becomes a vapour at normal room temperature. Usually, the evaporation rate is given in comparison to certain chemicals, such as butyl acetate, which evaporate fairly quickly. For example, the rate might be given as '0.5 (butyl acetate = 1)'. This means that, under specific conditions, 0.5 g of the material evaporates during the same time that 1 g of butyl acetate evaporates. Often, the evaporation rate is given only as greater or less than 1, which means the material evaporates faster or slower than the comparison chemical. In general, a hazardous material with a higher evaporation rate presents a greater hazard than a similar compound with a lower evaporation rate.

Excipient: Any largely inert substance added to a drug to give suitable consistency or form to the drug.

Excitotoxicity: Pathological process by which neurons are damaged and killed by the overactivation of receptors for the excitatory neurotransmitter glutamate, such as the N methyl D aspartate receptor (NMDA) and activated protein kinase (AMPA).

Excretion: A general term for removal of substances from the body.

Exothermic: A term used to characterise a chemical reaction that gives off heat as it proceeds.

Explosion data: Information on the explosive properties of a material. Quantitative explosion data are seldom available and is usually given in descriptive terms such as low, moderate, or high. The following types of information can be used to describe the explosive hazard of a chemical substance/material: (i) sensitivity to mechanical impact (this information indicates whether or not the chemical substance/material will burn or explode on shock (e.g. dropping a package) or friction (e.g. scooping up spilled chemical substance/material)) and (ii) sensitivity to static discharge (this information indicates how readily the material can be ignited by an electric spark).

Detailed information is available on the properties of commercial explosives. In Canada, the storage, transportation, and handling of commercial explosives are strictly regulated under the Explosives Act and Transport of Dangerous Goods Act. Commercial explosives are not regulated by the CPRs [e.g. not part of the WHMIS]. Under the U.S. OSHA Hazcom Standard, a chemical is identified as explosive if it causes a sudden, almost instantaneous, release of pressure, gas, and heat when subjected to sudden shock, pressure, or high temperature.

Explosion hazard: A chemical substance that is likely to blow up.

Explosives: Substances that cause a sudden, almost instantaneous, release of pressure, gas, and heat when subjected to sudden shock, pressure, or high temperature. An explosive material is a reactive substance that contains a great amount of potential energy that can produce an explosion if released suddenly, usually accompanied by the production of heat, light, sound, and pressure. Most commercial explosives are organic compounds such as nitroglycerin, trinitrotoluene (TNT), HMX, and nitrocellulose. Explosives are classified as low or high explosives according to their rates of burn: low explosives burn rapidly, while high explosives detonate. While these definitions are distinct, the problem of precisely measuring rapid decomposition makes practical classification of explosives difficult.

Explosive limits (chemical substance): The amounts of vapour in air that form explosive mixtures. These limits are expressed as lower and upper values and give the range of vapour concentrations in air that will explode if an ignition source is present.

Exposure: (i) Any kind of contact with a chemical substance by swallowing, breathing, or touching. (ii) Radiation or pollutants that come into contact with the body and present a potential health threat. The most common routes of exposure are through the skin and mouth or by inhalation.

Exposure assessment: Identifying the ways in which chemicals may reach individuals (e.g. by breathing), estimating how much of a chemical an individual is likely to be exposed to, and estimating the number of individuals likely to be exposed.

Exposure limits: The concentration of a substance in the workplace to which most workers can be exposed during a normal daily and weekly work schedule without adverse effects. An exposure limit is the concentration of a chemical in the workplace air to which most people can be exposed without experiencing harmful effects. Exposure limits should not be taken as sharp dividing lines between safe and unsafe exposures. It is possible for a chemical to cause health effects, in some people, at concentrations lower than the exposure limit. Exposure limits have different names and different meanings depending on who developed them and whether or not they are legal limits. For example, TLVs are exposure guidelines developed by the ACGIH. They have been adopted by many Canadian governments as their legal limits. Permissible exposure limits (PELs) are legal exposure limits in the United States. Sometimes, a manufacturer will recommend an exposure limit for a material. Exposure limits have not been set for many chemicals for many different reasons. For example, there may not be enough information available to set an exposure limit. Therefore, the absence of an exposure limit does not necessarily mean the material is not harmful.

In general, there are three different types of exposure limits: (a) TWA exposure limit is the TWA concentration of a chemical in air for a normal 8 h workday and 40 h workweek to which nearly all workers may be exposed day after day without harmful effects. TWA means that the average concentration has been calculated using the duration of exposure to different concentrations of the chemical during a specific time period. In this way, higher and lower exposures are averaged over the day or week. (b) Short-term exposure limit (STEL) is the average concentration to which workers can be exposed for a short period (usually 15 min) without experiencing irritation, long-term or irreversible tissue damage, or reduced alertness. The number of times the concentration reaches the STEL and the amount of time between these occurrences can also be restricted. (c) Ceiling (C) exposure limit is the concentration that should not be exceeded at any time.

Exposure potential: An estimate of the total dose of a chemical received by an exposed organism (e.g. a person) or by a population, not just via one pathway or medium but from all likely pathways.

Extinguishing media: Agents that can put out fires involving the material. The common extinguishing agents are water, carbon dioxide, dry chemical, 'alcohol' foam, and halogenated gases (halons). It is important to know which extinguishers can be used so they can be made available at the worksite. It is also important to know which agents cannot be used since an incorrect extinguisher may not work or may create a more hazardous situation. If several materials are involved in a fire, an extinguisher effective for all of the materials should be used.

Eukaryote: An organism (microorganism, plant, or animal) whose cells contain a membrane-bound nucleus and other membrane-bound organelles (compare prokaryote).

Eutrophication: A phenomena with the accumulation of nutrients in a lake or landlocked body of water. This occurs naturally over many years but has recently been accelerated by fertiliser run-off from farms and sewage input. Algal blooms result and their decay removes dissolved gen, eliminating aerobic organisms such as fish and may cause accumulation of sulphide in the water.

Eye hazard: A chemical substance that causes damage to or disease in the eyes.

EZ-Don facepiece harness: Provides quick, simple facepiece donning. A quick tug on the bottom two straps secures a snug fit. Takes seconds to don mask in an emergency.

FACOSH: The Federal Advisory Council on Occupational Safety and Health (FACOSH) is a source for federal agency safety and health programmes and policies. The work of the Advisory Council uses the expertise of its members and provides assistance to the Secretary of Labor and OSHA and leadership to federal agencies, in an effort to reduce the number of worker injuries and illnesses in the federal government. Advises the Secretary of Labor on appropriate policies and initiatives to enhance occupational safety and health in the federal sector and actively endorse these policies and initiatives.

FDA (U.S. Food and Drug Administration): Is involved in regulation of pesticides in the United States, particularly enforcement of tolerances in food and feed products.

Fecundity: (i) Ability to produce offspring frequently and in large numbers; (ii) in demography, the physiological ability to reproduce (iii) and the ability to produce offspring within a given period of time.

Federal Advisory Council on Occupational Safety and Health (FACOSH): Council established by Executive Order 11612 on July 26, 1971, to advise the Secretary of Labor on matters relating to the occupational safety and health of federal employees. FACOSH is a soe urce for federal agency safety and health programs and policies.

Federal Food, Drug, and Cosmetic Act (FFDCA): The regulatory act (P.L. 75–717, June 25, 1938) is the basic authority intended to ensure that foods are pure and wholesome, safe to eat, and produced under sanitary conditions; that drugs and devices are safe and effective for their intended uses; that cosmetics are safe and made from appropriate ingredients; and that all labelling and packaging is truthful, informative, and not deceptive. The FDA is primarily responsible for enforcing the FFDCA, although the U.S. DA also has some enforcement responsibility. The EPA establishes limits for concentrations of pesticide residues on food under this act.

Federal Hazardous Substances Act (FHSA): The FHSA (15 U.S.C 1261–1278), administered by the CPSC, requires that certain household products that are 'hazardous substances' bear cautionary labelling to alert consumers to potential hazards that those products present and inform them of the measures they need to protect themselves from those hazards. Any chemical substance that is toxic, corrosive, flammable or combustible, an irritant, a strong sensitiser, or that generates pressure through decomposition, heat, or other means requires labelling, if the product may cause substantial personal injury or substantial illness during or as a proximate result of any customary or reasonable foreseeable handling or use, including reasonable foreseeable ingestion by children.

Federal Register: A federal publication containing current presidential orders or directives, agency regulations, proposed agency rules, notices, and other documents that are required by statute to be published for wide public distribution. The *Federal Register* is published each federal working day. The U.S. DA publishes its rules, notices, and other documents in the *Federal Register*. Final regulations are organised by agency and programmes in the CFR.

Feromone (pheromone): Chemical substance used in olfactory communication between organisms of the same species eliciting a change in sexual or social behaviour.

Fertility: Ability to conceive and to produce offspring: for litter-bearing species, the number of offspring per litter is used as a measure of fertility. The reduced fertility is sometimes referred to as subfertility.

Fetus/Foetus: An organism in the later stages of development prior to birth. In humans, it is the unborn child from the end of the 2nd month of pregnancy to birth.

Fibrosis: Abnormal formation of fibrous tissue.

FID: Flame ionisation detector.

FIFRA: The FIFRA was enacted in 25 June 1947. The act instructs the EPA to regulate (i) the registration of all pesticides used in the United States, (ii) the licensing of pesticide applicators; (iii) the re-registration of all pesticide products; (iv) the storage, transportation, disposal, and recall of all pesticide products; and (v) the federal law regulating most pesticides and their uses.

First aid: Emergency care given immediately to an injured person. The purpose of first aid is to minimise injury and future disability. In serious cases, first aid may be necessary to keep the victim alive.

First-order process: A chemical process where the rate of reaction is directly proportional to the amount of chemical present.

First-pass effect: A chemical alteration resulting from biotransformation of a xenobiotic before it reaches the systemic circulation. Such biotransformation by the liver is referred to as a hepatic first-pass effect.

Flammable, flammability: Able to ignite and burn readily. Flammability is the ability of a material to ignite and burn readily. (See also Combustible.) Under the Canadian CPRs [part of WHMIS] and the U.S. Hazcom Standard, there are specific technical criteria for identifying flammable materials. (*See* Flammable aerosol, Flammable gas, Flammable liquid, Flammable solid, and Reactive flammable material.)

There are closely related criteria for the classification of certain flammable materials under the Canadian Transportation of Dangerous Goods (TDG) regulations and the U.S. DOT regulations. (See TDG Flammability Classification.) In Canada, local, provincial, and national fire codes also classify and regulate the use of flammable materials in workplaces.

(i) Any material that can be ignited easily and that will burn rapidly, (ii) as defined in the FHSA regulations at 16 CFR § 1500.3(c)(6) (ii), is a substance having a flashpoint above 20°F (−6.7°C) and below 100°F (37.8°C). An extremely flammable substance, as defined in the FHSA regulations at 16 CFR § 1500.3(c)(6) (iii), is any substance with a flashpoint at or below 20°F (−6.7°C).

Flammable aerosol: Under the Canadian CPRs, a material is identified as a flammable aerosol if it is packaged in an aerosol container that can release a flammable material. A flammable aerosol is hazardous because it may form a torch (explosive ignition of the spray) or because a fire fuelled by the flammable aerosol may flash back.

Flammable gas: A gas that can ignite readily and burn rapidly or explosively. Under the Canadian CPRs and under the U.S. Hazcom Standard, there are certain technical criteria for the identification of materials as flammable gases for the purposes of each regulation. Flammable gases can be extremely hazardous in the workplace. (a) If the gas accumulates so that its lower explosive limit (LEL) is reached and if there is a source of ignition, an explosion may occur. (b) If there is inadequate ventilation, flammable gases can travel a considerable distance to a source of ignition and flash back to the source of the gas.

Flammable liquid: A liquid that gives off a vapour that can be readily ignited at normal working temperatures. Under the Canadian CPRs, a flammable liquid is a liquid with a flashpoint (using a closed-cup test) below 37.8°C (100°F). The U.S. Hazcom Standard uses a similar, but not identical, definition.

Flammable liquids can be extremely hazardous in the workplace '(a) If there is inadequate ventilation, vapours can travel considerable distances to a source of ignition and flash back to the flammable liquid. (b) It may be difficult to extinguish a burning flammable liquid with water because water may not be able to cool the liquid below its flashpoint'.

Flammable solid: A material that can ignite readily and burn vigorously and persistently. There are certain technical criteria in the Canadian CPRs [part of the WHMIS] and in the U.S. OSHA Hazcom Standard for the identification of flammable solids for the purposes of each regulation. These criteria are based on ease of ignition and rate of burning. Flammable solids may be hazardous because heat from friction (e.g. surfaces rubbing together) or heat from processing may cause a fire. Flammable solids in the form of a dust or powder may be particularly hazardous because they may explode if ignited.

Flashback: Occurs when a trail of flammable gas, vapour, or aerosol is ignited by a distant spark, flame, or other source of ignition. The flame then travels back along the trail of gas, vapour, or aerosol to its source. A serious fire or explosion could result.

Flashpoint: (i) The lowest temperature at which evaporation of a substance produces enough vapour to form an ignitable mixture with air. (ii) The minimum temperature at which a liquid or a solid produces a vapour near its surface sufficient to form an ignitable mixture with the air; the lower the flashpoint, the easier it is to ignite the material.

Fluorosis (fluoridosis): Adverse effects of fluoride, as in dental or skeletal fluorosis.

Fly ash: (i) One of the residues generated in the combustion of coal. Fly ash is generally captured from the chimneys of coal-fired power plants. Small solid ash particles from the noncombustible portion of coal fuel. (ii) Fly ash contains environmental toxins in significant amounts. These include arsenic, barium, beryllium, boron, cadmium, chromium, chromium VI, cobalt, copper, fluorine, lead, manganese, nickel, selenium, strontium, thallium, vanadium, and zinc.

Foci: A small group of cells occurring in an organ and distinguishable, either in appearance or histochemically, from the surrounding tissue.

Foetotoxic: Harmful to the fetus/foetus.

Foetotoxicity: Toxicity to the foetus. Foetotoxicity describes the ability of a substance to harm the foetus (*See also* Embryotoxicity, Teratogenicity, and Reproductive effects).

Foetus: (i) The developing mammal or other viviparous vertebrate, after the embryonic stage and before birth. (ii) A term used to describe a developing human infant from approximately the third month of pregnancy until delivery. (iii) This term in medicine is applied to the young of mammals when fully developed in the womb. In human beings, this stage is reached after about 3 months of pregnancy. Prior to this, the developing mammal is at the embryo stage.

Food additive: (i) Any chemical substance or mixture of substances other than the basic foodstuff present in a food as a result of any phase of production, processing, packaging, storage, transport, or handling. The U.S. DA allows food additives in meat, poultry, and egg products only after they have received FDA safety approval. Food additives are regulated under the authority of the FFDCA and are subject to the Delaney Clause. (ii) Addition results, or may be reasonably expected to result (directly or indirectly), in the substance or its by-products becoming a component of, or otherwise affecting, the characteristics of the food to which it is added. Food additive does not include 'contaminants' or chemical substances added to food for maintaining or improving nutritional qualities.

Food allergy: Hypersensitivity reaction to chemical substances in the diet to which an individual has previously been sensitised.

Foodborne illnesses: Illnesses caused by pathogens that enter the human body through foods.

Food chain: (i) Sequence of organisms in an ecosystem each of which uses the next, lower member of the sequence as a food source. The food web, pyramid-shaped structure illustrating the feeding order in nature wherein each organism feeds upon the next lowest creature and feeding relationships between predators and prey wherein animals and plants get food in an ecosystem. (ii) Sequence of transfer of matter and energy in the form of food from organism to organism in ascending or descending trophic levels. (iii) A sequence of organisms, each of which uses the next, lower member of the sequence as a food source.

Food Residue Exposure Assessment Section (FREAS): The regulatory section/unit that evaluates every submission where a product could come in contact with food, including field crops, meat and dairy products, and processed foods. These evaluations are conducted in order to set the maximum residue limits (MRLs) for pesticides on food, both domestic and imported under the *Food and Drug Act*. Dietary risk assessments (DRAs) are also carried out to assess the potential daily intake of pesticide residues from all possible food sources. DRAs take into account the different eating patterns of infants, toddlers, children, adolescents, and adults and so include a detailed evaluation of the foods and drinks that infants and children consume in quantity such as fruits and fruit juices, milk, and soya products.

Food web: The network of interconnected food chains in an ecosystem.

Foot and mouth disease (FMD): A highly contagious viral disease of cattle and swine, as well as sheep, goats, deer, and other cloven-hoofed ruminants. Although rarely transmissible to humans, FMD is devastating to livestock and has critical economic consequences with potentially severe losses in the production and marketing of meat and milk.

Frameshift mutation: Such a mutation is a change in the structure of DNA causing the transcription of genetic information into RNA to be completely altered because the start point for reading has been changed. In other words, the reading frame for transcription has been altered.

Fugacity: The tendency for a substance to move from one environmental compartment to another. Originally the term was applied to the tendency of a gas to expand or escape related to its pressure in the system being studied.

Fugitive emissions: Air pollutants released to the air other than those from stacks or vents; typically small releases from leaks in plant equipment such as valves, pump seals, flanges, and sampling connections.

Full-face mask: Pertains to gas masks. A full-face mask protects the wearer's eyes, face, lungs, etc., from contamination.

Fumes: Fumes are very small, airborne, solid particles formed by the cooling of a hot vapour. For example, a hot zinc vapour may form when zinc-coated steel is welded. The vapour then condenses to form fine zinc fume as soon as it contacts the cool surrounding air. Fumes are smaller than dusts and are more easily breathed into the lungs.

Fumigants: Chemical substances that is vapourised in order to kill or repel pests.

Fumigation: Application of pesticides in the form of vapours or fumes.

Fungicide: A pesticidal chemical substance used to control or destroy fungi on food or grain crops.

Fungus: Type of plant that has no leaves, flowers, or roots. Both words, funguses and fungi, are the plural of fungus.

Galvanising: Application of a protective layer of zinc to a metal, chiefly steel, to prevent or inhibit corrosion.

Gamete: Reproductive cell (either sperm or egg) containing a haploid set of *chromosomes*.

Gametocide: Substance intended to kill gametes.

Gamma ray: A type of electromagnetic radiation with high energy and penetrating power and are released by atoms of radioactive elements. As it has a high penetrating power, gamma rays are used for radiation therapy to treat cancer.

Gas: The fluid having neither independent shape nor volume but tending to expand indefinitely. The word is often used to denote anaesthetics, combustibles (gasoline), poisonous materials, etc., whether liquid or solids at ordinary temperatures.

Gas mask: A device that fits snugly against the face to protect the wearer from breathing NBC agents. Attached filtering device is interchangeable.

Gastroenteritis: Inflammation of the stomach and intestine.

Gastrointestinal tract: (i) Pertaining or communicating with the stomach and intestine. (ii) The organ system of the body that includes the stomach and intestines.

Gavage: Administration of materials directly into the stomach by oesophageal intubation.

Gene: (i) The fundamental unit of heredity. (ii) Length of DNA that encodes a functional product, which may be a polypeptide or an RNA.

Gene amplification: Occurrence of extra copies of a gene; with respect to a plasmid, an increase in the number of plasmid copies per cell, which may be induced by a specific treatment. In the case of tumour cells, spontaneous gene amplification occurs frequently.

Gene expression: Transcriptional activation of a gene so that its functional product is produced.

General emergency: An event predicted, in progress, or having occurred that result in one or more of the following situations: (1) actual or imminent catastrophic reduction of facility safety or security systems with potential for the release of large quantities of hazardous materials (radiological or non-radiological) to the environment (the radiation dose from any release of radioactive material or a concentration in air from any release of other hazardous material is expected to exceed the applicable Protective Action Guide or Emergency Response Planning Guideline at or beyond the site boundary) and (2) actual or likely catastrophic failures in safety or security systems threatening the integrity of a nuclear weapon, component, or test device that may adversely impact the health and safety of workers and the public (*See* DOE O 151.1).

Gene therapy: Introduction of genetic material into an individual, or the modification of the individual's genetic material, in order to achieve a therapeutic or prophylactic objective.

General ventilation: As used in an MSDS, also known as dilution ventilation, the removal of contaminated air from the general area and the bringing in of clean air. This dilutes the amount of contaminant in the work environment. General ventilation is usually suggested for nonhazardous materials. (See also Mechanical ventilation, Local exhaust ventilation, and ventilation.)

Genetic epidemiology: Study of the correlations between phenotypic trends and genetic variation across population groups and the application of the results of such a study to control of health problems.

Gene map: Map showing the positions in the genome of *genes* or other genetic markers, either relative to each other or as a physical map of absolute distances.

Generally regarded as safe (GRAS): Phrase used to describe the U.S. FDA philosophy that justifies approval of food additives that may not meet the usual test criteria for safety but have been used extensively and have not demonstrated to cause any harm to consumers.

Genetically modified organism (GMO): Bacterium, plant, or animal whose DNA has been deliberately altered.

Genetic polymorphism: Existence of interindividual differences in DNA sequence coding for one specific gene, giving rise to different functional and (or) morphological traits.

Genetic susceptibility: Predisposition to a particular disease or sensitivity to a substance due to the presence of a specific allele or combination of alleles in an individual's genome.

Genetic toxicology: Study of chemically or physically induced changes to the structure of DNA, including epigenetic phenomena or mutations that may or may not be heritable.

Genome: (i) Complete set of chromosomal and extrachromosomal genes of an organism, a cell, an organelle, or a virus, that is, the complete DNA component of an organism. (ii) Genome includes both the DNA present in the chromosomes and that in subcellular organelles (e.g. mitochondria or chloroplasts). It also includes the RNA genomes of some viruses.

Genomics: (i) Science of using DNA- and RNA-based technologies to demonstrate alterations in genes expression. (ii) Method providing information on the consequences for gene expression of interactions of the organism with environmental stress, xenobiotics, etc. (iii) Capable of causing a change to the structure of the genome.

Genotype: (i) Genetic constitution of an organism as revealed by genetic or molecular analysis; the complete set of genes possessed by a particular organism, cell, organelle, or virus. (ii) The genetic identity of an individual that does not show as outward characteristics. The genotype refers to the pair of alleles for a given region of the genome that an individual carries.

Geo-nets: 3D structures composed of overlaying and intertwined parallel strands that create high-capacity flow channels. They are produced by the extrusion of high-density polyethylene (HDPE) and are therefore resistant to hazardous chemical pollutants and biological agents normally present in the living environment.

GERD: Gastro-esophageal reflux disease.

Germ-free animal: Axenic animal, meaning animal grown under sterile conditions in the period of postnatal development: such animals are usually obtained by Caesarean operation and kept in special sterile boxes in which there are no viable microorganisms (sterile air, food, and water are supplied).

Germ-line cell: Cell with a haploid chromosome content. *Note*: In animals, the germ-line cells are the sperm or egg (synonym gamete); in plants, the pollen cell or the ovum.

Germinal aplasia: Complete failure of gonad development.

Gingivitis: An inflammatory process limited to the mucosal epithelial tissue surrounding the cervical portion of the teeth and the alveolar processes. Gingivitis has been classified by clinical appearance like ulcerative, haemorrhagic, necrotising, and purulent. The aetiology includes drug-induced, hormonal, nutritional, infectious, plaque-induced, and duration. The most common type of gingivitis is a chronic form induced by plaque.

Glomerular: Pertaining to a tuft or cluster, as of a plexus of capillary blood vessels or nerve fibres, especially referring to the capillaries of the glomerular of the kidney.

GHS (Globally Harmonised System): GHS of Classification and Labelling of Chemicals developed by the United Nations. It is intended that GHS be adopted worldwide. GHS addresses the classification of chemicals by types of hazard (health, fire, reactivity, environmental) and proposes harmonised Hazcom elements (labels and SDSs).

Glomerulus: Tuft or a cluster, as of a plexus of capillary blood vessels or nerve fibres, for example, capillaries of the filtration apparatus of the kidney.

Glomerular filtration: Formation of an ultrafiltrate of the blood occurring in the *glomerulus* of the kidney.

Glomerular filtration rate: Volume of ultrafiltrate formed in the kidney tubules from the blood passing through the glomerular capillaries divided by time of filtration.

Glue sniffing: Solvent abuse using plastic cement or other solvent-based adhesives.

Glycome: Description of the complete set of carbohydrates and their functions in a living organism.

Glycomics (glycobiology): Global study of the structure and function of carbohydrates, especially oligosaccharides (short chains of sugars) in a living organism.

Gnotobiota: Specifically and entirely known microfauna and microflora of a specially reared laboratory animal.

Gnotobiont: Specially reared laboratory animal whose microflora and microfauna are specifically known in their entirety.

Goitre: Noncancerous enlargement of the thyroid gland, visible as a swelling at the front of the neck, that is often associated with iodine deficiency.

Goitrogen: Any substance (such as thiouracil) that induces the formation of a goitre.

Gonadotropic: Pertaining to effects on sex glands and on the systems that regulate them.

Good agricultural practice (GAP) in the use of pesticides: Nationally authorised safe uses of pesticides under actual conditions necessary for effective and reliable pest control. (ii) GAP encompasses a range of levels of pesticide applications up to the highest authorised use, applied in a manner that leaves a residue that is the smallest amount practicable. Authorised safe uses include nationally registered or recommended uses that take into account public and occupational *health* and environmental safety considerations. Actual conditions include any stage in the production, storage, transport, distribution, and processing of food commodities and animal feed.

Good laboratory practice (GLP) principles: (i) Fundamental rules incorporated in the Organisation for Economic Co-operation and Development (OECD) guidelines and national regulations concerned with the process of effective organisation and the conditions under which laboratory studies are properly planned, performed, monitored, recorded, and reported. (ii) For further information, refer to Directive 2004/10/EC, OECD Guidelines on Good Laboratory Practice part 1.

Good manufacturing practice (GMP) principles: Fundamental rules incorporated in national regulations concerned with the process of effective organisation of production and ensuring standards of defined quality at all stages of production, distribution, and marketing. Minimisation of waste and its proper disposal are part of this process.

Grab sample: (i) A single sample collected at a particular time and place that represents the composition of the water, air, or soil only at that time and place. (ii) A sample taken from the environment or from an effluent or other waste on a one-time basis. A grab sample should be collected over a short period of time, not to exceed 15 min.

Graded effect: Effect in which severity is related continuously to dose. Increases in dose rate or exposure level thus have a steadily increasing (sometimes linear) effect upon the severity of symptoms. (ii) This is an effect that can usually be measured on a continuous scale of intensity or severity and its magnitude related directly to dose.

Graminicide: A pesticide/herbicide used for the control of weedy grasses (Gramineae).

Granuloma: Granular growth or tumour, usually of lymphoid and epithelial cells.

Ground treatment of plants: Dusting or spraying of plants with pesticides by hand, by special machines, or by apparatus fixed to tractors or driven by them.

Group: A family of elements with similar chemical properties, represented by a vertical column in the periodic table.

Ground wire: A conductor leading from electrical equipment to a low resistance connection with the earth.

Guideline for exposure limits: Scientifically judged quantitative value (a concentration or number) of an environmental constituent that ensures aesthetically pleasing air, water, or food and from which no adverse effect is expected concerning noncarcinogenic end points, or that gives an acceptably low estimate of lifetime cancer risk from those substances that are proven human carcinogen or carcinogens with at least limited evidence of human carcinogenicity.

Guideline value: Quantitative measure (a concentration or a number) of a constituent of an environmental medium that ensures aesthetically pleasing air, water, or food and does not result in a significant risk to the user.

Guides to air quality: Sets of atmospheric concentrations and exposure times that are associated with specific effects of varying degrees of pollution on man, animals, vegetation, and the environment in general.

Guides to environmental quality: Sets of concentrations, numbers and exposure times that are associated with the specific effects of factors in environmental media on man, animals, vegetation, and the environment in general.

Guinea pig maximisation test (Magnusson and Kligman test): A widely used skin test for screening possible contact allergens: considered to be a useful method to identify likely moderate and strong sensitisers in humans.

Haem: The part of certain molecules that contains iron. The haem part of haemoglobin is the substance inside RBCs that binds to oxygen in the lungs and carries it to the tissues.

Haematemesis: Vomiting of blood.

Haematoma: (i) A pool of clotted or partially clotted blood in an organ, tissue, or body space, usually caused by a broken blood vessel. (ii) Localised accumulation of blood, usually clotted, in an organ, space, or tissue, due to a failure of the wall of a blood vessel.

Haematopoiesis: The formation of new blood cells.

Haematopoietic agents: Chemical substances that act on the blood or haematopoietic system, decrease haemoglobin function, deprive the body tissues of oxygen, blood in the urine.

Haematopoietic cells: Blood-forming cells produced in the bone marrow.

Haematopoietic toxin: A chemical substance that damages the blood or blood production. Blood toxins can prevent the blood from carrying oxygen to cells.

Haemoglobin: The substance inside RBCs that binds to oxygen in the lungs and carries it to the tissues.

Haemophilia: Group of hereditary disorders in which affected individuals fail to make enough of certain proteins needed to form blood clots.

Haemoptysis: Coughing or spitting up blood from the respiratory tract.

Haemorrhage: Loss of blood from damaged blood vessels. A haemorrhage may be internal or external and usually involves a lot of bleeding in a short time.

Haemosiderin: The name of the iron-protein molecule that is a source of iron for haemoglobin synthesis and other processes requiring iron.

Hair follicle: A shaft or opening on the surface of the skin through which hair grows.

Half-life ($t_{1/2}$): Time required for the concentration of a reactant in a given reaction to reach a value that is the arithmetic mean of its initial and final (equilibrium) values. For a reactant that is entirely consumed, it is the time taken for the reactant concentration to fall to one half of its initial value. The half-life of a reaction has meaning only in special cases, for example, (i) for a first-order reaction, the half-life of the reactant may be called the half-life of the reaction, and (ii) for a reaction involving more than one reactant, with the concentrations of the reactants in their stoichiometric ratios, the half-life of each reactant is the same and may be called the half-life of the reaction.

Half-face mask: Pertains to gas masks. A half-face mask does not protect the eyes, upper face, and forehead from biological or chemical agents. Can be used with the proper filter to protect the wearer from viruses.

Hallucination: A sight, sound, smell, taste, or touch that a person believes to be real but is not real. Hallucinations can be caused by nervous system disease, certain drugs, or mental disorders.

Hamartoma: A benign (not cancerous) growth made up of an abnormal mixture of cells and tissues normally found in the area of the body where the growth occurs.

Hand-foot syndrome: A condition marked by pain, swelling, numbness, tingling, or redness of the hands or feet. It sometimes occurs as a side effect of certain anticancer drugs. Hand-foot syndrome is also called palmar-plantar erythrodysaesthesia.

Handling and storage: Handle in a well-ventilated area. Maintain positive pressure in interior of occupied buildings during exterior application and close fresh air intakes in area of application. Avoid inhaling vapours and mists and getting in eyes, on skin, or on clothing. Wash hands and other contaminated areas thoroughly with soap and water after handling this product and before eating or smoking. Wash contaminated clothing and equipment thoroughly before reuse.

Haploid: Pertaining to a cell containing only one set of chromosomes.

Hard palate: The front, bony part of the roof of the mouth.

Harmful substance: A chemical substance that on following contact with an organism/individual can cause ill health or adverse effects either at the time of exposure or later in the life of the present and future generations.

Harmful: Chemical substances causing or capable of causing harmful effects.

Hazard: (i) Anything that has the ability to cause injury or for the potential to cause injury. Hazard is the potential for harmful effects. Hazardous means potentially harmful. The hazards of a material are evaluated by examining the properties of the material, such as toxicity, flammability, and chemical reactivity, as well as how the material is used. How a material is used can vary greatly from workplace to workplace and, therefore, so can the hazard. The hazard associated with a potentially toxic chemical substance is a function of its toxicity and the potential for exposure to the chemical substance. The probability of exposure to the chemical substance is a risk factor. (ii) A hazard is a situation that poses a level of threat to life, health, property, and the environment.

Hazard area: (i) Any place that is being in use with toxic chemical substances and hazardous materials. (ii) Any area in close proximity to fumigation enclosure into which lethal amount of fumigant may escape to cause health hazard.

Hazard assessment: (i) Determination of factors controlling the likely effects of a hazard such as the dose–effect and dose–response relationships, variations in target susceptibility, and mechanism of toxicity. (ii) The identification and characterisation of hazardous materials specific to a facility/site, analyses of potential accidents or events, and evaluation of potential consequences. The hazard assessment also includes a determination of the size of the geographic area surrounding the site, known as the emergency planning zone (EPZ), within which special planning and preparedness activities are required to reduce the potential health and safety impacts from an event involving hazardous materials. The hazard assessment provides the technical basis for the hazardous materials programme.

Hazard Characteristic Codes (HCC): The HCC classify chemical substances/materials by their primary hazard characteristic for the safe segregation and storage of hazardous materials. There are 10 basic classes defined by letters A–Z, and each of these is broken down into additional subclasses.

Hazard characterisation: A description of the potential adverse health effects attributable to a specific environmental agent, the mechanisms by which agents exert their toxic effects, and the associated dose, route, duration, and timing of exposure.

Hazard Communication Standards: An established policy of the OSHA that sets guidelines for Hazcom. The hazardous chemical safety (HCS) emphasises labelling, MSDS, and training. The U.S. OSHA regulation that details requirements for MSDSs and labelling (29 CRF Part 1910.1200).

Hazard evaluation: Establishment of a qualitative or quantitative relationship between hazard and benefit, involving the complex process of determining the significance of the identified hazard and balancing this against identifiable benefit.

Hazard identification: The first step in the risk assessment process. This step includes the identification of a chemical of concern and its potential adverse effects.

Hazard indicator: A quantitative measurement of a chemical's hazard. The site includes hazard indicators for numerous end points, including human health, ecological health, and combined human and ecological health. These are based on different combinations of factors, such as toxicity, persistence, and exposure potential.

Hazard warning: Any words, pictures, symbols, or combination thereof appearing on a label that convey the hazards of the chemical(s) in the container.

Hazard zone: One of four levels of hazard (hazard zones A through D) assigned to gases, as specified in safety regulations based on the LC_{50} value. (Refer to §173.116(a), 173.133(a).)

Hazardous chemical substance: A substance that has the potential to harm the health of persons/users and/or workers. Hazardous chemical substance can be a single chemical substance/entity or a mixture.

Hazardous combustion products: Chemicals that may be formed when a material burns. These chemicals may be toxic, may be flammable, or may have other hazards. The chemicals released and their amounts vary, depending upon conditions such as the temperature and the amount of air (or more specifically, oxygen) available. The combustion chemicals may be quite different from those formed by heating the same material during processing (thermal decomposition products). It is important to know which chemicals are formed by hazardous combustion in order to plan the response to a fire involving the material.

Hazardous decomposition products: Formed when a material decomposes (breaks down) because it is unstable or reacts with common materials such as water or oxygen (in air). This information should be considered when planning storage and handling procedures.

Hazardous ingredient: Under the Canadian Hazardous Products Act, a chemical must be listed in the Hazardous Ingredients Section of an MSDS if (i) it meets the criteria for a controlled product, (ii) it is on the Ingredient Disclosure List, (iii) there is no toxicological information available, or (iv) the supplier has reason to believe it might be hazardous.

Hazardous waste: Any kind of waste material that poses a risk to human health or the environment. Hazardous waste requires special types of storage and disposal to make it harmless or less dangerous.

HAZCHEM: Acronym for 'HAZardous CHEMicals'. The HAZCHEM emergency action.

HDPE (high-density polyethylene): Polyethylene that is more rigid and harder than lower-density materials. It also has higher tensile strength four times that of LDPE. It is three times better in compressive strength. Poly-Hi's HDPE meets FDA requirements for direct food-contact applications. It is also accepted by the U.S. DA, the NSF, and the Canadian Department of Agriculture.

Health: A condition describing the (i) state of complete physical, mental, and social well-being, and not merely the absence of disease or infirmity; (ii) a state of dynamic balance in which an individual's or a group's capacity to cope with the circumstances of living is at an optimal level; (iii) a state characterised by anatomical, physiological, and psychological integrity and ability to perform personally valued family, work, and community roles; (iv) an ability to deal with physical, biological, psychological, and social stress; (v) a feeling of well-being and freedom from the risk of disease and untimely death; and (vi) a sustainable steady state in which humans and other living organisms can coexist indefinitely (when discussed in the context of ecology).

Health advisory level (HAL): A non-regulatory agency in the United States dealing on health-based reference level of traces of chemical substances (usually in ppm/mg/L) in drinking water at which there are no adverse health risks when ingested over various periods of time.

Health-based exposure limit: Maximum concentration or intensity of exposure that can be tolerated without significant effect (based on only scientific and not economic evidence concerning exposure levels and associated health effects).

Health hazard: (i) A chemical substance for which there is statistically significant evidence based on at least one study conducted in accordance with established scientific principles that acute or chronic health effects may occur in exposed employees. Chemicals covered by this definition include carcinogens, toxic or highly toxic agents, reproductive toxins, irritants, corrosives, sensitisers, hepatotoxins, nephrotoxins, neurotoxins, that which act on the haematopoietic system, and agents that damage the lungs, skin, eyes, or mucous membranes. (ii) A chemical substance that is capable of causing an acute reaction, a chronic effect, or both. Health hazards can affect the whole body or a particular organ. Health hazard should have statistically significant evidence based on at least one study conducted in accordance with established scientific principles that acute or chronic health effects may occur in exposed employees.

Health surveillance: Periodic medico-physiological examinations of exposed workers with the objective of protecting health and preventing occupationally related disease.

Heavy metal: A term used commonly in the toxicological literature but having no generally agreed meaning, sometimes even applied to non-metals, and therefore a source of confusion and to be avoided. The term 'metal' is adequate without the qualifying adjective but may be misleading since it implies a solid material when toxicological concern is mostly for the ionic form or another chemical species.

HECD: Hall electron capture detector.

Henry's law constant: A measure of the tendency of a chemical to evaporate from a solution in water. It indicates solubility of a gas in a liquid solution at a constant temperature is proportional to the partial pressure of the gas above the solution.

Heparin: A chemical substance that slows the formation of blood clots. Heparin is made by the liver, lungs, and other tissues in the body and can also be made in the laboratory. Heparin may be injected into muscle or blood to prevent or break up blood clots. It is a type of anticoagulant.

Hepatectomy: Removal of the liver in part or all by surgery.

Hepatitis: Disease of the liver causing inflammation. Symptoms include an enlarged liver, fever, nausea, vomiting, abdominal pain, and dark urine.

Hepatitis C: Inflammation of the liver due to the hepatitis C virus (HCV), which is usually spread by blood transfusion, haemodialysis, and needlesticks. HCV causes most transfusion-associated hepatitis, and the damage it does to the liver can lead to cirrhosis and cancer. Transmission of the virus by sexual contact is rare. At least half of HCV patients develop chronic hepatitis C infection. Diagnosis is by blood test. Treatment is via antiviral drugs. Chronic hepatitis C may be treated with interferon sometimes in combination with antivirals. There is no vaccine for hepatitis C. Previously known as non-A, non-B hepatitis.

Hepatocyte: In histology terms, a parenchymal liver cell.

Hepatotoxic: Pertaining to anything that is harmful to the liver.

Hepatotoxin: A chemical substance that is poisonous to the liver cells and produce liver damage, jaundice, and liver enlargement.

Herbicide: A chemical substance used to kill herbs – plants.

Hernia: The bulging of an internal organ through a weak area or tear in the muscle or other tissue that holds it in place. Most hernias occur in the abdomen.

High blood pressure (hypertension): Blood pressure of 140/90 or higher. High blood pressure usually has no symptoms. It can harm the arteries and cause an increase in the risk of stroke, heart attack, kidney failure, and blindness.

Highly toxic substance: As defined by OSHA (Appendix A of 29 CFR 1910.1200) and in the FHSA regulations at 16 CFR § 1500.3(b)(6)(i), a substance with either (a) an LD_{50} of 50 mg/kg or less of body weight administered orally to rats, (b) an LD_{50} of 200 mg/kg or less of body weight when administered continuously on the bare skin of rabbits for 24 h or less, or (c) an LC_{50} in air of 200 ppm by volume or less of gas or vapour, or 2 mg/L by volume or less of mist or dust, when exposed to continuous inhalation for one hour or less to rats.

HIPS (high-impact styrene): A cost-effective material with good impact resistance and can be easily coloured. It extrudes with a matte uniform finish. HIPS is used for indoor applications, and if necessary, UV inhibitors can be added making it fairly weatherable. HIPS is translucent in its natural state and translucent to opaque when coloured.

Histochemistry: The branch of science concerned with the chemistry of microscopically visible structures. Chemical processes used to make cell structure microscopically visible.

Histology: (i) The branch of science related to the study of the microanatomy of tissues and their cellular structure. (ii) The study of tissues and cells under a microscope.

Histopathology: (i) The branch of science concerned with the study of microscopic changes in diseased tissues. (ii) Microscopic pathological study of the anatomy and cell structure of tissues in disease to reveal abnormal or adverse structural changes.

HIV (human immunodeficiency virus): HIV-1 is a retrovirus that is recognised as the causative agent of AIDS.

HIV positive: Infected with the HIV, the cause of AIDS.

HNPCC: Hereditary nonpolyposis colon cancer.

Hodgkin's disease: A cancer of the immune system that is marked by the presence of a type of cell called the Reed–Sternberg cell. The two major types of Hodgkin's disease are classical Hodgkin's lymphoma and nodular lymphocyte-predominant Hodgkin's lymphoma. Symptoms include the painless enlargement of lymph nodes, spleen, or other immune tissue. Other symptoms include fever, weight loss, fatigue, or night sweats. Also called Hodgkin's lymphoma.

Homeostasis: (i) A state of balance among all the body systems needed for the body to survive and function correctly. In homeostasis, body levels of acid, blood pressure, blood sugar, electrolytes, energy, hormones, oxygen, proteins, and temperature are constantly adjusted to respond to changes inside and outside the body, to keep them at a normal level. (ii) In medicine and biology, this term is applied to the inherent tendency in an organism towards maintenance of physiological and psychological stability.

Hormone: Chemical substance formed in one organ or part of the body and carried in the blood to another organ or part where it selectively alters functional activity.

Hormone: (i) A chemical substance, usually a peptide or steroid, produced by one tissue and conveyed by the bloodstream to another to affect physiological activity, such as growth or metabolism; (ii) a chemical substance released by one or more cells that affects cells in other parts of the organism/animal/humans. (iii) One of the chemical messengers produced by endocrine glands whose secretions are liberated directly into the bloodstream and transported to a distant part or parts of the body, where they exert a specific effect for the benefit of the body as a whole.

Hormone therapy (hormone treatment): Treatment that adds, blocks, or removes hormones. For certain conditions (such as diabetes or menopause), hormones are given to adjust low hormone levels. To slow or stop the growth of certain cancers (such as prostate and breast cancer), synthetic hormones or other drugs may be given to block the body's natural hormones. Sometimes surgery is needed to remove the gland that makes a certain hormone.

Household waste: Any kind of solid waste (such as garbage, trash, and sanitary waste from septic tanks) derived from households (including single and multiple residences, hotels and motels, bunkhouses, ranger stations, crew quarters, campgrounds, picnic grounds, and day-use recreation areas). This term is not applicable to consolidated shipments of household hazardous materials transported from collection centres. A collection centre is a central location where household waste is collected.

Human ecology: Interrelationship between humans and the entire environment – physical, biological, socio-economic, and cultural – including the interrelationships between individual humans or groups of humans and other human groups or groups of other species.

Hycar aprons: Nitrile blend provides reliable abrasion and cut resistance for longer protection against oils, fats, chemicals, acids, and grease. These rubber protective aprons can withstand the deteriorating effects of animal fats and greases while remaining flexible in cold environments.

Hydrocephalus: The abnormal build-up of CSF in the ventricles of the brain.

Hydrogen peroxide: Chemical substance used in bleaches, dyes, cleansers, antiseptics, and disinfectants. In a concentrated form, it is toxic and irritating to tissues.

Hydrolysis: Chemical reaction of a substance with water, usually resulting in the formation of one or more new compounds.

Hydronephrosis: Pathological chronic enlargement of the collecting channels of a kidney, leading to compression and eventual destruction of kidney tissue and diminishing kidney function.

Hydrophilic: Describing the character of a substance, material, molecular entity, or group of atoms that has an affinity for water.

Hydrophobic: (i) Denoting the property of attracting or associating with water molecules; characteristic of polar or charged molecules. (ii) Describing the character of a substance, material, molecular entity, or group of atoms that is insoluble or confers insolubility in water or resistance to wetting or hydration. The term denotes to a molecule or side group, tending to dissolve readily in organic solvents, but not in water, resisting wetting, not containing polar groups or subgroups.

Hydrosphere: A broad term for the water above, on, or in the Earth's crust, including oceans, seas, lakes, ground water, and atmospheric moisture.

Hygiene: Science of health and its preservation.

Hyperbilirubinemia: Excessive concentration of bilirubin in the blood.

Hypercalcaemia: Excessive concentration of calcium in the blood.

Hyperaemia: Excessive amount of blood in any part of the body.

Hyperglycaemia: (i) Excessive concentration of glucose in the blood. (ii) Higher than normal amount of glucose (a type of sugar) in the blood. Hyperglycaemia can be a sign of diabetes or other conditions.

Hyperkalaemia: Excessive concentration of potassium in the blood.

Hyperkerotosis: A condition marked by thickening of the outer layer of the skin, which is made of keratin (a tough, protective protein). It can result from normal use (corns, calluses), chronic inflammation (eczema), or genetic disorders.

Hypernatraemia: Excessive concentration of sodium in the blood.

Hyperparathyroidism: Abnormally increased parathyroid gland activity that affects, and is affected by, plasma calcium concentration.

Hyperplasia: Abnormal multiplication or increase in the number of normal cells in a tissue or organ.

Hypersensitivity: (i) A state of altered reactivity in which the body reacts with an exaggerated immune response to what is perceived as a foreign substance. (ii) State in which an individual reacts with allergic effects following exposure to a certain substance (allergen) after having been exposed previously to the same (*Note*: most common chemical-induced allergies are type I (IgE-mediated) and type IV (cell-mediated) hypersensitivity).

Hypersusceptibility: Excessive reaction following exposure to a given amount or concentration of a substance as compared with the large majority of other exposed subjects.

Hypertension: Persistently high blood pressure in the arteries or in a circuit, for example, pulmonary hypertension or hepatic portal hypertension.

Hypertrophy: Excessive growth in bulk of a tissue or organ through increase in size but not in number of the constituent cells.

Hypocalcaemia: Abnormally low calcium concentration in the blood.

Hypoglycaemia: Abnormally low blood sugar.

Hypothalamus: The area of the brain that controls body temperature, hunger, and thirst.

Hypothyroidism: Too little thyroid hormone. Symptoms include weight gain, constipation, dry skin, and sensitivity to the cold.

Hypotriglyceridaemia: In medicine, the term describes the situation of decreased blood triglyceride content.

Hypoxaemia: A state deficient in oxygenation of the blood.

Hypoxia: (i) A state abnormally low in dioxygen content or tension. (ii) Deficiency of dioxygen in the inspired air, in blood, or in tissues, short of anoxia.

IACUC: Institutional Animal Care and Use Committee.

IARC (International Agency for Research on Cancer): Organization that evaluates information on the carcinogenicity of chemicals, groups of chemicals, and chemicals associated with certain industrial processes. The IARC has published lists of chemicals that are generally recognised as human carcinogens, probable human carcinogens, or carcinogens in animal tests.

IBS (irritable bowel syndrome): A disorder of the intestines commonly marked by abdominal pain, bloating, and changes in a person's bowel habits. This may include diarrhoea or constipation, or both, with one occurring after the other. Also called as irritable colon, mucus colitis, and spastic colon.

ICBN: International Code of Botanical Nomenclature.

ICD: International Classification of Diseases.

ICZN: International Code of Zoological Nomenclature.

Ideopathic: To describe a disease of unknown cause.

IDLH (immediately dangerous to life or health): For the purposes of respirator selection, the NIOSH defines the IDLH concentration as the airborne concentration that poses a threat of exposure to airborne contaminants when that exposure is likely to cause death or immediate or delayed permanent adverse health effects or prevent escape from such an environment. The purpose of establishing an IDLH exposure concentration is to ensure that the worker can escape from a given contaminated environment in the event of failure of the respiratory protection equipment. In the event of failure of respiratory protective equipment, every effort should be made to exit immediately.

Ignitable: Capable of bursting into flames; ignitable chemical substances pose a fire hazard.

Immediately dangerous to life or health concentration (IDLH): A regulatory value defined as the maximum exposure concentration of a chemical substance in the workplace from which one could escape within 30 min without any escape-impairing symptoms or any irreversible health effects. This value should be referred to in respirator selection.

Immune complex: Product of an antigen–antibody reaction that may also contain components of the complement system.

Immune function: Production and action of cells that fight disease or infection.

Immune response: (i) The general reaction of the body to substances that are foreign or treated as foreign. It may take various forms, for example, antibody production, cell-mediated immunity, immunological tolerance, or hypersensitivity (allergy). (ii) The activity of the immune system against foreign substances (antigens).

Immune system: (i) An integrated network of organs, glands, and tissues that has evolved to protect the body from foreign substances, including bacteria, viruses, and other infection-causing parasites and pathogens. The immune system may produce hypersensitivity reactions that, in the extreme, can be fatal. If the immune system misidentifies normal body components as foreign, this leads to autoimmune disorders, such as lupus, in which the body destroys its own constituents. (ii) The complex group of organs and cells that defends the body against infections and other diseases.

Immunisation: (i) The technique used to cause an immune response that results in resistance to a specific disease, especially an infectious disease. (ii) An individual, when immunised, is protected from getting a particular disease caused by a particular biological agent.

Immunity: The condition of being protected against an infectious disease. Immunity can be caused by a vaccine, by previous infection with the same agent, or by transfer of immune substances from another person or animal.

Immunoassay: (i) A test that uses the binding of antibodies to antigens to identify and measure certain substances. Immunoassays may be used to diagnose disease. Also, test results can provide information about a disease that may help in planning treatment, for example, when oestrogen receptors are measured in breast cancer. (ii) Ligand-binding assay that uses a specific antigen or antibody, capable of binding to the analyte, to identify and quantify substances. The antibody can be linked to a radioisotope (radioimmunoassay, RIA) or to an enzyme that catalyses an easily monitored reaction (enzyme-linked immunosorbent assay, ELISA) or to a highly fluorescent compound by which the location of an antigen can be visualised.

Immunochemistry: The branch of science related to molecular aspects of immunology, the chemistry of antigens, antibodies, and their relationship to each other.

Immunodeficiency: The decreased ability of the body to fight infections and other diseases.

Immunoglobulin: (i) Family of closely related glycoproteins capable of acting as antibodies and present in plasma and tissue fluids; immunoglobulin E (IgE) is the source of antibody in type I hypersensitivity (allergic) reactions. (ii) A protein (Ig) that acts as an antibody. Immunoglobulins are made by B cells and plasma cells. An immunoglobulin is a type of glycoprotein with two heavy chains and two light chains.

Immunomodulation: Modification of the functioning of the immune system by the action of a substance that increases or reduces the ability to produce antibodies.

Immunosuppression: (i) In medicine, this term is applied to inhibition of the normal response of the immune system to an antigen. A decrease in the functional capacity of the immune response may be due to (a) the inhibition of the normal response of the immune system to an antigen or (b) prevention, by chemical or biological means, of the production of an antibody to an antigen by inhibition of the processes of transcription, translation, or formation of tertiary structure. (ii) Suppression of the body's immune system and its ability to fight infections and other diseases. Immunosuppression may be deliberately induced with drugs, as in preparation for bone marrow or other organ transplantation, to prevent rejection of the donor tissue. It may also result from certain diseases such as AIDS or lymphoma or from anticancer drugs.

Immunotherapy: Treatment to boost or restore the ability of the immune system to fight cancer, infections, and other diseases. Also used to lessen certain side effects that may be caused by some cancer treatments. Agents used in immunotherapy include monoclonal antibodies, growth factors, and vaccines. These agents may also have a direct antitumour effect.

Immunotoxic: Pertaining to any chemical substance harmful to the immune system.

Impervious: A term used to describe protective gloves and other protective clothing. If a material is impervious to a chemical, then that chemical cannot readily penetrate through the material or damage the material. Different materials are impervious (resistant) to different chemicals. No single material is impervious to all chemicals. If an MSDS recommends wearing impervious gloves, you need to know the type of material from which the gloves should be made.

Implant: A substance or object that is put in the body as a prosthesis or for treatment or diagnosis.

Implantation: Attachment of the fertilised ovum (blastocyst) to the endometrium and its subsequent embedding in the compact layer, occurring 6 or 7 days after fertilisation of the ovum.

Improvement actions: Actions taken to improve the efficiency of operations based on a good work practice or an innovative approach.

Inadequate evidence: When, because of major qualitative or quantitative limitations, the studies cannot be interpreted as showing either the presence or absence of a carcinogenic effect. This indicates that one of two conditions prevailed: (a) there are few pertinent data or (b) the available studies, while showing evidence of association, do not exclude chance, bias, or confounding.

Inadequate study: When carcinogenicity is not demonstrated because of major qualitative or quantitative limitations and the studies cannot be interpreted as valid for showing either the presence or absence of a carcinogenic effect.

Incineration: A method of treating solid, liquid, or gaseous wastes by burning.

Incision: A cut made in the body to perform surgery.

Incompatible materials: Chemical substances/materials that can react to cause a fire, explosion, or violent reaction or lead to the evolution of flammable gases or otherwise lead to injury to people or danger to property. Incompatible chemical substances/materials can react with the product or with components of the product and may (i) destroy the structure or function of a product; (ii) cause a fire, explosion, or violent reaction; or (iii) cause the release of hazardous chemicals.

Incontinence: Inability to control the flow of urine from the bladder (urinary incontinence) or the escape of stool from the rectum (faecal incontinence).

Indication: In medicine, a sign, symptom, or medical condition that leads to the recommendation of a treatment, test, or procedure.

Induction: Increase in the rate of synthesis of an enzyme in response to the action of an inducer or environmental conditions.

Inert ingredient: Anything other than the active ingredient of a product. It may be a solvent, colorant, filler, or dispersing agent. In some cases, inert ingredients may be hazardous.

Infection: (i) The growth of a parasitic organism within the body. (A parasitic organism is one that lives on or in another organism and draws its nourishment therefrom.) A person with an infection has another organism (a 'germ') growing within his or her system drawing its nourishment from the affected individual. (ii) Invasion and multiplication of germs in the body. Infections can occur in any part of the body and can spread throughout the body. The germs may be bacteria, viruses, yeast, or fungi. They can cause a fever and other problems, depending on where the infection occurs. When the body's natural defence system is strong, it can often fight the germs and prevent infection. Some cancer treatments can weaken the natural defence system.

Inferior vena cava: The large vein that empties into the heart. It carries blood from the legs and feet and from organs in the abdomen and pelvis.

Inflammation: Reaction of the body to injury or to infectious, allergic, or chemical irritation. Inflammation is marked by swelling, redness, heat, and/or pain because of dilation of the blood vessels accompanied by loss of plasma and leucocytes (WBCs) into the tissues. This is a protective reaction to injury, disease, or irritation of the tissues. Inflammation involves an influx of blood or other fluid into bodily tissue or organs in reaction to injury, disease, and/or foreign substances.

Infusion: A method of putting fluids, including drugs, into the bloodstream. Also called intravenous infusion. (ii) Therapeutic administration of a fluid other than blood, usually saline as a solution into a vein.

Injection: The introduction of chemicals into the body through puncture.

Ingestion: (i) Process of taking food and drink into the body by mouth. (ii) Taking a substance into the body by mouth and swallowing it. (iii) A process of swallowing such as eating or drinking. Chemical substances can get into or onto food, drink, utensils, cigarettes, or hands where they can then be ingested.

Inhalation: (i) Act of drawing in air, vapour, or gas and any suspended particulates into the lung. (ii) A process of breathing. Once inhaled, the contaminants get deposited in the lungs, taken into the blood, or both; (iii) taking a chemical substance/material into the body by breathing it in.

Inhaler: A device for giving medicines in the form of a spray that is inhaled (breathed in) through the nose or mouth. Inhalers are used to treat medical problems such as bronchitis, angina, emphysema, and asthma. They are also used to help relieve symptoms that occur when a person is trying to quit smoking.

Initial protective action zone: The area downwind from an incident in which persons may become incapacitated and unable to take protective action and/or incur serious or irreversible health effects.

Initiation: The first stage of carcinogenesis.

Initiator: Any agent that starts the process of tumour formation, usually by action on the genetic material, is called an initiator.

Injury (harm or hurt): A term applied in medicine to damage inflicted upon oneself by an external agent. The injury may be accidental or deliberate, as with a needlestick injury.

Inoculations: A series of injections given to humans to protect them against diseases caused by biological agents.

Insecticide: Any chemical substance used to kill insects (*See* Classification of pesticides, pp. 12–15).

Insomnia: Difficulty in going to sleep or getting enough sleep.

Integrated pest management (IPM): (i) A systematic plan that brings together different pest control strategies into one programme that is economically sound and that minimises environmental problems. (ii) The careful consideration of all available pest control techniques and subsequent integration of appropriate measures that discourage the development of pest populations and keep pesticides and other interventions to levels that are economically justified and reduce or minimise risks to human health and the environment. IPM emphasises the growth of a healthy crop with the least possible disruption to agroecosystems and encourages natural pest control mechanisms.

Integrated Risk Information System (IRIS): An electronic database that contains EPA's latest descriptive and quantitative regulatory information about chemical constituents. Files on chemicals maintained in IRIS contain information related to both noncarcinogenic and carcinogenic health effects.

Interferon: A naturally occurring chemical substance that interferes with the ability of viruses to reproduce. Interferon also boosts the immune system.

International Agency for Research on Cancer (IARC): An agency of the WHO that publishes *IARC Monographs on the Evaluation of the Carcinogenic Risk of Chemicals to Humans*. This publication documents reviews of information on chemicals and determinations of the cancer risk of chemicals.

Intestine: The long, tube-shaped organ in the abdomen that completes the process of digestion. The intestine has two parts, the small intestine and the large intestine. Also called bowel.

Interstitial pneumonia: Chronic form of pneumonia involving increase of the interstitial tissue and decrease of the functional lung tissue.

Intoxication: (i) Poisoning – pathological process with clinical signs and symptoms caused by a substance of exogenous or endogenous origin. (ii) Drunkenness following consumption of beverages containing ethanol or other compounds affecting the CNS.

In vitro: A term applied to describe a study carried out in isolation from the living organism in an experimental system (applied to studies of biological functions or processes to contrast with in vivo studies).

In vivo: The term used to describe any study carried out within the living organism as in contrast with 'in vitro'.

Intermediate duration exposure: Contact with a substance that occurs for more than 14 days and less than a year.

IPE: Intermediate potential of exposure.

Iris: The coloured tissue at the front of the eye that contains the pupil in the centre. The iris helps control the size of the pupil to let more or less light into the eye.

Iron: An important mineral the body needs to make haemoglobin, a substance in the blood that carries oxygen from the lungs to tissues throughout the body. Iron is also an important part of many other proteins and enzymes needed by the body for normal growth and development. It is found in red meat, fish, poultry, lentils, beans, and foods with iron added, such as cereal.

Irradiation: The use of high-energy radiation from x-rays, gamma rays, neutrons, protons, and other sources to kill cancer cells and shrink tumours. Radiation may come from a machine outside the body (external-beam radiation therapy), or it may come from radioactive material placed in the body near cancer cells (internal radiation therapy). Systemic irradiation uses a radioactive substance, such as a radiolabelled monoclonal antibody, that travels in the blood to tissues throughout the body.

Irritant: (i) Any chemical substance that causes inflammation following immediate, prolonged, or repeated contact with skin or mucous membrane. (ii) A chemical substance that causes a reversible inflammatory effect on living tissue by chemical action at the site of contact.

Irritancy/irritation: The ability of a hazardous chemical substance material to irritate the skin, eyes, nose, throat, or any other part of the body that it contacts. Signs and symptoms of irritation include tearing in the eyes and reddening, swelling, itching, and pain of the affected part of the body. Irritancy is often described as mild, moderate, or severe, depending on the degree of irritation caused by a specific amount of the hazardous chemical substance/material. Irritancy may also be described by a number on a scale of 0–4, where 0 (zero) indicates no irritation and 4 means severe irritation. Irritancy is usually determined in animal experiments. The CPRs (WHMIS) and the U.S. OSHA Hazcom Standard describe technical criteria for identifying the hazardous chemical substance materials that are skin or eye irritants for the purposes of each regulation.

Ischaemia: The deficiency of blood supply to any part of the body, relative to its local requirements. Ischaemia may cause tissue damage due to the lack of oxygen and nutrients.

ISCN: International System for Human Cytogenetic Nomenclature.

ISEA: International Safety Equipment Association.

Islets of Langerhans cell: A pancreatic cell that produces hormones (e.g. insulin and glucagon) that are secreted into the bloodstream. These hormones help control the level of glucose (sugar) in the blood. It is also called endocrine pancreas cell and islet cell.

Isoamyl acetate (banana oil) test: A test used by Safety Central and throughout the gas mask industry to detect leaks in the mask or faulty filters. If the wearer can smell a 'banana' odour, the mask is either (a) not fitted properly or (b) has a used and ineffective filter (canister).

Isotonic: Denoting a liquid exerting the same osmotic pressure or chemical potential of water (water potential) as another liquid with which it is being compared.

Itai-itai disease: Illness (renal osteomalacia) observed in the Toyama prefecture of Japan, resulting from the ingestion of cadmium-contaminated rice. The damage occurred to the renal and skeleto-articular systems, the latter being very painful ('itai' means 'ouch' in Japanese and refers to the intense pain caused by the condition).

IUPAC: The International Union of Pure and Applied Chemistry.

IV: Abbreviation for intravenous (used in relation to administration of drugs or other substances).

Jaundice: (i) A pathological condition in which the skin and the whites of the eyes become yellow, urine darkens, and the colour of stool becomes lighter than normal. Jaundice occurs when the liver is not working properly or when a bile duct is blocked. (ii) A condition characterised by deposition of bile pigment in the skin and mucous membranes, including the conjunctivae, resulting in yellow appearance of the patient or animal.

Joint Information Center: A facility jointly operated by the Department of Energy (DOE), DOE contractor, and state, tribal, and local governments to coordinate the release of accurate and timely information to the public during and after an emergency.

Karyotype: The particular chromosome complement of an individual or a related group of individuals, as defined by both the number and morphology of the chromosomes, usually in mitotic metaphase, and arranged by pairs according to the standard classification.

Kava kava: A herb native to islands in the South Pacific. Substances taken from the root have been used in some cultures to relieve stress, anxiety, tension, sleeplessness, and problems of menopause. Kava kava may increase the effect of alcohol and of certain drugs used to treat anxiety and depression. The U.S. FDA advises users that kava kava may cause severe liver damage. The scientific name is *Piper methysticum*. Kava kava is also called intoxicating pepper, rauschpfeffer, tonga, and yangona.

Keloid: A thick, irregular scar caused by excessive tissue growth at the site of an incision or wound.

Keratoacanthoma: A rapidly growing, dome-shaped skin tumour that usually occurs on sun-exposed areas of the body, especially around the head and neck. Keratoacanthoma occurs more often in males. Although in most patients it goes away on its own, in a few patients it comes back. Rarely, it may spread to other parts of the body.

Ketone: A type of chemical substance used in perfumes, paints, and solvents and found in essential oils (scented liquid taken from plants). Ketones are also made by the body when there is not enough insulin.

Ketosis: Pathological increase in the production of ketone bodies, for example, following blockage or failure of carbohydrate metabolism.

Kidney: (i) One of a pair of glandular organs located in the right and left side of the abdomen that eliminates/clears poisonous chemical substances from the blood, regulate acid concentration, and maintain water balance in the body by excreting urine. The kidneys are part of the urinary tract. The urine then passes through connecting tubes called 'ureters' into the bladder. The bladder stores the urine until it is released during urination. (ii) A pair of glandular organs in the dorsal region of the vertebrate abdominal cavity, functioning to maintain proper water and electrolyte balance, regulate acid–base concentration, filter the blood of metabolic wastes that are then excreted as urine, and also play a role in blood pressure regulation. (iii) Kidneys are situated on the sides of the spine that filter blood and eliminate metabolism waste in urine through the ureters.

Kg: Abbreviation for kilogram.

Kidney failure: A health condition in which the kidneys stop working and are not able to remove waste and extra water from the blood or keep body chemicals in balance. Acute or severe kidney failure happens suddenly (e.g. after an injury) and may be treated and cured. Chronic kidney failure develops over many years, may be caused by conditions like high blood pressure or diabetes, and cannot be cured. Chronic kidney failure may lead to total and long-lasting kidney failure, called end-stage renal disease (ESRD). A person in ESRD needs dialysis (the process of cleaning the blood by passing it through a membrane or filter) or a kidney transplant.

Kidney function test: A biochemical test in which blood or urine samples are checked for the amounts of certain substances released by the kidneys. A higher- or lower-than-normal amount of a substance can be a sign that the kidneys are not working the way they should, and also called renal function test.

Klinefelter syndrome: A genetic disorder in males caused by having one or more extra X chromosomes. Males with this disorder may have larger than normal breasts, a lack of facial and body hair, a rounded body type, and small testicles. They may learn to speak much later than other children and may have difficulty learning to read and write. Klinefelter syndrome increases the risk of developing extragonadal germ cell tumours and breast cancer.

Known human carcinogen: A chemical substance for which there is sufficient evidence of a cause and effect relationship between exposure to the material and cancer in humans.

Labial mucosa: The inner lining of the lips.

Lacrimal gland: A gland that secretes tears. The lacrimal glands are found in the upper, outer part of each eye socket.

Lacrimation: (i) Secretion and discharge of tears. (ii) Excessive production of tears when the eye is exposed to an irritant.

Lacrimator: A chemical substance that irritates the eyes and causes the production of tears.

Laparoscope: A thin, tube-like instrument used to look at tissues and organs inside the abdomen. A laparoscope has a light and a lens for viewing and may have a tool to remove tissue.

Large intestine: The long, tube-like organ that is connected to the small intestine at one end and the anus at the other. The large intestine has four parts: caecum, colon, rectum, and anal canal. Partly digested food moves through the caecum into the colon, where water and some nutrients and electrolytes are removed. The remaining material, solid waste called stool, moves through the colon, is stored in the rectum, and leaves the body through the anal canal and anus.

Larvicide: A chemical substance intended to kill larval life stage of an insect. Larvicides may be contact poisons, stomach poisons, growth regulators, or biological control agents.

Laryngitis: Inflammation of the larynx.

Larynx (voice box): The area of the respiratory tract and throat containing the vocal cords and used for breathing, swallowing, and talking.

Lassitude: A feeling of tiredness, weakness, and lack of interest in daily activities.

Latent period: The period suggesting (i) a delay between exposure to a harmful substance and the manifestations of a disease or other adverse effects and (ii) a period from disease initiation to disease detection.

Latent: Hidden, dormant, inactive.

Lavage: Irrigation or washing out of a hollow organ or cavity such as the stomach, the intestine, or the lungs.

Laxative (purgative): Substance that causes evacuation of the intestinal contents.

LC_{50} (median lethal concentration$_{50}$): The concentration of a material in air that causes the death of 50% (one-half) of a group of test animals. The material is inhaled over a set period of time, usually 1 or 4 h. The LC_{50} helps determine the short-term poisoning potential of a material. The concentration of a chemical substance that kills 50% of a sample population, typically expressed in mass per unit volume of air.

LCn: The abbreviation to describe the exposure concentration of a toxic chemical substance lethal to n per cent of a test population.

LD_{50} (median lethal dose 50): The amount of a material, given all at once, which causes the death of 50% (one-half) of a group of test animals. The LD_{50} can be determined for any route of entry, but dermal (applied to skin) and oral (given by mouth) LD_{50}s are most common. The LD_{50} is one measure of the short-term poisoning potential of a material. The amount of a chemical substance that kills 50% of a sample population, typically expressed as milligrams per kilogram of body weight.

LDn: The abbreviation to describe the dose of a toxicant lethal to n% of a test population.

LD_{LO}: Abbreviation for lowest lethal dose tested.

LDPE (low-density polyethylene): The first of the polyethylenes to be developed. It is an excellent material in electrical and chemical uses in low heat applications.

Lesion: States of health conditions describing (i) area of pathologically altered tissue; (ii) injury or wound; (iii) infected patch of skin.

Lethal concentration (LC): Concentration of a substance in an environmental medium that causes death following a certain period of exposure.

Lethal dose (LD): Amount of a substance or physical agent (e.g. radiation) that causes death when taken into the body.

Lethal: Deadly; fatal; causing death.

Lethargy: A health condition marked by drowsiness and an unusual lack of energy and mental alertness. It can be caused by many things, including illness, injury, or drugs.

Leucocyte (white blood cell/WBC): A type of immune cell. Most leucocytes are made in the bone marrow and are found in the blood and lymph tissue. Leucocytes help the body fight infections and other diseases. Granulocytes, monocytes, and lymphocytes are leucocytes.

Leukaemia: (i) Literally, 'white blood'. An abnormal increase in the number of leucocytes (WBCs) in the tissues of the body. It is a form of cancer, and there are many individual types of leukaemia that are defined by which kinds of WBCs are affected.

(ii) Cancer that starts in blood-forming tissue such as the bone marrow and causes large numbers of blood cells to be produced and enter the bloodstream. (iii) Progressive, malignant disease of the blood-forming organs, characterised by distorted proliferation and development of leucocytes and their precursors in the bone marrow and blood.

Leukopaenia: A condition in which there is a lower-than-normal number of leucocytes (WBCs) in the blood.

Libido: Sexual desire or sex drive and sexual urge.

Limit test: Acute toxicity test in which, if no ill effects occur at a pre-selected maximum dose, no further testing at greater exposure levels is required.

Limited evidence: The evidence of carcinogenicity when data suggest a carcinogenic effect but are limited because (a) the studies involve a single species, strain, or experiment; or (b) the experiments are restricted by inadequate dosage levels, inadequate duration of exposure to the agent, inadequate period of follow-up, poor survival, too few animals, or inadequate reporting; or (c) the neoplasms produced often occur spontaneously and, in the past, have been difficult to classify as malignant by histological criteria alone. This indicates that a causal interpretation is credible but that alternative explanations, such as chance, bias, or confounding, could not adequately be excluded.

LIMS (Laboratory Information Management Systems): Software that helps biological and chemical laboratories handle data generation, information management, and data archiving.

Lipophilic: Having an affinity for fat and high lipid solubility. *Note*: This is a physico-chemical property that describes a partitioning equilibrium of solute molecules between water and an immiscible organic solvent, favouring the latter, and that correlates with bioaccumulation.

Lipophobic: Having a low affinity for fat and a high affinity for water.

Liposome: (i) Artificially formed lipid droplet, small enough to form a relatively stable suspension in aqueous media, useful in membrane transport studies and in drug delivery. (ii) Lipid droplet in the endoplasmic reticulum of a fatty liver.

Liquid: Any chemical substance/material other than an elevated temperature material, with a melting point or initial melting point of 20°C (68°F) or lower at a standard pressure of 101.3 kPa (14.7 psia).

Liquid phase: A chemical substance/material that meets the definition of liquid when evaluated at the higher temperature at which it is offered for transportation or at which it is transported.

Liver: The largest glandular vital organ of the body in the upper abdomen. Liver has many functions, for instance, it produces substances that break down fats, helps in digestion, converts glucose to glycogen, produces urea (the main substance of urine), makes certain amino acids (the building blocks of proteins), filters harmful substances from the blood, and is also responsible for producing cholesterol.

Liver and gastrointestinal toxicity: Adverse effects to the structure and/or function of the liver, gall bladder, or gastrointestinal tract caused by exposure to a toxic chemical. The liver is frequently subject to chemical-induced injury because of its role as the body's principal site of metabolism. Chemicals that damage the liver can cause diseases such as hepatitis, jaundice, cirrhosis, and cancer.

Liver disease: Any disorder of the liver. The liver is a large organ in the upper right abdomen that aids in digestion and removes waste products from the blood.

Liver nodule: A medical term to describe any small node or aggregation of cells within the liver.

Liver transplant: Surgery to remove a diseased liver and replace it with a healthy liver (or part of one) from a donor.

Local anaesthesia: A temporary loss of feeling in one small area of the body caused by special drugs or other substances called anaesthetics. The patient stays awake but has no feeling in the area of the body treated with the anaesthetics.

Local exhaust ventilation: The removal of contaminated air directly at its source. This type of ventilation can help reduce worker exposure to airborne materials more effectively than general ventilation. This is because it does not allow the hazardous chemical substance/material to enter the work environment. It is usually recommended for hazardous airborne materials.

Locus: In a genomic context, the position on a chromosome. It may, therefore, refer to a marker, a gene, or any other landmark that can be described.

LOEL: Lowest-observed-effect level.

Long-term exposure: Repeated exposure by the oral, dermal, or inhalation route for more than 30 days, up to approximately 10% of the life span in humans (more than 30 days up to approximately 90 days in typically used laboratory animal species).

Lowest effective dose (LED): Lowest dose of a chemical substance inducing a specified effect in a specified fraction of exposed individuals.

Lowest-observed-effect level (LOEL): Lowest concentration or amount of a substance (dose), found by experiment or observation, that causes any alteration in morphology, functional capacity, growth, development, or life span of target organisms distinguishable from normal (control) organisms of the same species and strain under the same defined conditions of exposure.

LPE: Lowest potential for exposure.

Lumbar puncture: A procedure in which a thin needle called a spinal needle is put into the lower part of the spinal column to collect CSF or to give drugs.

Lung: One of a pair of organs in the chest that supplies the body with oxygen and removes carbon dioxide from the body. Lungs are sac-like structures where gas exchange occurs with the blood.

Lupus: A chronic, inflammatory, connective tissue disease that can affect the joints and many organs, including the skin, heart, lungs, kidneys, and nervous system. It can cause many different symptoms.

Lymph gland: A rounded mass of lymphatic tissue that is surrounded by a capsule of connective tissue. Lymph glands filter lymph (lymphatic fluid), and they store lymphocytes (WBCs). They are located along lymphatic vessels.

Lymphocyte: A type of immune cell that is made in the bone marrow and is found in the blood and in lymph tissue. The two main types of lymphocytes are B lymphocytes and T lymphocytes. B lymphocytes make antibodies, and T lymphocytes help kill tumour cells and help control immune responses. A lymphocyte is a type of WBC.

Lymphoma: General term comprising tumours and conditions allied to tumours arising from some or all of the cells of lymphoid tissue.

Lysimeter: Laboratory column of selected representative soil or a protected monolith of undisturbed field soil with facilities for sampling and monitoring the movement of water and chemical substances.

Lysis: The breakdown of a cell caused by damage to its plasma (outer) membrane. It can be caused by chemical or physical means (e.g. strong detergents or high-energy sound waves) or by infection with a strain virus that can lyse cells.

Lysosome: (i) A sac-like compartment inside a cell that has enzymes that can break down cellular components that need to be destroyed. (ii) Membrane-bound cytoplasmic organelle containing hydrolytic enzymes. The release of these enzymes from lysosomes damaged by xenobiotics can cause autolysis of the cell.

Macrophages: (i) WBCs within tissues, produced by the division of monocytes. (ii) Macrophages are large phagocytic cells found in connective tissue, especially in areas of inflammation.

Macroscopic (gross) pathology: The studies to know the diseased tissue changes that are visible to the naked eye.

Macular degeneration: A health condition in which there is a slow breakdown of cells in the centre of the retina (the light-sensitive layers of nerve tissue at the back of the eye). This blocks vision in the centre of the eye and can cause problems with activities such as reading and driving. Macular degeneration is most often seen in people who are over the age of 50.

Malaise: A kind of vague feeling of bodily discomfort.

Malignant: (i) A medical term used to describe a severe and progressively worsening disease – cancer. (ii) An adjective term to describe cells in a cancerous growth.

Mania: Emotional disorder (mental illness) characterised by an expansive and elated state (euphoria), rapid speech, distorted ideas, decreased need for sleep, distractibility, grandiosity, and poor judgment and increased motor activity.

Margin of safety (MOS): Ratio of the NOAEL to the theoretical or estimated exposure dose (EED).

Marking: A descriptive name, identification number, instructions, cautions, weight, specification, or UN marks, or combinations thereof, required by this subchapter on outer packagings of hazardous chemical substances/materials.

Mass mean diameter (MMD): Diameter of a particle with a mass equal to the mean mass of all the particles in a population.

Mast cell: A type of WBC.

Material at risk (MAR): The MAR is the amount and type of material available to be acted on by a given physical stress. For facilities, processes, and activities, the MAR is a value representing some maximum quantity of material present or reasonably anticipated for the process or structure being analysed.

Material causing immediate and serious toxic effects: The Canadian CPRs describe technical criteria for identifying materials that cause immediate and serious toxic effects. These criteria use information such as the LD_{50} or LC_{50} for a material. Based on the specific information, a material may be identified as toxic or very toxic in the class D – poisonous and infectious material.

Material causing other toxic effects: The Canadian CPRs describe technical criteria for identifying materials that cause toxic effects such as skin or respiratory sensitisation, mutagenicity, and carcinogenicity. Based on the specific information, a material may be identified as toxic or very toxic in the class D – poisonous and infectious material.

Material safety data sheet (MSDS): (i) Compilation of information required under the U.S. OSHA Hazcom Standard on the identity of hazardous substances, health and physical hazards, exposure limits, and precautions. (ii) Mandatory information that must accompany almost every chemical substance in the workplace except for items like cleaning supplies. An MSDS includes details such as the hazards, precautions, and first-aid procedures associated with the chemical.

Maximum allowable concentration (admissible, acceptable concentration) (MAC): Regulatory value defining the concentration that if inhaled daily (in the case of people working for 8 h with a working week of 40 h, in the case of the general population 24 h) does not, in the present state of knowledge, appear capable of causing appreciable harm, however long delayed during the working life or during subsequent life or in subsequent generations.

Maximum allowable concentration (admissible, acceptable concentration) (MAC): The maximum exposure to a biologically active physical or chemical agent that is allowed during an 8 h period (a workday) in a population of workers, or during a 24 h period in the general population, which does not appear to cause appreciable harm, whether immediate or delayed for any period, in the target population.

Maximum average daily concentration of an atmospheric pollutant: The peak daily average concentration of an air pollutant. The highest of the average daily concentrations recorded at a definite point of measurement during a certain period of observation.

Maximum contaminant level (MCL): Under the Safe Drinking Water Act (Unites States), primary MCL is a regulatory concentration for drinking water that takes into account both adverse effects (including sensitive populations) and technological feasibility (including natural background levels): secondary MCL is a regulatory concentration based on 'welfare', such as taste and staining, rather than health, but also takes into account technical feasibility. *Note*: MCL Goals (MCLG) under the Safe Drinking Water Act do not consider feasibility and are zero for all human and animal carcinogens.

Maximum exposure limit (MEL): OEL legally defined under control of substance hazardous to health (COSHH) as the maximum concentration of an airborne substance, averaged over a reference period, to which employees may be exposed by inhalation under any circumstances, and set on the advice of the Health and Safety Commission (HSC) Advisory Committee on toxic chemical substances.

Maximum permissible concentration (MPC): Same as MAC.

Maximum permissible daily dose: Maximum daily dose of a chemical substance whose penetration into a human body during a lifetime will not cause diseases or health hazards that can be detected by current investigation methods and will not adversely affect future generations.

Maximum permissible level (MPL): Level, usually a combination of time and concentration, beyond which any exposure of humans to a chemical or physical agent in their immediate environment is unsafe.

Maximum residue limit (MRL): (i) The maximum concentration of residue accepted by the European Union (EU) and the Minimal Risk Levels (MRLs) according to Agency for Toxic Substances and Diseases Registry (ATSDR) estimated for individual pesticides or veterinary drug residues in various food commodities. They are based on GAP (pesticides) or good practice in the use of veterinary drugs in which the product has been used in an efficacious manner and appropriate withdrawal periods have been followed. They are expressed as either the parent compound or a metabolite that is, or is representative of, the residue of toxicological concern in the food commodity. MRLs are not based upon toxicological data, but crude estimates of their toxicological significance are usually made by comparing the ADI with a calculation of the total intake of the residue based on the MRLs and food intake data of these commodities for which MRLs have been established. (ii) The MRL is usually determined by measurement, following a number (in the order of 10) of field trials, where the crop has been treated according to GAP and an appropriate preharvest interval.

Maximum residue limit for pesticide residues (MRL): Maximum contents of a pesticide residue (expressed as mg/kg fresh weight) recommended by the CAC to be legally permitted in or on food commodities and animal feeds. The MRL values are based on data obtained following GAP, and foods derived from commodities that comply with the respective MRLs are intended to be toxicologically acceptable.

Maximum tolerated dose (MTD): The highest dose of a drug or treatment that does not cause unacceptable side effects. The MTD is determined in clinical trials by testing increasing doses on different groups of people until the highest dose with acceptable side effects is found.

Mechanical ventilation: The movement of air by mechanical means (e.g. a wall fan). There are two kinds of mechanical ventilation: general ventilation and local exhaust ventilation.

Median effective concentration (EC_{50}): Statistically derived median concentration of a chemical substance in an environmental medium expected to produce a certain effect in 50% of test organisms in a given population under a defined set of conditions. (*Note*: ECn refers to the median concentration that is effective in n% of the test population.)

Median effective dose (ED_{50}): Statistically derived median dose of a chemical or physical agent (radiation) expected to produce a certain effect in 50% of test organisms in a given population or to produce a half-maximal effect in a biological system under a defined set of conditions (*Note*: EDn refers to the median dose that is effective in n% of the test population).

Median lethal concentration (LC_{50}): Statistically derived median concentration of a substance in an environmental medium expected to kill 50% of organisms in a given population under a defined set of conditions.

Median lethal dose (LD_{50}): Statistically derived single dose of a chemical that can be expected to cause death in 50% of a given population of organisms under a defined set of experimental conditions. This figure has often been used to classify and compare toxicity among chemical substances, but its value for this purpose is doubtful.

Medication: (i) A drug or medicine. (ii) The administration of a drug or medicine.

Mediastinum: The area between the lungs. The organs in this area include the heart and its large blood vessels, the trachea, the oesophagus, the thymus, and the lymph nodes but not the lungs.

MEDLINE: National Library of Medicine's database for scientific publications.

Meiosis: (i) Process of 'reductive' cell division, occurring in the production of gametes, by means of which each daughter nucleus receives half the number of chromosome characteristic of the somatic cells of the species. (ii) A special form of cell division in which each daughter cell receives half the amount of DNA as the parent cell. Meiosis occurs during formation of egg and sperm cells in mammals.

Melanin: Pigment that gives colour to skin and eyes and helps protect it from damage by UV light.

Melanocyte: A cell in the skin and eyes that produces and contains the pigment called melanin.

Melting point: The temperature at which a solid material becomes a liquid. The freezing point is the temperature at which a liquid material becomes a solid. Usually one value or the other is given on the MSDS. It is important to know the freezing or melting point for storage and handling purposes. For example, a frozen or melted material may burst a container. As well, a change of physical state could alter the hazards of the material.

Member of the public: Persons who are not occupationally associated with the facility or operations, meaning persons whose assigned occupational duties do not require them to enter the operation site.

Mercurialism (mad hatter syndrome): Chronic poisoning because of exposure to mercury, often by breathing its vapour but also by skin absorption and, less commonly, by ingestion. The CNS damage usually predominates in mercury poisoning.

Mesothelioma: Malignant spreading tumour of the mesothelium of the pleura, pericardium, or peritoneum, arising as a result of the presence of asbestos fibres. It is diagnostic of exposure to asbestos.

Metabolic activation: Biotransformation of relatively inert chemical substances to biologically reactive metabolites.

Metabolic enzymes: Proteins that catalyse chemical transformations of body constituents and in more common usage of xenobiotics.

Metabolism: (i) The sum of (total of) the physical and chemical changes that take place in living organisms. These changes include both synthesis (anabolism) and breakdown (catabolism) of body constituents. In a narrower sense, the physical and chemical changes that take place in a given chemical substance within an organism. It includes the uptake and distribution within the body of chemical compounds, the changes (biotransformations) undergone by such chemical substances, and the elimination of the compounds and their metabolites. (ii) Metabolism is the whole range of biochemical processes that occur within the living organism/animal. The term is commonly used to refer specifically to the breakdown of food and its transformation into energy.

Metabolite: Intermediate or product resulting from metabolism.

Metamorphosis: The process by which insects change their body structures as they develop.

Metaplasia: Abnormal transformation of an adult, fully differentiated tissue of one kind into a differentiated tissue of another kind.

Metastasis: (i) The movement of bacteria or body cells, especially cancer cells, from one part of the body to another, resulting in change in location of a disease or of its symptoms from one part of the body to another. (ii) growth of pathogenic microorganisms or of abnormal cells distant from the site of their origin in the body.

Methemoglobin: Derivative of haemoglobin that is formed when the iron(II) in the haem porphyrin is oxidised to iron(III); this derivative cannot transport dioxygen.

mg/m³: Abbreviation for milligrams (mg) of a material per cubic metre (m^3) of air. It is a unit of metric measurement for concentration (weight/volume). The concentrations of any airborne chemical can be measured in mg/m^3, whether it is a solid, liquid, gas, or vapour.

Microalbuminuria: Chronic presence of albumin in slight excess in urine.

Micromercurialism: Early or subclinical effects of exposure to elemental mercury detected at the low exposure levels.

Micronucleus test: Test for mutagenicity in which animals are treated with a test agent after which time the frequency of micronucleated cells is determined; if a test group shows significantly increased levels of micronucleated cells compared to a control group, the chemical is considered capable of inducing chromosomal damage.

Microsome: Artefactual spherical particle, not present in the living cell, derived from pieces of the endoplasmic reticulum present in homogenates of tissues or cells. (*Note*: Microsomes sediment from such homogenates (usually the S9 fraction) when centrifuged at 100,000 g for 60 min: the microsomal fraction obtained in this way is often used as a source of monooxygenase enzymes.)

Military mask: A gas mask that is styled for use by the military. As effective as a civilian gas mask for saving the wearer's life from NBC agents. Styled for wear while using firearms, carrying large amounts of equipment on the body, etc.

Minamata disease: Neurological disease caused by methylmercury, first seen in subjects ingesting contaminated fish from Minamata Bay in Japan.

Mind: The important part of the human body system that thinks, reasons, perceives, wills, and feels. The mind now appears in no way separate from the brain. (In neuroscience, there is no duality between the mind and body.)

Minimal risk level (MRL): Estimate of the daily human exposure to a hazardous substance that is likely to be without appreciable risk of adverse non-cancer health effects over a specified duration of exposure: this substance-specific estimate is used by ATSDR health assessors to identify contaminants and potential health effects that may be of concern at hazardous waste sites.

Miscible: Able to be mixed. Two liquids are said to be miscible if they are partially or completely soluble in each other. Commonly, the term miscible is understood to mean that the two liquids are completely soluble in each other.

Mist: A collection of liquid droplets suspended in air. A mist can be formed when spraying or splashing a liquid. It can also be formed when a vapour condenses into liquid droplets in the air.

Mitigation: Actions taken to prevent or reduce the severity of harm from the release of chemical substances.

Mitosis: Process by which a cell nucleus divides into two daughter nuclei, each having the same genetic complement as the parent cell: nuclear division is usually followed by cell division.

Mixed-function oxidases: These are important sets of oxidising enzymes that are involved in the metabolism of many foreign compounds, giving products of different toxicity from the parent compound.

Mixture: A substance/material composed of more than one chemical compound or element.

mmHg: The abbreviation for millimetres (mm) of mercury (Hg). It is a common unit of measurement for the pressure exerted by gases such as air. Normal atmospheric pressure is 760 mmHg.

Mode: Means any of the following methods of transportation by rail, highway, air, or water; any method and route of exposure to chemical substances by oral (ingestion), dermal (skin absorption), and inhalation (respiratory).

Model organisms: Species of animals, plants, or other organisms used to study basic biological processes to provide insight into other organisms.

Moderate scenarios: Scenarios that could be initiated by a single individual using materials or tools readily available in the facility or small quantities of flammables.

Molecular formula: Chemical notation of the exact number of atoms of each element present in the smallest unit of the substance.

Molecule: The smallest unit of a chemical substance that can exist alone and retain the character of that chemical substance.

Molecular weight: A number showing how heavy one molecule (or unit) of a chemical is compared to the lightest element, hydrogen, which has a weight of 1. The molecular weight has various technical uses, such as calculating conversions from ppm to mg/m^3 in air.

Molluscicide: A chemical substance used to kill molluscs, class of organisms/animals of the phylum Mollusca.

Monoclonal: Pertaining to a specific protein from a single clone of cells, all molecules of this protein being the same.

Monoclonal antibody: Antibody produced by cloned cells derived from a single lymphocyte.

Mononuclear phagocyte system (MPS): The set of cells consisting of macrophages and their precursors (blood monocytes and their precursor cells in bone marrow). The term has been proposed to replace reticuloendothelial system (RES) that does not include all macrophages and does include other unrelated cell types.

Monooxygenase (mixed-function oxidase): Enzyme that catalyses reactions between an organic compound and molecular oxygen in which one atom of the oxygen molecule is incorporated into the organic compound and one atom is reduced to water; involved in the metabolism of many natural and foreign compounds giving both unreactive products and products of different or increased toxicity from that of the parent compound.

Morbidity: The elative incidence of a particular disease. In common clinical usage, any disease state, including diagnosis and complications, is referred to as morbidity.

Mordant: Chemical substance that fixes a dyestuff in or on a material by combining with the dye to form an insoluble compound, used to fix or intensify stains in a tissue or cell preparation.

Mortality: (i) The ratio of deaths in an area to the population of that area, within a particular period of time. (ii) The death rate in a population or locality.

MRID: Master record identification.

MS: Mass spectrometry.

MSDS: Acronym for material safety data sheet. The MSDS is a document that contains information on the potential health effects of exposure and how to work safely with the material it is written about. It is an essential starting point to a health and safety programme. It contains hazard evaluations on the use, storage, handling, and emergency procedures all related to the material. The CPR system requires an MSDS before the product or chemical substance can be used in the workplace. The CPRs are part of the WHMIS.

MSHA: Acronym for Mine Safety and Health Administration. MSHA is the U.S. government agency responsible for enforcing the health and safety regulations and standards for American miners.

Müller's fibres: The elongated neuroglial cells traversing all the layers of the retina, forming its principal supporting element.

Multigeneration study: (i) Toxicological studies conducted in at least three generations of the test organism/animal. The test animals are exposed (usually continuous) to a candidate chemical substance and evaluated. (ii) Toxicity test in which two to three generations of the test organism are exposed to the substance being assessed. (iii) Toxicity test in which only one generation is exposed and effects on subsequent generations are assessed.

Muscle fibre: Any of the cells of skeletal or cardiac muscle tissue. Skeletal muscle fibres are cylindrical multinucleate cells containing contracting myofibrils, across which run transverse striations. Cardiac muscle fibres have one or sometimes two nuclei, contain myofibrils, and are separated from one another by an intercalated disk; although striated, cardiac muscle fibres branch to form an interlacing network.

Musculoskeletal toxicity: Adverse effects to the structure and/or function of the muscles, bones, and joints caused by exposure to a toxic chemical. Exposures to coal dust and cadmium, for example, have been shown to cause adverse changes to the musculoskeletal system. Examples of musculoskeletal diseases that can be caused by exposure to toxic chemicals include the bone disorders, arthritis, fluorosis, and osteomalacia.

Mustard agents: Hazardous and highly toxic chemical substances such as blister agents, used in the 1980s by the Iraqis against the Kurds. Mustard agents are known to cause blindness. Problem may continue for 30–40 years after exposure.

Mutagen: (i) A chemical substance or physical agent that causes mutations in the organism/ animal. (ii) A chemical substance capable of changing genetic material in a cell.

Mutagenesis: (i) A process by which the genetic information of an organism/animal is changed in a stable manner, either in nature or experimentally by the use of chemical substances or radiation. (ii) A process of production of mutations possibly leading to transformation and carcinogenesis.

Mutagenic: (i) Capable of inducing mutation (used mainly of extracellular factors such as x-rays or chemical pollution). (ii) Causing mutations; mutagenic substances may also be carcinogenic.

Mutagenicity: (i) A change in the genetic material of a living organism, usually in a single gene, which can be passed on to future generations. (ii) The property of a physical, chemical, or biological agent to induce mutations in living tissue. Mutagenicity is included on MSDSs because it is an early indicator of potential hazard, and often there is very little other evidence available on possible carcinogenic or reproductive effects.

Mutation: (i) Any heritable change in genetic material. This may be a chemical transformation of an individual gene (a gene or point mutation), which alters its function. On the other hand, this change may involve a rearrangement, or a gain or loss of part of a chromosome, which may be microscopically visible. This is designated a chromosomal mutation. Most mutations are harmful. (ii) A permanent structural alteration in DNA. In most cases, DNA changes have either no effect or cause harm, but occasionally a mutation can improve an organism's chance of surviving, and the beneficial change is passed on to the organism's descendants. Typically, mutations are more rare than polymorphism in population samples because natural selection recognises their lower fitness and removes them from the population.

Mutual Acceptance of Data (MAD): The 1981 decision of OECD on MAD indicating that the data generated in the testing of chemicals in an OECD member country in accordance with OECD Test Guidelines and OECD Principles of GLP shall be accepted in other member countries for purposes of assessment and other uses relating to the protection of man and the environment.

Myalgia: An adverse health effect indicating muscle pain. The term is derived from the Greek words *myos* (muscle) and *algos* (pain). Myalgia can occur in an isolated muscle or in multiple muscles across the body. Myalgia could be acute, short term or chronic, long term. There are a number of possible causes. Myalgia that involves extreme soreness and inhibition of movement in the neck, accompanied by fever, can be very serious.

Mycotoxin: Toxin produced by a fungus, for example, aflatoxins, tricothecenes, ochratoxin, and patulin.

Mydriasis: Extreme dilation of the pupil of the eye, either as a result of normal physiological response or in response to a chemical exposure.

Myelosuppression: Reduction of bone marrow activity leading to a lower concentration of platelets, red cells, and white cells in the blood.

Nanogram: A measure of weight. One nanogram weighs a billion times less than 1 g and almost a trillion times less than a pound.

Nanoparticle: Microscopic particle whose size is measured in nanometres, often restricted to so-called nanosized particles (NSPs; <100 nm in aerodynamic diameter), also called ultrafine particles.

Nanotechnology: (i) The field of research that deals with the engineering and creation of things from materials that are less than 100 nm (one billionth of a metre) in size, especially single atoms or molecules. Nanotechnology is being studied in the detection, diagnosis, and treatment of cancer. (ii) Scientific discipline involving the study of the actual or potential danger presented by the harmful effects of nanoparticles on living organisms and ecosystems, of the relationship of such harmful effects to exposure, and of the mechanisms of action, diagnosis, prevention, and treatment of intoxications.

Narcotic: (i) A chemical substance used to treat moderate to severe pain. Narcotics are like opiates such as morphine and codeine, but are not made from opium. They bind to opioid receptors in the CNS. Narcotics are now called opioids. (ii) A drug that causes insensibility or stupor. A narcotic induces narcosis, from the Greek 'narke' for 'numbness or torpor'. (iii) A drug such as marijuana that is subject to regulatory restrictions comparable to those for addictive narcotics.

Nasogastric: Describes the passage from the nose to the stomach. For example, a nasogastric tube is inserted through the nose, down the throat and oesophagus, and into the stomach.

Nasopharynx: The upper part of the throat behind the nose. An opening on each side of the nasopharynx leading into the ear.

National Cancer Institute (NCI): A part of the National Institutes of Health (NIH) of the United States, DHHS is the federal government's principal agency for cancer research. NCI conducts, coordinates, and funds cancer research, training, health information dissemination, and other programmes with respect to the cause, diagnosis, prevention, and treatment of cancer.

National Fire Protection Association (NFPA): An organisation that provides information about fire protection and prevention and developed a standard outlining a hazard warning labelling system that rates hazard(s) of a material during a fire (health, flammability, and reactivity hazards).

National Institute for Occupational Safety and Health (NIOSH): U.S. federal agency of the CDC that investigates and evaluates potential hazards in the workplace. The NIOSH is also responsible for conducting research and providing recommendations for the prevention of work-related illness and injuries.

National Institutes of Health (NIH): A U.S. federal agency that conducts biomedical research in its own laboratories; supports the research of non-federal scientists in universities, medical schools, hospitals, and research institutions throughout the country and abroad; helps in the training of research investigators; and fosters communication of medical information.

National Toxicology Program (NTP): U.S. federal interagency programme that coordinates toxicological testing programmes, develops and validates improved testing methods, and provides toxicological evaluations on substances of public health concern.

Nausea: (i) A feeling of sickness or discomfort in the stomach that may come with an urge to vomit. Nausea is a side effect of some types of cancer therapy. (ii) Nausea is the urge to vomit. It can be brought by many causes including systemic illnesses such as influenza, medications, pain, and inner ear disease.

NBC: Acronym for nuclear, biological, and chemical. Pertains to warfare and terrorist activities.

NCBI: National Center for Biotechnology Information.

Nebuliser: A device that produces a fine spray or mist often used for breathing treatments in patients with respiratory diseases. A device used to turn liquid into a fine spray. Nebulisers are commonly used for treatment of asthma, cystic fibrosis, and COPD.

Necrosis: (i) Mass death of areas of tissue surrounded by otherwise healthy tissue; (ii) the sum of morphological changes resulting from cell death by lysis and (or) enzymatic degradation, usually accompanied by inflammation and affecting groups of cells in a tissue.

Nematicide: A group of chemical substance used to kill nematodes, the parasites that infect agricultural crops and farm animals.

Nematodes: Organisms also known as roundworms that affect human health. Most parasitic roundworm diseases are transmitted to humans through soil. Cysticercosis, an infection caused by the larval form of the pork tapeworm, *Taenia solium*, is recognised as an increasingly important cause of severe neurological disease in developed countries because of eating undercooked pork, through contaminated water or food, or hand to mouth. Symptoms of *cysticercosis* include, but are not limited to, muscle pains, lumps under skin, and blurred vision. The symptoms typically occur months to years after the infection. Humans who ingest the *T. solium* eggs through contaminated food/undercooked pork become infected with *cysticercosis*, which later spreads to all of the human organs, typically the CNS.

Neoplasia: (i) Abnormal and uncontrolled cell growth. (ii) Formation of new and abnormal tissue as a tumour or growth by cell proliferation.

Neoplasm: (i) An abnormal mass of tissue that results when cells divide more than they should or do not die when they should. Neoplasms may be benign or malignant and also called tumour. Any new formation of tissue associated with disease such as a tumour.

Nephritis: Inflammation of the kidney, leading to kidney failure, usually accompanied by proteinuria, haematuria, oedema, and hypertension.

Nephrocalcinosis: A form of renal stone disease where the kidney tissue is characterised by foci of calcification in addition to numerous deposits of calcium phosphate and calcium oxalate.

Nephropathy: Any disease or abnormality of the kidney.

Nephrotoxic: (i) Chemical substances that are poisonous or damaging to the kidney. (ii) Any chemical substance harmful to the kidney.

Nephrotoxin: Agents that can cause toxic effects on the kidney. (ii) A toxic agent or chemical substance that inhibit, damages, or destroys the cells and/or tissues of the kidneys. The signs and symptoms of exposure to nephrotoxins include proteinuria – protein in the urine – oedema, kidney stones, and uraemia (an excess of urea in the blood, characterised by headache, nausea, and/or coma).

Neoprene: A synthetic rubber characterised by high resistance to oils, heat, or other substances.

Nerves: A bundle of fibres that receives and sends messages between the body and the brain. The messages are sent by chemical and electrical changes in the cells that make up the nerves.

Nerve cell (neuron): Type of cell that receives and sends messages from the body to the brain and back to the body. The messages are sent by a weak electrical current.

Nerve fibre: (i) Also known as axon, a slender process of a neuron, especially the prolonged axon that conducts nerve impulses away from the cell; classified as either afferent or efferent according to the direction the impulse flow and either myelinated or unmyelinated according to whether there is or is not a myelin sheath. (ii) A nerve fibre is a threadlike extension of a nerve cell and consists of an axon and myelin sheath (if present) in the nervous system.

Nervous system: The organised network of nerve tissue in the body. It includes the CNS (the brain and spinal cord), the peripheral nervous system (nerves that extend from the spinal cord to the rest of the body), and other nerve tissue.

Neural: Pertaining to a nerve or to the nerves.

Neuroglia (glial cell): Any of the cells that hold nerve cells in place and help them work the way they should. The types of neuroglia include oligodendrocytes, astrocytes, microglia, and ependymal cells.

Neuroleptic malignant syndrome (NMS): A life-threatening condition that may be caused by certain drugs used to treat mental illness, nausea, or vomiting. Symptoms include high fever, sweating, unstable blood pressure, confusion, and stiffness.

Neuron(e): Nerve cell, the morphological and functional unit of the central and peripheral nervous systems.

Neuropathy (peripheral neuropathy): (i) Any disease of the central or peripheral nervous system; (ii) nerve problem that causes pain, numbness, tingling, swelling, or muscle weakness in different parts of the body. It usually begins in the hands or feet and gets worse over time. Neuropathy may be caused by physical injury, infection, toxic substances, disease (such as cancer, diabetes, kidney failure, or malnutrition), or drugs, including anticancer drugs.

Neuropeptide: A member of a class of protein-like molecules made in the brain. Neuropeptides consist of short chains of amino acids, with some functioning as neurotransmitters and some functioning as hormones.

Neurotoxicity: (i) Adverse effects on the structure or function of the central and/or peripheral nervous system caused by exposure to a toxic chemical. Symptoms of neurotoxicity include muscle weakness, loss of sensation and motor control, tremors, cognitive alterations, and autonomic nervous system dysfunction. (ii) Able to produce chemically an adverse effect on the nervous system: such effects may be subdivided into two types: (1) CNS effects (including transient effects on mood or performance and presenile dementia such as Alzheimer's disease) and (2) peripheral nervous system effects (such as the inhibitory effects of organophosphorus compounds on synaptic transmission).

Neurotoxin: Agents that can cause toxic effects on the nervous system. A chemical substance that induces an adverse effect on the structure and/or function of the central and/or peripheral nervous system.

Neurotransmitter: A chemical that is made by nerve cells and used to communicate with other cells, including other nerve cells and muscle cells.

Neutrophil: A type of immune cell that is one of the first cell types to travel to the site of an infection. Neutrophils help fight infection by ingesting microorganisms and releasing enzymes that kill the microorganisms. A neutrophil is a type of WBC, a type of granulocyte, and a type of phagocyte.

NFPA: National Fire Protection Association (United States).

Niacin (nicotinic acid and vitamin B3): A nutrient in the vitamin B complex that the body needs in small amounts to function and stay healthy. Niacin helps some enzymes work properly and helps the skin, nerves, and digestive tract stay healthy. Niacin is found in many plant and animal products. It is water soluble (can dissolve in water) and must be taken in every day. Not enough niacin can cause a disease called pellagra (a condition marked by skin, nerve, and digestive disorders). A form of niacin is being studied in the prevention of skin and other types of cancer. Niacin may help to lower blood cholesterol.

Nicotine: An addictive, poisonous chemical found in tobacco. It can also be made in the laboratory. When it enters the body, nicotine causes an increased heart rate and increased use of oxygen by the heart and a sense of well-being and relaxation. It is also used as an insecticide.

NIDA: National Institute on Drug Abuse.

NIH: *See* National Institutes of Health.

NIOSH: (i) A U.S. federal agency that conducts research on occupational safety and health questions and makes recommendations to the federal OSHA about new standards for controlling toxic chemicals in the workplace. The NIOSH undertakes research and develops occupational health and safety standards and National Institute of Occupational and Safety Hazards. A U.S. government agency that puts their 'seal of approval' on items designed to protect the health and life of consumers. The NIOSH has the same 'standard of excellence' as Underwriters Laboratories (UL) does with electrical products.

Nitric acid: A toxic, corrosive, colourless liquid used to make fertilisers, dyes, explosives, and other chemicals.

NMR (nuclear magnetic resonance): A spectroscopic technique used for the determination of protein structure.

No-effect dose (subthreshold dose) (NED): Amount of a chemical substance that has no effect on the organism.

No evidence: When several adequate studies are available, which show that, within the limits of the tests used, the chemical is not carcinogenic.

No-observed-adverse-effect level (NOAEL): Greatest concentration or amount of a substance, found by experiment or observation, which causes no detectable adverse alteration of morphology, functional capacity, growth, development, or life span of the target organism under defined conditions of exposure.

No-observed-effect level (NOEL): The greatest concentration or amount of a chemical found by experiment or observation that causes no detectable adverse alteration of morphology, functional capacity, growth, development, or life span of the target organism. The maximum dose or ambient concentration that an organism can tolerate over a specified period of time without showing any detectable adverse effect and above which adverse effects are apparent. An exposure level at which there are no statistically or biologically significant increases in the frequency or severity of any effect between the exposed population and its appropriate control.

NOEL: No-observed-effect level.

Non-malignant (benign): Not cancerous and non-malignant tumours may grow larger but do not spread to other parts of the body.

Nonoccupational exposure: Environmental exposure outside the workplace to substances that are otherwise associated with particular work environments and/or activities and processes that occur there.

Nonprescription (over-the-counter, OTC): A medicine that can be bought without a prescription (doctor's order). Examples include analgesics (pain relievers) such as aspirin and acetaminophen.

NOS: Not otherwise specified.

NPRI: National Pollutant Release Inventory.

NTP: National Toxicology Program, part of the U.S. DHHS. The NTP has a large programme for testing the potential carcinogenicity of chemicals. It also does many other types of studies on short-term and long-term health effects.

Nucleus: (i) The structure in a cell that contains the chromosomes. The nucleus has a membrane around it and is where RNA is made from the DNA in the chromosomes. (ii) Compartment in the interphase eukaryotic cell bounded by a double membrane and containing the genomic DNA, with the associated functions of transcription and processing.

Nuisance dust, nuisance particulate: A term used historically by the ACGIH to describe airborne materials (solids and liquids), which have little harmful effect on the lungs and do not produce significant disease or harmful effects when exposures are kept under reasonable control. Nuisance particulates may also be called nuisance dusts. High levels of nuisance particulates in the air may reduce visibility and can get into the eyes, ears, and nose. Removal of this material by washing or rubbing may cause irritation.

Nutraceutical: Food or dietary supplement that is believed to provide health benefits.

Nutrient: A chemical compound (such as protein, fat, carbohydrate, vitamin, or mineral) contained in foods. These compounds are used by the body to function and grow.

Nutrition: (i) The science or practice of taking in and utilising foods. (ii) The taking in and use of food and other nourishing material by the body. Nutrition is a three-part process. First, food or drink is consumed. Second, the body breaks down the food or drink into nutrients. Third, the nutrients travel through the bloodstream to different parts of the body where they are used as 'fuel' and for many other purposes. To give the body proper nutrition, a person has to eat and drink enough of the foods that contain key nutrients.

Nystagmus: Involuntary, rapid, rhythmic movement (horizontal, vertical, rotary, mixed) of the eyeball, usually caused by a disorder of the labyrinth of the inner ear or a malfunction of the CNS.

Obese: Having an abnormally high, unhealthy amount of body fat.

Obesity: A condition marked by an abnormally high, unhealthy amount of body fat.

Occupational environment: Surrounding conditions at a workplace. The environment at a workplace.

Occupational exposure: Experience of substances, intensities of radiation, or other conditions while at work.

Occupational exposure assessment section (OEAS): The regulatory division/unit that conducts performs exposure assessments on all new active ingredients and all major new uses of a pesticide in order to determine how much exposure to a pesticide could occur in a typical day. These assessments take into account the different exposures that people could have to pesticides, such as those who work with the pesticides (formulators, applicators, and farmers) and bystanders (people working or living near where a pesticide is used). They also take into consideration the differing exposures that adults and children would have. Data considered include residues found in air and on indoor and outdoor surfaces following application in domestic, commercial, and agricultural situations. Routes and duration of exposure and the species tested in toxicity studies are also considered. Assessments of the effectiveness of personal protective equipment (PPE) are often performed, and wearing such equipment can be required as a condition of registration.

Occupational exposure limit (OEL): (i) Regulatory level of exposure to substances, intensities of radiation, or other conditions, specified appropriately in relevant government legislation or related codes of practice. (ii) A legally enforceable limit on the amount or concentration of a chemical substance to which workers may be exposed.

Occupational hazard: Danger to health, limb, or life that is inherent in, or is associated with, a particular occupation, industry, or work environment. Occupational hazards include risk of accident and of contracting occupational diseases.

Occupational hygiene: (i) The applied science concerned with the recognition, evaluation, and control of chemical, physical, and biological factors arising in or from the workplace, which may affect the health or well-being of those at work or in the community. (ii) Identification, recognition, evaluation, and control of biological, chemical, physical, and/or other occupational hazards.

Occupational medicine: Specialty devoted to the prevention and management of occupational injury, illness, and disability and the promotion of the health of workers, their families, and their communities.

Occupational Safety and Health Administration (OSHA): U.S. federal agency that develops and enforces occupational safety and health standards for all general as well as construction and maritime industries and businesses in the United States.

Occurrence (synonym: frequency): In epidemiology, a general term describing the frequency of a disease or other attribute or event in a population without distinguishing between incidence and prevalence.

Ocular: The adjective applied to anything pertaining to the eye.

Ocular albinism: For some individuals, albinism affects only their eyes. People with ocular albinism usually have blue eyes. In some cases, the iris (the coloured part of the eye) has very little colour and the person's eyes might look pink or reddish. This is caused by the blood vessels inside the eye showing through the iris. In some forms of ocular albinism, the hearing nerves may be affected and the person may develop hearing problems or deafness over time.

Odour: A sensation resulting from adequate chemical stimulation of the olfactory organ.

Odour threshold: The lowest concentration of a chemical substance in air that is detectable by smell. The odour threshold should only be regarded as an estimate. This is because odour thresholds are commonly determined under controlled laboratory conditions using people trained in odour recognition. As well, in the workplace, the ability to detect the odour of a chemical substance varies from person to person and depends on conditions such as the presence of other odorous materials. Odour cannot be used as a warning of unsafe conditions since workers may become used to the smell (adaptation), or the chemical may numb the sense of smell, a process called olfactory fatigue. However, if the odour threshold for a chemical substance is well below its exposure limit, odour can be used to warn off a problem with your respirator.

OECD (The Organisation of Economic Co-operation and Development): An international agency that supports programmes designed to facilitate trade and development. The OECD has published *Guidelines for Testing of Chemicals*. These guidelines contain recommended procedures for testing chemicals for toxic and environmental effects and for determining physical and chemical properties. The OECD is a Paris-based intergovernmental organisation with 29 member countries. The forum in which governments develop common solutions to various social problems, including issues related with the management of toxic chemical substances.

OEL: Occupational exposure limit.

Office for Human Research Protections (OHRP): The office within the U.S. DHHS that protects the rights, welfare, and well-being of people involved in clinical trials. It also makes sure that the research follows the law 45 CFR 46 (Protection of Human Subjects).

Ointment: A substance used on the skin to soothe or heal wounds, burns, rashes, scrapes, or other skin problems.

Olfactometer: Apparatus for testing the power of the sense of smell.
    Smell may affect emotion, behaviour, memory, and thought.

Oliguria: Excretion of a diminished amount of urine in relation to fluid intake.

Omentum: The parts of the body involved in sensing smell, including the nose and many parts of the brain. Smell may affect emotion, behaviour, memory, and thought.

Oncogene: (i) A gene that is a mutated (changed) from of a gene involved in normal cell growth. Oncogenes may cause the growth of cancer cells. Mutations in genes that become oncogenes can be inherited or caused by being exposed to substances in the environment that cause cancer. (ii) Gene that can cause neoplastic transformation of a cell. Oncogenes are slightly changed equivalents of normal genes known as proto-oncogenes. (iii) Genes carried by tumour viruses that are directly and solely responsible for the neoplastic transformation of host cells. Many oncogenes function after integration into the DNA of the host cell and some upregulate normal downstream host cell genes to cause neoplasia.

Oncogenesis: The process initiating and promoting the development of neoplasms/tumours through the action of biologic, chemical, or physical agents.

Oncogenic: Capable of producing tumours in animals, either benign (noncancerous) or malignant (cancerous).

Oncology: The scientific study of cancer.

Onset: In medicine, the first appearance of the signs or symptoms of an illness as, for example, the onset of rheumatoid arthritis. There is always an onset to a disease but never to the return to good health. The default setting is good health.

Opioids: Chemical substances (a type of alkaloid) used to treat moderate to severe pain. Opioids are like opiates, such as morphine and codeine, but are not made from opium. Opioids bind to opioid receptors in the CNS. Opioids are used to be called narcotics.

Ophthalmic: Pertaining to the eye.

OPPT (Office of Pollution Prevention and Toxics): The U.S. EPA-related management programmes under the Toxic Substances Control Act (TSCA) and the Pollution Prevention Act of 1990. Under these laws, the U.S. EPA evaluates new and existing chemicals and their risks and finds ways to prevent or reduce pollution before it gets into the environment. We also manage a variety of environmental stewardship programmes that encourage companies to reduce and prevent pollution.

Optic chiasma: A place in the brain where some of the optic nerve fibres coming from one eye cross optic nerve fibres from the other eye.

Oral cavity: The mouth. It includes the lips, the lining inside the cheeks and lips, the front two-thirds of the tongue, the upper and lower gums, the floor of the mouth under the tongue, the bony roof of the mouth, and the small area behind the wisdom teeth.

Organelle: (i) In cell biology, any structure that occurs in cells and that has a specialised function. (ii) Microstructure or separated compartment within a cell that has a specialised function, for example, ribosome, peroxisome, lysosome, Golgi apparatus, mitochondria (structures that make energy for the cell), lysosomes (sac-like containers filled with enzymes that digest and help recycle molecules in the cell), nucleus (a structure that contains the cell's chromosomes and is where RNA is made), and nucleolus (the most obvious and clearly differentiated nuclear sub-compartment, where ribosome biogenesis takes place, but it is becoming clear that the nucleolus also has non-ribosomal functions).

Organic peroxide: A type of reactive hazard that can catch fire on its own.

Organoleptic and organoleptic test: Use of an organ, especially a sense organ as of taste, smell, or sight and useful in nutritional technology.

Orthology: The term orthology describes genes in different species that derive from a common ancestor, that is, they are direct evolutionary counterparts.

OSH: Occupational safety and health.

OSHA (Occupational Safety and Health Administration): The branch of the U.S. government that sets and enforces occupational health and safety regulations. For example, the OSHA sets the legal exposure limits in the United States, which are called PELs. The OSHA also specifies what information must be given on labels and MSDSs for materials, which have been classified as hazardous using their criteria.

Osteomalacia: An adverse health condition marked by softening of the bones because of impaired mineralisation, with excess accumulation of osteoid. The symptoms include pain, tenderness, muscular weakness, anorexia, and loss of weight, resulting from deficiency of vitamin D and calcium.

Osteoporosis: Significant decrease in bone mass with increased porosity and increased tendency to fracture.

Over the counter: Medications/drugs that are sold without a prescription from a doctor.

Oxidative stress: (i) Adverse effects occurring when the generation of reactive oxygen species (ROS) in a system exceeds the system's ability to neutralise and eliminate them; excess ROS can damage a cell's lipids, protein, or DNA. (ii) A persistent *imbalance between antioxidants and pro-oxidants* in favour of the latter, resulting in (often) irreversible cellular damages.

Oxidiser: A chemical substance that causes the ignition of combustible materials without an external source of ignition; oxidisers can produce oxygen and therefore support combustion in an oxygen-free atmosphere.

Oxidising agent, oxidising material: Material that gives up oxygen easily or can readily oxidise other materials. Examples of oxidising agents are oxygen, chlorine, and peroxide compounds. These chemicals will support a fire and are highly reactive.

Under the Canadian CPRs and under the U.S. OSHA Hazcom Standard, there are specific criteria for the classification of materials as oxidising materials. The CPR is part of the national WHMIS.

P.o.: Abbreviation for per os, meaning oral administration.

Palpitation: (i) Unduly rapid or throbbing heartbeat that is noted by a patient; it may be regular or irregular. (ii) Undue awareness by a patient of a heartbeat that is otherwise normal.

Pancreas: A nodular organ in the abdomen that contains a mixture of endocrine glands and exocrine glands. The small endocrine portion consists of the islets of Langerhans secreting a number of hormones into the bloodstream. The large exocrine portion (exocrine pancreas) is a compound acinar gland that secretes several digestive enzymes into the pancreatic ductal system that empties into the duodenum.

Paracentric inversion: An inversion in which the breakpoints are confined to one arm of a chromosome; the inverted segment does not span the centromere.

Parakeratosis: In medicine, the imperfect formation of horn cells of the epidermis.

Paralogy: The relationship of homologous genes that arose by duplication of genes.

Paralysis: Loss or impairment of motor function.

Paranasal sinuses (nasal sinuses): Air-filled extensions of the respiratory part of the nasal cavity into the frontal, ethmoid, sphenoid, and maxillary cranial bones. They vary in size and form in different individuals and are lined by the ciliated mucous membranes of the nasal cavity.

Paraplegia: Severe or complete loss of motor function in the lower extremities and lower portions of the trunk. This condition is most often associated with spinal cord diseases, peripheral nervous system diseases, neuromuscular diseases, and muscular diseases and causes bilateral leg weakness.

Parkinson's disease: Any of several neurological conditions that resemble Parkinson's disease and that result from a deficiency or blockage of dopamine caused by degenerative disease, drugs, or toxins.

Parkinsonism: A group of disorders that feature impaired motor control characterised by bradykinesia, muscle rigidity, tremor, and postural instability. Parkinsonian diseases are generally divided into primary parkinsonism, secondary parkinsonism, and inherited forms. These conditions are associated with dysfunction of dopaminergic or closely related motor integration neuronal pathways in the basal ganglia.

Paraesthesias: Skin sensations, such as burning, numbness, itching, hyperaesthesia (increased sensitivity), or tingling, with no apparent physical cause (idiopathic). The most common locations of paraesthesias are the hands, arms, legs, and feet, although paraesthesias can occur anywhere on the body.

Parasympathetic: In medicine, pertaining to the parasympathetic nervous system that stimulates digestive secretions, slows the heart, constricts the pupils of the eyes, and dilates blood vessels or its disturbances.

Parenchyma(-al): In medicine, pertaining to a specific or functional component of a gland or organ.

Parenteral dosage: Method of introducing substances into an organism avoiding the gastrointestinal tract (subcutaneously, intravenously, intramuscularly, etc.).

Paresis: Slight or incomplete paralysis.

Particulate matter: A general term used in the context of pollution (i) to describe airborne solid or liquid particles of all sizes. The term aerosol is recommended to describe airborne particulate matter. (ii) Particles in air, usually of a defined size and specified as PMn where n is the maximum aerodynamic diameter (usually expressed in μm) of at least 50% of the particles.

Partition coefficient: The constant ratio that is found when a heterogeneous system of two phases is in equilibrium; the ratio of the concentrations (or strictly activities) of the same molecular species in the two phases is constant at constant temperature and pressure.

Patch test: Test for allergic sensitivity in which a suspected allergen is applied to the skin on a small surgical pad. Patch tests are also used to detect exposure to pesticides.

PE quoted: A polyethylene coating that provides lightweight industrial chemical protection. Not suggested for use with extreme chemicals.

PEEK (polyetheretherketone): A semi-crystalline thermoplastic. It is a very high-end exotic plastic with some of the best properties available of any thermoplastic.

PEL (permissible exposure limit): Legal limit in the United States set by the OSHA.

Peliosis: Purplish or brownish-red discoloration, easily visible through the epidermis, caused by haemorrhage into the tissues.

PEOSHA (Public Employees Occupational Safety and Health Act): A state law that sets PELs for New Jersey public employees.

Peptide: Any compound consisting of two or more amino acids, the building blocks of proteins. Peptides are combined to make proteins.

Peripheral neuropathy: A problem in peripheral nerve function (any part of the nervous system except the brain and spinal cord) that causes pain, numbness, tingling, swelling, and muscle weakness in various parts of the body. Neuropathies may be caused by physical injury; infection; toxic substances; disease such as cancer, diabetes, and kidney failure; malnutrition; or drugs such as anticancer drugs.

Peritoneum: The tissue that lines the abdominal wall and covers most of the organs in the abdomen.

Penetration resistance: Material resistance to liquid penetration is measured using ASTM F903 – the outside surface of the material in question is exposed to the test chemical for 1 h.

Percutaneous: The entry of a chemical substance through the skin following application on the skin.

Permeability: Ability or power to enter or pass through a cell membrane.

Permeation resistance: The ASTM F739 regulation is used to measure the permeation resistance of materials. Permeation is the molecular movement of chemicals through a material. If exposure to chemical vapours is a concern, these data should be analysed.

Permissible exposure limit (PEL): (i) Recommendation by the U.S. OSHA for a TWA concentration that must not be exceeded during any 8-h work shift of a 40 h working week. (ii) The legally enforceable maximum amount or concentration of a chemical that a worker may be exposed to under the OSHA regulations.

Peroxide former: A chemical substance that reacts with air or oxygen to form explosive peroxy compounds that are shock, pressure, or heat sensitive.

Peroxisome: The term in cell biology to describe the organelle present in the cytoplasm of eukaryotic cells and characterised by its content of catalase and other oxidative enzymes such as peroxidase.

Persistence: The attribution or the property of a chemical substance or the length of time that the substance remains in a particular environment before it is physically removed or chemically or biologically transformed.

Persistent inorganic pollutant (PIP): Inorganic chemical substance that is stable in the environment and liable to long-range transport, may bioaccumulate in human and animal tissue, and may have significant impacts on human health and the environment. For example, arsenides, fluorides, cadmium salts, and lead salts. It is important to remember that some inorganic chemicals, like crocidolite asbestos, are persistent in almost all circumstances, but others, like metal sulphides, are persistent only in unreactive environments; sulphides can generate hydrogen sulphide in a reducing environment or sulphates and sulphuric acid in oxidising environments. As with organic substances, persistence is often a function of environmental properties.

Persistent organic pollutant (POP): Organic chemical substance that is stable in the environment and liable to long-range transport, may bio-accumulate in human and animal tissue, and may have significant impacts on human health and the environment. The POPs include dioxin, PCBs, DDT, tributyltin oxide (TBTO), and many others. The Stockholm Convention on Persistent Organic Pollutants was adopted at a Conference of Plenipotentiaries held from 22 to 23 May 2001, in Stockholm, Sweden; by signing this convention, governments have agreed to take measures to eliminate or reduce the release of POPs into the environment.

Personal protective equipment (PPE): (i) All clothing and other work accessories designed to create a barrier against workplace hazards. Examples include safety goggles, blast shields, hard hats, hearing protectors, gloves, respirators, aprons, and work boots. (ii) Any clothing and/or equipment used to protect the head, torso, arms, hands, and feet from exposure to chemical, physical, or thermal hazards.

PPE is clothing or devices worn to help isolate a person from direct exposure to a hazardous material or situation. Recommended PPE is often listed on an MSDS. This can include protective clothing, respiratory protection, and eye protection. The use of PPE is the least preferred method of protection from hazardous exposures. It can be unreliable, and if it fails, the person can be left completely unprotected. This is why engineering controls are preferred. Sometimes, PPE may be needed along with engineering controls. For example, a ventilation system (an engineering control) reduces the inhalation hazard of a chemical, while gloves and an apron (PPE) reduce skin contact. In addition, PPE can be an important means of protection when engineering controls are not practical, for example, during an emergency or other temporary conditions such as maintenance operations.

Pest control operator: An individual/any person who undertakes pest control operations and includes the person or the firm or the company or organisation under whose control such a person is operating.

Pest Management Regulatory Agency (PMRA): The regulatory agency that conducts all pesticides that are carefully regulated in Canada through a programme of premarket scientific assessment, enforcement, education, and information dissemination. These activities are shared among federal, provincial/territorial, and municipal governments and are governed by various acts, regulations, guidelines, directives, and by-laws. Although it is a complex process, regulators at all levels work together towards the common goal – helping protect Canadians from any risks posed by pesticides and ensuring that pest control products do what they claim to on the label. The responsibilities of the PMRA include Pest Control Products Act (PCPA) regulations, pesticide registration and re-evaluation, human health and safety, environmental impact; value/efficacy assessment, alternative strategies, compliance, and enforcement.

Pest Management Regulatory Agency registration process: On submission for registration of a pesticide, the PMRA registration process involves and is empowered to evaluate the complete data of the pesticides for registration for health and environmental considerations, and for its value, including efficacy, it is first examined by screening officers of the PMRA's Submission Management and Information Division (SMID). The purpose of the initial screening is to make sure that submissions meet the format, content, and fee requirements of the agency before they are sent for detailed evaluation. The screening process ensures that only complete, accurate, and standardised submissions are brought forwards for assessment. To this end, the agency provides to industry detailed pre-submission guidance on administrative procedures and data requirements. In the screening unit, preliminary analyses of the studies are also carried out in order to determine if they are acceptable and whether they comply with international protocols.

Pesticide residue: Any chemical substance or mixture of chemical substances found in man or animals or in food and water following use of a pesticide: the term includes any specified derivatives, such as degradation and conversion products, metabolites, reaction products, and impurities considered to be of toxicological significance.

Pesticides: Chemical substances used to kill pests and minimise their impact on agriculture, health, and other human interests. Pesticides are often classified according to the organisms that they are used to control, for example, as fungicides, herbicides, insecticides, molluscicides, nematicides, rodenticides, and others.

Pests: Organisms that may adversely affect public health or attack agricultural crops, stored food grains, and other important materials of human use.

pH: Measure of the acidity or basicity (alkalinity) of a material when dissolved in water. It is expressed on a scale from 0 to 14. Roughly, pH can be divided into the following ranges: (a) pH 0–2 strongly acidic, (b) pH 3–5 weakly acidic, (c) pH 6–8 neutral, (d) pH 9–11 weakly basic, and (e) pH 12–14 strongly basic. According to the Controlled Products Regulations, chemical substances/materials with pH values of 0–2 or 11.5–14 may be classified corrosive. Corrosive chemical substances/materials must be stored and handled with great care and precautions.

Phagocytes: An immune system cell that can surround and kill microorganisms and remove dead cells. Phagocytes include macrophages.

Phagocytosis: (i) The ingestion of microorganisms, cells, and foreign particles by phagocytes, for example, phagocytic macrophages. (ii) Engulfing of microorganisms, other cells, and foreign particles by phagocytic cells (phagocytes). Process by which particulate material is endocytosed by a cell.

Pharmacist: A professional who fills prescriptions and, in the case of a compounding pharmacist, makes them. Pharmacists are familiar with medication ingredients, interactions, cautions, and hints.

Pharmacodynamics: Broadly, the science concerned with the study of the way in which xenobiotics exert their effects on living organisms. Such a study aims to define the fundamental physico-chemical processes that lead to the biological effect observed.

Pharmacogenomics (pharmacogenetics): The branch of pharmacology concerned with using DNA and amino acid sequence data to inform drug development and testing. The study of the influence and correlation of individual genetic variation with drug responses.

Pharmacokinetics: The science that describes quantitatively the uptake of drugs by the body, their biotransformation they undergo, their distribution, metabolism in the tissues, and the elimination of the drugs and their metabolites from the body. Both total amounts and tissue and organ concentrations are considered. The term 'toxicokinetics' is essentially the same term applied to xenobiotics other than drugs.

Pharmacology: The science that discusses the use and effects of drugs. Pharmacology has been subdivided into pharmacokinetics and pharmacodynamics.

Pharynx: Area including the throat, the part of the digestive tract between the oesophagus below and the mouth and nasal cavities above and in front.

Phase 1 reactions: A group of reactions that comprises every possible stage in the enzymic modification of a xenobiotic by oxidation, reduction, hydrolysis, hydroxylation, dehydrochlorination, and related reactions.

Phase 2 reactions: A group of reactions that comprises all reactions concerned with modification of a xenobiotic by conjugation. See conjugate.

Phase 0 clinical trial: A study of the pharmacodynamic and pharmacokinetic properties of a drug. In a Phase 0 trial, a limited number of doses and much lower doses of the drug are administered; therefore, there is less risk to the participant.

Phase I clinical trial: A test of a new biomedical or behavioural intervention in a small group of people (20–80) for the first time to determine the metabolism and pharmacological actions of the drug in humans, safety, side effects associated with increasing doses, and, if possible, early evidence of effectiveness. Phase I trials are closely monitored and may be conducted in patients or healthy volunteers.

Phase II clinical trial: A study of the biomedical or behavioural intervention in a large group of people (several hundred) to determine efficacy and to further evaluate safety. They include controlled clinical studies of effectiveness of a drug for a particular indication or indications in patients with the disease or condition under study and determination of common, short-term side effects and risks associated with the drug. Phase II studies are typically well controlled and closely monitored.

Phase III clinical trial: Expanded controlled and uncontrolled studies performed after preliminary evidence of drug effectiveness has been obtained. They are intended to gather additional information about effectiveness and safety needed to evaluate the overall benefit–risk relationship of the drug and to provide adequate basis for physician labelling. These studies usually include anywhere from several hundred to several thousand subjects.

Phase IV clinical trial: Postmarketing studies (generally randomised and controlled) carried out after licensure of a drug. These studies are designed to monitor effectiveness of an approved intervention in the general population and to collect information about any adverse effects associated with widespread use.

Phenotype: (i) The appearance or constitutional nature of an organism as contrasted with its genetic potential, the genotype. See genotype. (ii) The observable traits or characteristics of an organism, for example, hair colour, weight, or the presence or absence of a disease. Phenotypic traits are not necessarily genetic.

Pheromone: A hormone secreted by an animal, including insects, that stimulates others of the same species.

Phlebotomy: Incision of a vein for the drawing of blood. Other terms include venesection, venipuncture, venotomy, venous blood sampling, and incision of a vein for the drawing of blood.

Phlegm: Mucus that is coughed up from the lungs.

Phorias: Inability of one eye to attain binocular vision with the other because of imbalance of the muscles of the eyeball – also called heterotropia, squint.

Photophobia: Abnormal sensitivity to light. This may occur as a manifestation of eye diseases, migraine, subarachnoid haemorrhage, meningitis, and other disorders. Photophobia may also occur in association with depression and other mental disorders.

Photosensitisation: Sensitisation or heightened reactivity of the skin to sunlight, usually due to the action of certain drugs.

Phototoxicity: Adverse effects produced by exposure to light energy, especially those produced in the skin.

PIB: Piperonyl butoxide.

Pill: In pharmacy, a medicinal substance in a small round or oval mass meant to be swallowed. Pill often contains a filler material and a plastic substance such as lactose that permits the pill to be rolled by hand or machine into the desired form and may then be coated with a varnish-like substance.

Pinocytosis: Type of endocytosis in which soluble chemical substances are taken up by the cell and incorporated into vesicles for digestion.

Pituitary gland: The main endocrine gland that produces hormones that control other glands and many body functions, especially growth.

Plasma: The (i) fluid component of blood in which the blood cells and platelets are suspended; (ii) fluid component of semen produced by the accessory glands, the seminal vesicles, the prostate, and the bulbourethral glands; and (iii) cell substance outside the nucleus, that is, the cytoplasm.

Plasmid: Autonomous self-replicating extrachromosomal circular DNA molecule present in bacteria and yeast. The plasmids replicate autonomously each time a bacterium divides and are transmitted to the daughter cells. The DNA segments are commonly cloned using plasmid vectors.

Pleura: The thin layer of tissue covering the lungs and the wall of the chest cavity to protect and cushion the lungs. A small amount of fluid that acts as a lubricant allows the lungs to move smoothly in the chest cavity during breathing.

Ploidy: The number of sets of chromosomes present in an organism.

Plumbism: The chronic poisoning caused by absorption of lead or lead salts.

Pneumoconiosis: The fibrosis of the lungs that develops owing to (prolonged) inhalation of inorganic or organic dusts. Different causative factors are linked with different but specific types of pneumoconiosis. These include (i) anthracosis, from coal dust; (ii) asbestosis, from asbestos dust; (iii) byssinosis, from cotton dust; (iv) siderosis, from iron dust; (v) silicosis, from silica dust; and (vi) stannosis, from tin dust.

Pneumonia: An inflammatory infection that occurs in the lungs.

Pneumonitis: The term is used to describe the inflammation of the lung.

PNS: Peripheral nervous system.

Poison: A chemical substance or an agent that, taken into or formed within the organism, impairs the health of the organism or animal and may as well kill it.

Pollutant: An undesirable chemical substance as a solid, liquid, or gaseous condition or physical matter or sound and disturbs the normal environmental medium. The 'undesirability' of a pollutant is modulated with its concentration.

Low concentrations of most of the chemical substances are tolerable and/or even essential. (i) A primary pollutant is one emitted into the atmosphere, water, sediments, or soil from an identifiable source. (ii) A secondary pollutant is a pollutant formed by chemical reaction in the atmosphere, water, sediments, or soil.

Pollution: Introduction of pollutants into a solid, liquid, or gaseous environmental medium.

Polycarbonate: Type of plastic characterised by an excellent combination of toughness, transparency, heat and flame resistance, and dimensional stability. Polycarbonates are the material of choice for business equipment, lighting, signage, safety, and medical devices.

Polyclonal antibody: Antibody produced by a number of different cell types.

Polydipsia: A health condition indicating chronic excessive thirst.

Polymer: A natural or man-made material formed by combining units, called monomers, into long chains. The word polymer means many parts. Examples of polymers are starch (which has many sugar units), polyethylene (which has many ethylene units), and polystyrene (which has many styrene units). Most man-made polymers have low toxicity, low flammability, and low chemical reactivity. In these ways, polymers tend to be less hazardous than the chemicals (monomers) from which they are made.

Polymerise, polymerisation: The process of forming a polymer by combining large numbers of chemical units or monomers into long chains. Polymerisation can be used to make some useful materials. However, uncontrolled polymerisation can be extremely hazardous. Some polymerisation processes can release considerable heat, can generate enough pressure to burst a container, or can be explosive. Some chemicals can polymerise on their own without warning. Others can polymerise upon contact with water, air, or other common chemicals. Inhibitors are normally added to products to reduce or eliminate the possibility of uncontrolled polymerisation.

Polymorphism: A common variation in the sequence of DNA among individuals. Genetic variations occurring in more than 1% of the population would be considered useful polymorphisms for genetic linkage analysis.

Polypeptide: Linear polymer of amino acids connected by peptide bonds. Proteins are large polypeptides, and the two terms are commonly used interchangeably.

Polypropylene: A breathable material used for nonhazardous environments. Polypropylene (PP) provides protection against dry particulates, paint, and light chemicals.

Polysulphone: A high-performance amorphous resin and can withstand repeated 'autoclaving' cycles (steam sterilisation under high pressure) and is used extensively in medical applications.

Polyuria: A health condition indicating excessive production and discharge of urine.

POPs (persistent organic pollutants): Chemicals, chiefly compounds of carbon, that persist in the environment, bioaccumulate through the food chain, and pose a risk of causing adverse effects to human health and the environment.

Porphyria: Disturbance of porphyrin metabolism characterised by increased formation, accumulation, and excretion of porphyrins and their precursors.

Potentiation: The ability of one chemical substance to enhance the activity of another chemical substance to an extent greater than the simple summation of the two expected activities.

PP (polypropylene): Inexpensive material that offers a combination of outstanding physical, chemical, mechanical, thermal, and electrical properties not found in any other thermoplastic.

ppb: Parts per billion. Equivalent to 1×10^{-9}.

ppm: A unit of measure expressed as parts per million. Equivalent to 1×10^{-6}.

The abbreviation ppm stands for parts per million. It is a common unit of concentration of gases or vapour in air. For example, 1 ppm of a gas means that 1 unit of the gas is present for every 1 million units of air.

PPO (polyphenylene oxide and polystyrene): A rare example of a homogeneous mixture of two polymers. The compatibility of the two polymers in Noryl is caused by the presence of a benzene ring in the repeat units of both chains. Noryl has the unique ability to perform at a wide temperature range without sacrificing key properties. Noryl can remain stable under load from −40°F to +265°F.

Practical certainty: A concept involving the determination of a numerically specified low risk or socially acceptable risk that may be used in decision making where absolute certainty is not possible.

Precursor: A chemical substance from which another, usually more biologically active, chemical substance is formed.

Predicted environmental concentration: The estimated concentration of a chemical in an environmental compartment calculated from available information on its properties, its use and discharge patterns, and the quantities involved.

Preneoplastic: Before the formation of a tumour.

Prescription: A physician's order for the preparation and administration of a drug or device for a patient. A prescription has several parts. They include the superscription or heading with the symbol 'R' or 'Rx', which stands for the word recipe (meaning, in Latin, to take); the inscription, which contains the names and quantities of the ingredients; the subscription or directions for compounding the drug; and the signature, which is often preceded by the sign 's' standing for signa (Latin for mark), giving the directions to be marked on the container.

Primary: First or foremost in time or development. The primary teeth (the baby teeth) are those that come first. Primary may also refer to symptoms or a disease to which others are secondary.

Probability: The likelihood that something will happen. For example, a probability of less than .05 indicates that the probability of something occurring by chance alone is less than 5 in 100, or 5%. This level of probability is usually taken as the level of biological significance, so a higher incidence may be considered meaningful. The abbreviation for probability is P.

Procarcinogen: A chemical substance that has to be metabolised before it becomes a carcinogen.

Prodrug: Precursor converted to an active form of a drug within the body.

Prognosis: The expected course of a disease. The patient's chance of recovery. The prognosis predicts the outcome of a disease and therefore the future for the patient. His prognosis is grim, for example, while hers is good.

Prokaryote: An organism, for example, a mycoplasma, a blue-green alga, or a bacterium, whose cells contain no membrane-bound nucleus or other membranous organelles. See eukaryote.

Promoter: (i) An agent that increases tumour production by another substance when applied to susceptible organisms after their exposure to the first substance.

(ii) A term used in oncology to describe an agent that induces cancer when administered to an animal or human being who has been exposed to a cancer initiator.

Protein binding: The property of having a physico-chemical affinity for protein.

Proteinuria: Excretion of excessive amounts of protein (derived from blood plasma or kidney tubules) in the urine.

Pseudogene: A sequence of DNA that is very similar to a normal gene but that has been altered slightly so that it is not expressed. Such genes were probably once functional but, over time, acquired one or more mutations that rendered them incapable of producing a protein product.

Psychotropic drugs: Drugs or chemical substances that are known to exert an effect upon the mind and capable of modifying mental activity.

Public health surveillance: The ongoing systematic collection, analysis, and interpretation of data relating to public health.

Pulmonary: Pertaining to the lungs.

Pulmonary alveoli: Minute air-filled sacs in the vertebrate lung, thin walled and surrounded by blood vessels.

Pulmonary function: How well the lungs are working including expanding and contracting (inhaling and exhaling) and exchanging oxygen and carbon dioxide efficiently between the air (or other gases) within the lungs and the blood.

PVC aprons: Made from a thick 20 mil PVC material. Used in rigorous work environments. Recommended for use in aircraft production or battery manufacturing.

PVC or vinyl (polyvinyl chloride): A widely used moderately priced thermoplastic polymer material that provides good clarity. Rigid PVC can be used in both interior and exterior application. PVC is a tough, durable, versatile material available in clear to opaque colours with a moderately lustrous finish.

PVDF (polyvinylidene fluoride): A highly chemically resistant crystalline thermoplastic. It is used extensively in the chemical processing industry because of its properties.

Pyrexia: Condition in which the temperature of a human being or mammal is above.

Pyrogen: Any chemical substance that produces fever.

Pyrophoric: Defined in the U.S. OSHA Hazcom Standard as chemicals that will ignite spontaneously in air at a temperature of 130°F (54.4°C) or below. Regulatory definitions in other jurisdictions may differ.

Quality assurance: All planned and systematic actions necessary to provide adequate confidence that a product or service will satisfy given requirements for quality.

Quality control: Important inputs such as (i) operational techniques and activities that are used to fulfil requirements for quality and (ii) procedures incorporated in experimental protocols to reduce the possibility of error, especially human error. This is a requirement of GLP.

Quantitative structure–activity relationship (QSAR): Quantitative structure–biological activity model derived using regression analysis and containing parameters for physico-chemical constants, indicator variables, or theoretically calculated values.

Quantitative structure–metabolism relationship (QSMR): Quantitative association between the physico-chemical and (or) the structural properties of a substance and its metabolic behaviour.

Quarantine pest: A pest of potential economic and/or environmental importance to an area where it is not yet present or is present but not widely distributed and being officially controlled.

Quarantine treatment: Any kind of treatment that is applied for quarantine purpose for elimination of pest in accordance with phytosanitary regulations of the importing country.

Radiation therapy: Radioisotopes produce radiation and can be placed in or near the tumour or in the area near cancer cells. This type of radiation treatment is called internal radiation therapy, implant radiation, interstitial radiation, or brachytherapy. Systemic radiation therapy uses a radioactive substance, such as a radio-labelled monoclonal antibody, that circulates throughout the body. Also called radiotherapy, irradiation, and x-ray therapy.

Radiation toxicology: The scientific study involving research, education, prevention, and possible health hazards on exposure to radiation. Also the branch of science that discuss the dangers of radiation, and examine its myriad benefits, dangers of radiation and anxiety—nuclear energy and nuclear weapons, uranium, plutonium etc.

Radioactive material: A material whose nuclei spontaneously give off nuclear radiation.

Rate: (i) The measure of the frequency with which an event occurs in a defined population in a specified period of time. (ii) Rate at which this occurs.

Reactivity: The capacity of a substance to combine chemically with other substances.

Reactive flammable material: Defined by the Canadian CPRs as a material that is a dangerous fire risk because it can react readily with air or water. This category includes any material that is (i) spontaneously combustible, that is, a material that can react with air until enough heat builds up that it begins to burn, (ii) can react vigorously with air under normal conditions without actually catching fire, (iii) gives off dangerous quantities of flammable gas on reaction with water, or (iv) becomes spontaneously combustible when it contacts water or water vapour. Reactive flammable materials must be kept dry and isolated from oxygen (in air) or other oxidising agents. Therefore, they are often stored and handled in an atmosphere of unreactive gas, such as nitrogen or argon.

Readily biodegradable: Arbitrary classification of chemical substances that have passed certain specified screening tests for ultimate biodegradability; these tests are so stringent that such compounds will be rapidly and completely biodegraded in a wide variety of aerobic environments.

Recalcitrance: Ability of a substance to remain in a particular environment in an unchanged form.

Receptor: Molecular structure in or on a cell that specifically recognises and binds to a compound and acts as a physiological signal transducer or mediator of an effect.

Recombinant DNA technology: Methods involving the use of restriction enzymes to cleave DNA at specific sites, allowing sections of DNA molecules to be inserted into plasmid or other vectors and cloned in an appropriate host organism (e.g. a bacterial or yeast cell).

Recombinant DNA: DNA made by transplanting or splicing DNA into the DNA of host cells in such a way that the modified DNA can be replicated in the host cells in a normal fashion.

Recommended exposure level (REL): The highest allowable regulatory airborne concentration of a toxicant. This exposure concentration is not expected to injure workers. It may be expressed as a ceiling limit or as a TWA.

Recommended limit: Regulatory value of the maximum concentration of a potentially toxic chemical substance that is believed to be safe. Such limits often have no legal backing in which case a control or statutory guide level that should not be exceeded under any circumstances may be set.

Recording: A procedure, specified in national laws and regulations, for ensuring that the employer maintains information on (a) occupational accidents and diseases and (b) dangerous occurrences and incidents.

Recovery: A process leading to partial or complete restoration of a cell, tissue, organ, organism, or animal following its damage from exposure to a harmful chemical substance or agent.

RED: Reregistration eligibility decision.

Regulatory dose: U.S. EPA defined expected dose resulting from human exposure to a chemical substance at the level at which it is regulated in the environment.

Relative risk (RR): This term has different meanings and depends upon context. (i) Ratio of the risk of disease or death among the exposed to the risk among the unexposed: this usage is synonymous with 'risk ratio'. (ii) Alternately, the ratio of the cumulative incidence rate in the exposed to the cumulative incidence rate in the unexposed, that is, the cumulative incidence ratio. (iii) Sometimes used as a synonym for 'odds ratio'.

Remedy: A process that consistently helps treat or cure a disease. From the Latin 'remedium' meaning that heals again and again and as a medicine or therapy that relieves pain, cures disease, or corrects a disorder.

Renal: Pertaining to the kidneys.

Renal plasma flow: Volume of plasma passing through the kidneys in unit time.

Repellent: Any chemical substance used mainly to repel blood-sucking insects in order to protect man and animals. The term is also used for chemical substances that repel mammals, birds, rodents, mites, plant pests, and other organisms.

Reproductive effects: Problems in the reproductive process that may be caused by a substance. Possible reproductive effects include reduced fertility in the male or female, menstrual changes, miscarriage, embryotoxicity, foetotoxicity, teratogenicity, or harmful effects to the nursing infant from chemicals in breast milk. Most chemicals can cause reproductive effects if there is an extremely high exposure. In these cases, the exposed person would experience other noticeable signs and symptoms caused by the exposure. These signs and symptoms act as a warning of toxicity. Chemicals that cause reproductive effects in the absence of other significant harmful effects are regarded as true reproductive hazards. Very few workplace chemicals are known to be true reproductive hazards.

Reproductive toxicant: Chemical substance or preparation that produces non-heritable adverse effects on male and female reproductive function or capacity and on resultant progeny.

Reproductive toxicity: Adverse effects on sexual function and fertility in adult males and female, as well as developmental toxicity in the offspring (International Programme on Chemical Safety (IPCS) Environmental Health Criteria 225, Principles for Evaluating Health Risks to Reproduction Associated with Exposure to Chemicals).

Reproductive toxicology: The branch of scientific study to know the effects of chemical substances on the adult reproductive and neuroendocrine systems, embryo, foetus, neonate, and prepubertal mammal.

Residual: Something left behind. With residual disease, the disease has not been eradicated.

Respirable dust (respirable particles): Mass fraction of dust (particles) that penetrates to the unciliated airways of the lung (the alveolar region).

Respiratory toxicity: Adverse health effects on the structure or function of the respiratory system caused by exposure to a toxic chemical substance. Respiratory toxicants are known to produce a variety of acute and chronic pulmonary conditions, including local irritation, bronchitis, pulmonary oedema, emphysema, and cancer.

Reticuloendothelial system (RES): (i) A diffuse system of cells of varying lineage that include especially the macrophages and the phagocytic endothelial cells lining blood sinuses and that were originally grouped together because of their supposed phagocytic properties based on their ability to take up the vital dye trypan blue. (ii) This is the system with cells that have the ability to take up and retain certain dyes and particles ingested into a living animal. This term has generally been replaced by the term MPS. (iii) A system with a group of cells having the ability to take up and sequester inert particles and vital dyes, including macrophages and macrophage precursors; specialised endothelial cells lining the sinusoids of the liver, spleen, and bone marrow; and reticular cells of lymphatic tissue (macrophages) and bone marrow (fibroblasts).

Ribonucleic acid (RNA): The generic term for polynucleotides, similar to DNA but containing ribose in place of deoxyribose and uracil in place of thymine. These molecules are involved in the transfer of information from DNA, programming protein synthesis and maintaining ribosome structure. The four main types of RNA are heterogeneous nuclear RNA (hRNA), messenger RNA (mRNA), transfer RNA (tRNA), and ribosomal RNA (rRNA).

Risk: A combination of the likelihood of an occurrence of a hazardous event and the severity of injury or damage to the health of workers/people caused by the event.

Risk assessment: (i) The identification and quantification of the risk resulting from a specific use or occurrence of a chemical, taking into account the possible harmful effects on individual people or society using the chemical in the amount and manner proposed and all the possible routes of exposure. Quantification ideally requires the establishment of dose–effect and dose–response relationships in likely target individuals and populations. (ii) An organised process used to estimate the amount of risk of adverse human health effects from exposure to a toxic chemical substance and how likely or unlikely it is that the adverse effect will occur. How reliable and accurate this process is depends on the quantity and quality of the information that goes into the process. The four steps in a risk assessment of a toxic chemical substance are (i) hazard identification, (ii) dose–response assessment, (iii) exposure assessment, and (iv) risk characterisation. (v) Calculation of an individual's risk, employing appropriate mathematical equations, of having inherited a certain gene mutation, of developing a particular disorder, or of having a child with a certain disorder based upon analysis of multiple factors including family medical history and ethnic background.

Risk evaluation: The establishment of a qualitative or quantitative relationship between risks and benefits, involving the complex process of determining the significance of the identified hazards and estimated risks to those organisms or people concerned with or affected by them.

Risk management: The decision-making process involving considerations of political, social, economic, and engineering factors with relevant risk assessments relating to a potential hazard so as to develop, analyse, and compare regulatory options and to select the optimal regulatory response for safety from that hazard. Essentially risk management is the combination of three steps: risk evaluation, emission and exposure control, and risk monitoring.

Risk management: The process of actually trying to reduce risk, for instance, from a toxic chemical substance and/or of trying to keep it under control. Risk management involves not just taking action but also analysing and selecting among options and then evaluating their effect.

Risk: (i) The probability that damage to life, health, and/or the environment will occur as a result of a given hazard (such as exposure to a toxic chemical). Some risks can be measured or estimated in numerical terms, for example, one chance in a hundred. (ii) The term risk must not be confused with the term 'hazard'. It is most correctly applied to the predicted or actual frequency of occurrence of an adverse effect of a chemical or other hazard.

RNA (ribonucleic acid): A single-stranded nucleic acid, similar to DNA, but having a ribose sugar, instead of deoxyribose, and uracil instead of thymine as one of its bases.

Rodenticide: Any chemical substance used to kill rodents. See Pesticide.

RTECS: Registry of Toxic Effects of Chemical Substances.

S9 fraction: Supernatant fraction obtained from an organ (usually liver) homogenate by centrifuging at 9000 g for 20 min in a suitable medium; this fraction contains cytosol and microsomes.

Safety: The practical certainty that injury will not result from exposure to a hazard under defined conditions: in other words, the high probability that injury will not result. In the context of toxicology, the term safety denotes the high probability that injury will not result from exposure to a substance under defined conditions of quantity and manner of use, ideally controlled to minimise exposure.

Safety and Health Committee: A committee with representation of workers' safety and health representatives and employers' representatives, established and functioning at workplace level according to national laws, regulations, and practice.

Safety assessment/safety evaluation: The process of evaluating the safety or lack of safety of a chemical substance in the environment based upon its toxicity and current levels of human exposure.

Safety factors: Factors traditionally used in establishing public health and regulatory guidelines and standards based on toxicological data. These are based on NOELs in species of laboratory animals. These values are then reduced by a factor of ten to provide assurance that the animal data are protective of humans. In general, a tenfold factor is used to account for the possibility that humans are more sensitive than the animal species from which the data are obtained.

SARA: Superfund Amendments and Reauthorization Act of 1986 (United States).

SARS (severe acute respiratory syndrome): A respiratory disease in humans caused by the SARS coronavirus. The initial symptoms are flu-like and may include fever, myalgia, lethargy, cough, sore throat, gastrointestinal disorders, and several other nonspecific symptoms.

Sclerosis: Hardening of an organ or tissue, especially that due to excessive growth of fibrous tissue.

Screening: A test or tests, examination(s), or procedure(s) conducted to know the undetected abnormalities, unrecognised (incipient) diseases, or health defects. The pharmacological or toxicological screening consists of a specified set of procedures to which a series of chemical compounds are subjected to characterise pharmacological and toxicological properties and to establish dose–effect and dose–response relationships.

Secondary containment: An empty chemical-resistant container/dike placed under or around chemical storage containers for the purpose of containing a spill should the chemical container leak.

Sedative: A chemical substance that exerts a soothing or tranquillising effect to the organism or animal.

Selective pesticide: A pesticide that is effective only against certain species and that can control unwanted pests without serious injury to desirable species.

Sensitisation: The development, over time, of an allergic reaction to a chemical. The chemical may cause a mild response on the first few exposures, but as the allergy develops, the response becomes worse with subsequent exposures. Eventually, even short exposures to low concentrations can cause a very severe reaction. There are two different types of occupational sensitisation: skin and respiratory. Typical symptoms of skin sensitivity are swelling, redness, itching, pain, and blistering. Sensitisation of the respiratory system may result in symptoms similar to a severe asthmatic attack. These symptoms include wheezing, difficulty in breathing, chest tightness, coughing, and shortness of breath. Sensitisation indicates (i) the exposure to a substance (allergen) that provokes a response in the immune system such that disease symptoms will ensue on subsequent encounters with the same substance and (ii) the immune response whereby individuals become hypersensitive to chemical substances, pollen, dandruff, or other agents that make them develop a potentially harmful allergy when they are subsequently exposed to the sensitising material (allergen).

Serum (blood serum): Watery proteinaceous portion of the blood that remains after clotting.

SEV (severe): Severe erythema (beet redness) to slight eschar formation (injuries in depth) and severe oedema (raised more than 1 mm and extending beyond area of exposure) on the skin where the dose was applied.

SGOT: Serum glutamic–oxalic transaminase.

SGPT: Serum glutamic–pyruvic transaminase.

Short-term exposure: Repeated exposure by the oral, dermal, or inhalation route for more than 24 h, up to 30 days.

Short-term exposure limit (STEL): Defined by the ACGIH as the TWA airborne concentration to which workers may be exposed for periods up to 15 min, with no more than 4 such excursions per day and at least 60 min between them (*See also* TWA – Time-Weighted Average).

Siderosis: (i) The pneumoconiosis resulting from the inhalation of iron dust. (ii) Excess of iron in the urine, blood, or tissues, characterised by haemosiderin granules in urine and iron deposits in tissues.

SIDS (sudden infant death syndrome): A syndrome marked by the symptoms of sudden and unexplained death of an apparently healthy infant aged 1 month to 1 year.

Silicosis: Pneumoconiosis resulting from inhalation of silica dust.

SimHaz: Simple hazard tool (to identify those chemical substances that pose a high or low hazard).

Sister chromatid exchange (SCE): The process of producing reciprocal exchange of DNA, between the two DNA molecules of a replicating chromosome; (ii) the reciprocal exchange of chromatin between two replicated chromosomes that remain attached to each other until anaphase of mitosis; used as a measure of mutagenicity of substances that produces this effect.

Skeletal fluorosis: Osteosclerosis due to fluoride.

SKY (spectral karyotyping): A technique that allows for the visualisation of all of an organism's chromosomes together, each labelled with a different colour. This is achieved by using chromosome-specific, single-stranded DNA probes (each labelled with a different fluorophore) to hybridise or bind to the chromosomes of a cell, resulting in each chromosome being painted a different colour. This technique is useful for identifying chromosome abnormalities because it is easy to spot instances where a chromosome painted in one colour has a small piece of another chromosome, painted in a different colour, attached to it.

Soluble threshold limit concentration (STLC): Measurement used to determine the hazardous waste characterisation under California State regulations as outlined in Title 26 of the California Code of Regulations (CCR).

Solubility: The ability of a material to dissolve in water or another liquid. Solubility may be expressed as a ratio or may be described using words such as insoluble, very soluble, or miscible. Often, on an MSDS, the 'solubility' section describes solubility in water since water is the single most important industrial solvent. Solubility information is useful for planning spill clean-up and fire-fighting procedures.

Solvent: Chemical that can dissolve many different chemicals. Examples of common solvents are water, ethanol, acetone, hexane, and toluene. Solvent is a substance, usually a liquid, that acts as a dissolving agent or that is capable of dissolving another substance. In solutions of solids or gases in a liquid, the liquid is the solvent. In all other homogeneous mixtures like liquids, solids, or gases dissolved in liquids, solids in solids, and gases in gases, solvent is the component of the greatest amount.

Some evidence: When carcinogenicity is demonstrated by studies that are interpreted as showing a chemically related increased incidence of benign neoplasms, studies that exhibit marginal increases in neoplasms of several organs/tissues or studies that exhibit a slight increase in uncommon malignant or benign neoplasms.

Sorption: A physico-chemical process by which a fumigant is adsorbed or absorbed by the commodity. This may be reversible (adsorbed fumigant may be released on aeration of commodity) or irreversible (chemically bound by the commodity leading to residue of fumigant in treated commodity).

Specific gravity: The ratio of the density of a material to the density of water. The density of water is about 1 g/cc. Materials that are lighter than water (specific gravity less than 1.0) will float. Most materials have specific gravities exceeding 1.0, which means they are heavier than water and so will sink. Knowing the specific gravity is important for planning spill clean-up and fire-fighting procedures. For example, a light flammable liquid such as gasoline may spread and, if ignited, burn on top of a water surface.

Stability: The ability of a chemical substance/material to remain unchanged in the presence of heat, moisture, or air. An unstable chemical substance/material may decompose, polymerise, burn, or explode under normal environmental conditions. Any indication that the chemical substance/material is unstable gives warning that special handling and storage precautions may be necessary.

Statistical significance: The probability that a result is not likely to be due to chance alone. By convention, a difference between two groups is usually considered statistically significant if chance could explain it only 5% of the time or less. Study design considerations may influence the a priori choice of a different level of statistical significance

STEL: Short-term exposure limit.

Stimulant: A chemical or drug, such as caffeine or nicotine, that temporarily accelerates physiological activity.

STLC (soluble threshold limit concentration): An analysis to determine the amount of each analyte that is soluble in the 'waste extraction test', (W.E.T.) leachate.

Stochastic: Pertaining to any phenomenon obeying the laws of probability.

Stomach: The sac-shaped digestive organ that is located in the upper abdomen, under the ribs. The upper part of the stomach connects to the oesophagus, and the lower part leads into the small intestine.

Stoddard solvent: A chemical substance that is used to remove other chemicals and coatings from tools and other objects. Its common names are often 'safety solvent' and 'mineral spirits'.

STP: Standard temperature and pressure (0°C and 1 atm pressure).

Structure–activity relationship (SAR): A traditional practice of medicinal chemistry to modify the effect or the potency or activity of bioactive chemical compounds by modifying their chemical structure. Medical chemists use the chemical techniques of synthesis to insert new chemical groups into the biomedical compound and test the modifications in their biological effect. The study helps to identify and determine the chemical groups responsible for evoking a target biological effect in the organism/animal. The analysis of the dependence of the biological effects of a chemical upon its molecular structure produces a structure–activity relationship. Molecular structure and biological activity are correlated by observing the results of systematic structural modification on defined biological end points.

Subacute (subchronic) toxicity: The toxicity test to determine the adverse effects occurring as a result of repeated daily dosing of a chemical, or exposure to the chemical, for part of an organism's life span (usually not exceeding 10%). With experimental animals, the period of exposure may range from a few days to 6 months.

Subchronic exposure: Repeated exposure by the oral, dermal, or inhalation route for more than 30 days, up to approximately 10% of the life span in humans (more than 30 days up to approximately 90 days in typically used laboratory animal species).

Subchronic study: A toxicity study designed to measure effects from subchronic exposure to a chemical.

Sufficient evidence: Evidence of carcinogenicity when there is an increased incidence of malignant tumours: (a) in multiple species or strains, or (b) in multiple experiments (preferably with different routes of administration or using different dose levels), or (c) to an unusual degree with regard to incidence, site or type of tumour, or age at onset. Additional evidence may be provided by data on dose–response effects. This indicates that there is a causal relationship between the exposure and human cancer.

Suggested no adverse response level (SNARL): Regulatory value that defines the maximum dose or concentration that, on the basis of current knowledge, is likely to be tolerated by an organism without producing any adverse effect.

Suicidal: Pertaining to suicide. Taking of one's own life. As in a suicidal gesture, suicidal thought, or suicidal act. An 'online lifeline for suicidal undergrads' may help prevent college students from committing suicide.

Suicide: The act of causing one's own death. Suicide may be positive or negative and it may be direct or indirect. Suicide is a positive act when one takes one's own life.

Supervisor: A person responsible for the day-to-day planning, organisation, and control of a function/duty at the workplace.

Surfactant: A chemical substance that lowers surface tension.

Susceptible subgroups: Life stages, for example, the children or elderly, or to other segments of the population, for example, the asthmatics or immune-compromised, but are likely to be somewhat chemical-specific and may not be consistently defined in all cases.

Symptom: An indication that a person has a condition or disease. Some examples of symptoms are headache, fever, fatigue, nausea, vomiting, and pain.

Synapse: The place at which a nervous impulse passes from one neuron to another. Also include other terms such as neuron junction, neuron–neuron junction, and synaptic junction.

Syndrome: A set of signs and symptoms that tend to occur together and that reflect the presence of a particular disease or an increased chance of developing a particular disease.

Synergistic, synergism: The concept that exposure to more than one chemical can result in health effects greater than expected when the effects of exposure to each chemical are added together. Very simply, it is like saying $1 + 1 = 3$. When chemicals are synergistic, the potential hazards of the chemicals should be re-evaluated, taking their synergistic properties into consideration. Synergism is the adverse effect or risk from two or more chemical substances interacting with each other that is greater than what it would be if each chemical substance acts separately. A synergistic effect is the any effect of two chemical substances acting together that is greater than the simple sum of their effects when acting alone: such chemical substances are said to show synergism.

Synonyms: Alternative names for the same chemical. For example, methanol and methyl hydrate are synonyms for methyl alcohol. Synonyms may help in locating additional information on a chemical.

Systemic: Affecting many or all body systems or organs; not localised in one spot or area.

Systemic effect: A response to chemical exposure that affects the whole body. Systemic illnesses may cause symptoms in one or two areas, but the whole body is affected.

Systemic injection: The acute systemic injection test looks for potential toxic effects as a result of a single-dose injection in mice. Mice will be injected intraperitoneally or intravenously and observed for signs of toxicity immediately after injection up to 72 h post-injection.

Systemic toxic effects: Toxic effects produced by chemical substances are generally categorised according to the site of the toxic effect. In some cases, the effect may occur at only one site. This site is referred to as the specific target organ. In other cases, toxic effects may occur at multiple sites. This is referred as systemic toxicity. Following are types of systemic toxicity:

Systemic toxicity: Systemic toxicity (acute) evaluates the potential adverse effects of medical devices on the body's organs and tissues that are remote from the site of contact. Depending on the type of device being tested, topical, inhalation, intravenous, intraperitoneal, or oral administration of extracts or implantation of the device in the animal is observed for toxicity. There are four categories: acute (24 h), subacute (14–28 days), subchronic (90 days or 10% of an animal's life span), and chronic (anything longer).This is outlined in ISO 10993–11.

Tablespoon: An old-fashioned but convenient household measure of capacity. A tablespoon holds about 3 teaspoons, each containing about 5 cc, so a tablespoon = about 15 cc of fluid.

Tachycardia: Abnormally fast heartbeat. Excessive rapidity in the action of the heart, usually with a heart rate above 100 beats/min.

Tachypnoea: Abnormally fast breathing.

Taeniacide: A chemical substance intended to kill tapeworms.

Target: The organism, organ, tissue, cell, or cell constituent that is subject to the action of an agent.

Target organ dose: The amount of a potentially toxic substance reaching the organ chiefly affected by that substance.

Target organ effects: A response to chemical exposure that affects a particular organ or system, such as the lungs or liver. Under the U.S. OSHA Hazcom Standard, chemical substances are identified as having target organ effects if there is statistically significant evidence of an acute or chronic health effect determined in a scientifically valid study. The following agents are normally included: (i) hepatotoxins, agents that damage the lungs (including irritants); (ii) chemical substances/agents that act on the haematopoietic system; (iii) neurotoxins; (iv) nephrotoxins; (v) reproductive toxins (mutagens, embryotoxins, teratogens); (vi) cutaneous hazards (chemical substances that normally damage/affect the dermal layer of the skin); (vii) and eye hazards (chemical substances that affect the eye or visual capacity). There are no maximum dose criteria for chronic toxicity studies (for details, refer to Controlled Products Rules and Regulations, Workplace Hazardous Materials Information System (WHMIS)).

TCLO: Lowest toxic airborne concentration tested.

TDG: Transportation of dangerous goods. In Canada, the transportation of potentially hazardous materials is regulated under the federal TDG Act and Regulations that is administered by Transport Canada. The TDG Act and Regulations set out criteria for the classification of materials as dangerous goods and state how these materials must be packaged and shipped.

TDI (tolerable daily intake): An estimate of the amount of a chemical substance in air, food, or drinking water that can be taken in daily over a lifetime without appreciable health risk. TDIs are calculated on the basis of laboratory toxicity data to which uncertainty factors are applied. TDIs are used for chemical substances that do not have a reason to be found in food in contrast to chemical substances such as additives, pesticide residues, or veterinary drugs found in foods.

TEAM: Total exposure assessment methodology.

Telomere: Structure that terminates the arm of a chromosome.

Temporary acceptable daily intake: Value for the ADI proposed for guidance when data are sufficient to conclude that use of the chemical substance is safe over the relatively short period of time required to generate and evaluate further safety data, but are insufficient to conclude that use of the chemical substance is safe over a lifetime.

Temporary safe reference action level (TSRAL): Regulatory value defining the inhalational exposure level in the workplace that is safe for a short time but that should be reduced as soon as possible or appropriate respiratory protection employed.

Tendon: Fibrous bands or cords of connective tissue at the ends of muscle fibres that serve to attach the muscles to bones and other structures.

Tensile strength: The force required to break a material apart by pulling it from opposing directions. Measured in pounds and is reported in two directions.

Teratogen: An agent that causes the production of physical defects in the developing embryo. A teratogen is a hazardous chemical substance that can cause birth defects. Teratogenic means able to cause birth defects. (i) A descriptor term applied to any chemical substance that can cause nonheritable birth defects. (ii) Agent that, when administered prenatally to the mother, induces permanent structural malformations or defects in the offspring.

Teratogenesis: The production of non-heritable birth defects.

Teratogenic: Anything including chemical substances that produces non-heritable birth defects is said to be teratogenic.

Teratogenicity: The ability of a chemical to cause birth defects. Teratogenicity results from a harmful effect to the embryo or the fetus/foetus. (i) Potential to cause the production of nonheritable structural malformations or defects in offspring.

Teratology: Study of malformations, monstrosities, or serious deviations from normal development in organisms.

Testing of chemicals: (i) In the discipline of toxicology, testing of chemical substances means procedures and evaluation of the therapeutic and potentially toxic effects of chemical substances by their application through relevant routes of exposure with appropriate organisms or animals or biological systems and to relate the effects to dose following application. (ii) In chemistry, qualitative or quantitative analysis by the application of one or more fixed methods and comparison of the results with established standards.

Therapeutic index: The ratio between toxic and therapeutic doses. The higher the ratio, the greater the safety of the therapeutic dose.

Therapy: Treatment of an illness, disease, or disability. Therapy may be scientifically proven to treat an illness or unproven to treat an illness. Unproven treatments are also called alternative therapy.

Thermolabile: Unstable when heated; specifically subject to loss of characteristic properties on being heated to or above 55°C (thermolabile enzymes and vitamins).

Thermoplastic: A polymer material that turns to liquid when heated and becomes solid when cooled. There are more than 40 types of thermoplastics, including acrylic, PP, polycarbonate, and polyethylene.

Thermal decomposition products: These chemicals may be toxic, may be flammable, or may have other hazards. The chemicals released and their amounts vary depending upon conditions such as the temperature. The thermal decomposition products may be quite different from the chemicals formed by burning the same material (hazardous combustion products). It is important to know which chemicals are formed by thermal decomposition because this information is used to plan ventilation requirements for processes where a material may be heated.

Threshold: (i) Dose or exposure concentration of a chemical substance below which a defined (adverse) effect will not occur. (ii) A level of chemical exposure below which there is no adverse effect and above which there is a significant toxicological effect.

Threshold limit value (TLV): (i) Term used by the ACGIH to express the recommended exposure limits of a chemical to which nearly all workers may be repeatedly exposed, day after day, without adverse effect. (ii) This is a guideline value defined by the ACGIH to establish the airborne concentration of a potentially toxic substance to which it is believed that healthy working adults may be exposed safely through a 40-h working week and a full working life. This concentration is measured as a TWA concentration. They are developed only as guidelines to assist in the control of health hazards and are not developed for use as legal standards; TLV is the concentration of a potentially toxic substance that should not be exceeded during any part of the working exposure (ACGIH). (i) Recommended limits proposed by the ACGIH to which most workers can be exposed without adverse effect. TLVs may be expressed as a TWA, as a STEL, or as a ceiling value (CL). (ii). The TLV of a chemical substance is a level to which it is believed a worker can be exposed day after day for a working lifetime without adverse health effects. (iii) Maximum concentration of gas, which a person can withstand when continuously exposed for a period of eight hours.

Thrombocytopenia: Decrease in the number of blood platelets.

Time-weighted average concentration (TWA): Regulatory value defining the concentration of a substance to which a person is exposed in ambient air, averaged over a period, usually 8 h. For a person exposed to 0.1 mg/m^3 for 6 h and 0.2 mg/m^3 for 2 h, the 8-h TWA is $(0.1 \times 6 + 0.2 \times 2)/8$ that equals 0.125 mg/m^3.

Thyroid hormones: Hormones secreted by the thyroid gland. The thyroid gland makes T3 (triiodothyronine) and T4 (thyroxine), which together are considered thyroid hormones. T3 and T4 have identical effects on cells. Thyroid hormone affects heart rate, blood pressure, body temperature, and weight. T3 and T4 are stored as thyroglobulin, which can be converted back into T3 and T4.

T-lymphocyte: Animal cell that possesses specific cell surface receptors through which it binds to foreign substances or organisms, or those which it identifies as foreign, and which initiates immune responses.

Tolerable daily intake (TDI): Values applied to chemical contaminants in food and drinking water. The presence of contaminants is unwanted and they have no useful function, differing from additives and residues where there is or was deliberate use resulting in their presence. TDIs are calculated on the basis of laboratory toxicity data with the application of uncertainty factors. A TDI is thus an estimate of the amount of a substance (contaminant) in food or drinking water that can be ingested daily over a lifetime without appreciable health risk.

Tolerance: (i) The ability to experience exposure to potentially harmful amounts of a substance without showing an adverse effect. (ii) Adaptive state characterised by diminished effects of a particular dose of a substance: the process leading to tolerance is called 'adaptation'. (iii) In food toxicology, dose that an individual can tolerate without showing an effect. (iv) Ability to experience exposure to potentially harmful amounts of a substance without showing an adverse effect. (v) Ability of an organism to survive in the presence of a toxic substance: increased tolerance may be acquired by adaptation to constant exposure. (vi) In immunology, state of specific immunological unresponsiveness.

Topical application: Medicines applied directly to the surface of the body.

Topical effect: Consequence of application of a substance, medicine, or toxic chemical (for testing) to the surface of the body that occurs at the point of application.

Total threshold limit concentration (TTLC): An analysis to determine the total concentration of each target analyte in a sample. The samples are analysed using published U.S. EPA methods. When any target analyte exceeds the TTLC limits, the waste is classified as hazardous and its waste code is determined by the compound(s) that failed TTLC.

Toxaemia (blood poisoning): (i) Condition in which the blood contains toxins produced by body cells at a local source of infection or derived from the growth of microorganisms. (ii) Pregnancy-related condition characterised by high blood pressure, swelling and fluid retention, and proteins in the urine.

Toxic: Pertaining to any chemical substance able to cause injury to living organisms/animals as a result of physico-chemical interaction. In short, toxic means able to cause harmful health effects.

Toxic concentration low (TCLo): The lowest concentration of substance in air to which humans or animals have been exposed for any given period of time, which has produced any toxic effect in humans or has produced a tumourigenic or reproductive effect in animals or humans.

Toxic dose: Amount of a chemical substance that produces intoxication without lethal outcome.

Toxic dose low (TDLo): The lowest dose of a substance introduced by any route other than inhalation, over any given period of time, to which humans or animals have been exposed, and reported to produce any nonsignificant toxic effects in humans or to produce non-significant tumourigenic or reproductive effects in animals or humans.

Toxic effects: Toxicity is complex with many influencing factors and dosage is the most important. Xenobiotics are chemical substances foreign to the body, causing many types of toxicity by a variety of mechanisms. Some chemical substance in itself is toxic and is referred to as 'parent' compound. Other chemical substances must be metabolised (chemically changed within the body) before they cause toxicity to the animal or humans.

Toxic substance: A chemical, physical, or biological agent that may cause an adverse effect or effects to biological systems. In general, as defined in the FHSA regulations at 16 CFR § 1500.3(b)(5), any substance (other than a radioactive substance) that has the capacity to produce personal injury or illness to man through ingestion, inhalation, or absorption through any surface of the body. This term is further defined by the OSHA and in the FHSA regulations.

Toxicant: Any chemical substance that is potentially toxic.

Toxicity: The ability of a substance to cause harmful health effects. Descriptions of toxicity (e.g. low, moderate, severe) depend on the amount needed to cause an effect or the severity of the effect. Under the Canadian CPRs and the U.S. OSHA Hazcom Standard, there are specific technical criteria for identifying a material as toxic for the purpose of each regulation. (See also Very Toxic and Highly Toxic.) The CPRs are part of the national WHMIS. The term 'toxicity' is used in two different senses: (i) capacity to cause injury to a living organism and (ii) adverse effects of a chemical on a living organism. See Adverse effect. The severity of toxicity produced by any chemical is directly proportional to the exposure concentration and the exposure time. This relationship varies with the developmental stage of an organism and with its physiological status. (iii) The extent, quality, or degree to which a chemical substance can be poisonous or harmful to humans or other living organisms.

Toxicity assessment: The process of defining the nature of injuries that may be caused to an organism by exposure to a given chemical and the exposure concentration and time dependence of the chemically induced injuries. The aim of the assessment is to establish safe exposure concentration limits in relation to possible time of exposure.

Toxicity test: Experimental study to evaluate the adverse effects of exposure of a living organism or animal to a chemical substance for a defined duration under defined conditions.

Toxicodynamics: The physiological mechanisms by which toxins/chemical substances are absorbed, distributed, metabolised, and excreted.

Toxicokinetics: Methodology that determines the relationship between the systemic exposure of a chemical substance in experimental animals and the related toxicity. It is used primarily for establishing relationships between exposures in toxicology experiments in experimental animals and the corresponding exposures in humans.

Toxicological data sheet: A document that provides important information about the toxicology of a chemical substance, its production, application, properties, methods of identification, storage, precautions, and disposal.

Toxicology: Branch of medicine dealing with the study of poisons and toxic chemical substances. The study includes understanding the chemical nature of poisons, their origin and preparation, their physiological action, tests to recognise them, the pathological changes that these chemical substances produce, the antidotes, and their recognition by post-mortem evidence.

Toxification: Metabolic conversion of a potentially toxic chemical substance to a product that becomes more toxic.

Toxin: A fairly complex and highly toxic organic substance produced by a living organism. A toxin is a chemical, physical, or biological agent that causes disease or some alteration of the normal structure and function of an organism. Usually refers to poisonous substance produced during the metabolism and growth of certain microorganisms and some higher plant and animal species, or any poisonous isomer, homologue, or derivative of such a substance. It causes either permanent or reversible injury to the health of a living thing on contact or absorption, typically by interacting with biological macromolecules such as enzymes and receptors. Onset of effects may be immediate or delayed, and impairments may be slight or severe.

TPE (thermoplastic elastomer): A flexible thermoplastic that can be used to replace rubber and flexible urethanes. Hytrel is an elastomer made by DuPont.

Trade name: A name under which a product is commercially known. Some materials are sold under common names, such as Stoddard solvent or degreaser, or internationally recognised trade names, like Varsol. Trade names are sometimes identified by symbols such as R or TM.

Trait: A specific characteristic of an organism. Traits can be determined by genes or the environment, or more commonly by interactions between them. The genetic contribution to a trait is called the genotype. The outward expression of the genotype is called the phenotype.

Transcription: A process by which the genetic information encoded in a linear sequence of nucleotides in one strand of DNA is copied into an exactly complementary sequence of RNA.

Transdermal skin patch: A medicated adhesive patch that is put on the skin to release a dose of medication through the skin into the bloodstream.

Transformation: The process that involves (i) alteration of a cell by incorporation of foreign genetic material and its subsequent expression in a new phenotype, (ii) conversion of cells growing normally to a state of rapid division in culture resembling that of a tumour, and (iii) chemical modification of substances in the environment.

Transformation (neoplastic): The conversion of normal cells into tumour cells. Frequently this is the result of a genetic change (mutagenesis) and the same term is used to describe the genetic modification of bacteria for use in biotechnology.

Transgene: Gene from one source that has been incorporated into the genome of another organism.

Transplant: The grafting of a tissue from one place to another; just as in botany, a bud from one plant might be grafted onto the stem of another. The transplanting of tissue can be from one part of the patient to another (autologous transplantation), as in the case of a skin graft using the patient's own skin, or from one patient to another (allogenic transplantation), as in the case of transplanting a donor kidney into a recipient.

Treatment: (i) Administration or application of a chemical substance/drug to an organism or an infected area to control the problem. (ii) A process to control or to eliminate the pest that may include fumigation, irradiation, hot-water, hot-air, vapour heat, and cold treatments.

TRI chemicals: A list of about 650 toxic chemical substances or chemical categories included in the Toxics Release Inventory (TRI). In general, TRI chemicals are ones the U.S. EPA has found that can be reasonably anticipated to cause acute or chronic adverse human health effects or adverse environmental effects. The U.S. TRI under Section 313 of the Emergency Planning and Community Right-To-Know Act of 1986 (EPCRA).

Triple rinse: Washing procedure required before properly disposing of plastic and metal pesticide containers.

Trohoc: An epidemiological study that starts with the outcome and looks backwards for the causes (the term is disapproved by the majority of epidemiologists).

Trophic level: The amount of energy in terms of food that an organism needs. Organisms not needing organic food, such as plants, are said to be on a low trophic level, whereas predator species needing food of high-energy content are said to be on a high trophic level. The trophic level indicates the level of the organism in the food chain.

TSCA: In theory, this 1976 law gave the U.S. EPA the power to test, regulate, and screen nearly all chemicals produced or imported into the United States.

TTLC (total threshold limit concentration): An analysis to determine the total concentration of each target analyte in a sample.

Tumour (neoplasm): Any growth of tissue forming an abnormal mass. Cells of a benign tumour will not spread and will not cause cancer. Cells of a malignant tumour can spread through the body and cause cancer.

Tumourigenic: Pertaining to chemical substances/physical agents that cause tumour formation.

TWA (time-weighted average): The concentration of a chemical contaminant averaged over a workday (usually 8 h long). TWA is measured in a workplace by sampling the breathing zone for the whole workday. (ii) The average concentration of a chemical substance in air over the total exposure period of time that is usually an eight-hour workday.

Tychem 9400: Tough, durable, tear-resistant material that provides excellent protection against a broad range of chemicals.

Tychem SL: A lightweight fabric providing effective and economical protection against a broad range of industrial chemicals, including those used in agriculture and petroleum markets.

Tyvek: A material that provides protection in all kinds of industrial applications. Provides an excellent barrier in light splash situations and dry particulates such as asbestos, lead dust, and radioactive dusts. Also provides protection in food processing and painting.

Tyvek QC: Polyethylene coated. Provides excellent lightweight splash protection from many acids and other liquid chemicals and pesticides.

Ubiquinone: A naturally occurring benzoquinone important in electron transport in mitochondrial membranes. Coenzyme Q functions as an endogenous antioxidant; deficiencies of this enzyme have been observed in patients with many different types of cancer, and limited studies have suggested that coenzyme Q may induce tumour regression in patients with breast cancer. This agent may have immunostimulatory effects.

UCM: Urographic contrast medium.

UDP: Uridine diphosphate.

Ulcer: (i) A lesion on the surface of the skin or a mucous surface, produced by the sloughing of inflammatory necrotic tissue. (ii) A kind of defect, often associated with inflammation, occurring locally or at the surface of an organ or tissue owing to sloughing of necrotic tissue.

Ultem (polyetherimide (PEI)): Ultem was an amorphous resin introduced by G.E. in the 1970s. Ultem can withstand multiple 'autoclaving' cycles (steam sterilisation under high pressure). Autoclaving is the primary process used in hospitals to sterilise surgical tools and instruments.

Ultrafine particle: Particle in air of aerodynamic diameter less than 100 nm.

Ultraviolet rays: Radiation from the sun that can be useful or potentially harmful. UV rays from one part of the spectrum (UVA) enhance plant life. UV rays from other parts of the spectrum (UVB) can cause skin cancer or other tissue damage. The ozone layer in the atmosphere partly shields us from UV rays reaching the Earth's surface.

Uncertainty: Uncertainty occurs because of a lack of knowledge. It is not the same as variability. For example, a risk assessor may be very certain that different people drink different amounts of water but may be uncertain about how much variability there is in water intakes within the population. Uncertainty can often be reduced by collecting more and better data, whereas variability is an inherent property of the population being evaluated. Variability can be better characterised with more data but it cannot be reduced or eliminated. Efforts to clearly distinguish between variability and uncertainty are important for both risk assessment and risk characterisation.

Uncertainty safety factor (UF): This term may be used in either of two ways depending upon the context: (i) mathematical expression of uncertainty applied to data that are used to protect populations from hazards that cannot be assessed with high precision and, (ii) with regard to food additives and contaminants, a factor applied to the NOEL to derive ADI (the NOEL is divided by the safety factor to calculate the ADI). The value of the safety factor depends on the nature of the toxic effect, the size and type of the population to be protected, and the quality of the toxicological information available.

Unhemmed aprons: Unhemmed aprons are used mainly with food processing, industrial maintenance, and other hygienic applications.

Unstable: Pertaining to a chemical substance that can easily catch fire.

Unstable (reactive): Under the U.S. OSHA Hazcom Standard, a chemical is identified as unstable (reactive) if in the pure state, or as produced or transported, it will vigorously polymerise, decompose, condense, or become self-reactive under conditions of shock, pressure, or temperature.

Uraemia: Excess in the blood of urea, creatinine, and other nitrogenous end products of protein and amino acid metabolism; also, the constellation of signs and symptoms of chronic renal failure.

Urate oxidase: The hepatic peroxisomal enzyme that catalyses the oxygen-mediated conversion of uric acid into allantoin.

Urea: A soluble weakly basic nitrogenous compound (CH_4N_2O) that is the chief solid component of mammalian urine and an end product of protein decomposition and that is administered intravenously as a diuretic drug – called also carbamide.

Urethane aprons: Lightweight, long lasting, and very economical where water splash is likely. Use of aprons is ideal for food processing and heavy industrial abrasion areas.

Uric acid: Waste product leftover from normal chemical processes in the body and found in the urine and blood. Abnormal build-up of uric acid in the body may cause a condition called gout. Increased levels of uric acid in the blood and urine can be a side effect of chemotherapy or radiation therapy.

Urticaria: Vascular reaction of the skin marked by the transient appearance of smooth, slightly elevated patches (wheals, hives) that are redder or paler than the surrounding skin and often attended by severe itching.

U.S. EPA (United States Environmental Protection Agency): An agency of the federal government of the United States charged to regulate chemicals and protect human health by safeguarding the natural environment: air, water, and land. The U.S. EPA conducts environmental assessment, research, and education. It has the primary responsibility for setting and enforcing national standards under a variety of environmental laws, in consultation with state, tribal, and local governments.

Uvea: Pigmented vascular coat of the eyeball, consisting of choroid, ciliary body, and iris.

Vacuole: A membrane-bound cell organelle. In animal cells, vacuoles are generally small and help sequester waste products. In plant cells, vacuoles help maintain water balance. Sometimes a single vacuole can take up most of the interior space of the plant cell.

Validity of a measurement: Expression of the degree to which a measurement measures what it purports to measure.

Vapour: The gaseous form of a material that is normally solid or liquid at room temperature and pressure. Evaporation is the process by which a liquid is changed into a vapour. Sublimation is the process by which a solid is changed directly into the vapour state. Most of the organic solvents evaporate and produce vapours.

Vapour density: The weight per unit volume of a pure gas or vapour. The vapour density is commonly given as the ratio of the density of the gas or vapour to the density of air. The density of air is given a value of 1. Light gases (density less than 1) such as helium rise in air. If there is inadequate ventilation, heavy gases and vapours (density greater than 1) can accumulate in low-lying areas such as pits and along floors.

Vapour pressure: A measure of the tendency of a material to form a vapour. The higher the vapour pressure, the higher the potential vapour concentration. In general, a material with a high vapour pressure is more likely to be an inhalation or fire hazard than a similar material with a lower vapour pressure.

Vasoconstriction: Decrease of the calibre of the blood vessels leading to a decreased blood flow.

Vasodilation: Increase in the calibre of the blood vessels, leading to an increased blood flow.

Vehicle: Chemical substance(s), solvents, water, or suspending agents used to formulate active ingredients for administration or use to conduct toxicological tests.

Venom: Animal toxin generally used for self-defence or predation and usually delivered by a bite or sting.

Ventilation: (i) The process of supplying fresh air to an enclosed space/workplace in order to refresh/remove/replace the existing atmosphere. Ventilation is commonly used to remove contaminants such as fumes, dusts, and vapours associated with chemical substances and to provide a healthy and safe working environment. Ventilation is an engineering control and can also be accomplished by natural means by opening a window or by mechanical means using fans or blowers. (ii) A process to exchange air between the ambient atmosphere and the lungs. (iii) The amount of air inhaled per day (in terms of physiology). (iv) The oxygenation of blood. Ventilation is the movement of air. One of the main purposes of ventilation is to remove contaminated air from the workplace. There are several different kinds of ventilation: general ventilation, local exhaust ventilation, mechanical ventilation, and natural ventilation.

Ventricular fibrillation: Irregular heartbeat characterised by uncoordinated contractions of the ventricle.

Vermicide: A chemical substance intended to kill intestinal worms (roundworms and others).

Vertigo (dizziness): An illusion of movement as if the external world were revolving around an individual or as if the individual were revolving in space.

Very toxic: Under the Controlled Products Rules and Regulations, there are specific technical criteria for identifying a very toxic material. There are specific criteria for short-term lethality, long-term toxicity, teratogenicity and embryotoxicity, reproductive toxicity, carcinogenicity, respiratory sensitisation, and mutagenicity. The CPRs are part of the national WHMIS. Under the U.S. OSHA Hazcom Standard, the corresponding term is 'highly toxic', which has a specific definition.

Vesicant: (i) a chemical substance that causes blisters on the skin.

Vesicle: In cell biology, a small bladderlike, membrane-bound sac containing aqueous solution or fat. In pathology, a blister-like elevation on the skin containing serous fluid.

Vibrating screens: Equipment that is used for separating particles by size. The circular vibrating screens work on circular motion. The vibrating screens are mainly used in coal dressing, metallurgy, mine, power station, water conservancy project, building industry, light, etc.

Vinyl aprons, hemmed: Made of high-quality virgin vinyl resistant to acids, alkalis, solvents, chemicals, oils, fats, grease, and salt. Provide reliable tear, abrasion, and puncture resistance. Used in food processing, meat packing, assembly, restaurant work, and industrial maintenance.

Vinyl chloride: A chemical substance used in producing some plastics that is believed to be oncogenic.

Virtually safe dose (VSD): Human exposure over a lifetime to a carcinogen that has been estimated, using mathematical modelling, to result in a very low incidence of cancer, somewhere between zero and a specified incidence, for example, one cancer in a million exposed people. (ii) The dose of a chemical substance corresponding to the level of risk determined and accepted by regulatory agencies; the dose-to-risk relationship is based on a chemical dose–response curve.

Virucide: Chemical substance used to control viruses.

Volatile organic chemicals (VOC): Any organic chemical having, at 293.15 K, a vapour pressure of 0.01 kPa or more or having a corresponding volatility under the particular condition of use. (i) An organic compound evaporates readily to the atmosphere. VOCs contribute significantly to photochemical smog production and certain health problems. (ii) Any organic chemical compound that has high enough vapour pressure under normal conditions to significantly vaporise and enter the atmosphere. (iii) The list of VOCs is huge and varied. Although ubiquitous in nature and modern industrial society, they may also be harmful or toxic. (iv) VOCs are produced (a) naturally through biological mechanisms – metabolism – (b) directly by the use of fossil fuels, namely, gasoline; or (c) indirectly by the automobile exhaust. VOCs are useful as fuels, solvents, scents, precursors, propellants, refrigerants, drugs, pesticides, and markers.

Volatile: Able to evaporate. Volatility is the ability of a material to evaporate. The term volatile is commonly understood to mean that a material evaporates easily. On an MSDS, volatility is commonly expressed as the '% volatile'. The per cent volatile can vary from 0% (none of the material will evaporate) to 100% (all of the material will evaporate if given enough time). If a chemical substance/product contains volatile ingredients, there may be a need for ventilation and other precautions to control vapour concentrations.

Volatility: (i) The tendency of a liquid to evaporate into a gas or vapour form. Organic solvents on inhalation are in the form of vapours. (ii) Volatility in the context of chemistry/physics/thermodynamics is a measure of the tendency of a substance to vaporise. It has also been defined as a measure of how readily a substance vaporises. At a given temperature, substances with higher vapour pressures will vaporise more readily than substances with a lower vapour pressure.

Vulnerable zone: An area over which the airborne concentration of a chemical substance is accidentally released and could reach the level of concern as a health hazard.

Waste (chemical waste): Any chemical substance that is discarded deliberately or otherwise disposed of on the assumption that it is of no further use to the primary user.

Wastewater: The spent or used water from a home, community, farm area, or industry that contains dissolved or suspended matter.

Wasting syndrome: A disease marked by weight loss and atrophy of muscular and other connective tissues that is not directly related to a decrease in food and water intake.

Water pollution: The presence where water sources are polluted/contaminated with harmful or objectionable material and damage the quality of water.

Water reactive: Under the U.S. OSHA Hazcom Standard, a chemical is identified as water reactive if it reacts with water to release a gas that is either flammable or presents a health hazard.

Water-reactive material: A chemical substance that reacts with water that could generate enough heat for the item to spontaneously combust or explode. The reaction may also release a gas that is either flammable or presents a health hazard.

Weight of evidence for toxicity: The extent to which the available biomedical/toxicological data support the hypothesis that a chemical substance causes a defined toxic effect in humans.

Wheezing: Breathing with a raspy or whistling sound; a sign of airway constriction or obstruction.

Withdrawal: Physical and psychological symptoms that follow the discontinuance of an addicting drug. The symptoms that have been associated with smoking cessation include cravings to smoke, irritability, anxiety, insomnia, fatigue, dizziness, inability to concentrate, increased appetite, headache, cough, sore throat, constipation, gas, dry mouth, sore tongue and/or gums, postnasal drip, and tightness in chest.

Withdrawal effect: The effect of adverse event following withdrawal from a person or animal of a drug to which they have been chronically exposed or on which they have become dependent, for example, drug abuse or drug addict.

WHMIS: Workplace Hazardous Materials Information System, Canadian programme designed to protect workers by providing them and their employers with vital information about hazardous materials. WHMIS is implemented by a series of federal, provincial, and territorial acts and regulations. One that is used frequently in preparing MSDSs is the CPRs.

World Health Organization (WHO): The directing and coordinating authority for health within the United Nations system. It is responsible for providing leadership on global health matters. The role of the WHO includes (i) providing leadership on matters critical to health and engaging in partnerships where joint action is needed; (ii) shaping the research agenda and stimulating the generation, translation, and dissemination of valuable knowledge; (iii) setting norms and standards and promoting and monitoring their implementation; (iv) articulating ethical and evidence-based policy options; (v) providing technical support, catalysing change, and building sustainable institutional capacity; (vi) and monitoring the health situation and assessing health trends. The WHO's constitution came into force on 7 April 1948. More than 8000 people from more than 150 countries work for the organisation in 147 country offices, 6 regional offices, and a headquarters in Geneva, Switzerland.

Worker: Any person who performs work, either regularly or temporarily, for an employer/organisation.

Workers' health surveillance: A generic term that covers procedures and investigations to assess workers' health in order to detect and identify any abnormality.

The results of surveillance should be used to protect and promote the health of the individual, collective health at the workplace, and the health of the exposed working population. Health assessment procedures may include, but are not limited to, medical examinations, biological monitoring, radiological examinations, questionnaires, and/or review of health records.

Work practice controls: Changes in work procedures such as written safety policies, rules, supervision, schedules, and training with the goal of reducing the duration, frequency, and severity of exposure to hazardous chemical substances/materials.

Xanthoma: A non-neoplastic disorder characterised by a localised collection of histiocytes containing lipid. Xanthomas usually occur in the skin and subcutaneous tissues, but occasionally they may involve the deep soft tissues.

x disease: Hyperkeratotic disease in cattle following exposures to CDDs, naphthalenes, and related compounds.

Xenobiotic metabolism: The sum of the physical and chemical changes that affect foreign substances in living organisms from uptake to excretion.

Xenobiotic: (i) A chemical substance that is not a natural component of the organism exposed to it. (ii) A substance with a chemical structure foreign to a given organism. These include drugs, foreign substances or compounds, or exogenous substances.

Zero air: Atmospheric air purified to contain less than 0.1 ppm total hydrocarbons.

Zoocide: A chemical substance intended to kill animals.

Zygote: (i) A fertilised egg resulting from the fusion of two gametes and (ii) a cell obtained as a result of complete or partial fusion of cells produced by meiosis.

Sources: U.S. National Library of Medicine, Bethesda, MD; U.S. Environmental Protection Agency (U.S. EPA), Integrated Risk Information System (IRIS), IRIS Glossary (updated 2011), U.S. EPA, Washington, DC, 2002; National Institutes of Health, Health and Human Services; NIOSH, Dictionary of Cancer Terms, NCI, Bethesda, MD, http://grants.nih.gov/grants/glossary.htm

Abbreviations

AA	Adverse action; Activation analysis; Ascorbic acid; Atomic absorption
AAEE	American Academy of Environmental Engineers
AAF	Acetylaminofluorene
AALAS	American Association for Laboratory Animal Science
AAPCC	American Association of Poison Control Centers
AAS	Atomic absorption spectrophotometry
ABS	Acrylonitrile–butadiene–styrene
ACBM	Asbestos-containing building material
ACE	Angiotensin-converting enzyme
ACL	Alternative concentration limit
ACM	Asbestos-containing material
ACGIH	American Conference of Governmental Industrial Hygienists
Ach	Acetylcholine
AChE	Acetylcholinesterase
AcP	Acid phosphatase
ACS	American Chemical Society
ACTH	Adrenocorticotropic hormone
ACWM	Asbestos-containing waste material
AD	Alzheimer's disease
ADD	Average daily dose
ADH	Alcohol dehydrogenase; Antidiuretic hormone
ADI	Acceptable daily intake (humans)
ADP	Adenosine diphosphate
AEGL	Acute Emergency Guideline Level, developed by the EPA
AEP	Auditory-evoked potential
AES	Atomic emission spectroscopy
AFB	Air Force Base
AFID	Alkali flame ionisation detection
Ah	Aromatic hydrocarbon
AHH	Aryl hydrocarbon hydroxylase
AHI	Animal Health Institute (United States)
AI	Active ingredient
AIDS	Acquired immune deficiency syndrome
AIHA	American Industrial Hygiene Association
AIN	Acute interstitial nephritis
ALA	Aminolevulinic acid
ALAD	2-Aminolevulinic acid dehydratase
ALAS	o-Aminolevulinic acid synthetase
ALAT	Alanine aminotransferase
ALC	Approximate lethal concentration
ALD	Approximate lethal dose
ALMS	Atomic line molecular spectrometry
ALT	Alanine aminotransferase
AM	Alveolar macrophages
AMP	Amperometric titration
AMPA	Aminomethylphosphonic acid

ANTU	Alpha-naphthylthiourea
AOC	Area of contamination
AP	Alkaline phosphatase
APDM	Aminopyrine-*N*-demethylase
APHA	American Public Health Association
APHIS	Animal and Plant Health Inspection Service (U.S. DA)
APP	Ammonium polyphosphate
AQG	Air quality guidelines
A&R	Air and radiation
ARAC	Acid Rain Advisory Committee
ARF	Acute renal failure
ARL	Acceptable or tolerable residue limit
ART	Alternate reproductive test
ASAT	Aspartate aminotransferase
ASTM	American Society for Testing and Materials
ASV	Anodic stripping voltammetry
ATA	Alimentary toxic aleukia
ATH	Alumina trihydrate
ATLA	Alternatives to Laboratory Animals
atm	Atmosphere
ATPase	Adenosine triphosphatase
ATSDR	Agency for Toxic Substances and Disease Registry (U.S. CDC)
AUC	Area under the curve
AWI	Acceptable or tolerable weekly intake
BA	2-Bromoacrolein
BAER	Brainstem auditory-evoked response
BAF	Bioaccumulation factor
BAL	Bronchoalveolar lavage
BaP	Benzo(a)pyrene
BBB	Blood–brain barrier
BBPP	Bis(2,3-dibromopropyl) phosphate
BCF	Bioconcentration factor
BEA	2-Bromoethylamine
BEI	Biological Exposure Indices, published by the ACGIH
BEN	Balkan endemic nephropathy
BGG	Bovine gamma globulin
BHA	Butylated hydroxyanisole
BHK	Baby hamster kidney
BIC	Butyl isocyanate
BMAA	Beta-*N*-methylamino-*L*-alanine
BMD	Benchmark dose (U.S. EPA)
BMF	Biomagnification factor
BMP	Best management practice
BNP	3-Nitropropionic acid
BOAA	Beta-*N*-oxalylamino-*L*-alanine
BOD	Biochemical oxygen demand; Biological oxygen demand
BP	Boiling point
BSA	Bovine serum albumin
BSI	British Standards Institute

BSO	Benzene-soluble organics
BSP	Bromosulphophthalein
BUB	2-(3-Butylureido)benzimidazole
BUN	Blood urea nitrogen
Bw	Body weight
C	Ceiling limit; the concentration that should not be exceeded during any part of the working exposure (ACGIH)
°C	Celsius
CA	Carbon anhydrase; Carbon absorber; Chrysanthemic acid
CAA	Clean Air Act
CAAA	Clean Air Act Amendments
CALLA	Common acute lymphoblastic leukaemia antigen
CAMEO	Computer-Aided Management of Emergency Operations, maintained by the NOAA
cAMP	Cyclic adenosine monophosphate
CAS	Chemical Abstracts Service
CASRN	Chemical Abstracts Service Registry Number
CAT	Customer Acceptance Test
CBER	Center for Biologics Evaluation and Research (U.S. FDA)
CBF	Cerebral blood flow
CC	Critical concentration
CCAA	Canadian Clean Air Act
CCFA	Codex Committee on Food Additives (see definition of Codex Alimentarius Commission)
CCINFO	Canadian Centre for Occupational Health and Safety, Toronto, Canada
CCOHS	Canadian Centre for Occupational Health and Safety
CCPR	Codex Committee on Pesticide Residues
CCTTE	Computerised listing of chemicals being tested for their toxicological effects
CD	Cluster of differentiation; Coulometric detection
CDC	Centers for Disease Control and Prevention (the United States)
CDD	Chlorodibenzodioxin; Chlorinated dibenzo-p-dioxin
CDER	Center for Drug Evaluation and Research (U.S. FDA)
CDF	Chlorodibenzofuran
CDHS	Comprehensive Data Handling System
cDNA	Complementary DNA
CE	Capillary electrophoresis
CEC	Commission of the European Communities
CEFIC	European Chemical Industry Council
CEFIC	European Council of Chemical Industry Federations
CEI	Compliance evaluation inspection
CEM	Continuous emission monitoring
CEPA	Canadian Environment Protection Act
CEQ	Council on Environmental Quality
CERCLA	Comprehensive Environmental Response Compensation Liability Act of 1980 (amended in 1984 and later)
CFC	Chlorofluorocarbon – an ozone-depleting refrigerant
CFM	Chlorofluoromethanes
CFR	Code of Federal Regulations

cGMP	Cyclic guanosine monophosphate
ChE	Cholinesterase
Chd	Child
CHIPS	Chemical Hazards Information Profiles (EPA)
CHMP	Committee for Human Medicinal Products
CHO	Chinese hamster ovary
CHRIS	Chemical Hazards Response Information System, maintained by the U.S. Coast Guard
Ci	Curie
CI	Confidence interval
CIN	Chronic interstitial nephritis
CIS	International Occupational Safety and Health Information Centre
CKSCC	Cystic keratinising squamous cell carcinoma
C&L	Classification and labelling
Cl_2CA	3-(2,2-Dichlorovinyl)-2,2-dimethyl-cyclopropane carboxylic acid
CLD	Chemiluminescence nitrogen detector
CLM	Chemiluminescence method
CLV	Ceiling value
CMA	Chemical Manufacturers Association (United States)
CMI	Cell-mediated immunity
CML	Cell-mediated lympholysis
CMRs	Carcinogens, mutagens, and substances toxic to reproduction
CNS	Central nervous system
CO-Hb	Carboxyhaemoglobin
COD	Chemical oxygen demand
CODEN	A unique six-letter character code derived from the American Society for Testing and Materials CODEN for periodical titles and the CAS Source Index
COI	Coinvestigator
COPD	Chronic obstructive pulmonary disease
COT	Committee on Toxicity (United Kingdom)
cP	Centipoise
CP	Coproporphyrins
CPA	Cyclopropane carboxylic acid
CPIA	2-(4-Chlorophenyl)isovaleric acid
CPK	Creatine phosphokinase
CPSC	Consumer Product Safety Commission (16 CFR)
CRMs	Certified reference materials
CS	Cervical spondylosis; Customer service
CSDS	Chemical safety data sheet
CSF	Cerebrospinal fluid
CSIN	Chemical Substances Information Network (U.S. EPA)
CSIR	Council for Scientific and Industrial Research (India)
CSM	Committee on Safety of Medicines (the United Kingdom)
CSR	Chemical safety report
CSTEE	Scientific Committee on Toxicology, Ecotoxicology and the Environment
CT	Critical temperature
CT	Computerised tomography
Cv	Coefficient of variation

CWA	Clean Water Act
2,4-D	2,4-Dichlorophenoxyacetic acid
DAB	4-Dimethylaminobenzene
DABA	L-2,4-Diaminobutyric acid
DBCP	1,2-Dibromo-3-chloropropane
DBP	Di-n-butyl phthalate; 2,3-Dibromopropanol
DBT	Dibutyltin
DCB	Dichlorobenzene
DCVC	S-(1,2-Dichlorovinyl)-L-cysteine
DDC	Diethyldithiocarbamate
DDT	Dichlorodiphenyltrichloroethane – a toxic pesticide
DEA	U.S. Drug Enforcement Administration
DEHP	Diethylhexyl phthalate
DEN	Diethylnitrosamine
DES	Diethylstilbestrol
DFB	Diflubenzuron
DFP	Diisopropylfluorophosphate (a delayed neurotoxic)
DHPN	N-Bis(2-hydroxypropyl) nitrosamine
DiBP	Di-iso-butyl phthalate
DIDT	5,6-Dihydro-3H-imidazo (2,1-C)-1,2,4-dithiazole-3-thione
DIT	Diiodotyrosine
DL	Detection limit
DMA	Dimethylamine
DMBA	Dimethylbenzathraline
DMF	Dimethylformamide
DMP	Dimethylphenol
DMSO	Dimethyl sulphoxide
DNA	Deoxyribonucleic acid
DNCB	Dinitrochlorobenzene
DNPH	2,4-Dinitrophenyl hydrazine
DO	Dissolved oxygen
DOC	Dissolved organic carbon
DOE	U.S. Department of Energy
DOT	U.S. Department of Transportation
DPASV	Differential pulse anodic stripping voltammetry
DPTA	Diaminopropanoltetraacetic acid
DT_{50}	Degradation time for 50% of a compound
DTPA	Diethylenetriamine pentaacetic acid
EB	Ethyl benzene
EBDC	Ethylene bisdithiocarbamate
EBK	Ethyl n-butyl ketone
EC	Effective concentration; Electron capture; Emulsifiable concentrate
EC_{50}	Median effective concentration
ECD	Electron capture detection
ECETOC	European Chemical Industry Ecology and Toxicology Centre
ECG	Electrocardiogram
ECMO	Extracorporeal membrane oxidation
ECOD	7-Ethoxycoumarin-o-deethylase
EDA	Ethylenediamine; Exploratory data analysis

EDB	1,2-Dibromoethane (ethylene dibromide)
EDCF	Endothelial-derived contracting factor
EDI	Ethylene diisothiocyanate
EDL	Electrode discharge lamp
EDTA	Ethylenediaminetetraacetic acid
2-EE	2-Ethoxyethanol
2-EEA	2-Ethoxyethyl acetate
EEC	Electroencephalographic
EEC	European Economic Community
EEG	Electroencephalogram
EEGL	Emergency exposure guidance level
EF	Emission factor
EGF	Epidermal growth factor
EH	Epoxide hydratase
EHC	Environmental Health Criteria; Environmental Health Committee
EHS	Extremely hazardous substance
EI	Electron impact
EIFAC	European Inland Fisheries Advisory Commission of the FAO
EINECS	European Inventory of Existing Commercial Chemical Substances
EIS	Emissions Inventory System
EIS	Environmental impact statement
EIS	Environmental Inventory System
EJ	Environmental justice
EL	Exposure level
ELISA	Enzyme-linked immunosorbent assay
EMDI	Estimated maximum daily intake
EMG	Electromyography
EMR	Environmental Management Report
EMSL	Environmental Monitoring Support Systems Laboratory
EMTD	Estimated maximum tolerated dose
EMTS	Environmental monitoring testing site
ENG	Electroneurography
ENL	Erythema nodosum leprosum
ENU	Ethylnitrosourea
EO	Ethylene oxide
EOC	Emergency Operating Center
EP	Electrophoresis
EPA	Environmental Protection Agency (United States)
EPDM	Ethylene propylene rubber
EPI	Exposure/potency index
EPI	Environmental Policy Institute
EPS	Expandable polystyrene
EQC	External quality control
EQT	Environmental quality target
ER	Ecosystem restoration
ERL	Environmental Research Laboratory
ERL	Extraneous residue limit
EROD	7-Ethoxyresurofin-o-deethylase
ERPG	Emergency Response Planning Guideline published by the AIHA

ESA	Endangered Species Act. Environmentally sensitive area
ESC	Endangered Species Committee
ESCAL	Electron spectroscopy for chemical analysis
ESH	Environmental safety and health
ESIS	European Chemical Substance Information System
ESR	Erythrocyte sedimentation rate
ESRD	End-stage renal disease
ETA	Electrothermal atomisation
ETD	Ethylene bisthiuram disulphide
ETG	Epidermal transglutaminase
ETS	Emissions Tracking System; Environmental tobacco smoke
ETU	Ethylenethiourea
ETV	Electrothermal vaporisation
EU	Ethyleneurea
EU	European Union (1994, new name of the European Economic Community)
F_0	Parental generation
F_1	Filial generation, first
F_2	Filial generation, second
FAA	Flameless atomic absorption
FACS	Fluorescence-activated cell sorter
FAD	Flavin adenine dinucleotide
FAO	Food and Agriculture Organization
FATES	FIFRA and TSCA Enforcement System
f.c.	Field capacity
FCAT	Freund's complete adjuvant test
FD	Fluorescence detection
FDA	Food and Drug Administration (the United States)
FE	Fugitive emissions
FEF	Forced expiratory flow
FEP	Free erythrocyte porphyrin
FEPCA	Federal Environmental Pesticide Control Act; enacted as amendments to the FIFRA
FEV	Forced expiratory volume
FF	Federal facilities
FFDCA	Federal Food, Drug, and Cosmetic Act
fg	Femtogram (10^{-15} g)
FI	Flame ionisation
FID	Flame ionisation detector
FIFRA	Federal Insecticide, Fungicide, and Rodenticide Act
FLP	Flashpoint
F/M	Food to microorganism ratio
FMG	Fumigant management plan
FMLP	Formyl methionyl leucyl phenylalanine
FMN	Flavin mononucleotide
FOB	Functional observational battery
FOIA	Freedom of Information Act
FONSI	Finding of no significant impact
FP	Fine particulate
FP	Freezing point

FPA	Federal Pesticide Act
FPD	Flame photometric detector
FQPA	Food Quality Protection Act
FR	Federal Register
FR	Flame retardant
FRA	Federal Register Act
FS	Fluorescence spectrophotometry
FSA	Food Security Act
FSC	Food Safety Council (United States)
FSD	Flame photometric detector selective for sulphur maintained by the U.S. Coast Guard
FSH	Follicle-stimulating hormone
FTIR	Fourier transform infrared
FVC	Forced vital capacity
FWPCA	Federal Water Pollution and Control Act (aka CWA); Federal Water Pollution and Control Administration
G	Gas
g	Gram
GA	Gallium; Gibberellic acid
GABA	Gamma-aminobutyric acid
GAG	Glycosaminoglycan
GALT	Gut-associated lymphoid tissue
GAP	Good agricultural practice
GBM	Glomerular basement membrane
GC	Gas chromatography
GC-ECD	Gas chromatography with electron capture detector
GC/MS	Gas chromatograph/mass spectrograph
GC-SIM	Gas chromatography with selected ion monitoring
GDH	Glutamate dehydrogenase
GDMS	Glow discharge mass spectrometry
GEMS	Global Environmental Monitoring System
GESAMP	Group of Experts for the Scientific Aspects of Marine Pollution
GFAAS	Graphite furnace atomic absorption spectrometry
GFR	Glomerular filtration rate
GGT	Gamma-glutamyltranspeptidase
GH	Growth hormone
GHS	Globally Harmonised System of Classification and Labelling of Chemicals
GI	Gastrointestinal
GIFAP	International Group of National Association of Manufacturers of Agrochemical Products
GLC	Gas–liquid chromatography
GLDH	Glutamate dehydrogenase
GLP	Good laboratory practice
cGMP	Cyclic guanosine monophosphate
GOT	Glutamic–oxaloacetic transaminase
GOEV	Guide to Occupational Exposure Values, published by the ACGIH
GPC	Gel permeation chromatography
GPMT	Guinea pig maximisation test
GPT	Glutamic–pyruvic transaminase

GS	Glutamine synthetase
GSH	Glutathione-SH
GST	Glutathione-S-transferase
GTB	Glomerular tubular balance
GV	Guidance value
GWP	Global-warming potential
h	Hour
ha	Hectare
Hb	Haemoglobin
HBCD	Hexabromocyclododecane
HBDH	Hydroxybutyric dehydrogenase
HCB	Hexachlorobenzene
HCBD	Hexachloro-1,3-butadiene
HCFC	Hydrochlorofluorocarbon
HCFH-22	Chlorodifluoromethane ($CHClF_2$)
HCG	Human chorionic gonadotropin
HCH	Hexachlorocyclohexane
HCS	OSHA Hazard Communication Standard
HDL	High-density lipoprotein
HDPE	High-density polyethylene
H&E	Haematoxylin and eosin
HEAL	Human exposure assessment location
HECD	Hall electron capture detector
HEV	High endothelial venule
HEX	Hexachlorocyclopentadiene
HGPRT	Hypoxanthine–guanine phosphoribosyltransferase
HIPS	High-impact polystyrene
HIV	Human immunodeficiency virus
HLB	Hydrophilic–lipophilic balance
HLV	Hygienic limit value
hnRNA	Heterogeneous nuclear RNA
HPCA	Human progenitor cell antigen
HPI	Cyclohexane-1,2-dicarboximide
HPLC	High-performance liquid chromatography
HPTLC	High-performance thin-layer chromatography
HQ	Hydroquinone
HS	Headspace
HSA	Heat-stable antigen; Human serum albumin
HSG	Health and Safety Guide
IARC	International Agency for Research on Cancer
IC	Ion chromatography
ICAM	Intercellular adhesion molecule
IC_{50}	Median inhibitory concentration
ICD	International Classification of Diseases
ICG	Indocyanine green
ICNIRP	International Commission on Non-Ionising Radiation Protection
ICP	Inductively coupled plasma
ICST	Isolated cold stress testing
i.d.	Internal diameter

IDMS	Isotope dilution mass spectrometry
IFCS	Intergovernmental Forum on Chemical Safety
IFN	Interferon
Ig	Immunoglobulin
IL	Interleukin
ILO	International Labour Organisation
ILSI	International Life Science Institute
im	Intramuscular
ip	Intraperitoneal
IPA	Isopropylamine
IPCS	International Programme on Chemical Safety
IQ	Intelligence quotient
IQC	Internal quality control
IR	Infrared
IRPTC	International Register of Potentially Toxic Chemicals
ISO	International Organization for Standardization
ISPESL	National Institute of Occupational Safety and Prevention (Italy)
ISRO	Indian Space Research Organisation (India)
IT	Isomeric transition
IU	International unit
IUCLID	International Uniform Chemical Information Database
IUPAC	International Union of Pure and Applied Chemistry
iv	Intravenous
JECFA	Joint FAO/WHO Expert Committee on Food Additives
JMPR	Joint FAO/WHO Meeting on Pesticide Residues
JRC	Joint Research Centre of the European Commission
Kcal	Kilocalorie
keV	Kiloelectron volt
K_{ow}	Octanol/water partition coefficient
LAP	Leucine aminopeptidase
LAQL	Lowest analytically quantifiable level
LC	Liquid chromatography
LC_{50}	Median lethal concentration
LD_{01}	Lethal dose for 1%
LD_{50}	Median lethal dose
LDH	Lactate dehydrogenase
LDL	Low-density lipoprotein
LDPE	Low-density polyethylene
LDQ	Lowest detectable quantity
LEI	Lifetime exposure intensity
LEV	Local exhaust ventilation
Lf	Limit flocculation
LFA	Lymphocyte function-related antigen
LFP	Lavage fluid protein
LH	Luteinizing hormone
LI	Labelling index
LIF	Laser-induced fluorescence; Leukaemia inhibitory factor
LLD	Lowest lethal dose
LMS	Linear multistage model

LMW	Low molecular weight
LOAEL	Lowest-observed-adverse-effect level
LOD	Limit of determination
LOEL	Lowest-observed-effect level
LRNI	Lower reference nutrient intake (the United Kingdom)
LSC	Liquid scintillation counter
LT_{50}	Median lethal time
LTP	Long-term potentiation
MAA	Methoxyacetic acid
MAC	Maximum allowable concentration
MAD	Maximum allowable deviation
MAFF	Ministry of Agriculture, Forestry and Fisheries (Japan)
MAK	Maximum workplace concentration (Maximale Arbeitsplatzkonzentration)
MALT	Mucosa-associated lymphoid tissue
MAM	Methylazoxymethanol
MAP	Mutagenic activity profile
MAOI	Monoamine oxidase inhibitor
MARC	Monitoring and Assessment Research Centre (the United Kingdom)
MARE	Monoclonal anti-rat immunoglobulin E
MAS	Molecular absorption spectrometry
MAT	Mean absorption time
MATC	Maximum acceptable toxicant concentration
MBC	Minimum bactericidal concentration
MBDE	Mass balance differential equation
MBK	Methyl n-butyl ketone
MBP	Monobutyl phthalate
MBT	Monobutyltin
MC	Methyl chloroform
3-MC	3-Methylcholanthrene
MCD	Microcoulometric detection
MCH	Mean cell haemoglobin
MCHC	Mean cell haemoglobin concentration
mCi	Millicurie
MCL	Melanotic cell lines
MCPA	4-Chloro-o-tolyoxyacetic acid
MCV	Mean cell volume
MDA	Malondialdehyde
MDI	Methylene-diphenyl diisocyanate
MDMA	Methylenedioxymethamphetamine
2-ME	2-Methoxyethanol
2-MEA	2-Methoxyethyl acetate
MeB_{12}	Methylcobalamin
MED	Minimum effective dose; Minimal erythemal dose
MEHP	Monoethylhexyl phthalate
MEK	Methyl ethyl ketone
MEL	Maximum exposure limit (UK)
mEq	Milliequivalent
MeV	Megaelectron volt
MFO	Mixed-function oxidase

MHC	Major histocompatibility complex
MHW	Ministry of Health and Welfare (Japan)
MIBK	Methyl isobutyl ketone
MIC	Minimal inhibitory concentration
MIT	Methyl isothiocyanate; Monoiodotyrosine
MLD	Minimum lethal dose
MLR	Mixed lymphocyte response assay
MLSS	Mixed liquor suspended solids
MMA	Methoxyacetic acid
MMAD	Mass median aerodynamic diameter
MMH	Monomethylhydrazine
mm Hg	millimetres of mercury
MMMF	Man-made mineral fibre
MNNG	N-methyl-N-nitro-N-nitrosoguanidine
MNU	N-methyl-N-nitrosourea
mPa	Millipascal (7.5×10^{-6} mmHg)
MPC	Maximum permissible limit
MQL	Minimum quantifiable limit
MRBIS	Mean running bias index score
MRL	Maximum residue limit
mRNA	Messenger RNA
MRVIS	Mean running variance index score
MS	Mass spectrometry
MSD	Mass selective detection
MSDS	Material safety data sheet
MSHA	Mine Safety and Health Administration (the United States)
MSW	Municipal solid waste
MTBE	Methyl tertiary butyl ether
MTC	Maximum tolerable or acceptable concentration
MTD	Maximum tolerated dose
MTE	Mild toxic encephalopathy
MTI	N-(Hydroxymethyl)-3,4,5,6-tetrahydrophthalamide
NAA	Neutron activation analysis
NAC	N-Acetyl cysteine
NAD	Nicotinamide adenine dinucleotide
NADP	Nicotinamide adenine dinucleotide phosphate
NADPH	Reduced nicotinamide adenine dinucleotide phosphate
NBU	N-Nitrosobutylurea
NCAM	Neural cell adhesion molecule
NCD	New Chemicals Database
NCI	National Cancer Institute (United States); Negative ion chemical ionisation
NCDOL	North Carolina, Department of Labor
NCV	Nerve conduction velocity
ND	Not detectable
NDDC	Sodium diethyldithiocarbamate
NDMA	Nitrosodimethylamine
NDMC	Sodium dimethyldithiocarbamate
NEQUAS	National External Quality Assessment Scheme (the United Kingdom)
NFT	Neurofibrillary tangle

ng	Nanogram (10^{-9} g)
NIOSH	National Institute for Occupational Safety and Health (the United States)
NK	Natural killer
nm	Nanometre
NMCL	Nonmelanotic cell lines
NMN	*N*-Methylnicotinamide
NMOR	*N*-Nitrosomorpholine
NMR	Nuclear magnetic resonance
NNM	*N*-Nitrosomorpholine
NO	Nitrogen oxide
NOAEL	No-observed-adverse-effect level
NOEC	No-observed-effect concentration
NOEL	No-observed-effect level
NOLC	No-observed lethal concentration
NPD	Nitrogen–phosphorus-sensitive detector
NPSH	Nonprotein sulphhydryl
NSAID	Nonsteroidal anti-inflammatory drugs
NSD	Nitrogen-selective detector
NSF	National Sanitation Foundation
NTA	Nitrilotriacetic acid
NTE	Neuropathy target esterase
NTEL	No-toxic-effect level
NTP	National Toxicology Program (the United States)
OCT	Ornithine carbamoyltransferase
ODC	Ornithine decarboxylase
ODP	Ozone-depletion potential
OECD	Organisation for Economic Co-operation and Development
OEL	Occupational exposure limit
OER	Oxygen enhancement ratio
OES	Occupational exposure standard; Optical emission spectrometry
OET	Open epicutaneous test
ONS	Oncology Nursing Society
OP	Organophosphate
OPIDN	Organophosphate-induced delayed neuropathy
OR	Odds ratio
OSC	Oil-enhanced suspension concentrate
OSHA	Occupational Safety and Health Administration (the United States)
OSOR	One substance, one registration
OVA	Ovalbumin
OZT	Oxazolidinethione
PA	Polyamides
PAA	Photon activation analysis
2-PAM	2-Pyridine aldoxime methochloride (pralidoxime chloride)
PAH	*p*-Aminohippurate; Polycyclic aromatic hydrocarbon
PALS	Periarteriolar lymphocyte sheath
PAN	Peroxyacetyl nitrate; Polyacrylonitrile
PAS	Periodic acid Schiff stain
PB	Phenobarbital
PBA	Phenoxybenzoic acid

PBalc	3-Phenoxybenzyl alcohol
PBald	3-Phenoxybenzaldehyde
PBB	Polybrominated biphenyl
PBDD	Polybrominated dibenzodioxin
PBDE	Polybrominated diphenyl ether
PBDF	Polybrominated dibenzofuran
PBI	Protein-bound iodine
PBPK	Physiologically based pharmacokinetics
PBT	Polybutylene terephthalate
PCA	Para-chloroaniline (4-chloroaniline); Passive cutaneous anaphylaxis
PCB	Polychlorinated biphenyl
PCBD	S-(1,2,3,4,4-Pentachloro-1,3-butadienyl)
PCDD	Polychlorinated dibenzodioxin
PCDD	Polychlorinated dibenzo-p-dioxin
PCDF	Polychlorinated dibenzofuran
PCDPE	Polychlorinated diphenylether
PCE	Polychromatic erythrocytes
PCOM	Phase contrast optical microscopy
PCPY	Polychlorinated pyrene
PCQ	Polychlorinated quaterphenyl
PCT	Porphyria cutanea tarda
PCV	Packed cell volume
PD	Plasma desorption
PDG	Phosphate-dependent glutaminase
PE	Polyethylene
PEC	Persistent, bioaccumulative, and toxic
PEF	Peak expiratory flow
PEG	Pneumoencephalography
PEL	Permissible exposure limit
PET	Polyethylene terephthalate
PEYLL	Predicted environmental concentration
PFC	Plaque-forming cell
pg	Picogram (10^{-12} g)
PG	Prostaglandin
pH	The negative logarithm of the hydrogen ion concentration
PHF	Paired helical filaments
PHS	Prostaglandin-H-synthetase
PIB	Piperonyl butoxide
PIC	Picrotoxin
PID	Photoionisation detection
PIXE	Proton-induced x-ray emission
pKa	The negative logarithm of the dissociation constant
PMBA	p-Methylbenzyl alcohol
PMN	Polymorphonuclear leucocyte
PMTDI	Provisional maximum tolerable daily intake
PMR	Proportional mortality rate
PMSG	Pregnant mare serum gonadotropin
PNS	Peripheral nervous system
pO$_2$	Plasma partial pressure (concentration) of oxygen

POCP	Photochemical ozone-creation potential
PoG	Proteoglycan
POPs	Persistent organic pollutants
POPs:	Predicted no-effect concentrations
POS	Psycho-organic syndrome
PP	Polypropylene
ppb	Parts per billion
ppm	Parts per million
PPORD	Predicted no-effect concentrations
ppt	Parts per trillion
PSD	Passive sampling device
PSPS	Pesticides Safety Precautions Scheme (the United Kingdom)
PT	Prothrombin time
PTH	Parathyroid hormone
PTT	Partial thromboplastin time
PTU	Propylenethiourea
PTWI	Provisional tolerable weekly intake
PTZ	Pentylenetetrazole
PVC	Polyvinyl chloride
PYR	Pyrene
QA	Quality assurance
QAP	Quality assurance programme
QC	Quality control
QSAR	Quantitative structure–activity relationship
RACB	Reproductive assessment by continuous breeding
RAST	Radioallergosorbent test
RBC	Red blood cell
RBP	Retinal binding protein
RBPS	Risk-based performance standards
RCRA	Resource Conservation and Recovery Act
RDA	Recommended dietary allowance (United States)
REL	Recommended exposure limit
RER	Rough endoplasmic reticulum
RIA	Radioimmuno assay
RIPs	Risk assessment report
RIPT	Repeat insult patch test
RMA	Reflex modification audiometry
RNA	Ribonucleic acid
RNI	Reference nutrient intake (UK)
ROC	Reactive organic carbon
RPN	Renal papillary necrosis
RR	Relative risk
RTECS	Registry of toxic effects of chemical substances
RUBISCO	Ribulose 1,5-bisphosphate carboxylase
S9	$9000 \times g$ supernatant
SAM	S-Adenosylmethionine
SAP	Serum alkaline phosphatase
SAR	Structure–activity relationship

SAX Number	Each chemical's identifying code as used in SAX's Dangerous Properties of Industrial Chemicals (SAX)
sc	Subcutaneous
SC	Suspension concentrate
SCE	Sister chromatid exchange
SCF	Stem cell factor
SCID	Severe combined immunodeficiency
SCOPE	Scientific Committee on Problems of the Environment of the International Council of Scientific Unions
SCUBA	Self-contained underwater breathing apparatus
SD	Standard deviation
SDAT	Senile dementia of Alzheimer's type
SE	Standard error
SEM	Standard error of the mean; Scanning electron microscopy
SER	Smooth endoplasmic reticulum
SFC	Supercritical fluid chromatography
SFS	Subjective facial sensation
SG	Specific gravity
SGOMSEC	Scientific Group on Methodologies for the Safety Evaluation of Chemicals
SGOT	Serum glutamic–oxaloacetic transaminase
SGPT	Serum glutamic–pyruvic transaminase
SHE	Syrian hamster embryo
SIM	Selected ion monitoring
SIMS	Secondary ion mass spectrometry
SLE	Systemic lupus erythematosus
SMA	Sequential multiple analyser
SMR	Standardised mortality ratio
sol	Soluble in water
SOP	Standard operating procedure
SPF	Specific pathogen-free
SPM	Suspended particulate matter
S-PMA	S-Phenyl-mercapturic acid
SRT	Simple reaction time
SSB	Single-strand breaks
STEL	Short-term exposure limit
SVAs	Site vulnerability assessments
2,4,5-T	2,4,5-Trichlorophenoxyacetic acid
TADI	Temporary acceptable daily intake
TAN	Tropical ataxic neuropathy
TAP	Trialkyl/aryl phosphate
TBBPA	Tetrabromobisphenol A
TBG	Thyroxine-binding globulin
TBP	Tributyl phosphate
TBPP	Tris(2,3-dibromopropyl) phosphate
TBT	Tributyltin
TBTO	Tributyltin oxide
TCA	Tricarboxylic acid cycle
TCD	Thermal conductivity detection
TCDD	2,3,7,8-Tetrachlorinated dibenzo-p-dioxin

TCDF	2,3,7,8-Tetrachlorinated dibenzofuran
TCE	1,1,1-Trichloroethylene
TCP	Trichlorophenol; Tricresyl phosphate
TCPP	Tris(1-chloro-2-propyl)phosphate
TCR	T cell receptor
TDC	Thermal conductivity detection
TDI	Tolerable daily intake; Toluene diisocyanate
TDLAS	Tuneable diode laser absorption spectrometry
TEA	Tetraethyl ammonium; Thermal energy analyser
TEAC	Tetraethylammonium chloride
TEAM	Total exposure assessment methodology
TEEL	Temporary emergency exposure limit
TEF	Toxicity equivalency factor
TEPP	Tetraethyl pyrophosphate
TGA	Thermogravimetric analysis
TH	Thyroid hormone
THF	Tetrahydrofolate
TI	Tolerable intake
TLC	Thin-layer chromatography
TLV	Threshold limit value
TMCP	Tri-meta-cresyl phosphate
TMDI	Theoretical maximum daily intake
TML	Tetramethyllead
TMRL	Temporary maximum residue limit
TMT	Trimethyltin
TNF	Tumour necrosis factor
TOCP	Tri-ortho-cresyl phosphate
TOD	Total oxygen demand
TPA	12-O-Tetradecanoylphorbol-13-acetate
TPCP	Tri-para-cresyl phosphate
TPI	3,4,5,6-Tetrahydrophthalimide
TPIA	3,4,5,6-Tetrahydrophthalic acid
TPN	Total parenteral nutrition
TPO	Thyroid peroxidase
TPP	Triphenyl phosphate
TPTA	Triphenyltin acetate
TPTH	Triphenyltin hydroxide
tRNA	Transfer ribonucleic acid
TRP	Tubular reabsorption of phosphate
TSCA	Toxic Substances Control Act
TSCA	Substance of very high concern
TSH	Thyroid-stimulating hormone (thyrotropin)
TSP	Total suspended particulate
TST	Temperature sensitivity
TT	Toxicity threshold
TWA	Time-weighted average
UCM	Urographic contrast medium
UCL	Upper confidence limit
UDP	Uridine diphosphate

UDPGA	UDP-glucuronic acid
UDPGT	UDP-glucuronosyltransferase
UDS	Unscheduled DNA synthesis
UEL	Upper explosive limit
UF	Uncertainty factor
ULV	Ultra-low volume
UNEP	United Nations Environment Programme
UNICEF	United Nations Children's Fund
U.S. ATSDR	U.S. Agency for Toxic Substance and Disease Registry
UV	Ultraviolet
UVB	Ultraviolet B
VC	Vinyl chloride
VCAM	Vascular cell adhesion molecule
VER	Visual-evoked response
VHH	Volatile halogenated hydrocarbon
VLA	Very late antigen
VLDL	Very-low-density lipoproteins
VOC	Volatile organic carbon compound
VP	Vapour pressure
VSD	Virtually safe dose
VTR	Vibration threshold
v/v	Volume per volume
WAIS	Wechsler Adult Intelligence Scale
WBC	White blood cell
WEEL	Workplace environmental exposure limit
WEEL	Workplace Environmental Exposure Level, published by the AIHA
WG	Water-dispersible granule
WHMIS	Canadian Workplace Hazardous Materials Identification System
WHO	World Health Organization
WISN	Warfarin-induced skin necrosis
WP	Wettable powder
w/v	Weight per volume
XRF	X-ray-generated atomic fluorescence
XRFS	X-ray fluorescence spectroscopy
ZPP	Zinc protoporphyrin
ZRL	Zero risk level

Index